Wendell H. Fleming

Stochastic Analysis, Control, Optimization and Applications

A Volume in Honor of W.H. Fleming

William M. McEneaney, G. George Yin, and Qing Zhang
Editors

Springer-Science+Business Media, LLC

William M. McEneaney
Department of Mathematics
North Carolina State University
Raleigh, NC 27695

G. George Yin
Department of Mathematics
Wayne State University
Detroit, MI 48202

Qing Zhang
Department of Mathematics
University of Georgia
Athens, GA 30602

Library of Congress Cataloging-in-Publication Data

Stochastic analysis, control, optimization and applications : a volume in honor of W.H. Fleming / William McEneaney, G. George Yin, Qing Zhang, [editors].
p. cm.
Includes bibliographical references and index.
ISBN 978-1-4612-7281-6 ISBN 978-1-4612-1784-8 (eBook)
DOI 10.1007/978-1-4612-1784-8
1. Stochastic analysis. 2. Control theory. I. Fleming, Wendell Helms, 1928- . II. McEneaney, William. III. Yin, George, 1954- . IV. Zhang, Qing, 1959-
QA274.2.S7718 1999
519.2—dc21 98-45990
CIP

AMS Subject Classifications: 90, 93

Printed on acid-free paper.

Originally published by Birkhäuser Boston in 1999
Softcover reprint of the hardcover 1st edition 1999

ISBN 978-1-4612-7281-6

Typeset by the editors in TEX.

9 8 7 6 5 4 3 2 1

Dedicated to
Wendell H. Fleming
On the Occasion of His 70th Birthday

Contents

PART II. PARTIAL DIFFERENTIAL EQUATIONS AND VISCOSITY SOLUTIONS

PART III. STOCHASTIC CONTROL, FILTERING AND PARAMETER ESTIMATION

Preface

In view of Professor Wendell Fleming's many fundamental contributions, his profound influence on the mathematical and systems theory communities, his service to the profession, and his dedication to mathematics, we have invited a number of leading experts in the fields of control, optimization, and stochastic systems to contribute to this volume in his honor on the occasion of his 70th birthday.

These papers focus on various aspects of stochastic analysis, control theory and optimization, and applications. They include authoritative expositions and surveys as well as research papers on recent and important issues. The papers are grouped according to the following four major themes: (1) large deviations, risk sensitive and H_∞ control, (2) partial differential equations and viscosity solutions, (3) stochastic control, filtering and parameter estimation, and (4) mathematical finance and other applications.

We express our deep gratitude to all of the authors for their invaluable contributions, and to the referees for their careful and timely reviews. We thank Harold Kushner for having graciously agreed to undertake the task of writing the foreword. Particular thanks go to H. Thomas Banks for his help, advice and suggestions during the entire preparation process, as well as for the generous support of the Center for Research in Scientific Computation. The assistance from the Birkhäuser professional staff is also greatly appreciated.

Raleigh, NC — William McEneaney
Detroit, MI — George Yin
Athens, GA — Qing Zhang

Foreword

Wendell Fleming has made fundamental contributions to many exciting areas in contemporary mathematics and applied mathematics. These include geometric measure theory and the calculus of variations, deterministic and stochastic differential games, stochastic control, mathematical population genetics and mathematical finance. He has been awarded the first SIAM Reid Prize, and (with Herbert Federer) the Steele Prize of the American Mathematical Society and was an invited or Plenary speaker at a substantial number of the main events in stochastic analysis, differential games and PDE's.

Fleming's diplomatic skills, toughness, as well as his moral strength and sensitivity to the concerns of others were made evident through two terms as Chairman of the Brown University Applied Mathematics Department, one as Chairman of the Mathematics Department, as Chairman of the Brown University Faculty Policy Group (which we all counted on to solve the most contentious problems during the period of turmoil in many American Universities), and in many other quite stressful positions which required considerable skill as well as sensitivity to and understanding of diverse points of view on emotionally charged issues. His service to the profession is equally large, and included service on many journal editorial boards, the editorial board of Mathematical Reviews, chairing the AMS Committee on Employment and the widely cited AFOSR Future Directions in Control Project. Whenever a job requires both considerable technical knowledge and human judgment, Fleming is high on the list of candidates.

Fleming has written or coauthored three classical and highly influential books. His 1977 book Functions of Several Variables, aimed at serious upper level undergraduates, might be the best on the subject. The book Deterministic and Stochastic Optimal Control, coauthored with Ray Rishel, remains the standard source for the pre-viscosity solution subject. The most recent book, Controlled Markov Processes and Viscosity Solutions, coauthored with Mete Soner, is the most comprehensive and most accessible account to date of the viscosity solution perspective in stochastic control.

After receiving his Ph.D. with L.C. Young at Wisconsin in 1951 on generalized surfaces in the calculus of variations, Fleming became a staff member of the RAND Corporation, where he stayed until 1955. During those years, RAND was one of the most creative and exciting centers for mathematical work in dynamic programming, game theory, operations research and related areas. The dynamic interaction between applications and mathematics and the intrinsic excitement of mathematical modeling and analysis were environmental fixtures there. Such interactions have remained focal points of Fleming's work since then, and enabled him to master the unique pure and applied mathematical features and intricacies of such diverse areas as mathematical population genetics and mathematical finance.

An eloquent and detailed description of many of his contributions up to 1992 can be found in the preface to volume 34, no. 2, of the SIAM Journal on Control and Optimization, which was dedicated to him on the occasion of his 65th birthday. We will try to avoid excessive overlap.

Fleming has been at the center of developments in stochastic control since the beginning. The major themes of calculus of variations, stochastic processes (SDE's) and stochastic control, PDE's and stochastic and deterministic differential games have dominated much of his work. Each topic contributes much to the others, and one often finds ideas from each topic elegantly and intricately interwoven in his works on the other topics. As prescient and influential as his papers are, his influence extends far beyond his written words. He has inspired or taught an entire generation of scholars in stochastic systems theory. Numerous young (and not so young) people have been inspired through sharing his enthusiasm and insights. The following remarks on a few selected topics give some flavor of Fleming's contributions to stochastic control.

The 1963 paper "Some Markovian Optimization Problems" is probably the first complete work on modern nonlinear stochastic control. It firmly established the subject, as well as some of the most basic techniques. Up to that time, mathematical understanding in the stochastic control community was essentially limited to discrete time and linear-quadratic-Gaussian problems. The general intuition was "linear" and there was little understanding of or experience with the Itô calculus. Well known and accomplished authors were making "second order term" errors. As significant as the paper was at that time, it stands out even more strongly now that the early history of the subject is clearer. Starting with the fact that appropriate functionals of diffusions satisfy parabolic equations, Fleming extended the approximation in policy space (APS) method and estimates for fully elliptic linear operators to show that the sequence of approximations converged to the true minimum value function, which satisfied the correct Hamilton-Jacobi-Bellman (HJB) equation. Subsequently, analogous APS methods were used for more general stochastic models. The elegant integration of diffusion process and PDE methods put the connection on the map for nonlinear problems, to the benefit of both fields. This paper was the first correct general formulation of the "verification theorem" for nonlinear stochastic control in continuous time. As usual with Fleming's papers, this one was remarkably complete at a very early stage in the development of the field, and was a major force in later attracting considerable mathematical attention to it.

In essentially the same time period, two fundamental papers on (deterministic) differential games were published. Apart from the actual mathematical results, they introduced powerful techniques of approximation and smoothing which are widely used to this day. By using discrete time approximations to the games, adding small noise and defining appropriate upper and lower values, Fleming showed that one could approximate the actual value of the game arbitrarily well. The use of the smoothing effects of the small noise and the limit and approximation arguments anticipated the introduction of the current viscosity methods. It made it very clear that one could define the value function for a deterministic control process via the limit of small noise problems, hence via the well defined limit of a sequence of classical PDE's.

Fleming's introduction of logarithmic transforms goes back to the 1970's. This seemingly simple idea has had impressive consequences in the theories of controlled and uncontrolled large deviations, nonlinear eigenvalue problems and nonlinear risk sensitive control. The idea was initially applied to asymptotic problems (in particular, to large deviations type problems) relating to diffusions with small noise intensity ϵ. With positive ϵ, the solution satisfies a linear PDE. By representing the solution as a functional of the associated diffusion process in the form of the expectation of an exponential function $e^{\lambda t/\epsilon}\psi(x)$, and then computing the PDE satisfied by $\epsilon \log \psi(x)$, one gets a nonlinear PDE that is the HJB equation for a stochastic control problem. Thus, one gets a variational (a stochastic control) problem *before* taking limits as $\epsilon \to 0$, as opposed to the usual methods of large deviations. Then stochastic control techniques are used to show that the limit as $\epsilon \to 0$ solves the HJB equation for a controlled deterministic limit process, which is the same variational problem that one would get via large deviations theory. An asymptotic series expansion can also be derived, taking us beyond what is possible with large deviations theory.

While nonlinear transforms were previously used to linearize nonlinear PDE's (say, Burgers' equation), the importance of the reverse process and its control interpretation was unexpected. In addition, the method is valuable even if ϵ does not go to zero, as shown by Fleming's later work on the relations between "ergodic" and variational problems and on risk sensitive control, and Sheu's work on jump processes.

With the log transform approach, Fleming essentially initiated the field of control of large deviations. If the initial diffusion model included a control function and a cost function, then the transformation led naturally to a stochastic differential game and the Isaacs equation. This log transform and control combination was effectively exploited in the subsequent work of Fleming and McEneaney on risk sensitive nonlinear control, which used these methods together with the theory of viscosity solutions, to put the subject on a solid mathematical foundation. Indeed, before that work, it was far from obvious what a nonlinear risk sensitive or robust theory should be like. Later, Fleming (with James) showed the connections between nonlinear risk sensitive and H_∞ control, and (with Hernández-Hernández) applied it to finite state processes. The log transformation idea was a key foundational element in the work of Dupuis and Ellis on the powerful weak convergence approach to the theory of large deviations.

Hopefully, these comments on a few selected topics will illustrate the tremendous impact of Wendell Fleming's work. It is a pleasure to dedicate this book to him on the occasion of his 70th birthday.

Providence, RI Harold J. Kushner

W.H. Fleming's Curriculum Vitae

University Professor of Applied Mathematics and Mathematics
Brown University, Providence, RI 02912

Born: 1928, Guthrie, Oklahoma
Ph.D., 1951, University of Wisconsin

Employment

Rand Corporation 1951–55
Purdue University 1955–58
Brown University 1958–Present
Chairman, Department of Mathematics, 1965–68
Chairman, Division of Applied Mathematics, 1982–85, 1991–94
University Professor 1991–95
University Professor Emeritus and Professor (Research) 1995–present

Visiting Positions

University of Wisconsin, 1954–54, 1962–63
Stanford University, 1968–69 and Summer 1977
Institut de Recherche d'Informatique et d'Automatique
(one month visits: 1969, 1973, 1974, 1977)
University of Genoa (one month: 1973)
M.I.T., Fall semester 1980
University of Minnesota IMA, Fall semester 1985
University of Minnesota, Ordway Visiting Professor, spring quarter 1993

Honors

NSF Senior Postdoctoral Fellow 1968–69.
Guggenheim Fellow 1976–77.
Invited Plenary speaker, 1982 International Congress of Mathematicians, in Warsaw. (Postponed until August 1983.)
Fermi Lecturer, Scuola Normale Superiore Pisa, 1986.
Steele Prize, American Mathematical Society, 1987.
Plenary Speaker IEEE Conference on Decision and Control 1988.
Doctor of Science, Honoris Causa, Purdue University, 1991.
Reid Prize, Society for Industrial and Applied Math., 1994.
American Academy of Arts and Sciences, 1995.

Books

1. *Functions of Several Variables*, Addison-Wesley, 1965, and 2nd ed., Springer-Verlag, 1977.

2. *Deterministic and Stochastic Optimal Control*, (with R.W. Rishel), Springer-Verlag, 1975.

3. *Advances in Filtering and Optimal Stochastic Control*, co-editor with L.G. Gorostiza), Springer Lecture Notes in Control and Information Sciences, No. 42 (1982).

4. *Recent Advances in Dynamic Programming*, (co-editor with I. Capuzzo-Dolcetta and T. Zolezzi), Springer Lecture Notes in Math, No. 1119, 1985.

5. *Controlled Markov Processes and Nonlinear Evolution Equations*, Accademia Nazionale dei Lincei, Scuola Normale Superiore, 1987.

6. *Stochastic Differential Systems, Stochastic Control Theory and Applications*, (co-editor with P.-L. Lions) IMA Volumes in Math. and Its Applications No. 10, Springer-Verlag, 1987.

7. *Controlled Markov Processes and Viscosity Solutions*, (with H. M. Soner) Springer-Verlag, 1992.

Research Publications

1. A Generalized Notion of Boundary, (with L. C. Young), *Trans. Amer. Math. Soc.*, **76** (1954) 457-484.

2. On a Class of Games over Function Space and Related Variational Problems, *Ann. Math.*, **60** (1954) 578-594.

3. Variational Problems with Constraints, (with R. Bellman and D.V. Widder). *Annali di Mat.* (4), **41** (1956) 301-323.

4. A Note on Differential Games of Prescribed Duration, *Contribution Theory of Games*, **3**, *Ann. Math. Studies*, No. 39, 407-412.

5. On Differential Games with an Integral Payoff, (with L.D. Berkovitz), *Contribution Theory of Games*, **3**, *Ann. Math. Studies*, No. 39, 413-435.

6. An Example in the Problem of Least Area, *Amer. Math. Soc.*, **7** (1956), 1063-1974.

7. Representations of Generalized Surfaces as Mixtures, (with L.C. Young), *Rendiconti Circolo Mat. Palermo* (2), **5** (1956), 117-144.

8. Generalized Surfaces with Prescribed Elementary Boundary, (with L.C. Young), *Rendiconti Circolo Mat. Palermo* (2), **5** (1956), 320-340.

9. Functions with Generalized Gradient and Generalized Surfaces, *Annali di Mat. Ser.* 4, **46** (1957), 93-104.

10. Nondegnerate Surfaces of Finite Topological Type, *Trans. Amer. Math. Soc.*, **90** (1959), 323-335.

11. Irreducible Generalized Surfaces, *Rivista Mat. Univ. Parma*, **8** (1957), 251-281.

12. Nondegenerate Surfaces and Fine-Cyclic Surfaces, *Duke Mathematical J.*, **26** (1959), 137-146.

13. Functions whose Partial Derivatives are Measures, *Illinois J. Math.*, **4** (1960), 452-478.

14. An Integral Formula for Total Gradient Variation, (with R. Rishel), *Archiv der Math.*, **11** (1960), 218-222.

15. Normal and Integral Currents, (with H. Federer), *Ann. Math.*, **72** (1960), 458-520.

16. The convergence Problem for Differential Games, *J. Math. Anal. Appl.*, **3** (1961), 102-116.

17. On the Oriented Plateau Problem, *Rendiconti Circolo Mat. Palermo* (2), **11**, (1962), 1-22.

18. Some Markovian Optimization Problems, *J. Math. Mechanics*, **12**, (1963), 131-140.

19. The Convergence Problem for Differential Games II, *Contributions to the Theory of Games*, 5, Princeton University Press, 1964.

20. The Cauchy Problem for Degenerate Quasilinear Parabolic Equations, *J. Math. Mechanics*, **13** (1964), 987-1008.

21. A Problem of Random Accelerations, Report No. 403, University of Wisconsin Math. Res. Center, 1963.

22. Flat Chains Over a Finite Coefficient Group, *Trans. Amer. Math. Soc.*, **121** (1966), 160-186.

23. Duality and a priori Estimates in Markovian Optimization Problems, *J. Math. Anal. Appl.* **16** (1966), 254-279; Erratum, ibid. **19** (1966), 204.

24. On the Existence of Optimal Stochastic Controls, (with M. Nisio), *J. Math. Mechanics*, **15** (1966), 777-794.

25. Stochastic Lagrange Multipliers, *Math. Theory Control, Proc. Symp., Univ. of Southern California*, 1967, Academic Press, New York (1967), 443.

26. Optimal Control of Partially Observable Diffusions, *SIAM J. Control*, **6** (1968), 194-214.

27. Optimal Continuous Parameter Stochastic Control, *SIAM Review* **11** (1969), 470-509.

28. The Cauchy Problem fo a Nonlinear First-Order Partial Differential Equation, *J. Differential Equations*, **5** (1969), 515-530.

29. Controlled Diffusions Under Polynomial Growth Conditions, in *Calculus of Variations and Control Theory*, A.V. Balakrishnan, ed., Academic Press, New York, 1969, 209-234.

30. Stochastic Control for Small Noise Intensities, *SIAM J. Control* **9** (1971), 473-517.

31. Stochastically Perturbed Dynamical Systems, Proc. Conf. on Stochastic Differential Equations, Edmonton, July 1972. *Rocky Mountain Math. J.*, **4** (1974), 407-433.

32. Some One Dimensional Migration Models In Population Genetics Theory, (with C.H. Su), *Theoretical Population Biology*, **5** (1974), 431-449.

33. Diffusion Processes in Population Biology, *Applied Probab.*, **7** (1975), 100-105.

34. A Selection-migration Model in Population Genetics, *J. Math. Biology*, **2** (1975), 219-233.

35. Generalized Solutions in Optimal Stochastic Control, *Proc. Second Kingston Conf. on Differential Games*, Marcel Dekker, 1977.

36. Exit Probabilities and Optimal Stochastic Control, *Appl. Math. Optim.*, **4** (1978), 329-346.

37. Equilibrium Distributions of Continuous Polygenic Traits, *SIAM J. Appl. Math.*, **36** (1979), 148-168.

38. Some Measure-valued Markov Processes in Population Genetics Theory, (with M. Viot), *Indiana Univ. Math. J.*, **28** (1979), 817-844.

39. Measure-valued Processes in the Control of 55 Partially Observable Stochastic Systems, *Appl. Math. Optim.*, **6** (1980), 271-285.

40. Optimal Exit Probabilities and Differential Games, (with C-P Tsai), *Appl. Math. Optim.*, **7** (1981).

41. Optimal Control for Partially Observed Diffusions, (with E. Pardoux), *SIAM J. Control Optim.*, **20** (1982), 261-285.

42. Nonlinear Semigroup for Controlled Partially Observed Diffusions, *SIAM J. Control Optim.*, **20** (1982), 286-301.

43. Stochastic Calculus of Variations and Mechanics, *J. Optim. Theory Appl.*, **41** (1983), 55-74.

44. Optimal control and Nonlinear Filtering for Nondegenerate Diffusion Processes, (with S.K. Mitter), *Stochastics*, **7** (1982), 63-77.

45. Logarithmic Transformations and Stochastic Control, in *Advances in Filtering and Optimal Stochastic Control*, Springer Lecture Notes Control Inform. Sci. No. 42 (1982), 131-142.

46. On Stochastic Relaxed Controls for Partially Observed Diffusions, (with M. Nisio), *Osaka Math. J.* **93** (1984), 71-108.

47. Optimal Control of Markov Processes, *Proc. Intern. Congress of Math.*, 1983 (invited Plenary Address).

48. A PDE Approach to Asymptotic Estimates for Optimal Exit Probabilities, (with P. E. Souganidis), *Annali della Scuola Normale Superiore Pisa*, Ser. IV **23** (1986), 171-192.

49. Stochastic Variational Formula for Fundamental Solutions of Parabolic PDE, (with S-J. Sheu), *Appl. Math. Optim.*, **13** (1985), 193-204.

50. A Stochastic Control Approach to Some Large Deviations Problems, *Proc. Conf. on Recent Advances in Dynamic Programming*, Rome, March, 1984. Springer Lecture Notes in Math. No. 1119, 52-66.

51. An Optimal Stochastic Production Planning Problem with Randomly Fluctuating Demand, (with S.P. Sethi and H.M. Soner), *SIAM J. Control Optim.*, **25** (1987), 1494-1502.

52. Asymptotic Series and the Method of Vanishing Viscosity, (with P.E. Souganidis), *Indiana Univ. Math. J.*, **35** (1986), 425-447.

53. Convex Duality Approach to the Optimal Control of Diffusions, (with D. Vermes), *SIAM J. Control Optim.*, **27** (1989), 876-907.

54. On the Existence of the Dominant Eigenvalue and its Application to the Large Deviation Properties of an Ergodic Markov Process, (with S. J. Sheu and H.M. Soner), *Stochastics*, **22** (1987), 187-199.

55. Value Functions for Two-Player, Zero-Sum Stochastic Differential Games, (with P. E. Souganidis), *Indiana Univ. Math J.*, **38** (1989), 293-312.

56. Asymptotic Expansions for Markov processes with Levy Generators, (with H.M. Soner), *Appl. Math. Optim.*, **19** (1989), 203-223.

57. Generalized solutions and convex duality in optimal control in Partial Differential Equations and the Calculus of Variations, (Ed. F. Colombini et al), Birkhauser, 1989, 461-472.

58. Piecewise monotone filtering with small observation noise, (with E. Pardoux), *SIAM J. Control Optim.*, **27** (1989), 1156-1181.

59. Piecewise monotone filtering in discrete time with small observation noise, (with D. Ji, P. Salame and Q. Zhang), *IEEE Trans. Automatic Control*, **36** (1991), 1181-1185.

60. An optimal investment/consumption model with borrowing, (with T. Zariphopoulou), *Math. Oper. Res.*, **16** (1991), 802-822.

61. Numerical method for an optimal investment/consumption problem, (with B. Fitzpatrick), *Math. Oper. Res.*, **16** (1991), 823-841.

62. Asymptotic series and exit time probabilities, (with M. James), *Ann. Probab.*, **20** (1992), 1369–1384.

63. Nonlinear filtering with small observation noise: piecewise monotone observations, (with Q. Zhang), in Stochastic Analysis, E. Merzbach, A. Schwartz and E. Mayer-Wolf Eds. Academic Press, (1991) 153-168.

64. Piecewise monotone filtering with small observation noise, (with Q. Zhang), *Proc. Joint US - France Workshop on Stochastic Analysis*, (Eds. I. Karatzas and D. Ocone) April 1991.

65. Optimal portfolio rebalancing with transactions costs, (with S. Grossman, J-L. Vila and T. Zariphopoulou), preprint.

66. Hedging in incomplete markets with HARA utility, (with D. Duffie, H. M. Soner and T. Zariphopoulou), *J. Economic Dynamics Control*, **21** (1997), 753-782.

67. Stochastic control and large deviations, Springer Lecture Notes in Computer Sciences, No. 653, 1992, 291-300. INRIA 25th Anniversary Conference Volume.

68. Risk sensitive control on an infinite time horizon, (with W. M. McEneaney), *SIAM J. Control Optim.* **33** (1995), 1881-1915.

69. Optimal investment models and risk sensitive stochastic control, in IMA Volumes in Math and Appl. No. 65, 1995, 75-88.

70. The risk-sensitive index and the H_2 and H_∞ norms for nonlinear systems, (with M. R. James), *Math. Control, Signals and Systems*, **8** (1995), 199-221.

71. Risk sensitive control of finite state machines on an infinite horizon I, (with D. Hernandez-Hernandez), *SIAM J Control Optim.*, **35** (1997), 1790-1810.

72. Some results and problems in risk sensitive stochastic control, *Computational and Appl. Math.*, **16** (1997), 99-115.

73. Asymptotics for the principal eigenvalue and eigenvector of a nearly first order operator with large potential, (with S.-J. Sheu), *Ann. Probab.*, **25** (1997), 1953-1994.

74. Risk-sensitive production planning of a stochastic manufacturing system, (with Q. Zhang), *SIAM J. Control Optim.*, **36** (1998), 1147-1170.

75. Risk sensitive control of finite state machines on an infinite horizon II, (with D. Hernandez-Hernandez), to appear in *SIAM J. Control Optim.*

76. Deterministic nonlinear filtering, to appear in*Annali Scuola Normale Superiore Pisa.*

77. A max-plus based algorithm for an HJB equation of nonlinear filtering, (with W.M. McEneaney), submitted to *SIAM J. Control Optim.*

78. Optimal long term growth rate of expected utility of wealth, (with S.-J. Sheu), submitted to *Ann. Appl. Probab.*

Invited Summary Papers Published in Conference Proceedings

1. Optimal Control of Diffusion Processes, in *Functional Analysis and Optimization*, E.R. Cainiello Ed., Academic Press, (1966), 68-84.

2. Some Problems of Optimal Stochastic Control, in *Stochastic Optimization and Control*, H.F. Karreman Ed., John Wiley, New York (1968), 59-64.

3. Optimal Continuous Parameter Stochastic Control, *Actes du Congress Int'l. des Math.* 1970, Gauthier-Villars, Park 3 (1971), 163-167.

4. Nonlinear Partial Differential Equations: Probabilistic and Game Theoretic Methods, *Proc. CIME Summer School "Problems in Nonlinear Analysis,"* Varenna, August 1970, 95-128.

5. Dynamical Systems with Small Stochastic Terms, in *Techniques of Optimization*, A.V. Balakrishnan, Ed., Academic Press (1972), 325-334.

6. Optimal Control of Diffusion Processes, in *Stochastic Differential Equations*, J.B. Keller, H.P. McKean, Eds., SIMA-AMS Proc., Vol. VI (1973,) 163-171.

7. Distributed Parameter Systems in Population Biology, in *Control Theory, Numerical Methods and Computer systems Modeling*, 179-191, A. Bensoussan and J.L. Lions, Eds., Lecture Notes in Econ. and Math. Systems, 107 (1975), Springer-Verlag.

8. Some Stochastic Systems Depending on Small Parameters, (with C.P. Tsai), *Proc. Intl. Symposium on Dynamical Systems at Brown University*, Academic Press (1976) 103-114.

9. Inclusion Probability and Optimal Stochastic Control, *IRIA Seminars Review*, 1977.

10. Optimal Inclusion Probability and Differential Games, (with C.P. Tsai), *IRIA Seminars Rview*, 1977.

11. Some Measure-valued Population Processes, (with M. Viot), in *Stochastic Analysis*, A. Friedman and M. Pinsky, Eds., Academic Press (1978), 97-108.

12. Optimal Control of Markov Diffusion Processes, *Proc. Joint Automatic Control Conference*, Vol. 1, October 1978, 355-358.

13. Large Deviations for Diffusions Depending on Small Parameters: A Stochastic Control Method, *Proc. 1st AFCET-SMF Symposium, Ecole Polytechnique*, September, 1978.

14. Minimum Exit Probabilities and Differential Games, (with C.P. Tsai), *Proc. 3rd Kingston Conference on Differential Games and Control Theory*, Marcel Dekker, 1979.

15. Partially Observed Stochastic Control systems, (with E. Pardoux), *Proc. 18th IEEE Conf. on Decision and Control*, December 1979, 163-165.

16. Control Optimo de Processos de Diffusion Markovians, (with O. Hernandez Lerma), *Conferencias sobre sistemas estocasticas, Ciencia*, **32** (1981), 39-55.

17. *Lecture Notes on Diffusion Approximation and Optimal Stochastic Control*, Clemson University, 1979.

18. Stochastic Control Under Partial Observations, *4th Intl. Conf. on Analysis and Optimization of Systems*, INRIA, France, December 1980.

19. Optimal Control and Nonlinear Filtering of Nondegenerate Diffusions, (with S.K. Mitter), *Proc. 20th IEEE Conf. on Decision and Control*, December 1981.

20. A Regular Perturbation Expansion in Nonlinear Filtering, (with R.W. McGwier), *Proc. 22nd IEEE Conf. on Decision and Control*, December 1983, 82-83.

21. A PDE Approach to Asymptotic Estimates for Optimal Exit Probabilities, (with P. E. Souganidis), Springer Lecture Notes in Control and Information Science, *Proc. IFIP Conf.*, Marseille, March 1984.

22. Asymptotic Series for Solutions to the Dynamic Programming Equation for Diffusions with Small Noise, (with P.E. Souganidis) *Proc. 24th IEEE Conf. on Decision and Control*, Ft. Lauderdale, Florida, December 11-13, Vol. 1, 1985.

23. A Stochastic Production Planning Problem with Random Demand, (with H.M. Soner), *Proc. 24th IEEE Conf. on Decision and Control*, Vol. 1, Ft. Lauderdale, Florida, December 11-13, 1985.

24. Generalized Solutions in the Optimal Control of Diffusions, (with D. Vermes), *Proc. IMA Workshop*, June 1986, IMA Vols. in Math. and Appl. No. 10, Springer-Verlag, 1987.

25. Two-Player, Zero-Sum Stochastic Differential Games, (with P. E. Souganidis), *Analyse Mathematique et Applications*, Gauthier-Villars, 1988.

26. Piecewise Linear Filtering With Small Observation Noise, (with D. Ji and E Pardoux), *Proc. 8th INRIA Conf. on Analysis & Optim. of Systems*, Springer Lecture Notes in Control and Info. Sci Vol. 111 (1988), 725-739.

27. Piecewise Monotone Filtering With Small Observation Noise, (with E. Pardoux), *Proc. 27th Conf. on Decision and Control*, Dec. 1988.

28. Numerical methods for optimal investment-consumption model, (with B. Fitzpatrick), *Proc. 29th Conf. on Decision and Control*, Dec. 1990.

29. Risk sensitive optimal control and differential games, (with W. M. McEneaney), *Proc. Conf. on Adaptive and Stochastic Control*, University of Kansas, 1991, Springer Lecture Notes on Control and Info. Sci. No. 184, 1992, 185-187.

30. Risk sensitive control with ergodic cost criteria, (with W. M. McEneaney), *Proc. 31st Conf. on Decision and Control*, Dec. 1992.

31. Numerical methods for infinite horizon risk sensitive stochastic control, (with J. Yang), *Proc. 33rd IEEE Conf. on Decision and Control*, Dec. 1994.

32. Risk sensitive and robust nonlinear filtering, (with W.M. McEneaney), *Proc. 36th IEEE conf. on Decision and Control*, Dec. 1997.

33. Risk sensitive control of finite state machines on an infinite horizon I, (with D. Hernandez-Hernandez), *Proc. 36th IEEE Conf. on Decision and Control*, Dec. 1997.

34. Risk sensitive production planning, (with Q. Zhang), *Proc. 37th IEEE Conf. on Decision and Control*, Dec. 1998.

Service to The Profession (since 1985)

- Member of Organizing Committee for the Year 1985–86 Program at IMA, University of Minnesota.
- Member of Organizing Committee for Workshop on PDE and Stochastic Control, Brown University, August 1987.
- Member of Organizing Committee for the year 1992–93 Program on Control at IMA.
- Chairman, Panel on Future Directions in Control Theory, 1986-88.
- Chairman, AFOSR Review Panel for Mathematical Sciences, 1986-1988.
- Mathematical Reviews Editorial Committee 1991-1996.
- Member of Organizing Committee for 1996 AMS-SIAM Summer Seminar in Applied Mathematics.

Contributors and Addresses

Paolo Albano
Dipartimento di Matematica
Università di Roma Tor Vergata
Via della Ricerca Scientifica
00133 Roma, Italy
albano@axp.mat.uniroma2.it

Rami Atar
Division of Applied Mathematics
Brown University
Providence, RI 02912
U.S.A.
rami@cfm.brown.edu

H. Thomas Banks
Center for Res. in Scientific Comp.
Box 8205
North Carolina State University
Harrelson Hall
Raleigh, NC 27695-8205

Martino Bardi
Dip. di Matematica P. e A.
Università di Padova
via Belzoni 7
35131 Padova, Italy
bardi@math.unipd.it.

Guy Barles
Lab. de Math. et Phys. Theorique
Faculte des Sciences et Techniques
Universite de Tours
Parc de Grandmont, 37200 Tours
France, barles@univ-tours.fr

Tamer Başar
Coordinated Science Laboratory
University of Illinois
1308 W. Main Street
Urbana, Illinois 61801-2307, U.S.A.
tbasar@isaacs.csl.uiuc.edu

Alain Bensoussan
Centre National d'Etudes
Spatiales, 2 place Maurice Quentin
75039 PARIS cedex 01,
France

Leonard D. Berkovitz
Department of Mathematics
Purdue University
West Lafayette, IN 47907, U.S.A.
brkld@math.purdue.edu

Amarjit Budhiraja
Department of Mathematics
University of Notre Dame
Notre Dame, IN 46556
Amarjit.S.Budhiraja.1@nd.edu

Jeff Butera
Department of Mathematics
Box 3430, Montlieu Avenue
High Point University
High Point, NC 27262-3598, U.S.A.

Piermarco Cannarsa
Dipartimento di Matematica
Università di Roma Tor Vergata
Via della Ricerca Scientifica
00133 Roma, Italy
cannarsa@axp.mat.uniroma2.it

P. L. Chow
Department of Mathematics
Wayne State University
Detroit, Michigan 48202
U.S.A.
plchow@math.wayne.edu

Stefano P. Coraluppi
ALPHATECH Inc.
50 Mall Road
Burlington, MA 01803, U.S.A.
stefano.coraluppi@alphatech.com

Mark H. A. Davis
Tokyo-Mitsubishi International Pl
6 Broadgate
London EC2M 2AA, U.K.
mark.davis@t-mi.com

Martin V. Day
Department of Mathematics
Virginia Tech
Blacksburg, VA 24061-0123, U.S.A.
day@math.vt.edu

Tyrone E. Duncan
Department of Mathematics
University of Kansas
Lawrence, KS 66045, U.S.A.
duncan@math.ukans.edu

Paul Dupuis
Division of Applied Mathematics
Brown University
Providence, RI 02912, U.S.A.
dupuis@cfm.brown.edu

Robert J. Elliott
Dept. of Mathematical Sciences
University of Alberta
Edmonton, Canada, T6G 2G1
relliott@gpu.srv.ualberta.ca

Maurizio Falcone
Dipartimento di Matematica
Università di Roma, "La Sapienza"
Piazzale Aldo Moro, 2
00185 Roma, Italy
FALCONE@caspur.it

Mohammad Farid
Control Engineering Res. Center
City University
Northampton Square
London EC1V 0HB, U.K.
m.farid@city.ac.uk

G. Ferreyra
Department of Mathematics
Louisiana State University
Baton Rouge, LA 70803
U.S.A.
ferreyra@math.lsu.edu

Ben G. Fitzpatrick
Tempest Technologies LLC
4435 Coldwater Canyon Ave.
Suite 205,
Studio City, CA 91604, U.S.A.
fitzpatrick@tempest-tech.com

Tiziana Giorgi
Department of Mathematics
McMaster University
Hamilton, ON Canada L8S 4K1
giorgi@icarus.math.McMaster.ca

Paola Goatin
SISSA-ISAS,via Beirut 2-4
34014 Trieste
Italy
goatin@sissa.it

Piyush Gupta
Coordinated Science Laboratory, and
Dept. of Elec. & Computer Eng.
Univ. of Illinois, Urbana-Champaign
Urbana, IL 61801-2307
piyush@decision.csl.uiuc.edu

Daniel Hernandez-Hernandez
Departamento de Matemáticas
CINVESTAV–IPN
Apartado postal 14-740
México D.F. 07000, Mexico
dher@math.cinvestav.mx

James E. Hodder
Business School
University of Wisconsin-Madison
Madison, WI 53706, U.S.A.
jhodder@bus.wisc.edu

Chi-Fu Huang
Long-Term Capital Management
Greenwich, Connecticut
U.S.A.
chuang@ltcm.com

Hitoshi Ishii
Department of Mathematics
Tokyo Metropolitan University
Minami-Ohsawa 1 - 1, Hachioji-shi
Tokyo 192-03, Japan
ishii@math.metro-u.ac.jp

Matthew R. James
Department of Engineering
Faculty of Eng. and Info. Tech.
Australian National University
Canberra, ACT 0200, Australia
Matthew.James@faceng.anu.edu.au

Hugues Julien
Centre National d'Etudes
Spatiales, 2 place Maurice Quentin
75039 Paris Cedex 01
France

Maciej Kocan
Centre for Math. and its Appl's.
Australian National University
Canberra, ACT 0200, Australia
kocan@maths.anu.edu.au

Arthur J. Krener
Department of Mathematics
Unversity of California, Davis
Davis, CA 95616-8633, U.S.A.
ajkrener@ucdavis.edu

Vikram Krishnamurthy
Dept. of Elec. and Electronic Eng.
University of Melbourne
Parkville, Victoria 3052, Australia
v.krishnamurthy@ee.mu.oz.au

P.R. Kumar
Coordinated Science Laboratory, and
Dept. of Electrical & Computer Eng.
Univ. of Illinois, Urbana-Champaign
Urbana, IL 61801-2307
prkumar@decision.csl.uiuc.edu

Harold J. Kushner
Division of Applied Mathematics
P.O. Box F
Brown University
Providence, RI 02912
U.S.A.

Steven I. Marcus
Electrical Engineering Department
and Institute for Systems Research
University of Maryland
College Park, Maryland 20742, U.S.A.
marcus@isr.umd.edu

Jose-Luis Menaldi
Wayne State University
Department of Mathematics
Detroit, MI 48202
U.S.A.
menaldi@math.wayne.edu

Toshio Mikami
Dept. of Math., Hokkaido Univ.
Sapporo 060
Japan
mikami@math.sci.hokudai.ac.jp

Carl Mueller
Dept. of Mathematics
University of Rochester
Rochester, NY 14627, U.S.A.
cmlr@troi.cc.rochester.edu

Hideo Nagai
Department of Mathematical Science
Graduate School of Engineering Sci.
Osaka University
Toyonaka 560, Osaka, Japan
nagai@sigmath.es.osaka-u.ac.jp

Makiko Nisio
Faculty of Engineering
Osaka Electro-Communication Univ.
Hatsu-cho, Neyagawa 572-0833
Japan
nishio@isc.osakac.ac.jp

Daniel Ocone
Department of Mathematics
Rutgers, State Univ. of New Jersey
110 Frelinghuysen Rd.
Piscataway, NJ 08854-8019, U.S.A.
ocone@mathrutgers.edu

Etienne Pardoux
LATP, UMR-CNRS 6632
CMI, Université de Provence
39, rue F. Joliot Curie
F-13453 Marseille Cedex 13, France
pardoux@gyptis.univ-mrs.fr

Bozenna Pasik-Duncan
Department of Mathematics
University of Kansas
Lawrence, KS 66045, U.S.A.
bozenna@kuhub.cc.ukans.edu

Laura K. Potter
Center for Res. in Scientific Comp.
North Carolina State University
Box 8205
Raleigh, NC 27695-8205, U.S.A.

Raymond Rishel
Department of Mathematics
University of Kentucky
Lexington, KY 40506
U.S.A.

Maurice Robin
European Org. for Nuclear Research
(CERN)
CH - 1211, Geneva 23
Switzerland

Elisabeth Rouy
Lab. de Math. et Phys. Theorique
Faculte des Sciences et Techniques
Universite de Tours
Parc de Grandmont
37200 Tours, France

Suresh P. Sethi
School of Management
University of Texas at Dallas
Richardson, TX 75-083-0688
U.S.A.
sethi@utdallas.edu

Shuenn-Jyi Sheu
Institute of Mathematics
Academia Sinica
Nankang, Taipei
Taiwan
MASHEU@ccvax.sinica.edu.tw

Pierpaolo Soravia
Dip. di Matematica, P. ed A.
Università di Padova
Via Belzoni, 7
35131 Padova, Italy
soravia@galileo.math.unipd.it

Panagiotis E. Souganidis
University of Wisconsin – Madison
Department of Mathematics
480 Lincoln Dr.
Madison, WI 53706, U.S.A.
souganid@math.wisc.edu

P. Sundar
Department of Mathematics
Louisiana State University
Baton Rouge, LA 70803
U.S.A.

Michael Taksar
Department of Applied Mathematics
State Univ. New York, Stony Brook
Stony Brook, NY, U.S.A.
taksar@ams.sunysb.edu

Frederi Viens
Department of Mathematics
University of North Texas
Denton, TX 76203-5118, U.S.A.
fviens@unt.edu

Alexander D. Wentzell
Deapartment of Mathematics
Tulane University
New Orleans, LA 70118, U.S.A.

Christian Wypasek
9316 Tewkesbury Ct.
Raleigh, NC 27615
U.S.A.

MingQing Xiao
Department of Mathematics
University of California
Davis, CA 95616
U.S.A.
xiao@math.ucdavis.edu

Thaleia Zariphopoulou
Dept. of Mathematics and Finance
975 University Ave., Grainger Hall
University of Wisconsin–Madison
Madison, WI 53706, U.S.A.
zariphop@math.wisc.edu

Ofer Zeitouni
Department of Electrical Engineering
Technion
Haifa 32000, Israel
zeitouni@ee.technion.ac.il

Hanqin Zhang
Institute of Applied Mathematics
Academia Sinica
Beijing, 100080
China

Yue Zhang
MARC/CTA/PE3/E, Bldg 503
Michelin North American Res. Center
515 Michelin Road
Greenville, SC 29605, U.S.A.

Steven H. Zhu
Citicorp Center, 19FL
2-3-14 Higashi-Shinagawa
Shinagawa-ku, Tokyo 140
Japan

Part I

Large Deviations, Risk Sensitive and H_∞ Control

1

Representations for Functionals of Hilbert Space Valued Diffusions

Amarjit Budhiraja
Paul Dupuis

ABSTRACT: This paper contains two main results. The first is a variational representation for the expectation of a measurable function of a Hilbert space valued Brownian motion, when the function is uniformly positive and bounded from above and the Brownian motion has a trace class covariance. This representation is then applied to derive the second main result, which is the large deviation principle for a class of Hilbert space valued diffusions with small noise.

1 Introduction

The theory of large deviations is one of the classical topics in probability and statistics. For historical background and fundamental results in this area we refer the reader to Varadhan [24], Deuschel and Stroock [7], Dembo and Zeitouni [6], Freidlin and Wentzell [13], Ellis [12]. In a recent book [10] a new methodology was introduced for the analysis of large deviation problems. A crucial ingredient of this approach is the representation of the expectations whose asymptotic behavior is to be analyzed by value functions (minimal cost functions) of associated optimal stochastic control problems. With the representation in hand, one can then use weak convergence methods to study the convergence properties of the value functions as the large deviation scaling parameter tends to its limit. When rewritten in terms of the original expectations, one then obtains the desired large deviation result.

In the approach to large deviations just described, the variational representation for the pre-limit expectations is the starting point of the analysis. The canonical example of such a representation is the following (Proposition 1.4.2 [10]). Let $(\mathcal{V}, \mathcal{A})$ be a measurable space, k a bounded measurable function mapping $\mathcal{V}$ into $\mathbb{R}$ and θ a probability measure on $\mathcal{V}$. Then

$$-\log \int_{\mathcal{V}} e^{-k} d\theta = \inf_{\gamma \in \mathcal{P}(\mathcal{V})} \left\{ R(\gamma \| \theta) + \int_{\mathcal{V}} k d\gamma \right\}, \tag{1}$$

where $\mathcal{P}(\mathcal{V})$ is the space of all probability measures on $(\mathcal{V}, \mathcal{A})$ and $R(\cdot\|\cdot)$ denotes the relative entropy function (see Section 3 for the definition of relative entropy). Although this representation is very general (*e.g.*, it requires no topological properties on the underlying space), it must be rewritten in a more convenient form that reflects the structure of the underlying large deviation problem before it can be used in the convergence analysis. Many such refinements and various applications of the representations to the study of large deviation problems are in [10].

The main interest in the present work is the study of large deviations for infinite dimensional stochastic differential equations. The area of infinite dimensional stochastic calculus is a rapidly growing one, and various excellent references are available ([5], [25], [14], [17]). The "driving noise" in this calculus is an infinite dimensional (typically a Hilbert space valued) Brownian motion, and for this reason the underlying measure that plays the role of θ in (1) is Wiener measure on an infinite dimensional space. One of the main steps in our study is thus the derivation of a useful variational representation for expectations of exponential functionals of such a Brownian motion.

In a recent paper Boué and Dupuis [2] obtained the following representation for a finite dimensional Brownian motion. Let f be a bounded Borel measurable function mapping $\mathcal{C}([0,T] : \mathbb{R}^n)$ (the space of $\mathbb{R}^n$-valued continuous functions on $[0,T]$) into $\mathbb{R}$. Let W be a $\mathbb{R}^n$-valued standard Brownian motion. Then

$$-\log E \exp\{-f(W)\} = \inf_{v \in \mathcal{A}} E \left(\frac{1}{2} \int_0^T \|v(s)\|_0^2 + f\left(W + \int_0^{\cdot} v(s) ds \right) \right), \tag{2}$$

where $\mathcal{A}$ is the space of square integrable progressively measurable processes. One of our main results (Theorem 3.6) is the extension of this representation to the case of infinite dimensional Wiener processes. Once a proper representation is available one can address a variety of large deviation questions for infinite dimensional stochastic differential equations. We illustrate the general methodology by studying the case of Hilbert space valued small noise diffusions in Section 4. The weakest conditions known to the authors on the drift and diffusion coefficients under which such diffusions have a unique strong solution are in Leha and Ritter [19]. We show under exactly these conditions that a large deviation principle (LDP) holds. As is the case for finite dimensional diffusions with such general coefficients contraction mapping arguments cannot be applied and the usual approach is via approximation (usually through a time discretization) of the original problem by a sequence of simpler problems. The LDP for each approximating problem is derived using a contraction mapping argument, and one finally obtains the LDP for the original problem by obtaining bounds on the approximation errors. In contrast, the approach presented here avoids such approximations and discretizations, which can be very useful in certain

circumstances.
Small noise asymptotics of other classes of infinite dimensional stochastic differential equations have been studied by various authors. For example, the case of stochastic evolution equations in Hilbert spaces has been studied in [5], [22], [4]. Some references for the related case of reaction-diffusion stochastic partial differential equations are [26], [23], [16]. Although these classes of processes do not satisfy the assumptions made in Section 4, the representation proved in Section 3 is still applicable and can be used to establish the corresponding LDP. Unfortunately, space limitations prohibit the treatment of these cases in the present work.
The paper is organized as follows. In Section 2 we recall some facts about Hilbert space valued Brownian motions and weak convergence criteria for probability measures on Hilbert spaces. Section 3 is devoted to the proof of the main representation theorem. We also prove in this section a lemma (Lemma 3.1) concerning tightness of a certain sequence of Hilbert space valued processes. This lemma is used several times in rest of the paper. Finally in Section 4 we present the LDP for small noise diffusions in a Hilbert space.
One of the motivations for the present work is the study of image matching problems from a Bayesian perspective (*cf.* [11] and the references cited therein). Under appropriate regularity conditions on the coefficients, Hilbert space valued diffusions provide a natural class of prior models for the deformation that takes the canonical image into the target image suggested by the data. One potential application of large deviation theorems such as the one presented in Section 4 is to establish that the solution to the variational problem studied in [11] is an approximate maximum a posteriori estimator of the true deformation in the small noise limit.

2 Preliminaries

Let $(\Omega, \mathcal{F}, \theta)$ be a probability space with an increasing family of right continuous θ-complete sigma fields $\{\mathcal{F}_t\}_{0\leq t\leq T}$. Let $(H, \langle\cdot,\cdot\rangle)$ be a real separable Hilbert space and let Q be a strictly positive, symmetric, trace class operator (*cf.* [9]) on H. The precise definition of an $H-$valued Wiener process with covariance Q can be found in [5, p. 87, 90]. For our purposes, the following two properties are essential:

1. For every nonzero $h \in H$, $\langle Qh, h\rangle^{-1/2}\langle W(t), h\rangle$ is a one dimensional standard $\mathcal{F}_t-$Wiener process.

2. For every $h \in H$, $W(t,h) \doteq \langle W(t), h\rangle$ is a $\mathcal{F}_t-$martingale.

Define an inner product on H as follows. For $h, k \in H$

$$\langle h, k\rangle_0 \doteq \langle Q^{-1/2}h, Q^{-1/2}k\rangle.$$

As a notational convenience we will use the symbol H_0 for the Hilbert space $(H, \langle\cdot,\cdot\rangle_0)$. Denote the norms in H and H_0 by $\|.\|$ and $\|.\|_0$, respectively. Since Q is trace class the identity mapping from H_0 to H is Hilbert-Schmidt. This Hilbert-Schmidt embedding of H_0 in H will play a central role in many of the arguments to follow. One consequence of the embedding is that if $v^{(n)}$ is a sequence in H_0 such that $v^{(n)} \to 0$ weakly in H_0, then $\|v^{(n)}\| \to 0$. For an excellent exposition of stochastic calculus with respect to an $H-$valued Wiener process we refer the reader to [5]. Other useful references are [20], [21], [17].

The following two theorems are crucial ingredients of the proofs in this paper. Although the first theorem is standard (the reader may refer to [5, Theorem 10.14]), the second requires some elementary modifications of standard arguments and we refer the reader to [3] for the proof.

Let $\{\mathcal{G}_t\}_{0\le t\le T}$ be the $\theta-$completion of the filtration generated by $\{W(s) : 0 \le s \le t\}_{0\le t\le T}$. Define $\mathcal{A}$ to be the class of H_0-valued $\mathcal{F}_t-$progressively measurable processes ϕ that satisfy

$$\theta\left\{\int_0^T \|\phi(s)\|_0^2 ds < \infty\right\} = 1. \tag{3}$$

Finally, let

$$\mathcal{A}^W \doteq \{\phi \in \mathcal{A} : \phi \text{ is } \mathcal{G}_t - \text{progressively measurable}\}.$$

We refer the reader to [5, Chapter 4] for the definition of stochastic integrals of elements of $\mathcal{A}$ with respect to W.

Theorem 2.1 *Let $\psi \in \mathcal{A}$ be such that*

$$E\left(\exp\left\{\int_0^T \langle\psi(s), dW(s)\rangle - \frac{1}{2}\int_0^T \|\psi(s)\|_0^2 ds\right\}\right) = 1. \tag{4}$$

Then the process

$$\tilde{W}(t) \doteq W(t) - \int_0^t \psi(s)ds,$$

$t \in [0,T]$, is a $Q-$ Wiener process with respect to $\{\mathcal{F}_t\}$ on $(\Omega, \mathcal{F}, \gamma)$, where γ is the probability measure defined by

$$\frac{d\gamma}{d\theta} = \exp\left\{\int_0^T \langle\psi(s), dW(s)\rangle - \frac{1}{2}\int_0^T \|\psi(s)\|_0^2 ds\right\}.$$

Theorem 2.2 *Let $(M(t), \mathcal{G}_t)$ be a real valued local martingale with right continuous paths having left limits. Then there exists $\phi \in \mathcal{A}^W$ such that for all $0 \le t \le T$*

$$M(t) = M(0) + \int_0^t \langle\phi(s), dW(s)\rangle, \qquad a.s.$$

To finish this section we record two results that will be used in later sections to prove tightness for a sequence of Hilbert space valued processes. The first of these results is due to Aldous (cf. [25]). Let $(\mathcal{E}, d)$ be a Polish space. We denote by $\mathcal{C}([0,T] : \mathcal{E})$ the Polish space of continuous maps from $[0,T]$ to $\mathcal{E}$ equipped with the uniform topology. Although the tightness criteria in [25] is phrased in terms of processes with values in $\mathcal{D}([0,T] : \mathcal{E})$, the result as stated for processes in $\mathcal{C}([0,T] : \mathcal{E})$ follows as an immediate consequence (see [1] Chapter 3, Section 18, pages 150-153).

Theorem 2.3 *Let $\{X^{(n)}\}$ be a sequence of processes with paths in $\mathcal{C}([0,T] : \mathcal{E})$. Suppose that $\{X^{(n)}(t)\}$ is tight for each rational $t \in [0,T]$ and that for any sequence of stopping times $\{\tau_n\}$ such that $\tau_n \leq T$ and any sequence of nonnegative numbers $\{\delta_n\}$ converging to zero as $n \to \infty$,*

$$d(X^{(n)}(\tau_n + \delta_n), X^{(n)}(\tau_n)) \to 0$$

in probability as $n \to \infty$. Then $\{X^{(n)}\}$ is tight.

The proof of the following theorem can be found in [17].

Theorem 2.4 *Let K be a separable Hilbert space and let $\{e_i\}$ be a complete orthonormal system (CONS) in K. Define P_N to be the projection operator with range equal to $\text{span}\{e_1, \cdots, e_N\}$. Let $\{\mu^{(n)}\}$ be a sequence of probability measures on $(K, \mathcal{B}(K))$. Then $\{\mu^{(n)}\}$ is tight if and only if*

1. *for all $N > 0$*

$$\lim_{A\to\infty} \sup_n \mu^{(n)} \left\{ x \in K : \max_{1\leq i\leq N} |\langle x, e_i\rangle| > A \right\} = 0,$$

2. *for any $\delta > 0$,*

$$\lim_{N\to\infty} \sup_n \mu^{(n)} \{x \in K : \|x - P_N(x)\|_K \geq \delta\} = 0.$$

3 The representation theorem

This section is devoted to the proof of a representation theorem. For a bounded operator A on H let $\|A\|_{op}$ denote its operator norm. We begin with the following lemma.

Lemma 3.1 *Let $\{v^{(n)}\}$ be a sequence of elements of $\mathcal{A}$ (cf. (3)). Assume that*

$$M \doteq \sup_n \int_0^T E\|v^{(n)}(s)\|_0^2 ds < \infty. \tag{5}$$

Then the sequence $\left\{\int_0^{\cdot} v^{(n)}(s)ds\right\}$ is tight in $\mathcal{C}([0,T] : H)$.

Proof: For $0 \leq t \leq T$ define $X^{(n)}(t) \doteq \int_0^t v^{(n)}(s)ds$, and let $\{\tau_n\}$ and $\{\delta_n\}$ as in Theorem 2.3. The Cauchy-Schwarz inequality and the observation that $\|h\| \leq \|Q\|_{op}^{\frac{1}{2}} \|h\|_0$ for $h \in H_0$ imply

$$\|X^{(n)}(\tau_n + \delta_n) - X^{(n)}(\tau_n)\| \leq \sqrt{\delta_n}\, \|Q\|_{op}^{\frac{1}{2}} \left(\int_0^T \|v^{(n)}(s)\|_0^2 ds \right)^{1/2}.$$

Thus by (5) $\|X^{(n)}(\tau_n + \delta_n) - X^{(n)}(\tau_n)\|$ converges to 0 in $L^2(\theta)$. It now suffices, in view of Theorem 2.3, to show that for each fixed $t \in [0,T]$ the sequence $\{X^{(n)}(t)\}$ is tight in H. We will verify conditions 1 and 2 of Theorem 2.4. Let $\{\lambda_k\}$ be the sequence of eigenvalues of Q and let $\{e_k\}$ be a CONS of corresponding eigenvectors. Denote by $\mathbb{N}_0$ the set of positive integers. In order to verify condition 1, it suffices to note that for $A > 0$ and any $n, N \in \mathbb{N}_0$,

$$\theta\left\{\max_{1\leq i\leq N} |\langle X^{(n)}(t), e_i\rangle| > A\right\} \leq \theta\left\{\sum_{i=1}^{N} \langle X^{(n)}(t), e_i\rangle^2 > A^2\right\} \leq \frac{TM\,\|Q\|_{op}}{A^2}.$$

For condition 2, observe that if $f^{(n)}(s) \doteq Q^{-\frac{1}{2}} v^{(n)}(s)$ then the Cauchy-Schwarz inequality implies

$$\begin{aligned}
\|X^{(n)}(t) - P_N(X^{(n)}(t))\|^2 &= \sum_{j=N+1}^{\infty} \left\langle \int_0^t v^{(n)}(s)ds, e_j \right\rangle^2 \\
&= \sum_{j=N+1}^{\infty} \left\langle \int_0^t f^{(n)}(s)ds, Q^{\frac{1}{2}} e_j \right\rangle^2 \\
&\leq T \int_0^T \|f^{(n)}(s)\|^2 ds \sum_{j=N}^{\infty} \lambda_i.
\end{aligned}$$

Observing finally that $\|f^{(n)}(s)\| = \|v^{(n)}(s)\|_0$ and recalling (5), condition 2 is verified by an application of Chebychev's inequality. ∎

The following lemma will be used in some of the tightness arguments in Sections 3 and 4.

Lemma 3.2 *Let $\{v^{(n)}\}$ be a sequence of elements of $\mathcal{A}$. Assume there is $M < \infty$ such that*

$$\sup_n \int_0^T \|v^{(n)}(s)\|_0^2 ds < M$$

a.s. Suppose further that $v^{(n)}$ converges in distribution to v with respect to the weak topology on $L^2([0,T] : H_0)$. Then $\int_0^\cdot v^{(n)}(s)ds$ converges in distribution to $\int_0^\cdot v(s)ds$ in $C([0,T] : H)$.

Proof: For $N \in \mathbb{N}_0$ define

$$S_N \doteq \left\{ u \in L^2([0,T] : H_0) : \int_0^T \|u(s)\|_0^2 ds \leq N \right\}. \tag{6}$$

One can endow S_N with the weak topology, in which case it is a Polish space (*cf.* [9]). The lemma then follows immediately on observing that the map $\tau : S_N \to C([0,T] : H)$ defined by $\tau(u) \doteq \int_0^{\cdot} u(s)ds$ is continuous. ■

The following lemma concerning measurable selections will be used in the proof of the main theorem below. We refer the reader to [3] for a proof.

Lemma 3.3 *Let E_1, E_2 be Polish spaces and let $f : E_1 \times E_2 \to \mathbb{R}$ be a bounded continuous function. Let K be a compact set in E_2. Define for each $x \in E_1$ the sets*

$$\Gamma_x^1 \doteq \left\{ y \in K : \inf_{y_0 \in K} f(x, y_0) = f(x, y) \right\},$$

$$\Gamma_x^2 \doteq \left\{ y \in K : \sup_{y_0 \in K} f(x, y_0) = f(x, y) \right\}$$

Then for $i = 1, 2$ there exist Borel measurable functions $g_i : E_1 \to E_2$ such that $g_i(x) \in \Gamma_x^i$ for all $x \in E_1$.

For probability measures θ_1, θ_2 on $(\Omega, \mathcal{F})$ we define the relative entropy of θ_1 with respect to θ_2 by

$$R(\theta_1 \| \theta_2) \doteq \int_\Omega \left(\log \frac{d\theta_1}{d\theta_2}(\omega) \right) \theta_1(d\omega)$$

whenever θ_1 is absolutely continuous with respect to θ_2 and $\log(\frac{d\theta_1}{d\theta_2})$ is θ_1-integrable. In all other cases set $R(\theta_1 \| \theta_2) \doteq \infty$. Define $\mathcal{A}_N \doteq \{v \in \mathcal{A} : v(\omega) \in S_N \ \theta - a.s.\}$.

Lemma 3.4 *Let $\{f^{(n)}\}$ be a uniformly bounded sequence of real valued measurable functions on $C([0,T] : H)$ converging to f a.s. θ. Then*

$$\inf_{v \in \mathcal{A}_N} E\left(\frac{1}{2} \int_0^T \|v(s)\|_0^2 + f^{(n)} \left(W + \int_0^{\cdot} v(s)ds \right) \right) \tag{7}$$

converges to

$$\inf_{v \in \mathcal{A}_N} E\left(\frac{1}{2} \int_0^T \|v(s)\|_0^2 + f \left(W + \int_0^{\cdot} v(s)ds \right) \right) \tag{8}$$

as $n \to \infty$.

Proof: Let $\epsilon > 0$ be arbitrary. For each $n \in I\!N_0$ pick an element $v^{(n),\epsilon}$ of $\mathcal{A}_N$ such that

$$E\left(\frac{1}{2}\int_0^T \|v^{(n),\epsilon}(s)\|_0^2 + f^{(n)}\left(W + \int_0^{\cdot} v^{(n),\epsilon}(s)ds\right)\right) \tag{9}$$

is at most ϵ larger than the infimum in (7). Since $\{v^{(n),\epsilon}, n \in I\!N_0\}$ is tight in S_N, we can pick a subsequence (relabeled by n) along which $(v^{(n),\epsilon}, W)$ converges weakly to (v^ϵ, W). Using Lemma 3.2 we have that $W + \int_0^{\cdot} v^{(n),\epsilon}(s)ds$ converges weakly as elements of $C([0,T] : H)$ to $W + \int_0^{\cdot} v^\epsilon(s)ds$.
We next claim that

$$E\left(f^{(n)}\left(W + \int_0^{\cdot} v^{(n),\epsilon}(s)ds\right)\right) \to E\left(f\left(W + \int_0^{\cdot} v^\epsilon(s)ds\right)\right).$$

This is a consequence of [2, Lemma 2.8(b)], which states that for the last display to hold it is sufficient that the relative entropies

$$R\left(\mathcal{L}_\theta\left(W + \int_0^{\cdot} v^{(n),\epsilon}(s)ds\right) \|\mathcal{L}_\theta(W)\right)$$

be uniformly bounded in n, where $\mathcal{L}_\theta(W)$ and $\mathcal{L}_\theta\left(W + \int_0^{\cdot} v^{(n),\epsilon}(s)ds\right)$ denote the measures induced on $C([0,T] : H)$ by W and $W + \int_0^{\cdot} v^{(n),\epsilon}(s)ds$, respectively. But this is immediate since these relative entropies equal $E\int_0^T \|v^{(n),\epsilon}(s)\|_0^2 ds \leq N$. Using the weak convergence of $v^{(n),\epsilon}$ to v^ϵ and Fatou's lemma, it follows that

$$\liminf_{n\to\infty} E\left(\frac{1}{2}\int_0^T \|v^{(n),\epsilon}(s)\|_0^2 ds\right) \geq E\left(\frac{1}{2}\int_0^T \|v^\epsilon(s)\|_0^2 ds\right).$$

Thus the limit inferior, as $n \to \infty$, of the expression in (7) is at least the expression in (8).
For the reverse inequality, pick an element v^ϵ of $\mathcal{A}_N$ such that

$$E\left(\frac{1}{2}\int_0^T \|v^\epsilon(s)\|_0^2 ds + f\left(W + \int_0^{\cdot} v^\epsilon(s)ds\right)\right) \tag{10}$$

is at most ϵ larger than the infimum in (8). Clearly the expression in (10) with f replaced by $f^{(n)}$ is at least the infimum in (7). However, this quantity also converges to the expression in (10) as $n \to \infty$. This proves the reverse inequality and hence the lemma. ■

Lemma 3.5 *Let f be a bounded continuous function mapping $C([0,T] : H)$ into $I\!R$.*

1. Let $\tilde{v} \in \mathcal{A}$ be such that

$$E\left(\exp\left\{\int_0^T \langle \tilde{v}(s), dW(s)\rangle - \frac{1}{2}\int_0^T \|\tilde{v}(s)\|_0^2 ds\right\}\right) = 1,$$

define $\tilde{W}(t) \doteq W(t) - \int_0^t \tilde{v}(s)ds$, and let $E^{\tilde{v}}$ denote expectation with respect to the measure $\gamma^{\tilde{v}}$ defined by

$$d\gamma^{\tilde{v}} \doteq \exp\left\{\int_0^T \langle \tilde{v}(s), dW(s)\rangle - \frac{1}{2}\int_0^T \|\tilde{v}(s)\|_0^2 ds\right\} d\theta.$$

Let $v_0 \in \mathcal{A}$ be an elementary process such that for some $M_0 \in (0,\infty)$ $\|v_0(s)\|_0 \le M_0$ a.s., for all $s \in [0,T]$. Then for every $\epsilon > 0$ there exist elementary processes $v_1, v_2 \in \mathcal{A}^W$ such that for $i = 1,2$ $\|v_i(s)\|_0 \le M_0$ for all $s \in [0,T]$, and

$$\begin{aligned} &E\left(\frac{1}{2}\int_0^T \|v_1(s)\|_0^2 ds + f\left(W + \int_0^\cdot v_1(s)ds\right)\right) - \epsilon \\ \le\ & E^{\tilde{v}}\left(\frac{1}{2}\int_0^T \|v_0(s)\|_0^2 ds + f\left(\tilde{W} + \int_0^\cdot v_0(s)ds\right)\right) \qquad (11) \\ \le\ & E\left(\frac{1}{2}\int_0^T \|v_2(s)\|_0^2 ds + f\left(W + \int_0^\cdot v_2(s)ds\right)\right) + \epsilon. \end{aligned}$$

2. Let $\mathcal{A}^{(b)}$ denote the subclass of $\mathcal{A}$ consisting of bounded elementary processes. Then

$$\begin{aligned} &\inf_{\tilde{v}\in\mathcal{A}^{(b)}} E^{\tilde{v}}\left(\frac{1}{2}\int_0^T \|\tilde{v}(s)\|_0^2 ds + f\left(\tilde{W} + \int_0^\cdot \tilde{v}(s)ds\right)\right) \\ =\ & \inf_{v\in\mathcal{A}} E\left(\frac{1}{2}\int_0^T \|v(s)\|_0^2 ds + f\left(W + \int_0^\cdot v(s)ds\right)\right). \end{aligned}$$

Proof: For the proof of part 1 we will use Lemma 3.3. We will only show the first inequality in (11) since the proof of the second inequality is similar, save that the corresponding supremization part of Lemma 3.3 is used instead. Suppose that the elementary process v_0 takes the form

$$v_0(s,\omega) \doteq X_0(\omega)\mathcal{I}_{\{0\}}(s) + \sum_{j=1}^{l} X_j(\omega)\mathcal{I}_{(t_j,t_{j+1}]}(s),$$

where $(s,\omega) \in [0,T]\times\Omega$, $0 = t_1 \le t_2 \ldots \le t_{l+1} = T$ and X_j are H–valued $\mathcal{F}_{t_j}$-measurable random variables satisfying $\|X_j(\omega)\|_0 \le M_0$ a.s. for all

$j \in \{0, \ldots l\}$. There is a continuous function $F_1 : H_0^{\otimes l+1} \to \mathbb{R}$ such that $F_1(X_0(\omega), \ldots, X_l(\omega)) = \frac{1}{2}\int_0^T \|v_0(s)\|_0^2 ds$, a.s. Now define for $j = 1, \ldots, l$ measurable maps $\tilde{Z}_j$ from Ω to $\mathcal{M}_j \doteq C([0, t_{j+1} - t_j], H)$ by

$$\tilde{Z}_j(\omega)(s) \doteq \tilde{W}(\omega)(s + t_j) - \tilde{W}(\omega)(t_j), \quad 0 \le s \le t_{j+1} - t_j.$$

From the continuity and boundedness of the map f it follows that there exists a continuous bounded map $F_2 : \left(H_0^{\otimes l+1} \times \left(\prod_{j=1}^{l} \mathcal{M}_j\right)\right) \to \mathbb{R}$ such that

$$f\left(\tilde{W} + \int_0^\cdot v_0(s)ds\right) = F_2(X_j, 0 \le j \le l; \tilde{Z}_j, 1 \le j \le l),$$

a.s. For $1 \le i \le l$, let $\mathbf{X}_i$ and $\tilde{\mathbf{Z}}_i$ denote the vectors $(X_0, \ldots, X_i)$ and $(\tilde{Z}_1, \ldots, \tilde{Z}_i)$, respectively. With this notation

$$E^{\bar{v}}\left(\frac{1}{2}\int_0^T \|v_0(s)\|_0^2 ds + f\left(\tilde{W} + \int_0^\cdot v_0(s)ds\right)\right) = E^{\bar{v}}\left(F_1(\mathbf{X}_l) + F_2(\mathbf{X}_l, \tilde{\mathbf{Z}}_l)\right). \tag{12}$$

We recall that every probability measure on a Polish space is tight. This implies there is a compact set $K_0 \subset H_0$ such that

$$\begin{aligned} &E^{\bar{v}}\left(F_1(\mathbf{X}_l) + F_2(\mathbf{X}_l, \tilde{\mathbf{Z}}_l)\right) \\ &\quad \ge E^{\bar{v}}\left(\mathcal{I}_{K_0^{\otimes l+1}}(\mathbf{X}_l)\left(F_1(\mathbf{X}_l) + F_2(\mathbf{X}_l, \tilde{\mathbf{Z}}_l)\right)\right) - \epsilon/4(l+1). \end{aligned}$$

Since $(\tilde{W}(t), \mathcal{F}_t)$ is a Wiener process under $\gamma^{\bar{v}}$, if $0 \le u_1 \le u_2 \le T$ then $\tilde{W}(u_2) - \tilde{W}(u_1)$ is independent of $\mathcal{F}_{u_1}$. Therefore, $\tilde{Z}_j$ is independent of $\left(\mathbf{X}_j, \tilde{\mathbf{Z}}_{j-1}\right)$ under $\gamma^{\bar{v}}$. Let μ_j denote the standard Wiener measure on $\mathcal{M}_j$ and let $F_2^{(1)}$ be the real valued continuous map on $\left(H_0^{\otimes l+1} \times \left(\prod_{j=1}^{l-1} \mathcal{M}_j\right)\right)$ obtained by integrating out $\tilde{Z}_l$ from F_2, i.e $F_2^{(1)}(y) \doteq \int F_2(y, z)\mu_l(dz)$, where $y \in \left(H_0^{\otimes l+1} \times \left(\prod_{j=1}^{l-1} \mathcal{M}_j\right)\right)$. Recalling that $\|X_l\|_0 \le M_0$, a.s., and applying Lemma 3.3 with $E_2 \doteq H_0$, $E_1 \doteq \left(H_0^{\otimes l} \times \left(\prod_{j=1}^{l-1} \mathcal{M}_j\right)\right)$, $K \doteq K_0 \cap \{x \in H_0 : \|x\|_0 \le M_0\}$ and $f \doteq F_1 + F_2^{(1)}$, we have that there exists a measurable function $h : \left(H_0^{\otimes l} \times \left(\prod_{j=1}^{l-1} \mathcal{M}_j\right)\right) \to H_0$ satisfying $\|h(\cdot)\|_0 \le M_0$ such that the right side of (12) is bounded below by

$$E^{\bar{v}}\left(F_1(\mathbf{X}_{l-1}, h(\mathbf{X}_{l-1}, \tilde{\mathbf{Z}}_{l-1})) + F_2^{(1)}(\mathbf{X}_{l-1}, h(\mathbf{X}_{l-1}, \tilde{\mathbf{Z}}_{l-1}), \tilde{\mathbf{Z}}_{l-1})\right) - \epsilon/2(l+1).$$

By subtracting an additional $\epsilon/2(l+1)$ from this lower bound, we can take h to be a continuous map via an application of [8, Theorem V.16a] and the dominated convergence theorem. We now iterate the above procedure l times to obtain the following inequality:

$$E^{\bar{v}}\left(F_1(\mathbf{X}_l) + F_2(\mathbf{X}_l, \tilde{\mathbf{Z}}_l)\right) \ge E^{\bar{v}}\left(F_1(\Gamma(\tilde{\mathbf{Z}}_l)) + F_2(\Gamma(\tilde{\mathbf{Z}}_l), \tilde{\mathbf{Z}}_l)\right) - \epsilon,$$

where

- $\Gamma : \prod_{j=1}^{l} \mathcal{M}_j \to H_0^{\otimes l+1}$ is continuous.
- $\Gamma(\mathbf{z}_l)$ can be written $(\Gamma_0, \Gamma_1(\mathbf{z}_1), \ldots \Gamma_l(\mathbf{z}_l))$, where $\mathbf{z}_i \doteq (z_1, \ldots, z_i) \in \prod_{j=1}^{i} \mathcal{M}_j$,
- Γ_0 is a non-random element of H_0 bounded in norm by M_0,
- for $i = 1, \ldots l$ $\Gamma_i : \prod_{j=1}^{i} \mathcal{M}_j \to H_0$ satisfies $\|\Gamma_i(\mathbf{u})\|_0 \leq M_0$ for $\mathbf{u} \in \prod_{j=1}^{i} \mathcal{M}_j$.

Now define for $j = 1, \ldots, l$ measurable maps Z_j from Ω to $\mathcal{M}_j$ by

$$Z_j(\omega)(s) \doteq W(\omega)(s) - W(\omega)(t_j), \quad t_j \leq s \leq t_{j+1}$$

and let $\mathbf{Z}_i \doteq (Z_1, \ldots, Z_i)$ for $i \in \{1, \ldots l\}$. Finally define

$$\overline{v}(s, \omega) \doteq \Gamma_0 \mathcal{I}_{\{0\}}(s) + \sum_{j=1}^{l} \Gamma_j(\mathbf{Z}_j(\omega)) \mathcal{I}_{(t_j, t_{j+1}]}(s).$$

Clearly $\overline{v}(s)$ is an elementary process in $\mathcal{A}^W$ satisfying $\|\overline{v}(s)\|_0 \leq M_0$ for each $s \in [0, T]$ and

$$\begin{aligned} & E^{\tilde{v}} \left(\frac{1}{2} \int_0^T \|v_0(s)\|_0^2 ds + f\left(\tilde{W} + \int_0^{\cdot} v_0(s) ds \right) \right) \\ & \geq \ E \left(\frac{1}{2} \int_0^T \|\overline{v}(s)\|_0^2 ds + f\left(W + \int_0^{\cdot} \overline{v}(s) ds \right) \right) - \epsilon. \end{aligned}$$

This proves part 1.
Next, taking $\tilde{v}$ in (11) to be a bounded elementary process and $v_0 = \tilde{v}$, we obtain

$$\begin{aligned} & \inf_{\tilde{v} \in \mathcal{A}^{(b)}} E^{\tilde{v}} \left(\frac{1}{2} \int_0^T \|\tilde{v}(s)\|_0^2 ds + f\left(\tilde{W} + \int_0^{\cdot} \tilde{v}(s) ds \right) \right) \\ & \geq \ \inf_{v \in \mathcal{A}^{W,(b)}} E \left(\frac{1}{2} \int_0^T \|v(s)\|_0^2 ds + f\left(W + \int_0^{\cdot} v(s) ds \right) \right), \end{aligned} \tag{13}$$

where $\mathcal{A}^{W,(b)}$ is the subclass of $\mathcal{A}^W$ of bounded elementary processes. Since elements of $\mathcal{A}^{W,(b)}$ are piecewise constant, for every $v \in \mathcal{A}^{W,(b)}$ we can recursively construct $\tilde{v} \in \mathcal{A}^{(b)}$ such that

$$\begin{aligned} & E^{\tilde{v}} \left(\frac{1}{2} \int_0^T \|\tilde{v}(s)\|_0^2 ds + f\left(\tilde{W} + \int_0^{\cdot} \tilde{v}(s) ds \right) \right) \\ & = \ E \left(\frac{1}{2} \int_0^T \|v(s)\|_0^2 ds + f\left(W + \int_0^{\cdot} v(s) ds \right) \right) \end{aligned} \tag{14}$$

(see [2, Theorem 3.1] for details). Combining (13), (14), we have that

$$\inf_{\tilde{v}\in\mathcal{A}^{(b)}} E^{\tilde{v}}\left(\frac{1}{2}\int_0^T \|\tilde{v}(s)\|_0^2 ds + f\left(\tilde{W} + \int_0^{\cdot} \tilde{v}(s)ds\right)\right)$$
$$= \inf_{v\in\mathcal{A}^{W,(b)}} E\left(\frac{1}{2}\int_0^T \|v(s)\|_0^2 ds + f\left(W + \int_0^{\cdot} v(s)ds\right)\right). \quad (15)$$

Next, taking $\tilde{v} \equiv 0$ in (11) and observing that $\mathcal{A}^{W,(b)} \subset \mathcal{A}^{(b)}$, it follows

$$\inf_{v\in\mathcal{A}^{W,(b)}} E\left(\frac{1}{2}\int_0^T \|v(s)\|_0^2 ds + f\left(W + \int_0^{\cdot} v(s)ds\right)\right)$$
$$= \inf_{v\in\mathcal{A}^{(b)}} E\left(\frac{1}{2}\int_0^T \|v(s)\|_0^2 ds + f\left(W + \int_0^{\cdot} v(s)ds\right)\right). \quad (16)$$

Now let $v \in \mathcal{A}$ be such that $E\{\int_0^T \|v(s)\|_0^2 ds\} < \infty$. Choose a sequence $\{v^n : n \in \mathbb{N}_0\}$ in $\mathcal{A}$ such that each v^n is a bounded elementary process, $\lim_{n\to\infty} E\left(\int_0^T \|v^n(s) - v(s)\|_0^2 ds\right) = 0$, and $\sup_{n\in\mathbb{N}_0} E\left(\int_0^T \|v^n(s)\|_0^2 ds\right) < 1 + E\left(\int_0^T \|v(s)\|_0^2 ds\right)$. Clearly $\int_0^t (v^n(s) - v(s))ds$ converges to zero in probability for each $t \in [0,T]$. Also, an application of Lemma 3.1 shows that $\{\int_0^{\cdot}(v^n(s)-v(s))ds\}$ is tight in $C([0,T] : H)$. Thus $(W, \int_0^{\cdot} v^n(s)ds)$ converges weakly to $(W, \int_0^{\cdot} v(s)ds)$, and therefore

$$\lim_{n\to\infty} E\left(\frac{1}{2}\int_0^T \|v^n(s)\|_0^2 ds + f\left(W + \int_0^{\cdot} v^n(s)ds\right)\right)$$
$$= E\left(\frac{1}{2}\int_0^T \|v(s)\|_0^2 ds + f\left(W + \int_0^{\cdot} v(s)ds\right)\right).$$

Using $\mathcal{A} \subset \mathcal{A}^{(b)}$, this proves that

$$\inf_{v\in\mathcal{A}^{(b)}} E\left(\frac{1}{2}\int_0^T \|v(s)\|_0^2 ds + f\left(W + \int_0^{\cdot} v(s)ds\right)\right)$$
$$= \inf_{v\in\mathcal{A}} E\left(\frac{1}{2}\int_0^T \|v(s)\|_0^2 ds + f\left(W + \int_0^{\cdot} v(s)ds\right)\right). \quad (17)$$

The proof of part 2 is completed by combining (15), (16) and (17). ■

We now present the main result of this section. Though in the theorem we take f to be a bounded function, it can be shown (as in [2]) that the representation continues to hold if f is bounded from above.

Theorem 3.6 *Let f be a bounded, Borel measurable function mapping $\mathcal{C}([0,T]:H)$ into $\mathbb{R}$. Then*

$$-\log E\exp\{-f(W)\} = \inf_{v\in\mathcal{A}} E\left(\frac{1}{2}\int_0^T \|v(s)\|_0^2 ds + f\left(W + \int_0^{\cdot} v(s)ds\right)\right) \tag{18}$$

Proof: The proof presented here is adapted from that of the finite dimensional case studied in [2]. We claim that it suffices to prove the result for f that are continuous. To see this, let $\{f^{(n)}\}$ be a sequence of real valued continuous functions on $\mathcal{C}([0,T],H)$ such that $\sup_{x,n}|f^{(n)}(x)| \leq \sup_x |f(x)|$ and $f^{(n)}$ converges to f $\theta-$a.s. An application of dominated convergence theorem shows that

$$-\log E\exp\{-f^{(n)}(W)\} \to -\log E\exp\{-f(W)\}.$$

For $\mathcal{B}\subset\mathcal{A}$ and g a bounded, Borel measurable function mapping $\mathcal{C}([0,T]:H)$ into $\mathbb{R}$ define

$$\Lambda(\mathcal{B},g) \doteq \inf_{v\in\mathcal{B}} E\left(\frac{1}{2}\int_0^T \|v(s)\|_0^2 ds + g\left(W + \int_0^{\cdot} v(s)ds\right)\right).$$

To prove the claim, we must show that $\Lambda(\mathcal{A},f^{(n)})$ converges to $\Lambda(\mathcal{A},f)$ as $n\to\infty$. Let $K \doteq \sup_x |f(x)|$ and $\overline{\mathcal{A}} \doteq \{v\in\mathcal{A} : E\left(\int_0^T \|v(s)\|_0^2 ds\right) \leq 4K\}$. Then clearly $\Lambda(\mathcal{A},f^{(n)})$ equals $\Lambda(\overline{\mathcal{A}},f^{(n)})$ and $\Lambda(\mathcal{A},f)$ equals $\Lambda(\overline{\mathcal{A}},f)$. Let $\epsilon>0$ be arbitrary. Choose $N\in\mathbb{N}_0$ such that $\frac{4K^2}{N}\leq\epsilon/2$. Fix $v\in\overline{\mathcal{A}}$ and define the stopping time $\tau_N \doteq \inf\{s\in[0,T] : \int_0^s \|v(s)\|_0^2 \geq N\}\wedge T$. Recall that $\mathcal{A}_N \doteq \{v\in\mathcal{A} : \int_0^T \|v(s)\|_0^2 ds \leq N, \theta - a.s.\}$. Let $v_N\in\mathcal{A}_N$ be defined as $v_N(s) \doteq v(s)\mathcal{I}_{[0,\tau_N]}(s)$, where $\mathcal{I}$ denotes the indicator function. We observe that

$$\begin{aligned}\Lambda(\mathcal{A}_N,f^{(n)}) &\leq E\left(\frac{1}{2}\int_0^T \|v_N(s)\|_0^2 ds + f^{(n)}\left(W+\int_0^{\cdot} v_N(s)ds\right)\right)\\ &\leq E\left(\frac{1}{2}\int_0^T \|v(s)\|_0^2 ds + f^{(n)}\left(W+\int_0^{\cdot} v(s)ds\right)\right)+\epsilon,\end{aligned}$$

where the second line in the display above follows since $v\in\overline{\mathcal{A}}$ implies that the probability of the set $\{\tau_N<T\}$ is at most $\frac{4K}{N}$. Taking the infimum over all $v\in\overline{\mathcal{A}}$ in the inequality above we have that

$$\Lambda(\mathcal{A},f^{(n)}) \leq \Lambda(\mathcal{A}_N,f^{(n)}) \leq \Lambda(\mathcal{A},f^{(n)})+\epsilon.$$

Exactly the same argument gives

$$\Lambda(\mathcal{A},f) \leq \Lambda(\mathcal{A}_N,f) \leq \Lambda(\mathcal{A},f)+\epsilon.$$

Finally, an application of Lemma 3.4 shows that $\Lambda(\mathcal{A}_N, f^{(n)})$ converges to $\Lambda(\mathcal{A}_N, f)$ as $n \to \infty$. This proves the claim.
Henceforth we will assume that f is continuous. We will prove that the left side of (18) is bounded above and below by the right side.

Proof of the upper bound: From Proposition 1.4.2 of [10] it follows that

$$-\log E\exp\{-f(W)\} = \inf_{\gamma\in\mathcal{P}(\Omega):\gamma\ll\theta}\{R(\gamma\|\theta) + E^{\gamma}(f(W))\}, \tag{19}$$

where $\mathcal{P}(\Omega)$ is the class of all probability measures on $(\Omega, \mathcal{F})$. Let $\tilde{v} \in \mathcal{A}$ be a bounded elementary process. Clearly, $(\int_0^t \langle\tilde{v}(s), dW(s)\rangle, \mathcal{F}_t)_{0\le t\le T}$ is a real valued continuous martingale with quadratic variation $\int_0^t \|\tilde{v}(s)\|_0^2 ds$ ([5, Section 4]). The boundedness assumption also implies that the expectation $E(\exp\{\int_0^T \|\tilde{v}(s)\|_0^2 ds\})$ is finite, and therefore Proposition 5.12 of [18] yields

$$E\left(\exp\left\{\int_0^T \langle\tilde{v}(s), dW(s)\rangle - \frac{1}{2}\int_0^T \|\tilde{v}(s)\|_0^2 ds\right\}\right) = 1.$$

By Theorem 2.1 $d\gamma^{\tilde{v}} \doteq \exp\left\{\int_0^T \langle\tilde{v}(s), dW(s)\rangle - \frac{1}{2}\int_0^T \|\tilde{v}(s)\|_0^2 ds\right\} d\theta$ is a probability measure and, under $\gamma^{\tilde{v}}$, $\tilde{W}(t) \doteq W(t) - \int_0^t \tilde{v}(s)ds$ is a Q–Wiener process. Next from the definition of the relative entropy function

$$R(\gamma^{\tilde{v}}\|\theta) \doteq E^{\tilde{v}}\left(\log\frac{d\gamma^{\tilde{v}}}{d\theta}\right) = E^{\tilde{v}}\left(\int_0^T \langle\tilde{v}(s), dW(s)\rangle - \frac{1}{2}\int_0^T \|\tilde{v}(s)\|_0^2 ds\right), \tag{20}$$

where $E^{\tilde{v}}$ denotes the expectation with respect to $\gamma^{\tilde{v}}$. The expression on the right side above equals $\frac{1}{2}E^{\tilde{v}}\left(\int_0^T \|\tilde{v}(s)\|_0^2 ds\right)$, and consequently

$$R(\gamma^{\tilde{v}}\|\theta) + E^{\tilde{v}}(f(W)) = E^{\tilde{v}}\left(\frac{1}{2}\int_0^T \|\tilde{v}(s)\|_0^2 ds + f\left(\tilde{W} + \int_0^{\cdot} \tilde{v}(s)ds\right)\right).$$

It follows now from (19) that

$$-\log E\exp\{-f(W)\} \le E^{\tilde{v}}\left(\frac{1}{2}\int_0^T \|\tilde{v}(s)\|_0^2 ds + f\left(\tilde{W} + \int_0^{\cdot} \tilde{v}(s)ds\right)\right) \tag{21}$$

for every $\tilde{v}$ as above. Therefore, an application of part 2 of Lemma 3.5 yields

$$\begin{aligned}
&-\log E\exp\{-f(W)\} \\
&\le \inf_{\tilde{v}\in\mathcal{A}^{(b)}} E^{\tilde{v}}\left(\frac{1}{2}\int_0^T \|\tilde{v}(s)\|_0^2 ds + f\left(\tilde{W} + \int_0^{\cdot} \tilde{v}(s)ds\right)\right) \\
&= \inf_{v\in\mathcal{A}} E\left(\frac{1}{2}\int_0^T \|v(s)\|_0^2 ds + f\left(W + \int_0^{\cdot} v(s)ds\right)\right).
\end{aligned}$$

Proof of the lower bound: From Proposition 1.4.2 of [10] we have that

$$-\log E \exp\{-f(W)\} = R(\gamma_0 \| \theta) + E^{\gamma_0}(f(W)), \tag{22}$$

where $\frac{d\gamma_0}{d\theta} \doteq c \exp\{-f(W)\}$ a.s. and c is the normalizing constant. Define $L(t) = E(\frac{d\gamma_0}{d\theta} | \mathcal{G}_t)$. Clearly $(L(t), \mathcal{G}_t)_{0 \le t \le T}$ is a right continuous martingale bounded above and below by $e^{2\|f\|_\infty}$ and $e^{-2\|f\|_\infty}$ respectively. It follows from Theorem 2.2 that there exists $u \in \mathcal{A}^W$ such that for all $0 \le t \le T$

$$L(t) = 1 + \int_0^t \langle u(s), dW(s) \rangle.$$

We can rewrite the last equality as

$$L(t) = 1 + \int_0^t \langle \tilde{v}(s) L(s), dW(s) \rangle,$$

where $\tilde{v}(t) \doteq u(t)/L(t)$. Since $L(t)$ is a real valued continuous nonnegative martingale with $L(0) \equiv 1$, we have ([15, Lemma 7.1.4]) that

$$L(t) = \exp\left(\int_0^t \langle \tilde{v}(s), dW(s) \rangle - \frac{1}{2} \int_0^t \|\tilde{v}(s)\|_0^2 ds \right).$$

It follows from Theorem 2.1 that under γ_0

$$\tilde{W} \doteq W - \int_0^\cdot \tilde{v}(s) ds$$

is a Brownian motion with covariance Q. Therefore

$$-\log E \exp\{-f(W)\} = E^{\tilde{v}} \left(\frac{1}{2} \int_0^T \|\tilde{v}(s)\|_0^2 ds + f\left(\tilde{W} + \int_0^\cdot \tilde{v}(s) ds \right) \right). \tag{23}$$

As in the proof of part 2 of Lemma 3.5 we can approximate $\tilde{v}$ by a sequence $\{\tilde{v}^n, n \in \mathbb{N}_0\}$ of bounded elementary processes in $\mathcal{A}$ such that $E^{\tilde{v}} \left(\int_0^T \|\tilde{v}^n(s)\|_0^2 ds \right) \le 1 + E^{\tilde{v}} \left(\int_0^T \|\tilde{v}(s)\|_0^2 ds \right)$ and $E^{\tilde{v}} \left(\int_0^T \|\tilde{v}^n(s) - \tilde{v}(s)\|_0^2 ds \right) \to 0$ as $n \to \infty$. It now follows, as in the proof of part 2 of Lemma 3.5, that

$$E^{\tilde{v}} \left(\frac{1}{2} \int_0^T \|\tilde{v}^n(s)\|_0^2 ds + f(\tilde{W} + \int_0^\cdot \tilde{v}^n(s) ds) \right)$$

converges to the right side of (23) as $n \to \infty$.
Let $\epsilon > 0$ be arbitrary. We have shown that there exists $M_0 < \infty$ and an elementary process $v_0 \in \mathcal{A}$ satisfying $\|v_0(s)\|_0 \le M_0$ for all $s \in [0, T]$ such that

$$-\log E \exp\{-f(W)\} \ge E^{\tilde{v}} \left(\frac{1}{2} \int_0^T \|v_0(s)\|_0^2 ds + f\left(\tilde{W} + \int_0^\cdot v_0(s) ds \right) \right) - \epsilon.$$

The proof is now completed by applying part 1 of Lemma 3.5. ■

4 Hilbert space valued diffusions

In this section we will state an LDP for Hilbert space valued diffusions. Due to space limitations many details have been omitted. We refer the interested reader to [3] for complete proofs. Let H and K be two separable Hilbert spaces. As before, let W be a $H-$valued Brownian motion with the trace class covariance Q. We are interested in diffusion processes that take values in K. Let $\mathcal{L}(H,K)$ be the space of all bounded linear operators from H to K and denote the operator norm for $A \in \mathcal{L}(H,K)$ by $||A||_{op}$. We begin with a result proved in [19] concerning existence and strong uniqueness of solutions to $K-$valued stochastic differential equations. Let $A : K \to K$ and $G : K \to \mathcal{L}(H,K)$ be measurable mappings. Throughout this section the following conditions will be assumed on these functions. No other conditions on the stochastic differential equation will be required.

1. For every $n \in \mathbb{N}_0$ there exists $L_n < \infty$ such that for all $x, y \in K$ satisfying $||x|| \le n$ and $||y|| \le n$,

$$||A(x) - A(y)|| + ||G(x) - G(y)||_{op} \le L_n||x - y||. \tag{24}$$

2. There exists $B < \infty$ such that for all $x \in K$

$$||G(x)||_{op}^2 \le B(1 + ||x||^2) \text{ and } \langle x, A(x)\rangle \le B(1 + ||x||^2). \tag{25}$$

Theorem 4.1 *The stochastic differential equation:*

$$dX(t) = A(X(t))dt + G(X(t))dW(t),\ X(0) = x \tag{26}$$

has a unique strong solution as a $K-$valued process that satisfies

$$\int_0^T ||G(X(s))||_{op}^2\, ds < \infty \text{ and } \int_0^T ||A(X(s))||ds < \infty \text{ a.s.}$$

Although [19] explicitly considers only the case $H = K$, the proof applies to the the more general setup of the theorem above with only notational changes.

The following proposition is a consequence of Theorems 4.1. and 2.1.

Proposition 4.2 *Let $v \in A$. Then there exists a unique $K-$valued progressively measurable process $\{X^v(t); 0 \le t \le T\}$ such that*

1.

$$\theta\left\{\int_0^T ||G(X^v(s))||_{op}^2\, ds + \int_0^T ||A(X^v(s))||ds < \infty\right\} = 1,$$

2. for every $t \in [0,T]$

$$X^v(t) = x + \int_0^t G(X^v(s))v(s)ds + \int_0^t A(X^v(s))ds + \int_0^t G(X^v(s))dW(s). \tag{27}$$

Proof: Consider first the case when $v \in \mathcal{A}$ is such that

$$E\left(\exp\left\{-\int_0^T \langle v(s), dW(s)\rangle - \frac{1}{2}\int_0^T \|v(s)\|_0^2 ds\right\}\right) = 1.$$

In this case from Theorem 2.1 $\tilde{W} \doteq W + \int_0^\cdot v(s)ds$ is a $Q-$Wiener process under the probability measure

$$d\gamma^v \doteq \exp\left\{-\int_0^T \langle v(s), dW(s)\rangle - \frac{1}{2}\int_0^T \|v(s)\|_0^2 ds\right\} d\theta.$$

Let X^v be the unique solution to (26) with W replaced by $\tilde{W}$ on $(\Omega, \mathcal{F}, \gamma^v)$. Clearly X^v solves (27) a.s. θ and conditions 1 and 2 in the proposition are automatically satisfied. This shows existence. Uniqueness is argued in a similar manner using change of measure techniques. Now let $v \in \mathcal{A}$ be arbitrary and define $\tau_n \doteq \inf\{s \in [0,T] : \int_0^s \|v(s)\|_0^2 > n\} \wedge T$, where the infimum is defined to be ∞ if the set over which the infimum is taken is empty. Clearly τ_n increases to T, and in fact for every $\omega \in \Omega$ outside a θ-null set there exists a finite number $M(\omega)$ such that $\tau_n(\omega) \equiv T$ for all $n \geq M(\omega)$. Define $v^n(s) \doteq v(s)\mathcal{I}_{[0,\tau_n]}(s)$ for $0 \leq s \leq T$, where $\mathcal{I}_{[a,b]}(s)$ is the indicator function of the interval $[a,b]$. From the case considered previously we know there exist processes $\{X^{n,v}(t)\}_{0\leq t\leq T}$ that satisfy (27) with v replace by v^n. Also, by the strong uniqueness, $X^{m,v}(t) = X^{n,v}(t)$ a.s. for all $t \leq \tau_n$ and $m \geq n$. Define $X^v(t) \doteq X^{n,v}(t)$ for $t \leq \tau_n$. Clearly $X^v(t)$ is defined for $0 \leq t \leq T$ a.s. and by its construction and strong uniqueness solves (27). We leave it to the reader to verify that condition 2 is satisfied by X^v. For uniqueness, it is enough to note that if X^v and Y^v solve (27), then a.s. $X^v(t) = Y^v(t)$ for $t \leq \tau_n$ for every $n \in \mathbb{N}_0$. ■

The next proposition follows from the representation obtained in Theorem 3.6 and the fact that since X is a strong solution to (26) there exists a Borel measurable map $h : C([0,T] : H) \to C([0,T] : K)$ such that $X(t) = \pi(t) \circ h(W)$ θ - a.s., where $\pi(t)$ is the coordinate mapping on $C([0,T] : K)$.

Proposition 4.3 *Let A and G be as in Theorem 4.1 and let X be the unique solution to (26). Then for any bounded Borel function $f : C([0,T] : H) \to R$*

$$-\log E\exp\{-f(X)\} = \inf_{v\in\mathcal{A}} E\left(\frac{1}{2}\int_0^T \|v(s)\|_0^2 ds + f(X^v)\right),$$

where X^v is the unique solution to (27).

The following is the main result of this section. Define

$$\mathcal{D}_f \doteq \left\{ v \in L^2([0,T]:H_0) : f(t) = x + \int_0^t A(f(s))ds + \int_0^t G(f(s))v(s)ds \right\}.$$

Theorem 4.4 *Let G and A be as in Theorem 4.1. Let $\{X^\epsilon\}$ solve the equation*

$$dX^\epsilon(t) = A(X^\epsilon(s))ds + \sqrt{\epsilon}G(X^\epsilon(s))dW(s),\ X^\epsilon(0) = x. \tag{28}$$

Then $\{X^\epsilon\}$ satisfies the Laplace principle in $C([0,T]:K)$ with rate function

$$I_x(f) \doteq \inf_{v \in \mathcal{D}_f} \left\{ \frac{1}{2} \int_0^T \|v(t)\|_0^2 dt \right\},$$

where the infimum over the empty set is taken to be ∞.

As noted previously, space limitations prohibit giving the proof of this theorem. However, it is worth noting that given the representation in Proposition 4.3 and the tightness and convergence reults in Lemmas 3.1 and 3.2, the proof is very similar to the finite dimensional case in [2]. In fact, the only significant difference is an additional localization argument needed to deal with the fact that A and G can be unbounded as functions on H. For full details, we refer to [3].

Acknowledgments: This research supported in part by the Office of Naval Research (N00014-96-1-0279 and N00014-96-1-0276), the National Science Foundation (NSF-DMS-9704426), and the Army Research Office (DAAH04-96-1-0075).

Bibliography

[1] P. Billingsley. *Convergence of probability measures.* J.Wiley and Sons, New York, 1968.

[2] M. Boué and P. Dupuis. A variational representation for certain functionals of Brownian motion. *Annals of Probability*, to appear.

[3] A. Budhiraja and P. Dupuis. Representations and large deviations for Hilbert space valued diffusions. *Lefschetz Center for Dynamical Systems, Technical Report*, 1998.

[4] P. Chow. Large deviation problem for some parabolic Îto equations. *Communications on pure and applied mathematics*, 45:97–120, 1992.

[5] G. Da Prato and J. Zabczyk. *Stochastic Equations in Infinite Dimensions.* Cambridge University Press, 1992.

[6] A. Dembo and O. Zeitouni. *Large Deviations Techniques and Applications.* Academic Press, San Diego, Calif., 1989.

[7] J-D. Deuschel and D. Stroock. *Large Deviations.* Academic Press, San Diego, Calif., 1989.

[8] J.L. Doob. *Measure Theory.* Springer-Verlag, New York, 1994.

[9] N. Dunford and J. Schwartz. *Linear Operators Parts I, II, III.* Interscience Publishers, John Wiley and Sons, Inc., 1958.

[10] P. Dupuis and R. S. Ellis. *A Weak Convergence Approach to the Theory of Large Deviations.* John Wiley and Sons, Inc., 1997.

[11] P. Dupuis, U. Grenander, and M. I. Miller. A variational formulation of a problem in image matching. *The Quarterly of Applied Mathematics.*, to appear.

[12] R. S. Ellis. *Entropy, Large Deviations and Statistical Mechanics.* Springer-Verlag, New York, 1985.

[13] M. I. Freidlin and A. D. Wentzell. *Random Perturbations of Dynamical Systems.* Springer-Verlag, New York, 1984.

[14] K. Itô. *Foundations of Stochastic Differential Equations in Infinite Dimensional Spaces.* SIAM, Philadelphia, 1984.

[15] G. Kallianpur. *Stochastic Filtering Theory.* Springer-Verlag, 1980.

[16] G. Kallianpur and J. Xiong. Large deviation principles for a class of stochastic reaction diffusion equations. *To appear in Annals of Probability.*

[17] G. Kallianpur and J. Xiong. *Stochastic Differential Equations in Infinite Dimensional Spaces.* Institute of Mathematical Statistics, 1996.

[18] I. Karatzas and S. Shreve. *Brownian Motion and Stochastic Calculus.* Springer-Verlag, 1991.

[19] G. Leha and G. Ritter. On diffusion processes and their semigroups in Hilbert spaces with an application to interacting stochastic systems. *Annals of Probability*, 12:1077–1112, 1984.

[20] M. Metivier. *Semimartingales.* Walter de Gruyter, 1982.

[21] M. Metivier and J. Pellaumail. *Stochastic Integration.* New York, 1980.

[22] S. Peszat. Large deviation principle for stochastic evolution equations. *Probability Theory and Related Fields*, 98:113–136, 1994.

[23] R. Sowers. Large deviations for a reaction diffusion equation with non-gaussian perturbations. *Annals of Probability*, 20:504–537, 1992.

[24] S. R. S. Varadhan. *Large Deviations and Applications.* SIAM, Philadelphia, 1984.

[25] J.B. Walsh. An introduction to stochastic partial differential equations. *Ecole d'ete de Probabilites de Saint-Flour XIV, Lecture Notes in Mathematics*, 266–443, 1986.

[26] A. D. Wentzell. Large deviation principle for stochastic evolution equations. *Probability Theory and Related Fields*, 98:113–136, 1994.

2

Risk-Sensitive, Minimax, and Mixed Risk-Neutral/Minimax Control of Markov Decision Processes

Stefano P. Coraluppi
Steven I. Marcus

ABSTRACT: This paper analyzes a connection between risk-sensitive and minimax criteria for discrete-time, finite-state Markov Decision Processes (MDPs). We synthesize optimal policies with respect to both criteria, both for finite horizon and discounted infinite horizon problems. A generalized decision-making framework is introduced, leading to stationary risk-sensitive and minimax optimal policies on the infinite horizon with discounted costs.
We introduce the mixed risk-neutral/minimax objective, and utilize results from risk-neutral and minimax control to derive an information state process and dynamic programming equations for the value function. We synthesize optimal control laws both on the finite and infinite horizon, and establish the effectiveness of the controller as a tool to trade off risk-neutral and minimax objectives.

KEYWORDS: Markov Decision Processes, Risk-Sensitive Control, Minimax Control, Mixed Risk-Neutral/Minimax Control

1 Introduction

In the classical, risk-neutral approach to stochastic control, one seeks to minimize the expected total cost (or average cost) incurred in the evolution of a dynamical system. Risk-sensitive control is a generalization of this approach whereby we consider higher order moments of the probability distribution for the total cost as well. In minimax control, one is interested in minimizing the worst-case behavior of a dynamical system.
An early formulation of the risk-sensitive control problem is due to [HM72]. In the LQG setting, the problem was first studied by [Jac73], where it was found that in the risk-sensitive setting, the certainty equivalence principle does not hold in its original form. Extensions to the partially observed setting include [Whi81] and [BS85]. A somewhat surprising result is that

the conditional distribution of the state given past observations does not constitute an information state.

A good survey of work in nonlinear risk-sensitive control is given by [McE96a] and [McE96b]. The partially-observed MDP setting has been studied in [BJam], where an information state and dynamic programming equations for the value function on the finite horizon are introduced. Structural results for the value function are due to [FGMar].

Early work in minimax control of stochastic systems includes [BR71], where the connection between stochastic and deterministic descriptions of uncertainty is addressed. In the LQG setting, a connection between risk-sensitive control and H_∞ control is established in [GD88]. The connection between minimax and robust control is explored in [BB95]. In [BJam], a finite-state robust control problem is studied as the small-noise limit of a particular risk-sensitive control problem. Further connections between risk-sensitive control and a particular minimax control problem are explored in an interesting recent work (see [PJD97]).

An interesting fact both in risk-sensitive and minimax control is that in general, on the infinite horizon and with stationary discounted costs, there does not exist a stationary optimal policy. This is the case in the finite-state MDP setting as well. Dynamic programming equations in the full state observations case are derived in [CS87]. Alternate approaches to risk-sensitive control which lead to stationary optimal policies are developed in [Por75], [KP78], and [Eag75]. An alternate approach in the LQG setting is developed in [HS95]. Average cost approaches, which also lead to optimal stationary policies on the infinite horizon, are pursued in [MFHCF97], [FHH(1)], [FHH(2)], [HHM96], [HHM97].

While risk-neutral and minimax controllers are limiting special cases of the risk-sensitive controller, we show that in general the risk-sensitive controller does not effectively trade off risk-neutral and minimax objectives. We introduce a mixed risk-neutral/minimax objective, solve the associated optimal control problem, and show that it does effectively trade off risk-neutral and minimax objectives, at the cost of increased controller complexity with respect to the risk-sensitive controller. Our mixed risk-neutral/minimax formulation parallels the mixed H_2/H_∞ criterion that has been introduced in the linear systems setting. See [ZGBD94] and [DZGB94] for details.

This paper is organized as follows. In Section 2 we discuss our results on risk-sensitive and minimax control. In Section 3, we define and address the mixed risk-neutral/minimax problem, and motivate its usefulness as a tool to trade off risk-neutral and minimax objectives. We note that throughout our presentation, proofs are omitted due to space constraints. For detailed proofs of our results and further discussion, the reader is referred to [Cor97].

2 Risk-Sensitive and Minimax Control

We consider the class of discrete-time MDPs with finite state space X, finite control space U, and finite observation space Y. We denote the cardinality of these spaces by $|X|$, $|U|$, and $|Y|$. The probability transition matrix $P(u)$ is defined by $P_{ij}(u) = pr(x_{k+1} = j | x_k = i, u_k = u)$, and the observation matrix $Q(u)$ is defined by $Q_{ij}(u) = pr(y_k = j | x_k = i, u_{k-1} = u)$. We define $c_k(x_k, u_k) \geq 0$ to be the (possibly discounted) cost incurred by the system at time $k \geq 0$, given that it is in state $x_k \in X$ and that control $u_k \in U$ is used. If there is a finite horizon size N, there is a terminal cost $c_N(x_N) \geq 0$. A partial sum of costs is denoted by $C_{i,N} = \sum_{k=i}^{k=N-1} c_k(x_k, u_k) + c_N(x_N)$. The vector of terminal costs is denoted by c_N.

The *risk-neutral* objective is given by

$$J(\mu, \pi_0) = E^{\mu,\pi_0}[\sum_k c_k(x, u)], \tag{1}$$

where μ is a non-anticipative policy and π_0 is the probability distribution on the states of the system at time $k = 0$. A policy or control law is a sequence of mappings from available information to control actions. Let us denote by M this set of (non-anticipative) policies. Good references for the risk-neutral control of MDPs include [KV86], [Ber95], and [Put94].

If the state is observed, that is $y_k = x_k$, there exists a Markov policy that is optimal. In the partially observed setting, the conditional distribution of the state given past observations is an information state. It is defined recursively as follows:

$$\pi_{k+1} = r(\pi_k, u_k, y_{k+1}) = \frac{\pi_k P(u_k) \bar{Q}(y_{k+1}, u_k)}{\pi_k P(u_k) \bar{Q}(y_{k+1}, u_k) \underline{1}} \tag{2}$$

where $\bar{Q}(\cdot, \cdot)$ is a diagonal matrix with $\bar{Q}_{ii}(y, u) = pr(y_{k+1} = y | x_{k+1} = i, u_k = u)$, and $\underline{1} = [1, \ldots, 1]'$. The information state at each time k is a $|X|$-dimensional vector belonging to the space Π, the unit simplex in $\Re_+^{|X|}$, where $\Re_+$ is the set of non-negative real numbers. The value function has the important properties that it is piecewise linear and concave in π_k.

The *risk-sensitive* objective is given by

$$J^\gamma(\mu, \pi_0) = \frac{1}{\gamma} \log E^{\mu,\pi_0}[\exp(\gamma \sum_k c_k(x, u))]. \tag{3}$$

For small γ, (3) takes the form

$$J^\gamma(\mu, \pi_0) \simeq E^{\mu,\pi_0}[\sum_k c_k(x, u)] + \frac{\gamma}{2} Var^{\mu,\pi_0}[\sum_k c_k(x, u)], \tag{4}$$

and in the limit $\gamma \to 0$, (3) reverts to the risk-neutral objective (1). The parameter γ allows one to incorporate an aversion or preference for risk,

or variability in the cost incurred in the system's evolution. For $\gamma > 0$, we are penalized for variability in the cost incurred, so we say that we have a *risk-averse* objective.
An equivalently objective to (3) is given by

$$\hat{J}^{\gamma}(\mu, \pi_0) = E^{\mu,\pi_0}[\exp(\gamma \sum_k c_k(x, u))]. \quad (5)$$

In [BJam], an information state process for the MDP with respect to criterion (5) is defined, satisfying the following recursion:

$$\sigma_0^{\gamma} = \pi_0, \quad (6)$$
$$\sigma_{k+1}^{\gamma} = |Y|\sigma_k^{\gamma} D^{\gamma}(k, u_k)\bar{Q}(y_{k+1}, u_k), \quad (7)$$

where

$$D_{ij}^{\gamma}(k, u) := P_{ij}(u) \exp(\gamma c_k(i, u)), \quad (8)$$

and $\bar{Q}(\cdot,\cdot)$ is a diagonal matrix with $\bar{Q}_{ii}(y, u_k) = pr(y_{k+1} = y | x_{k+1} = i, u_k = u)$. The information state belongs to the space $R_+^{|X|}$, where R_+ is the space of non-negative real numbers. On the finite horizon, the value function associated with this information state is given by

$$S_{k,N}^{\gamma}(\sigma) := \inf_{\mu \in M} E^{\dagger}[\sigma_N^{\gamma} \cdot \exp(\gamma c_N) | \sigma_k^{\gamma} = \sigma]. \quad (9)$$

where the *exp* operator is defined component-wise, M denotes the set of non-anticipative policies, and $\dagger$ denotes a reference probability measure, under which all observations $y \in Y$ are independent and equiprobable at every time k. Dynamic programming equations for (9) are given by

$$S_{N,N}^{\gamma}(\sigma^{\gamma}) = \sigma^{\gamma} \cdot \exp(\gamma c_N), \quad (10)$$
$$S_{k,N}^{\gamma}(\sigma^{\gamma}) = \min_{u \in U} E^{\dagger}[S_{k+1,N}^{\gamma}(|Y|\sigma^{\gamma} D^{\gamma}(k, u)\bar{Q}(y_{k+1}, u)]. \quad (11)$$

It has been shown in [FGMar] that $S_{k,N}^{\gamma}(\cdot)$ is a concave and piecewise-linear function. These structural properties together with a normalized information state can be exploited to develop an algorithm to synthesize an optimal policy, similar to the algorithm given in [SS73] for risk-neutral control (with a minor correction in [Lov89]). See [Cor97] for details.
The *minimax* objective is given by

$$\bar{J}(\mu, \pi_0) = \sup_{\omega \in \Omega^{\mu}} \sum_k c_k(x_k, u_k), \quad (12)$$

where Ω^{μ} is the set of trajectories of the form $(x_0, u_0, x_1, u_1, \ldots)$ that occur with non-zero probability under policy μ. Note that, with respect to the minimax objective, the probability with which each trajectory occurs under a fixed policy μ is significant only to the extent that it is zero or non-zero.

2.1 Finite Horizon Results

The following result will be useful in establishing a connection between the risk-sensitive and minimax criteria. Its proof is similar to that of the Varadhan-Laplace Lemma, given e.g. in [BJam].

Lemma 1 (Modified Varadhan-Laplace Lemma). Let F^γ, F be real valued functions defined on a finite set Ω, where for all $\omega \in \Omega$ we have $F(\omega) = \lim_{\gamma\to\infty} F^\gamma(\omega)$. Also, let $p(\omega)$ be a nonnegative real number $\forall \omega \in \Omega$, independent of γ. Then

$$\lim_{\gamma\to\infty} \frac{1}{\gamma} \log \sum_{\omega\in\Omega} p(\omega) \exp\left[\gamma F^\gamma(\omega)\right] = \max_{\omega\in\Omega, p(\omega)\neq 0} F(\omega). \quad \square \tag{13}$$

Using Lemma 1, it can be shown that on the finite horizon, $\lim_{\gamma\to\infty} J^\gamma(\cdot,\cdot) = \bar{J}(\cdot,\cdot)$. That is, the large-risk limit of the risk-sensitive objective is the minimax objective. Let us define a statistic for the MDP by

$$s_k := \lim_{\gamma\to\infty} \frac{1}{\gamma} \log \sigma_k^\gamma, \ \forall k, \tag{14}$$

where the *log* operator is defined component-wise. Again using Lemma 1, it can be shown that the statistic satisfies the following recursion, where by $s[x]$ we mean the xth component of vector s:

$$s_0[x] = \begin{cases} 0 & \text{if } \pi_0[x] \neq 0, \\ -\infty & \text{otherwise,} \end{cases} \tag{15}$$

$$s_{k+1}[x'] = f_k(s_k, u_k, y_{k+1}). \tag{16}$$

The function $f_k(\cdot,\cdot,\cdot)$ is given by

$$f_k(s_k, u_k, y_{k+1}) = \begin{cases} \max_{x\in\tilde{X}(x',u_k)} [s_k[x] + c_k(x, u_k)] & \text{if } \begin{array}{r} \tilde{X}(x', u_k) \neq \emptyset, \\ x' \in \tilde{Y}(y_{k+1}, u_k) \end{array} \\ -\infty & \text{otherwise,} \end{cases} \tag{17}$$

where $\tilde{X}(x', u_k)$ is the set of states at time k from which, using control $u_k \in U$, there is a nonzero probability that the state of the system at time $k+1$ will be x'; $\tilde{Y}(y_{k+1}, u_k)$ is the set of states at time $k+1$ that can result in observation y_{k+1} at time $k+1$, if the control at time k is u_k.
It can be shown that the statistic and the objective (12) on the finite horizon are related by the following:

$$\bar{J}(\mu, \pi_0) = \max_{y_1,\ldots,y_N} \max_{i\in X} s_N[i]. \tag{18}$$

This motivates the following definition for the value function:

$$W_{k,N}(s) := \min_{\mu\in M} \max_{y_{k+1},\ldots,y_N} \max_{i\in X} s_N[i], \text{ where } s_k = s. \tag{19}$$

Indeed, we have

$$W_{0,N}(s_0) = W_{0,N}(\lim_{\gamma\to\infty} \frac{1}{\gamma} \log \pi_0) = \min_{\mu\in M} \bar{J}(\mu, \pi_0). \tag{20}$$

The value function at time k can be thought of as the worst case total cost incurred in the system's evolution, given an information state at time k, and given that an optimal policy is used thereafter.
The following result establishes that the statistic satisfying (15), (16) is an information state, and that there exists an optimal separated policy that can be computed by using the dynamic programming equations for the value function (19). First, we introduce the following notation for the set of all information states. Define $\tilde{R}_+^{|X|} := \{R_+, -\infty\}^{|X|}$.

Theorem 1 (Minimax Finite Horizon Dynamic Programming). The value function satisfies the following, $\forall s \in \tilde{R}_+^{|X|}$:

$$W_{N,N}(s) = \max_{i\in X} s[i], \tag{21}$$

$$W_{k,N}(s) = \min_{u\in U} \max_{y\in Y} W_{k+1,N}(f(s,u,y)). \tag{22}$$

A policy that achieves the minimum in equations (21) and (22), also achieves the minimum in (19). Furthermore, the policy is separated and is optimal with respect to (12). □

In risk-neutral and risk-sensitive control, the determination of optimal policies for partially observed MDPs typically involves the use of structural results for the value function. See [Cor97] for details. Without such results, the minimization in (11) over a continuum of information states (the unit simplex), is intractable. In the minimax control setting, the situation is greatly simplified since, on the finite horizon, we need only consider a finite number of information states. At time $k = 0$, there are $2^{|X|} - 1$ values that the information state s_0 can take, corresponding to all possible subsets of X of feasible initial states. At time $k > 0$, in the worst case there are $(2^{|X|} - 1)(|U| \cdot |Y|)^k$ feasible information states. A possible scheme for determining optimal policies on the finite horizon is the following:

1. Generate all information states of interest.
2. Use the dynamic programming equations (21), (22) to find the optimal control at each state of interest.

2.2 The Infinite Horizon Case

One way to insure that the objectives (1), (3), and (12) are bounded on the infinite horizon is to introduce a discounted cost structure. That is, we set $c_k(\cdot,\cdot) = \beta^k c(\cdot,\cdot)$, where $0 < \beta < 1$. In [CS87] it is shown that the limit

$$\hat{S}_k^\gamma(x) := \lim_{N\to\infty} \hat{S}_{k,N}^\gamma(x) \tag{23}$$

exists, for all $x \in X$ and $\gamma > 0$, where $\hat{S}_{k,N}^\gamma(x)$, $x \in X$ is the value function in the case of full state observations. Furthermore, the infinite horizon value function can be characterized as follows:

$$\hat{S}_0^\gamma = \min_{u\in U}\{D^\gamma(0,u)\hat{S}_0^{\beta\gamma}\}, \tag{24}$$

where the minimum is taken separately for each component of the vector equation. Analogously, in the partially observed setting we have the following.

Theorem 2 (Risk-Sensitive Infinite Horizon Dynamic Programming). For all $\sigma \in \Re_+^{|X|}$, and $\gamma > 0$, define

$$S_k^\gamma(\sigma) := \lim_{N\to\infty} S_{k,N}^\gamma(\sigma), \tag{25}$$

where $S_{k,N}^\gamma$ is defined in (9). The limit in (25) exists, and

$$S_0^\gamma(\sigma) = \min_{u\in U} E^\dagger[S_0^{\beta\gamma}(|Y|\sigma D^\gamma(0,u)\bar{Q}(y_1,u))]. \ \square \tag{26}$$

Proceeding in a similar fashion for the minimax objective, we introduce the following infinite horizon value function:

$$W_k(s) := \lim_{N\to\infty} W_{k,N}(s). \tag{27}$$

We can verify that the limit in (27) is well-defined by recalling that $W_{k,N} = \lim_{\gamma\to\infty} \frac{1}{\gamma}\log S_{k,N}^\gamma(\exp(\gamma s))$, and $\lim_{N\to\infty} S_{k,N}$ is well-defined. Thus

$$W_k(s) = \lim_{\gamma\to\infty} \frac{1}{\gamma}\log S_k^\gamma(\exp(\gamma s)) \tag{28}$$

We can relate the value function to the criterion (12) by taking the limit in (20) as $N \to \infty$. We obtain:

$$W_0(s_0) = \inf_{\mu\in M} \bar{J}(\mu,\pi_0) \tag{29}$$

The following result characterizes the infinite horizon value function.

Theorem 3 (Minimax Infinite Horizon Dynamic Programming). The value function (27) satisfies the following, $\forall s \in \tilde{R}_+^{|X|}$:

$$W_0(s) = \min_{u\in U}\max_{y\in Y} \beta W_0\left(\frac{f_0(s,u,y)}{\beta}\right). \ \square \tag{30}$$

In risk-neutral control, with finite state and action spaces, there exists a stationary optimal policy. In the full state observations setting, this policy can be determined through policy or value iteration techniques. Unfortunately, both in the risk-sensitive and the minimax settings, in general there does not exist a stationary optimal policy. Thus, the optimal policies satisfying equations (26) and (30) are difficult to determine. Given a tolerance bound $\epsilon > 0$, we can consider the truncation of the infinite horizon to a finite horizon of $N = \max\{\lceil \xi \rceil, 1\}$, where $\xi = \log[(1-\beta)\epsilon / \| c \|] / \log \beta$, and $\| c \| := \max_{x \in X, u \in U} |c(x,u)|$. Both for risk-sensitive and minimax criteria, if we solve the finite horizon dynamic programming equations with horizon size N and no terminal cost, and use a fixed, arbitrary policy thereafter, the resulting objectives (3) and (12) are within ϵ of optimal. See ([Cor97]) for details.

2.3 A Generalized Decision-Making Framework

Motivated by the the lack of stationary optimal policies for discounted risk-sensitive and minimax criteria, and the complexity associated with solving the dynamic programming equations (10), (11) or (21), (22) for a large horizon N, we would like to formulate optimal risk-sensitive and minimax decision-making in a more general setting, leading to stationary discounted optimal policies on the infinite horizon. An additional motivation is provided by decision theorists, many of whom argue (see e.g. [EZ89]) that a normative theory for decision-making must lead to stationary optimal policies on the infinite horizon.

Assume that the state of the MDP is observed. On the finite horizon, the value function corresponding to the risk-sensitive criterion (3) can be defined as

$$s^{\gamma}_{k,N}(i) := \min_{\mu} \frac{1}{\gamma} \log E^{\mu}[\exp(\gamma C_{k,N}) | x_k = i], \; i \in X \tag{31}$$

Recall that $C_{k,N} = \sum_{j=k}^{N-1} c_j(x_j, u_j) + c_N(x_N)$. The dynamic programming equations for (31) are given by

$$s^{\gamma}_{k,N}(i) = \min_{u \in U}\{c_k(i,u) + \frac{1}{\gamma}\log[\sum_j P_{ij}(u)\exp(\gamma s^{\gamma}_{k+1,N}(j))]\}, \tag{32}$$

$$s^{\gamma}_{N,N}(i) = c_N(i). \tag{33}$$

In the small-risk limit, $\gamma \to 0$, (32), (33) revert to the usual risk-neutral dynamic programming equations. On the infinite horizon, we have

$$s^{\gamma}_{k}(i) = \min_{u \in U}\{c_k(i,u) + \frac{1}{\gamma}\log[\sum_j P_{ij}(u)\exp(\gamma s^{\gamma}_{k+1}(j))]\}, \; k = 0, \ldots . \tag{34}$$

If $c_k(\cdot,\cdot) = \beta^k c(\cdot,\cdot)$, it can be shown that time-shifted value functions are related as follows:

$$s^{\gamma}_{k+1}(\cdot) = \beta s^{\beta\gamma}_k(\cdot). \tag{35}$$

Equation (35) also reverts to a well-known relationship in the risk-neutral case:

$$s^0_{k+1}(\cdot) = \beta s^0_k(\cdot). \tag{36}$$

A more general set of optimality equations than (32), (33) can be defined as follows:

$$\begin{aligned} h^{\gamma}_{k,N}(i) &= \min_{u\in U}\{c_k(i,u) + \\ &\qquad \frac{\beta'}{\gamma}\log[\sum_j P_{ij}(u)\exp(\gamma\beta'' h^{\gamma}_{k+1,N}(j))]\}, \qquad (37) \\ h^{\gamma}_{N,N}(i) &= c_N(i). \qquad (38) \end{aligned}$$

An interpretation for these optimality equations is that the value function at time k equals the cost incurred at time k, plus a (possibly discounted) contribution accounting for future costs. Note that if we set $\beta' = \beta'' = 1$, we revert to the classical risk-sensitive dynamic programing equations. If we set $\beta = \beta'' = 1$, we obtain the formulation that has been studies in a series of papers including [Por75] and [KP78], which we refer to as the *Porteus* formulation. A similar formulation in the LQG setting has been proposed recently in [HS95]. If we set $\beta = \beta' = 1$, we obtain the formulation introduced in [Eag75], which we refer to as the *Eagle* formulation.
On the infinite horizon, setting $c_k(\cdot,\cdot) = \beta^k c(\cdot,\cdot)$, the generalized optimality equation is given by

$$h^{\gamma}_k(i) = \min_{u\in U}\{\beta^k c(i,u) + \frac{\beta'}{\gamma}\log[\sum_j P_{ij}(u)\exp(\gamma\beta'' h^{\gamma}_{k+1}(j))]\},\ k = 0,\ldots. \tag{39}$$

Once again we obtain the classical, Porteus, and Eagle formulations as special cases of (39). A key feature of the generalized formulation (39) is that it is sufficient for one of β, β', and β'' to be less than 1, provided the others are set to 1, to insure boundedness of the value function h^{γ}_k. Thus, by setting either β' or β'' to be less than one, we can set $\beta = 1$. It can then be shown that $h^{\gamma}_k(\cdot) = h^{\gamma}(\cdot)$, that is we have a time-invariant value function, and furthermore there is a stationary policy that achieves the minimum in (16). It can further be shown that policy and value iteration techniques can be used to synthesize an optimal policy. See [Cor97] for details, and for extensions to the partial state observations setting.
The nature of the discount factors β, β', and β'' can be better understood by considering the small-risk limit, $\gamma \to 0$, of (39). We obtain the following:

$$h^0_k(i) = \min_{u\in U}\{\beta^k c(i,u) + \beta'\beta''\sum_j P_{ij}(u)h^0_{k+1}(j)\},\ k = 0,\ldots. \tag{40}$$

Note that this optimality equation is more general than the risk-neutral dynamic programming equation. On the other hand, each of the three special cases of (39) that we have considered (classical, Porteus, Eagle) is equivalent to risk-neutral control in the small-risk limit.
A generalized minimax formulation is given by

$$\bar{h}_{k,N}(i) = \min_{u\in U}\{c_k(i,u) + \beta'\beta'' \max_{j\in\tilde{X}'(i,u)} \bar{h}_{k+1,N}(j)\}, \quad (41)$$

$$\bar{h}_{N,N}(i) = c_N(i), \quad (42)$$

where once again $\tilde{X}'(i,u)$ is the set of states that the system reaches in one transition with nonzero probability, given that it is in state i and control u is used. On the infinite horizon and with $c_k(\cdot,\cdot) = \beta^k c(\cdot,\cdot)$, the generalized minimax formulation is given by

$$\bar{h}_k(i) = \min_{u\in U}\{\beta^k c(i,u) + \beta'\beta'' \max_{j\in\tilde{X}'(i,u)} \bar{h}_{k+1}(j)\}. \quad (43)$$

It can be shown that the generalized minimax formulation is the large-risk limit of the generalized risk-sensitive formulation. It follows that when $\beta = 1$ and at least one of β', β'' is less than 1, once again the value function is time-invariant, and there exists a stationary optimal policy that can be determined by policy or value iteration techniques.
An interesting consequence of introducing the additional discount parameters β' and β'' in the risk-sensitive formulation is that, unlike (32), (33), the equations (37), (38) are not dynamic programming equations. By this we mean that, in general, a policy μ^* achieving the minimum on the *r.h.s.* of equations (37), (38) does not minimize a criterion of expected utility form. More precisely, in general there does not exist a $U : \Re_+ \rightarrow \Re_+$, such that the objective $E^\mu[U(\sum_k c_k(x_k,u_k))]$ is minimized by policy μ^*. The same comment applies to the infinite horizon optimality equation (39). This can be understood in light of the axiomatic foundation of Utility Theory (see e.g. [HS84]), and some dynamic extensions discussed in [KP78].

3 Mixed Risk-Neutral/Minimax Control

The approach for defining the mixed risk-neutral/minimax objective is the following. We let a bound be given on the worst-case cost incurred, as a function of the probability distribution on $x_0 \in X$. Subject to this bound, an optimal policy is one for which the expected cost incurred is minimized. Specifically, let $\eta(\cdot) : \tilde{\Re}_+^{|X|} \rightarrow \Re_+$ be given, such that $\eta(s_0) \geq \eta_0(s_0)$, $\forall s_0 \in \tilde{\Re}_+^{|X|}$, where $\eta_0(s_0) = \min_{\mu\in M} \bar{J}(\mu, s_0)$. Recall that s_0 depends on π_0 as given by (15). Given the functional dependence of $\bar{J}(\mu,\pi_0)$ on π_0, by a slight abuse of notation we may write $\bar{J}(\mu, s_0)$ instead.
We define $M(\eta(\cdot)) \subset M$ to be the set of feasible policies such that for each initial probability distribution $\pi_0 \in \Pi$, where Π denotes the unit simplex,

the worst-case cost incurred does not exceed $\eta(s_0)$. That is, for $\mu \in M(\eta(\cdot))$ and for $s_0 \in \tilde{\Re}_+^{|X|}$,

$$\bar{J}(\mu, s_0) = \max_{\omega \in \Omega_0, p^{\mu,\pi_0}(\omega) \neq 0} \sum_k c_k(x,u)(\omega) \leq \eta(s_0). \tag{44}$$

We seek a policy $\mu^\star$ that minimizes the risk-neutral objective subject to a constraint on the allowable worst-case cost. That is, given $\eta(\cdot) \geq \eta_0(\cdot)$, an optimal policy $\mu^\star$ is one which satisfies

$$J(\mu^\star, \pi_0) = \min_{\mu \in M(\eta(\cdot))} J(\mu, \pi_0), \tag{45}$$

for all $\pi_0 \in \Pi$. Again, recall that s_0 depends on π_0 as given by (15).

3.1 Finite Horizon Results

In the general, partially observed setting, we wish now to address the task of determining an optimal policy $\mu^\star$ as defined by (45), for a given $\eta(\cdot) > \eta_0(\cdot)$. We will need to introduce an appropriate sufficient statistic, as well as dynamic programming equations for the value function.

We introduce the following statistic which combines the risk-neutral and the minimax information states. This statistic will be our candidate information state (sufficient statistic). This statistic is given by $\{g_k\}, k = 0, 1, \ldots$, where $g_k := (\pi_k, s_k)$.

We now introduce a number of definitions. Let Ω_k be the set of trajectories of the system beginning at time k. That is, elements of Ω_k are of the form $(x_k, u_k, x_{k+1}, \ldots)$. Let $p^{\mu,g,k}(\omega), \omega \in \Omega_k$, denote the probability of trajectory ω given that the information state at time k is $g_k = g$. Let $M(\eta(s_0), g, k) \subset M$ be the set of policies such that

$$\max_{\omega \in \Omega_k, p^{\mu,g,k}(\omega) \neq 0} [s[x_k] + \sum_{l=k}^{N} c_l(x_l, u_l)](\omega) \leq \eta(s_0). \tag{46}$$

That is, a policy μ is in $M(\eta(s_0), g, k)$ if the worst case cost incurred, given that the information state is g at time k, is no greater than $\eta(s_0)$. We say that an information state g is *feasible* at time k with respect to $\eta(s_0)$ if $G(\eta(s_0), g, k) \neq \emptyset$. Let $U(\eta(s_0), g, k) \subset U$ be the set of feasible controls, that is for $u \in U(\eta(s_0), g, k)$, $\exists \mu \in M(\eta(s_0), g, k)$ such that $\mu_k(g) = u$. Define the value function $V^{\eta(s_0)}$ as follows:

$$V_{k,N}^{\eta(s_0)}(g) := \min_{\mu \in M(\eta,g,k)} E[\sum_{l=k}^{N} c_l(x_l, u_l) | g_k = g]. \tag{47}$$

In particular, we have

$$V_{0,N}^{\eta(s_0)}(g_0) = \min_{\mu \in M(\eta(s_0),g_0,0)} J(\mu, \pi_0) = J(\mu^\star, \pi_0), \tag{48}$$

using equation (45).

Theorem 4 (Dynamic Programming). The value function defined in equation (47) satisfies the following dynamic programming equations for all feasible g:

$$V_{N,N}^{\eta(s_0)}(g) = \pi \cdot c_N, \tag{49}$$

$$V_{k,N}^{\eta(s_0)}(g) = \min_{u \in U(\eta(s_0),g,k)} E[c_k(x,u) + V_{k+1,N}^{\eta(s_0)}(g_{k+1}) | g_k = g]. \tag{50}$$

Furthermore, a policy $\mu_s^\star$ that achieves the minimum in equations (49) and (50) also achieves the minimum in (47). The optimal separated policy $\mu_s^\star$ is optimal within the larger class $M(\eta(s_0), (\pi_0, s_0), 0)$ of all feasible policies. □

Note that for a given time k, the feasible information states g for which we are interested in the minimization in (50) will be uncountably infinite in general. Thus we need structural results for the value function to make the minimization tractable. The following two lemmas will be useful to address this.

Lemma 2 [Ast69]. Let $f_1(x)$ and $f_2(x)$ be concave functions. The function $f(x) = \min\{f_1(x), f_2(x)\}$ is also concave. □

Lemma 3 [Ast69]. Let the function $g : \Pi \to \Re$ be concave and let A be a linear transformation from Π into Π. Then the function $f : \Pi \to \Re$ defined by

$$f(x) = \| Ax \| \cdot g(\frac{Ax}{\| Ax \|}), \; x \in \Pi, \tag{51}$$

is also concave. □

Using these two lemmas, we can show the following.

Theorem 5 (Concavity). The value function $V_{k,N}^{\eta(s_0)}(g) = V_{k,N}^{\eta(s_0)}(\pi, s)$ is concave as a function of π. □

Theorem 6 (Piecewise Linearity). The value function $V_{k,N}^{\eta(s_0)}(g) = V_{k,N}^{\eta(s_0)}(\pi, s)$ is piecewise linear as a function of π. □

The determination of optimal policies on the finite horizon can be achieved by generalizing the methodology used for risk-neutral control. A key observation is that only a finite number of values of the minimax information state will be of interest. Thus a scheme for determining optimal finite horizon policies is the following:

1. Generate all minimax information states s_k of interest, for $k = 0, 1, \ldots$ Discard those information states such that the corresponding g will be infeasible.

2. Implement a backwards dynamic programming iteration using (49), (50). For each k, $0 \leq k < N$, we must consider states $g = (\pi, s)$ such that s is generated by step (1) and $\pi \in \Pi$. For each value s, a risk-neutral methodology can be utilized.

In the worst case, the number of minimax information states will increase polynomially in the size of the horizon as follows:

$$|s_k| = (2^{|X|} - 1)(|U| \cdot |Y|)^k. \tag{52}$$

Also, in the worst case, the number of vectors needed to represent the value function $V_{k,N}(s, \cdot)$ is given by

$$|U|^{(|Y|^{N-k}-1)/(|Y|-1)}. \tag{53}$$

This can be derived by noting that the number of vectors needed at time k, or $|A_k|$, increases as follows:

$$|A_k| \leq |U| \cdot |A_{k+1}|^{|Y|}. \tag{54}$$

Thus, the controller complexity at time k is bounded by the product of (52) and (53).
A slight reduction in the complexity of the algorithm can be obtained with the following observation. Our algorithm is such that at time k, we consider separately information states $g_k = (\pi, s)$ and $g'_k = (\pi, s')$, with corresponding bounds given by $\eta(s_0)$ and $\eta(s'_0)$, respectively. Note that if $\eta(s_0)\underline{1} - s_0 = \eta(s'_0)\underline{1} - s'_0$, we need not repeat the minimization in (50) both for $s_k = s$ and $s_k = s'$. This observation leads to a more efficient procedure to determine an optimal policy in many instances, though the worst-case complexity is the same.
In the special case where the state of the system is observed, we know that both in risk-neutral and in minimax control there is a Markov policy that is optimal. Unfortunately this is not the case for the mixed control problem. The information state process $g_k, k = 0, 1, \ldots$, cannot be simplified in this manner. Intuitively, this follows from the fact that at time k, the optimal policy depends not only on the state of the system but on the total accumulated cost up to time k.
The complexity of the mixed risk-neutral/minimax controller is greater than that of the risk-sensitive controller in general. In the fully observed setting, it is well known that Markov policies are optimal for the risk-sensitive criterion (see [HM72]). In the general, partially observed setting, it has been shown (see [Cor97]) that the complexity of the risk-sensitive controller is the same as the risk-neutral controller.

3.2 The Infinite Horizon Case

We derive dynamic programming equations characterizing the value function on the infinite horizon. Defining $V_k^{\eta(s_0)} := \lim_{N\to\infty} V_{k,N}^{\eta(s_0)}$, we obtain the following, using (50):

$$V_k^{\eta(s_0)}(g) = \min_{u\in U(\eta(s_0),g,k)} E[c_k(x,u) + V_{k+1}^{\eta(s_0)}(g_{k+1})]. \tag{55}$$

Assuming $c_k(\cdot,\cdot) = \beta^k c(\cdot,\cdot)$, time-shifted value functions can be related by observing that

$$V_{k+1}^{\eta(s_0)}(\pi, s) = \beta V_k^{\frac{\eta(s_0)}{\beta}}(\pi, \frac{s}{\beta}). \tag{56}$$

Combining (55) and (56), we obtain the following equation characterizing the value function:

$$\begin{aligned} V_k^{\eta(s_0)}(\pi, s) &= \min_{u\in U(\eta(s_0),(s,\pi),0)} E[c(x,u) + \\ &\qquad \beta V_k^{\frac{\eta(s_0)}{\beta}}(r(\pi,u,y_{k+1}), \frac{f_k(s,u,y_{k+1})}{\beta})]. \end{aligned} \tag{57}$$

This equation reverts to the risk-neutral dynamic programming equation as $\eta(s_0) \to \infty$, that is as we relax the constraint on worst-case cost.
On the infinite horizon the optimal policy will be non-stationary in general, as with the minimax control problem. This fact makes it difficult to directly utilize equation (57) in constructing an optimal policy. A near-optimal policy can be determined by considering an appropriate finite horizon approximation, as established by the following result.

Theorem 7 (Finite Horizon Approximation). Consider the MDP on the infinite horizon, with initial distribution π_0 on the states, and $\eta(\cdot) > \eta_0(\cdot)$. Let $\epsilon > 0$ be given. Then $\exists N > 0$ such that the policy $\hat{g}$ satisfies the following, $\forall \pi_0 \in \Pi$:

$$J(\hat{\mu}, \pi_0) - J(\mu^\star, \pi_0) < \epsilon, \tag{58}$$

$$\bar{J}(\hat{\mu}, \pi_0) < \eta(s_0) + \epsilon, \tag{59}$$

where $\mu^\star$ is an optimal mixed risk-neutral/minimax policy with robustness bound $\eta(\cdot)$, and $\hat{\mu}$ is an optimal mixed risk-neutral/minimax finite horizon policy on $(0, N-1)$, with $c_N = 0$ and robustness bound $\eta(\cdot)$. The policy $\hat{\mu}$ on $(N, \ldots)$ is arbitrary but fixed. □

In general, it is not possible to construct a near-optimal policy through a finite horizon approximation if we require the worst-case cost to be no greater than $\eta(s_0)$. That is, it is necessary to relax the bound on worst-case cost by ϵ, in order to achieve near-optimality in performance.

3.3 Control for Performance and Robustness

We will refer to the risk-neutral objective, which indicates expected total cost incurred, as a system's *performance*. Also, we will refer to the minimax objective, which indicates worst-case total cost incurred, as a system's *robustness*. When both objectives are of interest, we would like to utilize a family of controllers that provides a good way to trade off performance and robustness. In this section we will quantify what we mean by "good", and we will examine both the mixed risk-neutral/minimax and the risk-sensitive families of controllers in this light.

It is easy to see that the mixed risk-neutral/minimax controller has risk-neutral and minimax controllers as limiting cases. Specifically, as $\eta(s_0) \to \infty$ we have $M(\eta(s_0), g_0, 0) \to M$, so that $\lim_{\eta(s_0)\to\infty} V_{0,N}^{\eta(s_0)}(g_0) = V_{0,N}(\pi_0)$ using (48). That is, as we relax the constraint on worst-case behavior we recover the risk-neutral formulation. Similarly, as $\eta(s_0) \to \eta_0(s_0)$, the mixed risk-neutral/minimax controller will be an optimal minimax controller. In general, there may be more than one minimax controller, since there may be more than one policy achieving the robustness bound $\eta_0(s_0)$. As noted earlier, the risk-sensitive controller also has risk-neutral and minimax controllers as limiting cases, as $\gamma \to 0$ and $\gamma \to \infty$, respectively.

While both families of controllers provide a link between the risk-neutral and minimax objectives, this itself is not sufficient to motivate the use of either family to trade off performance and robustness. Additional properties of the families of controllers are required. We proceed by first introducing some terminology. For the purposes of this discussion, we will not distinguish between two policies for which the performance (1) and the robustness (12) are the same. The terminology that we introduce in this section is in part borrowed from the language of portfolio theory. See [Sha70] for details.

We say that a policy μ *dominates* another policy μ' if the performance and robustness characteristics of μ are both at least as good as those of μ', for all probability distributions $\pi_0 \in \Pi$ on the initial state x_0. We say that a policy is *efficient* if it is dominated by no policy other than itself. We say that a family of policies is *efficient* if each policy in the family is itself an efficient policy. We say that a family of policies is *complete* if it is efficient, and if every efficient policy belongs to the family. We say that a family of policies is *monotonic* in a parameter if, for each probability distributions $\pi_0 \in \Pi$ on the initial state x_0, a decrease (increase) in the parameter does not worsen performance, and an increase (decrease) in the parameter does not worsen robustness.

In order to effectively determine a policy which trades off performance and robustness as desired, one would like to identify a family of policies indexed by a parameter, that is both monotonic in the parameter and efficient. Then, one can search among this class of efficient policies, adjusting the parameter in a straighforward manner. If the family is also complete, one can achieve a more precise tradeoff than if it is not.

In the family of all risk-neutral optimal policies, there is exactly one that is efficient, the policy for which criterion (12) is smallest. Likewise, in the family of all minimax optimal policies, there is exactly one efficient policy, the policy for which criterion (1) is smallest. Other policies in these families, if they exist, are not efficient, though they are not dominated by any policy not in the respective family. Clearly then, the family of all risk-neutral optimal policies is only efficient if it consists of a single policy. The same is true of the family of all minimax optimal policies. Both families are complete if and only if there is a unique risk-neutral optimal policy, a unique minimax optimal policy, and these are the same.
By construction, for a given $\eta(\cdot) \geq \eta_0(\cdot)$, the mixed risk-neutral/minimax optimal policy is efficient. It follows immediately that the family of all mixed risk-neutral/minimax policies, $\{\mu^\eta, \eta(\cdot) \geq \eta_0(\cdot)\}$, is efficient. Furthermore, the family is complete. Indeed, let μ be any efficient policy, and let $\eta(\cdot)$ be its corresponding robustness. Since there is a mixed risk-neutral/minimax policy with threshold $\eta(\cdot)$, it follows that μ must be a mixed risk-neutral/minimax optimal policy. Finally, the family is monotonic in $\eta(\cdot)$. Indeed, as we increase η, we degrade the robustness characteristics and monotonically improve performance. This follows by observing (50) and noting that for $\eta_2(s_0) > \eta_1(s_0)$, $U(\eta_1(s_0), g, k) \subset U(\eta_2(s_0), g, k)$, $\forall g, k$. It follows that $V_{k,N}^{\eta_2(s_0)}(g) \leq V_{k,N}^{\eta_1(s_0)}(g)$.
Efficient policies are deterministic. Note that since the family of mixed risk-neutral/minimax optimal policies is a (complete) family of deterministic policies, it follows that every efficient policy is deterministic. Another simple way that this property of an efficient policy can be established is the following. Let μ_{nd} be a non-deterministic policy whereby with probability p we choose the (deterministic) policy μ_{d1}, and with probability $(1-p)$ we choose the (deterministic) policy μ_{d2}, $\mu_{d2} \neq \mu_{d1}$. We will show that μ_{nd} is not efficient. Since every non-deterministic policy can be expressed as a convex combination of deterministic policies, we will conclude that every efficient policy is deterministic. Let $\pi_0 \in \Pi$ be given. Let the performance under the two deterministic policies be p_{d1} and p_{d2} respectively, and let the robustness be r_{d1} and r_{d2}. The worst-case cost incurred under policy μ_{nd} will equal the greater of that for μ_{d1} and for μ_{d2}. That is,

$$r_{nd} = \max\{r_{d1}, r_{d2}\}. \tag{60}$$

The performance under policy μ_{nd} is given by

$$p_{nd} = p \cdot p_{d1} + (1-p) \cdot p_{d2}. \tag{61}$$

If $p_{d1} < p_{d2}$, we have $p_{d1} < p_{nd}$ and $r_{d1} \leq r_{nd}$, so that μ_{nd} is dominated by μ_1 and so is not efficient. Similarly, if $p_{d2} < p_{d1}$, we have $p_{d2} < p_{nd}$ and $r_{d2} \leq r_{nd}$, so that μ_{nd} is dominated by μ_1 and so is not efficient. If $p_{d1} = p_{d2}$, then since $\mu_{d1} \neq \mu_{d2}$, it must be that $r_{d1} < r_{d2}$ or $r_{d1} < r_{d2}$. Assume *w.l.o.g.* that $r_{d1} < r_{d2}$. Then $p_{d1} = p_{nd}$ and $r_{d1} < r_{nd}$, so again we conclude that μ_{nd} is not efficient. We conclude that every efficient policy is deterministic.

A Risk-Sensitive Example. The following example shows that, in general, the family $\{\mu^{\gamma}, \gamma > 0\}$ of risk-sensitive controllers is not efficient, and is not monotonic in γ. Consider a fully observed MDP evolving on a horizon of size $N = 1$, with state space $X = \{1, 2, 3\}$, and control space $U = \{1, 2, 3\}$. Let the probability transition matrices $P(u), u \in U$ be given by

$$P(u) = \begin{bmatrix} 0.5 - \theta(u) & 2\theta(u) & 0.5 - \theta(u) \\ 0.5 - \theta(u) & 2\theta(u) & 0.5 - \theta(u) \\ 0.5 - \theta(u) & 2\theta(u) & 0.5 - \theta(u) \end{bmatrix}, \tag{62}$$

where $0 \leq \theta(u) \leq 0.5$, $u \in U$. Let the cost at time 0 be given by $c_0(x, 1) = 0$, $c_0(x, 2) = \xi$, $c_0(x, 3) = 2\xi$, with $\xi > 0$, for $x \in X$. Let the terminal cost at time 1 be given by $c_1(1) = 0$, $c_1(2) = c$, and $c_1(3) = 2c$. In particular, set $\xi = 0.01$, $c = 1$, $\theta(1) = 0$, $\theta(2) = 0.49$, $\theta(3) = 0.5$.

It is easy to verify the following. The risk-neutral policy is to select action $u = 1$ at time 0, for any initial states $x \in X$. The minimax policy is to select action $u = 3$ at time 0, for any initial states $x \in X$. For $\gamma = 0.1$, the risk-sensitive policy is to select action $u = 2$ at time 0, for any initial states $x \in X$. The risk-sensitive policy with $\gamma = 0.1$ is dominated by the risk-neutral policy, showing that the family of risk-sensitive policies is not efficient and is not monotonic in γ.

4 Conclusions

This paper overviews a number of contributions to the literature on risk-sensitive and minimax control for finite state systems. Key results include a large-risk-limit connection between risk-sensitive and minimax control in the MDP setting, infinite horizon discounted dynamic programming equations for both risk-sensitive and minimax criteria, and a generalized framework for discounted optimal decision-making, allowing for controllers that retain risk-sensitivity without sacrificing stationarity on the infinite horizon.

In addition, the paper discusses a mixed risk-neutral/minimax objective. The optimal control problem is addressed by generalizing known results for risk-neutral and minimax control. On the infinite horizon, ϵ-optimal policies are constructed by considering a sufficiently large, finite horizon approximation. The mixed risk-neutral/minimax objective provides a family of controllers that can be used to effectively trade off performance and robustness in controller design.

Acknowledgments: This research was partially supported by the National Science Foundation under Grant EEC 9402384.

References

[Ast69] K. Astrom. Optimal control of markov processes with incomplete state information ii. the convexity of the loss function. *Journal of Mathematical Analysis and Applications*, 26:403–406, 1969.

[BB95] T. Basar and P. Bernhard. *H^∞-Optimal Control and Related Minimax Design Problems.* Birkhauser, 1995.

[Ber95] D. P. Bertsekas. *Dynamic Programming and Optimal Control.* Athena Scientific, 1995.

[BJam] J. S. Baras and M. R. James. Robust and risk-sensitive output feedback control for finite state machines and hidden markov models. *Journal of Mathematical Systems, Estimation, and Control*, to appear.

[BR71] D. P. Bertsekas and I. B. Rhodes. On the minimax feedback control of uncertain systems. In *Proc. IEEE Conference on Decision and Control*, 451–455, 1971.

[BS85] A. Bensoussan and J. H. Van Schuppen. Optimal control of partially observable stochastic systems with an exponential-of-integral performance index. *SIAM Journal on Control and Optimization*, 23(4):599–613, 1985.

[Cor97] S. P. Coraluppi. *Optimal Control of Markov Decision Processes for Performance and Robustness.* PhD thesis, University of Maryland, 1997.

[CS87] K. J. Chung and M. J. Sobel. Discounted mdp's: Distribution functions and exponential utility maximization. *SIAM Journal on Control and Optimization*, 25:49–62, 1987.

[DZGB94] J. Doyle, K. Zhou, K. Glover, and B. Bodenheimer. Mixed H_2 and H_∞ performance objectives II: optimal control. *IEEE Transactions on Automatic Control*, 39(8):1575–1587, 1994.

[Eag75] J. N. Eagle. *A Utility Criterion for the Markov Decision Process.* PhD thesis, Stanford University, 1975.

[EZ89] L. G. Epstein and S. E. Zin. Substitution, risk aversion, and the temporal behavior of consumption and asset returns: A theoretical framework. *Econometrica*, 57(4):937–969, 1989.

[FGMar] E. Fernández-Gaucherand and S. I. Marcus. Risk-sensitive optimal control of hidden markov models: Structural results. *IEEE Transactions on Automatic Control*, 42(10): 1418-1422, 1997.

[FHH(1)] W. H. Fleming and D. Hernández-Hernández. Risk-sensitive control of finite state machines on an infinite horizon II. Technical report, Division of Applied Mathematics, Brown University.

[FHH(2)] W. H. Fleming and D. Hernández-Hernández. Risk-sensitive control of finite state machines on an infinite horizon I. *SIAM Journal on Control and Optimization*, 35(5): 1790-1810, 1997.

[GD88] K. Glover and J. C. Doyle. State-space formulae for all stabilizing controllers that satisfy an H_∞-norm bound and relations to risk sensitivity. *Systems and Control Letters*, 11:167–172, 1988.

[HHM96] D. Hernández-Hernández and S. I. Marcus. Risk-sensitive control of markov processes in countable state space. *Systems and Control Letters*, 29:147–155, 1996.

[HHM97] D. Hernández-Hernández and S. I. Marcus. Existence of risk sensitive optimal stationary policies for controlled markov processes. *Applied Mathematics and Optimization*, to appear.

[HM72] R. A. Howard and J. E. Matheson. Risk-sensitive markov decision processes. *Management Science*, 18(7):356–369, 1972.

[HS84] D. P. Heyman and M. J. Sobel. *Stochastic Models in Operations Research, Vol. II: Stochastic Optimization*. McGraw-Hill, 1984.

[HS95] L. P. Hansen and T. J. Sargent. Discounted linear exponential quadratic gaussian control. *IEEE Transactions on Automatic Control*, 40:968–971, 1995.

[Jac73] D. H. Jacobson. Optimal stochastic linear systems with exponential performance criteria and their relation to deterministic differential games. *IEEE Transactions on Automatic Control*, 18(2):124–131, 1973.

[KP78] D. M. Kreps and E. L. Porteus. Temporal resolution of uncertainty and dynamic choice theory. *Econometrica*, 46(1):185–200, 1978.

[KV86] P. R. Kumar and P. Varaiya. *Stochastic Systems: Estimation, Identification, and Adaptive Control*. Prentice-Hall, 1986.

[Lov89] W. S. Lovejoy. A note on exact solution of partially observed markov decision processes. Technical report, Graduate School of Business, Stanford University, 1989.

[MFHCF97] S. Marcus, E. Fernandez-Gaucherand, D. Hernandez-Hernandez, S. Coraluppi, P. Fard. Risk-sensitive markov decision processes. *Systems and Control in the Twenty-First Century*, 263-279. C. I. Byrnes, et. al. (eds.), Birkhauser, 1997.

[McE96a] W. M. McEneaney. Risk-sensitive control of nonlinear systems. *SIAM Activity Group on Control and System Theory Newsletter*, 4(1), 1996.

[McE96b] W. M. McEneaney. Risk-sensitive control of nonlinear systems. *SIAM Activity Group on Control and System Theory Newsletter*, 4(2), 1996.

[PJD97] I. Peterson, M. James, and P. Dupuis. Minimax optimal control of stochastic uncertain systems with relative entropy constraints. In *Proc. IEEE Conference on Decision and Control*, San Diego, CA, Dec. 1997.

[Por75] E. Porteus. On the optimality of structured policies in countable stage decision processes. *Management Science*, 22(2):148–157, 1975.

[Put94] M. L. Puterman. *Markov Decision Processes: Discrete Stochastic Dynamic Programming.* John Wiley and Sons, 1994.

[Sha70] W. Sharpe. *Portfolio Theory and Capital Markets.* McGraw-Hill, 1970.

[SS73] R. D. Smallwood and E. J. Sondik. The optimal control of partially observable markov processes over a finite horizon. *Operations Research*, 21:1071–1088, 1973.

[Whi81] P. Whittle. Risk-sensitive linear/quadratic/gaussian control. *Advances in Applied Probability*, 13:764–777, 1981.

[ZGBD94] K. Zhou, K. Glover, B. Bodenheimer, and J. Doyle. Mixed H_2 and H_∞ performance objectives I: robust performance analysis. *IEEE Transactions on Automatic Control*, 39(8):1564–1574, 1994.

3

Partially Observed Control Problems with Multiplicative Cost

Daniel Hernández-Hernández

ABSTRACT: In this paper the risk sensitive optimal control problem of discrete time partially observed systems on an infinite horizon is considered. Defining an appropriate information state we formulate an equivalent control problem with completely observed state dynamics, which is solved using dynamic games and dynamic programming methods.

KEYWORDS: risk sensitive control; stochastic dynamic games; information state

1 Introduction

In this paper we consider risk sensitive control problems of discrete time partially observed systems on an infinite horizon. This problem is solved introducing an information state, and defining an equivalent control problem with completely observed state dynamics. Then, using dynamic programming methods, we obtain an optimal control, which is separated through the information state. The dynamic programming equation is a nonlinear eigenvalue problem, and the existence of solution is proved using the vanishing discount method for stochastic dynamic games. We refer to Fleming and Hernández-Hernández, where similar results are given for the discrete (finite state machine) case.

The use of game theory to solve this problem is not surprising, and it has been explored extensively in the study of risk sensitive control problems by several authors, e.g. Baras and James (1997), Fleming and McEneaney (1992),(1995), Dai Pra, Meneghini and Runggaldier (1996), Hernández-Hernández and Marcus (1996). It is also worth mention that most of the literature in this field has been inspired by Whittle's pioneer contributions (see Whittle (1990)). For linear systems and exponential of the sum of quadratic costs, the control problem has been studied by Jacobson (1973), while extensions to the partially observed setting are due to Bensoussan and Van Schuppen (1985). See also Marcus et al (1997), where some recent results on risk sensitive Markov decision processes are surveyed.

We begin Section 2 with the formulation of the problem, and introduction of the main ideas. In Section 3 the ergodic properties of the information state are studied for a noncontrolled system. Establishing some contractive properties (see Theorem 3.1) of the information state dynamics, it is possible to prove, under suitable conditions, the existence of a unique invariant measure, while the Law of Large Numbers gives an explicit formula for the rate of growth of the cost functional. For finite state space models similar results have been obtained by Furstenberg and Kifer (1983). Finally, in Section 4 we present the solution to the risk sensitive control problem.

Notation. Given a generic metric space $\mathcal{X}$, its Borel σ-algebra is denoted by $\mathcal{B}(\mathcal{X})$; $P(\mathcal{X})$ stands for the set of probability measures on $\mathcal{X}$, endowed with the weak topology, and $\mathcal{M}(\mathcal{X})$ denotes the set of signed measures on $\mathcal{X}$. Given $\eta \in \mathcal{M}(\mathcal{X})$, its total variation norm is denoted by $\|\eta\|$. Finally, $C(\mathcal{X})$ is the space of real valued continuous functions on $\mathcal{X}$.

2 Preliminaries

In this section we shall formulate a risk sensitive control problem for partially observed systems. Let $(\Omega, \mathcal{F})$ be an underlying measurable space, and $\{x_t\}$ be a Markov process with state space E, stationary transition kernel $P^u(x, dz)$, and initial distribution θ_0, where E is a compact separable metric space and u is a control parameter with $u \in U$ a compact metric space.

Let $\{w_t\}$ be i.i.d. random variables with values in $\mathbb{R}^d$ and define

$$y_t = h(x_t) + w_t, \quad t = 1, 2 \ldots,$$

where $h : E \to \mathbb{R}^d$. This process is measured and is called the observed process of the system. We denote by $\mathcal{U}$ the set of admissible control sequences $\widetilde{u} = \{u_t\}$, where u_t is a $U-$ valued random variable adapted to the observations σ-field $\mathcal{Y}^t = \sigma(y_1, \ldots, y_t)$, with $\mathcal{Y}^0 = \{\Omega, \emptyset\}$.

Given the cost per unit time function $l : E \times U \to \mathbb{R}$, for each admissible control $\widetilde{u} \in \mathcal{U}$, we define the cost funtional (to be minimized) by

$$J(\widetilde{u}) = \limsup_{T \to \infty} \frac{1}{T} \cdot \frac{1}{\gamma} \log E_{\theta_0}^{\widetilde{u}} \exp\{\gamma \sum_{t=0}^{T-1} l(x_t, u_t)\}, \tag{2.1}$$

and the partial observed control problem is to find $\widetilde{u}^* \in \mathcal{U}$ such that

$$J(\widetilde{u}^*) = \inf_{\widetilde{u} \in \mathcal{U}} J(\widetilde{u}) =: \Lambda.$$

Here $\gamma > 0$ is the risk averse parameter, and analogously to $E_{\theta_0}^{\widetilde{u}}$, $\mathcal{P}_{\theta_0}^{\widetilde{u}}$ denotes the probability measure induced by $\{x_t\}$, $\{y_t\}$ on Ω, given $\widetilde{u} \in \mathcal{U}$, and the initial distribution θ_0. Without loss of generality we shall set $\gamma = 1$. Throughout we make the following assumptions.

(A1) For each $u \in U$, $P^u(x, \cdot)$ is Feller.

(A2) h is continuous.

(A3) l is nonnegative, bounded and continuous.

(A4) The random variables $\{w_t\}$ have a strictly positive continuous density g.

(A5) There is $\nu \in P(E)$ and $p^u(x, z)$ such that

(a) $P^u(x, A) = \int_A p^u(x, z)\nu(dz)$
(b) $0 < c \leq p^u(x, z) \leq c' \quad \forall u \in U,\ x, z \in E$,
and let $\lambda := \frac{c}{c'}$.

Remark 2.1. When E is finite P^u is the matrix of transitions from x_t to x_{t+1} when control u_t is used. We note that if the components of P^u are bounded away from zero uniformly on $u \in U$ then Assumption (A5) is satisfied.

In order to solve the above control problem we shall introduce an information state $\theta \in P(E)$, and formulate an equivalent completely observed risk sensitive control problem with state variable θ. Moreover, using dynamic programming we shall find an optimal control, which is separated through the information state.
Let $\tilde{u} \in \mathcal{U}$ and define $\mathcal{G}^T \subset \mathcal{F}$ the σ-field $\sigma(x_0, \ldots, x_T, y_1, \ldots, y_T)$. Then (cf. James, Baras and Elliott (1994)), there exists a probability measure $\mathcal{P}^+$ on Ω such that $\mathcal{P}^+|_{\mathcal{G}^T}$ is equivalent to $\mathcal{P}^{\tilde{u}}|_{\mathcal{G}^T}$ with

$$\frac{d\mathcal{P}^{\tilde{u}}}{d\mathcal{P}^+}\Big|_{\mathcal{G}^T} = \overline{L}_T =: \prod_{t=1}^{T} \frac{r(x_t, y_t)}{g(y_t)},$$

where $r(x, y) := g(y - h(x))$. This follows from the Carathéodory Extension Theorem using the fact that $\overline{L}_T^{-1}$ is a $\mathcal{G}^T-$ martingale with mean one. Further, under $\mathcal{P}^+$, $\{y_t\}$ are i.i.d. with density g, independent of $x_{t'}$, $t' \leq t$, and x_t has the same transition kernel.
Now, for $A \in \mathcal{B}(E)$, define $\sigma_0 = \theta_0$, and

$$\begin{aligned}\sigma_T(A) &= E^+\{\overline{L}_T I_A(x_T)\exp\{\sum_{t=0}^{T-1} l(x_t, u_t)\}|\mathcal{Y}^T\} \\ &= \int\int \frac{1}{g(y_T)} I_A(z) r(z, y_T) e^{l(x, u_{T-1})} p^{u_{T-1}}(x, dz)\sigma_{T-1}(dx),\end{aligned}$$

and the information state $\theta_T \in P(E)$ by

$$\theta_T(A) = \frac{\sigma_T(A)}{|\sigma_T|}, \tag{2.2}$$

where $|\sigma_T| := \sigma_T(\Omega)$.
The recursions of θ_T and $|\sigma_T|$ are analogous to those given for the finite state model (see Fleming and Hernández-Hernández)). Next, given $y \in \mathbb{R}^d$, $u \in U$, $\theta \in P(E)$, we define the positive measure

$$F_1(\theta, y, u)(A) := \int\int \frac{r(z,y)}{g(y)} I_A(z) e^{l(x,u)} p^u(x,z)\nu(dz)\theta(dx),$$

and $F(\theta, y, u) \in P(E)$ by

$$F(\theta, y, u)(A) = \frac{F_1(\theta, y, u)(A)}{|F_1(\theta, y, u)|}.$$

Then,

$$\theta_{T+1}(A) = \frac{F_1(\theta_T, y_{T+1}, u_T)(A)}{|F_1(\theta_T, y_{T+1}, u_T)|}, \tag{2.3}$$

and

$$|\sigma_{T+1}| = |\sigma_T|\, |F_1(\theta_T, y_{T+1}, u_T)|. \tag{2.4}$$

Furthermore, defining $G(\theta, y, u) := \log |F_1(\theta, y, u)|$, we get

$$\begin{aligned} V(\theta_0, T; \widetilde{u}) &:= E^+ \exp\left[\sum_{t=0}^{T-1} l(x_t, u_t)\right] \overline{L}_T \\ &= E^+|\sigma_T| \qquad\qquad (2.5) \\ &= E^+ \exp\{\sum_{t=0}^{T-1} G(\theta_t, y_{t+1}, u_t)\}. \end{aligned}$$

Therefore, $J(\widetilde{u})$ (see (2.1)) can be expressed in terms of the completely observed information state θ_t by

$$J(\widetilde{u}) = \limsup_{T\to\infty} \frac{1}{T} \log E_\theta^+ \exp\{\sum_{t=1}^{T} G(\theta_{t-1}, y_t, u_{t-1})\}. \tag{2.6}$$

3 Ergodic behavior of the information state

Throughout this section we study the asymptotic behavior of $|\sigma_T|$ (see (2.4), and the ergodic properties of the information state θ_T for the noncontrolled system. Therefore, for easy of notation, we shall drop the parameter u from the definitions of P, F_1, F, etc.
The following result is fundamental to obtain the main theorems of this paper.
Theorem 3.1. *Let θ_t, $\widetilde{\theta}_t$ be the information states defined in (2.2) with initial distributions θ and $\widetilde{\theta}$ respectively. Then,*

$$||\theta_t - \widetilde{\theta}_t|| \le 2\lambda^{-1}(1-\lambda^2)^t||\theta - \widetilde{\theta}||, \quad \textit{for } t = 1, 2, \ldots, \tag{3.1}$$

with λ as in (A5).
For finite state space models there is a similar result, which is based in the theory of random products of matrices (see Kaijser(1978), Arapostathis and Marcus (1990)). However, for general spaces some technical details are different, and we are going to give the proof. Before that, we shall introduce some preliminary results.
Given $\theta \in P(E)$, $y \in \mathbb{R}^d$, define $\theta^* \in P(E)$ by

$$\theta^*(B) = \frac{\int_B \int r(z,y)e^{l(x)}p(x,z)\nu(dz)\theta(dx)}{\int\int r(z,y)e^{l(x)}p(x,z)\nu(dz)\theta(dx)}. \tag{3.2}$$

Then, it follows that

$$F(\theta,y)(A) = \int K_1 d\theta^*, \tag{3.3}$$

where

$$K_1(\cdot;y,A) = \frac{\int r(z,y)I_A(z)p(\cdot,z)\nu(dz)}{\int r(z,y)p(\cdot,z)\nu(dz)}.$$

Lemma 3.2. *Let θ, $\widetilde{\theta} \in P(E)$, and $y \in \mathbb{R}^d$. Then,*

$$\|F(\theta,y) - F(\widetilde{\theta},y)\| \le 2(1-\lambda^2)(\theta^* - \widetilde{\theta}^*)(E^+),$$

where E^+ is the positive set of the Hahn decomposition of $\theta^* - \widetilde{\theta}^*$.
Proof. Firstly, we recall that

$$\|F(\theta,y) - F(\widetilde{\theta},y)\| = 2 \sup_{A\in\mathcal{B}(E)} |F(\theta,y)(A) - F(\widetilde{\theta},y)(A)|.$$

Now, let $A \in \mathcal{B}(E)$ arbitrary, but fixed. Then, for any x, $x' \in E$,

$$\begin{aligned}\int r(z,y)I_A(z)p(x,z)\nu(dz) &\ge \lambda \int r(z,y)I_A(z)p(x',z)\nu(dz)\\ &= \lambda K_1(x';y,A)\int r(z,y)p(x',z)\nu(dz)\\ &\ge \lambda^2 K_1(x';y,A)\int r(z,y)p(x,z)\nu(dz).\end{aligned}$$

Therefore,

$$sp(K_1) := \sup_{x\in E} K_1(x;y,A) - \inf_{x'\in E} K_1(x';y,A) \le 1-\lambda^2, \tag{3.4}$$

and from (3.3) we get

$$\begin{aligned}F(\theta,y)(A) - F(\widetilde{\theta},y)(A) &\le sp(K_1)(\theta^* - \widetilde{\theta}^*)(E^+)\\ &= \frac{1}{2}sp(K_1)\|\theta^* - \widetilde{\theta}^*\|.\end{aligned}$$

Thus, from (3.4) and interchanging the roles of θ and $\widetilde{\theta}$ the lemma follows. □

Lemma 3.3. *Let θ^* and $\widetilde{\theta}^*$ be as in Lemma 3.2. Then, $(\theta^* - \widetilde{\theta}^*)(E^+) \leq \lambda^{-1}\|\theta - \widetilde{\theta}\|$.*

Proof. Given $y \in \mathbb{R}^d$, for each $x \in E$, define

$$\rho(x) = \int r(z,y)e^{l(x)}p(x,z)\nu(dz).$$

Hence, from (3.2), we obtain

$$\begin{aligned}
(\theta^* - \widetilde{\theta}^*)(E^+) &= \theta^*(E^+) - \widetilde{\theta}^*(E^+) \\
&= \frac{\int_{E^+}\rho d\theta}{\int \rho d\theta} - \frac{\int_{E^+}\rho d\widetilde{\theta}}{\int \rho d\widetilde{\theta}} \\
&= \left[\frac{\int_{E^+}\rho d\theta}{\int \rho d\theta} - \frac{\int_{E^+}\rho d\widetilde{\theta}}{\int \rho d\theta}\right] + \left[\frac{\int_{E^+}\rho d\widetilde{\theta}}{\int \rho d\theta} - \frac{\int_{E^+}\rho d\widetilde{\theta}}{\int \rho d\widetilde{\theta}}\right]. \qquad (3.5)
\end{aligned}$$

The first term in the r.h.s. yields,

$$\begin{aligned}
\frac{1}{\int \rho d\theta}\left[\int_{E^+}\rho d\theta - \int_{E^+}\rho d\widetilde{\theta}\right] &= \frac{1}{\int \rho d\theta}\int_{E^+}\rho(x)(\theta - \widetilde{\theta})(dx) \\
&= \frac{1}{\int \rho d\theta}\int_{E}\rho(x)(\theta - \widetilde{\theta})^+(dx) \\
&\leq \lambda^{-1}(\theta - \widetilde{\theta})^+(E) \qquad (3.6) \\
&= \frac{1}{2}\lambda^{-1}\|\theta - \widetilde{\theta}\|.
\end{aligned}$$

In the inequality (3.6) we have used the fact that for any probability measures $\vartheta, \widetilde{\vartheta} \in P(E)$,

$$\frac{\int \rho d\widetilde{\vartheta}}{\int \rho d\vartheta} \leq \lambda^{-1}. \qquad (3.7)$$

On the other hand, using again (3.7), the second term in (3.5) yields,

$$\begin{aligned}
\int_{E^+}\rho d\widetilde{\theta} \cdot \frac{\int \rho d\widetilde{\theta} - \int \rho d\theta}{\int \rho d\theta \int \rho d\widetilde{\theta}} &\leq \frac{\int \rho d(\widetilde{\theta} - \theta)^+}{\int \rho d\theta} \\
&\leq \lambda^{-1}(\widetilde{\theta} - \theta)^+(E) \\
&= \frac{1}{2}\lambda^{-1}\|\theta - \widetilde{\theta}\|. \qquad (3.8)
\end{aligned}$$

We complete the proof adding (3.6) and (3.8). □

Proof of Theorem 3.1. For $t = 1$, its proof follows from Lemmas 3.2 and

3.3. Now, let $y_1, \dots y_{t+1}$ be a sequence in $\mathbb{R}^d$, $A \in \mathcal{B}(E)$, and define

$$k_{t+1}(x_0; y_1, \dots, y_{t+1}, A) := \int\int \cdots \int r(x_{t+1}, y_{t+1}) e^{l(x_t)} I_A(x_{t+1}) \cdot$$

$$\cdot p(x_t, x_{t+1}) \nu(dx_{t+1}) r(x_t, y_t) e^{l(x_{t-1})} \cdot$$

$$\cdot p(x_{t-1}, x_t) \nu(dx_t) \cdots r(x_1, y_1) p(x_0, x_1) \nu(dx_1),$$

$$k_t(x_1; y_2, \dots, y_{t+1}, A) := \int\int \cdots \int r(x_{t+1}, y_{t+1}) e^{l(x_t)} I_A(x_{t+1}) \cdot$$

$$\cdot p(x_t, x_{t+1}) \nu(dx_{t+1}) r(x_t, y_t) e^{l(x_t)} \cdot$$

$$\cdot p(x_{t-1}, x_t) \nu(dx_t) \cdots r(x_2, y_2) e^{l(x_1)} p(x_1, x_2) \nu(dx_2),$$

and

$$K_{t+1}(x_0; y_1, \dots, y_{t+1}) := \frac{k_{t+1}(x_0; y_1, \dots, y_{t+1}, A)}{k_{t+1}(x_0; y_1, \dots, y_{t+1}, E)}.$$

Also, we define $\theta' \in P(E)$ by

$$\theta'(B) = \frac{\int_B \int \cdots \int r(x_{t+1}, y_{t+1}) e^{l(x_t)} p(x_t, x_{t+1}) \nu(dx_{t+1}) \cdots r(x_1, y_1) e^{l(x_0)}}{\int\int \cdots \int r(x_{t+1}, y_{t+1}) e^{l(x_t)} p(x_t, x_{t+1}) \nu(dx_{t+1}) \cdots r(x_1, y_1) e^{l(x_0)}} \cdot$$

$$\cdot \frac{p(x_0, x_1) \nu(dx_1) \theta(dx_0)}{p(x_0, x_1) \nu(dx_1) \theta(dx_0)}.$$

Then, it follows that

$$\theta_{t+1}(A) = \int K_{t+1} d\theta'. \tag{3.9}$$

Now, for each $x_0, x_0' \in E$, we have

$$K_{t+1}(x_0; y_1, \dots, y_{t+1}) - K_{t+1}(x_0'; y_1, \dots, y_{t+1}) =$$

$$= \frac{1}{k_{t+1}(x_0; y_1, \dots, y_{t+1}, E)} \cdot$$

$$\cdot \int k_t(x_1; y_2, \dots, y_{t+1}, A) \cdot r(x_1, y_1) p(x_0, x_1) \nu(dx_1)$$

$$- \frac{1}{k_{t+1}(x_0'; y_1, \dots, y_{t+1}, E)} \cdot$$

$$\cdot \int k_t(x_1; y_2, \dots, y_{t+1}, A) r(x_1, y_1) p(x_0', x_1) \nu(dx_1)$$

$$= \int \left\{ \frac{r(x_1, y_1) p(x_0, x_1)}{k_{t+1}(x_0; y_1, \dots, y_{t+1}, E)} \cdot k_t(x_1; y_2, \dots, y_{t+1}, E) - \right.$$

$$\left. - \frac{r(x_1, y_1) p(x_0', x_1)}{k_{t+1}(x_0', y_1, \dots, y_{t+1}, E)} \cdot k_t(x_1; y_2, \dots, y_{t+1}, E) \right\} \cdot$$

$$\cdot \frac{k_t(x_1; y_2, \dots, y_{t+1}, A)}{k_t(x_1; y_2, \dots, y_{t+1}, E)} \nu(dx_1). \tag{3.10}$$

However,

$$\int \frac{k_t(x_1; y_2, \ldots, y_{t+1}, E) r(x_1, y_1) p(x_0, x_1) \nu(dx_1)}{k_{t+1}(x_0; y_1, \ldots, y_{t+1}, E)} = 1 \tag{3.11}$$

Then, proceeding as in the proof of Lemma 3.2,

$$\frac{k_t(x_1; y_2, \ldots, y_{t+1}, E) r(x_1, y_1) p(x_0, x_1)}{k_{t+1}(x_0; y_1, \ldots, y_{t+1}, E)} \geq$$

$$\lambda^2 \frac{k_t(x_1; y_2, \ldots, y_{t+1}, E) r(x_1, y_1) p(x_0', x_1)}{k_{t+1}(x_0'; y_1 \ldots, y_{t+1}, E)} \tag{3.12}$$

Let $E' \subset E$ be the set of $x_1 \in E$ such that the integrand in (3.10) is nonnegative. Thus, (3.11) and (3.12) imply that

$$\begin{aligned} 0 &\leq \int_{E'} \left\{ \frac{r(x_1, y_1) p(x_0, x_1)}{k_{t+1}(x_0; y_1, \ldots, y_{t+1}, E)} \cdot k_t(x_1; y_2, \ldots, y_{t+1}, E) - \right. \\ &\quad \left. - \frac{r(x_1, y_1) p(x_0', x_1)}{k_{t+1}(x_0', y_1, \ldots, y_{t+1}, E)} \cdot k_t(x_1; y_2, \ldots, y_{t+1}, E) \right\} \nu(dx_1) \\ &= \int_{E \setminus E'} \left\{ \frac{r(x_1, y_1) p(x_0, x_1)}{k_{t+1}(x_0; y_1, \ldots, y_{t+1}, E)} \cdot k_t(x_1; y_2, \ldots, y_{t+1}, E) - \right. \\ &\quad \left. - \frac{r(x_1, y_1) p(x_0', x_1)}{k_{t+1}(x_0', y_1, \ldots, y_{t+1}, E)} \cdot k_t(x_1; y_2, \ldots, y_{t+1}, E) \right\} \nu(dx_1) \\ &\leq (1 - \lambda^2). \end{aligned}$$

Therefore,

$$\begin{aligned} spK_{t+1}(\cdot; y_1, \ldots, y_{t+1}) &\leq (1 - \lambda^2) sp\{ \frac{k_t(\cdot; y_2, \ldots, y_{t+1}, A)}{k_t(\cdot; y_2, \ldots, y_{t+1}, E)} \} \\ &= (1 - \lambda^2) spK_t(\cdot, y_2, \ldots, y_{t+1}) \\ &\leq (1 - \lambda^2)^{t+1}, \end{aligned}$$

by induction hypothesis. Finally, using (3.7) and (3.9) and paralleling Lemmas 3.2 and 3.3, we complete the proof.

□

Now, let $Q(\theta, \cdot)$ be the transition kernel of the information state, which in view of (2.3) it is given by

$$Q(\theta, B) = \int I_B(F(\theta, y)) g(y) dy. \tag{3.13}$$

We denote by $Q^t(\theta, \cdot)$ the distribution of θ_t with initial condition θ. Note that, from (A1)–(A5), it follows that θ_t is Feller.

Proposition 3.4. *(a) There exists a probability measure $\pi \in P(P(E))$ such that, for any $\theta \in P(E)$, $Q^t(\theta, \cdot)$ converges weakly to π, and it is the unique invariant measure of Q.*

(b) For every $\theta \in P(E)$,

$$\lim_{T\to\infty} \frac{1}{T} E_\theta \sum_{t=0}^{T-1} f(\theta_t) \to \int f d\pi \qquad \pi - \text{a.s.}$$

for every $f \in L_1(P(E), \mathcal{B}(P(E)), \pi)$.

Sketch of the proof. The first part follows immediatly from Lemmas 2.1, 2.2 in Kaijser (1978) and Theorem 3.1, while the second part is a straightforward application of the Ergodic Theorem (see e.g. Yosida (1980)). □
Now, since $\{y_t\}_{t=1}^{\infty}$ are i.i.d. random variables under P^+, we have that the process

$$z_0 = (\theta_0, y_1),\ z_1 = (\theta_1, y_2), \ldots, \text{ with values in } P(E) \times \mathbb{R}^d,$$

is a Markov process, and in view of Proposition 3.4.(a), it has a unique invariant measure ζ. Moreover, it is given by

$$\int f d\zeta = \int f(\theta, y) g(y) dy\ \pi(d\theta), \quad f \in C(P(E) \times \mathbb{R}^d).$$

On the other hand, from (2,4), we have that

$$\begin{aligned} \frac{1}{T}\sum_{t=0}^{T-1} G(\theta_t, y_{t+1}) &= \frac{1}{T}\sum_{t=0}^{T-1} \log |F_1(\theta_t, y_{t+1})| \\ &= \frac{1}{T}\sum_{t=0}^{T-1} \log \frac{|\sigma_{t+1}|}{|\sigma_t|} \\ &= \frac{1}{T} \log |\sigma_T|. \end{aligned}$$

The next proposition follows from the ergodicity of the process z_t and the Ergodic Theorem.

Proposition 3.5. *Let μ be the common distribution of y_t. Then, for each $\theta \in P(E)$,*

$$\frac{1}{T} E_{\theta,\mu} \sum_{t=0}^{T-1} G(z_t) \to \int\int G(\theta, y) g(y) dy\ \pi(d\theta) \quad \zeta - \text{a.s.}$$

provided $G \in L_1(P(E) \times \mathbb{R}^n, \mathcal{B}(P(E) \times \mathbb{R}^n), \zeta)$.

4 Risk sensitive control problem

In this section we shall present the solution to the risk sensitive control problem following Fleming and Hernández-Hernández. The approach we follow depends in a fundamental way on a duality result.

Let μ be the common distribution of y_t under $\mathcal{P}^+$, and define the relative entropy function $I(\cdot||\mu) : P(\mathbb{R}^d) \to \mathbb{R} \cup \{+\infty\}$ by

$$I(\eta||\mu) = \begin{cases} \int \log(\frac{d\eta}{d\mu}(x))\eta(dx) & \text{if } \eta << \mu \text{ and } \log \frac{d\eta}{d\mu} \in L_1(\eta) \\ +\infty & \text{otherwise} \end{cases}$$

We denote by $\Gamma \subset P(\mathbb{R}^d)$ the set of measures η such that $I(\eta||\mu)$ is finite. The next lemma establishes, using a Legendre-type transformation, the duality relationship between the relative entropy function and the logarithmic moment generating function.

Lemma 4.1 [Dupuis and Ellis (1997), Proposition II.4.2]. *Let ψ be a bounded function defined on $\mathbb{R}^d$. Then,*

$$\log \int e^{\psi(z)}\mu(dz) = \sup_{\eta \in P(\mathbb{R}^d)} \{\int \psi(z)\eta(dz) - I(\eta||\mu)\},$$

and the supremum is attained at the unique probability measure η^* defined by

$$\frac{d\eta^*}{d\mu} = \frac{e^\psi}{\int e^\psi d\mu}.$$

Stochastic dynamic game. The stochastic dynamics of the state θ_t for the controlled sequence (two players) is

$$\theta_{t+1} = F(\theta_t, y_{t+1}, u_t). \tag{4.1}$$

Here, the sequence $\{u_t\} \subset U$ represents the controls that the minimizer controller (Player 1) chooses at each time step, while the maximizer controller (Player 2) chooses a sequence $\{\eta_t\} \subset \Gamma$, where η_t is the distribution of y_{t+1}. The information available at each time t for both players is the history of observations $\{y_1, \ldots, y_t\}$. We shall make this precise using strategies.

Let $\mathcal{V} := \{\vec{u} = \{\overline{u}_t\} \mid \overline{u}_0 = u_0 \in U,\ \overline{u}_t : (\mathbb{R}^d)^t \times \Gamma^t \to U,\ t \geq 1\}$ be the set of strategies for Player 1, and let $\mathcal{W} = \{\vec{\eta} = \{\overline{\eta}_t\} \mid \overline{\eta}_0 = \eta_0(u_0) \in \Gamma,\ \overline{\eta}_t : (\mathbb{R}^d)^t \times U^t \to \Gamma,\ t \geq 1\}$ be the set of strategies for Player 2. Then, given the strategies $\vec{u}$, $\vec{\eta}$, we obtain the control sequences $\tilde{u} = \{\tilde{u}_t\}$, $\tilde{\eta} = \{\tilde{\eta}_t\}$, with $\tilde{u}_t$, $\tilde{\eta}_t$ $\mathcal{Y}^t$-adapted, determined by $\tilde{u}_0 = u_0$, $\tilde{\eta}_0 = \overline{\eta}_0(u_0)$, $\tilde{u}_1(y_1) = \overline{u}_1(y_1, \eta_1)$, $\tilde{\eta}_1(y_1) = \overline{\eta}_1(y_1, \tilde{u}_0, \tilde{u}_1(y_1)) \ldots$. Note that the Player 1 chooses first at each time step t (upper game). Moreover, $(\vec{u}, \vec{\eta})$ defines a probability measure $\mathcal{P}^{\vec{u},\vec{\eta}}$ on $\mathcal{Y}^\infty$, where $\mathcal{Y}^\infty$ is the σ-algebra generated by the "cylinder sets", in the following way:

$$P^{\vec{u},\vec{\eta}}(A_1, A_2, \ldots, A_n) = \int_{A_1}\int_{A_2} \cdots \int_{A_n} \tilde{\eta}_0(u_0)(dy_1)\tilde{\eta}_1(y_1)(dy_2) \ldots$$

$$\cdot \tilde{\eta}_{n-1}(y_1, \ldots, y_{n-1})(dy_n).$$

We denote by $E^{\vec{u},\vec{\eta}}$ the corresponding expectation operator. Finally, the reward per stage function is $K: P(E)\times U\times\Gamma \to \mathbb{R}$, defined by $K(\theta,u,\eta) := \int[G(\theta,y,u) - \log\frac{d\eta}{d\mu}(y)]\eta(dy)$.
Now, we shall introduce a sequence of infinite horizon discounted games. Let $\beta \in (0,1)$ the discount factor. Given $\theta \in P(E)$, $\vec{u} \in \mathcal{V}$ and $\vec{\eta} \in \mathcal{W}$, define the payoff functional

$$I_\beta(\theta,\vec{u},\vec{\eta}) = E^{\vec{u},\vec{\eta}} \sum_{t=0}^{\infty} \beta^t K(\theta_t,u_t,\eta_t).$$

Definition 4.2. When there exists a pair of strategies $\vec{u}^*, \vec{\eta}^*$ such that

$$I_\beta(\theta,\vec{u}^*,\vec{\eta}) \le I_\beta(\theta,\vec{u}^*,\vec{\eta}^*) \le I_\beta(\theta,\vec{u},\vec{\eta}^*) \text{ for all } \vec{u},\vec{\eta}, \tag{4.2}$$

the value $I_\beta(\theta,\vec{u}^*,\vec{\eta}^*)$ is called the upper value of the game, and $\vec{u}^*,\vec{\eta}^*$ are referred as a saddle point. The upper value function of this game is denoted by $V_\beta(\theta)$.
In addition to (A1)-(A5) we make the following additional assumption.

(A6) The mapping $(u,x,z) \to p^u(x,z)$ is continuous.

Lemma 4.3 [Fleming and Hernández-Hernández, Lemma 4.5]. *There is a unique continuous concave solution to the Isaacs equation*

$$\begin{aligned} V_\beta(\theta) \quad = \quad & \inf_{u\in U}\sup_{\eta\in\Gamma} \int \Big[\beta V_\beta(F(\theta,y,u)) \\ & + G(\theta,y,u) - \log(\frac{d\eta}{d\mu}(y))\Big]\,\eta(dy), \end{aligned} \tag{4.3}$$

and it is the upper value function defined above. Furthermore, the pair of strategies $\vec{u} = \{\overline{u}_t^*\}$, $\vec{\eta} = \{\overline{\eta}_t^*\}$, with $\overline{u}_t^*(y_1,\dots y_t,\eta_0\dots,\eta_{t-1}) = u^*(\theta_t)$ and $\overline{\eta}_t^*(y_1,\dots,y_t,u_0,\dots,u_t) = \eta^*(\theta_t,u_t)$, where

$$u^*(\theta) \in \operatorname*{argmin}_{u\in U} \int e^{\beta V_\beta(F(\theta,y,u))+G(\theta,y,u)}\mu(dy)$$

and

$$\eta^*[\theta,u](A) = \frac{\int_A e^{\beta V_\beta(F(\theta,y,u))+G(\theta,y,u)}\mu(dy)}{\int e^{\beta V_\beta(F(\theta,y,u))+G(\theta,y,u)}\mu(dy)}$$

is a saddle point.
Lemma 4.4. *Let $u \in U$ and $y \in \mathbb{R}^d$ fixed, but arbitrary. Then, for any $\theta,\ \tilde{\theta} \in P(E)$,*

$$|G(\theta,y,u) - G(\tilde{\theta},y,u)| \le \frac{1}{2}\lambda^{-1}\|\theta - \tilde{\theta}\|. \tag{4.4}$$

Proof. Let us define

$$\rho(x) := \int r(z,y)e^{l(x,u)}p^u(x,z)\nu(dz)$$

Then, from definition,

$$G(\theta,y,u)-G(\widetilde{\theta},y,u)=\log\int\rho d\theta-\log\int\rho d\widetilde{\theta},$$

and

$$\begin{aligned}
G(\theta,y,u)-G(\widetilde{\theta},y,u) &= \log\frac{\int\rho d\theta}{\int\rho d\widetilde{\theta}} \\
&= \log[\frac{\int\rho d\theta-\int\rho d\widetilde{\theta}}{\int\rho d\widetilde{\theta}}+1] \\
&= \log[\frac{\int\rho d(\theta-\widetilde{\theta})}{\int\rho d\widetilde{\theta}}+1] \\
&\leq \log[\frac{\int\rho d(\theta-\widetilde{\theta})^{+}}{\int\rho d\widetilde{\theta}}+1] \\
&\leq \log[\frac{1}{2}\lambda^{-1}\|\theta-\widetilde{\theta}\|+1] \quad \text{from (3.7)} \\
&\leq \frac{1}{2}\lambda^{-1}\|\theta-\widetilde{\theta}\|.
\end{aligned}$$

Interchanging the role of θ and $\widetilde{\theta}$ the proof is done.

□

The next result is analogous to Lemma 4.6 in Fleming and Hernández-Hernández, and its proof is based in the above lemma and Theorem 3.1.
Lemma 4.5. *For each $\beta\in(0,1)$ and $\theta,\widetilde{\theta}\in P(E)$,*
(a)

$$|(1-\beta)V_\beta(\theta)|\leq\|G\|$$

(b)

$$|V_\beta(\theta)-V_\beta(\widetilde{\theta})|\leq C\|\theta-\widetilde{\theta}\|.$$

for some constant C independent of β.
Now,we observe that, using Lemma 4.1, we can rewrite (4.3) as

$$e^{V_\beta(\theta)}=\inf_{u\in U}\int e^{\beta V_\beta(F(\theta,y,u))+G(\theta,y,u)}\mu(dy). \tag{4.5}$$

Since V_β is continuous and $P(E)$ is compact, there is $\theta_\beta\in\operatorname{argmin}V_\beta$, and defining $W_\beta(\cdot)=V_\beta(\cdot)-V_\beta(\theta_\beta)$, (4.5) is equivalent to

$$e^{(1-\beta)V_\beta(\theta_\beta)+W_\beta(\theta)}=\inf_{u\in U}\int e^{\beta W_\beta(F(\theta,y,u))+G(\theta,y,u)}\mu(dy). \tag{4.6}$$

If $\{W_\beta\}$ is an equicontinuous family in $C(P(E))$, we can prove, using also Lemma 4.6(a), the existence of a sequence $\beta_n\uparrow 1$ such that $W_{\beta_n}\to W$ as

$n \to \infty$ in $C(P(E))$, and $(1-\beta_n)V_{\beta_n}(\theta_{\beta_n}) \to \Lambda$. Therefore, letting $\beta_n \uparrow 1$ in (4.6), we get

$$\exp\{\Lambda + W(\theta)\} = \min_{u\in U}\left[E^{+}\exp\{G(\theta,y,u) + W(F(\theta,y,u))\}\right], \qquad (4.7)$$

which is the dynamic programming equation of the infinite horizon risk sensitive control problem.

The proof of the next lemma is based in Proposition 4.6 of Runggaldier and Stettner (1994). The author wishes to thank Prof. W. Runggaldier for providing him a copy of this book.

Lemma 4.6. *The family of functions* $\{W_\beta\}_{\beta\in(0,1)}$ *is equicontinuous.*

Proof. Let $\theta,\ \widetilde{\theta} \in P(E)$, and define

$$\tau(\theta,\widetilde{\theta}) = \inf_{u\in U}\ \inf_{A\in\mathcal{B}(E),\nu(A)>0} \frac{\int\int_A p^u(x,z)\nu(dz)\theta(dx)}{\int\int_A p^u(x,z)\nu(dz)\widetilde{\theta}(dx)}. \qquad (4.8)$$

Then, in view of (A5)-(A6), $\tau(\theta,\widetilde{\theta}) \to 1$ as $\tau(\widetilde{\theta},\theta) \to 1$ when θ converges weakly to $\widetilde{\theta}$. On the other hand, note that $\tau(\theta,\widetilde{\theta}) \le 1$ and $\tau(\widetilde{\theta},\theta) \le 1$. When the equality is satisfied the lemma follows in a straightforward way from (4.6). So, let us assume that $\tilde{\tau} := \tau(\theta,\widetilde{\theta})\tau(\widetilde{\theta},\theta) < 1$. Further, following Runggaldier and Stettner (1994), for each $u \in U$, $y \in \mathbb{R}^d$, we define $m^u(\theta,\widetilde{\theta},y) \in \mathcal{M}(E)$ by

$$m^u(\theta,\widetilde{\theta},y) = F(\widetilde{\theta},y,u) - \tau(\theta,\widetilde{\theta})\tau(\widetilde{\theta},\theta)F(\theta,y,u). \qquad (4.9)$$

Now, from (4.6), we have

$$\exp\{W_\beta(\theta) - W_\beta(\widetilde{\theta})\} \le$$

$$\le \sup_u \frac{\int\int[\int e^{\beta W_\beta(F(\theta,y,u))}r(z,y)dy]e^{l(x,u)}p^u(x,z)\nu(dz)\theta(dx)}{\int\int[\int e^{\beta W_\beta(F(\widetilde{\theta},y,u))}r(z,y)dy]e^{l(x,u)}p^u(x,z)\nu(dz)\widetilde{\theta}(dx)}$$

$$= \sup_u \Bigg[\frac{\int\int[\int e^{\beta W_\beta(F(\theta,y,u))}r(z,y)dy]e^{l(x,u)}}{\int\int\left[\int e^{\beta W_\beta(F(\widetilde{\theta},y,u))}r(z,y)dy\right] e^{l(x,u)}p^u(x,z)\nu(dz)\widetilde{\theta}(dx)}\cdot$$

$$\cdot[p^u(x,z)\nu(dz)\theta(dx) - \tilde{\tau}p^u(x,z)\nu(dz)\widetilde{\theta}(dx)] +$$

$$+ \frac{\tilde{\tau}\int\int\left[\int e^{\beta W_\beta(F(\theta,y,u))}r(z,y)dy\right] e^{l(x,u)}p^u(x,z)\nu(dz)\widetilde{\theta}(dx)}{\int\int\left[\int e^{\beta W_\beta(F(\widetilde{\theta},y,u))}r(z,y)dy\right] e^{l(x,u)}p^u(x,z)\nu(dz)\widetilde{\theta}(dx)}\Bigg].$$

Then, from Lemma 4.6, we observe that the first summand within the brackets is bounded above by $e^{2C+\|l\|}(1-\tau^2(\widetilde{\theta},\theta)\tau(\theta,\widetilde{\theta}))$, while the concavity of $e^{\beta W_\beta}$ and (4.9) imply that the second term is bounded above by one. Therefore,

$$W_\beta(\theta) - W_\beta(\widetilde{\theta}) \leq e^{2C+\|l\|}(1 - \tau^2(\widetilde{\theta}, \theta)\tau(\theta, \widetilde{\theta})),$$

and interchanging the role of θ and $\widetilde{\theta}$ we get the result.

□

We now state the main result of this section.

Theorem 4.7. *(a) There exists a solution* $\Lambda \in \mathbb{R}$ *and* $W \in C(P(E))$ *to the dynamic programming equation (4.7).*
(b) $\Lambda \leq J(\widetilde{u})$ *for all* $\widetilde{u} \in \mathcal{U}$.
(c) Let $\widetilde{u}^* = \{u_t^*\} \in \mathcal{U}$ *such that, for each* $t = 0, 1, \ldots$, $u_t^* = u^*(\theta_t)$, *where* $u^* : S \to U$ *is a function with*

$$u^*(\theta) \in \underset{u \in U}{\operatorname{argmin}}[E^+ \exp\{W(F(\theta, y, u)) + G(\theta, y, u)\}].$$

Then, $\Lambda = J(\tilde{u}^*)$.

Proof. The first part follows from Lemmas 4.5(a) and 4.6 applying the Arzela-Ascoli Theorem. The proof of parts (a) and (b) was given already in Theorem 4.2 of Fleming and Hernández-Hernández.

□

Acknowledgments: This research was supported by CONACYT grant 3115P-E9608.

References

Arapostathis A. and Marcus S.I. (1990). Analisys of an identification algorithm arising in the adaptive estimation of Markov chains, Math. of Control, Signals and Systems 3, pp. 1-29.

Baras J.S. and James M.R. (1997). Robust and risk-sensitive output feedback control for finite state machines and hidden Markov models, J. Math. Systems, Estimation and Control.

Bensoussan A. and Van Schuppen J.H. (1985). Optimal control of partially observable stochastic systems with exponential of integral performance index, SIAM J. Control Optim. 23, pp. 599-613.

Dai Pra P. Meneghini L. and Runggaldier W. (1996). Connections between stochastic control and dynamic games, Math of Control Signals and Systems 9, pp. 303-326.

Dupuis P. and Ellis R.S. (1997). *A Weak Convergence Approach to the Theory of Large Deviations*, John Wiley & Sons, New York.

Fleming W.H. and Hernández-Hernández D. Risk sensitive control of finite state machines on an infinite horizon II, SIAM J. Control Optim. (to appear).

Fleming W.H. and McEneaney W.M. (1995). Risk sensitive control on an infinite horizon, SIAM J. Control Optim. 33, pp. 1881-1915.

Fleming W.H. and McEneaney W.M. (1992). Risk sensitive and differential games, Lecture Notes in Control and Info. Sci. No. 184, Springer, pp. 185-197.

Furstenberg H. and Kesten H. (1960). Products of random matrices, Ann. Math. Statist. 31, pp. 457-469.

Furstenberg H. and Kifer Y. (1983). Random matrix products and measures on proyective spaces, Israel J. Math. 46, pp. 12-33.

Hernández-Hernández D. and Marcus S.I. (1996). Risk sensitive control of Markov processes in countable state space, Sys. Control Lett. 29, pp. 147-155. Corrigendum, Systems and Control Letters (to appear).

Jacobson D.H. (1973). Optimal stochastic linear systems with exponential performance criteria and their relation to deterministic differential games. IEEE Trans. Autom. Control 18 pp. 124-131.

James M.R., Baras J.S. and Elliott R. (1994). Risk sensitive control and dynamic games for partially observed discrete time nonlinear systems, IEEE Trans. Aut. Control 39 pp. 750-792.

Kaijser T. (1978). A limit theorem for Markov chains in compact metric spaces with applications to products of random matrices, Duke Mathematical J. 45 pp. 311-349.

Marcus S.I., et al (1997). Risk sensitive Markov decision processes, in "Systems and Control in the Twenty-First Century", C.I. Byrnes, et al (eds.) Boston, Birkhäuser.

Runggaldier W. and Stettner L. (1994). *Approximations of Discrete Time Partially Observed Control Problems*, Applied Mathematics Monographs. Gardini Editori e Stampatori.

Whittle P. (1990). *Risk Sensitive Optimal Control*, John Wiley, New York.

Yosida K. (1980). *Functional Analysis*, Springer-Verlag, Berlin.

Fleming W.H. and Hernández-Hernández D. [illegible]

Fleming W.H. and McEneaney W.M. (1995), Risk sensitive control on an infinite horizon, SIAM J. Control Optim. 33, pp. 1881-1915.

Fleming W.H. and McEneaney W.M. (1992), Risk sensitive optimal control and differential games, Lecture Notes in Control and Info. Sci. No. 184, Springer, pp. 185-197.

Furstenberg H. and Kesten H. (1960), Products of random matrices, Ann. Math. Statist. 31, pp. 457-469.

Furstenberg H. and Kifer Y. (1983), Random matrix products [illegible]

4

Nonlinear Semigroups for Partially Observed Risk-Sensitive Control and Minimax Games

Matthew R. James

ABSTRACT: This paper presents a heuristic discussion of Nisio semigroups and generators for partially observed risk-sensitive control and minimax games as well as a unified framework based on a space of generalized measures. These ideas are motivated by the current lack of understanding of nonlinear infinite dimensional dynamic programming PDEs, the practical need to accommodate "singular" information states, and the desire to develop a common mathematical framework useful for "robust control".

KEYWORDS: Risk-sensitive stochastic control, minimax differential games, nonlinear H_∞ control, robust control, Nisio semigroup and generator, large deviations, max-plus measures.

1 Introduction

In recent years there has been a renewed interest in stochastic control and deterministic game theory, fueled to a large extent by developments in the H_∞ approach to robust control. These developments show fundamental connections amoung H_∞ control, deterministic differential minimax games, and risk-sensitive stochastic control; see [2], [3], [5], [12], [15], [22], [25], [24], [27], [33], [34], plus many others. Perhaps the most challenging aspect of these developments, both from mathematical as well as practical points of view, is the problem of partial or incomplete state information. When perfect state measurements are not available, the measurements are passed through a filtering process and the outcome is used to determine the control value. For linear gaussian systems, this is achieved using a modified Kalman filter. In the general nonlinear case, a suitable information state is used. The technical difficulties arise because the information state is an infinite dimensional quantity, with dynamics described by PDEs. When dynamic programming methods are used, one is led to a dynamic programming PDE defined on an infinite dimensional space. PDEs at this level of abstraction are not well understood; [4], [20], [31], etc. The viscosity framework has been successfully used for a wide range of finite dimensional control problems, and infinite dimensional extensions ([31] or [20]) could be developed for the risk-sensitive problem, although there are substantial technical issues to resolve and this has not been done to my knowledge. A viscosity solution framework for the deterministic minimax games has been proposed in [27]; however it is believed that significant reformulation and new developments are required; these are largely open and difficult problems, and are important in part because such an infinite dimensional game type PDE underlies nonlinear H_∞ control.

The viscosity solution concept is important because it provides a means for proving existence and uniqueness results for nonlinear PDE even though the solutions are not smooth and the nonlinearities preclude standard integration-by-parts weak solution methods. Before the viscosity solution concept was discovered, numerous approaches were tried, with varying levels of generality and success. One very interesting approach was the Nisio semigroup, [32]. Here, attention is focused on a nonlinear semigroup which is naturally associated with a control problem, and properties of this semigroup and its generator are studied. In the paper [10], Fleming studied the Nisio semigroup and generator for a class of nonlinear partially observed stochastic control problems. This work prompted the author to consider the application of this approach to partially observed risk-sensitive stochastic control problems as well as to partially observed deterministic minimax differential games. Motivation for this comes from the current lack of understanding (discussed above) of the basic PDE underlying nonlinear H_∞ control and from the practical need to deal with information states which are in a sense singular, [16], [17], [18]. A further motivation is a desire to develop a unified theory of robust control which places risk-sensitive control and minimax games in a common mathematical framework.

This paper presents a heuristic discussion of Nisio semigroups and generators for partially observed risk-sensitive control and minimax games as well as a unified framework based on a space of generalized measures. This space of generalized measures includes conventional measures as well as max-plus type measures, [28], [6] (see also [30], [1], etc). The ideas presented here have not been fully developed and are the subject of current investigations, [17].

2 Risk-Sensitive Control

The output feedback risk-sensitive stochastic optimal control problem for the system

$$\begin{aligned} dx_t^\varepsilon &= b(x_t^\varepsilon, u_t)\,dt + \sqrt{\varepsilon}\,dw_t \\ dy_t^\varepsilon &= h(x_t^\varepsilon)\,dt + \sqrt{\varepsilon}\,dv_t \end{aligned} \tag{1}$$

is to minimize the functional

$$J^{\mu,\varepsilon}(u) = \mathbf{E}\left[\exp\frac{\mu}{\varepsilon}\left(\int_0^T L(x_s^\varepsilon, u_s)\, ds + \Phi(x_T^\varepsilon)\right)\right] \tag{2}$$

over the class $\mathcal{U}_{0,T}$ of control policies which are non-anticipating functionals of the observation process y. In (1), w and v are standard independent Wiener processes, and $b \in C_b^1(\mathbf{R}^n \times \mathbf{R}^m, \mathbf{R}^n)$, $h \in C_b^1(\mathbf{R}^n)$, $L \in C_b^1(\mathbf{R}^n \times \mathbf{R}^m)$, $\Phi \in C_b^1(\mathbf{R}^n)$, with $L \geq 0$, $\Phi \geq 0$.
The stochastic system (1) is assumed defined on a probability space $(\Omega, \mathcal{F}, \mathbf{P}^u)$, and we write $\mathcal{G}_t = \sigma(x_s, y_s; 0 \leq s \leq t)$ and $\mathcal{Y}_t = \sigma(y_s; 0 \leq s \leq t)$. There is an equivalent probability measure $\mathbf{P}^\dagger$ under which y is a standard Wiener process independent of the state process [7]. For $u \in \mathcal{U}_{0,T}$ we have

$$\frac{d\mathbf{P}^u}{d\mathbf{P}^\dagger}|_{\mathcal{G}_t} = \exp\left(-\tfrac{1}{\varepsilon}\left[\tfrac{1}{2}\int_0^t |h(x_s^\varepsilon)|^2\, ds - \int_0^t h(x_s^\varepsilon)\, dy_s^\varepsilon\right]\right).$$

Write Z_t^ε for the right hand side of this expression.
Following an idea of [3] for linear systems, we express $J^{\mu,\varepsilon}(u)$ in terms of a new "state" process and then solve the resulting optimal stochastic control problem with complete "state" information. To this end, we define an information state $\sigma_t^{\mu,\varepsilon}(dx)$ by

$$\langle \sigma_t^{\mu,\varepsilon}, \eta\rangle = \mathbf{E}^\dagger\left[\eta(x_t^\varepsilon) Z_t^\varepsilon \exp\left(\frac{\mu}{\varepsilon}\int_0^t L(x_s^\varepsilon, u_s)\, ds\right) | \mathcal{Y}_t\right], \tag{3}$$

for all test functions η, where

$$\langle \sigma, \eta\rangle = \int_{\mathbf{R}^n} \sigma(x)\eta(x)\, dx = \int_{\mathbf{R}^n} \eta(x)\sigma(dx)$$

is the usual inner product in $L_2(\mathbf{R}^n)$, or the integral of a function η against a measure σ. The evolution of the information state process is governed by the stochastic PDE

$$d\sigma_t^{\mu,\varepsilon} = \left(A^{\varepsilon, u_t *} + \frac{\mu}{\varepsilon} L^{u_t}\right)\sigma_t^{\mu,\varepsilon}\, dt + \tfrac{1}{\varepsilon} h \sigma_t^{\mu,\varepsilon}\, dy_t^\varepsilon, \tag{4}$$

where $A^{\varepsilon,u}\phi = \frac{\varepsilon}{2}\Delta\phi + b^u \cdot D\phi$, the asterisk denoting adjoint, $L^u = L(\cdot, u)$, and $b^u = b(\cdot, u)$. Using the information state process, one obtains the representation

$$J^{\mu,\varepsilon}(u) = \mathbf{E}^\dagger\left[\langle \sigma_T^{\mu,\varepsilon}, e^{\frac{\mu}{\varepsilon}\Phi}\rangle\right]. \tag{5}$$

Thanks to the representation (5), the problem now is to minimize the RHS of (5) with dynamics (4) which are completely observable. The dynamic programming value function is defined by

$$S^{\mu,\varepsilon}(\sigma, t) = \inf_{u \in \mathcal{U}_{0,t}} \mathbf{E}_\sigma^\dagger\left[\langle \sigma_t^{\mu,\varepsilon}, e^{\frac{\mu}{\varepsilon}\Phi}\rangle\right]. \tag{6}$$

Here, the notation indicates expectation given the initial condition $\sigma_0 = \sigma$. Formally, Hamilton-Jacobi-Bellman (HJB) equation for $S^{\mu,\varepsilon}$ is (c.f. [31], [21]):

$$\begin{aligned} \frac{\partial S^{\mu,\varepsilon}}{\partial t} &= \frac{1}{2\varepsilon} D^2 S^{\mu,\varepsilon}(h\sigma, h\sigma) + \inf_{u \in \mathbf{R}^m} \left\{ DS^{\mu,\varepsilon} \cdot \left(A^{\varepsilon,u\,*}\sigma + \frac{\mu}{\varepsilon} L^u \sigma \right) \right\} \\ S^{\mu,\varepsilon}(\sigma, 0) &= \langle \sigma, e^{\frac{\mu}{\varepsilon}\Phi} \rangle \end{aligned} \tag{7}$$

A verification theorem for this problem would say that if there exists a sufficiently smooth solution $S^{\mu,\varepsilon}$ to (7) and if $u^* \in \mathcal{U}_{0,T}$ is a policy such that $u_t^* = \bar{u}_t^*(\sigma_t^{\mu,\varepsilon})$, where $\bar{u}_t^*(\sigma)$ achieves the minimum in (7), then u^* is optimal.

In general, it is unlikely that the value function $S^{\mu,\varepsilon}$ will be smooth in a classical sense, and so one might hope to appeal to a generalized interpretation of the PDE (7). However, the viscosity solution theory for infinite dimensional PDEs of the type (7) is in its infancy [4], [31], [20], [21]. In [10], Fleming found the Nisio [32] semigroup associated with a partially observed stochastic control problem. This framework is valid quite generally. We now briefly sketch how this can apply to the risk-sensitive problem.

The information state $\sigma^{\mu,\varepsilon}$ is a measure on $\mathbf{R}^n$, and the value function $S^{\mu,\varepsilon}$ is a function defined on the space of all such measures $\mathcal{M} = \mathcal{M}(\mathbf{R}^n)$ (the topology of w^*-convergence was used in [10]). Let $\mathcal{F}(\mathcal{M})$ denote a suitable space of functions on $\mathcal{M}$ which includes the subset of functions of the form

$$\begin{aligned} \mathcal{D}_1 = \{ \psi \ : \ & \psi(\sigma) = g(\langle f_1, \sigma \rangle, \ldots, \langle f_J, \sigma \rangle), \\ & g \in C_b^\infty(\mathbf{R}^J), f_1, \ldots, f_J \in C_b^\infty(\mathbf{R}^n), J = 1, 2, \ldots \}. \end{aligned}$$

For $\psi \in \mathcal{F}(\mathcal{M})$, $\sigma \in \mathcal{M}$ and $t \geq 0$ define

$$\mathcal{S}_t^{\mu,\varepsilon} \psi(\sigma) = \inf_{u \in \mathcal{U}_{0,t}} \mathbf{E}_\sigma^\dagger \left[\psi(\sigma_t^{\mu,\varepsilon}) \right]. \tag{8}$$

The semigroup property for $\mathcal{S}_t^{\mu,\varepsilon}$ reads

$$\mathcal{S}_{t+s}^{\mu,\varepsilon} \psi = \mathcal{S}_s^{\mu,\varepsilon} \mathcal{S}_t^{\mu,\varepsilon} \psi.$$

The value function can be recovered via the choice $\psi(\sigma) = \langle \sigma, e^{\frac{\mu}{\varepsilon}\Phi} \rangle$:

$$S^{\mu,\varepsilon}(\sigma, t) = \mathcal{S}_t^{\mu,\varepsilon} \langle \sigma, e^{\frac{\mu}{\varepsilon}\Phi} \rangle.$$

The generator $\mathcal{L}^{\mu,\varepsilon}$ of $\mathcal{S}_t^{\mu,\varepsilon}$ is defined, when it exists, by

$$\mathcal{L}^{\mu,\varepsilon} \psi(\sigma) = \lim_{t \downarrow 0} \frac{\mathcal{S}_t^{\mu,\varepsilon} \psi(\sigma) - \psi(\sigma)}{t}. \tag{9}$$

The set $\mathcal{D}_1$ is a set of test functions on which the "generator" $\mathcal{L}^{\mu,\varepsilon}$ of $\mathcal{S}_t^{\mu,\varepsilon}$ can be explicitly represented. For fixed $u \in \mathbf{R}^m$ define, for $\psi \in \mathcal{D}_1$,

$$\begin{aligned}\mathcal{L}^{\mu,\varepsilon;u}\psi(\sigma) &= \tfrac{1}{2\varepsilon}\textstyle\sum_{j,k=1}^{J} g_{jk}(\langle f_1,\sigma\rangle,\ldots,\langle f_J,\sigma\rangle)\langle hf_j,\sigma\rangle\langle hf_k,\sigma\rangle \\ &\quad + \textstyle\sum_{j=1}^{J} g_j(\langle f_1,\sigma\rangle,\ldots,\langle f_J,\sigma\rangle)\langle A^{\varepsilon,u} + \tfrac{\mu}{\varepsilon}L^u,\sigma\rangle\end{aligned} \tag{10}$$

where g_j and g_{jk} denote first and second order partial derivatives of g. The generator $\mathcal{L}^{\mu,\varepsilon}$ is defined by

$$\mathcal{L}^{\mu,\varepsilon}\psi(\sigma) = \inf_{u\in\mathbf{R}^m} \mathcal{L}^{\mu,\varepsilon;u}\psi(\sigma).$$

3 Minimax Games

Consider the minimax game defined for the system

$$\begin{aligned}\dot{x}_t &= b(x_t,u_t) + w_t, \\ \dot{y}_t &= h(x_t) + v_t, \ y(0) = 0,\end{aligned} \tag{11}$$

where w and v are unknown square integrable disturbances on $[0,T]$, and the problem is to minimize the cost

$$\begin{aligned}J_p^\mu(u) &= \sup_{(w,v)\in L_2([0,T],\mathbf{R}^{n+1})} \Big\{ p(x(0)) + \int_0^T L(x_t,u_t) \\ &\quad -\tfrac{1}{2\mu}\left(|w_t|^2 + |v_t|^2\right) dt + \Phi(x_T)\Big\}\end{aligned} \tag{12}$$

over the class $\mathcal{U}_{0,T}$ of control policies which are non-anticipating functionals of the observation y.
The information state is defined by

$$\begin{aligned}p_t^\mu(x) &= \sup_{\xi\in C([0,T],\mathbf{R}^n)} \Big\{ p(\xi_0) + \int_0^t L(\xi_s,u_s) \\ &\quad -\tfrac{1}{\mu}\left(\tfrac{1}{2}|\dot{\xi} - b(\xi_s,u_s)|^2 + \tfrac{1}{2}|h(\xi_s)|^2 - h(\xi_s)\dot{y}_s\right) ds \ : \ \xi_t = x\Big\}.\end{aligned} \tag{13}$$

The cost function $J_p^\mu(u)$ can be represented in terms of the information state:

$$J_p^\mu(u) = \sup_{y\in C([0,T],\mathbf{R})} \left\{ (p_T^\mu, \Phi) - \tfrac{1}{2\mu}\int_0^T |\dot{y}_s|^2\, ds \ : \ p_0^\mu = p \right\}. \tag{14}$$

Here, we have used the "max-plus inner product"

$$(p,q) = \sup_{x\in\mathbf{R}^n} \{p(x) + q(x)\}.$$

The supremum over $y \in C([0,T],\mathbf{R})$ is defined to equal $-\infty$ if y is not absolutely continuous. Note that all solutions of (11) will produce abolutely continuous trajectories, and in view of the initial condition $y(0)=0$ there is a natural bijection between $y(\cdot)$ and $\dot{y}(\cdot)$.
The PDE satisfied by the information state p_t^μ is a nonlinear first-order Hamilton-Jacobi equation

$$\frac{\partial p^\mu}{\partial t} = \sup_{w\in\mathbf{R}^n}\left[-Dp^\mu(b^{u_t}+w)-\tfrac{1}{2\mu}|w|^2\right]+L^{u_t}-\tfrac{1}{\mu}\left[\tfrac{1}{2}|h|^2-h\dot{y}_t\right]. \quad (15)$$

The problem now is to minimize the RHS of (14) with dynamics (15) which are completely observable. The dynamic programming value function is given by

$$W^\mu(p,t)=\inf_{u\in\mathcal{U}_{0,t}}\sup_{y\in C([0,t],\mathbf{R})}\left\{(p_t^\mu,\Phi)-\tfrac{1}{2\mu}\int_0^t|\dot{y}_s|^2\,ds\ :\ p_0^\mu=p\right\}, \quad (16)$$

and formally satisfies the HJI equation

$$\begin{aligned}\frac{dW^\mu}{dt} &= \inf_{u\in\mathbf{R}^m}\sup_{\dot{y}\in\mathbf{R}}\left\{DW^\mu\cdot\left(\sup_{w\in\mathbf{R}^n}\left[-Dp^\mu(b^u+w)-\tfrac{1}{2\mu}|w|^2\right]+L^u\right.\right.\\ &\qquad\left.\left.-\tfrac{1}{\mu}\left[\tfrac{1}{2}|h|^2-h\dot{y}\right]\right)-\tfrac{1}{2\mu}|\dot{y}|^2\right\},\\ W^\mu(p,0) &= (p,\Phi).\end{aligned} \quad (17)$$

In this and similar equations, $\dot{y}\in\mathbf{R}$ is regarded as a dummy variable and no differentiation is implied.
The verification theorem would then say that if there exists a sufficiently smooth solution W^μ to the HJI equation (17) and if $u^*\in\mathcal{U}_{0,T}$ is a policy such that $u_t^*=\bar{u}_t^*(p_t^\mu)$, where $\bar{u}_t^*(p)$ achieves the minimum in (17), then u^* is optimal.
The infinite dimensional PDE (17) is not well understood at present. A definition of viscosity solution has been given in [27], although there is no known uniqueness theory.
We now describe how one might define a semigroup corresponding to this problem, by analogy with the stochastic case; [17]. Let $\mathcal{X}_e$ denote the space of upper semicontinuous functions $p:\mathbf{R}^n\to\mathbf{R}\cup\{-\infty\}$ which are bounded above. Let $\mathcal{F}(\mathcal{X}_e)$ denote a suitable space of functions defined on $\mathcal{X}_e$ which includes the subset of functions

$$\begin{aligned}\mathcal{D}_2=\{\psi\ :\ &\psi(p)=g((f_1,p),\ldots,(f_J,p)),\\ &g\in C_b^\infty(\mathbf{R}^J),f_1,\ldots,f_J\in C_b^\infty(\mathbf{R}^n),J=1,2,\ldots\}.\end{aligned}$$

For $p\in\mathcal{X}_e$, $\psi\in\mathcal{F}(\mathcal{X}_e)$ and $t\geq 0$ define

$$\mathcal{W}_t^\mu\psi(p)=\inf_{u\in\mathcal{U}_{0,t}}\sup_{y\in C([0,t],\mathbf{R})}\left\{\psi(p_t^\mu)-\tfrac{1}{2\mu}\int_0^t|\dot{y}_s|^2\,ds\ :\ p_0=p\right\}. \quad (18)$$

The semigroup property for $\mathcal{W}_t^\mu$ reads

$$\mathcal{W}_{t+s}^\mu \psi = \mathcal{W}_s^\mu \mathcal{W}_t^\mu \psi,$$

and the value function can be recovered via the choice $\psi(p) = (p, \Phi)$:

$$W^\mu(\sigma, t) = \mathcal{W}_t^\mu(p, \Phi).$$

The generator $\mathcal{L}^\mu$ of $\mathcal{W}_t^\mu$ is defined, when it exists, by

$$\mathcal{L}^\mu \psi(p) = \lim_{t \downarrow 0} \frac{\mathcal{W}_t^\mu \psi(p) - \psi(p)}{t}. \tag{19}$$

By analogy with the stochastic case, we would like to define the generator on the set $\mathcal{D}_2$. To motivate this, consider the function $g(p) = (p, f)$, where $f \in C_b(\mathbf{R}^n)$, and $p \in \mathcal{X}_e$. Assume that p satisfies a tightness (coercivity) condition such as $p(x) \leq -c_1|x|^2 + c_2$, where $c_1 > 0$ and $c_2 \geq 0$. Let $\bar{x}_t \in \operatorname{argmax}_x\{p(x) + t\phi(x)\}$, where $\phi \in C_b(\mathbf{R}^n)$. Then $\bar{x}_t$ exists for each $t > 0$ and $\lim_{t\downarrow 0} \bar{x}_t \in \operatorname{argmax}_x\{p(x) + f(x)\}$ (at least through a subsequence). Then

$$\begin{aligned}\limsup_{t\downarrow 0} \frac{g(p+t\phi) - g(p)}{t} &\leq \limsup_{t\downarrow 0} \phi(\bar{x}_t) \\ &\leq \sup\left\{\phi(\bar{x}) \ : \ \bar{x} \in \operatorname*{argmax}_{x\in\mathbf{R}^n}\{p(x) + f(x)\}\right\}.\end{aligned}$$

Also, for any $\bar{x} \in \operatorname{argmax}_x\{p(x) + f(x)\}$,

$$\liminf_{t\downarrow 0} \frac{g(p+t\phi) - g(p)}{t} \geq \phi(\bar{x}).$$

Therefore we have the Gateaux-type derivative

$$\nabla_p g(p)[\phi] = \lim_{t\downarrow 0} \frac{g(p+t\phi) - g(p)}{t} = \sup\left\{\phi(\bar{x}) \ : \ \bar{x} \in \operatorname*{argmax}_{x\in\mathbf{R}^n}\{p(x) + f(x)\}\right\}.$$

This defines an operator which acts on functions ϕ by maximizing over the set of maximizers of the function $p + f$.
This leads to the following. For fixed $u \in \mathbf{R}^m$, $\dot{y} \in \mathbf{R}^p$ (think of $\dot{y}$ as a dummy variable), $\psi \in \mathcal{D}_2$, define

$$\begin{aligned}\mathcal{L}^{\mu;u,\dot{y}}\psi(p) = {}& \textstyle\sum_{j=1}^J g_j((f_1,p),\ldots,(f_J,p)) \\ & \sup\Big\{ \sup_{w\in\mathbf{R}^n}[Df_j(\bar{x})(b(\bar{x},u) + w) - \tfrac{1}{2\mu}|w|^2] + L(\bar{x},u) \\ & -\tfrac{1}{\mu}\left[\tfrac{1}{2}|h(\bar{x})|^2 - h(\bar{x})\dot{y}_t\right] \\ & \quad : \ \bar{x} \in \operatorname*{argmax}_{x\in\mathbf{R}^n}\{p(x) + f_j(x)\}\Big\} - \tfrac{1}{2\mu}|\dot{y}|^2.\end{aligned} \tag{20}$$

Then one has

$$\mathcal{L}^{\mu}\psi(p) = \inf_{u\in\mathbf{R}^m} \sup_{\dot{y}\in\mathbf{R}^p} \mathcal{L}^{\mu;u,\dot{y}}\psi(p)$$

for the "generator" corresponding to $\mathcal{W}_t^{\mu}$.

Note that the risk-sensitive generator $\mathcal{L}^{\mu,\varepsilon}$ is defined in terms of integrals on $\mathbf{R}^n$ (recall (10)), whereas the minimax generator $\mathcal{L}^{\mu}$ is defined in terms of maximization over (a subset of) $\mathbf{R}^n$. Further, in both cases the underlying differential operators act on the test functions f_j. This permits consideration of cases where σ does not possess a smooth density, or when p is not a smooth, or even everywhere finite, function. Such singular cases are important in H_∞ control [17]. An important example is

$$p = \delta_M + \breve{p},$$

where δ_M ($M \subset \mathbf{R}^m$) is the "max-plus delta function"

$$\delta_M(x) = \begin{cases} 0 & \text{if } x \in M \\ -\infty & \text{if } x \notin M \end{cases}$$

and $\breve{p}$ is a real-valued function defined on M.

4 Small Noise Limits and a Common Framework

The deterministic minimax differential game is related to the risk-sensitive stochastic control problem via large deviation type small noise limits. For the nonlinear full state feedback case, this connection was established in [12], [25]. For linear systems, the relationship is more direct [22], [33]. In the nonlinear output feedback case, we need to evaluate two limits; one for the information state, and one for the value function, [24], [23], [34].

First, note that the L_2 inner product and the sup-pairing are related by

$$(p, q) = \lim_{\varepsilon\to 0} \frac{\varepsilon}{\mu} \log\langle e^{\frac{\mu}{\varepsilon}p}, e^{\frac{\mu}{\varepsilon}q}\rangle.$$

The large deviation-type limit for the information states

$$\lim_{\varepsilon\to 0} \frac{\varepsilon}{\mu} \log \sigma_t^{\mu,\varepsilon}(x) \;=\; p_t^{\mu}(x) \tag{21}$$

can be proven (in this formula $\sigma_t^{\mu,\varepsilon}(x)$ is interpreted as a density) using small noise optimal control or viscosity methods, [9], [14], [19], [26]. The corresponding limit result for the value function is

$$\lim_{\varepsilon\to 0} \frac{\varepsilon}{\mu} \log S^{\mu,\varepsilon}(e^{\frac{\mu}{\varepsilon}p}, t) \;=\; W^{\mu}(p,t). \tag{22}$$

These limit results have been proven in the discrete-time case [23], and complete results in the continuous-time case are not yet available, although some results have been obtained. Much of the technical difficulty is due to the current lack of knowledge concerning the infinite dimensional PDE (17), and the nature of the limits (21), (22). In particular, the limits involve functions defined on different spaces. In this section we adopt a variant of the framework used by [28] (see also [6]) and propose a means of unifying risk-sensitive *stochastic* control and *deterministic* differential games. In this framework, large deviation limits are interpreted as weak convergence limits in an enlarged space of generalized measures.

We construct a space $\hat{Q}_1 = \hat{Q}_1(\mathbf{R}^n)$ of generalized measures as follows. $\hat{Q}_1$ contains all measures $\sigma = \sigma^{\mu,\varepsilon}$ with integral (A is a Borel subset of $\mathbf{R}^n$)

$$\tilde{\int}_A f d\sigma^{\mu,\varepsilon} = \left(\int_A |f|^{\frac{\mu}{\varepsilon}} d\sigma^{\mu,\varepsilon}\right)^{\frac{\varepsilon}{\mu}} \tag{23}$$

and all max-plus measures $\sigma = \sigma_p$ ($p \in \mathcal{X}_e$) with integral

$$\tilde{\int}_A f d\sigma_p = \sup_{x \in A} |f(x)| e^{p(x)}. \tag{24}$$

The limit

$$\lim_{n\to\infty} \tilde{\int} f d\sigma^n = \tilde{\int} f d\sigma \tag{25}$$

(for all suitable test functions f) can be used to characterize weak convergence $\sigma^n \Longrightarrow \sigma$ as $n \to \infty$ on the space $\hat{Q}_1$. Therefore if $\sigma^{\mu,\varepsilon}$ is a sequence of measures satisfying the large deviations limit

$$\lim_{\varepsilon\to 0} \frac{\varepsilon}{\mu} \log \int e^{\frac{\mu}{\varepsilon}\phi(x)} \sigma^{\mu,\varepsilon}(dx) = \sup_{x\in\mathbf{R}^n} \{\phi(x) + p^\mu(x)\} = (\phi, p^\mu) \tag{26}$$

then equivalently

$$\lim_{\varepsilon\to 0} \tilde{\int} e^\phi d\sigma^{\mu,\varepsilon} = \tilde{\int} e^\phi d\sigma_{p^\mu} \tag{27}$$

($\phi \in C_b(\mathbf{R}^n)$). This is the weak convergence interpretation of the large deviations limit. Note that

$$\tilde{\int} e^\phi d\sigma_{p^\mu} = e^{(\phi, p^\mu)}.$$

These integrals can be extended to functionals F on the path space $C([0,T])$ as follows. In the risk-sensitive case, with $\sigma = \sigma^{\mu,\varepsilon}$, define

$$\tilde{\mathbf{E}}^\dagger_{\sigma^{\mu,\varepsilon}}[F(\sigma^{\mu,\varepsilon}(\cdot), y(\cdot))] = \left(\mathbf{E}^\dagger_{\sigma^{\mu,\varepsilon}}[(F(\sigma^{\mu,\varepsilon}(\cdot), y(\cdot)))^{\frac{\mu}{\varepsilon}}]\right)^{\frac{\varepsilon}{\mu}}, \tag{28}$$

or in the minimax case

$$\tilde{\mathbf{E}}^{\dagger}_{\sigma_{p^\mu}}[F(p^\mu_\cdot, y(\cdot))] = \sup_{y\in C([0,t],\mathbf{R})} \left\{ F(p^\mu_\cdot, y(\cdot)) \exp(-\tfrac{1}{2\mu}\int_0^t |\dot{y}_s|^2\, ds) \ : \ p^\mu_0 = p^\mu \right\}. \tag{29}$$

In formula (28), $\sigma^{\mu,\varepsilon}$ is a sequence satisfying the large deviations limit, and $\sigma^{\mu,\varepsilon}_\cdot$ denotes the corresponding risk-sensitive information state trajectory with initial condition $\sigma^{\mu,\varepsilon}_0 = \sigma^{\mu,\varepsilon}$. Similarly, in (29), $p^\mu_\cdot$ is the minimax information state trajectory initialized at $p^\mu_0 = p^\mu$.

We are now in a position to define a semigroup which generalizes both $\mathcal{S}^{\mu,\varepsilon}_t$ and $\mathcal{W}^\mu_t$ in a unified manner. Let $\mathcal{F}(\hat{\mathcal{Q}}_1)$ denote a suitable space of functions on $\hat{\mathcal{Q}}_1$ containing

$$\mathcal{D}_3 = \{\psi \ : \ \psi(\sigma) = g(\tilde{\int} f_1 d\sigma, \ldots, \tilde{\int} f_J d\sigma),$$

$$g \in C^\infty_b(\mathbf{R}^J), f_1, \ldots, f_J \in C^\infty_b(\mathbf{R}^n), J = 1, 2, \ldots\}.$$

This set includes $\mathcal{D}_1$ and $\mathcal{D}_2$ embedded in the obvious ways.

Define the operator, for $\varepsilon \geq 0$, by

$$\tilde{\mathcal{S}}^{\mu,\varepsilon}_t \psi(\sigma) = \inf_{u\in\tilde{\mathcal{U}}_{0,t}} \tilde{\mathbf{E}}^{\dagger}[\psi(\sigma_t)]. \tag{30}$$

This is the generalized semigroup.

If $\sigma^{\mu,\varepsilon} \Longrightarrow \sigma_{p^\mu}$ as $\varepsilon \to 0$, and if ψ is continuous with respect to this convergence (e.g. $\psi \in \mathcal{D}_3$), we have

$$\lim_{\varepsilon\to 0} \tilde{\mathcal{S}}^{\mu,\varepsilon}_t \psi(\sigma^{\mu,\varepsilon}) = \tilde{\mathcal{S}}^{\mu,0}_t \psi(\sigma_{p^\mu}). \tag{31}$$

This provides a unified interpretation of the limits (21), (22).

For $\psi \in \mathcal{D}_3$, in the risk-sensitive case, with $\sigma = \sigma^{\mu,\varepsilon}$, we have

$$\tilde{\mathcal{S}}^{\mu,\varepsilon}_t \psi(\sigma^{\mu,\varepsilon}) = \left(\mathcal{S}^{\mu,\varepsilon}_t (g(\langle f_1^{\frac{\mu}{\varepsilon}}, \sigma\rangle^{\frac{\varepsilon}{\mu}}, \ldots, \langle f_J^{\frac{\mu}{\varepsilon}}, \sigma\rangle^{\frac{\varepsilon}{\mu}}))^{\frac{\mu}{\varepsilon}}_{\sigma=\sigma^{\mu,\varepsilon}} \right)^{\frac{\varepsilon}{\mu}} \tag{32}$$

while in the minimax case, with $\sigma = \sigma_{p^\mu}$,

$$\tilde{\mathcal{S}}^{\mu,0}_t \psi(\sigma_{p^\mu}) = \inf_{u\in\mathcal{U}_{0,t}} \sup_{y\in C([0,t],\mathbf{R})} \{ g((f_1, e^{p^\mu_t}), \ldots, (f_J, e^{p^\mu_t})) \exp(-\tfrac{1}{2\mu}\int_0^t |\dot{y}_s|^2\, ds)\}. \tag{33}$$

This can written, using the change of variables

$$g = \exp\zeta\log, \quad f_j = \exp\phi_j, \ j = 1, \ldots, J, \tag{34}$$

as

$$\tilde{\mathcal{S}}^{\mu,0}_t \psi(\sigma_{p^\mu}) = \exp\left(\mathcal{W}^\mu_t \zeta((\phi_1, p), \ldots, (\phi_J, p))\right). \tag{35}$$

This discussion reveals an interesting unification of risk-sensitive stochastic control and deterministic minimax games at the semigroup level. We turn now to a consideration of the generators. Again one has a unified view, but it is not as complete as the semigroup perspective. The key issue is how to relate the generators $\mathcal{L}^{\mu,\varepsilon}$, $\mathcal{L}^{\mu}$ through a weakly convergent large deviation sequence $\sigma^{\mu,\varepsilon} \Longrightarrow \sigma_{p^{\mu}}$.

The generator for $\tilde{S}_t^{\mu}$ is defined by

$$\tilde{\mathcal{L}}^{\mu,\varepsilon}\psi(\sigma) = \lim_{t\downarrow 0}\frac{\tilde{S}_t^{\mu,\varepsilon}\psi(\sigma) - \psi(\sigma)}{t}. \tag{36}$$

This expression can be evaluated, in the risk-sensitive case, with $\sigma = \sigma^{\mu,\varepsilon}$, when $\psi \in \mathcal{D}_3$. This involves the use of formula (10) and the chain rule. To relate this with the minimax game, fix $u \in \mathbf{R}^m$ and note that

$$\tilde{\mathcal{L}}^{\mu,\varepsilon;u}\psi(\sigma^{\mu,\varepsilon}) = \frac{d}{dt}\tilde{S}_t^{\mu,\varepsilon;u}\psi(\sigma^{\mu,\varepsilon})|_{t=0},$$

and in view of (35) we make use of the change of variables (34) with $J = 1$, $f_1 = f = e^{\phi}$, and consider

$$\begin{aligned}
&\tfrac{d}{dt}\log \tilde{S}_t^{\mu,\varepsilon;u}\psi(\sigma^{\mu,\varepsilon})|_{t=0}\\
&= \frac{\langle he^{\frac{\mu}{\varepsilon}\phi}, \sigma^{\mu,\varepsilon}\rangle^2}{2\mu\langle e^{\frac{\mu}{\varepsilon}\phi}, \sigma^{\mu,\varepsilon}\rangle^2}[(\zeta'(\log \tilde{\int} e^{\phi} d\sigma^{\mu,\varepsilon}))^2\\
&\quad + \tfrac{\varepsilon}{\mu}\zeta''(\log \tilde{\int} e^{\phi} d\sigma^{\mu,\varepsilon}) - \zeta'(\log \tilde{\int} e^{\phi} d\sigma^{\mu,\varepsilon})]\\
&\quad + \zeta'(\log \tilde{\int} e^{\phi} d\sigma^{\mu,\varepsilon})\frac{\langle(\frac{\varepsilon}{2}\Delta\phi + \frac{\mu}{2}|D\phi|^2 + b^u\cdot D\phi + L^u)e^{\frac{\mu}{\varepsilon}\phi}, \sigma^{\mu,\varepsilon}\rangle}{\langle e^{\frac{\mu}{\varepsilon}\phi}, \sigma^{\mu,\varepsilon}\rangle}.
\end{aligned} \tag{37}$$

Then as $\varepsilon \to 0$ the RHS tends to

$$\begin{aligned}
\tilde{\mathcal{L}}^{\mu,0;u}\zeta((\phi, p^{\mu})) &\triangleq \tfrac{1}{2\mu}(\bar{h}^{\mu})^2[(\zeta'((\phi,p^{\mu})))^2 - \zeta'((\phi,p^{\mu}))]\\
&\quad + \zeta'((\phi,p^{\mu}))\int(\frac{\mu}{2}|D\phi|^2 + b^u\cdot D\phi + L^u)d\bar{\sigma}^{\mu}
\end{aligned} \tag{38}$$

where

$$\bar{h}^{\mu} = \int h d\bar{\sigma}^{\mu}$$

and $\bar{\sigma}^{\mu}$ is a probability measure supported on the set $\operatorname{argmax}_x\{\phi(x) + p^{\mu}(x)\}$ arising as the limit

$$\lim_{\varepsilon\to 0}\frac{\int \kappa e^{\frac{\mu}{\varepsilon}\phi} d\sigma^{\mu,\varepsilon}}{\int e^{\frac{\mu}{\varepsilon}\phi} d\sigma^{\mu,\varepsilon}} = \int \kappa d\bar{\sigma}^{\mu}.$$

Using the representation (c.f. [8]),

$$\tfrac{1}{2}(\bar{h}^\mu \zeta')^2 = \sup_{\dot{y}} \{ \zeta' \bar{h}^\mu \dot{y} - \tfrac{1}{2}|\dot{y}|^2 \}$$

we can rewrite (38) as

$$\begin{aligned} &\tilde{\mathcal{L}}^{\mu,0;u}\zeta((\phi, p^\mu)) \\ &= \sup_{\dot{y}} \Big\{ \zeta'((\phi, p^\mu))[\int (\frac{\mu}{2}|D\phi|^2 + b^u \cdot D\phi + L^u \\ &-\frac{1}{\mu}(\tfrac{1}{2}(\bar{h}^\mu)^2 - h\dot{y}))d\bar{\sigma}^\mu] - \frac{1}{2\mu}|\dot{y}|^2 \Big\}. \end{aligned} \tag{39}$$

There is a close resemblance between this formula and the expression (20) for the generator $\mathcal{L}^{\mu;u}$ of the minimax game semigroup $\mathcal{W}_t^{\mu;u}$ (with u fixed):

$$\begin{aligned} &\mathcal{L}^{\mu;u}\zeta((\phi, p^\mu)) \\ &= \sup_{\dot{y}} \Big\{ \zeta'((\phi, p)) \sup \Big[\frac{\mu}{2}|D\phi|^2 + D\phi(\bar{x}) \cdot b(\bar{x}, u) + L(\bar{x}, u) \\ &\quad -\tfrac{1}{\mu}\left[\tfrac{1}{2}|h(\bar{x})|^2 - h(\bar{x})\dot{y}_t\right] \\ &\quad : \ \bar{x} \in \operatorname*{argmax}_{x \in \mathbf{R}^n} \{p(x) + f_j(x)\} \Big] - \tfrac{1}{2\mu}|\dot{y}|^2 \Big\} \end{aligned} \tag{40}$$

(the supremum over w in (20) has been evaluated).
The expressions (39) and (40) are not in general the same. There is at least one instance when they agree; viz., when $\operatorname{argmax}_x \{\phi(x) + p^\mu(x)\} = \{\bar{x}\}$ is unique, in which case both (39) and (40) reduce to

$$\begin{aligned} &\sup_{\dot{y}} \Big\{ \zeta'((\phi, p^\mu))[\frac{\mu}{2}|D\phi(\bar{x})|^2 + b(\bar{x}, u) \cdot D\phi(\bar{x}) + L(\bar{x}, u) \\ &\quad -\tfrac{1}{\mu}(\tfrac{1}{2}(h(\bar{x}))^2 - h(\bar{x})\dot{y})] - \tfrac{1}{2\mu}|\dot{y}|^2 \Big\}. \end{aligned} \tag{41}$$

In H_∞ control, there is an important class of examples corresponding to uniqueness of $\operatorname{argmax}_x\{V(x)+p^\mu(x)\} = \{\bar{x}\}$, where V is the state feedback minimax value function, called certainty equivalence examples, [2] (in this case, the value function is given by $W^\mu(p,t) = (p, V)$).
It is interesting to note that in [23] the small noise limit was formally evaluated with $\sigma_t^{\mu,\varepsilon}$ interpreted as densities, and after applying the logarithmic transformations to the PDE (7) and cancelling numerator and denominator $\sigma^{\mu,\varepsilon}$ terms (c.f. (37)) the PDE (17) was obtained.

Appendix - State Feedback

In this appendix we consider briefly the state feedback case, see also [12], [13], [25].

Consider the stochastic system

$$dx_t^\varepsilon = b(x_t^\varepsilon, u_t)\,dt + \sqrt{\varepsilon}\,dw_t \tag{42}$$

with risk-sensitive cost (2) to be minimized over the class of causal state feedback controllers. The semigroup is defined by

$$S_t^{\mu,\varepsilon} f(x) = \inf_{u_{state}} \mathbf{E}_x \left[e^{\int_0^t L(x_s^\varepsilon, u_s)ds} f(x_t^\varepsilon) \right] \tag{43}$$

with fixed-control generator

$$\mathcal{L}^{\mu,\varepsilon} f(x) = A^{\varepsilon;u} f(x) + L(x,u). \tag{44}$$

This expression is well-defined for all $f \in C^\infty(\mathbf{R}^n)$ by Ito's rule.
The minimax game defined for the system

$$\dot{x}_t = b(x_t, u_t) + w_t, \tag{45}$$

with the objective of minimizing the cost (12) (with $p \equiv 0$) over the class of all causal state feedback controllers. The semigroup is

$$\mathcal{W}_t^\mu \phi(x) = \inf_{u_{state}} \sup_{w \in L_2[0,t]} \left\{ \phi(x_t) + \int_0^t [L(x_s, u_s) - \tfrac{1}{2\mu}|w_s|^2]ds \ : \ x_0 = x \right\}. \tag{46}$$

The fixed-control generator is defined by

$$\begin{aligned} \mathcal{L}^{\mu;u}\phi(x) &= \tfrac{\mu}{2}|D\phi(x)|^2 + b(x,u)\cdot D\phi(x) + L(x,u) \\ &= \sup_w \{ D\phi(x)\cdot(b(x,u)+w) - \frac{1}{2\mu}|w|^2 + L(x,u) \} \end{aligned} \tag{47}$$

This generator is well-defined for all $\phi \in C^\infty(\mathbf{R}^n)$ by the method of characteristics for first-order PDE.
We now show how these are related via a large deviations limits as in §4. Define the generalized semigroup as follows: in the risk-sensitive case write

$$\tilde{S}_t^{\mu,\varepsilon} f(x) = \left(\inf_{u_{state}} \mathbf{E}_{x_\varepsilon} \left[e^{\frac{\mu}{\varepsilon}\int_0^t L(x_s^\varepsilon, u_s)ds} |f(x_t^\varepsilon)|^{\frac{\mu}{\varepsilon}} \right] \right)^{\frac{\varepsilon}{\mu}} \tag{48}$$

and in the minimax game case write

$$\tilde{S}_t^{\mu,0} f(x) = \inf_{u_{state}} \sup_{w \in L_2[0,t]} \left\{ |f(x_t)| e^{\int_0^t [L(x_s,u_s) - \frac{1}{2\mu}|w_s|^2]ds} \ : \ x_0 = x \right\}. \tag{49}$$

Now the probability measures for the stochastic system (42) obey the large deviations principle and so we have the convergence

$$\lim_{\varepsilon \to 0} \tilde{S}_t^{\mu,\varepsilon} f(x) = \tilde{S}_t^{\mu,0} f(x) = \exp\left(\mathcal{W}_t^\mu \log f(x)\right). \tag{50}$$

To relate the fixed-control generators, use (44) and the change of variables

$$f = \exp \phi \tag{51}$$

and calculate

$$\frac{d}{dt} \log \tilde{S}_t^{\mu,\varepsilon;u} e^{\phi(x)}|_{t=0} = \frac{\varepsilon}{2}\Delta\phi(x) + \frac{\mu}{2}|D\phi(x)|^2 + b(x,u) \cdot D\phi(x) + L(x,u). \tag{52}$$

As $\varepsilon \to 0$ this tends to

$$\tilde{\mathcal{L}}^{\mu,\varepsilon}\phi(x) \triangleq \frac{\mu}{2}|D\phi(x)|^2 + b(x,u) \cdot D\phi(x) + L(x,u) \tag{53}$$

which is just (47).

Acknowledgments: This research was supported in part by the Australian Research Council.

Bibliography

[1] M. Akian, J.P. Quadrat and M. Viot, *Bellman Processes*, in: Lecture Notes in Control & Info. Science, 199, 1994, Springer Verlag, New York.

[2] T. Basar and P. Bernhard, "H^{∞}-Optimal Control and Related Minimax Design Problems: A Dynamic Game Approach", Second Edition, Birkhauser, Boston, 1995.

[3] A. Bensoussan and J.H. van Schuppen, *Optimal Control of Partially Observable Stochastic Systems with an Exponential-of-Integral Performance Index*, SIAM J. Control Optim., **23** (1985) 599-613.

[4] M.G. Crandall and P.L. Lions, *Hamilton-Jacobi Equations in Infinite Dimensions*, J. Func. Anal., Part I, Uniqueness of Solutions, 62, 339-396; Part II, Existence of Solutions, 65, 1986, 368-405; Part III, 68, 1986, 214-247; Part IV, Unbounded Linear Terms, 90, 1990, 137-283; Part V, B-Continuous Solutions, 97, 1991, 417-465.

[5] J.C. Doyle, K. Glover, P.P. Khargonekar and B.A. Francis, *State-Space Solutions to Standard H_2 and H_∞ Control Problems*, IEEE Trans. Aut. Control, **AC-34** (8) (1989) 831-847.

[6] P. Dupuis and R.S. Ellis, "A Weak Convergence Approach to the Theory of Large Deviations", Wiley, New York, 1997.

[7] R.J. Elliott, "Stochastic Calculus and Applications", Springer Verlag, New York, 1982.

[8] W.H. Fleming, *The Cauchy Problem for a Nonlinear First Order Differential Equation*, J. Diff. Equations, 5 (1969) 555-530.

[9] W.H. Fleming, *Stochastic Control for Small Noise Intensities*, SIAM J. Control Optim., 9, (1971), 473-517.

[10] W.H. Fleming, *Nonlinear Semigroup for Controlled Partially Observed Diffusions*, SIAM J. Control Optim., 20(2) 1982, 286-301.

[11] W.H. Fleming and E. Pardoux, *Optimal Control for Partially Observed Diffusions*, SIAM J. Control Optim., **20** (2) (1982) 261-285.

[12] W.H. Fleming and W.M. McEneaney, *Risk Sensitive Optimal Control and Differential Games*, Proc. Conf. on Adaptive and Stochastic Control, Univ. of Kansas, Springer Verlag, 1991.

[13] W.H. Fleming and W.M. McEneaney, *Risk-Sensitive Control on an Infinite Time Horizon*, SIAM J. Control Optim., 33 (1995), 1881-1915.

[14] W.H. Fleming and H.M. Soner, "Controlled Markov Processes and Viscosity Solutions", Springer-Verlag, New York, 1993.

[15] W.H. Fleming and M.R. James, *The Risk-Sensitive Index and the H_2 and H_∞ Norms for Nonlinear Systems*, J. Mathematics of Control, Signals, and Systems, 8 (1996) 199-221.

[16] J.W. Helton and M.R. James, *Reduction of Controller Complexity in Nonlinear H_∞ Control*, 34rd IEEE CDC, New Orleans, 1995.

[17] J.W. Helton and M.R. James, "A General Framework for Extending H_∞ Control to Nonlinear Systems", monograph, to appear, SIAM, 1998.

[18] J.W. Helton, M.R. James and W.M. McEneaney, *Nonlinear Control: The Joys of Having Extra Sensors*, submitted to 1998 IEEE CDC.

[19] O. Hijab, "Minimum Energy Estimation", PhD Thesis, UC Berkeley, 1980.

[20] O. Hijab, *Partially Observed Control of Markov Processes III*, Annals of Probability, 18, 1990, 1099-1125.

[21] O. Hijab, *Partially Observed Control for Markov Processes*, Proc. Conf. on Applied Stochastic Analysis, Rutgers University, Springer Verlag, 1991 (I. Karatzas and D. Ocone, eds.).

[22] D.H. Jacobson, *Optimal Stochastic Linear Systems with Exponential Performance Criteria and Their Relation to Deterministic Differential Games*, IEEE Trans. Aut. Control, **AC-18** (2) (1973) 124-131.

[23] M.R. James, J.S. Baras and R.J. Elliott, *Risk-Sensitive Control and Dynamic Games for Partially Observed Discrete-Time Nonlinear Systems*, IEEE Trans. Automatic Control, 39 (1994), 780-792.

[24] M.R. James, J.S. Baras and R.J. Elliott, *Output Feedback Risk-Sensitive Control and Differential Games for Continuous-Time Nonlinear Systems*, 32nd IEEE CDC, San Antonio, Dec. 1993, 3357-3360.

[25] M.R. James, *Asymptotic Analysis of Nonlinear Stochastic Risk-Sensitive Control and Differential Games*, Math. Control, Signals, and Systems, **5**(4) (1992) 401-417.

[26] M.R. James and J.S. Baras, *Nonlinear Filtering and Large Deviations: A PDE-Control Theoretic Approach*, Stochastics, **23**(3) (1988) 391-412.

[27] M.R. James and J.S. Baras, *Partially Observed Differential Games, Infinite Dimensional HJI Equations, and Nonlinear H_∞ Control*, SIAM J. Control Optim., 1996.

[28] T. Jiang, *The Metric of Large Deviation Convergence*, preprint, 1995.

[29] P.R. Kumar and P. Varaiya, "Stochastic Systems: Estimation, Identification, and Adaptive Control", Prentice-Hall, Englewood Cliffs, 1986.

[30] G.L. Litvinov and V.P. Maslov, *Correspondence Principle for Idempotent Calculus and Some Computer Applications*, preprint, 1995.

[31] P.L. Lions, *Viscosity Solutions of Fully Nonlinear Second Order Equations and Optimal Stochastic Control in Infinite Dimensions, Part II: Optimal Control of Zakai's Equation*, CEREMADE Report 8825, 1989.

[32] M. Nisio, *On a Nonlinear Semigroup Attached to Optimal Stochastic Control*, Publ. RIMS Kyoto Univ. 13 (1976), 513-537.

[33] P. Whittle, *Risk-Sensitive Linear Quadratic Gaussian Control*, Adv. Appl. Prob., **13** (1981) 764-777.

[34] P. Whittle, *A Risk-Sensitive Maximum Principle: The Case of Imperfect State Observation*, IEEE Trans. Aut. Control, **36** (1991) 793-801.

5

Nonlinear, Dissipative, Infinite Dimensional Systems

Maciej Kocan
Pierpaolo Soravia

ABSTRACT: In this paper we study and characterize dissipative, infinite dimensional, nonlinear and unbounded systems. We show that the system is dissipative with respect to a given supply rate and a function V is a storage function for the system if and only if V satisfies a suitable partial differential inequality in the sense of viscosity solutions. We also characterize by a representation formula (as the available storage) the minimal among all solutions of such a partial differential inequality. Applications to nonlinear $\mathcal{H}_\infty$ control are also discussed.

KEYWORDS: Evolution equations, distributed parameter systems, nonlinear systems, dissipative systems, viscosity solutions, nonlinear $\mathcal{H}_\infty$ control

1 Introduction

In this paper we study the controlled evolution equation

$$\dot{y}(t) + Ay(t) \ni f(y(t), w(t)), \quad t \geq 0, \quad y(0) = x, \tag{1.1}$$

where A is a maximal monotone operator in a real Hilbert space H, possibly nonlinear and multivalued, with domain $D(A) \subset H$ (in particular, $x \in \overline{D(A)}$). The function f on the right hand side of (1.1), $f\colon \overline{D(A)} \times W \to H$, where W is another real Hilbert space (or a closed subset of a Hilbert space), is at least continuous, see precise assumptions in the next section. Solutions of (1.1) are then meant in the mild sense and $y(\cdot) \equiv y(\cdot, x, w)$ indicates the unique solution of (1.1). The evolution equation (1.1) is affected by the unknown deterministic disturbance $w(\cdot) \in L^2_{\text{loc}}(0, +\infty; W)$.
We are given a function $h\colon \overline{D(A)} \times W \to \mathbf{R}$, $-h$ is usually called the supply rate, and the goal we want to achieve is a characterization of dissipative systems according to the following definition.

Definition 1.1 *The system (1.1) is said to be dissipative with respect to the supply rate $-h$ if there is a nonnegative function $V\colon \overline{D(A)} \to [0, +\infty)$*

such that for all $t \geq 0$ *and* $w \in L^2_{\mathrm{loc}}(0, +\infty; W)$

$$V(y(t,x,w)) + \int_0^t h(y(s,x,w), w(s))ds \leq V(x). \tag{1.2}$$

A function V satisfying Definition 1.1 is called a storage function. The definition of dissipative system was first formulated and studied by Willems (1972), see also Moylan (1974) and Hill and Moylan (1976).
In the general framework of infinite dimensional, unbounded and nonlinear systems, we will show that a nonnegative lower semicontinuous function V is a storage function if and only if it satisfies the following partial differential inequality

$$\langle Ax, DV(x)\rangle + \inf_{w \in W} \Big\{ -\langle f(x,w), DV(x)\rangle - h(x,w)\Big\} \geq 0, \quad x \in \overline{D(A)}. \tag{1.3}$$

Of course, the meaning of the differential inequality (1.3) has to be interpreted, and (1.3) may have nonsmooth solutions, which are of interest. Solutions of (1.3) are going to be meant in the sense of the recent theory of viscosity solutions for unbounded, nonlinear Hamilton-Jacobi equations, see Tataru (1992), (1994) and Crandall and Lions (1994), as applied to equations without comparison principle, see Kocan and Soravia (1998). Of course, a given system is not necessarily dissipative, and finite solutions to (1.3) may not exist. The counterpart of the results we prove, in the case of finite dimensional systems, was discussed in James (1993).
Dissipative systems have found an important application in nonlinear $\mathcal{H}_\infty$ control theory, when one wants to test if the "$\mathcal{H}_\infty$ norm" of a given feedback control satisfies a prescribed bound. We refer the reader to the last section for a discussion about this application. In general, when such a feedback control is not available, the discussion of the $\mathcal{H}_\infty$ control problem becomes much more complicated, since the problem is related to the study of a suitably defined differential game. For more complete results on this issue in the case of nonlinear, infinite dimensional systems, as well as more references on this side of the problem, we refer the reader to the forthcoming paper Kocan and Soravia (1997), see also Soravia (1996) and (1998) and the references therein.

2 Dissipative systems

In this section we describe our main results. Before doing so, we state the main assumptions that we need in the following.
The operator A is maximal monotone in a real Hilbert space H, with domain $D(A) \subset H$. It is well known that $-A$ generates a semigroup of contractions on $\overline{D(A)}$, which we will denote by $S(t)$. For notational simplicity we always assume that $(0,0) \in A$; it follows that $\|S(t)x\| \leq \|x\|$ for

all $x \in \overline{D(A)}$. We always use the same notation $\|x\|$ to indicate the norm of the element x in an appropriate vector space. Moreover, P will denote the projection in H onto $\overline{D(A)}$.
W is another real Hilbert space and the unknown disturbance $w(\cdot)$ in (1.1) is an element of $L^2_{\text{loc}} \equiv L^2_{\text{loc}}(0, +\infty; W)$.
We assume that there are $\gamma > 0$, $L > 0$, $p \in [1,2]$ and $q \in [1,2)$ satisfying $pq \leq 2$ such that the function f: $\overline{D(A)} \times W \to H$ in the dynamics (1.1) and the supply rate $-h$: $\overline{D(A)} \times W \to \mathbf{R}$ satisfy

$$f \text{ is continuous,} \quad h \text{ is lower semicontinuous,} \tag{2.1}$$

and

$$\|f(x,w) - f(z,w)\| \leq L\|x - z\|, \ \|f(x,w)\| \leq L(1 + \|x\| + \|w\|^q), \tag{2.2}$$
$$-L(1 + \|x\|^p + \|w\|^2) \leq h(x,w) \leq L(1 + \|x\|^p) - \gamma^2\|w\|^2, \tag{2.3}$$

for all $x, z \in \overline{D(A)}$ and $w \in W$. A special case interesting for the applications, in particular in $\mathcal{H}_\infty$ control, where the assumptions above are satisfied, is when $q = 1$ and $h(x,w) = g(x) - \gamma^2\|w\|^2$. Here, g is lower semicontinuous with at most quadratic growth. Our current framework does not allow for a discussion of passive systems, i.e. systems that are dissipative with respect to the supply rate $-h(x,w) = \langle x, w\rangle$, where now $H = W$.
Note that the dissipation inequality of (1.2) can be rewritten as

$$V(x) = \sup_{w \in L^2_{\text{loc}}} \sup_{t \geq 0} \left\{ \int_0^t h(y(s,x,w), w(s))ds + V(y(t,x,w)) \right\}, \tag{2.4}$$

since one of the two inequalities in (2.4) is obvious by choosing $t = 0$ in the sup operation. We will show that this property characterizes the viscosity solutions of (1.3), as we define next.
In order to define viscosity solutions of (1.3), we first need to define precisely the class of test functions that we are going to use. We will write Lip(X) and $C^1(X)$ for the spaces of all real-valued Lipschitz continuous and continuously Fréchet differentiable, respectively, functions on the space X. Given $\psi \in \text{Lip}(X)$, $L(\psi)$ will denote the best Lipschitz constant for ψ.

Definition 2.1 *We say that $\Phi = \varphi + \psi \in C^1(H) + Lip(H)$ is a supertest function if $\varphi(Px) \geq \varphi(x)$ for all $x \in H$. Subtest functions are defined by reversing the inequality above.*

Then we need to define appropriately how to interpret the term "$\langle Ax, DV\rangle$" in (1.3). Note that it does not make sense directly unless $x \in D(A)$.

Definition 2.2 *For a given function Φ: $\overline{D(A)} \to \mathbf{R}$ and $\hat{x} \in \overline{D(A)}$, we define*

$$D_A^+\Phi(\hat{x}) = \limsup_{\substack{\overline{D(A)} \ni x \to \hat{x} \\ h \to 0+}} \tfrac{1}{h}\left(\Phi(x) - \Phi(S(h)x)\right).$$

Similarly, $D_A^-\Phi$ is defined by replacing lim sup *by* lim inf *above.*

We can now define viscosity solutions of (1.3).

Definition 2.3 *A lower semicontinuous function $V\colon \overline{D(A)} \to \mathbf{R}$ is a viscosity solution of the partial differential inequality (1.3) if for all supertest functions $\Phi = \varphi + \psi$, if $V - \Phi$ has relative minimum at $\hat{x} \in \overline{D(A)}$, then*

$$D_A^+\Phi(\hat{x})+\inf_{w\in W}\left\{-\langle f(\hat{x},w),D\varphi(\hat{x})\rangle+L(\psi)\|f(\hat{x},w)\|-h(\hat{x},w)\right\}\geq 0. \quad (2.5)$$

We refer the reader to Crandall and Lions (1994) and Kocan and Soravia (1998) (in particular Remark 3.7) and the references therein for more details and an extensive discussion of the definition of viscosity solution in our situation. Note that the notion of viscosity solution of (1.3) discussed in Crandall and Lions (1994) is slightly different (i.e. stronger) than ours.
In this paper, we prove the following results.

Theorem 2.4 *Assume that (2.1), (2.2) and (2.3) hold. If $V\colon \overline{D(A)} \to [0,+\infty)$ is a storage function, then its lower semicontinuous envelope V_* is a viscosity solution of (1.3).*

In the previous statement, the lower semicontinuous envelope of a function $V\colon \overline{D(A)} \to \mathbf{R}$ is defined as $V_*(\hat{x}) = \liminf_{\overline{D(A)}\ni x\to\hat{x}} V(x)$. Note that $V \equiv V_*$ if and only if V is lower semicontinuous.

Theorem 2.5 *Assume that (2.1), (2.2) and (2.3) hold. If $V\colon \overline{D(A)} \to [0,+\infty)$ is a viscosity solution of (1.3), then it is a storage function. In particular, if V is a storage function, V_* is also a storage function.*

We immediately prove the following consequence of the two theorems above.

Corollary 2.6 *Assume that (2.1), (2.2) and (2.3) hold. The system (1.1) is dissipative with respect to the supply rate $-h$ if and only if there is a nonnegative viscosity solution of the partial differential inequality (1.3). In this case, the available storage function*

$$V_a(x) = \sup_{w\in L^2_{\text{loc}}}\sup_{t\geq 0}\int_0^t h(y(s,x,w),w(s))ds \quad (2.6)$$

is lower semicontinuous, the minimal viscosity solution of (1.3) and the minimal storage function.

The available storage function defined in the previous statement is helpful in practice since, as it is clear from its definition, the system is dissipative if and only if V_a is finite.

Remark 2.7 *If the available storage function V_a is not only finite but of at most quadratic growth, then it satisfies in $\overline{D(A)}$ not only the differential* *inequality (1.3) but also the following variational inequality*

$$\min\Big\{\langle Ax, DV(x)\rangle + \inf_{w\in W}\{-\langle f(x,w), DV(x)\rangle - h(x,w)\},\ V(x)\Big\} = 0. \tag{2.7}$$

When solutions of (2.7) are meant in the viscosity sense, this fact is not much harder to prove than Theorem 2.4. However, we will not prove this statement in full detail, see instead Kocan and Soravia (1997) and a similar proof in Section 3.

Proof of Corollary 2.6. The first statement is an obvious consequence of the two theorems above. We only need to show the properties of V_a. We first show that V_a is a storage function; then it is clearly the minimal one. By definition

$$V_a(x) \geq \int_0^t h(y(s,x,w), w(s))ds \tag{2.8}$$

for all $w \in L^2_{\rm loc}$ and $t \geq 0$. Fix $T \geq 0$ and $w \in L^2(0,T;W)$. Then for all possible extensions of w, by patching an element of $L^2_{\rm loc}$ we have

$$V_a(x) \geq \int_0^T h(y(s,x,w), w(s))ds + \int_T^{T+t} h(y(s,x,w), w(s))ds = \\ \int_0^T h(y(s,x,w), w(s))ds + \int_0^t h(y(s, y(T,x,w), w(\cdot+T)), w(s+T))ds$$

for all $t \geq 0$. The conclusion follows from the definition of $V_a(y(T,x,w))$. Now, by Theorem 2.5, if U is a nonnegative supersolution of (1.3), it is a storage function, hence it satisfies (1.2). U being nonnegative, it follows from (1.2) that $U \geq V_a$.
On the other hand, since V_a is a storage function, by Theorem 2.4, V_{a*} solves the differential inequality (1.2), hence the previous argument can be applied to V_{a*} and we get $V_{a*} \geq V_a$, which shows that V_a is lower semicontinuous. △
Storage functions can be used as Liapunov functions to prove local stability of the undisturbed system – meaning that $w(\cdot) \equiv 0$ in (1.1) – at certain points. This is important for applications, for instance to $\mathcal{H}_\infty$ control.
For notational simplicity in the following we will always consider stability at the origin, but this role can be played by any point in $\overline{D(A)}$.

Definition 2.8 *We say that the undisturbed system*

$$\dot{y}(t) + Ay(t) \ni f(y(t), 0), \quad t \geq 0, \tag{2.9}$$

is locally stable at 0 *if for every* $\epsilon > 0$ *there is* $\delta > 0$ *such that for every* $x \in \overline{D(A)}$ *satisfying* $\|x\| \leq \delta$

$$\|y(t,x,0)\| \leq \epsilon$$

for all $t \geq 0$. *The undisturbed system is locally asymptotically stable at* 0 *if moreover*

$$\lim_{t\to+\infty} y(t,x,0) = 0$$

for all $\|x\| \leq \delta$.

Recall that $(0,0) \in A$; in particular $S(\cdot)0 \equiv 0$. If in addition $f(0,0) = 0$ then 0 is an equilibrium of (2.9). The following discussion shows that stability of this equilibrium can be deduced from the existence of certain storage functions.
Suppose that

$$h(x,0) \geq 0 \tag{2.10}$$

for all $x \in \overline{D(A)}$. If V is a storage function then by definition we have

$$V(x) \geq V(y(t,x,0))$$

for all $t \geq 0$. Therefore the sublevel sets of V:

$$\mathcal{V}_\epsilon = \left\{x \in \overline{D(A)}:\ V(x) \leq \epsilon\right\},$$

$\epsilon > 0$, are invariant for (2.9).
Suppose now that V vanishes and is continuous at the origin and moreover V is positive definite, meaning that there is a strictly increasing, continuous function ω: $[0,\infty) \to [0,\infty)$ satisfying $\omega(0) = 0$ such that $V(x) \geq \omega(\|x\|)$ for all $x \in \overline{D(A)}$. It follows that the family $\{\mathcal{V}_\epsilon\}_{\epsilon>0}$ is a base of neighborhoods of 0 in $\overline{D(A)}$. Indeed, by the assumed continuity of V at 0 and $V(0) = 0$, every $\mathcal{V}_\epsilon$ is a neighborhood of 0. Let $\delta > 0$. If $x \in \overline{D(A)}$ and $\|x\| > \delta$ then $V(x) > \omega(\delta)$ and therefore

$$\mathcal{V}_{\omega(\delta)} \subset \left\{x \in \overline{D(A)}:\ \|x\| \leq \delta\right\}.$$

Since every $\mathcal{V}_\epsilon$ is invariant for (2.9), local stability at 0 follows.
If moreover $h(x,0)$ is also positive definite, for simplicity with respect to the same function ω, then the undisturbed system is asymptotically stable at 0. Indeed, take any $x \in \overline{D(A)}$. Since

$$\int_0^{+\infty} h(y(s,x,0),0)ds \leq V(x) < \infty,$$

there is a sequence $t_n \to \infty$ such that $h(y(t_n,x,0),0) \to 0$, and h positive definite yields that $y(t_n,x,0) \to 0$. Thus for any fixed $\epsilon > 0$, for all sufficiently large n we have $y(t_n,x,0) \in \mathcal{V}_\epsilon$. By invariance of $\mathcal{V}_\epsilon$ with respect to (2.9) we conclude that $y(t,x,0) \in \mathcal{V}_\epsilon$ for all sufficiently large t. Since $\{\mathcal{V}_\epsilon\}_\epsilon$ is a base of open neighborhoods of 0, asymptotic stability follows. We obtained the following result.

Proposition 2.9 *Assume that (2.2) holds. If the system (1.1) is dissipative with respect to the supply rate $-h$ satisfying (2.10) and has a positive definite storage function vanishing and continuous at the origin, then the undisturbed system (2.9) is locally stable at the origin. If moreover $h(x,0)$ is also positive definite, then the undisturbed system is locally asymptotically stable at* 0.

The proof of Proposition 2.9 shows that the origin is in fact globally attractive and that, with a little more technical effort, V locally positive definite is enough to conclude. Of course, coupling Proposition 2.9 with Theorem 2.5, local stability of (2.10) can be deduced from existence of viscosity solutions of (1.3) possessing additional properties.

3 Proof of Theorem 2.4

Before proving the Theorem, observe that by (2.2) and Gronwall's inequality

$$\|y(t,x,w) - y(t,z,w)\| \le \|x-z\|e^{Lt} \tag{3.1}$$

for all $x, z \in \overline{D(A)}$, $w \in L^2_{\text{loc}}$ and $t \ge 0$. Similarly,

$$\|y(t,x,w) - S(t)x\| \le L\Big(t + t\|x\| + \int_0^t \|w\|^q\Big)e^{Lt} \tag{3.2}$$

for all $x \in \overline{D(A)}$, $w \in L^2_{\text{loc}}$ and $t \ge 0$. From (3.2) it follows that $y(t,x,w) \to x$ as $t \to 0^+$, uniformly in all w bounded in $L^2(0,t;W)$.

Proof of Theorem 2.4. Let V be a storage function; we want to show that V_* solves (1.3). By definition, let $\Phi = \varphi + \psi$ be a Lipschitz continuous supertest function (it is not restrictive to assume φ to be globally Lipschitz) and assume that $V_* - \Phi$ attains a local minimum at $\hat{x} \in \overline{D(A)}$, such that $V_*(\hat{x}) = \Phi(\hat{x})$. If $D_A^+\Phi(\hat{x}) = +\infty$ there is nothing to prove. Otherwise we argue by contradiction and suppose that there are $\theta > 0$ and $w^* \in W$ such that

$$D_A^+\Phi(\hat{x}) - \langle f(\hat{x},w^*), D\varphi(\hat{x})\rangle + L(\psi)\|f(\hat{x},w^*)\| - h(\hat{x},w^*) < -2\theta. \tag{3.3}$$

First we show that

$$D_A^+\Phi(\hat{x}) > -\infty. \tag{3.4}$$

Choose $x_n \in \overline{D(A)}$ such that $x_n \to \hat{x}$ and $V(x_n) \to V_*(\hat{x})$. Since $y(t,x_n,0) \to \hat{x}$ as $n \to \infty$ and $t \to 0^+$, by the minimum property of Φ, (1.2), (3.2) and (2.3) we get for all small $t > 0$

$$\begin{aligned}
\Phi(S(t)\hat{x}) - V(x_n) &\le \Phi(S(t)\hat{x}) - \Phi(y(t,x_n,0)) + V(y(t,x_n,0)) - V(x_n) \\
&\le L(\Phi)\big(\|y(t,x_n,0) - S(t)\hat{x}\|\big) - \textstyle\int_0^t h(y(s,x_n,0),0)ds \\
&\le L(\Phi)\big(Lt(1+\|x_n\|)e^{Lt} + \|x_n - \hat{x}\|\big) + L\textstyle\int_0^t (1+\|y(s,x_n,0)\|^p)ds \\
&\le M\,(t + \|x_n - \hat{x}\|),
\end{aligned}$$

where the constant M does not depend on n and all sufficiently small $t > 0$. As $n \to +\infty$ we then get $\Phi(S(t)\hat{x}) - \Phi(\hat{x}) \le O(t)$ as $t \to 0^+$, which yields (3.4).

For $s \in [0,t]$ put $y_n(s) = y(s, x_n, w^*)$. From (3.3), (2.2) and (2.1) it follows that, if t is sufficiently small, then

$$D_A^+\Phi(\hat{x}) - \langle v(s), D\varphi(y_n(s))\rangle + L(\psi)\|v(s)\| - h(y_n(s), w^*) < -\theta$$

for all $s \in [0,t]$, where we denoted $v(s) = f(y_n(s), w^*)$. The previous step allows us to integrate from 0 to t, using Tataru's result, see Tataru (1994) and Corollary 4.8 in Kocan, Soravia and Swiech (1997), which yields (note that it works just like formally except for the term $o(t)$)

$$t\theta \leq \int_0^t h(y_n(s), w^*)ds + \Phi(y_n(t)) - \Phi(x_n) + o(t)$$

as $t \to 0^+$, uniformly for all n. By the local minimum property of Φ we obtain for all n

$$t\theta \leq \int_0^t h(y_n(s), w^*)ds + V(y_n(t)) - \Phi(x_n) + o(t)$$

as $t \to 0^+$. From (1.2) we then get $t\theta \leq V(x_n) - \Phi(x_n) + o(t)$, and letting $n \to +\infty$ we therefore obtain $t\theta \leq o(t)$ and a contradiction when choosing t sufficiently small. Δ

4 Auxiliary value function

In this section we introduce an approximating value function which will play a crucial role in the proof of our main result Theorem 2.5. Throughout the rest of this section it is assumed that f is continuous, (2.2) holds, $\lambda > 0$ is fixed and $U: H \to \mathbf{R}$ is a fixed bounded Lipschitz continuous function. We also suppose that $g: H \times W \to \mathbf{R}$ is a fixed function such that for some $Q > 0$

$$-Q(1 + \|x\|^2) \leq g(x, w) \leq Q \tag{4.1}$$

for all $(x, w) \in H \times W$, and such that for every $R > 0$ there is $L_R > 0$ such that

$$|g(x, w) - g(z, v)| \leq L_R(\|x - z\| + \|w - v\|) \tag{4.2}$$

for all $x, z \in H$ satisfying $\|x\|, \|z\| \leq R$ and all $u, v \in W$. Functions U and g are approximations of a nonnegative supersolution of (1.3) and of $h(x, w) + L\|w\|^2$, respectively, arising in the proof of Theorem 2.5, see Section 5.

Let

$$G(x, w) = g(x, w) - L\|w\|^2.$$

Define $\rho: \mathbf{R} \to (0, \pi)$ as $\rho(t) = \frac{\pi}{2} + \tan^{-1}(t)$ and note that ρ is strictly increasing and 1-Lipschitz continuous on $\mathbf{R}$. Define also $W: \overline{D(A)} \times \mathbf{R} \to (0, \pi)$ according to

$$W(x, r) = \rho(U(x) + r). \tag{4.3}$$

We will consider an auxiliary value function $V^\lambda \equiv V^\lambda_{W,G}$ defined, for $(x,r) \in \overline{D(A)} \times \mathbf{R}$, by

$$V^\lambda(x,r) = \sup_{w \in L^2_{\mathrm{loc}}} \sup_{t \geq 0} e^{-\lambda t} W(y(t,x,w), r + r(t,x,w)), \tag{4.4}$$

where

$$r(t,x,w) = \int_0^t G(y(s,x,w), w(s))ds. \tag{4.5}$$

Choosing $t = 0$ in (4.4) shows that $V^\lambda \geq W > 0$. We begin by establishing some regularity properties of the auxiliary value function V^λ.

Lemma 4.1 *Suppose that $V^\lambda(x,r) > \delta > 0$ and $r < K$. Then there are $T = T(\delta)$ and $M = M(\delta, K)$ such that*

$$V^\lambda(x,r) = \sup_{0 \leq t \leq T} \sup \left\{ e^{-\lambda t} W(y(t,x,w), r + r(t,x,w)) \colon \int_0^T \|w\|^2 \leq M \right\}.$$

Proof. Since $\rho < \pi$, only t satisfying $\delta \leq e^{-\lambda t}\pi$ are relevant, which determines T. Further, for $t \leq T$

$$W(y(t,x,w), r + r(t,x,w)) \leq \rho\Big(\|U\|_{L^\infty} + K + TQ - L \int_0^t \|w\|^2 \Big),$$

and thus M is determined from $\delta \leq \rho(\|U\|_{L^\infty} + K + TQ - LM)$. △
Using Lemma 4.1, (3.1) and (3.2) and properties of ρ one easily proves the following result.

Lemma 4.2 *1. $V^\lambda(x,r) \to 0$ as $r \to -\infty$, uniformly in $x \in \overline{D(A)}$.*

2. If $V^\lambda(x_1,r_1) > V^\lambda(x_2,r_2) > \delta > 0$, where $r_1, \|x_1\|, \|x_2\| \leq K$, then there is $N = N(\delta,K) > 0$ such that $V^\lambda(x_1,r_1) - V^\lambda(x_2,r_2) \leq N\|x_1 - x_2\| + r_1 - r_2$. In particular, V^λ is Lipschitz continuous in a neighborhood of every point.

3. Suppose that $V^\lambda(x,r) > \delta > 0$ and $r < K$. Then there is $\nu = \nu(\delta,K) > 0$ such that $\liminf_{h \to 0+} \frac{1}{h}(V^\lambda(x,r+h) - V^\lambda(x,r)) > \nu$.

The dynamic programming principle for the auxiliary value function can be written as follows.

Lemma 4.3 *If $V^\lambda(\hat{x},\hat{r}) > W(\hat{x},\hat{r}) \geq \delta > 0$ then there are $\varepsilon > 0$ and $M = M(\delta) > 0$ such that if $\|x - \hat{x}\| + |r - \hat{r}| + t < \varepsilon$ then*

$$V^\lambda(x,r) \leq \sup \left\{ e^{-\lambda t} V^\lambda(y(t,x,w), r + r(t,x,w)) \colon \int_0^t \|w\|^2 \leq M \right\}.$$

For more details concerning the standard proofs of the two previous lemmas, we refer to Lemma 4.3 in Soravia (1996) or Proposition 6.4 in Kocan and Soravia (1997) for similar arguments.
Given Lemma 4.3, one expects $V = V^\lambda$ to be a subsolution of

$$\lambda V + \langle Ax, D_x V\rangle + \inf_{w\in W}\Big\{ - \langle f(x,w), D_x V\rangle - G(x,w) D_r V\Big\} = 0 \quad (4.6)$$

on $\big\{(x,r) \in \overline{D(A)} \times \mathbf{R}\colon\ V^\lambda(x,r) > W(x,r)\big\}$. Recalling (2.2), note that the Hamiltonian in (4.6) is not necessarily finite unless $D_r V > 0$. However, by Lemma 4.2, $D_r V^\lambda > 0$ in the viscosity sense. As a consequence, V^λ indeed does solve (4.6) if solutions are understood in an appropriate sense, as we define below. Observe that the variable r does not appear in the unbounded term $\langle Ax, D_x V\rangle$, thus equation (4.6) is an example of an unbounded equation in separated variables, as studied in Kocan and Swiech (1995).

Definition 4.4 *An upper semicontinuous function $V\colon \overline{D(A)} \times \mathbf{R} \to \mathbf{R}$ is a viscosity subsolution of (4.6) in $\Omega \subset \overline{D(A)} \times \mathbf{R}$ if for every subtest function Φ as in Definition 2.1, $\eta \in C^1(\mathbf{R})$ and local maximum $(\hat{x},\hat{r}) \in \Omega$ of $V(x,r) - \Phi(x) - \eta(r)$ relative to Ω we have $\eta'(\hat{r}) > 0$ and*

$$\begin{aligned} \lambda V(\hat{x},\hat{r}) + D_A^- \Phi(\hat{x}) + \inf\nolimits_{w\in W} \Big\{ &- \langle f(\hat{x},w), D\varphi(\hat{x})\rangle \\ &- L(\psi)\|f(\hat{x},w)\| - G(\hat{x},w)\eta'(\hat{r})\Big\} \le 0. \end{aligned} \quad (4.7)$$

Supersolutions of (4.6) are defined analogously, by replacing upper semicontinuity by lower semicontinuity, subtest function by supertest function, local maximum by local minimum, and by reversing the inequality in (4.7), after changing $D_A^-\Phi$ to $D_A^+\Phi$ and $-L(\psi)$ to $L(\psi)$.

Proposition 4.5 *The auxiliary function V^λ is a viscosity subsolution (in the sense of Definition 4.4) of (4.6) in $\Omega = \big\{(x,r) \in \overline{D(A)}\times\mathbf{R}\colon\ V^\lambda(x,r) > W(x,r)\big\}$.*

The following simple fact is needed in the proof of Proposition 4.5.

Lemma 4.6 *Suppose that $w \in L^2_{\mathrm{loc}}(0,\infty;\mathbf{R})$, $q \in [1,2)$ and $M > 0$. Then there is $K = K(q,M) > 0$ such that for all $t \ge 0$*

$$M\int_0^t |w|^q - \int_0^t |w|^2 \le Kt.$$

Proof of Lemma 4.6. Denoting $W = \int_0^t |w|^2$, by Hölder's inequality

$$M\int_0^t |w|^q - \int_0^t |w|^2 \le M W^{\frac{q}{2}} t^{1-\frac{q}{2}} - W = t\Big(M\left(\tfrac{W}{t}\right)^{\frac{q}{2}} - \tfrac{W}{t}\Big);$$

now let $K = \max_{r \geq 0} \left(Mr^{q/2} - r\right)$. $\triangle$

Proof of Proposition 4.5. Let $\Phi = \varphi + \psi$ be a subtest function as in Definition 2.1, $\eta \in C^1(\mathbf{R})$ and suppose that $V^\lambda(x,r) - \Phi(x) - \eta(r)$ has a local maximum at $(\hat{x},\hat{r}) \in \overline{D(A)} \times \mathbf{R}$, where $V^\lambda(\hat{x},\hat{r}) > W(\hat{x},\hat{r})$. By Lemma 4.2 $\eta'(\hat{r}) > 0$. First we will show that

$$D_A^- \Phi(\hat{x}) < \infty. \tag{4.8}$$

Without loss of generality we can assume that $\Phi \in \mathrm{Lip}(H)$. Let M and ϵ be as in Lemma 4.3 and suppose that $t \in (0,\epsilon)$, $\|w\|_{L^2(0,t;W)} \leq M$ and $V^\lambda(y(t), r(t)) > W(\hat{x},\hat{r})$, where we denoted $y(t) = y(t,\hat{x},w)$ and $r(t) = \hat{r} + \int_0^t G(y(s),w(s))ds$. From Lemma 4.2 either $V^\lambda(y(t),r(t)) \leq V^\lambda(S(t)\hat{x},\hat{r})$ or

$$V^\lambda(y(t),r(t)) - V^\lambda(S(t)\hat{x},\hat{r}) \leq N\|y(t) - S(t)\hat{x}\| + r(t) - \hat{r}$$

for some universal $N > 0$, and from (3.2) and Lemma 4.6, in both cases we get

$$V^\lambda(y(t),r(t)) \leq V^\lambda(S(t)\hat{x},\hat{r}) + O(t)$$

as $t \to 0^+$, uniformly in $\|w\|_{L^2(0,t;W)} \leq M$. Therefore Lemma 4.3 yields $V^\lambda(\hat{x},\hat{r}) \leq V^\lambda(S(t)\hat{x},\hat{r}) + O(t)$, and by the maximum property of $V^\lambda - \Phi$, $\Phi(\hat{x}) - \Phi(S(t)\hat{x}) \leq O(t)$, which gives (4.8).

Once (4.8) is established, the rest of the proof of Proposition 4.5 is similar to the proof of Theorem 2.4 and we omit it. $\triangle$

5 Proof of Theorem 2.5

We are ready to proceed with the proof of our main result Theorem 2.5. The main tool for proving optimality principles of type (2.4) is comparison principle for viscosity solutions, see e.g. Tataru (1991) for the first result in this direction for equations with nonlinear A. However, comparison principle for (1.3) clearly fails: for instance, one can modify V by any constant without affecting (1.3). Further difficulties are due to the fact that h may change sign. Thus to apply the known techniques of viscosity solutions some approximations are required.

Suppose that U is a nonnegative solution of (1.3). We start introducing a change of variables. Recall that ρ: $\mathbf{R} \to (0,\pi)$ has been defined in Section 4 by $\rho(t) = \frac{\pi}{2} + \tan^{-1}(t)$ and define W: $\overline{D(A)} \times \mathbf{R} \to (0,\pi)$ as in (4.3). A formal computation suggests that W ought to be a solution of

$$\langle Ax, D_x W\rangle + \inf_{w \in W} \Big\{ - \langle f(x,w), D_x W\rangle - h(x,w) D_r W \Big\} \geq 0 \tag{5.1}$$

in $\overline{D(A)} \times \mathbf{R}$, where viscosity solutions of (5.1) are defined similarly as in Definition 4.4. We do not show explicitly that after the change of variables

(5.1) is satisfied since this simple proof parallels that of Lemma 3.8 in Kocan and Soravia (1998).

In addition to approximating the equation (1.3), we will approximate the function U by more regular functions. We construct a nondecreasing sequence $\{U_n\}_n$ of bounded, Lipschitz continuous functions from H to $\mathbf{R}$ such that $U = \sup_n U_n$ on $\overline{D(A)}$. This procedure is well-known and can be accomplished by means of inf-convolutions, combined with an appropriate cut-off technique. For instance

$$U_n(x) = \inf_{y \in \overline{D(A)}} \{ \min(U(y), n) + n\|x - y\|^2 \} \tag{5.2}$$

has the desired properties.

Approximations for h are more complicated. Let $g(x, w) = h(x, w) + L\|w\|^2$; note that by (2.3) we have

$$-L(1 + \|x\|^p) \le g(x, w). \tag{5.3}$$

For $x \in H$ and $w \in W$ define

$$g_n(x, w) = \inf_{y \in \overline{D(A)}, v \in W} \{ \min(g(y, v), n) + n \left(\|x - y\|^2 + \|w - v\|^2 \right) \}. \tag{5.4}$$

By (5.3), for sufficiently large n (determined by L) g_n is finite. Clearly the sequence of functions $\{g_n\}_n$ is nondecreasing. By standard results on inf-convolutions, see e.g. Attouch and Azé (1993), we obtain that g_n converge to g on $\overline{D(A)} \times W$, so that $g = \sup_n g_n$, and each g_n is Lipschitz continuous in its both variables on every set of the form $\{x \in H \colon \|x\| \le R\} \times W$, $R > 0$. Moreover, for n sufficiently large we have

$$-2L(1 + \|x\|^p) \le g_n(x, w) \le n \tag{5.5}$$

for all $(x, w) \in H \times W$. Now put $h_n(x, w) = g_n(x, w) - L\|w\|^2$. It follows that the sequence h_n is nondecreasing and approximates h from below, i.e. $h = \sup_n h_n$ on $\overline{D(A)} \times W$.

Proof of Theorem 2.5. Suppose that $U \colon \overline{D(A)} \to [0, +\infty)$ is a supersolution of (1.3). By the discussion above, W given by (4.3) is a viscosity solution of (5.1).

Fix $\lambda > 0$. Let $U_n \colon H \to \mathbf{R}$ be given by (5.2) and $g_n \colon H \times W \to \mathbf{R}$ by (5.4). For every n, let $W_n(x, r) = \rho(U_n(x) + r)$ and denote $V_n \equiv V^{\lambda}_{W_n, h_n}$, where V^{λ} is the auxiliary value function from Section 4. By Proposition 4.5, V_n is a continuous solution of

$$\lambda V_n + \langle Ax, D_x V_n \rangle + \inf_{w \in W} \Big\{ -\langle f(x, w), D_x V_n \rangle - h_n(x, w) D_r V_n \Big\} \le 0 \tag{5.6}$$

on $\big\{(x, r) \in \overline{D(A)} \times \mathbf{R} \colon V_n(x, r) > W_n(x, r)\big\}$.

We will show that

$$V_n \le W \quad \text{on } \overline{D(A)} \times \mathbf{R}. \tag{5.7}$$

We argue by contradiction and suppose that $V_n - W > 0$ somewhere in $\overline{D(A)} \times \mathbf{R}$. For $\alpha, \beta, \delta, \epsilon > 0$, $x, y \in \overline{D(A)}$ and $r, s \in \mathbf{R}$ let

$$\Phi(x,y,r,s) = V_n(x,r) - W(y,s) - \tfrac{\alpha}{2}\|x-y\|^2 - \tfrac{\beta}{2}(r-s)^2 \\ -\delta(\|x\|^2 + \|y\|^2) - \epsilon(r^2 + s^2).$$

If δ and ϵ are sufficiently small then $\sup \Phi \geq 2\tau > 0$, where τ does not depend on any of the parameters. By the perturbed optimization result in Kocan and Swiech (1996) one can find $\hat{x}, \hat{y} \in \overline{D(A)}$, $\hat{r}, \hat{s} \in \mathbf{R}$ and $a, b \in \mathbf{R}$ such that $|a|, |b| < \varepsilon$, $\Phi(\hat{x}, \hat{y}, \hat{r}, \hat{s}) \geq \sup \Phi - \varepsilon$ and the map

$$\Phi(x,y,r,s) - \varepsilon d(x,\hat{x}) - \varepsilon d(y,\hat{y}) - ar - bs$$

has a strict global maximum at $(\hat{x}, \hat{y}, \hat{r}, \hat{s})$. The function d that we are using above is the so-called Tataru distance, see Tataru (1992), (1994) and Crandall and Lions (1994), namely

$$d(x,y) = \inf_{t \geq 0} \{t + \|x - S(t)y\|\},$$

which does not define a metric, but yet a version of the Ekeland ε–variational principle with d in place of a usual distance function holds. Note that $\Phi(\hat{x}, \hat{y}, \hat{r}, \hat{s}) \geq \sup \Phi - \varepsilon \geq 2\tau - \varepsilon \geq \tau$, for all sufficiently small ε, and then

$$V_n(\hat{x},\hat{r}) \geq W(\hat{y},\hat{s}) + \tfrac{\alpha}{2}\|\hat{x}-\hat{y}\|^2 + \tfrac{\beta}{2}|\hat{r}-\hat{s}|^2 + \delta(\|x\|^2 + \|y\|^2) + \varepsilon(|\hat{r}|^2 + |\hat{s}|^2) + \tau. \tag{5.8}$$

Since V_n is bounded from above by π, (5.8) gives

$$\|\hat{x} - \hat{y}\| \leq \sqrt{2\pi/\alpha}, \quad |\hat{r} - \hat{s}| \leq \sqrt{2\pi/\beta}. \tag{5.9}$$

Recall that $V_n \geq W_n$ and first suppose that $V_n(\hat{x}, \hat{r}) = W_n(\hat{x}, \hat{r})$. Since $W \geq W_n$ for every n, from (5.8)

$$\tau \leq V_n(\hat{x},\hat{r}) - W(\hat{y},\hat{s}) \leq W_n(\hat{x},\hat{r}) - W_n(\hat{y},\hat{s}) \leq L(W_n)(\|\hat{x}-\hat{y}\| + |\hat{r}-\hat{s}|),$$

which, due to (5.9), cannot happen if α and β are sufficiently big. Thus we can assume that $V_n(\hat{x}, \hat{r}) > W_n(\hat{x}, \hat{r})$ for all choices of parameters.
From (5.8) we obtain that $V_n(\hat{x}, \hat{r}) \geq \tau$ and $W(\hat{y}, \hat{s}) \leq \pi - \tau$, and Lemma 4.2 and the properties of U and ρ, respectively, give that $\hat{r}$ is bounded from below and $\hat{s}$ is bounded from above, uniformly in all parameters, and now (5.9) shows that for any β fixed, $\hat{r}$ and $\hat{s}$ are bounded, uniformly in all other parameters. Now Lemma 4.2 together with the fact that $V_n(\hat{x}, r) - \frac{\beta}{2}(r - \hat{s})^2 - \epsilon r^2 - ar$ has a maximum at $r = \hat{r}$ yield that

$$\beta(\hat{r} - \hat{s}) + 2\epsilon\hat{r} + a \geq \nu = \nu(\beta) > 0. \tag{5.10}$$

In particular, we can apply in (5.6) the definition of viscosity solution at $(\hat{x}, \hat{r})$ by the maximum property of $(\hat{x}, \hat{y}, \hat{r}, \hat{s})$. Since $W(\hat{y}, s) + \frac{\beta}{2}(\hat{r} - s)^2 + \varepsilon s^2 + bs$ has a minimum point at $\hat{s}$, then

$$\beta(\hat{r} - \hat{s}) - 2\varepsilon\hat{s} - b > 0$$

and the definition of viscosity solution can be applied to (5.1) at $(\hat{y},\hat{s})$ as well. Therefore, by adapting Doubling Theorem 3.1 in Crandall and Lions (1994) we obtain (note that the contributions of the unbounded terms drop out)

$$\begin{aligned}\lambda V_n(\hat{x},\hat{r}) \leq \inf\nolimits_{w\in W} \Big\{ &- \langle f(\hat{y},w), \alpha(\hat{x}-\hat{y}) - 2\delta\hat{y}\rangle + \varepsilon\|f(\hat{y},w)\| \\ &-h(\hat{y},w)(\beta(\hat{r}-\hat{s}) - 2\varepsilon\hat{s} - b)\Big\} \\ - \inf\nolimits_{w\in W} \Big\{ &- \langle f(\hat{x},w), \alpha(\hat{x}-\hat{y}) + 2\delta\hat{x}\rangle - \varepsilon\|f(\hat{x},w)\| \\ &-h_n(\hat{x},w)(\beta(\hat{r}-\hat{s}) + 2\varepsilon\hat{r} + a)\Big\} + 2\varepsilon.\end{aligned} \tag{5.11}$$

From (5.8) we also get

$$\epsilon|\hat{r}|, \epsilon|\hat{s}| \leq \sqrt{\epsilon\pi}, \quad \delta\|\hat{x}\|, \delta\|\hat{y}\| \leq \sqrt{\delta\pi}. \tag{5.12}$$

In particular, by (5.10) and (5.12) we have that, for any fixed β,

$$\beta(\hat{r}-\hat{s}) > 0 \tag{5.13}$$

for all sufficiently small δ, ϵ.
For $(x,p,e,s) \in \overline{D(A)} \times H \times \mathbf{R} \times (0,+\infty)$ define

$$F_n(x,p,e,s) = \inf_{w\in W} \big\{ - \langle f(x,w), p\rangle + \|f(x,w)\|e - h_n(x,w)s\big\}. \tag{5.14}$$

Note that for every $R > 0$ there is $C = C(R)$ such that if $\|p\|, |e|, s < R$ and $s > \frac{1}{R}$ then $\inf_{w\in W}$ in (5.14) can be replaced by the infimum over $\big\{w \in W\colon \|w\| \leq C(1+\|x\|^{p/2})\big\}$. This follows since, by definition, $-h_n(x,w) = L\|w\|^2 - g_n(x,w)$ and g_n satisfies (5.5), and by (2.2) (see Lemma 5.4 in Kocan and Soravia (1998) for a similar computation).
From (2.3) we have $-h(x,w) \geq \gamma^2\|w\|^2 - L(1+\|x\|^p)$ and then the same is true for F which is defined as F_n in (5.14) but with h replacing h_n.
Thus (5.10), (5.12) and (5.9) show that for α,β fixed, $\delta < 1$ and sufficiently small ε, all $w \in W$ relevant for the inf's on the right hand side of (5.11) are uniformly bounded by $C\big(1+\|\hat{x}\|^{p/2}+\|\hat{y}\|^{p/2}\big)$, where $C = C(\alpha,\beta)$, and then taking lim sup as $\varepsilon \to 0^+$ in (5.11) we arrive at

$$\begin{aligned}\lambda\tau \leq \limsup\nolimits_{\varepsilon\to 0^+} \Big(&F(\hat{y}, \alpha(\hat{x}-\hat{y}) - 2\delta\hat{y}, 0, \beta(\hat{r}-\hat{s})) \\ &-F_n(\hat{x}, \alpha(\hat{x}-\hat{y}) + 2\delta\hat{x}, 0, \beta(\hat{r}-\hat{s}))\Big).\end{aligned} \tag{5.15}$$

Recalling (5.4), we have that

$$h_n(\hat{x},w) \leq h(\hat{y},w) + n\|\hat{x}-\hat{y}\|^2.$$

Therefore from (5.15), (2.2) and (5.13) we get

$$\lambda\tau \leq \limsup_{\varepsilon\to 0+} \Big(L\alpha\|\hat{x}-\hat{y}\|^2 + n\beta|\hat{r}-\hat{s}|\|\hat{x}-\hat{y}\|^2 + \\ 2\delta \sup\Big\{ |\langle f(\hat{y},w),\hat{y}\rangle| + |\langle f(\hat{x},w),\hat{x}\rangle| : \|w\| \leq C(1+\|\hat{x}\|^{p/2}+\|\hat{y}\|^{p/2})\Big\}\Big). \tag{5.16}$$

Recall that by standard arguments, see e.g. the proof of Theorem 4.1 in Ishii (1992), from (5.8) we have

$$\limsup_{\beta\to\infty} \limsup_{\alpha\to\infty} \limsup_{\delta\to 0+} \limsup_{\varepsilon\to 0+} \left(\alpha\|\hat{x}-\hat{y}\|^2 + \delta(\|\hat{x}\|^2+\|\hat{y}\|^2)\right) = 0.$$

Now taking the iterated lim sup in (5.16), first as $\delta \to 0^+$, while recalling (5.9), (2.2) and $pq \leq 2$, then as $\alpha \to \infty$, and finally as $\beta \to \infty$, we arrive at a contradiction. Thus (5.7) is proved.

So far we showed that for every $\lambda > 0$ and $n \geq 1$

$$W(x,r) \geq \sup_{t\geq 0} \sup_{w\in L^2_{\text{loc}}} e^{-\lambda t} W_n(y(t,x,w), r + r(t,x,w)) \tag{5.17}$$

for all $x \in \overline{D(A)}$ and $r \in \mathbf{R}$. Fix $t > 0$ and $w \in L^2(0,t;W)$. From (5.17), letting $\lambda \to 0^+$ we obtain that

$$W(x,r) \geq W_n(y(t,x,w), r + r(t,x,w)),$$

which implies, since ρ is increasing, for all $n \geq 1$ and $t > 0$

$$U(x) \geq \int_0^t h_n(y(s,x,w), w(s))ds + U_n(y(t,x,w)), \tag{5.18}$$

for all $w \in L^2(0,t;W)$. Letting $n \to \infty$ in (5.18), by (5.5) and the Monotone Convergence Theorem we get that for all $x \in \overline{D(A)}$, $t > 0$, and $w \in L^2(0,t;W)$

$$U(x) \geq \int_0^t h(y(s,x,w), w(s))ds + U(y(t,x,w)),$$

which proves that U is a storage function and concludes the proof. △

6 Applications to $\mathcal{H}_\infty$ control

In this section we recall the formulation of state feedback $\mathcal{H}_\infty$ suboptimal control problem and its relationship with dissipative systems. Suppose we are now given the new system

$$\dot{y} + Ay \ni f(y,w) + B(y)u, \quad t \geq 0, \quad y(0) = x \in \overline{D(A)}, \tag{6.1}$$

where the only difference compared to (1.1) consists in the fact that the right hand side has an additional control term. Namely, U is another real Hilbert space and the operator-valued function $B\colon H \to \mathcal{L}(U,H)$ is assumed to be Lipschitz continuous, where $\mathcal{L}(U,H)$ is the vector space of all linear and bounded operators from U to H equipped with the operator norm. In (6.1) we are free to choose a function $u(\cdot)$ taking its values in U.
Given a lower semicontinuous cost function $g\colon \overline{D(A)} \to [0,+\infty)$ and a constant $\gamma > 0$ (the disturbance attenuation level), the goal to achieve is to find a feedback control $u\colon \overline{D(A)} \to U$ such that the closed loop system

$$\dot{y} + Ay \ni f(y,w) + B(y)u(y), \quad t \geq 0, \quad y(0) = x \tag{6.2}$$

is well-defined (i.e., it has a unique mild solution $y(\cdot, x, w)$), it is locally asymptotically stable at the origin when $w \equiv 0$, and the following gain condition is satisfied:

$$\int_0^t (g(y(s,x,w)) + \|u(y(s,x,w))\|^2)ds \leq \gamma^2 \int_0^t \|w(s)\|^2 ds + V(x) \tag{6.3}$$

for all $x \in \overline{D(A)}$, $w \in L^2_{\text{loc}}$ and $t \geq 0$, where $V\colon \overline{D(A)} \to [0,+\infty)$ is a function vanishing and continuous at the origin.
If $u(\cdot)$ is a feedback that makes the closed loop system well-defined, then the discussion in Section 2 leads to the fact that the main point that we really need to determine is whether system (6.2) with supply rate

$$-h(x,w) = \gamma^2\|w\|^2 - g(x) - \|u(x)\|^2 \tag{6.4}$$

is dissipative or not. Therefore the question can be decided by our theorems in the previous sections.
This procedure applies for example if we can find a C^1 function $V\colon \overline{D(A)} \to [0,+\infty)$ with Lipschitz continuous Fréchet derivative DV, solving in classical sense (which we explain below) the partial differential inequality

$$\begin{array}{c}\langle Ax, DV(x)\rangle + \inf_{w\in W}\Big\{ -\langle f(x,w), DV(x)\rangle + \gamma^2\|w\|^2\Big\} \\ +\frac{1}{4}\|[B(x)]^* DV(x)\|^2 \geq g(x)\end{array} \tag{6.5}$$

in $\overline{D(A)}$. In this case, in fact, the well-known choice for the feedback is

$$u(x) = -\tfrac{1}{2}[B(x)]^* DV(x).$$

By observing that with this choice we have

$$\tfrac{1}{4}\|[B(x)]^* DV(x)\|^2 = -\langle Bu(x), DV(x)\rangle - \|u(x)\|^2,$$

Theorem 2.5 can be applied and we obtain that V is a storage function for the closed loop system with respect to the supply rate in (6.4).

Below we give an example where the above discussion can be carried out in full detail, despite the nonlinear unbounded term A in the dynamics. Often however, a smooth solution of (6.5) may not exist, or may not be easy to determine, and (6.5) has to be interpreted as the dynamic programming equation of a differential game. In this case, the technical details to approach the $\mathcal{H}_\infty$ problem become sensibly heavier and will be presented elsewhere, see Kocan and Soravia (1997). Existence of a finite nonnegative solution of the partial differential inequality (6.5) can be also shown to be necessary to solve the $\mathcal{H}_\infty$ suboptimal problem.
We denote

$$F(x,p) = \inf_{w\in W} \{ -\langle f(x,w),p\rangle + \gamma^2\|w\|^2\},$$

so that $F\colon \overline{D(A)} \times H \to \mathbf{R}$. Classical solvability of (6.5) means that V is Fréchet differentiable at every point $x \in D(A)$ with derivative $DV(x) \in H$ and

$$\langle y, DV(x)\rangle + F(x, DV(x)) + \tfrac{1}{4}\|[B(x)]^* DV(x)\|^2 \geq g(x)$$

for every $y \in Ax$. By Proposition 6.3 in Kocan, Soravia and Swiech (1997) classical solutions of (6.5) are viscosity solutions in the sense of Crandall and Lions (1994), and therefore viscosity solutions in our sense, Definition 2.3.
We look for classical solutions of (6.5) of the special form $V(x) = K\|x\|^2$, where K is a positive constant. Note that V vanishes at the origin and it is continuous and positive definite.
Since $(0,0) \in A$ and A is monotone, we have $\langle y,x\rangle \geq 0$ for every $(x,y) \in A$ and thus to verify that $V \in C^1$ is a classical (and therefore viscosity) solution of (6.5) it is sufficient for V to satisfy pointwise

$$F(x, DV(x)) + \tfrac{1}{4}\|[B(x)]^* DV(x)\|^2 \geq g(x) \tag{6.6}$$

for all $x \in \overline{D(A)}$. This now becomes a standard task.
In the following, we suppose that

$$0 \leq g(x) \leq M\|x\|^2, \tag{6.7}$$

and assume that, for any constant $K > 0$, we have

$$\inf\left\{\tfrac{1}{\|x\|^2}F(x,2Kx)\colon 0 \neq x \in \overline{D(A)}\right\} \geq -\tfrac{a}{\rho(\gamma)}K^2 - 2bK, \tag{6.8}$$

where a, b are independent of K and $\rho(\cdot)$ is a positive function such that $\lim_{\gamma\to+\infty} \rho(\gamma) = +\infty$. The last condition holds, for instance, if $f(x,w) = f_1(x) + Cw$, where $C\colon W \to H$ is linear and bounded, f_1 is Lipschitz continuous and $f_1(0) = 0$.
In our case (6.6) reads

$$F(x, 2Kx) + K^2\|[B(x)]^* x\|^2 \geq g(x),$$

and it is certainly satisfied if we can find $K > 0$ such that

$$\left(\inf_{\|x\|=1} \|[B(x)]^* x\|^2 - \tfrac{a}{\rho(\gamma)}\right) K^2 - 2bK - M \geq 0. \tag{6.9}$$

We then need to assume that we have enough control on the system, namely that

$$\inf_{\|x\|=1} \|[B(x)]^* x\| > 0. \tag{6.10}$$

Therefore, for sufficiently large γ, i.e.

$$\rho(\gamma) \inf_{\|x\|=1} \|[B(x)]^* x\|^2 > a, \tag{6.11}$$

we get that the algebraic equation (6.9) has a positive solution K. Moreover note that

$$h(x, 0) = g(x) + \|u(x)\|^2 = g(x) + K^2 \|[B(x)]^* x\|^2$$

is then positive definite if either g is positive definite or $B(x) \equiv B$ is a linear and bounded operator from U to H satisfying (6.10).
We obtained the following result.

Theorem 6.1 *Consider system (6.1) and assume that (2.2), (6.7), (6.8) and (6.10) hold. Suppose moreover that either g is positive definite or $B(\cdot) \equiv B$: $U \to H$ is linear and bounded. Then, for any disturbance attenuation level $\gamma > 0$ satisfying (6.11), the $\mathcal{H}_\infty$ control problem for the nonlinear system (6.1) has a feedback solution given by $u(x) = -K[B(x)]^* x$, where $K > 0$ solves the algebraic equation (6.9). The feedback is linear when $B(\cdot)$ is constant.*

Acknowledgments: The first author was supported by the Australian Research Council. The second author was supported by Carnegie Mellon University, CNR-GNAFA and Università di Padova.

References

Attouch, H. and Azé, D. (1993) *Approximation and regularization of arbitrary functions in Hilbert spaces by the Lasry-Lions method*, Ann. Inst. H. Poincaré Anal. Non Linéaire **10**, 289–312.

Crandall, M.G. and Lions, P.L. (1994). *Hamilton-Jacobi equations in infinite dimensions, part VI. Nonlinear A and Tataru's method refined*, in Evolution Equations, Control Theory and Biomathematics, Clement, P. and Lumer, G. eds., Lect. Notes Pure Appl. Math. **155**, Marcel Dekker, 51–89.

Hill, D.J. and Moylan, P.J. (1976). *The stability of nonlinear, dissipative systems*, IEEE Trans. Automat. Control **21**, 708–711.

Ishii, H. (1992) *Viscosity solutions for a class of Hamilton-Jacobi equations in Hilbert spaces*, J. Func. Anal. **105**, 301–341.

James, M.R. (1993). *A partial differential inequality for dissipative, nonlinear systems*, Systems Control Lett. **21**, 315–320.

Kocan, M. and Soravia, P. (1997). *Differential games and nonlinear $\mathcal{H}_\infty$ control in infinite dimensions*, Mathematics research report, Australian National University **MRR 020-97**.

Kocan, M. and Soravia, P. (1998). *A viscosity approach to infinite-dimensional Hamilton-Jacobi equations arising in optimal control with state constraints*, SIAM J. Control Optim. **36**, 1348–1375.

Kocan, M., Soravia, P. and Swiech, A. (1997). *On differential games for infinite dimensional systems with nonlinear, unbounded operators*, J. Math. Anal. Appl. **21**, 395–423.

Kocan, M. and Swiech, A. (1995). *Second order unbounded parabolic equations in separated form*, Studia Math. **115**, 291–310.

Kocan, M. and Swiech, A. (1996). *Perturbed optimization on product spaces*, Nonlin. Anal. Theory Methods Appl. **26**, 81–90.

Moylan, P.J. (1974). *Implications of passivity in a class of nonlinear systems*, IEEE Trans. Automat. Control **19**, 373–381.

Soravia, P. (1996). *$\mathcal{H}_\infty$ control of nonlinear systems: differential games and viscosity solutions*, SIAM J. Control Optim. **34**, 1071–1097.

Soravia, P. (1998). *Nonlinear $\mathcal{H}_\infty$ control*, in Bardi, M. and Capuzzo-Dolcetta, I., Optimal control and viscosity solutions of Hamilton-Jacobi equations, Birkhäuser.

Tataru, D. (1991). *On the equivalence between the dynamic programming principle and the dynamic programming equation*, in Estimation and control of distributed parameter systems, Desch, W. et al. eds., International Series of Numerical Mathematics, **100**, Birkhäuser Verlag, 331–340.

Tataru, D. (1992). *Viscosity solutions for Hamilton-Jacobi equations with unbounded nonlinear terms*, J. Math. Anal. Appl. **163**, 345–392.

Tataru, D. (1994). *Viscosity solutions for Hamilton-Jacobi equations with unbounded nonlinear term: a simplified approach*, J. Diff. Equations **111**, 123–146.

Willems, J.C. (1972). *Dissipative dynamical systems part I: general theory*, Arch. Rat. Mech. Anal. **45**, 321–351.

6

Singular Limits of Bellman Equations of Ergodic Type Related to Risk-Sensitive Control

Hideo Nagai

ABSTRACT: We take up a particular class of Bellman-Isaacs equations of ergodic type related to risk-sensitive control, whose solutions satisfying a kind of semi-boundedness condition are unique up to additive constants. To study their singular limits we prove dynamic programing principle for ergodic control. Although the limit equations have in general infinitely many viscosity solutions, by giving variational representations, we characterize the singular limits among those solutions.

KEYWORDS: Bellman-Isaacs equation of ergodic type, risk-sensitive control, singular limit, dynamic programming principle.

1 Introduction

Let us consider a stochastic control problem with a risk-sensitive, averaging cost criterion:

$$\zeta_\epsilon = \liminf_{T\to\infty} \frac{\epsilon}{T\theta} \log E[e^{\frac{\theta}{\epsilon}\int_0^T \Phi(X_s,z_s)ds}], \tag{1.1}$$

where X_s is a controlled process governed by the stochastic differential equation

$$dX_t = \sqrt{\epsilon}\sigma(X_t)dB_t + \beta(X_t, z_t)dt,\ X_0 = x, \tag{1.2}$$

z_t is a control process and B_t is a standard $\mathcal{F}_t$ - Brownian motion process defined on a filtered probability space $(\Omega, \mathcal{F}, P; \mathcal{F}_t)$. The constant θ is called risk sensitive parameter. The minimum value $\chi_\epsilon = \inf_{z_\cdot} \zeta_\epsilon$ is considered to be characterized by the following Bellman equation of ergodic type:

$$\chi_\epsilon = \frac{\epsilon}{2}a^{ij}D_{ij}v_\epsilon + \frac{\theta}{2}a^{ij}D_iv_\epsilon D_jv_\epsilon + Q(x, \nabla v_\epsilon), \tag{1.3}$$

where $Q(x,p) = \inf_z\{\beta^i(x,z)p_i + \Phi(x,z)\}$, $a = \sigma\sigma^*$, and it is regarded as the following Bellman-Isaacs equation

$$\chi_\epsilon = \frac{\epsilon}{2}a^{ij}D_{ij}v_\epsilon + \inf_z\{\beta^i(x,z)D_iv_\epsilon + \Phi(x,z)\} + \sup_{y\in R^N}\{y^iD_iv_\epsilon - \frac{1}{2\theta}a_{ij}^{-1}y^iy^j\}.$$

Note that here we have utilized summation convention and σ^* stands for transposed matrix of σ. It is important to change to such a form from the viewpoint considering the relationships between risk sensitive control and H_∞ control. In fact, when letting ϵ tend to 0, we have the limit equation under suitable conditions

$$\chi_0 = \inf_z\{\beta(x,z)^iD_iv + \Phi(x,z)\} + \sup_{y\in R^N}\{y^iD_iv - \frac{1}{2\theta}a_{ij}^{-1}y^iy^j\} \tag{1.4}$$

and the pair of the constant χ_0 and the function v is considered to be a solution of the equation and to be connected to a differential game relevant to nonlinear H_∞ control (cf. Fleming and McEneaney(1995) and references therein).

In the present paper, we shall make some remarks on the passage to the limit as $\epsilon \to 0$ of the solution of (1.3). Note that under some conditions the solution of (1.3) with a certain semi-boundedness condition is seen to be unique up to additive constants to v_ϵ and it is a stabilizing solution (cf. Bensoussan and Frehse (1992) and Nagai (1996)). However, there exist some cases that such uniqueness does not hold for the limit equation (1.4). Indeed, we have an example such that (1.4) has infitely many viscosity solutions satisfying such semi-boundedness condition as we shall show below. Therefore we need in general to pick up a particular solution of (1.4) as the limit of the unique solution of (1.3), distinguishing from others. One may consider the solution must be the one relevant to the differential game connected to H_∞ control, although it is not attempted to verify it here. In place, by taking up a particular case such that (1.4) may have infinitely many solutions, we specify the solution of (1.4) obtained as the limit of the solution of (1.3) by giving a kind of variational representation.

Let us explain more precisely. We specialize to the case where $\sigma(x) \equiv I$, $\beta(x,z) = -\nabla U + z$, $\Phi(x,z) = V(x) + \frac{1}{2}|z|^2$, and then (1.3) reads

$$\chi_\epsilon = \frac{\epsilon}{2}\Delta v_\epsilon - \nabla U \cdot \nabla v_\epsilon - \frac{1-\theta}{2}|\nabla v_\epsilon|^2 + V(x). \tag{1.3$'$}$$

Taking up functions U and V with general growth order, equation (1.3)' has been obtained from a Bellman equation of discounted type as the limit as the discount factor tends to 0 (cf. Kaise and Nagai (1998) and section 2). We note that the pair $(v_\epsilon, \chi_\epsilon)$ of a function v_ϵ and a constant χ_ϵ is considered to be a solution in such equation as (1.3)'. By taking a transformation,

$$w_\epsilon = (1-\theta)v_\epsilon + U \tag{1.5}$$

we have the following equation

$$\frac{\epsilon}{2}\Delta w_\epsilon - \frac{1}{2}|\nabla w_\epsilon|^2 + \hat{V}_\epsilon = (1-\theta)\chi_\epsilon, \tag{1.6}$$

where

$$\hat{V}_\epsilon = \frac{1}{2}|\nabla U|^2 + (1-\theta)V - \frac{\epsilon}{2}\Delta U \tag{1.7}$$

and we can see that the solution v_ϵ of (1.3)' such that $(1-\theta)v_\epsilon + U$ is bounded from below is unique up to additive constants provided that $\hat{V}_\epsilon(x) \to \infty$, as $|x| \to \infty$ (cf. Bensoussan and Frehse (1992), Nagai (1996)). We shall consider the unique solution of (1.3)' in such a sense and study its asymptotic behavor as $\epsilon \to 0$. The reason why such solution is specialized can be seen by looking at the following example.

Example 1. Let us set $U(x) = \frac{1}{2}x^*Ax$ and $V(x) = \frac{1}{2}x^*Rx$, then the solution v_ϵ of (1.3)' are given by the formula

$$v_\epsilon = \frac{1}{2}x^*Px, \quad \chi_\epsilon = \mathrm{tr}P, \tag{1.8}$$

using a solution P of the following algebraic Riccati equation:

$$-AP - PA - (1-\theta)P^2 + R = 0 \tag{1.9}$$

provided that (1.9) has a solution. Let P^+ be a stabilizing solution of (1.9), namely such solution that $-A-(1-\theta)P^+$ is negative definite. Then,denoting the corresponding solution of (1.3)' to P^+ by v_ϵ^+,

$$(1-\theta)v_\epsilon^+ + U = \frac{1-\theta}{2}x^*P^+x + \frac{1}{2}x^*Ax$$

turns out to be bounded from below.

For general U and V we have no more Riccati equation and we pick up such solution of (1.3)' that $(1-\theta)v_\epsilon + U$ is bounded from below. Now, under suitable conditions we have obtained (cf. Kaise and Nagai (1998)) the limit equation of (1.3)' as $\epsilon \to 0$:

$$-\nabla U \cdot \nabla v - \frac{1-\theta}{2}|\nabla v|^2 + V = \frac{1}{1-\theta}\min \hat{V}, \tag{1.10}$$

where

$$\hat{V} = \frac{1}{2}|\nabla U|^2 + (1-\theta)V.$$

We have actually shown that a subsequence of $\{v_\epsilon\}$ of the solution of (1.3)' such that $(1-\theta)v_\epsilon + U$ is bounded from below converges to a viscosity solution v of (1.10) such that $(1-\theta)v + U$ is bounded from below. However, the solution of (1.10) satisfying such lower boundedness condition is not unique unless $\hat{V}(x)$ has a single minimum (cf. Ishii, Nagai and Teramoto (1996)). Actually (1.10) can have infinitely many viscosity solutions satisfying such lower boundedness condition as we see in the following example.

Example 2. We consider the case where $U \equiv 0$, $V(x) \geq 0$ and simplify (1.10) as

$$|\nabla v|^2 = V. \tag{1.11}$$

Let $N = 1$ and $V(x) = x^2(x-\alpha)^2$, $\alpha > 0$. Then

$$u_1 = \begin{cases} -\frac{1}{3}x^3 + \frac{\alpha}{2}x^2, & x \leq 0 \\ \frac{1}{3}x^3 - \frac{\alpha}{2}x^2, & 0 < x \end{cases}$$

and

$$u_2 = \begin{cases} -\frac{1}{3}x^3 + \frac{\alpha}{2}x^2, & x \leq \alpha \\ \frac{1}{3}x^3 - \frac{\alpha}{2}x^2 + \frac{\alpha^3}{3}, & \alpha < x \end{cases}$$

are both classical solutions bounded from below of (1.11) with $v(0) = 0$. Let $u_c^*(x) = u_1(x) + c$, $0 \leq c \leq \frac{\alpha^3}{3}$ and p be the intersection point of u_2 and u_c^*. We define a function $v_c(x)$ as $v_c(x) = u_2(x)$, $x \leq p$; $v_c(x) = u_c^*(x), x \geq p$. Then $v_c(x)$ turns out to be a viscosity solution bounded from below of (1.11).

To pick up the limit function of v_ϵ among the solutions of (1.10) in such a case that it may have infinitely many solutions, we give the limit the variational representation:

$$\begin{aligned} v(x) &= \inf_{x_0=x, x_\infty=a} \{\int_0^\infty (V(x_s) + \frac{1}{2\kappa}|\dot{x}_s + \nabla U(x_s)|^2) ds\} \\ &\wedge \inf_{x_0=x, x_\infty=-a} \{\int_0^\infty (V(x_s) + \frac{1}{2\kappa}|\dot{x}_s + \nabla U(x_s)|^2) ds\}, \end{aligned}$$

in the case of $\kappa = 1 - \theta > 0$, and if $\kappa < 0$ we prove the representation where inf is replaced by sup and $\wedge$ by $\vee$ in the above holds for the limit, by using dynamic programming principle for ergodic control which is shown in more general setting in section 3. Here a and $-a$ are the minimum points of U and V. We note that such kind of representation has been given by Fleming and Sheu (1998) in the case where U and V have a single minimum. Finally we give a remark on the size of risk sensitive parameter, which is relevant to disturbance attenuation level in H_∞ control. It is prescribed by the condition $\hat{V}_\epsilon \to \infty$, $|x| \to \infty$. So, in the case that $U(x) \sim \frac{c_1}{2}|x|^{2l}$, $V(x) \sim \frac{c_2}{2}|x|^{2m}$, $|x| \gg 1$ $l, m \geq 1$, $c_1, c_2 > 0$, the condition is satisfied if either $\theta < 1$, or $1 < \theta, 2l > m+1$, or $\theta < 1 + \frac{c_1^2 l^2}{c_2}, 2l = m+1$ holds.

2 Preliminary results

In the present section we summarize the results obtained in Kaise and Nagai (1998) mentioned in Introduction. We deduce equation (1.1) from the following Bellman equation of discounted type

$$\alpha v_\alpha = \frac{\epsilon}{2}\Delta v_\alpha - \nabla U \cdot \nabla v_\alpha - \frac{1-\theta}{2}|\nabla v_\alpha|^2 + V \tag{2.1}$$

To prove the existence of the solution of (2.1) we take a transformation:

$$u_\alpha = e^{\frac{\theta-1}{\epsilon}(v_\alpha - \frac{1}{\theta-1}U)} \tag{2.2}$$

and obtain the following equation

$$-\frac{\epsilon}{2}\Delta u_\epsilon + \frac{1}{\epsilon}\tilde{V}_\alpha u_\alpha = -\alpha u_\alpha \log u_\alpha, \tag{2.3}$$

where

$$\tilde{V}_\alpha = \frac{1}{2}|\nabla U|^2 + (1-\theta)V - \frac{\epsilon}{2}\Delta U + \alpha U.$$

Let us assume that

$$U \in W^{3,\infty}_{loc}, \quad V \in W^{1,\infty}_{loc} \tag{2.4}$$

$$\frac{1}{2}|\nabla U|^2 + (1-\theta)V - \frac{1}{2}\Delta U \to \infty, \ |x| \to \infty \tag{2.5}$$

$$U,\ V \geq 0, \text{ and } U(x) + c|x|^2 \text{ is convex for } \exists c > 0 \tag{2.6}$$

$$e^{-\delta\sqrt{\bar{w}}} \in L^2(R^N), \quad \exists\delta > 0, \tag{2.7}$$

where $\bar{w} = w - \min w$, $w = \frac{1}{2}|\nabla U|^2 + (1-\theta)V - \frac{1}{2}\Delta U$. Then we have the following theorem.

Theorem 2.1 (Bensoussan-Nagai) *Under assumptions (2.4)-(2.7) there exists a unique solution $u_\alpha \in H^1_{\bar{V}_\alpha}$ of (2.3) for $0 < \alpha < \frac{1}{\delta}$, where*

$$H^1_{\bar{V}_\alpha} = \{f; \int \{|\nabla f|^2 + (\bar{V}_\alpha + 1)\}f^2 dx < \infty\}, \ \bar{V}_\alpha = \tilde{V}_\alpha - \min \tilde{V}_\alpha.$$

Remark. We give a comment on the above assumptions. In the case where $U \equiv 0$ (2.5) and (2.6) means that $\theta < 1$, $V \geq 0$, $V(x) \to \infty, |x| \to \infty$, which is nothing but the case considered in Bensoussan and Nagai (1991). As far as the existence and uniqueness of the solution of (2.3) the proof can be reduced to the case. In view of a risk-sensitive control problem (2.5) has a meaning to relax the assumption on the size of risk-sensitive parameter θ assumed in Nagai (1996) by adding a drift coefficint ∇U having a stabilizing effect.

By setting $v_\alpha = \frac{\epsilon}{\theta-1}\log u_\alpha + \frac{1}{\theta-1}U$, we obtain a solution of (2.1). From the solution we can derive a solution of (1.1) in the following way.

Theorem 2.2 (Kaise-Nagai) *Besides the above conditions we assume that*

$$U(x) \leq M\hat{V}(x) \equiv M(\frac{1}{2}|\nabla U|^2 + (1-\theta)V), \quad x \notin B(0,R), \; \exists R, M > 0, \tag{2.8}$$

then

$$\lim_{\alpha\to 0} \alpha v_\alpha = \frac{\epsilon}{1-\theta}\lambda_1^\epsilon + \frac{1}{1-\theta}\min \hat{V}_\epsilon \equiv \chi_\epsilon \tag{2.9}$$

uniformly on each compact set and

$$\lim_{\alpha\to 0}(v_\alpha - \frac{\epsilon}{\theta-1}\int \psi_\alpha^2(\log u_\alpha - \log\psi_\alpha)dx) = \frac{\epsilon}{\theta-1}\log\psi_\epsilon + \frac{U}{\theta-1} \equiv v_\epsilon \tag{2.10}$$

in H^1_{loc} *and uniformly on each compact set, where* ψ_α *(resp.* ψ_ϵ*) is the principal eigenfunction of the schrödinger operator* $-\frac{\epsilon}{2}\Delta + \frac{1}{\epsilon}\tilde{V}_\alpha$ *(resp.* $-\frac{\epsilon}{2}\Delta + \frac{1}{\epsilon}\hat{V}_\epsilon$*) and* λ_1^ϵ *is the eigenvalue corresponding to* ψ_ϵ*.*

Here we note that under our assumptions $\hat{V}_\epsilon(x) \to \infty$, $|x| \to \infty$, and we can see the operator $-\frac{\epsilon}{2}\Delta + \frac{1}{\epsilon}\hat{V}_\epsilon$ on $L^2(R^N)$ has a discrete spectrum

$$0 \leq \lambda_1^\epsilon < \lambda_2^\epsilon \leq \lambda_3^\epsilon \leq ....,$$

and the principal eigenvalue λ_1^ϵ is simple. Furthermore, the principal eigenfunction ψ_ϵ satisfies the estimates

$$0 < \psi_\epsilon \leq c_1 e^{-c_2|x|}, \quad c_1, c_2 > 0. \tag{2.11}$$

Thus we obtain the solution $(v_\epsilon, \chi_\epsilon)$ of (1.1) defined by the formula (2.9) and (2.10), which turns out to satisfy that $(1-\theta)v_\epsilon + U$ is bounded from below because of estimates (2.11). It is easy to see that $\lim_{\epsilon\to 0}\chi_\epsilon = \frac{1}{1-\theta}\min\hat{V}$ (cf.Kaise and Nagai(1998)) and moreover we have the following theorem.

Theorem 2.3 (Kaise-Nagai) *Under the assumptions of Theorem 2.2 there exists a subsequence* $\{v_{\epsilon_n}\}$ *of* $\{v_\epsilon\}$ *such that* v_{ϵ_n} *converges unifromly on each compact set to a function* v*, which is a viscosity solution of the following equation (2.12) such that* $(1-\theta)v + U$ *is bounded from below:*

$$\nabla U \cdot \nabla v + \frac{1-\theta}{2}|\nabla v|^2 - V + \frac{1}{1-\theta}\min\hat{V} = 0 \tag{2.12}$$

3 Dynamic programming principle for ergodic control

To study more precisely asymptotic behaviour of the solution v_ϵ we shall prove dynamic programming principle for ergodic control in the present section. We assume that there exists $R > 0$ such that

$$\frac{1}{2}|\nabla U|^2 + (1-\theta)V - \frac{1}{2}\Delta U \geq c_0|x|^{2m}, \; c_0, m > 0, \; x \notin B(0,R) \tag{3.1}$$

$$-x \cdot \nabla U \leq c_1(|x|^2 + 1), \quad c_1 > 0, \tag{3.2}$$

$$|\nabla U|^2 \leq c_2(1 + |x|^{2m}), \; |\nabla V| \leq c_2(1 + |x|^{2m-1}), c_2 > 0 \tag{3.3}$$

besides assumptions (2.4) and (2.6). Let $\kappa = 1 - \theta$ and set

$$w_\epsilon = \kappa v_\epsilon + U, \tag{3.4}$$

then we have

$$\frac{\epsilon}{2}\Delta w_\epsilon - \frac{1}{2}|\nabla w_\epsilon|^2 + \hat{V}_\epsilon = \kappa \chi_\epsilon \tag{3.5}$$

We first show the following lemma.

Lemma 3.1 *Assume (3.1)-(3.3), (2.4) and (2.6). Let w_ϵ be a solution bounded from below, then there exists $c > 0$ such that*

$$\inf_{|x|=r} w_\epsilon(x) \geq c r^{m+1}, \tag{3.6}$$

for sufficiently large r.

Proof. We can assume that $w_\epsilon \geq 0$ without loss of generality. Set

$$\psi(x) = \begin{cases} c_r(1 - \frac{4|x-x_r|^2}{r^2}), & x \in B(x_r, \frac{r}{2}) \\ 0 & x \notin B(x_r, \frac{r}{2}), \end{cases}$$

where x_r is any point with $|x_r| = r$ and c_r is a constant determined later. Setting $S(x) = w_\epsilon(x) - \psi(x)$, we have $S(x) \geq 0, \; x \in \partial B(x_r, \frac{r}{2})$. Furthermore we have

$$\begin{aligned} &\frac{\epsilon}{2}\Delta S - \frac{1}{2}(\nabla w_\epsilon + \nabla\psi) \cdot \nabla S \\ = \; &\frac{\epsilon}{2}\Delta w_\epsilon - \frac{\epsilon}{2}\Delta\psi - \frac{1}{2}|\nabla w_\epsilon|^2 + \frac{1}{2}|\nabla\psi|^2 \\ = \; &-\hat{V}_\epsilon + \kappa\chi_\epsilon - \frac{4\epsilon N}{r^2}c_r + \frac{32c_r^2}{r^4}|x - x_r|^2 \\ \leq \; &-\frac{c_0}{2^{2m+2}}r^{2m} + \kappa\chi_\epsilon + \frac{8c_r^2}{r^2}. \end{aligned}$$

Thus, if we set $c_r = \frac{\sqrt{c_0}}{2^{m+3}}r^{m+1}$, then we can see that

$$\frac{\epsilon}{2}\Delta S - \frac{1}{2}(\nabla w_\epsilon + \nabla\psi) \cdot \nabla S < 0$$

for sufficiently large r, namely S is superharmonic in $B(x_r, \frac{r}{2})$ for large r. Therefore by maximum principle we see that $S(x) \geq 0, \; x \in B(x_r, \frac{r}{2})$. In particular $S(x_r) \geq 0$, which means that $w_\epsilon(x_r) \geq c_r$.

The following estimate is also useful.

Lemma 3.2 *Let v_ϵ be a solution of (1.1) such that $\kappa v_\epsilon + U$ is bounded from below. Then, under the assumptions of Lemma 3.1 we have*

$$|\nabla v_\epsilon|^2 \le c(|x|^{2m}+1), \tag{3.6}$$

where c is a constant independent of ϵ.

Prrof. Let us set $G = |\nabla v_\epsilon|^2$, then

$$\begin{aligned}
&-\frac{\epsilon}{2}\Delta G + (\nabla U + \kappa \nabla v_\epsilon)\cdot\nabla G \\
&\quad = -\epsilon D_{ij}v_\epsilon D_{ij}v_\epsilon - D_{ij}D_iv_\epsilon D_jv_\epsilon + D_iv_\epsilon D_iV \\
&\quad \le -\frac{\epsilon}{N}(\Delta v_\epsilon)^2 + c|\nabla v_\epsilon|^2 + \nabla v_\epsilon\cdot\nabla V,
\end{aligned}$$

where $D_i = \frac{\partial}{\partial x_i}$, $D_{ij} = \frac{\partial^2}{\partial x_i \partial x_j}$ and summation convention is utilized. Take a point $x_0 \in B(0,R)$ and a cut-off function φ in $B(x_0,R)$ satisfying

$$0 \le \varphi(x) \le \varphi(x_0) = 1, \quad |\nabla\varphi|^2 \le \frac{k_1}{R}\varphi, \quad |\delta\varphi| \le \frac{2k_2}{R}, \; k_1, k_2 > 0.$$

Let x be a maximum point of φG in $B(x_0,R)$, then at x we have

$$\begin{aligned}
0 &\le -\frac{\epsilon}{2}\Delta(\varphi G) + (\nabla U + \kappa\nabla v_\epsilon)\cdot\nabla(\varphi G) \\
&= \varphi(-\frac{\epsilon}{2}\Delta G + (\nabla U + \kappa\nabla v_\epsilon)\cdot\nabla G) + \epsilon\frac{|\nabla\varphi|^2}{\varphi}G - \frac{\epsilon}{2}G\Delta\varphi \\
&\quad + G(\nabla U + \kappa\nabla v_\epsilon)\cdot\nabla\varphi \\
&\le \varphi(-\frac{\epsilon}{N}(\Delta v_\epsilon)^2 + c|\nabla v_\epsilon|^2 + \nabla v_\epsilon\cdot\nabla V) + \epsilon\frac{|\nabla\varphi|^2}{\varphi}G - \frac{\epsilon}{2}G\Delta\varphi \\
&\quad + G(|\nabla U| + |\kappa||\nabla v_\epsilon|)k_1\sqrt{\varphi} \\
&= -\frac{\epsilon}{N}(\frac{1}{\epsilon}\nabla U\cdot\nabla v_\epsilon + \frac{\kappa}{2\epsilon}|\nabla v_\epsilon|^2 + \frac{1}{\epsilon}(\chi_\epsilon - V))^2 + c\varphi G + \varphi\sqrt{G}|\nabla V| \\
&\quad + \frac{\epsilon(k_1+k_2)}{R}G + k_1\sqrt{\varphi}|\nabla U|G + k_1|\kappa|\sqrt{\varphi}G^{\frac{3}{2}} \\
&= -\frac{\kappa^2\varphi}{4N\epsilon}|\nabla v_\epsilon|^4 - \frac{\epsilon\varphi}{N}(\frac{1}{\epsilon}\nabla U\cdot\nabla v_\epsilon + \frac{1}{\epsilon}(\chi_\epsilon - V))^2 - \frac{\kappa\varphi}{N\epsilon}\nabla U\cdot\nabla v_\epsilon|\nabla v_\epsilon|^2 \\
&\quad - \frac{\kappa\varphi}{N\epsilon}(\chi_\epsilon - V)|\nabla v_\epsilon|^2 + c\varphi G + \varphi\sqrt{G}|\nabla V| + \frac{\epsilon}{R}(k_1+k_2)G \\
&\quad + k_1\sqrt{\varphi}|\nabla U|G + k_1|\kappa|\sqrt{\varphi}G^{\frac{3}{2}} \\
&\le -\frac{\kappa^2\varphi}{4N\epsilon}G^2 + (\frac{\varphi}{N\epsilon}|\nabla U| + k_1\sqrt{\varphi})|\kappa|G^{\frac{3}{2}} \\
&\quad + (\frac{|\kappa|}{N\epsilon}|\chi_e - V| + c + \frac{\epsilon}{R}(k_1+k_2) + k_1|\nabla U|)G + \varphi|\nabla V|\sqrt{G}
\end{aligned}$$

Thus we obtain

$$0 \le -\frac{\kappa^2}{4N\epsilon}\varphi G^{\frac{3}{2}} + (\frac{|\nabla U|}{N\epsilon} + k_1)|\kappa|\sqrt{\varphi}G + \varphi|\nabla V|$$

$$+\left(\frac{|\kappa|}{N\epsilon}|\chi_\epsilon - V| + \frac{\epsilon}{R}(k_1+k_2) + k_1|\nabla U|\right)G^{\frac{1}{2}}.$$

When $|\nabla V| \le G$ we have

$$\begin{aligned} 0 \ &\le\ -\frac{\kappa^2}{4N}\varphi G + \left(\frac{|\kappa||\nabla U|}{N} + \epsilon k_1|\kappa| + \epsilon\right)\sqrt{\varphi G} \\ &\quad + \frac{|\kappa|}{N}|\chi_\epsilon - V| + c\epsilon + \frac{\epsilon^2}{R}(k_1+k_2) + k_1|\nabla U|\epsilon. \end{aligned}$$

Hence, setting $X = \sqrt{\varphi G}$, $A = \frac{\kappa^2}{4N}$, , $B = \frac{|\kappa||\nabla U|}{N} + k_1|\kappa| + 1$, $C = \frac{|\kappa|}{N}|\chi_\epsilon - V| + C + \frac{k_1+k_2}{R} + k_1|\nabla U|$, we obtain

$$AX^2 - BX - C \le 0,$$

from which we have

$$\begin{aligned} X^2 \ &\le\ \frac{B^2}{A^2} + \frac{2C}{A} = \frac{16N^2}{\kappa^4}\left(\frac{|\kappa||\nabla U|}{N} + k_1|\kappa| + 1\right)^2 \\ &\quad + \frac{8N}{\kappa^2}\left(\frac{\kappa}{N}|\chi_\epsilon - V| + c + \frac{k_1+k_2}{R} + k_1|\nabla U|\right) \end{aligned}$$

Including the case where $G \le |\nabla V|$ we conclude that

$$\begin{aligned} \varphi G \ &\le\ \frac{16N^2}{\kappa^4}\left(\frac{|\kappa||\nabla U|}{N}\right. \\ &\quad \left. + k_1|\kappa| + 1\right)^2 + \frac{8N}{\kappa^2}\left(\frac{\kappa}{N}|\chi_\epsilon - V| + c + \frac{k_1+k_2}{R} + k_1|\nabla U|\right) + |\nabla V|. \end{aligned}$$

Since $G(x_0) = \varphi(x_0)G(x_0) \le \varphi(x)G(x)$ and $\chi_\epsilon \to \chi_0$ as $\epsilon \to 0$ we see that

$$|\nabla v_\epsilon(x_0)| \le c'(1 + R^{2m}), \quad |x_0| \le R.$$

Now let v_ϵ be a solution of (1.1) such that $\kappa v_\epsilon + U$ is bounded from below and X_t be a solution of the following stochastic differential equation:

$$dX_t = \sqrt{\epsilon}dB_t - \nabla U(X_t)dt - \kappa\nabla v_\epsilon(X_t)dt, \ \ X_0 = x, \tag{3.7}$$

then we have the following lemma.

Lemma 3.3 *Assume the assumptions of Lemma 3.1. Then, for each t we have*

$$v_\epsilon(x) = E_x\left[\int_0^t \{V(X_s) + \frac{\kappa}{2}|\nabla v_\epsilon(X_s)|^2\}ds + v_\epsilon(X_t)\right] - \chi_\epsilon t. \tag{3.8}$$

Proof. We first note that (3.7) has a local solution since ∇U and ∇v_ϵ belong to $W^{2,p}_{loc}$ for $\forall p > 1$ and therefore those are locally Lipshitz because of the Sobolev embedding theorem. Let us set

$$\tau_R = \begin{cases} \inf\{s; |X_s| \ge R\}, & \text{if } \{\ \} \ne \phi \\ \infty, & \text{if } \{\ \} = \phi. \end{cases}$$

Then by Ito's formula we have

$$\begin{aligned}
&w_\epsilon(X_{t\wedge\tau_R}) - w_\epsilon(X_0)\\
&= \sqrt{\epsilon}\int_0^{t\wedge\tau_R} \nabla w_\epsilon(X_s)dB_s + \int_0^{t\wedge\tau_R} (\frac{\epsilon}{2}\Delta w_\epsilon(X_s) - |\nabla w_\epsilon|^2(X_s))ds\\
&= -\int_0^{t\wedge\tau_R} \{\hat{V}_\epsilon(X_s) + \frac{1}{2}|\nabla w_\epsilon|^2(X_s)\}ds\\
&+ \sqrt{\epsilon}\int_0^{t\wedge\tau_R} \nabla w_\epsilon(X_s)dB_s + \chi_\epsilon t\wedge\tau_R.
\end{aligned}$$

Therefore we have

$$E_x[w_\epsilon(X_{t\wedge\tau_R})] \leq w_\epsilon(x) + \chi_\epsilon t + M \tag{3.9}$$

for some constant M. By Lemma 3.1 $\inf_{|x|=R} w_\epsilon(x) \to \infty$, $R \to \infty$ and therefore from (3.9) it follows that

$$P_x(\tau_R < t) \to 0, \quad R \to \infty. \tag{3.10}$$

Thus we see that our solution X_t is a global one and moreover we have

$$E_x[\int_0^t \{\hat{V}(X_s) + \frac{1}{2}|\nabla w_\epsilon(X_s)|^2\}ds + w_\epsilon(X_t)] \leq w_\epsilon(x) + \chi_\epsilon t.$$

Thus, using Lemma 3.1, we see that

$$E_x[\int_0^t |X_s|^{2m}ds + |X_t|^{m+1}] < \infty. \tag{3.11}$$

Aplpying Ito's formula to v_ϵ we have

$$v_\epsilon(X_t) - v_\epsilon(x) = \sqrt{\epsilon}\int_0^t \nabla v_\epsilon(X_s)dB_s - \int_0^t \{\frac{\kappa}{2}|\nabla v_\epsilon(X_s)|^2 + V(X_s)\}ds + \chi_\epsilon t.$$

$\sqrt{\epsilon}\int_0^t \nabla v_\epsilon(X_s)dB_s$ is a martingale and $E_x[|v_\epsilon(X_t)|] < \infty$ by Lemma 3.2 and (3.11) we obtain

$$v_\epsilon(x) = E_x[\int_0^t \{\frac{\kappa}{2}|\nabla v_\epsilon(X_s)|^2 + V(X_s)\}ds + v_\epsilon(X_t)] - \chi_\epsilon t.$$

We are now prepared to prove Dynamic programming principle for ergodic control. Let $\mathcal{A}$ be the set of all progressively measurable proxess $\{z_s\}$ such that $\int_0^t |z_s|ds < \infty$, a.s. $\forall t$ and

$$E_x[\int_0^t |X_s|^{2m}ds + |X_t|^{m+1}] < \infty, \quad \forall t,$$

where X_t is a solution of

$$X_t = x + \sqrt{\epsilon}B_t - \int_0^t \nabla U(X_s)ds + \int_0^t z_s ds.$$

Proposition 3.1 *Let v_ϵ be a solution of (1.1) such that $\kappa v_\epsilon + U$ is bounded from below. If $\kappa > 0$, then we have*

$$v_\epsilon(x) = \inf_{\mathcal{A}} E_x[\int_0^t \{V(X_s) + \frac{1}{2\kappa}|z_s|^2\}ds + v_\epsilon(X_t)] - \chi_\epsilon t, \quad \forall t. \tag{3.12}$$

Furthermore, if $\kappa < 0$ (3.12) replacing inf *by* sup *holds.*

Proof. We give the proof in the case where $\kappa > 0$. For $\{z_s\} \in \mathcal{A}$ we have

$$\begin{aligned} v_\epsilon(X_t) - v_\epsilon(x) &= \sqrt{\epsilon}\int_0^t \nabla v_\epsilon(X_s)dB_s + \int_0^t (-\nabla U(X_s) + z_s) \cdot \nabla v_\epsilon(X_s)ds \\ &\quad + \frac{\epsilon}{2}\int_0^t \Delta v_\epsilon(X_s)ds \\ &\geq -\int_0^t (V(X_s) + \frac{1}{2\kappa}|z_s|^2\}ds + \chi_\epsilon t + \sqrt{\epsilon}\int_0^t \nabla v_\epsilon(X_s)dB_s. \end{aligned}$$

Since $\int_0^t \nabla v_\epsilon(X_s)dB_s$ is a martingale because of Lemma 3.2 we obtain

$$v_\epsilon(x) \leq E_x[\int_0^t \{V(X_s) + \frac{1}{2\kappa}|z_s|^2\}ds + v_\epsilon(X_t)] - \chi_\epsilon t, \quad \forall t. \tag{3.13}$$

From (3.13) along with Lemma 3.3 we conclude our present proposition.

The following theorem is the main result in the present section.

Theorem 3.1 *Let us assume the assumption of Lemma 3.1 and $m \geq 1$. We set a limit function $v(x)$ of a subsequence $\{v_{\epsilon_n}(x)\}$ of $\{v_\epsilon(x)\}$. Then, if $\kappa > 0$ we have*

$$v(x) = \inf_{\mathcal{A}_0}[\int_0^t \{V(x_s) + \frac{1}{2\kappa}|z_s|^2\}ds + v(x_t) - \chi_0 t], \quad \forall t, \tag{3.14}$$

where

$$x_t = x - \int_0^t \nabla U(x_s)ds + \int_0^t z_s ds$$

and

$$\mathcal{A}_0 = \{z_s \,;\, z_s : \textit{measurable}, \int_0^t |z_s|^{2m} ds < \infty, \forall t\}$$

Furthermore, if $\kappa < 0$, (3.14) replacing inf *by* sup *holds.*

To prove the theorem we need some lemmas.

Lemma 3.4 *Let $z_. \in \mathcal{A}_0$ and Y_t be a solution of*

$$Y_t = x + \sqrt{\epsilon}B_t - \int_0^t \nabla U(Y_s)ds + \int_0^t z_s ds, \tag{3.15}$$

then we have

$$E[|Y_t|^{2m}] \le ce^{ct}(|x|^{2m} + \int_0^t |z_s|^{2m} ds + ct),$$

where c is a constant independent of ϵ.

Proof. By Ito's formula

$$\begin{aligned}
&|Y_t|^{2m} - |x|^{2m} \\
&= 2\sqrt{\epsilon}m \int_0^t |Y_s|^{2(m-1)} Y_s^* dB_s - 2m \int_0^t |Y_s|^{2(m-1)} Y_s \cdot \nabla U(Y_s) ds \\
&+ 2m \int_0^t |Y_s|^{2(m-1)} z_s \cdot Y_s ds \\
&+ \epsilon \int_0^t \{Nm|Y_s|^{2(m-1)} + 2m(m-1)|Y_s|^{2(m-1)}\} ds \\
&\le 2\sqrt{\epsilon}m \int_0^t |Y_s|^{2(m-1)} Y_s^* dB_s + c\int_0^t (|Y_s|^{2m} + |Y_s|^{2(m-1)}) ds \\
&+ 2m \int_0^t (\frac{1}{2m}|z_s|^{2m} + \frac{2m-1}{2m}|Y_s|^{2m}) ds + c\epsilon \int_0^t |Y_s|^{2(m-1)} ds \\
&\le 2\sqrt{\epsilon}m \int_0^t |Y_s|^{2(m-1)} Y_s^* dB_s + \int_0^t \{c_1|Y_s|^{2m} + c_2 + |z_s|^{2m}\} ds.
\end{aligned}$$

Setting

$$\sigma \equiv \sigma_R \equiv \sigma_R^\epsilon = \begin{cases} \inf\{s; |Y_s| \ge R\}, & \text{if } \{\} \ne \phi \\ \infty, & \text{if } = \phi, \end{cases}$$

$$E_x[|Y_{t\wedge\sigma}|^{2m}] \le |x|^{2m} + \int_0^t |z_s|^{2m} ds + c_1 \int_0^t E_x[|Y_{s\wedge\sigma}|^{2m}] ds + c_2 t.$$

Set $\eta(t) = E[|Y_{t\wedge\sigma}|^{2m}]$, and $\alpha(t) = |x|^{2m} + \int_0^t |z_s|^{2m} ds + c_2 t$, then we have

$$\eta(t) \le \alpha(t) + c_1 \int_0^t \eta(s) ds.$$

Therefore, using Gronwall's inequality, we obtain

$$\begin{aligned}
\eta(t) &\le \alpha(t) + c_1 \int_0^t \alpha(s) e^{c_1(t-s)} ds \\
&\le c_1 e^{c_1 t} \alpha(t).
\end{aligned}$$

It is easy to see that $P(\sigma_R < t) \to 0$, $R \to \infty$ and we conclude our present lemma.

Lemma 3.5 *Let* $Y_t^\epsilon \equiv Y_t$ *be as above and* y_t *a solution of*

$$y_t = x - \int_0^t \nabla U(y_s) ds + \int_0^t z_s ds,$$

then we have

$$P(|Y_t^\epsilon - y_t| \to 0, \ \epsilon \to 0) = 1,, \ \forall t,$$

by taking a subsequence if necessary.

Proof. Take $T > 0$. We shall show first $P(\sigma_R^\epsilon \leq T) \to 0$ as $\epsilon \to 0$, taking R sufficintly large. In the same way as the proof of Lemma 3.4, by Ito's formula we have

$$\begin{aligned} |Y_t|^2 - |x|^2 &\leq 2\sqrt{\epsilon}\int_0^t Y_s^* dB_s + c\int_0^t |Y_s|^2 ds \\ &\quad + c\int_0^t |z_s|^2 ds + c\epsilon t. \end{aligned}$$

Set $\alpha_\epsilon(t) = |x|^2 + c\int_0^t |z_s|^2 ds + c\epsilon t + 2\sqrt{\epsilon}\sup_{0\leq s\leq t} |\int_0^s Y_r^* dB_r|$ and $\xi_\epsilon(t) = \sup_{0\leq s\leq t} |Y_s|^2$ and we have

$$\xi_\epsilon(t) \leq \alpha_\epsilon(t) + c\int_0^t \xi_\epsilon(s)ds.$$

Therefore we obtain

$$\xi_\epsilon(t) \leq \alpha_\epsilon(t) + c\int_0^t \alpha_\epsilon(s)e^{c(t-s)}ds.$$

by Gronwall's inequality. Hence

$$P[\xi_\epsilon(T) \geq R] \leq P[\alpha_\epsilon(T) \geq \frac{R}{2}] + P[c\int_0^T \alpha_\epsilon(s)e^{c(T-s)}ds \geq \frac{R}{2}]. \tag{3.16}$$

When taking R sufficintly large we see that the righthand side of (3.16) tends to 0 as $\epsilon \to 0$ by using Doob's inequality and Lemma 3.4. Thus we obtain

$$P[\sigma_R^\epsilon \leq T] = P[\xi_\epsilon(T) \geq R] \to 0, \quad \epsilon \to 0.$$

Since

$$|Y_t^\epsilon - y_t| \leq \sqrt{\epsilon}|B_t| + \int_0^t (|\nabla U(Y_s) - \nabla U(y_s)|)ds$$

Arguments using Gronwall's inequality imply that

$$P(|Y_t^\epsilon - y_t| \to 0, \ \epsilon \to 0) = 1, \quad t < T,$$

by taking a subsequence if necessary.

Lemma 3.6 *Let x_t^ϵ be a solution of*

$$x_t^\epsilon = x - \int_0^t \nabla U(x_s^\epsilon)ds - \kappa\int_0^t \nabla v_\epsilon(x_s^\epsilon)ds, \tag{3.17}$$

then there exists $\hat{z}_s \in L^2(0,T)$, $\forall T$ *such that* $|x_t^\epsilon - x_t| \to 0$, $\epsilon \to 0$, $\forall t$ *taking a subsequence if necessary, where*

$$x_t = x - \int_0^t \nabla U(x_s)ds + \int_0^t \hat{z}_s ds. \tag{3.18}$$

Moreover, $\hat{z}_s \in L^p(0,T)$, $\forall p > 1$.

Proof. We first note that there exists a compact set K such that

$$x_t^\epsilon \in K, \quad \forall t,\ \epsilon \leq 1. \tag{3.19}$$

In fact, since $x_t^\epsilon = x - \int_0^t \nabla w_\epsilon(x_s^\epsilon)ds$ we have

$$w_\epsilon(x_t^\epsilon) - w_\epsilon(x) = -\int_0^t |\nabla w_\epsilon|^2(x_s^\epsilon)ds$$

and therefore $w_\epsilon(x_t^\epsilon) \leq w_\epsilon(x)$. We fix as $w_\epsilon(0) = 0$. Then on account of Lemma 3.1 $w_\epsilon(x) \to \infty$, $|x| \to \infty$ uniformly with respect to ϵ and $w_\epsilon(x) = \kappa v_\epsilon(x) + U(x) \leq C(x)$ for some $C(x)$ independent of ϵ because of Lemma 3.2. So we see (3.19). From (3.19) and Lemma 3.2 we have

$$\int_0^T |\dot{x}_s^\epsilon|^2 ds = \int_0^T |\nabla w_\epsilon(x_s^\epsilon)|^2 ds \leq M_K T$$

for some constant M_K. Therefore we see that there exists $\xi_s \in L^2(0,T)$ and $\dot{x}_s^\epsilon$ converges weakly in $L^2(0,T)$ by taking a subsequence if necessary. Define x_t by $x_t = x + \int_0^t \xi_s ds$ and then we have

$$|x_t^\epsilon - x_t| = |\int_0^t (\dot{x}_s^\epsilon - \xi_s) \cdot 1_{(0,t)}(s)e ds| \to 0, \quad \epsilon \to 0$$

for all $t < T$, where $e = (1,1,...1)^* \in R^N$. Since $\nabla U(x_t^\epsilon) \to \nabla U(x_t)$, $\forall t$ boundedly,by setting $\hat{z}_t = \xi_t + \nabla U(x_t)$, we obtain our lemma. Since $\dot{x}_s^\epsilon$ is also L^p bounded for all $p > 1$ we see that $\hat{z}_s \in L^p$.

Proof of Theorem 3.1. We only give the proof of the theorem for $\kappa > 0$. The proof for $\kappa < 0$ is similar. We first note that $\mathcal{A}_0 \subset \mathcal{A}$ by virtue of Lemma 3.4 since $2m \geq m + 1$. Therefore by Proposition 3.1, for the solution Y_t of (3.15) corresponding to $z_. \in \mathcal{A}_0$ we have

$$v_\epsilon(x) \leq E_x[\int_0^t (V(Y_s^\epsilon) + \frac{1}{2\kappa}|z_s|^2)ds + v_\epsilon(Y_t^\epsilon)] - \chi_\epsilon t, \quad \forall t \tag{3.20}$$

We shall show that $V(Y_s^\epsilon)$ and $|v_\epsilon(Y_t^\epsilon)|$ are uniformly integrable and so $\lim_{\epsilon\to 0} E_x[V(Y_t^\epsilon)] = V(y_t)$ and $\lim_{\epsilon\to 0} E_x[v_\epsilon(Y_t^\epsilon)] = v(y_t)$. For that we need only to prove that $|Y_t^\epsilon|^{2m}$ is uniformly integrable. Set $\Lambda_\epsilon(t) = |x|^{2m} +$

$c_2 t + \int_0^t |z_s|^{2m} ds + \sup_{0 \le s \le t} 2\sqrt{\epsilon} m | \int_0^s |Y_r|^{2m-1} Y_r^* dB_r|$. Then, in a similar way to the proof of Lemma 3.4 we have

$$|Y_t|^{2m} \le \Lambda_\epsilon(t) + c_1 \int_0^t |Y_s|^{2m} ds,$$

from which we obtain

$$|Y_t|^{2m} \le \Lambda_\epsilon(t) + \int_0^t e^{c_1(t-s)} \Lambda_\epsilon(s) ds \le c_1 e^{c_1 t} \Lambda_\epsilon(t).$$

Thus we have

$$\begin{aligned} &E[|Y_t|^{2m}; |Y_t|^{2m} \ge R] \\ &\quad \le c_1 e^{c_1 t}(c_2 t + \int_0^t |z_s|^{2m} ds + |x|^{2m}) P[|Y_t|^{2m} \ge R] \\ &\quad + E[\sup_{0 \le s \le t} 2\sqrt{\epsilon} m | \int_0^s |Y_r|^{2(m-1)} Y_r^* dB_r | 1_{\{|Y_t^\epsilon| \ge R\}}]. \end{aligned}$$

On account of Hörder's inequality, Burkholder-Davis-Gundy's inequality, Chebyshev's inequality and Lemma 3.4 we see that

$$\sup_{0 \le \epsilon \le 1} E[|Y_t|^{2m}; |Y_t|^{2m} \ge R] \to 0, \quad R \to \infty.$$

Now we can easily see that

$$v(x) \le \int_0^t (V(y_s) + \frac{1}{2\kappa}|z_s|^2) ds + v(y_t) - \chi_0 t, \quad \forall t \tag{3.21}$$

for $z_s \in \mathcal{A}_0$ by using Lemma 3.5 and 3.1.

Let X_t be a solution of (3.7) and w_ϵ a nonnegative solution of (1.5) normalized as $w_\epsilon(0) = c_1$. We note that $|w_\epsilon(x)| \le c_2$, $x \in B(0,r)$ and $\inf_{|x|=R} |w_\epsilon(x)| \ge c_3 R^{m+1}$ for sufficintly large R, where c_2, c_3 are positive constant independent of ϵ. We further note that $\hat{V}_\epsilon(x) \ge -c_4$ because of our assumptions. We take R so large that

$$\frac{c_2 + |\chi_\epsilon| T + c_4 T}{c_3 R^{m+1}} < \frac{1}{2}.$$

It is posible since $\chi_\epsilon \to \chi_0$, $\epsilon \to 0$. Set

$$\tau_R^\epsilon = \begin{cases} \inf\{s; |X_s^\epsilon| \ge R\}, & \text{if } \{\} \ne \phi \\ \infty, & \text{if } \{\} = \phi, \end{cases}$$

then, in the same way as the proof of Lemma 3.3 we obtain by Ito's formula

$$w_\epsilon(X_{T \wedge \tau_R^\epsilon}) \le w_\epsilon(x) + \sqrt{\epsilon} \int_0^{T \wedge \tau_R^\epsilon} \nabla w_\epsilon(X_s) dB_s + \chi_\epsilon T + c_4 T.$$

Therefore we have

$$c_3R^{m+1}1_{\{\tau_R^\epsilon<T\}} \le c_2 + |\chi_\epsilon|T + c_4T + \sqrt{\epsilon}\int_0^{T\wedge\tau_R^\epsilon} \nabla w_\epsilon(X_s)dB_s. \quad (3.22)$$

Since $E[\int_0^T |\nabla w_\epsilon(X_s)|^2 ds]$ is bounded by a constant independent of ϵ we see that $\sqrt{\epsilon}\int_0^T \nabla w_\epsilon(X_s)dB_s \to 0$, a.s. by taking a subsequence if necessary. Thus from (3.22) it follows that $1_{\{\tau_R^\epsilon \le T\}} \to 0$, a.s. by taking a subsequence if necessary. For such sequence we see that

$$|X_{t\wedge\tau_R^\epsilon} - x_t^\epsilon| \to 0, \quad \epsilon \to 0 \quad (3.23)$$

uniformly on $t \le T$, a.s.. In fact, on the set $\{\tau_R^\epsilon > T\}$ we have

$$\sup_{0\le t\le T} |X_t - x_t^\epsilon| \le \int_0^t L_R \sup_{0\le r\le s} |X_r - x_r^\epsilon| ds + \sqrt{\epsilon} \sup_{0\le t\le T} |B_t|.$$

Standard arguments using Gronwall's inequality implies (3.23). Hence, by virtue of Lemma 3.6 we see that

$$P[|X_{t\wedge\tau_R^\epsilon}^\epsilon - x_t| \to 0,\ \epsilon \to 0,\ \forall t] = 1,$$

for x_t defined by (3.18). Set

$$\hat{z}_s^\epsilon = \begin{cases} -\kappa\nabla v_\epsilon(X_s), & s \le \tau_R^\epsilon \\ 0, & \tau_R^\epsilon < s, \end{cases}$$

then for $\eta_s \in L^2(0,T)$

$$\begin{aligned} |\int_0^T \hat{z}_s^\epsilon \cdot \eta_s ds| &\le |\int_0^{T\wedge\tau_R^\epsilon} (\kappa\nabla v_\epsilon(X_s) - \kappa v_\epsilon(x_s^\epsilon)) \cdot \eta_s ds| \\ &\quad + |\kappa \int_0^T \nabla v_\epsilon(x_s^\epsilon) \cdot \eta_s ds| \\ &\le L_R(\int_0^{T\wedge\tau_R^\epsilon} |X_s - x_s^\epsilon|^2 ds)^{\frac{1}{2}} (\int_0^T |\eta_s|^2 ds)^{\frac{1}{2}} \\ &\quad + |\kappa \int_0^T \nabla v_\epsilon(x_s^\epsilon) \cdot \eta_s ds| \end{aligned}$$

Thereofore we see that

$$P(\hat{z}_s^\epsilon \to \hat{z}_s, \text{ weakly in } L^2(0,T)) = 1.$$

We moreover see that

$$\int_0^T |\hat{z}_s|^2 ds \le \liminf_{\epsilon\to 0} \int_0^T |\hat{z}_s^\epsilon|^2 ds, \text{ a.s..} \quad (3.24)$$

It is easy to see in the same way as the proof of Lemma 3.3 that

$$v_\epsilon(x) = E_x[\int_0^{t\wedge\tau_R^\epsilon} \{V(X_s)+\frac{\kappa}{2}|\nabla v_\epsilon(X_s)|^2\}ds+v_\epsilon(X_{t\wedge\tau_R^\epsilon})-\chi_\epsilon t\wedge\tau_R^\epsilon]. \quad (3.25)$$

From (3.24) and (3.25) we obtain

$$\int_0^t (V(x_s) + \frac{1}{2\kappa}|\hat{z}_s|^2)ds + v(x_t) - \chi_0 t \leq v(x). \quad (3.26)$$

We conclude our theorem for $\kappa > 0$ from (3.19) and (3.26).

4 Representation of the limit

In the present section we assume in the case of $\kappa > 0$ that

$$V(a) = V(-a) = 0, \ \nabla U(a) = \nabla U(-a) = 0, \ \exists a \neq 0 \quad (4.1)$$

$$U, V \text{ are symmetric with respect to } \{x; x\cdot a = 0\} \quad (4.2)$$

$$V(x) > 0, \ x \neq a, \ -a. \quad (4.3)$$

In the case of $\kappa < 0$ we replace (4.3) by the following (4.3)'

$$\frac{1}{2}|\nabla U|^2(x) + \kappa V(x) > 0, U(x) > 0 = U(a) = U(-a), \ x \neq a, -a. \quad (4.3)'$$

Theorem 4.1 *Besides the assumptions of Theorem 3.1 we assume (4.1)-(4.3) and* $\kappa \equiv 1-\theta > 0$, *then limit function* $v(x)$ *of* $v_{\epsilon_n}(x)$ *normalized as* $v_{\epsilon_n}(a) = 0$ *satisfies*

$$\begin{aligned} v(x) &= \inf_{x_0=x, x_\infty=a} \{\int_0^\infty (V(x_s) + \frac{1}{2\kappa}|\dot{x}_s + \nabla U(x_s)|^2)ds\} \\ &\wedge \inf_{x_0=x, x_\infty=-a} \{\int_0^\infty (V(x_s) + \frac{1}{2\kappa}|\dot{x}_s + \nabla U(x_s)|^2)ds\} \end{aligned}$$

In the case of $\kappa < 0$, *we assume (4.1),(4.2) and (4.3)' besides the condition of Theorem 3.1. Then the above representation holds by replacing* inf *by* sup *and* $\wedge$ *by* $\vee$ *in the righthand side.*

Corollary 4.1 *Under the assumptions of Theorem 4.1 the solution* v_ϵ *of (1.1) with* $v_\epsilon(a) = 0$ *such that* $\kappa v_\epsilon + U$ *is bounded from below converges uniformly on each compact set to the function* $v(x)$ *defined by the variational representation in the above theorem as* $\epsilon \to 0$, *which is a viscosity solution of (1.9).*

Proof. In the case of $\kappa > 0$. We first note that

$$\chi_0 = \frac{1}{\kappa}\min \hat{V} = \frac{1}{\kappa}\min(\frac{1}{2}|\nabla U|^2 + \kappa V) = 0.$$

Moreover x_t defined by (3.18) satisties that $x_t \in K$, $\forall t$ for some compact set K because of (3.19). Therefore by (3.26) we see that

$$|v(x_t)| \leq M, \ \forall t$$

for some constant M. Thus in view of (3.26) we have

$$\limsup_{t\to\infty} V(x_t) = 0 \tag{4.4}$$

In fact, we assume that $\limsup_{t\to\infty} V(x_t) = c > 0$. Then there exists t_n such that $t_n + 1 \leq t_{n+1}$ and that $V(x_{t_n}) \geq c' > 0$, $\forall n$.

$$\begin{aligned} |x_t - x_{t_n}| &\leq \int_{t_n}^t |\nabla U(x_s)|ds + \int_{t_n}^t |\hat{z}_s|ds \\ &\leq c_1|t-t_n| + (\int_{t_n}^t |\hat{z}_s|^2 ds)^{\frac{1}{2}}|t-t_n|^{\frac{1}{2}} \end{aligned}$$

Because of (3.26) we have $\int_0^\infty |\hat{z}_s|^2 ds < \infty$. Therefore

$$|x_t - x_{t_n}| \leq c_2|t-t_n|^{\frac{1}{2}}.$$

For $\forall \eta > 0$ there exists δ such that $|V(x_t) - V(x_{t_n})| < \eta$, $|t - t_n| < \delta$ and therefore $V(x_t) \geq c' - \eta > 0$. Hence

$$\int_0^\infty V(x_s)ds \geq \sum_{n=1}^\infty \int_{t_n}^{t_n+\delta} V(x_s)ds = \infty,$$

which contradicts with (3.26).
From (4.4) it follows that $\lim_{t\to\infty} x_t = a$, or $-a$ by virtue of (4.3). Therefore $\lim_{t\to\infty} v(x_t) = 0$ since $v_\epsilon(a) = 0$ and also $v_\epsilon(-a) = 0$ because of (4.2). Hence we obtain

$$\int_0^\infty (V(x_s) + \frac{1}{2\kappa}|\hat{z}_s|^2)ds \leq v(x).$$

So we have

$$\begin{aligned} v(x) &\geq \inf_{x_0=x, x_\infty=a}\{\int_0^\infty (V(x_s) + \frac{1}{2\kappa}|\dot{x}_s + \nabla U(x_s)|^2)ds\} \\ &\wedge \inf_{x_0=x, x_\infty=-a}\{\int_0^\infty (V(x_s) + \frac{1}{2\kappa}|\dot{x}_s + \nabla U(x_s)|^2)ds\}. \end{aligned}$$

Converse inequality can be easily obtained from (3.14).

The case $\kappa < 0$. We have $\chi_0 = 0$. Let x_t be a solution of (3.18). Then

$$\begin{aligned} U(x_t) - U(x) &= -\int_0^t |\nabla U|^2(x_s)ds + \int_0^t \hat{z}_s \cdot \nabla U(x_s)ds \\ &\leq -\frac{1}{2}\int_0^t |\nabla U|^2(x_s)ds + \frac{1}{2}\int_0^t |\hat{z}_s|^2 ds. \end{aligned}$$

Therefore

$$\frac{1}{2\kappa}\int_0^t |\hat{z}_s|^2 ds + \int_0^t V(x_s)ds + \frac{1}{2\kappa}(U(x) - U(x_t))$$
$$\leq \frac{1}{\kappa}\int_0^t (\frac{1}{2}|\nabla U|^2(x_s) + \kappa V(x_s))ds.$$

Then, $\forall \delta > 0$ there exists t_0 such that

$$x_{t_0} \in B(a, \delta) \cup B(-a, \delta).$$

Indeed, if it does not hold

$$\frac{1}{\kappa}\int_0^t (\frac{1}{2}|\nabla U|^2(x_s) + \kappa V(x_s))ds \to -\infty, \quad t \to \infty$$

since $\frac{1}{2}|\nabla U|^2(x) + \kappa V(x) > \exists c > 0$ for $x \in B(a, \delta)^c \cap B(-a, \delta)^c$. Then

$$v(x) + \frac{1}{2\kappa}(U(x) - U(x_t))$$
$$\leq \frac{1}{2\kappa}\int_0^t |\hat{z}_s|^2 ds + \int_0^t V(x_s)ds + v(x_t) + \frac{1}{2\kappa}(U(x) - U(x_t))$$
$$\leq \frac{1}{\kappa}\int_0^t (\frac{1}{2}|\nabla U|^2(x_s) + \kappa V(x_s))ds.$$

Thus we have $v(x) + \frac{1}{2\kappa}(U(x) - U(x_t)) \to -\infty$, $t \to \infty$, which is a contradiction.
Set

$$\bar{z}_t = \begin{cases} \hat{z}_t, & t \leq t_0 \\ 0, & t > t_0 \end{cases}$$

and consider a solution $\bar{x}_t$ of

$$\bar{x}_t = x - \int_0^t \nabla U(\bar{x}_s)ds + \int_0^t \bar{z}_s ds.$$

Since $U(x) > U(a) = U(-a) = 0$, $x \neq a, -a$ we have $\lim_{t\to\infty} \bar{x}_t = a$, or $-a$. We note that $\forall \eta > 0$ there exists δ_1 such that$|v(x)| \leq \eta$, $x \in B(a, \delta_1) \cup B(-a, \delta_1)$. We can assume $\delta < \delta_1$. Then we have

$$\begin{aligned} v(x) &\leq \frac{1}{2\kappa}\int_0^{t_0} |\hat{z}_s|^2 ds + \int_0^{t_0} V(x_s)ds + v(x_{t_0}) \\ &\leq \frac{1}{2\kappa}\int_0^t |\bar{z}_s|^2 ds + \int_0^t V(\bar{x}_s)ds + v(x_{t_0}), \quad t \geq t_0. \end{aligned}$$

Thus we obtain

$$v(x) - \eta \le \frac{1}{2\kappa}\int_0^\infty |\bar{z}_s|^2 ds + \int_0^\infty V(\bar{x}_s)ds, \quad \forall \eta > 0,$$

which implies

$$\begin{aligned} v(x) \quad \le \quad & \sup_{x_0=x, x_\infty=a} \{\int_0^\infty (V(x_s) + \frac{1}{2\kappa}|\dot{x}_s + \nabla U(x_s)|^2) ds\} \\ & \vee \sup_{x_0=x, x_\infty=-a} \{\int_0^\infty (V(x_s) + \frac{1}{2\kappa}|\dot{x}_s + \nabla U(x_s)|^2) ds\}. \end{aligned}$$

Converse inequality is easily seen and we conclude our present theorem.

Remark. Fleming and Sheu (1998) have obtained such representation for $v(x)$ as Theorem 4.1 in the case where $\frac{1}{2}|\nabla U|^2 + \kappa V$ has a single minimum, namely $a = 0$.

References

A. Bensoussan and J. Frehse (1992) On Bellman equations of ergodic control in R^N, J. Reine Angew. Math.,vol. 429, pp.125-160

A. Bensoussan and H. Nagai (1991) An ergodic control problem arising from the principal eigenfunction of an elliptic operator, Math. Soc. Japan 43 , pp.49-65.

W. H. Fleming and W. M. McEneaney (1995) *Risk-sensitive control on an infinite time horizon*, SIAM J. Control and Optimization, vol.33 , pp.1881-1915.

W. H. Fleming and S.-J. Sheu (1998) Asymptotics for the first eigenvalue and eigenfunction of a nearly first order operator with large potential, to appear in Ann. of Prob.

H. Ishii, H. Nagai and F. Teramoto (1996) A singular limit on risk sensitive control and semi-classical analysis, Proceedings of the 7th Japan-Russia Symp. on Prob. and Math. Stat.pp.164-173, World Scientific.

H. Kaise and H. Nagai (1998) Bellman equations of ergodic type related to risk-sensitive control and their singular limits, Asymptotic Analysis, vol. 16, pp.347-362

H. Nagai (1996) Bellman equation of risk-sensitive control, SIAM J. Control and Optimization, vol. 34, pp.74-101.

7

Game Approach to Risk Sensitive Control for Stochastic Evolution Systems

Makiko Nisio

ABSTRACT: The relations between risk sensitive control problems and dynamic games are considered for stochastic evolution systems. We obtain the game representation of the logarithmic transformation of value function, without passing to small noise limit, and derive a speed of convergence of value function when the parameter of noise tends to 0.

KEYWORDS: risk sensitive control, viscosity solution, dynamic game, stochastic evolution system.

1 Introduction

The relationships between risk sensitive control problems and dynamic games are considered in various ways. They mostly show that small noise limits are related to differential games via Legendre type transformations. Significant early contributions in this subject are due to Jacobson(1973) and Whittle(1981), where they dealt with linear Gaussian dynamics. For controlled diffusion processes, Bensoussan and Nagai(1997) and Fleming and McEneaney (1992) investigated finite time horizon case and in (1995) infinite time horizon one. Recently Dai Pra, Menghini and Runggaldier(1996) introduced the dual games for wider classes, using duality relation between free energy of bounded functions and relative entropy of probability measures, without passing to small noise limits.
Motivated by their works for finite dimensional cases, we will here study the same proplems for stochastic evolution syatems. In Nisio(1996), we dealt with the system governed by stochastic differential equation (1) on Hilbert space H and the risk sensitive criterion (2) below with $q = 0$ and a bounded function f. Denoting by ν^ε the logarithmic transformation of value function, we customarily evaluated the continuity of ν^ε and showed the compactness of $\{ \nu^\varepsilon, \varepsilon > 0 \}$. Moreover we proved that its small noise limit coincided with the value function of deterministic differential game, via viscosity solution arguments. Since ν^ε satisfies Isaacs equation (see (4)

with $q = 0$), it is formally related to a stochastic differential game, where the control region of one player is H. But the continuity of (upper or lower) value function is delicate, because H is not bounded. In Nisio(to appear) we constructed a stochastic differential game which gave a game representation of ν^ε. Moreover this representation provided the speed of convergence of ν^ε, when ε tends to 0.

In this paper we treat general f and q with (A6) below. For the study of small noise asymptotics, we will use a different method, namely the game representation of ν^ε, since the evaluation of ν^ε, uniformly in ε, is difficult.

Let W_k, $k = 1, 2, \ldots$, be one dimensional independent Brownian motions. D denotes a bounded convex open domain of $\mathbf{R}^n$ with smooth boundary. Let $\mathbf{U}$ be a compact convex subset of $L^2(D, \mathbf{R}^M)$ and $\mathbf{B}(\mathbf{U})$ its Borel field. A process taking values in $\mathbf{U}$ is called an admissible control, if it is Brownian adapted and right continuous paths with left limits. Let us put H^k=Sobolev space $H_0^k(D)$, $\|\cdot\|_k$= its norm and $H = H^0$, $\|\cdot\| = \|\cdot\|_0$ for simplicity. For $\varepsilon > 0$, when a controller applies an admissible control U, the system ξ is governed by the following stochastic differential equation on H,

$$d\xi(t) = [A\xi(t) + \delta(\xi(t), U(t))]dt + \sqrt{\varepsilon} dM(t), \quad for \quad 0 < t < T, \tag{1}$$

$$with\ initial\ condition, \quad \xi(0) = \eta \quad (\in H),$$

where A; $H^1 \to H^{-1}$ is a uniformly elliptic differential operator

$$A\zeta = \sum_{ij=1}^n \frac{\partial}{\partial x_i}(a^{ij}(x)\frac{\partial \zeta}{\partial x_j}) + \sum_{i=1}^n r^i(x)\frac{\partial \zeta}{\partial x_i} - c(x)\zeta, \quad for \quad \zeta \in H^1,$$

δ; $H \times \mathbf{U} \to H$, *and* $M(t)$ is an H valued random noise having the form,

$$dM(t) = \sum_{i=1}^\infty \sqrt{m_i} e_i dW_i, \quad \text{with summable } m_i \text{ and an orthonormal}$$

base $e_i (\in C_0^\infty(D))$, $i = 1, 2, \cdots$. Precise formulation and assumptions are given in Section 2.

For f; $H \times \mathbf{U} \to \mathbf{R}^1$ and q; $H \to \mathbf{R}^1$ (see (A6)), the risk sensitive criterion $\mathcal{G}^\varepsilon$ is defined by

$$\mathcal{G}^\varepsilon(t, \eta, U) = E \exp \frac{1}{\varepsilon}[\int_0^t f(\xi(s), U(s))ds + q(\xi(t))] \tag{2}$$

with the solution ξ of (1), and the controller wants to minimize $\mathcal{G}^\varepsilon$.

In order to study a small noise asymptotics of value function, we will deal with the logarithmic transformation, (see, Chapter VI in Fleming and Soner (1993)),

$$\nu^\varepsilon(t, \eta) = \varepsilon \log \inf_{U \in \mathcal{U}} \mathcal{G}^\varepsilon(t, \eta, U)$$

where $\mathcal{U}$ is the set of admissible controls. Using a standard technique, we can see that ν^ε satisfies the following Hamilton-Jacobi-Bellman equation in viscosity sense (see Definition 2),

$$0 = \frac{\partial \nu}{\partial t}(t,\eta) - \langle A^* \partial\nu(t,\eta), \eta\rangle - \inf_{u \in \mathbf{U}} [\, \frac{\varepsilon}{2} trace(\Lambda^2 \partial^2 \nu(t,\eta))$$

$$+\langle \partial\nu(t,\eta), \delta(\eta,u)\rangle + \frac{1}{2}||\Lambda\partial\nu(t,\eta)||^2 + f(\eta,u)\,],$$

$$for \quad (t,\eta) \in (0,T) \times H, \tag{3}$$

$$with\ initial\ condition \quad \nu(0,\eta) = q(\eta)$$

where A^* = adjoint of A, $\langle \cdot, \cdot \rangle$ = duality pair between H^{-1} and H^1 and a linear transformation Λ on H is defined by $\Lambda e_i = \sqrt{m_i} e_i$, $i = 1, 2, \ldots$ This equation (3) is equivalent to the upper Isaacs equation (4) by Legendre transformation.

$$0 = \frac{\partial \nu}{\partial t}(t,\eta) - \langle A^* \partial\nu(t,\eta), \eta\rangle - \inf_{u \in \mathbf{U}} \sup_{\zeta \in H} \mathbf{H}^\varepsilon(\partial^2\nu(t,\eta), \nu(t,\eta), \eta, u, \zeta),$$

$$for \quad (t,\eta) \in (0,T) \times H, \tag{4}$$

$$with\ initial\ condition \quad \nu(0,\eta) = q(\eta)$$

where

$$\mathbf{H}^\varepsilon(Z,p,\eta,u,\zeta)$$
$$= \frac{\varepsilon}{2} trace(\Lambda^2 Z) + \langle p, \delta(\eta,u)\rangle + \langle \Lambda p, \Lambda\zeta\rangle - \frac{1}{2}||\Lambda\zeta||^2 + f(\eta,u).$$

Thus our main step is to show that $\nu^\varepsilon(t,\eta)$ coincides with the upper value function (in Elliott-Kalton sense), say V^ε, of stochastic differential game associated with (4), via viscosity solution arguments (Theorem 3). Moreover we will study the existence of limit of V^ε, when ε tends to 0, and the speed of convergence (Theorem 5). The limit function V^0 turns out to be the unique viscosity solution of (4) with $\varepsilon = 0$, and coincides with the upper value function of deterministic differential game. Since the equation (4) satisfies Isaacs condition (see (21)), we will remark that our game has the value (Proposition 5).

In Section 2, we study some regularity properties of solutions of stochastic differential equations on H. (e.g. Da Prato and Zabczyk(1992), Rozovskii(1990)). In particular, the time continuity will be discussed. We construct the game associated with (4) (see Theorem 1) and study its properties (Proposition 3) in Section 3 and risk sensitive control problems and its small noise limit will be treated in Sections 4 and 5 respectively.

2 Preliminaries

Let (Ω, F, F_θ, P) be a canonical coordinate space with an infinite dimensional standard Wiener measure P , namely Ω is the path space $\{\omega \in C([0,T], \mathbf{R}^{\mathbf{N}});\ \omega(0) = 0\}$ endowed with the usual product topology, where $\mathbf{N}$ denotes the set of natural numbers, and the coordinate function $W_k(t,\omega) = \omega_k(t)$, $k = 1, 2, \ldots,$ are independent one dimensional Brownian motions. F_θ denotes the σ-field generated by $\{\omega_k(s);\ s \leq \theta, k = 1, 2, \ldots\}$ and $F = F_T$.
Let us assume the following conditions (A1)~(A3) on A,
(A1) a^{ij} and r^i are *in* $C^3(\bar{D})$
(A2) $n \times n$ matrix $(a^{ij}(x))$ is uniformly positive definite, say

$$\sum_{ij=1}^{n} a^{ij}(x)t_i t_j \geq \lambda_0 |t|^2 \qquad \textit{for}\ \ t = (t_1, \ldots, t_n),\ \ \textit{with}\ \lambda_0 > 0$$

(A3) $c(x)$ is a non-negative continuous function on $\bar{D}$.
From (A2) and (A3), it follows that $-A$ is coercive, say

$$\langle -A\zeta, \zeta\rangle \geq \lambda\|\zeta\|_1^2 - \mu\|\zeta\|^2, \qquad \textit{with}\ \ \lambda > 0.$$

The operator $B;\ H \to H^2$, defined by

$$B = [\, I - (A - \sum_{i=1}^{n} r^i(x)\frac{\partial}{\partial x_i})\,]^{-1} \qquad \textit{with boundary value}\ 0$$

is a compact operator on H and satisfies the following structual condition,

$$\langle -A^* B\phi, \phi\rangle \geq \frac{1}{2}\|\phi\|^2 - p|\phi|_B^2, \quad \textit{with}\ \ p \geq 0,$$

where $|\phi|_B^2 = \langle B\phi, \phi\rangle$ and $|\cdot|_B$ is called B-norm (see Crandall and Lions (1991)), which is equivalent to $\|\cdot\|_{-1}$-norm, by (A2).
Putting $|\cdot| =$ norm of $L^2(D, \mathbf{R}^M)$, we assume (A4) and (A5),
(A4) $\delta;\ H \times \mathbf{U} \to H$ is bounded and Lipschitz continuous, say

$$\hat{\delta} = \sup_{\zeta u}\|\delta(\zeta, u)\| \ \text{ and } \ \|\delta(\xi, u) - \delta(\zeta, v)\| \leq L(\|\xi - \zeta\| + |u - v|)$$

(A5) $m_i \geq 0$, $i = 1, 2, \ldots,$ is summable, say $m = \sum_{i=1}^{\infty} m_i$.
Denoting by $M^2(0,T;H^k)$ the subset of $L^2((0,T) \times \Omega; H^k)$ consisting of F_t -progressively measurable processes, we deal with a stochastic evolution equation (5), for $F \in M^2(0,T;H)$,

$$d\xi(t) = (A\xi(t) + \delta(\xi(t), U(t)) + F(t))dt + \sqrt{\varepsilon}dM(t), \qquad \textit{for}\ \ 0 < t < T, \quad (5)$$

$$\textit{with initial condition}\ \ \xi(0) = \eta\ (\in H).$$

We will here define a solution according to Pardoux(1979), Rozovskii(1990).

Definition 1. $\xi \in M^2(0,T;H^1)$ is called a solution of the above equation, if $\xi \in C([0,T];H)$ a.s. and, for any t and smooth function ϕ with support in D,

$$\langle \xi(t),\phi\rangle = \langle \eta,\phi\rangle + \int_0^t \langle A\xi(s),\phi\rangle + \langle \delta(\xi(s),U(s)),\phi\rangle + \langle F(s),\phi\rangle\ ds$$

$$+\sqrt{\varepsilon}\sum_{i=1}^{\infty}\int_0^t \sqrt{m_i}\langle e_i,\phi\rangle dW_i(s). \qquad a.s..$$

Employing a successive approximation, we can obtain a unique solution $\xi(t,\eta)=\xi^{\varepsilon}(t,\eta,F,U)$ with following properties

$$E(\sup_{s\le t}\|\xi(s,\eta)\|^2 + \int_0^t \|\xi(s,\eta)\|_1^2 ds) \le K_1(t+\|\eta\|^2 + \int_0^t E\|F(s)\|^2 ds) \quad (6)$$

$$E(\sup_{s\le t}|\,\xi(s,\eta)\,|_B^2 + \int_0^t \|\xi(s,\eta)\|^2 ds) \le K_1(t+|\,\eta\,|_B^2 + \int_0^t E|\,F(s)\,|_B^2 ds) \quad (7)$$

where K_1 is independent of t,η,F,U and ε. Moreover, since the dynamics of $\xi(s,\eta)-\xi(s,\tilde\eta)$ does not depend on F and dM, we can easily see

$$\sup_{t\le T}\|\xi(s,\eta)-\xi(s,\tilde\eta)\|^2 + \int_0^T \|\xi(s,\eta)-\xi(s,\tilde\eta)\|_1^2 ds \le K_2\|\eta-\tilde\eta\|^2, \quad (8)$$

$$\sup_{t\le T}|\,\xi(s,\eta)-\xi(s,\tilde\eta)\,|_B^2 + \int_0^T \|\xi(s,\eta)-\xi(s,\tilde\eta)\|^2 ds \le K_2|\,\eta-\tilde\eta\,|_B^2, \quad (9)$$

with a constant K_2 independent of η,U,F,ε and a probability parameter ω.

Proof. We will show only (7) , because we can see the other parts in Rozovskii(1990) (Theorem 4 in Section 3.1) and Nisio(1998) (see Propositions 2.1 and 2.2). Dropping η, for simplicity, we consider

$$\begin{aligned}
d|\xi(t)|_B^2 &= 2\langle B\xi(t),d\xi(t)\rangle + |d\xi(t)|_B^2\\
&= 2(\langle A^*B\xi(t),\xi(t)\rangle + 2\langle B\xi(t),\delta(\xi(t),U(t))+F(t)\rangle)dt\\
&\quad +\varepsilon\sum_{i=1}^{\infty} m_i\langle Be_i,e_i\rangle dt + 2\sqrt{\varepsilon}\langle B\xi(t),dM(t)\rangle\\
&\le (-\|\xi(t)\|^2 + 2p|\xi(t)|_B^2 + 2|\xi(t)|_B^2 + |\delta(\xi(t),U(t))|_B^2)dt\\
&\quad +(|F(t)|_B^2 + m\varepsilon)dt + 2\sqrt{\varepsilon}\langle B\xi(t),dM(t)\rangle.
\end{aligned}$$

Hence integrating from 0 to t, we get the following three evaluations.

$$\begin{aligned}
Ee^{-2(p+1)t}|\xi(t)|_B^2 &\le |\eta|_B^2 + \int_0^t (\hat\delta^2 + E|F(\theta)|_B^2 + \varepsilon m)d\theta\\
&\le k(t+|\eta|_B^2 + \int_0^t E|F(\theta)|_B^2 d\theta),
\end{aligned}$$

$$E\int_0^t \|\xi(\theta)\|^2 d\theta \le -E|\xi(t)|_B^2 + |\eta|_B^2 + (2p+2)\int_0^t E|\xi(\theta)|_B^2 d\theta$$
$$+(\hat{\delta}^2 + m\varepsilon)t + \int_0^t E|F(\theta)|_B^2 d\theta,$$

$$\sup_{\theta\le t} |\xi(\theta)|_B^2 \le (2p+2)\int_0^t E|\xi(\theta)|_B^2 d\theta + (\hat{\delta}^2 + m\varepsilon)t$$
$$+\int_0^t E|F(\theta)|_B^2 d\theta + 2\sqrt{\varepsilon}\sup_{\theta\le t}\int_0^\theta \langle B\xi(s), dM(s)\rangle.$$

Applying a martingale inequality (see (3.2), P.110 in Ikeda and Watanabe(1981)) to the last term, we have

$$(E\sup_{\theta\le t}\int_0^\theta \langle B\xi(s), dM(s)\rangle)^2 \le 4\sum_{i=1}^\infty m_i E\int_0^t \langle B\xi(s), e_i\rangle^2 ds \tag{10}$$
$$\le 4m\int_0^t E|\xi(s)|_B^2 ds.$$

Now noting three evaluations above, we complete the proof of (7).

Proposition 1.

$$E\sup_{\theta\le\tau} |\,\xi(s+\theta,\eta) - \xi(s,\eta)\,|_B^2 \le K_3\tau[1 + E\|\xi(s,\eta)\|^2$$
$$+\int_0^\tau E|\,F(s+\theta)\,|_B^2 d\theta] + K_4(\varepsilon\tau\int_0^\tau E|\,F(s+\theta)\,|_B^2 d\theta)^{\frac{1}{2}}$$

with K_3 and K_4 independent of s, η, F, U and ε.

Proof. Putting $\zeta(\theta) = \xi(s+\theta,\eta) - \xi(s,\eta)$, we see that $d\zeta(\theta) = d_\theta\xi(s+\theta,\eta)$ and

$$d|\,\zeta(\theta)\,|_B^2 = 2\langle B\zeta(\theta), d\zeta(\theta)\rangle + |\,d\zeta(\theta)\,|_B^2$$
$$= \Big(2\langle A^*B\zeta(\theta), \zeta(\theta)\rangle + 2\langle A^*B\zeta(\theta), \xi(s,\eta)\rangle$$
$$+2\langle B\zeta(\theta), \delta(\xi(s+\theta,\eta), U(s+\theta))\rangle + 2\langle B\zeta(\theta), F(s+\theta)\rangle\Big)\, d\theta$$
$$+\varepsilon\sum_{i=1}^\infty m_i|\,e_i\,|_B^2 d\theta + 2\sqrt{\varepsilon}\langle B\zeta(\theta), dM(s+\theta)\rangle$$

Noting the following evaluations,

$$1st\ \ term\ \le -\|\zeta(\theta)\|^2 + 2p|\,\zeta(\theta)\,|_B^2$$

$$2nd\ \ term\ \le 2|\,A^*B\,|\|\zeta(\theta)\|\|\xi(s,\eta)\| \le \|\zeta(\theta)\|^2 + |\,A^*B\,|^2\|\xi(s,\eta)\|^2$$

where $|\cdot|$ denotes the operator norm,

$$3rd\ \ term \le |\zeta(\theta)|_B^2 + \hat{\delta}^2 \quad and\ \ 4th\ \ term \le |\zeta(\theta)|_B^2 + |F(s+\theta)|_B^2,$$

we get

$$\exp[-2(p+1)\theta]E|\zeta(\theta)|_B^2 \le (|A^*B|^2 E\|\xi(s,\eta)\|^2 + \hat{\delta}^2 + \varepsilon m)\theta + \int_0^\theta \exp[-2(p+1)h]E|F(s+h)|_B^2 dh. \tag{11}$$

Next we replace the above evaluation of 4th term by the following one,

$$\int_0^\tau |\langle B\zeta(\theta), F(s+\theta)\rangle| d\theta \le \sup_{\theta\le\tau} |\zeta(\theta)|_B \int_0^\tau |F(s+\theta)|_B d\theta$$
$$\le \frac{1}{8} \sup_{\theta\le\tau} |\zeta(\theta)|_B^2 + 2(\int_0^\tau |F(s+\theta)|_B d\theta)^2,$$

and similally for the 3rd term,

$$\int_0^\tau |\langle B\zeta(\theta), \delta(\xi(s+\theta), U(s+\theta))| d\theta$$

$$\le \sup_{\theta\le\tau} |\zeta(\theta)|_B \int_0^\tau |\delta(\xi(s+\theta), U(s+\theta))|_B d\theta \le \frac{1}{8} \sup |\zeta(\theta)|_B^2 + 2\hat{\delta}^2\tau^2.$$

Then we obtain

$$|\zeta(\tau)|_B^2 \le \int_0^\tau (-\|\zeta(\theta)\|^2 + 2p|\zeta(\theta)|_B^2 + \|\zeta(\theta)\|^2 + |A^*B|^2 \|\xi(s,\eta)\|^2) d\theta$$
$$+4(\int_0^\tau |F(s+\theta)|_B d\theta)^2 + 4\hat{\delta}^2\tau^2 + \varepsilon m\tau$$
$$+2\sqrt{\varepsilon} \sup_{\theta\le\tau} \int_0^\theta \langle B\zeta(h), dM(s+h)\rangle dh + \frac{1}{2} \sup_{\theta\le\tau} |\zeta(\theta)|_B^2.$$

Since the right hand side is incresing in τ , we can replace the left hand side by $\sup_{s\le\tau} |\zeta(s)|_B^2$. This yields

$$\frac{1}{2} \sup_{\theta\le t} |\zeta(\theta)|_B^2 \le 2p \int_0^t |\zeta(\theta)|_B^2 d\theta + (|A^*B|^2 \|\xi(s,\eta)\|^2 + 4t\hat{\delta}^2 + \varepsilon m)t$$
$$+4(\int_0^t |F(s+\theta)|_B d\theta)^2 + 2\sqrt{\varepsilon} \sup_{\theta\le t} \int_0^\theta \langle B\zeta(h), dM(s+h)\rangle.$$

Now applying the martingale inequality (10) to the last term with (11), we can conclude Proposition 1.

Proposition 2. Suppose $\sup_{\theta\leq t}\|F(\theta)\| \leq c$, with probability 1. Then

$$E\sup_{\theta\leq t}\|\xi(\theta,\eta)-e^{\theta A}\eta\|^4 \leq K_5(1+c^4)t^2$$

with constant K_5 independent of η, F, U and ε.

Proof Let us consider the system $\zeta(\theta)=\xi(\theta,\eta)-e^{tA}\eta$. Then, from the coercive condition of $-A$ and (5), it follows

$$e^{-(2+\mu)t}\|\zeta(t)\|^2 \leq K_1(1+c^2)t+2\sqrt{\varepsilon}\int_0^t \langle e^{-(2+\mu)h}\zeta(h), dM(h)\rangle.$$

Taking the square of both sides and using a martingale inequality we can complete the proof.

3 Stochastic differential games

Here we will study stochastic differential games associated with the upper Isaacs equation (4). We assume the condition (A6), besides (A1)~(A5).

(A6) $f;\ H\times U \rightarrow [0,\infty)$ is Lipschitz continuous, and $q;\ H \rightarrow [0,\infty)$ is B-Lipschitz continuous, say
$|f(\xi,u)-f(\zeta,v)| \leq \kappa(\|\xi-\zeta\|+|u-v|)$ and
$|q(\xi)-q(\zeta)| \leq \kappa|\xi-\zeta|_B$.

Let us put $Y_N=\{\Lambda\zeta;\ \|\zeta\|\leq N\}$. Then Y_N is compact, because of (A5). $\mathcal{Y}_N$ denotes the set of processes taking values in Y_N, which are F_t-progressively measurable and right continuous paths with left limits. $Y=\bigcup_{N=1}^{\infty}Y_N$ and $\mathcal{Y}=\bigcup_{N=1}^{\infty}\mathcal{Y}_N$ are called a control region and the set of admissible controls of opponent respectively.
When players apply admissible controles Y and U respectively, the system $X(t)=X^{\varepsilon}(t,\eta,Y,U)$ is governed by the stochastic evolution equation ,

$$dX(t)=[AX(t)+\delta(X(t),U(t))+Y(t)]dt+\sqrt{\varepsilon}dM(t),$$
$$for\quad 0<t<T, \tag{12}$$
$$with\ initial\ condition,\quad X(0)=\eta\ (\in H),$$

and the pay-off J^{ε} is given by

$$J^{\varepsilon}(t,\eta,Y,U)=E\int_0^t f(X(s),U(s))-\frac{1}{2}\|Y(s)\|^2ds+q(X(t)). \tag{13}$$

The opponent wants to maximize J^{ε} and the controller tries to minimize it.

An admissible strategy γ of controller is a mapping $\gamma;\ \mathcal{Y} \to \mathcal{U}$, which is $(\mathbf{B}[0,T] \times F, \mathbf{B}(\mathbf{U}))$ -measurable and non-anticipative, namely if $P(Y(s) = \tilde{Y}(s)) = 1$ *for* $s < t$, *then* $P(\gamma Y(t) = \gamma\tilde{Y}(t)) = 1$. $\mathcal{R}$ denotes the set of admissible strategies of controller. $\mathcal{A}_N$ denotes the set of mappings from $\mathcal{U}$ into $\mathcal{Y}_N$ with non-anticipative and measurable property. Putting $\mathcal{A} = \bigcup_{N=1}^{\infty} \mathcal{A}_N$, we call $\alpha \in \mathcal{A}$ an admissible strategy of opponent.

The upper and lower value functions (in Elliott-Kalton sense) are defined as follows,

$$\textit{upper value function}\quad V^\varepsilon(t,\eta) = \sup_{\alpha\in\mathcal{A}} \inf_{U\in\mathcal{U}} J^\varepsilon(t,\eta,\alpha,U)$$

$$\textit{lower value function}\quad v^\varepsilon(t,\eta) = \inf_{\gamma\in\mathcal{R}} \sup_{Y\in\mathcal{Y}} J^\varepsilon(t,\eta,Y,\gamma)$$

where $J^\varepsilon(t,\eta,\alpha,U) = J^\varepsilon(t,\eta,\alpha U,U)$ and $J^\varepsilon(t,\eta,Y,\gamma) = J^\varepsilon(t,\eta,Y,\gamma Y)$. Recalling (7), we shall note

$$J^\varepsilon(t,\eta,Y,U) \le k[1 + t + |\eta|_B + (\int_0^t E\|Y(s)\|^2 ds)^{\frac{1}{2}}]$$

$$-\frac{1}{2}\int_0^t E\|Y(s)\|^2 ds, \tag{14}$$

where k depends on K_1 and κ. Since the function $p(x) = a + kx - \frac{1}{2}x^2$ attains its maximum at $x = k$ and $p(x) < 0$ for $x > k + (2a + k^2)^{\frac{1}{2}}$, we have

$$J^\varepsilon(t,\eta,Y,U) < 0 \qquad \textit{for}\quad E\int_0^t \|Y(s)\|^2 ds > 4k(k + 1 + T + |\eta|_B).$$

Therefore putting $r(c) = 4k(k + 1 + T + c)$ and

$$\mathcal{A}(t,c) = \{\alpha \in \mathcal{A},\ E\int_0^t \|\alpha U(s)\|^2 ds \le r(c) \quad \textit{for any } U\},$$

we can replace $\mathcal{A}$ in the definition of upper value function by $\mathcal{A}(t,|\eta|_B)$. The proof is following. For $\alpha \notin \mathcal{A}(t,|\eta|_B)$, there is $\hat{U}$ such that $E\int_0^t \|\alpha\hat{U}(s)\|^2 ds > r(|\eta|_B)$. Hence $\inf_{U\in\mathcal{U}} J^\varepsilon(t,\eta,\alpha,U)$ $\le J^\varepsilon(t,\eta,\alpha,\hat{U}) < 0$. But, by the positivity of f and q, $J^\varepsilon(t,\eta,0,U) \ge 0$, for any U. This means that 0 strategy in $\mathcal{A}(t,|\eta|_B)$ is better than α.

Let us put

$$V_N^\varepsilon(t,\eta) = \sup_{\alpha\in\mathcal{A}_N} \inf_{U\in\mathcal{U}} J^\varepsilon(t,\eta,\alpha,U) \qquad \textit{and} \qquad \mathcal{A}_N(t,c) = \mathcal{A}_N \cap \mathcal{A}(t,c).$$

then V_N^ε satisfies the following properties.

Proposition 3

$$
\begin{aligned}
&\text{(i)} && V_N^\varepsilon(t,\eta) = \sup_{\alpha \in \mathcal{A}_N(t, |\eta|_B)} \inf_{U \in \mathcal{U}} J^\varepsilon(t,\eta,\alpha,U) \\
&\text{(ii)} && 0 \le V_N^\varepsilon(t,\eta) \le c_1(1+|\eta|_B) \\
&\text{(iii)} && |V_N^\varepsilon(t,\eta) - V_N^\varepsilon(t,\tilde{\eta})| \le c_2|\eta-\tilde{\eta}|_B \\
&\text{(iv)} && |V_N^\varepsilon(s+t,\eta) - V_N^\varepsilon(t,\eta)| \le c_3\sqrt{s}(1+\|\eta\|) + c_4(\varepsilon s)^{\frac{1}{4}}, \\
& && \quad \textit{for} \quad 0<t<t+s<T,
\end{aligned}
$$

where $c_i, i = 1,2,3,4,$ are independent of ε, N, t and η.

Proof. we will prove only (iv), because the others are easy.
For $\hat{\varepsilon} > 0$, we take an $\hat{\varepsilon}$-optimal strategy $\hat{\alpha} \in \mathcal{A}_N(s+t, |\eta|_B)$ such that

$$V_N^\varepsilon(s+t,\eta) < \inf_{U \in \mathcal{U}} J^\varepsilon(s+t,\eta,\hat{\alpha},U) + \hat{\varepsilon}.$$

Then, for $\hat{U}$ with the following property

$$\inf_{U \in \mathcal{U}} J^\varepsilon(t,\eta,\hat{\alpha},U) > J^\varepsilon(t,\eta,\hat{\alpha},\hat{U}) - \hat{\varepsilon},$$

we get

$$
\begin{aligned}
V_N^\varepsilon(s+t,\eta) - V_N^\varepsilon(t,\eta) &< \inf_{U \in \mathcal{U}} J^\varepsilon(s+t,\eta,\hat{\alpha},U) - \inf_{U \in \mathcal{U}} J^\varepsilon(t,\eta,\hat{\alpha},U) + \hat{\varepsilon} \\
&\le J^\varepsilon(s+t,\eta,\hat{\alpha},\hat{U}) - J^\varepsilon(t,\eta,\hat{\alpha},\hat{U}) + 2\hat{\varepsilon} \\
&= E\int_t^{s+t} f(\hat{X}(\theta),\hat{U}(\theta)) - \frac{1}{2}\|\hat{\alpha}\hat{U}(\theta)\|^2 d\theta + q(\hat{X}(s+t)) - q(\hat{X}(t)) + 2\hat{\varepsilon}
\end{aligned}
$$

where $\hat{X}(t) = X^\varepsilon(t,\eta,\hat{\alpha},\hat{U})$. Now (6) and Proposition 1 derive

$$V_N^\varepsilon(s+t,\eta) - V_N^\varepsilon(t,\eta) \le c_3\sqrt{s}(1+\|\eta\|) + c_4(\varepsilon s)^{\frac{1}{4}},$$

$$\textit{for} \quad 0<t<t+s<T, \tag{15}$$

because of $\hat{\alpha} \in \mathcal{A}_N(s+t, |\eta|_B)$.
For the opposite inequality, we take $\hat{\varepsilon}$ -optimal strategy $\tilde{\alpha} \in \mathcal{A}_N(t, |\eta|_B)$ such that $V_N^\varepsilon(t,\eta) < \inf_{U \in \mathcal{U}} J^\varepsilon(t,\eta,\tilde{\alpha},U) + \hat{\varepsilon}$ and define α^* by $\alpha^* U(\theta) = \tilde{\alpha}U(\theta)$ *on* $[0,t]$, $\alpha^* U(\theta) = 0$ on (t,T). Then we get

$$
\begin{aligned}
V_N^\varepsilon(s+t,\eta) - V_N^\varepsilon(t,\eta) &> \inf_{U \in \mathcal{U}} J^\varepsilon(s+t,\eta,\alpha^*,U) - \inf_{U \in \mathcal{U}} J^\varepsilon(t,\eta,\alpha^*,U) - \hat{\varepsilon} \\
&\ge \inf_{U \in \mathcal{U}} E[q(X^\varepsilon(s+t,\eta,\alpha^*,U)) - q(X^\varepsilon(t,\eta,\alpha^*,U))] - \hat{\varepsilon}.
\end{aligned}
$$

Again noting Proposition 1 and the definition of α^*, we have the opposite inequality of (15), which concludes (iv).

Now Proposition 3 yields

Proposition 4 As $N \to \infty$, V_N^ϵ is increasing to V^ϵ uniformly on any bounded set of $[0,T] \times H$. Moreover V^ϵ satisfies the same properties as (i)∼(iv) in Proposition 3.

Next we will recall below the definition of viscosity solution of (16),

$$0 = \frac{\partial W}{\partial t}(t,\eta) - \langle A^* \partial W(t,\eta), \eta\rangle + \mathbf{F}(\partial^2 W(t,\eta), \partial W(t,\eta), \eta),$$
$$\textit{for} \quad (t,\eta) \in (0,T) \times H, \tag{16}$$
$$\textit{with initial condition} \quad W(0,\cdot) = q(\cdot).$$

$\Phi \in C^{12}((0,T) \times H)$ is called a test function, if
(i). Φ is weakly sequentially lower semi-continuous and bounded from below,
and
(ii). $\partial\Phi(t,\eta) \in H^2$ and $A^*\partial\Phi$ are continuous.
$g \in C^2(H)$ is called radial, if $g(\eta) = \tilde{g}(\|\eta\|)$ with $\tilde{g} \in C^2[0,\infty)$ increasing from 0 to ∞.
Noting $-A+\mu I$ is dissipative, we have Defiition 2 according to Crandall and Lions(1990).

Definition 2. $W \in C([0,T] \times H)$ is called a subsolution (resp. supersolution) of (16), if $W(0) = q$ and the following condition (C1) (resp. (C2)) holds for any test function Φ and radial function g.
(C1). If $W - \Phi - g$ has a local maximum at $(\hat{t},\hat{\eta}) \in (0,T) \times H$, then

$$\frac{\partial \Phi}{\partial t}(\hat{t},\hat{\eta}) - \langle A^*\partial\Phi(\hat{t},\hat{\eta}), \hat{\eta}\rangle + \mathbf{F}(\partial^2(\Phi+g)(\hat{t},\hat{\eta}), \partial(\Phi+g)(\hat{t},\hat{\eta}), \hat{\eta})$$
$$\leq \mu\tilde{g}'(\|\hat{\eta}\|)\|\hat{\eta}\|.$$

(C2). If $W + \Phi + g$ has a local minimum at $(\hat{t},\hat{\eta}) \in (0,T) \times H$, then

$$-\frac{\partial \Phi}{\partial t}(\hat{t},\hat{\eta}) + \langle A^*\partial\Phi(\hat{t},\hat{\eta}), \hat{\eta}\rangle + \mathbf{F}(-\partial^2(\Phi+g)(\hat{t},\hat{\eta}), -\partial(\Phi+g)(\hat{t},\hat{\eta}), \hat{\eta})$$
$$\geq -\mu\tilde{g}'(\|\hat{\eta}\|)\|\hat{\eta}\|.$$

W is called a viscosity solution, if it is both a sub- and super-solution.
Putting $H_B = H$ equipped with B-norm and

$$\begin{aligned} & UC_\eta^+([0,T] \times H_B) \\ = \; & \{w \in C([0,T] \times H_B);\ w(t,\cdot) \text{ is non-negative and uniformly} \\ & \text{continuous in } H_B, \text{uniformly in } t\}, \end{aligned}$$

we have Theorem.

Theorem 1 V^ε satisfies the dynamic programming principle and is the unique non-negative viscosity solution of the upper Isaacs equation (4), in $UC^+_\eta([0,T]\times H_B)$.

Proof. We divide the proof into five parts. For Y_N and bounded f and q, Theorem is given in Nisio (Theorem 4.2, to appear), using similar arguments as Fleming and Souganidis(1989). So, we will sketch the outline of this case in steps 1~3.
Step 1. When α satisfies the following additional property (R) below we call it r-strategy of opponent.
(R) For any $t\in(0,T)$ and $U\in\mathcal{U}$, the mapping $(\theta,\omega)\to\alpha(U_{\omega^-})(\theta,\omega^+)$ is $(\mathbf{B}[0,T-t]\times F,\mathbf{B}(Y))$-measurable, where $\omega^-=\omega_t^-$: stopped path $\omega^+=\omega_t^+$: shifted path and $U_{\omega^-}(\theta,\omega^+)=U(t+\theta,\omega)$.
Replacing admissible strategies by r-strategies, we define the r-upper value function W_N^ε by

$$W_N^\varepsilon(t,\eta)=\sup_{\alpha\in\mathrm{A}_N}\inf_{U\in\mathcal{U}}J^\varepsilon(t,\eta,\alpha,U)$$

where A_N is the set of r-strategies in $\mathcal{A}_N$. Again using B-norm, we can prove the super-optimality dynamic programming principle by a rather standard argument. Hence W_N^ε turns out to be a supersolution of the upper Isaacs equation (17)

$$0=\frac{\partial W}{\partial t}(t,\eta)-\langle A^*\partial W(t,\eta),\eta\rangle$$
$$-\inf_{u\in U}\sup_{y\in \mathrm{Y}_N}\mathbf{H}^\varepsilon(\partial^2W(t,\eta),\partial W(t,\eta),\eta,u,y). \tag{17}$$

Step 2. We construct the unique viscosity solution of (17), using the following semi-discrete approximation. Putting $\Delta=2^{-j}T$, $j=1,2,\ldots$, we call U Δ-step, *if* $U(t)=u$ *for* $t\in[0,\Delta)$ and $U(t)=U(j\Delta)$ for $t\in[j\Delta,j\Delta+\Delta)$. $\mathcal{U}_j$ denotes the set of Δ-step admissible controls of controller. A strategy γ is called Δ-step, if $\gamma Y\in\mathcal{U}_j$ and $\gamma Y(t)$ does not depend on Y, for $t\in[0,\Delta)$. $\mathcal{R}_j$ denotes their collection.
Let us define G_{jN}^ε by

$$G_{jN}^\varepsilon(t,\eta)=\inf_{\gamma\in\mathcal{R}_j}\sup_{Y\in\mathcal{Y}_N}J^\varepsilon(t,\eta,Y,\gamma).$$

Then we can see that G_{jN}^ε is decreasing, as $j\to\infty$, and its limit, say G_N^ε, satisfies the dynamic programming principle and turns out to be the unique viscosity solution of (17) in $UC^+_\eta([0,T]\times H_B)$.
Step 3. Now the comparison theorem in Świech(1994) yields

$$V_N^\varepsilon(t,\eta)\geq W_N^\varepsilon(t,\eta)\geq G_N^\varepsilon(t,\eta). \tag{18}$$

On the other hand, for any $\alpha \in \mathcal{A}_N$ and $\gamma \in \mathcal{R}_j$, there exist $\hat{Y} \in \mathcal{Y}_N$ and $\hat{U} \in \mathcal{U}_j$ such that

$$\alpha\hat{U} = \hat{Y} \quad \text{and} \quad \gamma\hat{Y} = \hat{U}.$$

Hence we get

$$\sup_{Y \in \mathcal{Y}_N} J^\varepsilon(t,\eta,Y,\gamma) \geq J^\varepsilon(t,\eta,\hat{Y},\gamma) \geq J^\varepsilon(t,\eta,\alpha,\hat{U}) \geq \inf_{U \in \mathcal{U}_j} J^\varepsilon(t,\eta,\alpha,U)$$
$$\geq \inf_{U \in \mathcal{U}} J^\varepsilon(t,\eta,\alpha,U)$$

This means

$$G^\varepsilon_{jN}(t,\eta) \geq V^\varepsilon_N(t,\eta).$$

Letting j tend to ∞, we obtain the opposite inequality of (18), which yields that V^ε_N coincides with G^ε_N. Hence V^ε_N satisfies the dynamic programming principle and is the unique viscosity solution of (17).

Step 4. For general f and q, with (A6), we put $f_k = \min(f,k)$ and $q_k = \min(q,k)$. Replacing f and q by f_k and q_k, we define the upper value function W^ε_{Nk}, in the same way as V^ε_N. Then W^ε_{Nk} satisfies the dynammic programming principle and is increasing to V^ε_N uniformly on any bounded set of $[0,T] \times H$, as $k \to \infty$. Thus from the definition of $\mathcal{A}_N$ and (6), it follows that V^ε_N satisfies the dynammic programming principle,

$$V^\varepsilon_N(s+t,\eta) = \sup_{\alpha \in \mathcal{A}_N} \inf_{U \in \mathcal{U}} E[\int_0^s f(X(\theta),U(\theta)) - \frac{1}{2}\|\alpha U(\theta)\|^2 d\theta$$
$$+V^\varepsilon_N(t,X(s))], \tag{19}$$

We can show that V^ε_N is a viscosity solution of (17), using the same method as in Nisio (1998.Theorem 4.1), because of compactness of Y_N. But we need a slighltly careful evaluation, since V^ε_N is not bounded.

Suppose $V^\varepsilon_N - \Phi - g$ has a local maximum at $(\hat{t},\hat{\eta})$. Then, for any $\hat{\varepsilon} > 0$, there is $d = d(\hat{\varepsilon}) > 0$ such that
$V^\varepsilon_N(\hat{t},\hat{\eta}) - \Phi(\hat{t},\hat{\eta}) - g(\hat{\eta}) \geq V^\varepsilon_N(t,\eta) - \Phi(t,\eta) - g(\eta)$
and
$|F(\hat{t},\hat{\eta}) - F(t,\eta)| < \hat{\varepsilon}$ *for* $F = \Phi, g, \frac{\partial\Phi}{\partial t}, \partial\Phi, \partial g, A^*\partial\Phi, \partial^2\Phi, \partial^2 g$
hold, whenever $|t-\hat{t}| < d$ and $\|\eta - \hat{\eta}\| < d$. Let us put $\tau =$ the exit time from the ball with radius d and center $\hat{\eta}$, and $X(t) \equiv X(t,\hat{\eta},\alpha,U)$. Let us fix $\hat{\theta}$ such that $\|e^{hA}\hat{\eta} - \hat{\eta}\| < \frac{d}{2}$ for $h < \hat{\theta}$. Then Proposition 2 derives for $\alpha \in \mathcal{A}_N$,

$$P(\tau < s) = P(\sup_{h<s}\|X(h) - \hat{\eta}\| > d) \leq P(\sup_{h<s}\|X(h) - e^{hA}\hat{\eta}\| > \frac{d}{2})$$
$$\leq k_1(1+N^4)s^2 d^{-4} \quad \textit{for} \quad s < \hat{\theta}$$

Moreover Proposition 1 and (iii) and (iv) of Proposition 3 yield

$$\begin{aligned} & |E(V_N^\varepsilon(\hat{t}-\theta, X(\theta)) - V_N^\varepsilon(\hat{t},\hat{\eta}); \tau < \theta)| \\ \le\ & k_1[E(|X(\theta)-\hat{\eta}|_B; \tau<\theta) + (\sqrt{\theta}(1+\|\hat{\eta}\|) + \sqrt[4]{\theta})P(\tau<\theta))] \\ \le\ & k_2\theta^{\frac{5}{4}} d^{-2}, \end{aligned}$$

because of the inequality: $E(Z;\tau<\theta) \le \sqrt{EZ^2P(\tau<\theta)}$.
For the other part, $E(V_N^\varepsilon(\hat{t}-\theta, X(\theta)) - V_N^\varepsilon(\hat{t},\hat{\eta}); \tau \ge \theta)$, we apply the same calculations as (4.3) in Nisio(1998), and get that V_N^ε is a subsolution of (17). Since we can show that V_N^ε is a supersolution by a similar calculation, V_N^ε is the unique viscosity solution of (17).
Step 5. Since V^ε is B-continuous, local maximum (resp. minimum) can be replaced by strictly local maximum (resp. minimum). Suppose that $V^\varepsilon - \Phi - g$ has a strictly local maximum at $(\hat{t},\hat{\eta})$. Then, by Proposition 4, we can take a local maximum point $(\hat{t}_N, \hat{\eta}_N)$ of $V_N^\varepsilon - \Phi - g$, which converges to $(\hat{t},\hat{\eta})$ in $[0,T]\times H$. Hence we get, putting $n = \sup_{p=1}^\infty \|\partial\Phi(\hat{t}_p,\hat{\eta}_p)\|$

$$\sup_{y\in Y_N} (\langle \Lambda\partial\Phi(\hat{t}_N,\hat{\eta}_N), y\rangle - \frac{1}{2}\|y\|^2) = \frac{1}{2}\|\Lambda\partial\Phi(\hat{t}_N,\hat{\eta}_N)\|^2, \quad for \quad N \ge n.$$

As $N\to\infty$, this derives that V^ε is a subsolution of (3), which is equivalent to (4). Since we can show that V^ε is a supersolution in the same way, V^ε is a unique solution in $UC_\eta^+([0,T]\times H_B)$.
Let us fix k arbitrally. From (19), we see

$$\begin{aligned} V_N^\varepsilon(s+t,\eta) \ge \sup_{\alpha\in\mathcal{A}_k} \inf_{U\in\mathcal{U}} E[\int_0^s f(X(\theta),U(\theta)) - \frac{1}{2}\|\alpha U(\theta)\|^2 ds \\ +V_N^\varepsilon(t,X(s))], \qquad for \quad N\ge k. \end{aligned} \tag{20}$$

On the other hand, using (7) and Proposition 3, we get for $\alpha\in\mathcal{A}_k$ and $U\in\mathcal{U}$

$$E(|V_N^\varepsilon(t,X(s)) - V^\varepsilon(t,X(s))|; |X(s)|_B \ge k^{-3}) \le ck^{-1}$$

with c independent of $N, U$and α. Moreover from Proposition 4, it follows that $E(|V_N^\varepsilon(t,X(s)) - V^\varepsilon(t,X(s))|; |X(s)|_B \le k^{-3})$ tends to 0 uniformly in U and α, as $N\to\infty$. From these evaluations, V^ε satisfies (20) with error term ck^{-1}. Letting k tend to ∞, we get the super-optimality condition. Since the sub-optimality condition is clear by (19) and Proposition 4, we can complete the proof.
Considering the lower value function v^ε we can show that v^ε belongs to $UC_\eta^+([0,T]\times H_B)$ and satisfies the lower Isaacs equation,

$$0 = \frac{\partial v}{\partial t}(t,\eta) - \langle A^*\partial v(t,\eta),\eta\rangle - \sup_{\zeta\in H}\inf_{u\in U} \mathbf{H}^\varepsilon(\partial^2 v(t,\eta), v(t,\eta),\eta,u,\zeta),$$

$$for \quad (t,\eta) \in (0,T) \times H$$

$$with\ initial\ condition \quad v(0,\eta) = q(\eta).$$

Since $\mathbf{H}^\varepsilon$ satisfies the Isaacs condition, namely

$$\sup_{\zeta \in H} \inf_{u \in U} \mathbf{H}^\varepsilon(Z,p,\eta,u,\zeta) = \inf_{u \in U} \sup_{\zeta \in H} \mathbf{H}^\varepsilon(Z,p,\eta,u,\zeta), \tag{21}$$

the lower value function turns out to be a viscosity solution of (4) and coincides with V^ε.

Proposition 5 Our stochastic differential game has the value.

4 Stochastic risk sensitive control

When a controller applies an admissible control $U \in \mathcal{U}$, we have the stochastic system ξ^ε of (1) and the risk sensitive criterion $\mathcal{G}^\varepsilon$ of (2). The controller wants to minimize $\mathcal{G}^\varepsilon$ and its value function γ^ε is defined by

$$\gamma^\varepsilon(t,\eta) = \inf_{U \in \mathcal{U}} \mathcal{G}^\varepsilon(t,\eta,U).$$

First of all, we remark the integrability of exponential criterion.

Proposition 6. For any positive constant z

$$E \exp \frac{z}{\varepsilon}[\int_0^t \|\xi(s)\| ds + |\xi(t)|_B] \le \exp \frac{z}{\varepsilon} c(1 + |\eta|_B), \quad for \ \ t \le T$$

holds, where c is independent of t, U, ε *and* z, and $\xi(s)$ = solution of (1).

Proof. Let us consider Gaussian system ζ, governed by the following equation

$$d\zeta(t) = A\zeta(t)dt + \sqrt{\varepsilon} dM(t), \qquad for \qquad 0 < t < T,$$

$$with\ initial\ condition, \quad \zeta(0) = \eta \ (\in H).$$

Putting $g_k(\zeta) = z \min(\|\zeta\|, k)$, $p_k(\zeta) = z \min(|\zeta|_B, k)$ and

$$I_k^\varepsilon(t,\eta) = E \exp \frac{1}{\varepsilon}[\int_0^t g_k(\zeta(s))ds + p_k(\zeta(t))],$$

we can show that I_k^ε satisfies the following parabolic equation on H in viscosity sense, by Itô's formula,

$$0 = \frac{\partial I}{\partial t}(t,\eta) - \langle A^* \partial I(t,\eta), \eta \rangle - \frac{\varepsilon}{2} trace(\Lambda^2 \partial^2 I(t,\eta)) - \frac{1}{\varepsilon} g_k(\eta) I(t,\eta),$$

$$for \quad (t,\eta) \in (0,T) \times H$$

$$with\ initial\ condition, \quad I(0,\eta) = \exp \frac{1}{\varepsilon} p_k(\eta).$$

Therefore its logarithmic transformation $\mathcal{I}_k^\varepsilon$ satisfies the equation

$$0 = \frac{\partial \mathcal{I}}{\partial t}(t,\eta) - \langle A^* \partial \mathcal{I}(t,\eta), \eta \rangle - \frac{\varepsilon}{2} trace(\Lambda^2 \partial^2 \mathcal{I}(t,\eta))$$

$$- \sup_{y \in H} (\langle \partial \mathcal{I}(t,\eta), \Lambda y \rangle - \frac{1}{2} ||\Lambda y||^2) + g_k(\eta), \quad for \quad (t,\eta) \in (0,T) \times H$$

$$with\ initial\ condition, \quad \mathcal{I}(0,\eta) = p_k(\eta).$$

Since $\mathcal{I}_k^\varepsilon$ also belongs to $UC_\eta^+([0,T] \times H_B)$ by (9) and Proposition 1, $\mathcal{I}_k^\varepsilon$ is related to stochasic differential game (12) and (13) with δ=0. Therefore (ii) of Proposition 3 yields

$$0 \le \mathcal{I}_k^\varepsilon(t,\eta) \le c_1 z(1 + |\eta|_B)$$

with c_1 independent of ε, t, η, z and k. Thus the monotone convergence theorem derives, as $k \to \infty$,

$$\varepsilon \log E \exp \frac{z}{\varepsilon} [\int_0^t ||\zeta(s)|| ds + |\zeta(t)|_B] \le c_1 z(1 + |\eta|_B). \tag{22}$$

On the other hand, $Z(t) = \xi(t) - \zeta(t)$ moves according to the following evolution equation

$$\frac{\partial Z}{\partial t}(t) = AZ(t) + \delta(\xi(t), U(t)) \qquad for \quad 0 < t < T.$$

Again we see

$$\frac{d}{dt} |Z(t)|_B^2 = 2\langle A^* BZ(t), Z(t) \rangle + 2\langle BZ(t), \delta(\xi(t), U(t)) \rangle$$

$$\le - \| Z(t) \|^2 + (2p+1) |Z(t)|_B^2 + |\delta(\xi(t), U(t))|_B^2$$

by the structural condition. Hence we get

$$e^{-(2p+1)t} |Z(t)|_B^2 \le \hat{\delta}^2 t$$

and

$$\int_0^t \| Z(s) \|^2 ds \le -|Z(t)|_B^2 + (2p+1) \int_0^t |Z(s)|_B^2 ds + \hat{\delta}^2 t.$$

Noting $(\int_0^t \| Z(s) \| ds)^2 \le t \int_0^t \| Z(s) \|^2 ds$, we have

$$\int_0^t ||Z(s)|| ds + |Z(t)|_B \le c_2 \qquad for \quad 0 < t < T,$$

with c_2 independent of U and ω. Now recalling (22), we can complete the proof.
Appealing to Propositions 1 and 2, we get

Theorem 2. The value function is the unique viscosity solution of the following Hamilton-Jacobi-Bellman equation, in $UC_\eta^+([0,T] \times H_B)$,

$$0 = \frac{\partial \gamma}{\partial t}(t,\eta) - \langle A^* \partial \gamma(t,\eta), \eta \rangle - \inf_{u \in U} [\frac{\varepsilon}{2} trace(\Lambda^2 \partial^2 \gamma(t,\eta))$$

$$+ \langle \partial \gamma(t,\eta), \delta(\eta,u) \rangle + \frac{1}{\varepsilon} f(\eta,u)\gamma(t,\eta)], \qquad for \quad (t,\eta) \in (0,T) \times H$$

$$with\ initial\ condition, \quad \gamma(0,\eta) = \exp \frac{1}{\varepsilon} q(\eta).$$

Therefore its logarithmic transformation ν^ε is the unique viscosity solution of (3), which is equivalent to (4) by Legendre transformation. Noting Theorem 1 and Proposition 5, we have a game representation of ν^ε.

Theorem 3. ν^ε coincides with the value function V^ε of stochastic differential game.

5 Small noise limit

Let us consider a differential game of (12) and (13) with $\varepsilon = 0$. Namely the (stochastic) system $X(t) = X^0(t,\eta,Y,U)$ is governed by the evolution equation with random cofficients

$$dX(t) = [AX(t) + \delta(X(t),U(t)) + Y(t)]dt, \qquad for \qquad 0 < t < T,$$

$$with\ initial\ condition, \quad X(0) = \eta\ (\in H),$$

So replacing J^ε by J^0, we define the upper (resp. lower) value function V^0 (resp. v^0).

Theorem 4

$$|V^\varepsilon(t,\eta) - V^0(t,\eta)| \leq c\sqrt{\varepsilon t}, \qquad for \qquad t < T,$$

with c independent of η.

Proof. Putting $\zeta(t) = X^\varepsilon(t) - X^0(t)$, we have the following equation,

$$d\zeta(t) = [A\zeta(t) + \delta(X^\varepsilon(t),U(t)) - \delta(X^0(t),U(t))]dt + \sqrt{\varepsilon} dM(t)$$

$$with\ initial\ condition, \qquad \zeta(0) = 0.$$

So, again we have

$$
\begin{aligned}
d|\zeta(\theta)|_B^2 &= \Big(2\langle A^*B\zeta(\theta),\zeta(\theta)\rangle + 2\langle B\zeta(\theta),\delta(X^\varepsilon(\theta),U(\theta)) \\
&\quad -\delta(X^0(\theta),U(\theta))\rangle + m\varepsilon\Big)d\theta + 2\sqrt{\varepsilon}\langle B\zeta(\theta),dM(\theta)\rangle \\
&\le \Big(-\|\zeta(\theta)\|^2 + 2p|\zeta(\theta)|_B^2 + 2L|\zeta(\theta)|_B\|\zeta(\theta)\| + m\varepsilon\Big)d\theta \\
&\quad +2\sqrt{\varepsilon}\langle B\zeta(\theta),dM(\theta)\rangle.
\end{aligned}
$$

Hence we see

$$E|\zeta(t)|_B^2 \le tm\varepsilon\exp[2(p+L^2)t]$$

and

$$E\int_0^t \|\zeta(s)\|^2 ds \le -2E|\zeta(t)|_B^2 + 4(p+L^2)\int_0^t |\zeta(s)|_B^2 ds + 2\varepsilon tm.$$

Therefore

$$E(\int_0^t \|\zeta(s)\|^2 ds + |\zeta(t)|_B^2) \le \varepsilon tk$$

with k independent of U and α. Now recalling the definitions of J^ε and J^0 and (A6), we complete the proof.
Since V^ε satisfies (ii), (iii) and (iv) of Proposition 3, V^0 has the same ones and is in $UC_\eta^+([0,T]\times H_B)$. Although V^0 is the small noise limit and the upper value function of differential game, this game is not deterministic one, because an admissible control is stochastic. Now we will show that V^0 coincides with the upper value function of deterministic differential game, via viscosity solution technique. From Theorem 4, we can easily see the following proposition, using the routine.

Proposition 7. V^0 is the unique viscosity solution of (4) with $\varepsilon = 0$ in $UC_\eta^+([0,T]\times H_B)$.

Next we will remark that V^0 is the upper value function of (deterministic) differential game on H. Replacing stochastic concepts by deterministic ones, we put

$$\mathsf{Y}_N = L^2([0,T],Y_N), \quad \mathsf{Y} = \bigcup_{N=1}^\infty \mathsf{Y}_N, \quad \mathsf{U} = L^2([0,T],\mathbf{U}),$$

$\mathsf{A}_N = \{\mathsf{a};\mathsf{U}\to\mathsf{Y}_N,$ measurable and non-anticipative$\}$, $\mathsf{A} = \bigcup_{N=1}^\infty \mathsf{A}_N$,
$\mathsf{R} = \{\mathsf{r};\mathsf{Y}\to\mathsf{U},$ measurable and non-anticipative$\}$.
a (resp. r) is called an admissible strategy of player I (resp. II).

When players I and II apply $Y \in \mathsf{Y}$ and $U \in \mathsf{U}$ respectively, the system moves according to the evolution equation,

$$\frac{\partial X}{\partial t}(t) = AX(t) + \delta(X(t), U(t)) + Y(t) \tag{23}$$

$$\textit{with initial condition,} \qquad X(0) = \eta \ (\in H),$$

and the pay-off J is given by

$$\mathsf{J}(t, \eta, Y, U) = \int_0^t f(X(s), U(s)) - \frac{1}{2}\|Y(s)\|^2 ds + q(X(t)). \tag{24}$$

Then the upper value function V, defined by

$$\mathsf{V}(t, \eta) = \sup_{\mathsf{a} \in \mathsf{A}} \inf_{U \in \mathsf{U}} \mathsf{J}(t, \eta, \mathsf{a}U, U),$$

turns out to be the unique viscosity solution of (4) with $\varepsilon = 0$ in $UC^+_\eta([0,T] \times H_B)$. Therefore V coincides with V^0, by Proposition 7.

From Isaacs condition (21), it follows that V^0 coincides with the lower value function v^0 and V becomes the value function of the game. Now recalling Theorems 3 and 4, we get

Theorem 5 As $\varepsilon \to 0$, ν^ε converges uniformly to the value function of deterministic differential game with (23) and (24). That is

$$|\, \nu^\varepsilon(t, \eta) - \mathsf{V}(t, \eta)\,| \le c\sqrt{\varepsilon t}, \qquad \textit{for} \qquad t < T.$$

References

Bensoussan, A. and Nagai, H. (1997). Min-Max characterization of a small noise limit on risk-sensitive control, *SIAM. J.Contol Optim.* **35**, 1093-1115.

Crandall, M. G. and Lions, P. L. (1990). Hamilton-Jacobi equations in infinite dimensions, Part 4. *J.Func.Anal.* **90**, 237-283.

Crandall, M. G. and Lions, P. L. (1991). Hamilton-Jacobi equations in infinite dimensions, Part 5. *J.Func.Anal.* **91**, 417-465.

Da Prato, G. and Zabczyk, J. (1992). *Stochastic Equations in Infinite Dimensions,* Cambridge Univ. Press, Cambridge.

Dai Pra, P. Meneghini, L. and Runggaldier, W. J. (1996). Connections between stochastic control and dynamic games. *Math.Control Signals Systems,* **9**, 303-326.

Fleming, W. H. and McEneaney, W. M. (1992). Risk sensitive control and differential games. *Proc. Stochastic Theory and Adaptive Control,* Springer-Verlag, New York, 185-197.

Fleming, W. H. and McEneaney, W. M. (1995). Risk sensitive control on an infinite time horizon. *SIAM J.Control Optim.* **33**, 1881-1915.

Fleming, W. H. and Soner, H. M. (1993). *Controlled Markov Processes and Viscosity Solutions,* Springer-Verlag, Berlin.

Fleming, W. H. and Souganidis, P. E. (1989). On the existence of value functions of two-player, zero-sum stochastic differential games. *Indiana Univ.Math.J.* **38**, 293-314.

Ikeda, N. and Watanabe, S. (1981). *Stochastic Differential Equations and Diffusion Processes,* North-Holland Publ.- Kodansha.

Jacobson, D. H. (1973). Optimal stochastic linear systems with exponential performance criteria and their relation to deterministic differential games. *IEEE Trans.Automat Control,* **18**, 124-131.

Nisio, M. (1996). On sensitive control and differential games in infinite dimensional space. *Itô's Stochastic Calculus and Probability Theory,* ed. by N.Ikeda et al. Springer-Verlag, Tokyo, 281-292.

Nisio, M. (1998). On infinite dimensional stochastic differential games. *Osaka J.Math.* **35**, 15-33.

Nisio, M. On value functions of stochastic differential games in infinite dimensions and its application to sensitive control. *to appear in Osaka J.Math.*.

Pardoux, E. (1979). Stochastic partial differential equations and filtering of diffusion processes. *Stochastic,* **3**, 127-167.

Rozovskii, B. (1990). *Stochastic Evolution Systems,* Kluwer Acad. Publ. Dordrecht.

Święch, A. (1994). Unbounded second order partial differential equations in infinite dimensional Hilbert spaces. *Comm.P.D.E.* **19**,1999-2036.

Whittle, P. (1981). Risk-sensitive-linear-quadratic-gaussian control. *Adv. Appl.Proba.* **13**, 764-777.

8

On the Solutions of the Equation Arising from the Singular Limit of Some Eigen Problems

Shuenn-Jyi Sheu
Alexander D. Wentzell

ABSTRACT: We consider a family of matrices $A^\varepsilon = \left(A^\varepsilon(i,j)\right)_{i,j=1}^{M}$, indexed by $\varepsilon > 0$, having each element the asymptotics $A^\varepsilon(i,j) \approx \exp\,(\varepsilon^{-1} V(i,j))$. We study the limiting behavior of the principal eigenvalue μ^ε and the corresponding normalized eigenvector φ^ε for A^ε. Under some conditions, we can write a limiting equation for $\lambda^\varepsilon = \varepsilon \log \mu^\varepsilon, W^\varepsilon(i) = \varepsilon \log \varphi^\varepsilon(i)$ as $\varepsilon \to 0$. We study it solutions. We show also by some examples that the limit of W^ε, when it exists, may depend on the higher order asymptotics of $A^\varepsilon(i,j)$.

KEYWORDS: Matrix, Principal eigenvalue, Normalized eigenvector, Singular limit.

1 Introduction

In this paper we consider a family of matrices $A^\varepsilon = \left(A^\varepsilon(i,j)\right)_{i,j=1}^{M}$ with nonegative elements indexed by $\varepsilon > 0$. Each element has the asymptotics

$$A^\varepsilon(i,j) \approx \exp(\frac{1}{\varepsilon} V(i,j)), \quad \text{as} \quad \varepsilon \to 0$$

(the precise meaning will be given at the beginning of Section 2). Under suitable conditions, for each small ε, A^ε has the principal eigenvalue μ^ε (the positive eigenvalue which has the largest absolute value among all eigenvalues). The eigenvector corresponding to μ^ε is positive and is unique after a suitable normalization. This is denoted by φ^ε. We are interested in the asymptotic behavior of $\mu^\varepsilon, \varphi^\varepsilon$. See Freidlin and Wentzell(1984), Wentzell(1972), for the related studies.
In general, $\mu^\varepsilon, \varphi^\varepsilon$ are analyzed through the transformation

$$\lambda^\varepsilon = \varepsilon \log \mu^\varepsilon, \quad W^\varepsilon(i) = \varepsilon \log \varphi^\varepsilon(i).$$

In Section 2, the limiting equation for $\lambda^\varepsilon, W^\varepsilon(\cdot)$ is given. See (2.4). We also

give the conditions used in this paper. Then we show that λ^ε converges to λ^0. An expression for λ^0 will be given. After giving some elementary properties for the solutions of the equation (2.4), we introduce an equivalence relation associated to the problem. This concept will be shown to be important for describing the solution of (2.4).
In Section 3, we give a representation for the solutions of (2.4) using $\mathcal{E}$, the family of equivalence classes defined in Section 3. In particular, we show that the solution of (2.4) is unique if $\mathcal{E}$ has only one element.
In Section 4, we consider two classes of examples in which the solution of (2.4) is not unique, but in the examples of the first class the limit of W^ε exists, and for the second class it does not. The limit(and whether it exists)depends on the higher order asymptotics of $A^\varepsilon(i,j)$.
We now mention that the asymptotics of the principal eigenvalue and of the corresponding normalized eigenfunction for the operators of form

$$L^\varepsilon f = (\varepsilon L_1 + L_2)f + \frac{1}{\varepsilon} V f$$

are closely related to the problem studied here. Here $\varepsilon L_1 + L_2$ is the generator of a continuous time Markov process on a space $\mathcal{M}$ and V is a function on $\mathcal{M}$. This can be considered as the continuous time counterpart of the problem. This was studied in different contexts. Consider $\mathcal{M} = R^d$, $L_1 = \Delta$ the Laplacian operator, and $L_2 = 0$. By considerations from physics, it is interesting to know the exact order of the split for the first two eigenvalues $\lambda_1^\varepsilon < \lambda_2^\varepsilon$ of L^ε. The problem can be solved if one knows the behavior of λ_1^ε and the normalized eigenfunction φ_1^ε corresponding to λ_1^ε. See Harrell(1978, 1980), Jona-Lasinio, Martinelli and Scoppola(1981), Simon(1985). The limits of $\mu_1^\varepsilon = \varepsilon\lambda_1^\varepsilon$, $W_1^\varepsilon = \varepsilon \log \varphi_1^\varepsilon$ satisfy the equation

$$\frac{1}{2}|\nabla W|^2 + V = \mu.$$

The equation needs to be interpreted in a generalized sense. Only viscosity solutions are considered. See Fleming and Soner(1992). Similar to (2.4), the solution of this equation in general lacks uniqueness. However, μ is determined uniquely under some general conditions.
Another example is from the theory of risk sensitive control. This is to consider the exponential growth rate of the expectation

$$E[\exp(\frac{\mu}{\varepsilon}\int_0^T \ell(x_t^{\varepsilon,u}, u_t)dt)]$$

which is assumed to be of order $\exp(\frac{\mu}{\varepsilon}\lambda_u^{\varepsilon,\mu}T)$ as T tends to ∞. The goal is to minimize $\lambda_u^{\varepsilon,\mu}$ for all *admissible* control processes u_t. For each u_t, $x_t^{\varepsilon,u}$ is the process generated by $L^{\varepsilon,u}$,

$$L^{\varepsilon,u} f(x) = \frac{\varepsilon}{2}\Delta f(x) + b(x,u)\nabla f(x).$$

Here μ is the risk parameter. Let $\lambda^{\varepsilon,\mu}$ be the minimal cost. The dynamic programming equation is

$$\frac{\mu}{\varepsilon}\lambda^{\varepsilon,\mu}\varphi^{\varepsilon,\mu} = \min_u[L^{\varepsilon,u}\varphi^{\varepsilon,\mu} + \frac{\mu}{\varepsilon}\ell^u\varphi^{\varepsilon,\mu}], \tag{1.1}$$

$\ell^u(x) = \ell(x,u)$. Under some quite general conditions, $\lambda^{\varepsilon,\mu}$ converges to λ^μ as ε tends to 0. It is also important to know the behavior of $\varphi^{\varepsilon,\mu}$ since it can be used to obtain an optimal control for the problem by taking *argument* in the right hand of (1.1). The limits of $\lambda^{\varepsilon,\mu}$, $W^{\varepsilon,\mu} = \varepsilon\mu^{-1}\log\varphi^{\varepsilon,\mu}$ satisfy the equation

$$\lambda = \min_u[\frac{\mu}{2}|\nabla W|^2 + b^u \cdot \nabla W + \ell^u], \tag{1.2}$$

$b^u(x) = b(x,u)$. See Fleming and McEneaney(1995). Again, the viscosity solutions are considered for this equation. Because of the lack of uniqueness for the the solution of (1.2), the limiting behavior of $W^{\varepsilon,\mu}$ is not well understood. See Fleming and Hernandez-Hernandez(1997) for a model closely related to ours. See also other studies in Fleming and James(1995), Fleming and Sheu,related to this problem.
The arguement developed in this paper can not be directly applied to these problems. They are very interesting and need further study.

2 Preliminary

Let $\mathcal{M} = \{1, ..., M\}$ be a finite set, $\mathcal{A}^\varepsilon = (A^\varepsilon(i,j))_{i,j=1}^M$ be a matrix for each $\varepsilon > 0$ such that

$$A^\varepsilon(i,j) = \exp(\frac{1}{\varepsilon}V^\varepsilon(i,j)), \quad i,j = 1, ..., M.$$

We assume the following conditions.

$$\begin{aligned} &\text{(i)} \quad \lim_{\varepsilon\to 0} V^\varepsilon(i,j) = V(i,j) \quad \text{for all} \quad i,j, \\ &\text{(ii)} \quad -\infty < V(i,j) < \infty \quad \text{for all} \quad i,j. \end{aligned} \tag{2.1}$$

The condition (2.1)(ii) can be relaxed for the discussion made here. We need only some kind of "irreducibility" condition. More details are left to the interested reader.
Under conditions (2.1), it is known for small ε that there is a unique eigenvalue μ^ε of $\mathcal{A}^\varepsilon$ with the folowing properties: μ^ε is positive; Any eigenvalue of $\mathcal{A}^\varepsilon$ is in the set $\{\mu; |\mu| \le \mu^\varepsilon\}$; Let φ^ε be an eigenvector of $\mathcal{A}^\varepsilon$ corresponding to μ^ε, that is,

$$\mu^\varepsilon\varphi^\varepsilon(i) = \sum_j A^\varepsilon(i,j)\varphi^\varepsilon(j) \tag{2.2}$$

for all i. Then φ^ε is positive and unique if we impose the condition $\varphi^\varepsilon(i_0) = 1$, where $i_0 \in \mathcal{M}$ is a fixed element. We shall call μ^ε the principal eigenvalue of $\mathcal{A}^\varepsilon$.

In everything that follows we assume $\varphi^\varepsilon(i_0) = 1$.

We denote

$$\lambda^\varepsilon = \varepsilon \log \mu^\varepsilon, \quad W^\varepsilon(i) = \varepsilon \log \varphi^\varepsilon(i). \tag{2.3}$$

The following result is an easy consequence of the definition. We shall provide a short proof for the convenience of the reader.

Lemma 2.1 Assume conditions in (2.1). Then the following hold.

(i) $\limsup_{\varepsilon\to 0} |\lambda^\varepsilon| \le \max_{i,j}\{|V(i,j)|\}$

(ii) $\limsup_{\varepsilon\to 0} |W^\varepsilon(i)| \le \max_{i,j}\{|V(i,j) - V(i_0,j)|\}$

Proof. For each $\varepsilon > 0$, let i^ε be a point in $\mathcal{M}$ at which the maximum of $W^\varepsilon(\cdot)$ is attained. From (2.1),

$$\mu^\varepsilon \varphi^\varepsilon(i^\varepsilon) \le \sum_j A^\varepsilon(i^\varepsilon, j)\varphi^\varepsilon(i^\varepsilon),$$

that is,

$$\mu^\varepsilon \le \sum_j A^\varepsilon(i^\varepsilon, j).$$

Then it is easy to see that

$$\limsup_{\varepsilon\to 0} \lambda^\varepsilon \le \max_{i,j}\{V(i,j)\}.$$

If we consider the minimum of $W^\varepsilon(\cdot)$ instead, we can also show that

$$\liminf_{\varepsilon\to 0} \lambda^\varepsilon \ge \min_{i,j}\{V(i,j)\}.$$

These two imply (i).

To prove (ii), from (2.2)

$$\begin{aligned} \mu^\varepsilon \varphi^\varepsilon(i) &= \textstyle\sum_j A^\varepsilon(i,j)\varphi^\varepsilon(j) \\ &\le \max_j \tfrac{A^\varepsilon(i,j)}{A^\varepsilon(i_0,j)} \textstyle\sum_j A^\varepsilon(i_0,j)\varphi^\varepsilon(j) \\ &= \max_j \tfrac{A^\varepsilon(i,j)}{A^\varepsilon(i_0,j)} \mu^\varepsilon \varphi^\varepsilon(i_0). \end{aligned}$$

Then we have

$$\limsup_{\varepsilon\to 0} W^\varepsilon(i) \le \max_j\{V(i,j) - V(i_0,j)\}.$$

Similarly, we have

$$\limsup_{\varepsilon\to 0}(-W^\varepsilon(i)) \le \max_j\{V(i_0,j) - V(i,j)\}.$$

These two imply (ii). This completes the proof.

By Lemma 2.1, we may consider the limits of $\lambda^\varepsilon, W^\varepsilon(\cdot)$ along a sequence $\varepsilon = \varepsilon_n \to 0$. Let $\lambda^0, W^0(\cdot)$ be such limits. We can see easily from (2.2) that they satisfy the equation

$$\lambda^0 + W^0(i) = \max_j\{V(i,j) + W^0(j)\} \tag{2.4}$$

for any $i = 1, ..., M$. The following questions naturally arise. What are the solutions of (2.4) ? Is the solution unique ? Do $\lambda^\varepsilon, W^\varepsilon(\cdot)$ converge as $\varepsilon \to 0$? We shall give some answers to these questions in the rest of the paper.
In the following, $\lambda^0, W^0(\cdot)$ denote a solution of (2.4) no matter whether they are limits of $\lambda^\varepsilon, W^\varepsilon(\cdot)$ or not. We shall present some interesting properties for $\lambda^0, W^0(\cdot)$ in the rest of this section.

Lemma 2.2 The following relations hold.

(i) $\lambda^0 = \sup_{\{i_k\}}\left\{\limsup_{N\to\infty}\frac{1}{N}\sum_{k=1}^N V(i_k,i_{k+1})\right\},$

(ii) $\lambda^0 = \max\left\{\frac{1}{N}(V(i_1,i_2) + \cdots + V(i_N,i_1)) : i_1, ..., i_N,\ N = 1, 2, \cdots\right\}.$

Therefore, λ^0 is uniquely determined by $V(\cdot,\cdot)$. We then have

$$\lim_{\varepsilon\to 0}\lambda^\varepsilon = \lambda^0,$$

with λ^0 defined by (i) or (ii).
Proof. To prove (i), for any $\{i_k\}$, by (2.4), we have

$$N\lambda^0 + W^0(i_1) \ge V(i_1,i_2) + \cdots + V(i_N,i_{N+1}) + W^0(i_{N+1}).$$

Dividing this by N, letting $N\to\infty$, we get

$$\lambda^0 \ge \limsup_{N\to\infty}\frac{1}{N}\sum_{k=1}^N V(i_k,i_{k+1}).$$

Therefore, we have

$$\lambda^0 \ge \sup_{\{i_k\}}\left\{\limsup_{N\to\infty}\frac{1}{N}\sum_{k=1}^N V(i_k,i_{k+1})\right\}.$$

We need to prove the reverse inequality, hence the equality. Starting from any i_1 we can choose $i_2, i_3, \cdots$ step by step such that

$$\lambda^0 + W^0(i_k) = V(i_k, i_{k+1}) + W^0(i_{k+1}), \quad k = 1, 2, \cdots. \tag{2.5}$$

This is possible by (2.4). Then

$$N\lambda^0 + W^0(i_1) = V(i_1, i_2) + \cdots + V(i_N, i_{N+1}) + W^0(i_{N+1}).$$

Therefore,

$$\lambda^0 = \lim_{N\to\infty} \frac{1}{N} \sum_{k=1}^{N} V(i_k, i_{k+1}).$$

This proves (i).
For (ii), comparing the right hand side of (i) and (ii) , using the result (i), we have

$$\lambda^0 \geq \max\left\{ \frac{1}{N}(V(i_1, i_2) + \cdots + V(i_N, i_1)) : i_1, ..., i_N, N = 1, 2, 3, \cdots \right\}.$$

To prove the opposite inequality, we again choose $i_1, i_2, \cdots$ satisfying (2.5). Since $\{i_1, i_2, \cdots\} \subset \mathcal{M}$, there are $N_1 < N_2$ such that $i_{N_1} = i_{N_2}$. Let $N = N_2 - N_1$. Then

$$N\lambda^0 = V(i_{N_1}, i_{N_1+1}) + \cdots + V(i_{N_2-1}, i_{N_2}).$$

Following from these, we have (ii).

Lemma 2.3 Assume $\{i_1, ..., i_N\}$ is such that

$$\lambda^0 = \frac{1}{N}(V(i_1, i_2) + \cdots + V(i_{N-1}, i_N) + V(i_N, i_1)), \tag{2.6}$$

that is, at $\{i_1, ..., i_N\}$ the maximum in (ii) of Lemma 2.2 is attained. Then

$$\lambda^0 + W^0(i_k) = V(i_k, i_{k+1}) + W^0(i_{k+1}), \quad k = 1, 2, ..., N. \tag{2.7}$$

Here we denote $i_{N+1} = i_1$.
Proof. By (2.4),

$$\begin{aligned} \lambda^0 + W^0(i_k) &= \max_j \{V(i_k, j) + W^0(j)\} \\ &\geq V(i_k, i_{k+1}) + W^0(i_{k+1}), \quad k = 1, 2, ..., N. \end{aligned} \tag{2.8}$$

Adding these up,

$$N\lambda^0 + \sum_{k=1}^{N} W^0(i_k) \geq V(i_1, i_2) + \cdots + V(i_N, i_1) + \sum_{k=1}^{N} W^0(i_k).$$

This becomes an equality by the condition (2.6), which is possible only if we have equality in (2.8) for every k. This completes the proof of the result.

Remark 2.4 Lemma 2.3 states that if at $\{i_1, ..., i_N\}$ the maximum in (ii) of Lemma 2.2 is attained, then $W^0(i_k)$ is determined uniquely by $W^0(i_1)$ through the relation (2.7). This is a motivation for introducing an equivalence relation which can help us to describe the solutions of (2.4). We say $i, j \in \mathcal{M}$ are equivalent if there exist $i_1, ..., i_N$ such that (2.6) holds and $i, j \in \{i_1, ..., i_N\}$. A consequence of the above remark is that the values of $W^0(\cdot)$ determines each other in each equivalence class. This is made precise in the next proposition.

Lemma 2.5 If $i \neq j$ are in the same equivalence class, then

$$W^0(i) = \tilde{V}(i, j) + W^0(j),$$

where

$$\tilde{V}(i, j) = \max\{(V(i_1, i_2) - \lambda^0) + \cdots + (V(i_N, i_{N+1}) - \lambda^0)\},$$

and the maximum is taken over all $i_1, ..., i_N$ and $N > 0$ such that $i_1 = i, i_{N+1} = j$.

Proof. By the condition on i, j, there are

$$j_1 = i, j_2, ..., j_n = j, j_{n+1}, ..., j_m$$

such that

$$\lambda^0 = \frac{1}{m}(V(j_1, j_2) + \cdots + V(j_m, j_1)),$$

and

$$W^0(i) = (V(j_1, j_2) - \lambda^0) + \cdots + (V(j_{n-1}, j_n) - \lambda^0) + W^0(j).$$

The latter follows from (2.7). We shall prove that for any $\{i_1, i_2, ..., i_{N-1}, i_N\}$ with $i_1 = i, i_N = j$ we have

$$\begin{aligned}(V(i_1, i_2) - \lambda^0) + \cdots + (V(i_{N-1}, i_N) - \lambda^0)\\ \leq (V(j_1, j_2) - \lambda^0) + \cdots + (V(j_{n-1}, j_n) - \lambda^0).\end{aligned} \tag{2.9}$$

Then our result follows from this.
Suppose that (2.9) does not hold for some $i_1, ..., i_N$. Then

$$\begin{aligned}&V(i_1, i_2) + \cdots + V(i_{N-1}, i_N) + V(j_n, j_{n+1}) + \cdots + V(j_m, j_1)\\ > \ &V(j_1, j_2) + \cdots + V(j_m, j_1) + (N - n)\lambda^0\\ = \ &(N - n + m)\lambda^0.\end{aligned}$$

Since $i_N = j_n, i_1 = j_1$, the above inequality contradicts the result of Lemma 2.2. This ends the proof.

3 Solutions of the limiting equation

Let $\mathcal{E}$ denote the set of equivalence classes defined in Remark 2.4 which contain more than one element or contain only one element i with $V(i,i) = \lambda^0$. The elements of $\mathcal{E}$ will be denoted by $E_1, E_2, \ldots$. We will describe the solution of (2.4) with the help of $\mathcal{E}$. To do this, we choose, for each E, an arbitrary $i_E \in E$. Define, for $E_1, E_2 \in \mathcal{E}$,

$$\tilde{V}(E_1, E_2) = \max\{(V(i_1, i_2) - \lambda^0) + \cdots + (V(i_N, i_{N+1}) - \lambda^0)\}.$$

The maximum is taken over all $\{i_1, ..., i_{N+1}\}, N > 0$, such that $i_1 = i_{E_1}, i_{N+1} = i_{E_2}$ and i_k is not in E for $E \neq E_1, E_2$.
For any $i \in \mathcal{M}$ and $E_1 \in \mathcal{E}$, define

$$\tilde{V}(i, E_1) = \max\{(V(i, i_1) - \lambda^0) + \cdots + (V(i_N, i_{E_1}) - \lambda^0)\}.$$

The maximum is taken over all $\{i_1, ..., i_N\}$ such that i_k does not belong to any $E \in \mathcal{E}$, $E \neq E_1$. We have the following relation

$$\tilde{V}(i, E_1) = \tilde{V}(i, i_{E_1}), \quad \text{if } \ i \in E_1$$

(for the notation $\tilde{V}(i,j)$, see Lemma 2.5). It is also easy to see that for $E_1 \neq E_2$ and $i \in E_1$

$$\tilde{V}(i, E_2) \leq \tilde{V}(i, i_{E_1}) + \tilde{V}(E_1, E_2)$$

since we can show that the right hand side is equal to

$$\max\{(V(i, i_1) - \lambda^0) + \cdots + (V(i_N, i_{E_2}) - \lambda^0)\}.$$

The maximum is taken over all $\{i_1, ..., i_N\}$ such that i_k does not belong to any $E \in \mathcal{E}$, $E \neq E_1, E_2$.
We consider the following equation for the unknown function $W(i_E), E \in \mathcal{E}$:

$$W(i_E) = \max_{\tilde{E} \in \mathcal{E}}\{\tilde{V}(E, \tilde{E}) + W(i_{\tilde{E}})\}, \tag{3.1}$$

for $E \in \mathcal{E}$. Let $W(i_E), E \in \mathcal{E}$, be a solution of (3.1). Let us extend this function to all points $i \in \mathcal{M}$. If $i \in E, E \in \mathcal{E}$,

$$W(i) = \tilde{V}(i, i_E) + W(i_E). \tag{3.2}$$

If i is not in $\bigcup_{E \in \mathcal{E}} E$,

$$W(i) = \max_{E}\{\tilde{V}(i, E) + W(i_E)\}. \tag{3.3}$$

We now state one of our main results in this section.

Theorem 3.1 Assume $W(\cdot)$ satisfies (3.1)-(3.3). Then $W(\cdot)$ is a solution of (2.4). Conversely, every solution of (2.4) satisfies (3.1)-(3.3).
Proof. Let $W(\cdot)$ satisfy (3.1)-(3.3). We need to prove

$$W(i) = \max_j \{V(i,j) - \lambda^0 + W(j)\} \tag{3.4}$$

for all i.
First assume i is not in $\bigcup_{E\in\mathcal{E}} E$. Let $E_1 \in \mathcal{E}$. By definition

$$\begin{aligned}\tilde{V}(i, E_1) = \quad & \max\{\max\{V(i,j) - \lambda^0 + \tilde{V}(j, E_1); j \text{ is not in } \textstyle\bigcup_{E\in\mathcal{E}} E\}, \\ & \max\{V(i,j) - \lambda^0 + \tilde{V}(j, i_{E_1}); j \in E_1\}\}.\end{aligned}$$

By (3.2), (3.3) and the above relation, it is easy to show that (3.4) holds for this i.
Now assume $i \in E_1$ for some $E_1 \in \mathcal{E}$. By (3.2) and (3.3),

$$\begin{aligned}&\max\nolimits_j \{V(i,j) - \lambda^0 + W(j)\} \\ &= \max\{\max\{V(i,j) - \lambda^0 + W(j); j \text{ is not in } \textstyle\bigcup_{E\in\mathcal{E}} E\}, \\ &\qquad \max\{V(i,j) - \lambda^0 + W(j); j \in \textstyle\bigcup_{E\in\mathcal{E}} E\}\},\end{aligned}$$

$$\begin{aligned}&\max\{V(i,j) - \lambda^0 + W(j); j \text{ is not in } \textstyle\bigcup_{E\in\mathcal{E}} E\} \\ &= \max\{\max\{V(i,j) - \lambda^0 + \tilde{V}(j,E); j \text{ is not in } \textstyle\bigcup_{E\in\mathcal{E}} E\} + W(i_E); E \in \mathcal{E}\},\end{aligned}$$

$$\begin{aligned}&\max\{V(i,j) - \lambda^0 + W(j); j \in \textstyle\bigcup_{E\in\mathcal{E}} E\} \\ &= \max\{V(i,j) - \lambda^0 + \tilde{V}(j, i_E) + W(i_E); j \in E, E \in \mathcal{E}\}.\end{aligned}$$

We then have

$$\begin{aligned}&\max\nolimits_j \{V(i,j) - \lambda^0 + W(j)\} \\ = \quad &\max\{\max\{\max\{V(i,j) - \lambda^0 + \tilde{V}(j,E); j \text{ is not in } \textstyle\bigcup_{E\in\mathcal{E}} E\}, \\ &\max\{V(i,j) - \lambda^0 + \tilde{V}(j, i_E); j \in E\}\} + W(i_E); E \in \mathcal{E}\}.\end{aligned}$$

If $E = E_1$, then

$$\max\{V(i,j) - \lambda^0 + \tilde{V}(j, i_E); j \in E\} = \tilde{V}(i, i_{E_1}),$$

see the proof of Lemma 2.5. Also

$$\max\{V(i,j) - \lambda^0 + \tilde{V}(j,E); j \text{ is not in } \bigcup_{E\in\mathcal{E}} E\} \le \tilde{V}(i, i_{E_1}).$$

Therefore,

$$\max\{\max\{V(i,j)-\lambda^0+\tilde{V}(j,E); j \text{ is not in } \textstyle\bigcup_{E\in\mathcal{E}} E\},$$

$$\max\{V(i,j)-\lambda^0+\tilde{V}(j,i_E); j\in E\}\}+W(i_E)$$

$$=\tilde{V}(i,i_{E_1})+W(i_{E_1})=W(i).$$

Now let $E\in\mathcal{E}, E\neq E_1$,

$$\max\{\max\{V(i,j)-\lambda^0+\tilde{V}(j,E); j \text{ is not in } \textstyle\bigcup_{E\in\mathcal{E}} E\},$$

$$\max\{V(i,j)-\lambda^0+\tilde{V}(j,i_E); j\in E\}\}$$

$$= \max\{(V(i,i_1)-\lambda^0)+\cdots+(V(i_N,i_E)-\lambda^0)\},$$

where the maximum on the right hand side is taken over all $\{i_1,...,i_N\}$ such that i_k is not in $\tilde{E}$ for $\tilde{E}\neq E$. This last value is equal to $\tilde{V}(i,E)$. By a remark before (3.1), this is less than $\tilde{V}(i,i_{E_1})+\tilde{V}(E_1,E)$. But

$$\max\{\tilde{V}(E_1,E)+W(i_E); E\in\mathcal{E}\}=W(i_{E_1})$$

by (3.1). We conclude that

$$\max_j\{V(i,j)-\lambda^0+W(j)\}$$

$$= \tilde{V}(i,i_{E_1})+W(i_{E_1})$$

$$= W(i).$$

This completes the proof of the first statement of the theorem.
To prove the second statement, let $W(\cdot)$ be a solution of (2.4). The equality (3.2) follows from Lemma 2.5. We shall prove (3.1)(the proof of (3.3) is similar). By repeatly using (2.4), it is not difficult to see

$$W(i_E)\geq\max_{\tilde{E}\in\mathcal{E}}\{\tilde{V}(E,\tilde{E})+W(i_{\tilde{E}})\}.$$

It remains to prove the opposite inequality. By (2.4), there are $i_1=i_E$, $i_2,i_3,\cdots$ such that

$$W(i_k)=V(i_k,i_{k+1})-\lambda^0+W(i_{k+1})$$

for all k. It is easy to see that there is $k>1$ such that $i_k\in\bigcup_{E\in\mathcal{E}} E$. Let N be the first such k and denote E_1 the one contains i_N. Then

$$W(i_E)=(V(i_1,i_2)-\lambda^0)+\cdots+(V(i_{N-1},i_N)-\lambda^0)+W(i_N)$$

$$=(V(i_1,i_2)-\lambda^0)+\cdots+(V(i_{N-1},i_N)-\lambda^0)+\tilde{V}(i_N,i_{E_1})+W(i_{E_1})$$

$$\leq\tilde{V}(E,E_1)+W(i_{E_1}).$$

That is,

$$W(i_E) \leq \max_{\tilde{E}\in\mathcal{E}}\{\tilde{V}(E,\tilde{E}) + W(\tilde{E})\}.$$

We have completed the proof of the theorem.

The following result is immediate from Theorem 3.1.

Corollary 3.2 Assume $\mathcal{E}$ has only one element. Then (2.4) has a unique solution with $W^0(i_0) = 0$. Therefore, $W^\varepsilon(\cdot)$ converges as $\varepsilon \to 0$.

Since $W(\cdot)$ above is determined by $\{W(i_E), E \in \mathcal{E}\}$, it follows from Theorem 3.1 that solving (2.4) is equivalent to solving (3.1). The following result on the structure of the solution of (3.1) in turn gives an interesting property for the solution of (2.4).

Theorem 3.3 Assume $\mathcal{E}$ has more than one element. Then the set of solutions of (3.1) is convex with at least one interior point.

Proof. Since $\tilde{V}(E,E) = 0$ for all $E \in \mathcal{E}$, (3.1) is equivalent to

$$W(i_E) \geq \max\{\tilde{V}(E,\tilde{E}) + W(i_{\tilde{E}}); \tilde{E} \neq E\}.$$

This implies that the set of solutions of (3.1) is convex. In what follows, we prove that this set has an interior point.

We first observe that $\tilde{V}(E_1,E_2), E_1,E_2 \in \mathcal{E}$, satisfies the following properties.

$$\begin{aligned} &\text{(i)} \quad \tilde{V}(E,E) = 0 \quad \text{for all} \quad E, \\ &\text{(ii)} \quad \text{if} \quad \{E_1,E_2,\cdots,E_N\} \ \text{has more than one element, then} \\ &\qquad \tilde{V}(E_1,E_2) + \cdots + \tilde{V}(E_N,E_1) < 0. \end{aligned} \tag{3.5}$$

We choose another $\hat{V}(i,j)$ satisfying the following properties.

$$\begin{aligned} &\text{(i)} \quad \hat{V}(E,E) = 0 \quad \text{for all} \quad E, \\ &\text{(ii)} \quad \text{if} \quad \{E_1,E_2,\cdots,E_N\} \ \text{has more than one element, then} \\ &\qquad \hat{V}(E_1,E_2) + \cdots + \hat{V}(E_N,E_1) < 0 \\ &\text{(iii)} \quad \tilde{V}(E_1,E_2) < \hat{V}(E_1,E_2), \quad E_1 \neq E_2. \end{aligned} \tag{3.6}$$

The existence of such $\hat{V}$ is an easy consequence of (3.5).

We consider the equation

$$W(i_E) = \max\{\hat{V}(E,\tilde{E}) + W(i_{\tilde{E}}); \tilde{E} \in \mathcal{E}\}.$$

It follows from the discussion in Section 2 that this equation has a solution. Let $W(\cdot)$ be a solution. By (3.6)(iii), $W(\cdot)$ is also a solution of (3.1). We show that this solution is in the interior of the set of solutions of (3.1). By the condition,

$$W(i_E) \geq \hat{V}(E, \hat{E}) + W(i_{\hat{E}})$$

for $E \neq \tilde{E}$. This implies

$$W(i_E) \geq \tilde{V}(E, \hat{E}) + W(i_{\hat{E}}) + 2\delta,$$

$$2\delta = \min\{\hat{V}(E_1, E_2) - \tilde{V}(E_1, E_2); E_1 \neq E_2\}.$$

Let $\tilde{W}(\cdot)$ be such that

$$|\tilde{W}(i_E) - W(i_E)| < \delta, \quad E \in \mathcal{E}.$$

Then

$$\tilde{W}(i_E) \geq \tilde{V}(E, \tilde{E}) + \tilde{W}(i_{\tilde{E}})$$

for $E \neq \tilde{E}$. Therefore, $\tilde{W}$ is also a solution. This completes the proof.

Corollary 3.4 If $\mathcal{E}$ has more than one element, then the solution of (2.4) satisfying $W^0(i_0) = 0$ is not unique.

Example 3.5 The following model is considered in [FH]. Let $\mathcal{M} = \{1, ..., M\}$ and Ω be finite sets. The functions $f(\cdot), b(\cdot,\cdot), \theta(\cdot)$ are defined on $\mathcal{M}, \mathcal{M} \times \Omega, \Omega$ respectively. There are $i_\phi \in \mathcal{M}, \omega_\phi \in \Omega$ such that

$$f(i_\phi) = i_\phi, \; b(i, \omega_\phi) = f(i)$$

and

$$\begin{cases} \theta(\omega_\phi) = 0 \\ \theta(\omega) > 0, \quad \omega \neq \omega_\phi. \end{cases}$$

The function $\ell(\cdot)$ defined on $\mathcal{M}$ satisfies

$$\begin{cases} \ell(i_\phi) = 0 \\ \ell(i) > 0, \quad i \neq i_\phi. \end{cases}$$

Define the function $V_0 : \mathcal{M} \times \mathcal{M} \to R \cup \{\infty\}$ by

$$V_0(i, j) = \min\{\theta(\omega); j = b(i, \omega)\}.$$

If for any i, j there is ω such that $j = b(i, \omega)$, then (2.1)(ii) holds for $V(\cdot,\cdot) = V_0(\cdot,\cdot)$. We define the stochastic matrix $P_\varepsilon(i, j)$ by

$$P_\varepsilon(i, j) = \frac{1}{Z_\varepsilon(i)} \exp(-\frac{V_0(i, j)}{\varepsilon}),$$

$Z_\varepsilon(i)$ being the normalizing constant. This generates a Markov chain x_t on $\mathcal{M}$. Let

$$A^\varepsilon(i, j) = \exp(\frac{\mu}{\varepsilon}\ell(i)) P_\varepsilon(i, j).$$

Then the principal eigenvalue μ^ε determines the exponential growth rate of the expectation

$$E[\exp(\frac{\mu}{\varepsilon}\sum_{t=0}^{T}\ell(x_t))].$$

as $T \to \infty$. It is easy to see that

$$V(i,j) = V^{(\mu)}(i,j) = \mu\ell(i) - V_0(i,j) \quad \text{for all} \quad i,j.$$

Then $\lambda^0 = \lambda^{(\mu)}, W^0(i) = W^{(\mu)}(i)$ depend on μ. The quantity $\lambda^{(\mu)}$ is non-decreasing in μ. There is μ^* such that $\lambda^{(\mu)} = 0$ if and only if $\mu \le \mu^*$. It is not difficult to see that for $\mu < \mu^*$, $\mathcal{E} = \{\{i_\phi\}\}$. Therefore, $W^\varepsilon(i)$ converges as $\varepsilon \to 0$. Moreover, by using Lemma 2.2, we can show that

$$\mu^* = \min\{\frac{V_0(i_1,i_2) + \cdots + V_0(i_{N-1},i_N) + V_0(i_N,i_1)}{\ell(i_1) + \cdots + \ell(i_N)}\}$$

where the minimum is taken over all $\{i_1, ..., i_N\} \ne \{i_\phi\}$. For $\mu = \mu^*$, there is an equivalence class that contains elements other than i_ϕ. However, it does not rule out the possibility that (2.4) has a unique solution, hence the convergence of W^ε.

4 The existence of the limit: examples

In Section 3, we show that if $\mathcal{E}$ consists of only one element, then $W^\varepsilon(i)$ converges as $\varepsilon \to 0$. However, if $\mathcal{E}$ has more than one element, then any limit (along a sequence) of W^ε is a solution of (2.4) and (2.4) has multiple solutions. In the latter case, it remains a question whether W^ε converges as $\varepsilon \to 0$. We will not give the answer to this question in great generality. Instead, we shall consider two classes of examples in which the solution of (2.4) is not unique; in the example of the first class the limit of W^ε exist, for the second class it does not. We find that the limit of W^ε depends not only on $V(\cdot,\cdot)$, the limit of V^ε, but also on the higher order asymptotics of V^ε. We also remark that, due to Theorem 3.1, it seems that the general case can be reduced to the model studied here.
Assume that each $\{i\}, i \in \mathcal{M}$, is an equivalence class, that is,

$$\mathcal{E} = \{\{i\}; i = 1, ..., M\}. \tag{4.1}$$

The condition (4.1) is equivalent to

$$\lambda^0 = V(i,i) \quad \text{for all} \quad i \tag{4.2}$$

and

$$(V(i_1,i_2) - \lambda^0) + \cdots + (V(i_N,i_1) - \lambda^0) < 0 \tag{4.3}$$

if $\{i_1, ..., i_N\}$ contains more than one element.
We also assume that

$$V^\varepsilon(i,j) = V(i,j) + \varepsilon a^\varepsilon(i,j), \tag{4.4}$$

$$\lim_{\varepsilon\to 0} a^\varepsilon(i,j) = a(i,j)$$

and

$$a(1,1) > a(i,i), \quad \text{for} \quad i \neq 1.$$

Let $W^\varepsilon(\cdot), \lambda^\varepsilon$ be defined as in (2.2), (2.3).

Theorem 4.1 Assume the conditions (4.1),(4.4). Then W^ε has a limit as $\varepsilon \to 0$, which we will denote by W^0. There is a unique $B < 0$ such that

$$\begin{aligned} \lambda^0 + B + W^0(1) &= \max_{j\neq 1}\{V(1,j) + W^0(j)\} \\ \lambda^0 + W^0(i) &= \max_{j\neq i}\{V(i,j) + W^0(j)\}, \quad i \neq 1. \end{aligned} \tag{4.5}$$

The solution of this equation is unique.
Proof. From equation (2.2), we have

$$(\mu^\varepsilon - e^{\frac{V^\varepsilon(i,i)}{\varepsilon}})\varphi^\varepsilon(i) = \sum_{j\neq i} e^{\frac{V^\varepsilon(i,j)}{\varepsilon}}\varphi^\varepsilon(j).$$

Denote

$$\mu^\varepsilon - e^{\frac{V^\varepsilon(i,i)}{\varepsilon}} = e^{\frac{\alpha^\varepsilon(i)}{\varepsilon}}.$$

As in Section 2, we can show that $|\alpha^\varepsilon(i)|$ are bounded. Let $\alpha^0(i)$ be a limit of $\alpha^\varepsilon(i)$ and $W^0(i)$ of $W^\varepsilon(i)$ along some sequence $\varepsilon = \varepsilon_n \to 0$. Then

$$\alpha^0(i) + W^0(i) = \max_{j\neq i}\{V(i,j) + W^0(j)\}.$$

We first show that $\alpha^0(1) < \lambda^0$ holds. Suppose to the contrary that $\alpha^0(1) \geq \lambda^0$. Then we must have $\alpha^0(1) = \lambda^0$ since $\exp(\varepsilon^{-1}\alpha^\varepsilon(1)) < \mu^\varepsilon$, which implies $\alpha^0(1) \leq \lambda^0$ and hence $\alpha^0(1) = \lambda^0$. From

$$e^{\frac{\alpha^\varepsilon(i)}{\varepsilon}} = \mu^\varepsilon - e^{\frac{V^\varepsilon(1,1)}{\varepsilon}} + e^{\frac{V^\varepsilon(1,1)}{\varepsilon}} - e^{\frac{V^\varepsilon(i,i)}{\varepsilon}} \tag{4.6}$$

and the conditions, it is easy to see that $\alpha^0(i) = \lambda^0$ for all i. That is,

$$\lambda^0 + W^0(i) = \max_{j\neq i}\{V(i,j) + W^0(j)\}.$$

A simple argument shows that this contradicts (4.3). Therefore, we must have $\alpha^0(1) < \lambda^0$.
Using (4.6) again and (4.2), we can prove that $\alpha^0(i) = \lambda^0$ for $i \neq 1$. This proves (4.5) with $B = \alpha^0(1) - \lambda^0$.

We now show that there is a unique B such that (4.5) has a solution W^0. We shall also show that the solution is unique. For each $B \leq 0$, define

$$V_B(i,j) = \begin{cases} V(i,j) & \text{if } i \neq 1 \\ V(1,j) - B & \text{if } i = 1. \end{cases}$$

Consider the equation

$$\lambda_B + W^0(i) = \max_{j \neq i}\{V_B(i,j) + W^0(j)\}.$$

It is easy to see that λ_B is continuous and nonincreasing in B. By the property (4.3), $\lambda_B < \lambda^0$ for $B = 0$. Also, we can show that $\lambda_B \to \infty$ as $B \to -\infty$. Therefore, there is B^* such that $\lambda_{B^*} = \lambda^0$. Consider $B \leq B^*$. Using the property (4.3), it is not difficult to show that any equivalence class defined by $\{V_B(\cdot,\cdot)\}$ must contain 1. Then λ_B must be a decreasing function in B, $B \leq B^*$. This also proves that B such that $\lambda_B = \lambda^0$ is unique. With this B, (4.5) has a unique solution since $\mathcal{E}$ generated by $\{V_B(\cdot,\cdot)\}$ has only one element.
We conclude that W^ε has a unique limit as $\varepsilon \to 0$, hence it converges. The limit satisfies (4.5). This completes the proof of the theorem.

Now let us consider another example. Let $\big(a(i,j)\big)$ and $\big(a'(i,j)\big)$ be two matrices, $a(1,1) > a(i,i)$ for $i \neq 1$, and $a'(2,2) > a'(i,i)$ for $i \neq 2$. Suppose $V^\varepsilon(i,j)$ is given by the formula (4.4), and $\lim_{n\to\infty} a^{\varepsilon_n}(i,j) = a(i,j)$, $\lim_{n\to\infty} a^{\varepsilon'_n}(i,j) = a'(i,j)$ for two different sequences $\varepsilon_n, \varepsilon'_n \to 0$. Then, for some i, $\lim_{\varepsilon\to 0} W^\varepsilon(i)$ does not exist: the partial limit is equal to $W^0(i)$ along the sequence ε_n, and to $W^{0\prime}(i)$ along ε'_n, where W^0 is the solution of (4.5), while $W^{0\prime}$ is the solution of the same system with 1 changed to 2 (and $B < 0$ changed to a $B' < 0$).

Acknowledgments: Research of the first author was performed in part while he was visiting the Division of Applied Mathematics, Brown University, and the research was partially supported by NSC grant 86-2115-001-008. Research of the second author was performed in part while visiting Academia Sinica, Taipei.

References

Fleming, W. H. and Hernandez-Hernandez, D. (1997) Risk-sensitive control of finite state machines on an infinite horizon I, SIAM J. Control Optim. 35, 1790-1810.

Fleming, and M. R. James, M. R. (11995) The risk-sensitive index and the H_2 and H_∞ norms for nonlinear systems, Math. Control Signals Systems 8, 199-221.

Fleming, W. H. and McEneaney, W. M. (1995) Risk-sensitive control on an infinite time horizon, SIAM J. Control Optim. 33, 1881-1915 .

Fleming, W. H. and S. J. Sheu, S. J. Asymptotics for the principal eigenvalue and eigenfunction of a nearly first order operator with large potential, to appear in Ann. Probab.

Fleming, W. H. and Soner, H. M. (1992) *Controlled Markov Processes and Viscosity Solutions*, Springer-Verlag, New York.

Freidlin, M. I. and Wentzell, A. D. (1984) *Random Perturbations of Dynamical Systems* , Springer-Verlag, New York.

Harrell, E. M. (1978) On the rate of asymptotic eigenvalue degeneracy, Commun. Math. Phys. 60 , 73-95.

Harrell,E. M. (1980) Double wells, Commun. Math. Phys. 75, 239-261.

Jona-Lasinio, G., Martinelli, F. and Scoppola, E. (1981) New approach to the semiclassical limit of quantum mechanics, Commun. Math. Phys. 80, 223-254.

Simon, B. (1985) Semiclassical analysis of low lying eigenvalues. IV. The flea on the elephant, J. Functional Analy. 63, 123-136.

Wentzell, A. D. (1972) On the asymptotics of eigenvalues of matrices with elements of order $\exp(-\frac{V_{ij}}{\epsilon})$, Soviet Math. Dokl.13, 65-68.

9

Nonlinear H^∞ Controller Design via Viscosity Supersolutions of the Isaacs Equation

MingQing Xiao
Tamer Başar

ABSTRACT: We consider the solvability of Hamilton-Jacobi-Isaacs equations that arise in finite and infinite-horizon nonlinear H^∞ control problems where the system is affine in the control and the disturbance, while the cost function is not necessarily continuous in the state and the control. In each case we prove the existence of viscosity supersolutions under the assumption that the value function is finite, and provide a method for constructing H^∞ disturbance attenuating feedback controllers by using only the viscosity supersolution. We also obtain a result on global asymptotic stability of the closed-loop system under the H^∞ controller and the worst-case disturbance.

KEYWORDS: Viscosity supersolution, Isaacs equation, H^∞ disturbance attenuating feedback controller, H^∞ control.

1 Introduction

Consider the nonlinear H^∞ control problem with cost functional

$$J_\gamma(t,x:u,w) = \frac{1}{2}\int_t^{t_f} \{g(x(s)) + h(u(s)) - \gamma^2|w(s)|^2\}ds + \psi(x(t_f)) \quad (1)$$

where $x(\cdot)$ is the solution of the input-affine system:

$$\begin{aligned} \dot{x}(s) &= A(x(s)) + B_1(x(s))u(s) + B_2(x(s))w(s), \quad t \le s \le t_f \quad (2) \\ x(t) &= x. \end{aligned}$$

Here, $u(\cdot)$ is the control variable with values in $\mathcal{R}^p$, w is a reference trajectory or disturbance with values in $\mathcal{R}^l$, and x is the state vector with values in $\mathcal{R}^n$. We assume that A, B_1, B_2 are matrix functions on $\mathcal{R}^n$, of appropriate sizes, $g : \mathcal{R}^n \to [0,+\infty)$, and $h : \mathcal{R}^l \to [0,+\infty)$; more precise conditions on A, B_1, B_2, g, h and ψ will be given in section 2. If $t_f = +\infty$,

we set $\psi \equiv 0$. The Hamilton-Jacobi-Isaacs (HJI) equations associated with this problem in the finite and infinite horizons are (cf. Fleming and Soner (1992) or Başar and Bernhard (1995)), respectively:
(I) *Finite horizon*, i.e. $t_f < +\infty$:

$$\frac{\partial V(s,x)}{\partial s} + H(s,x,\nabla_x V(s,x)) = 0 \qquad (3)$$
$$V(t_f, x) = \psi(x)$$

where

$$H(s,x,p) = \inf_u \sup_w \{[A(x) + B_1(x)u + B_2(x)w]^T p + g(x) + h(u) - \gamma^2|w|^2\} \qquad (4)$$

(II) *Infinite horizon*, i.e. $t_f = +\infty$:

$$H(x, \nabla_x V(x)) = 0 \qquad (5)$$

where

$$H(x,p) = \inf_u \sup_w \{[A(x) + B_1(x)u + B_2(x)w]^T p + g(x) + h(u) - \gamma^2|w|^2\} \qquad (6)$$

Remark 1.1. Note that in both finite and infinite horizon cases, the Isaacs' condition is satisfied since the control u and disturbance w appear separately in system (2) and cost function (1).

As standard in nonlinear H^∞ control, we are interested in, for a given open subset $\mathcal{O} \subset \mathcal{R}^n$, finding a state feedback controller $u(t) = \mu(x(t))$ such that system (2) has L_2-gain for some prescribed $\gamma > 0$; that is for any $w \in \mathcal{W}$ the following holds, with $x(0) = x_0 \in \mathcal{O}$:
Finite-horizon case:

$$\int_0^{t_f} \{g(x(s)) + h(u(s))\}ds + \psi(x(t_f)) \le \gamma^2 \int_0^{t_f} |w(s)|^2 ds + \varphi(x_0) \qquad (7)$$

Infinite-horizon case:

$$\int_0^t \{g(x(s)) + h(u(s))\}ds \le \gamma^2 \int_0^t |w(s)|^2 ds + \varphi(x_0) \quad \forall t > 0 \qquad (8)$$

where $\varphi : \mathcal{O} \subset \mathcal{R}^n \to [0,\infty)$ is a nonnegative function.
Inequalities (7) and (8) are related to the existence of an energy or storage function (e.g. see James (1993)) $\Theta(\cdot,\cdot) : \mathcal{R}_+ \times \mathcal{R}^n \mapsto \mathcal{R}^1 \cup \{\infty\}$. That is for a fixed measurable control $\mu(\cdot)$ and any measurable disturbance $w(\cdot)$, if for any $t_2 \ge t_1 \ge 0$, Θ satisfies

$$\Theta(t_2, x(t_2,\mu(\cdot),w(\cdot))) - \Theta(t_1, x(t_1,\mu(\cdot),w(\cdot))) \le \int_{t_1}^{t_2} (\gamma^2|w(s)|^2 - (g(x(s,\mu(s),w(s))) + h(\mu(s))))ds \qquad (9)$$

where $x(\cdot, \mu(\cdot), w(\cdot))$ denotes the solution of (2) corresponding to $\mu(\cdot)$ and $w(\cdot)$, then Θ is called a *storage function* associated with the controller $\mu(\cdot)$. Note that if Θ is nonnegative and $\Theta(0, x_0) = \varphi(x_0)$, (9) becomes (8), and in the finite horizon case, moreover if $\Theta(t_f, x(t_f)) = \psi(x(t_f))$, (9) becomes (7).

The nonlinear H^∞ control problem as formulated above has been intensively studied in recent years, with one focus of attention being the existence of solutions to the HJI equation. It is already known that for nonlinear problems, the HJI equation may not admit a classical solution (e.g. Fleming and Soner (1992), Başar and Bernhard (1995), McEneaney (1995), Soravia (1996)), and hence one has to direct attention to weaker notions of a solution, and in particular the viscosity solution. Viscosity solutions have been investigated during the last decade in various contexts, including optimal control, differential games, and control of systems governed by partial differential equations. In the context of nonlinear H^∞ control problems, several authors have studied the existence of value or storage functions under various assumptions. Among these are James (1993), who used continuous storage (or energy) functions associated with nonlinear systems for the study of dissipative systems originally studied by Willems (1973) (see also Hill and Moylan (1976)), and Soravia (1996), who showed that for a wide class of nonlinear systems the smallest nonnegative continuous viscosity solution is the available storage. Most of this prior work has dealt with continuous viscosity solutions.

In this paper, we consider a wider class of storage functions associated with the nonlinear system (2), which are only *lower semicontinuous*. In fact, in various applications, to require the storage function (or value function) to be continuous could be quite restrictive. In the recent work of McEneaney (1995), the continuity of the value function requires that either the control space be compact or additional assumptions be imposed on the cost function. Simple examples can be devised to show that the value function need not be continuous; one such example is given later in section 5 (see Example 5.2) where the value function is only lower semicontinuous. Accordingly, one objective of this paper is to establish conditions under which the HJI equation of the nonlinear H^∞ control problem of the type formulated above admits only a viscosity supersolution. A second objective is to establish connections between such viscosity supersolutions and feedback controller design.

The remainder of the paper is structured as follows. Section 2 provides assumptions and the definition of viscosity supersolution. In section 3, we study the nonlinear system (2) with cost function (1) as a deterministic differential game, with the controller representing the minimizing player and the disturbance representing the maximizing player. We show that if value function is finite then it will be a viscosity supersolution of the corresponding HJI equation. Furthermore, for a given viscosity supersolution of the HJI equation, under appropriate assumptions, we present a way to construct a nonlinear H^∞ disturbance attenuating feedback controller by using only the viscosity supersolution. As an application, in section 4 we discuss the case in which the control is constrained. Section 5 provides two examples to illustrate the results of the previous sections.

2 Preliminaries and Assumptions

We open this section by recalling the definition of viscosity solutions of the nonlinear, first-order, partial differential equations (3), (5) in the lower semicontinuous case. We take Ω as an open set in $\mathcal{R}^n$.

Definition 1 *The lower (respectively, upper) semicontinuous function* $V : [0, t_f] \times \Omega \to \mathcal{R}$ *is a viscosity supersolution (respectively, subsolution) of*

$$\begin{aligned} &\frac{\partial V}{\partial t}(t,x) + H(x, \nabla_x V(t,x)) = 0 \ \textit{in} \quad (0,t_f] \times \Omega, \\ &V(t_f, x) = \psi(x) \qquad\qquad \textit{on} \ \{t_f\} \times \Omega \end{aligned} \tag{10}$$

if whenever $\phi \in C^1(\overline{Q}_T)$ $(Q_{t_f} = \Omega \times (0, t_f])$ *is such that* $V - \phi$ *attains a local minimum (respectively, local maximum), we have*

$$\phi_t(t_0, x_0) + H(t_0, x_0, \phi_x(t_0, x_0)) \leq 0 \tag{11}$$

(respectively,

$$\phi_t(t_0, x_0) + H(t_0, x_0, \phi_x(t_0, x_0)) \geq 0 \tag{12}$$

)
V is a viscosity solution of (10) if V is both a viscosity supersolution and a viscosity subsolution.

In the case of $t_f = \infty$, viscosity supersolution and viscosity subsolution are defined similarly, with $\partial V/\partial t$ and ϕ_t not appearing in (10), (11), and (12).

Assumptions

A1. There exists a constant $K_a > 0$ such that

$$|A(x) - A(y)| \leq K_a |x - y| \qquad \forall x, y \in \mathcal{R}^n$$

A2. There exists a constant $b_1 > 0$ such that

$$|B_1(x) - B_1(y)| \leq b_1 |x - y| \qquad \forall x, y \in \mathcal{R}^n$$

A3. There exists a constant $b_2 > 0$ such that

$$|B_2(x) - B_2(y)| \leq b_2 |x - y| \qquad \forall x, y \in \mathcal{R}^n$$

A4. $g : \mathcal{R}^n \to [0, +\infty]$ is lower semicontinuous, and for any sequence $\{x_n\} \subset \mathcal{R}^n$, $\{g(x_n)\}$ bounded implies that $\{x_n\}$ is bounded.
A5. $h : \mathcal{R}^p \to [0, +\infty]$ is a proper convex function with $h(0) = 0$, and there exits a nondecreasing continuous function $\omega : [0, +\infty) \to [0, +\infty)$ with properties $\omega(0) = 0, \lim_{r \to +\infty} \omega(r^2)/r = +\infty$ such that

$$h(u) \geq \omega(|u|^2).$$

A6. $\psi : \mathcal{R}^n \to \mathcal{R}^+$ is convex, locally Lipschitz.

Throughout this paper, we adopt the following notation:

$$\mathcal{U} = L^2_{loc}(\mathcal{R}^+, \mathcal{R}^p), \qquad \mathcal{W} = L^2_{loc}(\mathcal{R}^+, \mathcal{R}^l),$$

$h^*(x) = \sup_{y \in \mathcal{R}^n} \{x^T y - h(y)\}$ is the conjugate function of h, and ∂h is the subdifferential of h, defined as

$$\partial h(x) = \{y \in \mathcal{R}^n : h(x) \leq h(z) + y^T(x - z), \quad \forall z \in \mathcal{R}^n\}$$

Remark 2.1. By using the notation of the conjugate function of h, the HJI equation (3) can equivalently be written as

$$\begin{aligned} &\frac{\partial V_\gamma}{\partial t}(t,x) + g(x) + A(x)^T \nabla_x V_\gamma(t,x) - h^*(-B_1(x)^T \nabla_x V_\gamma(t,x)) \\ &\qquad + \frac{1}{4\gamma^2} \nabla_x V_\gamma(t,x)^T B_2(x) B_2(x)^T \nabla_x V_\gamma(t,x) = 0 \qquad (13) \\ &V_\gamma(t_f, x) = \psi(x) \end{aligned}$$

and the HJI equation (5) becomes equivalent to

$$\begin{aligned} &g(x) + A(x)^T \nabla_x V_\gamma(x) - h^*(-B_1(x)^T \nabla_x V_\gamma(x)) \\ &\qquad + \frac{1}{4\gamma^2} \nabla_x V_\gamma(x)^T B_2(x) B_2(x)^T \nabla_x V_\gamma(x) = 0 \qquad (14) \end{aligned}$$

3 Nonlinear H^∞ Control

Definition 2 *Let $\mathcal{O}$ be an open subset of $\mathcal{R}^n$. An admissible γ-attenuation feedback control for system (2) is a (nonlinear, multivalued) mapping F : $[0, t_f] \times \mathcal{R}^n \to \mathcal{R}^p (= U)$ such that $u \in Fx$ and*

1. *for every $w \in L^2([0, t_f]; \mathcal{R}^l)(= \mathcal{W})$ and $x_0 \in \mathcal{O}$, the following differential inclusion*

$$\begin{aligned} &\dot{x}(s) \in A(x(s)) + B_1(x(s))F(x(s)) + B_2(x(s))w(s), \qquad (15) \\ &x(0) = x_0 \end{aligned}$$

has at least one solution on $[0, t_f]$;

2. *there exists a nonnegative function $\varphi : \mathcal{R}^n \to \mathcal{R}^+$ associated with this feedback control such that for any $w \in \mathcal{W}$ we have*

$$\int_0^{t_f} \{g(x(s))+h(u(s))\}ds+\psi(x(t_f)) \leq \gamma^2 \int_0^{t_f} |w(s)|^2 ds+\varphi(x_0) \quad (16)$$

Remark 3.1. In the infinite-horizon case, (16) holds for every $t_f > 0$, and with $\psi \equiv 0$.
Remark 3.2. Solutions to differential inclusions (15) are understood in the Carathéodory sense, i.e., absolutely continuous functions verifying (15) almost everywhere.
Remark 3.3. If a feedback controller satisfies (1) of Definition 2 only, we call this controller an *admissible feedback control* for system (2).

3.1 The Finite-Horizon Case

We denote by $\mathcal{M}_{CL}$ the class of admissible feedback controls for system (2) and define the value function:

$$V_\gamma(t,x) = \inf_{\mathcal{M}_{CL}} \sup_{\mathcal{W}} J_\gamma(t,x;\mu,w) \quad (17)$$

Theorem 3.1. *Suppose that state feedback control*

$$\mu(x(t)) = \partial h^*\big(- B_1(x(t))^T \hat{p}(t,x(t))\big)$$

is an admissible feedback control, where

$$(\cdot,\hat{p}) \in D^- V_\gamma(t,x) = \big\{(q(t,x),p(t,x)) \in \mathcal{R}^{n+1} : \liminf_{(s,y)\to(t,x)} \frac{V_\gamma(s,y) - V_\gamma(t,x) - q(s-t) - p(y-x)}{|s-t|+|y-x|} \geq 0\}$$

Then, for this controller to be an admissible γ-attenuation feedback control, it is necessary that the HJI equation (13) have a nonnegative viscosity supersolution V_γ with property $dom(V_\gamma) \supset \overline{dom(g)}$.

Remark 3.4. $D^- V_\gamma(t,x)$ is convex for every $(t,x) \in [0,t_f]\times\mathcal{R}^n$. Moreover if V_γ is differentiable at (t,x) we have

$$D^- V_\gamma(t,x) = \Big\{\Big(\frac{\partial}{\partial t} V_\gamma(t,x), D_x V_\gamma(t,x)\Big)\Big\}$$

(cf. Fleming and Soner (1992)) and the feedback controller becomes

$$\mu(x(t)) = \partial h^*\big(- B_1(x(t))^T D_x V_\gamma(t,x(t))\big) \quad (18)$$

Proof of Theorem 3.1. Since the feedback control μ is an admissible γ-attenuation feedback control, the value function V_γ defined in (17) is finite, and it is also lower semicontinuous by assumptions (A4) and (A5). Let $\phi \in C^1([0,t_f]\times\mathcal{R}^n)$ be such that $V_\gamma - \phi$ achieves a local minimum at $(t_0, x_0) \in [0,t_f]\times\mathcal{R}^n$, and $x(t)$ be the trajectory of (2) with $x(t_0) = x_0, t \geq t_0$, under the given controller and the fixed $w(t) = \frac{1}{2\gamma^2}B_2(x_0)^T\phi_x(t,x_0)$. Thus for $t - t_0$ sufficiently small, we have

$$V_\gamma(t_0,x_0) - \phi(t_0,x_0) \leq V_\gamma(t,x(t)) - \phi(t,x(t)) \tag{19}$$

Thus, (19) yields $\forall w \in \mathcal{W}$

$$\frac{\phi(t,x(t)) - \phi(t_0,x_0)}{t-t_0} \leq \frac{1}{t-t_0}\int_{t_0}^{t}(\gamma^2|w(s)|^2 - g(x(s)) - h(u(s)))ds \tag{20}$$

Letting $t \downarrow 0$, (20) yields

$$\begin{aligned}\phi_t(t_0,x_0) + \phi_x(t_0,x_0)^T(A(x_0) + B_1(x_0)u(t_0) + B_2(x_0)w(t_0))\\ \leq \gamma^2|w(t_0)|^2 - g(x_0) - h(u(t_0))\end{aligned} \tag{21}$$

Note that (21) can be rewritten as

$$\begin{aligned}&\phi_t(t_0,x_0) + g(x_0) + A(x_0)^T\phi_x(t_0,x_0) + (B_1(x_0)u(t_0))^T\phi_x(t_0,x_0)\\ &-(B_2(x_0)u)^T\phi_x(t_0,x_0) + \frac{1}{4\gamma^2}\phi_x(t_0,x_0)^TB_2(x_0)B_2(x_0)^T\phi_x(t_0,x_0)\\ &\quad +h(u(t_0)) - \gamma^2|w(t_0) - \frac{1}{2\gamma^2}B_2(x_0)^T\phi_x(t_0,x_0)|^2 \leq 0\end{aligned}$$

Since

$$h(u) + h^*(-B_1^T\phi_x) \geq (-B_1u)^T\phi_x, \tag{22}$$

we thus arrive at

$$\begin{aligned}&\phi_t(t_0,x_0) + g(x_0) + A(x_0)^T\phi_x(t_0,x_0) - h^*(-B_1^T(x_0)\phi_x)\\ &\quad + \frac{1}{4\gamma^2}\phi_x(t_0,x_0)^TB_2(x_0)B_2(x_0)^T\phi_x(t_0,x_0) \leq 0\end{aligned}$$

and this completes the proof of Theorem 3.1. □

We are more interested in this question of the construction of a γ-attenuation feedback controller for system (2) for a given viscosity supersolution of (13). The answer to the question is provided in the following theorem.

Theorem 3.2. *Suppose that the following four conditions hold:*

1. *HJI equation (13) has a nonnegative viscosity supersolution V_γ with the property that for any $t \in [0,t_f]$, $dom(V_\gamma(t,\cdot)) \supset \overline{dom(g)}$;*
2. *There exists a pair $(\hat{q},\hat{p}) \in D^-V_\gamma$ such that $\hat{p}(t_f,x) \geq \psi(x), x \in \Omega$;*

3. The feedback controller

$$\mu(x(t)) = \partial h^*\big(- B_1(x(t))^T \hat{p}(t, x(t))\big)$$

is an admissible feedback control;

4. There exists a function $\varphi : [0, t_f] \times \Omega \to \mathcal{R}$ with $\nabla\varphi$ existing a.e. *on Ω, such that*

$$\varphi \geq \psi \quad \forall (t, x) \in [0, t_f] \times dom(V_\gamma), \quad \nabla\varphi = (\hat{p}, \hat{q}) \quad a.e.\ x \in dom(V_\gamma).$$

Then, the state feedback controller μ is an admissible γ-attenuation feedback control, i.e. we have

$$\int_0^{t_f} \{g(x(s)) + h(\mu(x(s)))\} ds + \psi(x(t_f)) \leq \gamma^2 \int_0^t |w(s)|^2 ds + \varphi(x_0)$$

for any $w \in \mathcal{W}$ and $x_0 \in \mathcal{O}$.

In order to prove Theorem 3.2, we first establish Lemma 3.3 below.

Lemma 3.3. *Assume that $V_\gamma(x, t_f) \geq \psi(x), x \in \Omega$. V_γ is a viscosity supersolution of (13) with $(t, x) \in [0, t_f] \times \Omega$ if and only if*

$$\begin{aligned} &\hat{q}(t, x) + g(x) + A(x)^T \hat{p}(t, x) - h^*(-B_1(x)^T \hat{p}(t, x)) \\ &\quad + \frac{1}{4\gamma^2} \hat{p}(t, x)^T B_2(x) B_2(x)^T \hat{p}(t, x) \leq 0 \end{aligned} \tag{23}$$

for any $(\hat{q}, \hat{p}) \in D^- V_\gamma(t, x)$, $(t, x) \in [0, t_f] \times \Omega$.

Proof: Suppose that (23) holds. Let $\phi \in C^1([0, t_f] \times \mathcal{R}^n)$ be such that (t_0, x_0) is a local minimizer of $V_\gamma - \phi$ with $V_\gamma(t_0, x_0) = \phi(t_0, x_0), (t_0, x_0) \in [0, t_f] \times \Omega$. Since $\phi \in C^1([0, t_f] \times \mathcal{R}^n)$, it gives

$$\begin{aligned} \phi(t, x) = \phi(t_0, x_0) + \phi_t(t_0, x_0)(t - t_0) + \phi_x(t_0, x_0)(x - x_0) \\ + o(|t - t_0|) + o(|x - x_0|) \end{aligned}$$

Hence there exists a $\delta > 0$ such that

$$\begin{aligned} V(t, x) \geq \phi(t_0, x_0) + \phi_t(t_0, x_0)(t - t_0) + \phi_x(t_0, x_0)(x - x_0) \\ + o(|t - t_0|) + o(|x - x_0|) \end{aligned}$$

with $|t - t_0| + |x - x_0| < \delta$. Let $\hat{p} = \phi_x(t_0, x_0), \hat{q} = \phi_t(t_0, x_0)$. According to the definition of $D^- V_\gamma$ (given in Theorem 3.1) and (23), we have

$$\begin{aligned} &\phi_t(t_0, x_0) + g(x_0) + A(x_0)^T \phi_x(t_0, x_0) - h^*(-B_1(x_0)^T \phi_x(t_0, x_0)) \\ &\quad + \frac{1}{4\gamma^2} \phi_x(t_0, x_0)^T B_2(x_0) B_2(x_0)^T \phi_x(t_0, x_0) \leq 0 \end{aligned} \tag{24}$$

and hence V_γ is a viscosity supersolution of (13) with $(t, x) \in [0, t_f] \times \Omega$.

Conversely, suppose that V_γ is a viscosity supersolution of (13). Let $(\hat{q}, \hat{p}) \in D^- V_\gamma(t,x), (t,x) \in [0,t_f] \times \Omega$. As (s,y) closes to (t,s) sufficiently, we have

$$V_\gamma(s,y) \geq V_\gamma(t,x) + q(s-t) + p(y-x) + o(|s-t|) + o(|y-s|)$$

Define the test function

$$\phi(s,y) := V_\gamma(t,x) + q(s-t) + p(y-x) + g_1(|s-t|) + g_2(|y-s|)$$

where $g_i : (0,\infty) \to (0,\infty), i = 1,2$ are nondecreasing functions such that

$$g_i(r) = o(r), \qquad Dg_i(r)\big|_{r=0} = 0$$

(For the construction of these functions, see Fleming and Soner (1993).) Therefore (t,x) is in fact a strict minimizer of $V - \phi$, and $\phi_x(t,x) = \hat{p}, \phi_t(t,x) = \hat{q}$, and (23) holds by the definition of viscosity supersolution. □

Proof of Theorem 3.2. By Lemma 3.3 and hypothesis (3), we have

$$\begin{aligned}&\varphi_t(t,x) + g(x) + A(x)^T\varphi_x(t,x) - h^*(-B_1(x)^T\varphi_x(t,x))\\ &\quad+\frac{1}{4\gamma^2}\varphi_x(t,x)^T B_2(x)B_2(x)^T\varphi_x(t,x) \leq 0 \quad a.e.\ x \in dom(V_\gamma)\end{aligned}$$

By hypotheses (2) and (3), and the fact that

$$h(\mu) + h^*(-B_1^T\varphi_x) = (-B_1\mu)^T\varphi_x,$$

we have

$$\begin{aligned}&\frac{d\varphi(t,x(t))}{dt} = \varphi_t + \Big(A(x(t)) + B_1(x(t))\mu(x(t)) + B_2(x(t))w(t)\Big)^T\varphi_x\\ &\leq -g(x(t)) + h^*(-B_1^T\varphi_x) + \big(B_1(x(t))\mu(x(t)) + B_2(x(t))w(t)\big)^T\varphi_x\\ &\qquad+\frac{1}{4\gamma^2}\varphi_x(t,x)^T B_2(x)B_2(x)^T\varphi_x(t,x)\\ &\leq -g(x(t)) - h(\mu(x(t))) + \gamma^2|w(t)|^2\end{aligned}$$

where $x(t)$ satisfies

$$\dot{x} = A(x) + B_1(x)\mu(x) + B_2(x)w, \qquad x(0) = x_0$$

Therefore

$$\varphi(x(t_f)) + \int_0^{t_f}\{g(x) + h(\mu(x)) - \gamma^2|w|^2\}ds \leq \varphi(x_0), \quad \forall t \in [0,t_f]$$

Since $\varphi \geq \psi$, this completes the proof of Theorem 3.2. □

3.2 The Infinite-Horizon Case

In this case, the value function is

$$V_\gamma(x) = \inf_{\mathcal{M}_{CL}} \sup_{\mathcal{W}} \sup_{t_f<\infty} J_\gamma(0, x; u, w) \tag{25}$$

where J_γ is defined by (1), with implicit dependence on t_f. As a counterpart of Theorem 3.1, we state the following theorem without proof.

Theorem 3.4. *There exists an admissible* γ*-attenuation feedback control*

$$\mu(x(t)) = \partial h^*\big(-B_1(x(t))^T\hat{p}(x(t))\big)$$

where

$$\hat{p} \in D^- V_\gamma(x) = \Big\{p \in \mathcal{R}^n, \liminf_{y\to x} \frac{V_\gamma(y) - V_\gamma(x) - p(y-x)}{|y-x|} \geq 0\Big\}$$

for system (2) only if the HJI equation (14) admits a supersolution V_γ *with property* $dom(V_\gamma) \supset \overline{dom(g)}$.

Lemma 3.5. V_γ *is a supersolution of (14) with* $x \in \Omega$ *if and only if*

$$g(x) + A(x)^T p(x) - h^*(-B_1(x)^T p(x)) + \frac{1}{4\gamma^2} p(x)^T B_2(x) B_2(x)^T p(x) \leq 0$$

holds for any $p \in D^- V_\gamma(x)$, $x \in \Omega$.

Proof: The proof is similar to the one of Lemma 3.3. □

The following theorem now provides a sufficient condition for stabilization of the nonlinear system (2).

Theorem 3.6. *Assume that*

1. *HJI equation (14) has a nonnegative supersolution* V_γ *with property*

$$dom(V_\gamma) \supset \overline{dom(g)}.$$

2. *The state feedback controller*

$$\mu(x) = \partial h^*(-B_1(x)^T p(x)) \tag{26}$$

is an admissible state feedback control.

3. *There exists a function* $\varphi : \Omega \to \mathcal{R}$ *with* $\nabla\varphi$ *existing a.e. on* Ω, *such that*

$$\varphi \geq 0 \quad x \in dom(V_\gamma), \quad \nabla_x \varphi = p, \quad a.e.\ x \in dom(V) \subset \mathcal{R}^n \tag{27}$$

4. *$g(x(\cdot)) \in L^1(\mathcal{R}^+;\mathcal{R})$ implies $x \in L^2(\mathcal{R}^+;\mathcal{R}^n)$.*

Then with the feedback controller (26), the worst-case system

$$\dot{x}^* = Ax^* + B_1(x^*)\partial h^*(-B_1(x^*)^T p(x^*)) + \frac{1}{\gamma^2} B_2(x^*)B_2(x^*)^T p(x^*) \quad (28)$$

is globally asymptotically stable, i.e.

$$x^* \in C(\mathcal{R}^+;\mathcal{R}^n) \cap L^2(\mathcal{R}^+;\mathcal{R}^n); \qquad \lim_{t\to\infty} x^*(t) = 0.$$

Remark 3.5. Assumption (2) of Theorem 3.6 says that system (28) is dissipative with respect to the supply rate

$$\gamma^2 B_2(x^*)B_2(x^*)^T p(x^*) - g(x^*) - h(\mu(x^*(t))).$$

Proof of Theorem 3.6. By Lemma 3.5, and hypothesis (3) of the theorem, we have

$$\begin{aligned} &g(x) + A(x)^T \nabla_x \varphi(x) - h^*(-B_1(x)^T \nabla_x \varphi(x)) \\ &+\frac{1}{4\gamma^2} \nabla_x \varphi(x)^T B_2(x) B_2(x)^T \nabla_x \varphi(x) \le 0 \quad a.e.\ x \in dom(V_\gamma) \end{aligned} \quad (29)$$

This inequality implies that

$$\begin{aligned} \frac{d\varphi(x^*(t))}{dt} \le\ & \gamma^2 p(x^*(t))^T B_2(x^*(t)) B_2(x^*(t))^T p(x^*(t)) \\ & -g(x^*(t)) - h(\mu(x^*(t))) \end{aligned} \quad (30)$$

This yields for any $t > 0$

$$\begin{aligned} \varphi(x_0) - \varphi(x^*(t)) \ge \int_0^t & \{g(x^*(s)) + h(\mu(x^*(s))) \\ & -\gamma^2 p(x^*(s))^T B_2(x^*(s)) B_2(x^*(s))^T p(x^*(s))\} ds, \end{aligned} \quad (31)$$

and hypothesis (3) and (31) yield

$$V(x^*(t)) + \int_0^t L(s, x^*, p) ds \le \varphi(x_0), \qquad \forall t > 0 \quad (32)$$

where

$$\begin{aligned} L(s, x^*, p) = &\ g(x^*(s)) + h(\mu(x^*(s))) \\ & -\gamma^2 p(x^*(s))^T B_2(x^*(s)) B_2(x^*(s))^T p(x^*(s)). \end{aligned} \quad (33)$$

Thus from (32) we get

$$\int_0^t g(x^*(s)) ds \le C, \qquad \forall t > 0$$

Hypothesis (2) implies that $x^* \in C(\mathcal{R}^+, \mathcal{R}^n)$; hence with hypothesis (4) we have

$$x^* \in C(\mathcal{R}^+, \mathcal{R}^n) \cap L^2(\mathcal{R}^+, \mathcal{R}^n) \quad \text{and} \quad \lim_{t\to\infty} x^*(t) = 0$$

as claimed. □

Theorem 3.7. *Under the assumptions of Theorem 3.6, for any $w \in \mathcal{W}$ and system trajectory generated by*

$$\dot{x}(t) = Ax(t) + B_1(x(t))\partial h^*(-B_1(x(t))^T p(x(t))) + B_2(x(t))w,$$
$$x(0) = x_0$$

we have for any $t > 0$

$$\int_0^t \{g(x(s)) + h(\mu(x(s)))\}ds \le \gamma^2 \int_0^t |w(s)|^2 ds + \varphi(x_0) \tag{34}$$

Proof: As in the proof of Theorem 3.2, we have

$$g(x) + A(x)^T \nabla_x \varphi(x) - h^*(-B_1(x)^T \nabla_x \varphi(x))$$
$$+\frac{1}{4\gamma^2} \nabla_x \varphi(x)^T B_2(x) B_2(x)^T \nabla_x \varphi(x) \le 0 \quad a.e.\ x \in dom(V)$$

which yields

$$\frac{d\varphi(x(t))}{dt} \le -\gamma^2 |w - \frac{1}{2\gamma^2} B_2(x(t))^T \nabla_x \varphi(x(t))|^2$$
$$+\gamma^2 |w(t)|^2 - g(x(t)) - h(\mu(x(t))) \tag{35}$$

Since $\varphi > 0$, integrating (35), we can arrive at (34). □

Corollary 3.8. *Suppose that (14) has a solution V_γ of class C^1 and the state feedback controller*

$$\mu(x) = \partial h^*(-B_1(x)^T \nabla_x V_\gamma)$$

is an admissible γ-attenuation state feedback controller. Then V_γ is a storage function.

Proof: With control μ, we have the following system

$$\dot{x} = Ax + B_1(x)\partial h^*(-B_1(x)^T \nabla_x V_\gamma) + B_2(x)w, \quad x(0) = x_0.$$

Using (14), we get

$$\frac{d}{dt} V_\gamma(x(t)) = (\nabla_x V_\gamma)^T \dot{x}(t)$$
$$= (\nabla_x V_\gamma)^T (A(x(t)) + B_1(x(t))\partial h^*(-B_1(x(t))^T \nabla_x V) + B_2(x(t))w(t))$$
$$= -g(x(t)) - h(\mu(x(t))) + \gamma^2 |w(t)|^2 - \gamma^2 |w(t) - \tfrac{1}{2\gamma^2} B_2^T(x(t)) \nabla_x V_\gamma|^2$$
$$\le -g(x(t)) - h(\mu(x(t))) + \gamma^2 |w(t)|^2.$$

Integrating the above inequality on any interval $[t_1, t_2]$, we have

$$V(x(t_2)) \le V_\gamma(x(t_1)) + \int_{t_1}^{t_2} \{\gamma^2 |w(t)|^2 - g(x(t)) - h(\mu(x(t)))\}dt$$

which shows that V_γ is a storage function. □

4 An Application to Problems with Control Constraints

In this section, we consider system (2) with control constraints. Let U_0 be a closed convex subset of control space U. Then, (2) can be written as

$$\dot{x}(s) = A(x(s)) + B_1(x(s))u(s) + B_2(x(s))w(s), u(t) \in U_0 \quad (36)$$
$$x(t) = x.$$

Let I_{U_0} be the indicator function for the set U_0, i.e.

$$I_{U_0}(u) = \begin{cases} 0 & \text{if } u \in U_0 \\ +\infty & \text{if } u \notin U_0 \end{cases}$$

We consider, for the finite t_f, the cost function

$$J_\gamma(t, x; u, w) = \int_t^{t_f} \{g(x(s)) + |u(s)|^2 - \gamma^2|w(s)|^2\}ds + \psi(x(t_f)) \quad (37)$$

and if $t_f = +\infty$, we set $\psi \equiv 0$. In order to incorporate this problem into an unconstrained optimization, we modify (37) into

$$J_\gamma(t, x; u, w) = \int_t^{t_f} \{g(x(s)) + |u(s)|^2 + I_{U_0}(u) - \gamma^2|w(s)|^2\}ds + \psi(x(t_f)).$$

Clearly, if $u \notin U_0$, $J(t, x; u, w) \equiv +\infty$. Define $P_{U_0} : U \to U_0$ by $P_{U_0}u = u_0 \in U_0$, where u_0 is defined by

$$\inf_{v \in U_0} |v - u| = |u_0 - u|$$

If we let $h(u) = |u|^2 + I_{U_0}(u)$ then the conjugate function of h is given by

$$h^*(u) = \frac{1}{4}(|u|^2 - |u - P_{U_0}u|^2).$$

We can now directly apply results of Theorems 3.2 and 3.6, for the finite- and infinite-horizon cases, respectively, to lead to the following:

Theorem 4.1. *Assume that:*

1. *The HJI equation*

$$\begin{aligned} &V_{\gamma t}(t, x) + g(x) + A(x)^T \nabla_x V_\gamma(t, x) \\ &\quad -2(-B_1(x)^T \nabla_x V_\gamma(t, x))^T P_{U_0}((-B_1(x)^T \nabla_x V_\gamma(t, x)) \\ &\quad +(P_{U_0}(-B_1(x)^T \nabla_x V_\gamma(t, x)))^T P_{U_0}((-B_1(x)^T \nabla_x V_\gamma(t, x))) \\ &\quad\quad +\frac{1}{4\gamma^2} \nabla_x V_\gamma(t, x)^T B_2(x) B_2(x)^T \nabla_x V_\gamma(t, x) = 0 \\ &V_\gamma(t_f, x) = \psi(x) \end{aligned}$$

admits a nonnegative viscosity supersolution V_γ with the property that $dom(V_\gamma) \supset \overline{dom(g)}$;

2. *There exists a pair* $(\hat{p}, \hat{q}) \in D^- V_\gamma$ *such that* $\hat{p}(t_f, x) \geq \psi(x)$;

3. *The state feedback controller*

$$\mu(x) = P_{U_0}(-B_1(x)^T \hat{p}(t, x))$$

is an admissible feedback control (see Definition 3);

4. *There exists a function* $\varphi : [0, t_f] \times \overline{domV_\gamma} \to \mathcal{R}$ *with a.e. existing* $\nabla\varphi$ *such that*

$$\varphi \geq \psi \quad \forall (t, x) \in [0, t_f] \times dom(V_\gamma), \nabla\varphi = (\hat{p}, \hat{q}) \quad a.e.\ x \in dom(V_\gamma).$$

Then, the state feedback controller μ *is a* γ*-attenuation feedback control, i.e. we have* $\forall w \in \mathcal{W}, x \in \mathcal{O}$,

$$\int_0^{t_f} \{g(x(t)) + h(\mu(x(t)))\} dt + \psi(x(t_f)) \leq \gamma^2 \int_0^{t_f} |w(t)|^2 dt + \varphi(x)$$

Theorem 4.2. *Suppose that:*

1. *The following HJI equation*

$$\begin{aligned} g(x) + A(x)^T \nabla_x V_\gamma(x) \\ +2(-B_1(x)^T \nabla_x V_\gamma(x))^T P_{U_0}(-B_1(x)^T \nabla_x V_\gamma(x)) \\ +P_{U_0}(-B_1(x)^T \nabla_x V_\gamma(x))^T P_{U_0}(-B_1(x)^T \nabla_x V_\gamma(x))) \\ +\frac{1}{4\gamma^2} \nabla_x V_\gamma(x)^T B_2(x) B_2(x)^T \nabla_x V_\gamma(x) = 0 \end{aligned}$$

has a nonnegative viscosity supersolution V_γ *with the property that* $dom(V_\gamma) \supset \overline{dom(g)}$;

2. *The feedback controller*

$$\mu(x) = P_{U_0}(-B_1(x)^T \hat{p}(x))$$

is an admissible feedback control;

3. *There exists a function* $\varphi : \overline{domV_\gamma} \to \mathcal{R}$ *with a.e. existing* $\nabla_x\varphi$ *such that*

$$\varphi \geq 0 \quad x \in dom(V_\gamma), \quad \nabla_x\varphi = \hat{p}, \quad a.e.\ x \in dom(V_\gamma) \subset \mathcal{R}^n.$$

Then, μ *is a* γ*-attenuation feedback controller. Moreover, the system associated with this control* μ *and worst-case disturbance,*

$$\dot{x} = A(x) + B_1(x) P_{U_0}(-B_1(x)^T \hat{p}(x)) + \frac{1}{\gamma^2} B_2(x) B_2^T(x) \hat{p}(x),$$

is asymptotically stable for $x_0 \in \mathcal{O}$.

Remark 4.1. *In the case of $U_0 = \{0\}$, the HJI equations in Theorems 4.1 and 4.2 reduce to HJB equations with only a maximization (for the disturbance) appearing in the Hamiltonians.*

Now let us consider the case when system (36) is a time-invariant linear system

$$\begin{aligned} \dot{x}(t) &= A^0 x(t) + B_1^0 u(t) + B_2^0 w(t), \quad u(t) \in U_0 \\ x(0) &= x_0 \end{aligned} \tag{38}$$

Suppose that $g(x) = x^T Q x, \psi(x) \equiv 0$, U_0 is a subspace of U, and the HJI equation in Theorem 4.1 has a nonnegative viscosity supersolution V_γ (it is in fact a viscosity solution). It is not difficult to see that V_γ in fact is a (unique) classical solution and $V_\gamma(t,x) = x^T Z_\gamma^{t_f}(t) x$, where $Z_\gamma^{t_f}$ satisfies

$$\begin{aligned} &\dot{Z}_\gamma^{t_f}(t) + A^T Z_\gamma^{t_f}(t) + Z_\gamma(t) A - 2 Z_\gamma^{t_f}(t) B_1 P_{U_0} B_1^T Z_\gamma^{t_f}(t) \\ &\qquad + Z_\gamma^{t_f}(t)(B_1 P_{U_0}^T P_{U_0} B_1^T + \frac{1}{\gamma^2} B_2 B_2^T) Z_\gamma^{t_f}(t) + Q = 0 \\ &Z_\gamma^{t_f}(t_f) = 0 \end{aligned} \tag{39}$$

If there exists a constant C such that

$$Z_\gamma^{t_f}(t) \le C \qquad t \in [0, t_f], \quad \forall t_f > 0,$$

then $\lim_{t_f \to \infty} Z_\gamma^{t_f}(t)$ exists for all t, and it is independent of t. Define

$$Z_\gamma^\infty = \lim_{t_f \to \infty} Z_\gamma^{t_f}(t).$$

Then $V_\gamma(x) = x^T Z_\gamma^\infty x$ is the minimal, nonnegative solution of (5). More precisely, Z_γ^∞ is the minimal, nonnegative-definite solution of

$$\begin{aligned} A^T Z_\gamma + Z_\gamma A - 2 Z_\gamma B_1 P_{U_0} B_1^T Z_\gamma + Z_\gamma (B_1 P_{U_0}^T P_{U_0} B_1^T \\ + \tfrac{1}{\gamma^2} B_2 B_2^T) Z_\gamma + Q = 0 \end{aligned} \tag{40}$$

In the finite-horizon case, the optimal feedback control and the worst disturbance are given, respectively, by

$$\mu(x(t)) = -P_{U_0} B_1^T Z_\gamma^{t_f}(t) x(t) \tag{41}$$

$$\nu(x(t)) = \gamma^{-2} B_2^T Z_\gamma^{t_f}(t) x(t) \tag{42}$$

while in the infinite-horizon case they are

$$\mu(x(t)) = -P_{U_0} B_1^T Z_\gamma^\infty x(t) \tag{43}$$

$$\nu(x(t)) = \gamma^{-2} B_2^T Z_\gamma^\infty x(t) \tag{44}$$

with $\lim_{t \to \infty} x(t) = 0$ by Theorem 3.6. Some further analysis on this problem leads to the following result:

Theorem 4.3. *Let $x_0 = 0$ and*

$$\rho(u,w) = \sup_{t>0} \frac{\int_0^t \{|x(s)|_Q^2 + |u(s)|^2 + I_{U_0}(u(s))\}ds}{\int_0^t |w(s)|^2 ds}. \tag{45}$$

Given $\gamma > 0$,

$$\sup_w \inf_u \rho(u,w) < \gamma^2$$

if and only if (40) admits a minimal, nonnegative, continuous viscosity solution. In this case the admissible γ-attenuation feedback controller is given by (41). Moreover

$$\sup_w \inf_u \rho(u,w) < \gamma^2 \Longrightarrow \inf_{\mathcal{M}_{CL}} \sup_w \rho(u,w) = \sup_w \inf_u \rho(u,w) < \gamma^2.$$

Proof: The proof is similar to that of Corollary 4.2 of Xiao and Başar (1996), and is not included here due to space limitations.

Remark 4.2. In the case $U_0 = U$, the generalized Riccati equation (39) becomes

$$\dot{Z}_\gamma(t) + A^T Z_\gamma(t) + Z_\gamma(t)A - Z_\gamma(t)(B_1B_1^T - \frac{1}{\gamma^2}B_2B_2^T)Z_\gamma(t) + Q = 0$$
$$Z_\gamma(t_f) = 0$$

and the generalized algebraic Riccati equation (40) becomes

$$A^T Z_\gamma + Z_\gamma A - Z_\gamma(B_1B_1^T - \frac{1}{\gamma^2}B_2B_2^T)Z_\gamma + Q = 0$$

Both equations are consistent with the known results (for this special case of standard LQ H^∞ control problem); see, e.g. Başar and Bernhard (1995).

5 Examples

1. Bilinear dynamics with control constraints.
Consider the bilinear (scalar) dynamics

$$\dot{x}(t) = (u(t) + w(t))x(t), \quad x(0) = x_0 \tag{46}$$

and the cost function

$$J_\gamma(x; u, w) = \int_0^\infty \{x^2(t) + u^2(t) + I_{U_0}(u(t)) - \gamma^2 w^2(t)\}dt$$

Note that in this case there is trivially an admissible feedback controller. The associated HJI equation is

$$x^2 - h^*(-V_x x) + \frac{1}{4\gamma^2}V_x^2 x^2 = 0 \tag{47}$$

and

$$h^*(u) = \frac{1}{4}(u^2 - (u - P_{U_0}u)^2)$$

First let us look at the particular control constraint $U_0 = \{0\}$, which corresponds to the control free case. In this case $h^* \equiv 0$ and (47) becomes

$$x^2 + \frac{1}{4\gamma^2}V_x^2 x^2 = 0 \tag{48}$$

Clearly, for any $x \neq 0$ (48) has no viscosity supersolution. Suppose that the control constraint is $u \geq 0$, and thus $U_0 = [0, +\infty)$. Then the HJI equation is

$$x^2 + \frac{1}{4}\{(V_x x + P_{U_0}(-V_x x))^2 - (V_x x)^2\} + \frac{1}{4\gamma^2}V_x^2 x^2 = 0 \tag{49}$$

It is easy to verify that (49) has a viscosity supersolution if and only if $x \leq 0$, which means that in order for the attenuation level to be $\gamma > 1$, the initial condition must be negative. In this case the viscosity supersolution of (49) is

$$V_\gamma(x) = -\frac{2\gamma}{\sqrt{\gamma^2 - 1}}x. \tag{50}$$

The feedback control is given by

$$\mu^\infty(x(t)) \quad = \quad \begin{cases} 0 & \text{if} \quad x(t) \geq 0 \\ -\frac{\gamma}{\sqrt{\gamma^2-1}}x(t) & \text{if} \quad x(t) \leq 0 \end{cases}$$

Now we change the constraint to $-a \leq u \leq a$, i.e. $U_0 = [-a, a]$ where $a > 0$. When $\gamma > 1$ the HJI equation (49) has a viscosity supersolution on the interval $-\frac{a}{2}\sqrt{\gamma^2 - 1} \leq x \leq \frac{a}{2}\sqrt{\gamma^2 - 1}$:

$$V_\gamma(x) \quad = \quad \begin{cases} \frac{2\gamma}{\sqrt{\gamma^2-1}}|x| & \text{if} \quad |x| \leq a\frac{\sqrt{\gamma^2-1}}{2\gamma} \\ G(x) & \text{if} \quad x \geq a\frac{\sqrt{\gamma^2-1}}{2\gamma} \\ H(x) & \text{if} \quad x \leq -a\frac{\sqrt{\gamma^2-1}}{2\gamma} \end{cases}$$

where

$$G'(x) = \frac{a\gamma^2 - \gamma\sqrt{a^2\gamma^2 - a^2 - 4x^2}}{x}, \quad H'(x) = \frac{-a\gamma^2 - \gamma\sqrt{a^2\gamma^2 - a^2 - 4x^2}}{x}$$

The controller and the disturbance obtained using this viscosity solution are:

$$\mu_\gamma^\infty(x) = \begin{cases} -\frac{\gamma}{\sqrt{\gamma^2-1}}|x| \\ -\frac{1}{2}\left(a\gamma^2 - \gamma\sqrt{a^2\gamma^2 - a^2 - 4x^2}\right) \\ \frac{1}{2}\left(a\gamma^2 - \gamma\sqrt{a^2\gamma^2 - a^2 - 4x^2}\right) \end{cases}$$

$$\nu_\gamma^\infty(x) = \begin{cases} -\frac{1}{\gamma\sqrt{\gamma^2-1}}|x| \\ \frac{1}{2\gamma}\left(a\gamma - \sqrt{a^2\gamma^2 - a^2 - 4x^2}\right) \\ -\frac{1}{2\gamma}\left(a\gamma - \sqrt{a^2\gamma^2 - a^2 - 4x^2}\right) \end{cases}$$

in the corresponding intervals $[-a\frac{\sqrt{\gamma^2-1}}{2\gamma}, a\frac{\sqrt{\gamma^2-1}}{2\gamma}]$, $[a\frac{\sqrt{\gamma^2-1}}{2\gamma}, a\frac{\sqrt{\gamma^2-1}}{2}]$, and $[-a\frac{\sqrt{\gamma^2-1}}{2}, -a\frac{\sqrt{\gamma^2-1}}{2\gamma}]$. It is not difficult to see that the worst-case system dynamics described by

$$\dot{x}^*(t) = (\mu^\infty(x^*(t)) + \nu^\infty(x^*(t)))x^*(t)$$

has a unique solution

$$x^* \in C(\mathcal{R}^+;\mathcal{R}) \cap L^2(\mathcal{R}^+;\mathcal{R}); \qquad \lim_{t\to\infty} x^*(t) = 0$$

for any $x_0 \in [-\frac{a}{2}\sqrt{\gamma^2-1}, \frac{a}{2}\sqrt{\gamma^2-1}]$.

Remark 5.1. In this example we can see the connections between the initial condition, control constraint, and the desired attenuation level. Increasing γ results in a larger range of values of initial conditions, while making the control more restrained will limit the initial data.

2. Value function only lower semicontinuous.
Here we still consider the bilinear dynamics of Example 5.1, but the cost function is now taken as:

$$J_\gamma(x;u,w) = \int_0^\infty \{g(x(t)) + u^2(t) - \gamma^2 w^2(t)\}dt$$

where

$$g(x) = \begin{cases} (2x+1)^2 & \text{if} \quad x > 1 \\ x^2 & \text{if} \quad x \le 1 \end{cases}$$

Note that g is lower semicontinuous. The corresponding HJI equation is

$$g(x) - \frac{1}{4}(1 - \frac{1}{\gamma^2})V_x^2 x^2 = 0 \tag{51}$$

When $\gamma > 1$, the viscosity supersolution of (51) is

$$V_\gamma(x) = \begin{cases} \frac{2\gamma}{\sqrt{\gamma^2-1}}(2+\ln x) & \text{if} \quad x > 1 \\ \frac{2\gamma}{\sqrt{\gamma^2-1}}|x| & \text{if} \quad x \leq 1 \end{cases}$$

which is lower semicontinuous. The controller obtained using this viscosity supersolution is

$$\mu_\gamma^\infty(x) = \begin{cases} -\frac{2\gamma}{\sqrt{\gamma^2-1}} & \text{if} \quad x > 1 \\ -\frac{\gamma}{\sqrt{\gamma^2-1}}|x| & \text{if} \quad x \leq 1 \end{cases}$$

The worst disturbance is

$$\nu_\gamma^\infty(x) = \begin{cases} \frac{2}{\gamma\sqrt{\gamma^2-1}} & \text{if} \quad x > 1 \\ \frac{1}{\gamma\sqrt{\gamma^2-1}}|x| & \text{if} \quad x \leq 1 \end{cases}$$

It is now easy to see that the corresponding worst-case system trajectory

$$\dot{x}(t) = (\mu_\gamma^\infty(x(t)) + \nu_\gamma^\infty(x(t)))x(t)$$

is globally asymptotically stable.

6 Concluding Remarks

In this paper, we have studied the existence of viscosity supersolutions to HJI equations that arise in a class of nonlinear H^∞ control problems. We have shown that if the min-max value of a general cost function (not necessarily continuous) is finite, lower semicontinuous value functions (or more general lower semicontinuous storage functions) exist. The notion of an admissible γ-attenuation state feedback controller is introduced, and it is shown that a nonlinear H^∞ controller can be constructed by using only such a viscosity supersolution. It can furthermore be shown that such a controller is admissible when the value function is convex. If the convexity property does not hold, however, then verification of the existence of such a controller stands out as a challenging problem for the future. A possible approach here is to use the Continuous Selection Theorem or the Measurable Selection Theorem (see Deimling (1988)). Another approach is to use the Viability Theorem (see Aubin and Frankowska (1990)) in which case additional assumptions may be needed. Moreover one can establish a the relationship between storage functions, and viscosity and epicontingent supersolutions in the lower semicontinuous case (see Frankowska and Quincampoix (1996)).

Acknowledgments: This research was supported in part by the U.S. Department of Energy under Grant DE-FG-02-97-ER-13939, and in part by the National Science Foundation under Grant 93-12807.

References

Aubin, J. and Frankowska, H. (1990) *Set-Valued Analysis*, Birkhäuser, Boston, MA.

Başar, T. and Bernhard, P. (1995). *H^∞-Optimal Control and Related Minimax Design Problems*, Birkhäuser, Boston, MA.

Deimling, K. (1988) *Nonlinear function analysis*, Springer-Verlag.

Frankowska, H. and Quincampoix, M. (1996) *Dissipative control systems and disturbance attenuation for nonlinear H^∞-problem*, Preprints of 7th International Symposium on Differential Games and Applications, Kanagawa, Japan.

Fleming, W. and Soner, H. (1992). *Controlled Markov Processes and Viscosity Solutions*, Springer-Verlag.

Hill, D. and Moylan, P. (1976). *The stability of nonlinear dissipative systems*, IEEE Trans. Automat. Control **25**, 708-711.

James, M. (1993). *A partial differential inequality for dissipative nonlinear systems*, Systems & Control Letters **21**, 315-326.

Lions, P. (1982). *Generalized solution of Hamilton-Jacobi equations*, Pitman Publishing Inc.

McEneaney, W. (1995). *Uniqueness for viscosity solutions of nonstationary Hamilton-Jacobi-Bellman equations under some a priori conditions (with applications)*, SIAM J. Control and Optimization **33**, 1560-1576.

McEneaney, W. (1995). *Robust control and differential games on a finite time horizon*, Math. Control Signals Systems **8**, 138-166.

Soravia, P. (1996). *H_∞ control of nonlinear systems: differential games and viscosity solutions*, SIAM J. Control and Optimization **34**, 1071-1097.

Xiao, M. and Başar, T. (1996). *H^∞ control of a class of infinite-dimensional linear systems with nonlinear outputs*, Preprints of 7th International Symposium on Differential Games and Applications, Kanagawa, Japan.

Willems, J. (1973). *Dissipative dynamical systems*, Arch. Rat. Mech. Anal. **45**, 321-351.

Part II

Partial Differential Equations and Viscosity Solutions

10

Singularities of Semiconcave Functions in Banach Spaces

Paolo Albano
Piermarco Cannarsa

ABSTRACT: We study the structure of the set of points of nondifferentiability of semiconcave functions defined on an infinite dimensional space. We show that the singular set is $\infty - 1$ countably rectifiable. Moreover, we give a sufficient condition for a singular point to be a propagation point of the singular set. As an application we consider a Mayer Optimal Control problem for an evolution equation of parabolic type.

KEYWORDS: convex functions, Banach spaces, nonsmooth analysis, singularities, optimal control.

1 Introduction

The notion of semiconcave function $g : \Omega \to \mathbb{R}$, where Ω is an open subset of $\mathbb{R}^N$, is a natural extension of the notion of concave function. In fact, a special form of this property can be given saying that g admits a local representation as the sum of a concave function with a smooth function.
Semiconcave functions arise naturally in optimal control as this property is typical for the value function, see e.g. Bardi and Capuzzo Dolcetta (1997) and the references therein. Semiconcavity can also be used to guarantee uniqueness of weak solutions to first order partial differential equations, see Kruzhkov (1975) and Lions (1982).
With minor modifications, the definition of semiconcave function can be generalized to infinite dimensions, where Ω is an open subset of a Banach space X. In this case the theory is motivated by applications to the optimal control of partial differential equations, see e.g. Li and Yong (1995).
It is well known that, for a semiconcave function g on a Banach space X, a natural replacement of the gradient is the superdifferential D^+g. In fact, $D^+g(x)$ is a non–empty bounded closed convex subset of X^* for every point x. In this paper we study the set of points at which D^+g is multi–valued, the so–called *singularities* of g, that we denote by $\Sigma(g)$.
In the first place, we introduce a hierarchy in the singular set $\Sigma(g)$ according

to the dimension of the superdifferential D^+g. More precisely, using the notion of rectifiability that is recalled in section 2, we show that the set of points x such that $D^+g(x)$ contains an n–dimensional ball is $\infty - n$ rectifiable. Moreover, in the above statement, one can switch n and $\infty - n$ to obtain a complementary rectifiability result. Here, our approach is similar to the one developed by Zaíček (1978) for monotone operators, and by Alberti, Ambrosio and Cannarsa (1992) for semiconcave functions in finite dimensions. Finer results in the more restrictive class of concave functions were obtained by Zaíček (1979), Veselý (1986,1987), and Alberti (1994). Then, in section 4, we study the problem of the *propagation* of singularities for a semiconcave function g. In a finite dimensional setting, this problem was treated by Ambrosio, Cannarsa and Soner (1993) using geometric measure theory. Since their approach seems hard to extend to infinite dimensions, here we have to develop a somewhat new technique. It is interesting to point out that this method yields a result that seems to be new even in finite dimensions. In fact, for any point $x_0 \in \Sigma(g)$ we construct a Lipschitz singular arc $x(t)$ for g, such that $x(0) = x_0$ and $x(t) \neq x_0$ for $t \neq 0$. This construction is possible under suitable conditions on D^+g that are easy to check in applications.
We conclude this paper with an example of a Mayer optimal control problem in infinite dimensions. A semiconcavity result for the value function V of this problem, and of more general control systems, was obtained by Cannarsa and Frankowska (1996). Therefore, we can easily apply the abstract results of section 3 to the set $\Sigma(V)$. We note that the singular points of V are starting points of multiple optimal trajectories for the control problem. Moreover, using the fact that V satisfies the Hamilton–Jacobi–Bellman equation, we are able to check that the assumptions of our propagation theorem hold for the value function V.

2 Preliminaries

Throughout this paper we denote by $\mathbb{N} = \{1, 2, ...\}$ the set of natural numbers. Let X be a *real separable* Banach space with *separable dual* X^*, and let

$$E = \{e_j\}_{j \in \mathbb{N}}$$

be a dense subset of X. Then, X is an Asplund space (see e.g. Giles (1982)) and, in particular, X possesses an equivalent norm that is Fréchet differentiable on $X \setminus \{0\}$. We denote by $|\cdot|$ such a norm, by $\langle .,. \rangle$ the duality pairing between X and X^*, and by $|\cdot|_*$ the dual norm

$$|p|_* = \sup_{|x|=1} |\langle p, x \rangle|, \quad \forall p \in X^* .$$

If W is another Banach space, then we denote by $\mathcal{L}(X; W)$ the Banach space of all bounded linear operators $L : X \to W$, and by $||\cdot||$ the standard

norm in $\mathcal{L}(X;W)$.
For any $R > 0$ and $x \in X$ we set

$$B_R(x) = \{y \in X \,:\, |x - y| < R\}$$

and we abbreviate $B_R = B_R(0)$.
Let $\Omega \subset X$ be an open set. We denote by $C(\Omega;W)$ the space of all continuous functions $g : \Omega \to W$, and by $C^1(\Omega;W)$ the space of all functions $g \in C(\Omega;W)$ that are Fréchet differentiable at every point $x \in \Omega$ with continuous derivative, Dg. Furthermore, for any $\theta \in [0,1]$, $C^{1,\theta}(\Omega;W)$ stands for the set of all functions $g \in C^1(\Omega;W)$ with Hölder continuous derivative, that is

$$||Dg(x) - Dg(z)|| \leq C|x - z|^\theta \quad \forall x, z \in \Omega,$$

for some constant $C > 0$. Also, we define $C^{1,\theta}_{loc}(\Omega;W)$ to be the space of all functions $g \in C^1(\Omega;W)$ such that $g \in C^{1,\theta}(B_R(x);W)$ for all points $x \in \Omega$ and some spherical neighbourhood $B_R(x) \subset \Omega$. Finally, in the above notation, we drop the arrival set when $W = \mathbb{R}$.
We now recall the definition of a semiconcave function in Ω, a notion that is classical in finite dimensions and that was applied to infinite dimensional problems in Lions (1988) and Cannarsa (1989). We also introduce some special classes of such functions. A function $\omega : [0,\infty[\to [0,\infty[$ is called a *modulus* if ω is nondecreasing and $\omega(r) \to 0$ as $r \downarrow 0$.

Definition 2.1 *A function $g \in C(\Omega)$ is said to be* semiconcave *if a modulus ω exists such that, for any point $x_0 \in \Omega$ and some ball $B_R(x_0) \subset \Omega$,*

$$\lambda g(x) + (1-\lambda)g(y) - g(\lambda x + (1-\lambda)y) \leq C\lambda(1-\lambda)\omega(|x-y|)|x-y|$$

for all $\lambda[0,1]$, all $x, y \in B_R(x_0)$ and some constant $C = C(x_0, R) \geq 0$. We denote by $SC(\Omega)$ the set of all semiconcave functions $g : \Omega \to \mathbb{R}$. Moreover, for any $\alpha \in]0,1]$ we define $SC^\alpha(\Omega)$ to be the set of all functions $g \in SC(\Omega)$ with modulus $\omega(r) = r^\alpha$. Furthermore, we denote by $SC^\alpha_w(\Omega)$ the set of all functions $g \in SC^\alpha(\Omega)$ that are sequentially weakly upper semicontinuous in Ω.

Semiconcave functions share many properties with concave functions. For instance, it is easy to show that any semiconcave function in Ω is locally Lipschitz continuous. This fact can be proved arguing as in the finite dimensional case, see e.g. Fleming (1977) for $g \in SC^1(\Omega)$ and Alberti, Ambrosio and Cannarsa (1992) for a general g. Consequently, g is Fréchet differentiable on a dense set by a result of Preiss (1990), and the gradient of g, Dg, is locally bounded. Now, let us denote by $D_*g(x)$ the set of all points $p \in X^*$ for which a sequence $\{x_k\}_{k\in\mathbb{N}}$ exists in Ω so that

(i) $\quad x = \lim_{k\to\infty} x_k,$

(ii) $\quad g$ is Fréchet differentiable at x_k,

(iii) $Dg(x_k)$ weakly-* converges to p as $k \to \infty$.

In view of the above remarks we have that $D_*g(x) \neq \emptyset$ for any $x \in \Omega$. Just like the concave case, semiconcave functions possess a natural notion of generalized gradient, given by the *superdifferential*

$$D^+g(x) = \left\{ p \in X^* \ : \ \limsup_{y \to x} \frac{g(y) - g(x) - \langle p, y - x \rangle}{|y - x|} \leq 0 \right\}, \quad \forall x \in \Omega.$$

Actually, a similar generalization of the gradient is the *subdifferential* of g defined as

$$D^-g(x) = \left\{ p \in X^* \ : \ \liminf_{y \to x} \frac{g(y) - g(x) - \langle p, y - x \rangle}{|y - x|} \geq 0 \right\}.$$

However, for a semiconcave function, the superdifferential is much more interesting than the subdifferential. Indeed, in view of the proposition below, D^+g is non–empty at every point. Therefore, either D^-g is empty, or g turns out to be differentiable.

Proposition 2.2 *Let $g \in SC(\Omega)$ and let $x_0 \in \Omega$. Then $D^+g(x_0)$ is non–empty since*

$$D^+g(x_0) = \overline{\mathrm{co}}\, D_*g(x_0),$$

where $\overline{\mathrm{co}}$ denotes the closed convex hull. Moreover, there exists a ball $B_R(x_0)$ and a constant $C \geq 0$ such that

$$g(y) - g(x) - \langle p, y - x \rangle \leq C\omega(|y - x|)|y - x| \tag{1}$$

for all points $y, x \in B_R(x_0)$ and all functionals $p \in D^+g(x)$. Furthermore,

$$\langle p - q, y - x \rangle \leq 2C\omega(|y - x|)|y - x| \tag{2}$$

for all $y, x \in B_R(x_0)$ and all covectors $p \in D^+g(y)$, and $q \in D^+g(x)$.

For the proof of the above proposition the reader is referred to Cannarsa (1989) for semiconcave functions with a linear modulus ω, the argument in the general case being similar.
We now need to recall some known facts about topological complements. Let Y be a subspace of X. Then another subspace Z exists so that every element $x \in X$ can be uniquely represented as

$$x = x_Y + x_Z, \quad x_Y \in Y, x_Z \in Z.$$

In this case, we say that X is the *direct sum* of Y and Z, or $X = Y \oplus Z$, and we call such a subspace Z a *complement* of Y. Moreover, if the projections $\pi_Z : X \to Z$ and $\pi_Y : X \to Y$ are continuous, then we call Z a *topological complement* of Y. If $\dim Z < \infty$, we say that

$$\operatorname{codim} Y = \dim Z.$$

A subspace Y is said to be of codimension n (codim$Y = n$), if a subspace $Z \subset Y$ exists such that $X = Y \oplus Z$ and dim$Z = n$. Similar definitions will be used for subspaces of X^* as well.
For any $n \geq 1$ we denote by $\mathcal{F}_n(E)$ the family of all linear subspaces S of X, that are generated by n linearly independent elements of E. For any given $S \in \mathcal{F}_n(E)$, we set

$$S^{\perp} = \{p \in X^* : \langle p, x \rangle = 0, \forall x \in S\}.$$

The next two lemmas can be easily derived from the above definitions. We give a proof for the reader's convenience.

Lemma 2.3 *Let P be an n–dimensional subspace of X^*. Then, a topological complement $S^{\perp}$ of P exists for some $S \in \mathcal{F}_n(E)$, i.e.*

$$X^* = P \oplus S^{\perp} \tag{3}$$

where the projections $\pi_P : X^* \to P$ and $\pi_{S^{\perp}} : X^* \to S^{\perp}$ are continuous with respect to the norm $|\cdot|_*$ defined above.

Remark 2.4 Consider the quotient space $X^*/S^{\perp}$, of X^* modulo the equivalence relation

$$p, q \in X^*, \quad p \sim q \quad \Longleftrightarrow \langle p - q, x \rangle = 0\,, \quad \forall x \in S.$$

For any $p \in X^*$, we denote by $[p]_s = p + S^{\perp}$ the equivalence class of p modulo the above relation. From (3) it follows that dim $X^*/S^{\perp} = n$. Consequently, $X^*/S^{\perp}$ is separable. In the sequel, with any $S \in \mathcal{F}_n(E)$, we shall associate a dense countable subset $\{V_n(S)\}_{n \in \mathbb{N}}$ of $X^*/S^{\perp}$ without paying attention to the way this set is constructed.

Proof of Lemma 2.3 – In this proof we shall denote by $p(x)$ the value of a functional $p \in X^*$ at a point $x \in X$, instead of using the heavier notation $\langle p, x \rangle$. Let $p_1, \ldots, p_n \in X^*$ be such that

$$P = \text{ span } \{p_1, \ldots, p_n\}.$$

The crucial step of the proof is the construction of linearly independent vectors $e_{k_1}, \ldots, e_{k_n} \in E$ and covectors $q_1, \ldots q_n \in X^*$ such that

$$\text{span } \{p_1, \ldots, p_n\} = \text{ span } \{q_1, \ldots, q_n\} \quad \text{and} \quad q_i(e_{k_j}) = \delta_{ij} \tag{4}$$

where δ_{ij} is the Kronecker symbol. For this purpose, we proceed by induction on n. Let $n = 1$. Since E is dense in X, an element $e_{k_1} \in E$ exists such that $p_1(e_{k_1}) \neq 0$. Then

$$q_1(x) := \frac{p_1(x)}{p_1(e_{k_1})} \quad (x \in X)$$

satisfies (4). Now, suppose that our assertion holds for n and let $P =$ span $\{p_1, \ldots, p_n, p_{n+1}\}$. Then, without loss of generality, we can assume the existence of linearly independent vectors $e_{k_1}, \ldots, e_{k_n} \in E$ such that

$$p_i(e_{k_j}) = \delta_{ij} \text{ for } i, j = 1, \ldots, n.$$

Defining

$$\tilde{p}_{n+1}(x) := p_{n+1}(x) - \sum_{j=1}^{n} p_{n+1}(e_{k_j}) p_j(x), \quad x \in X,$$

we have that $\tilde{p}_{n+1}(e_{k_j}) = 0$ for $j = 1, \ldots, n$. Moreover, since E is dense in X, a vector $e_{k_{n+1}} \in E$ exists such that $\tilde{p}_{n+1}(e_{k_{n+1}}) \neq 0$. Therefore, taking

$$q_{n+1}(x) := \frac{\tilde{p}_{n+1}(x)}{\tilde{p}_{n+1}(e_{k_{n+1}})}, \quad x \in X,$$

we conclude that

$$q_{n+1}(e_{k_j}) = \delta_{n+1,j}.$$

Consequently, $e_{k_1}, \ldots, e_{k_{n+1}}$ are linearly independent. To obtain (4) for $n+1$, it then suffices to set

$$q_i(x) = p_i(x) - p_i(e_{k_{n+1}}) q_{n+1}(x), \quad i = 1, \ldots, n, \quad x \in X.$$

Next, taking $e_{k_1}, \ldots, e_{k_n} \in E$ and $q_1, \ldots q_n \in X^*$ as in (4), let us show that

$$X^* = \text{ span } \{q_1, \ldots q_n\} + (\text{span } \{e_{k_1}, \ldots, e_{k_n}\})^{\perp}.$$

Representing any $x \in X$ as

$$x = \sum_{i=1}^{n} q_i(x) e_{k_i} + \left(x - \sum_{i=1}^{n} q_i(x) e_{k_i} \right),$$

we deduce that any $h \in X^*$ can be split in the sum $h = h_1 + h_2$, where

$$h_1(x) = \sum_{i=1}^{n} q_i(x) h(e_{k_i}), \quad h_2(x) = h\left(x - \sum_{i=1}^{n} q_i(x) e_{k_i} \right).$$

Since $h_1 \in$ span $\{q_1, \ldots q_n\}$ and $h_2 \in$ (span $\{e_{k_1}, \ldots, e_{k_n}\})^{\perp}$, we has thus obtained (3). Finally, the continuity of $\pi_P : h \mapsto h_1$ and $\pi_{S^\perp} : h \mapsto h_2$ easily follows from the above construction. ∎

The proof of the following lemma is similar and will be omitted.

Lemma 2.5 *Let $E^* = \{e_j^* : j \in N\}$ be a dense subset of X^*. Then, for any subspace $P \subset X^*$ of codimension n, one can find a topological complement $U \in \mathcal{F}_n(E^*)$.*

We now proceed to recall the notion of rectifiable sets (see e.g. Federer (1969)). Let $n \in \mathbb{N}$ be fixed.

Definition 2.6 *We will say that a set $\Sigma \subset X$ is n* rectifiable *iff a bounded set $A \subset \mathbb{R}^n$ and a Lipschitz function, f, exist so that*

$$f : A \to \Sigma$$

is surjective. Moreover, we will say that a set $\Sigma \subset X$ is countably n rectifiable *iff*

$$\Sigma = \bigcup_{j \in \mathbb{N}} Q_j \tag{5}$$

where the sets $Q_j \subset X$, $j \in \mathbb{N}$, are n rectifiable.

By analogy with the previous definition, we will say that $\Sigma \subset X$ is $\infty - n$ rectifiable if a subspace $Y \subset X$, of codimension n, and a bounded set $A \subset Y$ exist so that Σ is the image of A under a Lipschitz map f. Similarly, we will say that Σ is countably $\infty - n$ rectifiable if it can be represented as in (5), for some family of $\infty - n$ rectifiable sets $\{Q_j\}$.

3 Singular Sets of Semiconcave Functions

In this section we study the singularities of a semiconcave function on an open set $\Omega \subset X$. Let P be a subspace of X^* and $r > 0$. We denote by $B_r^P \subset P$ the P–ball with radius r and center at 0. We say that B is a *ball of dimension* $n \in \mathbb{N}$ in X^* if there exist a subspace $P \subset X^*$, with $\dim P = n$, a covector $p \in X^*$, and a radius $r > 0$ such that

$$B = p + B_r^P \,. \tag{6}$$

Similarly, we say that B is a *ball of codimension* n in X^* if (6) holds for a subspace P such that $\operatorname{codim} P = n$. Let $\Omega \subset X$ be an open set and $g : \Omega \to \mathbb{R}$ be a semiconcave function. We introduce the following singular sets of g.

Definition 3.1 *The set of all points $x \in \Omega$ such that $D^+g(x)$ fails to be a singleton is called the* singular set *of g and is denoted by $\Sigma(g)$. The points of $\Sigma(g)$ are the* singular points *of g. Moreover, for any $n \in \mathbb{N}$, we denote by $\Sigma_n(g)$ (resp. $\Sigma_{\infty-n}(g)$) the set of all points $x \in \Omega$ such that $D^+g(x)$ contains a ball of dimension n (resp. codimension n).*

Notice that the above notation is somewhat redundant as $\Sigma(g) = \Sigma_1(g)$. The following theorem extends the result of Alberti, Ambrosio and Cannarsa (1992) from finite to infinite dimensions, and the results of Zaíček (1978) from concave to semiconcave functions.

Theorem 3.2 *Let $g \in SC(\Omega)$, and fix $n \in \mathbb{N}$. Then,*
(i) $\Sigma_n(g)$ is countably $\infty - n$ rectifiable;
(ii) $\Sigma_{\infty-n}(g)$ is countably n rectifiable.

For the proof of Theorem 3.2 we need the following lemma. We recall that $E = \{e_j\}_{j\in\mathbb{N}}$ is dense subset of X.

Lemma 3.3 *Let $g \in SC(\Omega)$, and $\rho > 0$ be a real number. Then for any $j \in \mathbb{N}$ there exist radii $R_j > 0$ and constants $C_j \geq 0$ such that*

$$\text{(i)} \quad \Omega \subset \bigcup_{j\in\mathbb{N}} B_{R_j}(e_j)$$

$$\text{(ii)} \quad g(x) + g(y) - 2g\left(\frac{x+y}{2}\right) \leq C_j|x-y|\omega(|x-y|), \quad \forall x, y \in B_{R_j}(e_j)$$

$$\text{(iii)} \quad C_j\omega(2R_j) \leq \rho$$

Proof of Lemma – Let j be a given natural number. We denote by $\mathcal{A}_j$ the set of all numbers $r \in]0,1]$ for which a constant $C \geq 0$ exists such that

$$\begin{cases} g(x) + g(y) - 2g\left(\frac{x+y}{2}\right) \leq C|x-y|\omega(|x-y|) & \forall x, y \in B_r(e_j) \\ C\omega(2r) \leq \rho. \end{cases} \tag{7}$$

Note that from the semiconcavity of g it follows that $\mathcal{A}_j \neq \emptyset$. Define

$$R_j := \frac{\sup \mathcal{A}_j}{2}$$

and denote by C_j the infimum of all constants $C \geq 0$ satisfying (7) with $r = R_j$. Let now $x \in \Omega$ be fixed. Then, from the semiconcavity of g we derive the existence of $r_x \in]0,1]$ and $C_x \geq 0$ such that

$$\begin{cases} g(z) + g(y) - 2g\left(\frac{z+y}{2}\right) \leq C_x|z-y|\omega(|z-y|), & z, y \in B_{r_x}(x) \\ C_x\omega(2r_x) \leq \rho \end{cases}$$

From the density of E in X we obtain that $k \in \mathbb{N}$ exists such that $|e_k - x| < r_x/3$. Then $B_{2r_x/3}(e_k) \subset B_{r_x}(x)$, and so $2r_x/3 \in \mathcal{A}_k$, which yields in turn $r_x/3 \leq R_k$. Therefore, by the choise of e_k, $x \in B_{R_k}(e_k)$ and the proof is complete. ■

Proof of Theorem 3.2 – To describe the idea underneath the proof, let us suppose that X is a Hilbert space and identify $X^* = X$. Then, for any fixed $x_0 \in \Sigma_n(g)$ we have that

$$p_0 + B_r^{P_0} \subset D^+g(x_0)$$

for some $p_0 \in X$ and some n–dimensional subspace $P_0 \subset X$. Suppose further that a sequence $\{x_k\} \subset \Sigma_n(g) \setminus \{x_0\}$ exists such that

$$x_k \to x_0 \qquad \text{as } k \to \infty$$

$$\frac{x_k - x_0}{|x_k - x_0|} \to \theta$$

$$p_0 + B_r^{P_0} \subset D^+ g(x_k).$$

Then, by (2), we have that

$$\langle p, x_k - x_0 \rangle \leq C|x_k - x_0|\omega(|x_k - x_0|)$$

for any $p \in B_r^{P_0}$. Therefore,

$$\langle p, \theta \rangle \leq 0, \qquad \forall p \in B_r^{P_0},$$

and, since this is also true for $-p$, we conclude that $\theta \perp P_0$. Consequently, near x_0, $\Sigma_n(g)$ should be orthogonal to P_0. So, one can try to describe $\Sigma_n(g)$, near x_0, as the graph of a Lipschitz map from $P_0^\perp$ to P_0.
Turning this simple geometric argument into a rigourous proof requires some technical work. This is the reason why we split our reasoning in three steps: in the first one we introduce a family $\{Q(r, S, V)\}$ of subset of Ω; in the second one we show that $Q(r, S, V)$ is $\infty - n$ rectifiable; in the last step we show that $\Sigma_n(g)$ can be covered by countably many sets Q_j of this family.

STEP 1: DEFINITION OF $Q(r, S, V)$.
Let $r \in \mathbb{Q}$, $r > 0$, $S \in \mathcal{F}_n(E)$ and $V \in X^*/S^\perp$. We denote by $Q(r, S, V)$ the set of all points $x \in \Omega$ for which one can find an n-dimensional subspace $P \subset X^*$ and two covectors $p, q \in X^*$ such that

$$\begin{array}{ll} \text{(a)} & X^* = P \oplus S^\perp \\ \text{(b)} & p + B_r^P \subset D^+ g(x) \\ \text{(c)} & |p|_* < \dfrac{1}{r} \\ \text{(d)} & \|\pi_P\| < \dfrac{1}{r} \\ \text{(e)} & q + S^\perp = V \\ \text{(f)} & q - p \in P, \quad |q - p|_* < \dfrac{r}{2} \end{array}$$

Notice that (b) and (f) imply that $q \in D^+ g(x)$.

STEP 2: RECTIFIABILITY OF $Q(r, S, V)$.
Let constants R_j, C_j be as in Lemma 3.3, with $\rho = \frac{r^2}{16}$. Since

$$Q(r, S, V) = \bigcup_{j \in \mathbb{N}} Q(r, S, V) \cap B_{R_j}(e_j)$$

it sufficies to show that each set

$$Q_j := Q(r, S, V) \cap B_{R_j}(e_j), \qquad j \in \mathbb{N}$$

is $\infty - n$ rectifiable. For this purpose, let Z be a complement of S, so that any vector $x \in X$ admits a unique representation as

$$x = x_s + x_z\,, \qquad x_s \in S\,,\, x_z \in Z\,.$$

We claim that the the projection π_Z is one–to–one on Q_j. Indeed, let $x, y \in Q_j$, and let p_x, q_x, P_x and p_y, q_y, P_y be associated with x and y according to Step 1. Let now $h \in X^*$ be such that

$$\langle h, y_s - x_s \rangle > \frac{|y_s - x_s|}{2}\,, \qquad |h|_* \leq 1\,, \tag{8}$$

(the existence of h is a standard consequence of the Hahn–Banach theorem). By Lemma 2.3, h can be represented as

$$h = h_1 + h_2 \qquad (h_1 \in P_y,\ h_2 \in S^\perp)\,.$$

From (d) above we deduce that $|\frac{r^2}{2}h_1|_* \leq \frac{r}{2}|h|_* \leq \frac{r}{2}$. Consequently, by (d) and (f),

$$q_y + \frac{r^2}{2}h_1 = p_y + (q_y - p_y) + \frac{r^2}{2}h_1 \in p_y + B_r^{P_y}\,,$$

so that, in light of (b), $q_y + \frac{r^2}{2}h_1 \in D^+g(y)$. Therefore, (2) and (iii) of Lemma 3.3 yield

$$\langle q_x - q_y - \frac{r^2}{2}h_1, x - y\rangle \leq 2C_j\omega(|x-y|)|x-y| \leq \frac{r^2}{8}(|x_z - y_z| + |x_s - y_s|)\,. \tag{9}$$

Moreover, $q_x - q_y \in S^\perp$ in view of (e), and so, recalling (8),

$$\begin{aligned} &\langle q_x - q_y - \tfrac{r^2}{2}h_1, x - y\rangle \\ &= \langle q_x - q_y - \tfrac{r^2}{2}h_1, x_z - y_z\rangle + \tfrac{r^2}{2}\langle h_1, y_s - x_s\rangle \\ &> -|q_x - q_y - \tfrac{r^2}{2}h_1|_*|x_z - y_z| + \tfrac{r^2}{4}|x_s - y_s|\,. \end{aligned} \tag{10}$$

Furthermore, properties (c), (d), (f) imply that

$$\begin{aligned} |q_x - q_y - \tfrac{r^2}{2}h_1|_* &\leq |q_x - p_x|_* + |p_x|_* + |p_y|_* + |q_y - p_y|_* + \tfrac{r^2}{2}|h_1|_* \\ &< \tfrac{3}{2}r + \tfrac{2}{r}\,. \end{aligned}$$

Then, from (9) and (10) we conclude that, for any $x, y \in Q_j$,

$$\frac{r^2}{8}|x_s - y_s| \leq (\frac{r^2}{8} + \frac{3}{2}r + \frac{2}{r})|x_z - y_z|\,.$$

From the last inequality it follows that π_Z is one–to–one on Q_j, and that π_Z^{-1} is Lipschitz on $\pi_Z(Q_j)$. Since $\pi_Z(Q_j) \subset Z$ is bounded and $\operatorname{codim} Z = n$, we have thus proved that Q_j is $\infty - n$ rectifiable.

STEP 3: A COVERING OF $\Sigma_n(g)$.
Let $x \in \Sigma_n(g)$. Then, by the very definition of $\Sigma_n(g)$, one can find a rational number $r > 0$, a covector $p \in X^*$, and an n-dimensional subspace $P \subset X^*$ so that

$$p + B_r^P \subset D^+g(x) \text{ and } |p|_* < \frac{1}{r}.$$

Moreover, by Lemma 2.3 P has a topological complement $S^\perp$; equivalently, a subspace $S \in \mathcal{F}_n(E)$ exists such that, after possibly reducing r,

$$X^* = P \oplus S^\perp, \qquad \|\pi_P\| < \frac{1}{r}.$$

Now, let $\{V_j(S)\}_{j\in\mathbb{N}}$ be a dense subset of $X^*/S^\perp$, chosen as in Remark 2.4, and recall that $p \mapsto [p]_S$ denotes the canonical projection of X^* onto $X^*/S^\perp$. Then, an element $V_{j_0}(S) = [\tilde{q}]_S$, $\tilde{q} \in X^*$, exists so that

$$|p - \tilde{q}|_* < \frac{r^2}{2}.$$

We claim that $x \in Q(r, S, V_{j_0}(S))$. Indeed, let us define $q = \tilde{q}_1 + p_2$, where $\tilde{q}_1$ and p_2 are obtained from the representations

$$p = p_1 + p_2, \qquad (p_1 \in P,\ p_2 \in S^\perp), \qquad \tilde{q} = \tilde{q}_1 + \tilde{q}_2, \qquad (\tilde{q}_1 \in P,\ \tilde{q}_2 \in S^\perp).$$

Then, $[q]_S = V_{j_0}(S)$, and $q - p = \tilde{q}_1 - p_1 \in P$. Moreover,

$$|q - p|_* = |\pi_P(\tilde{q} - p)|_* \leq \frac{1}{r}|\tilde{q} - p|_* < \frac{r}{2},$$

and our claim follows. Finally, as $\mathcal{F}_n(E)$ is countable, we have shown that

$$\{Q(r, S, V_j(S)) \ : r \in \mathbb{Q}^+,\ S \in \mathcal{F}_n(E),\ j \in \mathbb{N}\}$$

is a countable covering of $\Sigma_n(g)$.
We have thus proved the part of our conclusion that concerns $\Sigma_n(g)$. The analogous assertion concerning $\Sigma_{\infty-n}(g)$ can be shown in a similar way using Lemma 2.5 instead of Lemma 2.3. ∎

Remark 3.4 We note that the above proof makes no use of the separability of X^*. Therefore, part (i) of the conclusion of Theorem 3.2 holds under weaker assumptions. However, we do need the separability of X^* for (ii).

4 Propagation of singularities

In this section we focus our attention on the behaviour of a semiconcave function g about a singular point x_0. In particular, we are interested in giving conditions to ensure that a singularity propagates in a suitable sense. In the finite dimensional case, such a problem was studied by Ambrosio, Cannarsa and Soner (1993). A first step for the development of a theory of propagation of singularities in infinite dimensions is represented by the theorem below, that is new even in finite dimensions. We will restrict our set–up assuming that X be a Hilbert space.

Theorem 4.1 *Let X be a Hilbert space, let $g \in SC^1(\Omega)$, and let $x_0 \in \Sigma(g)$. Let*

$$p_0 \in D^+g(x_0) \setminus D_*g(x_0)\,, \tag{12}$$

and suppose that, for some vector $q \in X \setminus \{0\}$ and some $T_0 > 0$,

$$p_0 + tq \notin D^+g(x_0)\,, \quad \forall 0 < t \leq T_0\,. \tag{13}$$

Then, a number $0 < T \leq T_0$ and two Lipschitz arcs $x, p : [0, T] \to X$ exist so that

$$\begin{aligned} &\text{(i)} \quad \langle x(t) - x_0, q\rangle < 0, \ \ \forall 0 < t \leq T \\ &\text{(ii)} \quad (x(0), p(0)) = (x_0, p_0) \\ &\text{(iii)} \quad (x(t), p(t)) \in \Sigma(g) \times D^+g(x(t)), \quad \forall t \in [0, T]\,. \end{aligned} \tag{14}$$

A point $x_0 \in \Sigma(g)$ for which x and p exist satisfying (14) will be called a *propagation point* of $\Sigma(g)$.

Proof – First of all, we recall that, by Proposition 2.2, a number $R > 0$ and a constant $C \geq 0$ exist such that

$$g(x) - g(x_0) - \langle p_0, x - x_0\rangle - C|x - x_0|^2 \leq 0 \tag{15}$$

and

$$\langle p_x - p_y, x - y\rangle \leq 2C|x - y|^2\,, \tag{16}$$

for all points $x, y \in \overline{B}_R(x_0)$ and all vectors $p_x \in D^+g(x), p_y \in D^+g(y)$. Now, define

$$\phi(x) = g(x) - g(x_0) - (C + 1)|x - x_0|^2 \quad (x \in \Omega)\,.$$

It is easy to see $\phi \in SC^1(\Omega)$ and $D^+\phi(x_0) = D^+g(x_0)$. Then, by assumption (13), for any $t \in]0, T_0]$ a point $y_t \in B_R(x_0)$ exists such that

$$\psi_t(y_t) = \phi(y_t) - \langle p_0 + tq, y_t - x_0\rangle > 0\,. \tag{17}$$

Now, define, for $t \in]0, T_0]$,

$$\psi_t(x) = \phi(x) - \langle p_0 + tq, x - x_0\rangle\,, \quad x \in \overline{B}_R(x_0)\,.$$

We note that ψ_t is weakly upper semicontinuous. So, ψ_t has a maximum at some point $x(t) \in \overline{B}_R(x_0)$. Moreover, $\psi_t(x(t)) > 0$ in view of (17). Now, (15), with $x = x(t)$, implies (i) since

$$0 < \psi_t(x(t)) \leq -t\langle q, x(t) - x_0\rangle - |x(t) - x_0|^2 ,$$

as well as the estimate

$$0 < |x(t) - x_0| < t|q| , \qquad \forall t \in]0, T_0] . \tag{18}$$

Moreover, $0 \in D^+\psi_t(x(t))$ for $0 < |t| \leq \min\{T_0, R\}$ since, by (18), $x(t)$ lies in the interior of $B_R(x_0)$. Thus, for any such t,

$$p(t) = p_0 + tq + 2(C+1)[x(t) - x_0] \in D^+g(x(t)) . \tag{19}$$

Incidentally, we note that, defining $x(0)$ and $p(0)$ as in (ii), the above equation is satisfied for $t = 0$ as well.

We claim that $0 < T \leq \min\{T_0, R\}$ exists such that $x(t) \in \Sigma(g)$ for any $t \in [0, T]$. In fact, let g be differentiable at $x(t_k)$, for some sequence $t_k \downarrow 0$.Then, by (19) and (18),

$$\lim_{k\to\infty} Dg(x(t_k)) = \lim_{k\to\infty} \{p_0 + t_k q + 2(C+1)[x(t_k) - x_0]\} = p_0 ,$$

and so $p_0 \in D_*g(x_0)$, in contradiction with the choice of p_0. This proves our claim and (iii).

We now proceed to show that $x(\cdot)$ is Lipschitz continuous on $[0, T]$. Let $0 \leq s < t \leq T$. Subtracting from (19) the same equation evaluated at s, we obtain

$$(t-s)q + 2(C+1)[x(t) - x(s)] = p(t) - p(s) . \tag{20}$$

Hence, taking the scalar product with $x(t) - x(s)$, we have

$$\begin{aligned}(t-s)\langle q, x(t) - x(s)\rangle + 2(C+1)|x(t) - x(s)|^2 \\ = \langle p(t) - p(s), x(t) - x(s)\rangle \leq 2C|x(t) - x(s)|^2 ,\end{aligned}$$

where we have also used estimate (16). Therefore, by an easy computation, $|x(t) - x(s)| \leq |q|\frac{t-s}{2}$ as desired. Moreover, (20) implies that $p(\cdot)$ is Lipschitz continuous on $[0, T]$ as well. This ends our proof. ■

Remark 4.2 **(a)** Assumption (12) cannot be dropped. This fact can be easily checked looking at behaviour of the concave function of two real variables $g(x, y) = -\sqrt{x^2 + y^4}$ about the origin.

(b) Assumptions (12) and (13) colud be rephrased requiring that the topological boundary of $D^+g(x_0)$ be not entirely covered by points of $D_*g(x_0)$. On the other hand, such a covering exists for the concave function $g(x) = -|x|$ ($x \in \mathbb{R}$) at $x_0 = 0$, which gives another example of an isolated singularity.

(c) Notice that (i) forces, in particular, $x(t) \neq x_0$ for any $t \in]0, T]$.

The same technique used in the proof of the theorem above yields the following propagation result in the space $SC^{\alpha}_{w}(\Omega)$, $\alpha \in]0,1[$. In this case, however, we are not able to construct any singular arc.

Theorem 4.3 *Let X be a Hilbert space, let $g \in SC^{\alpha}_{w}(\Omega)$, and let $x_0 \in \Sigma(g)$. Let*

$$p_0 \in D^+g(x_0) \setminus D_*g(x_0)\,,$$

and suppose that, for some vector $q \in X \setminus \{0\}$ and some $T_0 > 0$,

$$p_0 + tq \notin D^+g(x_0)\,, \quad \forall 0 < t \leq T_0\,.$$

Then, x_0 is a cluster point for the singular set $\Sigma(g)$, i.e. a sequence $\{x_n\}$ exists such that $x_n \in \Sigma(g) \setminus \{x_0\}$, for all $n \in \mathbb{N}$, and $\lim_{n\to\infty} x_n = x_0$.

5 Application to Optimal Control

In this section we describe an application of the general results on semiconcave functions to a simple example of Mayer problem in optimal control. Let X be a real Hilbert space, and let $r, T > 0$ be given. For any $(s,x) \in [0,T] \times X$ consider a control system governed by the *state equation*

$$\begin{cases} y'(t) = Ay(t) + u(t), & t \in [s,T] \\ y(s) = x. \end{cases} \tag{21}$$

In this section we impose the following assumptions on the data of equation (21):

$$\begin{cases} i) & A : D(A) \subset X \to X \text{ is the infinitesimal generator of an} \\ & \text{analytic semigroup }, e^{tA}, \text{ such that } \|e^{tA}\| \leq e^{-\omega t} \text{ for} \\ & \text{all } t \geq 0 \text{ and some } \omega > 0\,; \\ ii) & e^{tA} \text{ is compact in } X \text{ for any } t > 0; \\ iii) & u : [s,T] \to \overline{B}_r \text{ is a measurable function, called } \textit{control}. \end{cases} \tag{22}$$

Remark 5.1 We note that, under our assumption (22) for any $\alpha \in [0,1]$ the fractional power of $-A$, $(-A)^{\alpha}$ is well defined (see Pazy (1983)). Moreover, we will denote by $D(-A)^{\alpha}$ the domain of $(-A)^{\alpha}$.

For every control u, we denote by $y(\cdot; s, x, u)$ the mild solution

$$y(t) = e^{(t-s)A}x + \int_s^t e^{(t-\sigma)A}u(\sigma)d\sigma$$

of problem (21).

Let now $g : X \to \mathbf{R}$ be a given continuous function. We are interested in the Mayer optimal control problem below:

$$\text{minimize } g(y(T; s, x, u)) \text{ over all controls } u. \tag{23}$$

A control u at which the above minimum is attained is said to be *optimal* in (s, x) and the corresponding solution of (21) is called an *optimal trajectory*. We will impose the following assumptions on g:

$$\begin{cases} i) & g \in C^1(X); \\ & \\ ii) & \text{for any } R > 0 \text{ a constant } C_R > 0 \text{ exists such that} \\ & g(x) + g(y) - 2g\left(\frac{x+y}{2}\right) \leq C_R|x-y|^2 \quad \forall x, y \in B_R\,. \end{cases} \tag{24}$$

The example below describes a simple model problem that fits in the above abstract framework. For more details and examples the reader is referred to the recent monography Li and Yong (1995).

Example 5.2 Let us consider a system whose state $y(t, \xi)$ is described by the parabolic equation

$$\begin{array}{ll} \frac{\partial y}{\partial t}(t,\xi) = \frac{\partial^2 y}{\partial^2 \xi}(t,\xi) + u(t,\xi)\,, & (t,\xi) \in]s,T[\times]0,\pi[\\ y(t,0) = y(t,\pi) = 0\,, & t \in]s,T[\\ y(s,\xi) = x(\xi)\,, & \xi \in]0,\pi[\end{array} \tag{25}$$

where $x \in L^2(0,\pi)$ and $\int_0^\pi |u(t,\xi)|^2 d\xi \leq r^2$ for a.e. $t \in [s,T]$. The above equation is related, for instance, to the evolution of the temperature distribution in a uniform rod whose end points are kept at 0 temperature. Let us consider the problem of minimizing the functional

$$g(y(T,\cdot)) := \int_0^\pi [|y(T,\xi)|^2 - |y(T,\xi)|^{3/2}]d\xi \tag{26}$$

over all trajectory–control pairs $\{y, u\}$ of equation (25). This problem can be put in the abstract form (21)–(23) taking $X = L^2(0,\pi)$ and

$$D(A) = H^2(0,\pi) \cap H_0^1(0,\pi)\,; \quad Ax = \frac{d^2x}{d\xi^2}.$$

Moreover, we note that the function g given by (26) satisfies (24).

Remark 5.3 In this paper we have considered a linear state equation just for simplicity. All the result we obtain in this section hold for semilinear control systems

$$\begin{cases} y'(t) = Ay(t) + f(t, y(t), u(t)), & t \in [s,T] \\ y(s) = x \end{cases}$$

under suitable smoothness assumptions on the nonlinear term f.

The Dynamic Programming approach to problem (21)–(23) is based on the properties of the *value function* V, defined as

$$V(s,x) = \inf_u g(y(T;s,x,u)).$$

We recall below some of these properties, that are more relevant for the application of our abstract results. The following proposition is proved in Cannarsa and Frankowska (1996).

Proposition 5.4 *Assume* (22) *and* (24), *and let* $R > \frac{1}{T}$ *and* $\alpha \in [0,1[$ *be fixed. Then, a constant* $C > 0$ *exists such that*

$$|V(s_1,x_1) - V(s_0,x_0)| \leq C[|s_1 - s_0| + |(-A)^{-\alpha}(x_1 - x_0)|]$$

and

$$V(s_1,x_1) + V(s_0,x_0) - 2V\left(\frac{s_1+s_0}{2}, \frac{x_1+x_0}{2}\right) \leq C(|s_1 - s_0| + |x_1 - x_0|)^2$$

for all $s_1, s_0 \in [0, T - \frac{1}{R}]$ *and all* $x_1, x_0 \in B_R$. *In particular,*

$$V \in SC^1(]0,T[\times X).$$

From the above proposition and Theorem 3.2 we immediately obtain the following

Corollary 5.5 *Let* V *be the value function defined above. Then,*
(i) $\Sigma_n(V)$ *is countably* $\infty - n$ *rectifiable;*
(ii) $\Sigma_{\infty-n}(V)$ *is countably* n *rectifiable.*

We now turn to the application of Theorem 4.1, concerning propagation of singularities, which requires conditions (12) and (13) to be checked. For this purpose, we will use the so-called Dynamic Programming (or Hamilton–Jacobi–Bellman) equation of problem (21)–(23)

$$\begin{cases} V_s + \langle V_x, Ax\rangle = r|V_x| \\ V(T,x) = g(x). \end{cases}$$

In order to give a precise sense to the above equation, let us recall another result from Cannarsa and Frankowska (1996).

Lemma 5.6 *Assume* (22) *and* (24). *Then, for every* $\alpha \in]0,1[$ *and every* $(s,x) \in [0,T] \times D((-A)^{1-\alpha})$,

$$p_s - \langle (-A^*)^\alpha p_x, (-A)^{1-\alpha} x\rangle = r|p_x|$$

for all $(p_s, p_x) \in D_* V(s,x)$.

The main result of this section requires the notion of propagation point of the set $\Sigma(V)$ introduced in section 4.

Theorem 5.7 *Assume* (22) *and* (24). *Let* $(s,x) \in \Sigma(V)$ *be a singular point such that* $x \in D((-A)^{1-\alpha})$ *for some* $0 < \alpha \leq 1$. *Then* (s,x) *is a propagation point of the singular set* $\Sigma(V)$.

In order to apply Theorem 4.1 we have to show that

$$D^+V(s,x) \setminus D_*V(s,x) \neq \emptyset. \tag{27}$$

For this purpose we need the following optimality conditions.

Lemma 5.8 *Assume* (22) *and* (24). *Let* $(s,x) \in [0,T] \times X$ *and fix a vector* $(p_s, p_x) \in D_*V(s,x)$ *with* $p_x \neq 0$. *Then the Hamiltonian system*

$$\begin{cases} y'(t) = Ay(t) + r\,\dfrac{p(t)}{|p(t)|}\,, & y(s) = x \\ p'(t) = -A^*p(t)\,, & p(s) = p_x\,, \qquad p(T) = -Dg(y(T)) \end{cases} \tag{28}$$

has a unique solution (y,p). *Moreover, y is an optimal trajectory for problem* (21)–(23) *in* (s,x).

Most of the results contained in the above lemma are well known. As for the optimality of y, this is a special case of a result obtained by Cannarsa and Frankowska (1996).
We are now able to show that V satisfies condition (27).

Lemma 5.9 *Let* $(s,x) \in [0,T] \times D((-A)^\alpha)$ *be a singular point of* V. *Then*

$$D^+V(s,x) \setminus D_*V(s,x) \neq \emptyset.$$

Proof – We argue by contradiction assuming that

$$D^+V(s,x) = D_*V(s,x).$$

Then Lemma 5.6 yields

$$p_s - \langle (-A^*)^\alpha p_x, (-A)^{1-\alpha}x \rangle = r|p_x|\,, \quad \forall (p_s,p_x) \in D^+V(s,x)\,. \tag{29}$$

We claim that the set $D^+V(s,x)$ is a segment. Indeed, let (p_s,p_x) and (p_s,p'_x) be any two points of $D^+V(s,x)$ and define $p_x^\lambda = \lambda p_x + (1-\lambda)p'_x$, $\lambda \in [0,1]$. Since $D^+V(s,x)$ is a convex set, $(p_s,p_x^\lambda) \in D^+V(s,x)$ for any $\lambda \in [0,1]$. Now, equation (29) implies that

$$\begin{aligned} p_s &= \langle (-A^*)^\alpha p_x^\lambda, (-A)^{1-\alpha}x \rangle + r|p_x^\lambda| \\ p_s &= \langle (-A^*)^\alpha p_x, (-A)^{1-\alpha}x \rangle + r|p_x| \\ p_s &= \langle (-A^*)^\alpha p'_x, (-A)^{1-\alpha}x \rangle + r|p'_x|\,. \end{aligned}$$

Hence,

$$|p_x^\lambda| = \lambda|p'_x| + (1-\lambda)|p_x|\,,$$

and so $p'_x = \rho p_x$ for some $\rho > 0$. Moreover, $\rho = 1$ as, from equation (29),

$$\langle(-A^*)^\alpha p_x, (-A)^{1-\alpha}x\rangle + r|p_x| = \rho\langle(-A^*)^\alpha p_x, (-A)^{1-\alpha}x\rangle + \rho r|p_x| .$$

We have thus proved that for any number p_s there exists only one vector p_x such that $(p_s, p_x) \in D^+V(s,x)$. In view of the convexity of $D^+V(s,x)$ our claim follows.
Now, let us suppose that

$$D^+V(s,x) \supset \{\lambda(p_s,p_x) + (1-\lambda)(p'_s,p'_x) \ : \ \lambda \in [0,1]\}$$

for some (p_s, p_x) and (p'_s, p'_x) in $D^+V(s,x)\setminus\{0\}$. Then, using equation (29) as in the first part of this proof, we conclude that

$$(p'_s, p'_x) = \mu(p_s, p_x)$$

for some $\mu \in]0,1]$.
We claim that $\mu = 1$. Indeed, let $(y(\cdot), p(\cdot))$ and $(\overline{y}(\cdot), \overline{p}(\cdot))$ be the solutions of equation (28) with initial conditions $(y(s), p(s)) = (x, p_x)$ and $(\overline{y}(s), \overline{p}(s)) = (x, \mu p_x)$, respectively. Then, $\overline{p}(\cdot) = \mu p(\cdot)$ and $\overline{y}(\cdot) = y(\cdot)$, as the first equation in (28) is homogeneus of degree 0 with respect to p. Now, recalling the transversality condition in (28), we obtain

$$p(T) = -Dg(y(T)) = \overline{p}(T) = \mu p(T) .$$

Therefore, $\mu = 1$ as claimed, and $D^+V(s,x) = \{(p_s, p_x)\}$.
To conclude our proof it is enough to note that $D^+V(s,x) = \{(p_s, p_x)\}$ implies that $(s,x) \notin \Sigma(V)$, in contradiction with our assumption. ∎

Proof of Theorem 5.7 – We have to construct vectors p^0 and q satisfying the assumptions of Theorem 4.1.
From Lemma 5.9 we derive the existence of a vector

$$p^1 \in D^+V(s,x) \setminus D_*V(s,x) .$$

Clearly, (13) is satisfied if

$$(p^1_s + t, p^1_x) \notin D^+V(s,x), \quad \forall t > 0, \tag{30}$$

or

$$(p^1_s - t, p^1_x) \notin D^+V(s,x), \quad \forall t > 0. \tag{31}$$

In fact, one can take $p^1 = p^0$, and $q = (1,0)$ (resp. $q = (-1,0)$) if (30) (resp. (31)) holds true.
Suppose, on the other hand, that positive numbers $t_+, t_- > 0$ exist so that $(p^1_s \pm t_\pm, p^1_x) \in D^+V(s,x)$, and define

$$p^+_s = \max\{p_s \in \mathbb{R} \ : \ (p_s, p^1_x) \in D^+V(s,x)\} ,$$
$$p^-_s = \min\{p_s \in \mathbb{R} \ : \ (p_s, p^1_x) \in D^+V(s,x)\} .$$

We note that $p_s^+ > p_s^-$, and $(p_s^{\pm}, p_x^1) \in D^+V(s,x)$, as the superdifferential is a closed set.
We claim that at least one of these two points lies in $D^+V(s,x)\setminus D_*V(s,x)$. For if $(p_s^{\pm}, p_x^1) \in D_*V(s,x)$, then Lemma 5.6 would imply that $p_s^+ = p_s^-$, a contradiction.
We have thus found a vector, say (p_s^+, p_x^1), such that $(p_s^+, p_x^1) \in D^+V(s,x)\setminus D_*V(s,x)$. Moreover, (p_s^+, p_x^1) satisfies (30) by construction. Therefore, Theorem 4.1 can be applied once again taking $p^0 := (p_s^+, p_x^1)$ and $q = (1,0)$, and the proof is complete. ■

References

Alberti, G. (1994). On the structure of singular sets of convex functions, *Calc. Var. Partial Differential Equations* **2**, 17-27.

Alberti, G., Ambrosio, L. and Cannarsa, P. (1992). On the singularities of convex functions, *Manuscripta Math.* **76**, 421-435.

Ambrosio, L., Cannarsa, P. and Soner, H.M. (1993). On the propagation of singularities of semi–convex functions, *Annali Scuola Norm. Sup. Pisa Scienze Fisiche e Matematiche* - Serie IV **20**, 597-616.

Bardi, M. and Capuzzo Dolcetta, I. (1997). *Optimal control and viscosity solutions of Hamilton–Jacobi–Bellman equations*, Birkhäuser, Boston.

Cannarsa, P. (1989). Regularity properties of solutions to Hamilton Jacobi equations in infinite dimensions an nonlinear optimal control, *Differential and Integral Equations* **2** , 479-493.

Cannarsa, P. and Frankowska, H. (1992). Value functions and optimality conditions for semilinear control problems, *Appl. Math. Optim.* **26**, 139-169.

Cannarsa, P. and Frankowska, H. (1996). Value functions and optimality conditions for semilinear control problems II: parabolic case, *Appl. Math. Optim.* **33**, 1-33.

Cannarsa, P. and Tessitore, M. E. (1997). On the behaviour of the value function of a Mayer optimal control problem along optimal trajectories, preprint.

Clarke, F.H. (1983). *Optimization and nonsmooth analysis*, Wiley & Sons, New York.

Federer, H. (1969). *Geometric measure theory* , Springer, New York.

Fleming, W.H. (1977). *Functions of several variables* , Springer, New York.

Giles, J.R. (1982). *Convex analysis with application in differentiation of convex functions*, Pitman, Boston.

Kruzhkov, S.N. (1975). Generalized solutions of Hamilton-Jacobi equations of the eikonal type I, *Math. USSR Sb.* **27**, 406-445.

Li, X. and Yong, J. (1995). *Optimal control theory for infinite dimensional systems*, Birkhäuser, Boston.

Lions, P.L. (1988). Viscosity solutions of fully non-linear second-order equations and optimal stochastic control in infinite dimensions. Part I: the case of bounded stochastic evolutions, *Acta Math.* **161**, 243-278.

Lions, P.L. (1982). *Generalized solutions of Hamilton-Jacobi equations*, Pitman, Boston.

Pazy, A. (1983). *Semigroups of linear operators and applications to partial differential equations*, Springer-Verlag, New York-Heidelberg- Berlin.

Preiss, D. (1990). Differentiability of Lipschitz functions on Banach spaces, *J. Funct. Anal.* **91**, 312-345.

Veselý, L. (1986). On the multiplicity points of monotone operators on separable Banach spaces, *Commentationes Mathematicae Universitatis Carolinae* **27**, 551-570.

Veselý, L. (1987). On the multiplicity points of monotone operators on separable Banach spaces II, *Commentationes Mathematicae Universitatis Carolinae* **28**, 295-299.

Zaíček, L. (1978). On the points of multiplicity of monotone operators, *Commentationes Mathematicae Universitatis Carolinae* **19**, 179-189.

Zaíček, L. (1979). On the differentiation of convex functions in finite and infinite dimensional spaces, *Czechoslovak Mathematical Journal* **29**, 340-348.

11

Invariant Sets for Controlled Degenerate Diffusions: A Viscosity Solutions Approach

Martino Bardi
Paola Goatin

ABSTRACT: We study invariance and viability properties of a closed set for the trajectories of either a controlled diffusion process or a controlled deterministic system with disturbances. We use the value functions associated to suitable optimal control problems or differential games and analyze the related Dynamic Programming equation within the theory of viscosity solutions.

KEYWORDS: degenerate diffusion; invariance; viability; stochastic control; differential games; viscosity solutions; Hamilton-Jacobi-Bellman equations; nonsmooth analysis.

1 Introduction

Consider the controlled Ito stochastic differential equation in $\mathbb{R}^N$

$$(SDE)\begin{cases} dX_t = \sigma^{\alpha_t}(X_t)dB_t + f^{\alpha_t}(X_t)dt, & t > 0, \\ X_0 = x. \end{cases} \tag{1.1}$$

where B_t is an M-dimensional Brownian motion and α_t is the control taking values in a given set A. A set K is *invariant* for (SDE) if for all initial points $x \in K$ and all admissible controls α, the trajectory X_t of (SDE) remains in K for all $t > 0$ almost surely. One of the main results of this paper is a characterization of the closed invariant sets for (SDE) under general assumptions on the data. Of course this problem is interesting when the diffusion process described by (SDE) is degenerate, namely, the matrix $\sigma\sigma^T$ is merely positive semidefinite (in the nondegenerate case the only invariant set is the whole $\mathbb{R}^N$). From now on we will assume $K \subseteq \mathbb{R}^N$ is a closed set.
Let us recall the classical theorem of Nagumo (1942) that solves this problem in the deterministic case (i.e., $\sigma \equiv 0$) and without controls, that is, for

the ODE

$$\begin{cases} X_t' = f(X_t), & t > 0, \\ X_0 = x. \end{cases}$$

In this case K *is invariant if and only if* $f(x) \in T_K(x)$ *for all* $x \in \partial K$, where $T_K(x)$ is the Bouligand-Severi contingent cone to K at x (see Sect. 4 for the definition; $T_K(x)$ coincides with the tangent space if K is a smooth manifold near x). An equivalent characterization can be given in terms of the cone of generalized interior normal vectors, or *1st order normal cone*,

$$\mathcal{N}_K^1(x) := \left\{ p \in \mathbb{R}^N : \; p \cdot (y - x) \geq o(|y - x|) \; \text{ as } \; K \ni y \to x \right\}.$$

This is the positive polar cone of $T_K(x)$, and it reduces to the half line generated by the interior normal to K at x if K is the closure of an open set with smooth boundary and $x \in \partial K$. We can reformulate Nagumo's theorem as follows:

K *is invariant if and only if* $f(x) \cdot p \geq 0$ *for all* $p \in \mathcal{N}_K^1(x)$ *and all* $x \in \partial K$.

For deterministic systems with control, that is,

$$\begin{cases} X_t' = f(X_t, \alpha_t), & t > 0, \\ X_0 = x. \end{cases} \tag{1.2}$$

the property characterizing the invariance of a closed set K becomes

$$f(x, a) \cdot p \geq 0 \quad \forall\, a \in A, \; p \in \mathcal{N}_K^1(x), \; x \in \partial K. \tag{1.3}$$

The corresponding result for the controlled diffusion process (SDE) requires the following *2nd order normal cone*

$$\begin{aligned} \mathcal{N}_K^2(x) \; := \; & \{(p, Y) \in \mathbb{R}^N \times S(N) : \; \text{for } K \ni y \to x \\ & p \cdot (y - x) + \frac{1}{2}(y - x) \cdot Y(y - x) \geq o(|y - x|^2)\}, \end{aligned}$$

where $S(N)$ is the set of symmetric $N \times N$ matrices. The relation between the 1st and the 2nd order normal cones is transparent, in particular $p \in \mathcal{N}_K^1(x)$ if $(p, Y) \in \mathcal{N}_K^2(x)$. Here is the Invariance Theorem for (SDE):

all the trajectories of (SDE) *starting in* K *remain forever in* K *a.s. if and only if*

$$f^\alpha(x) \cdot p + \frac{1}{2} trace(\sigma^\alpha(x) \sigma^\alpha(x)^T Y) \geq 0 \quad \forall \alpha \in A, \; (p, Y) \in \mathcal{N}_K^2(x), \; x \in \partial K. \tag{1.4}$$

This result is better understood if K is the closure of an open set whose boundary is twice differentiable at the point $x \in \partial K$. Let us consider for simplicity an uncontrolled equation of the form

$$dX_t = f(X_t)dt + \sigma(X_t)dB_t$$

where B_t is a one-dimensional Brownian motion and σ is a vector field. In this case the condition (1.4) becomes

$$\sigma(x) \cdot \vec{n}(x) = 0 \text{ and } \frac{1}{2} S(P\sigma(x)) \geq f(x) \cdot \vec{n}(x), \tag{1.5}$$

where $\vec{n}(x)$ is the exterior normal to K, P is the orthogonal projection on the tangent space to K at x, S is the second fundamental form of ∂K at x (oriented with $\vec{n}(x)$). This says that the diffusion vector σ is tangential to ∂K and the component of the drift pointing outward K is smaller than a quantity depending on the curvature matrix of ∂K and the diffusion vector. This is the stochastic generalization of the condition $f(x) \cdot \vec{n}(x) \leq 0$ in the deterministic case.

Previous sufficient conditions of invariance of closed sets for uncontrolled diffusions can be found in the books of Friedman (1976) (for K closure of a smooth open set) and of Ikeda and Watanabe (1981) (for K smooth manifold and stochastic differential equations in the Stratonovich sense). Aubin and Da Prato (1990) (1995) characterized the invariance of random closed sets by means of a notion of stochastic contingent set, see also Gautier and Thibault (1993) and Milian (1993) (1995) (1997).

Our approach is completely different from the preceding. We start from the observation that K is invariant if and only if the first exit time $\hat{t}_x(\alpha_.)$ of the trajectory of (SDE) from K is $+\infty$ for all $x \in K$ and all admissible controls $\alpha_.$ almost surely. Then we consider the value function

$$v(x) := \inf_{\alpha_.} E(1 - e^{-\hat{t}_x(\alpha_.)}), \tag{1.6}$$

and note that

$$K \text{ is invariant} \iff v(x) = 1 \ \forall x \in K.$$

A Hamilton-Jacobi-Bellman equation is associated to the stochastic control problem of (1.6) via Dynamic Programming, and even if the value function is not continuous in general, it satisfies such nonlinear PDE in the viscosity sense in the interior of K, and a generalized Dirichlet boundary condition on ∂K, see P.L. Lions (1983a) (1983b), Ishii (1989), Ishii and Lions (1990). We proved in Bardi and Goatin (1997) that v is indeed the maximal subsolution of a suitable boundary value problem (BVP) for the HJB equation, see Sect. 2. Thus we can establish if K is invariant by checking whether the constant 1 is the maximal subsolution of (BVP), and this leads naturally to the condition (1.4) involving the 2nd order normal cone.

The approach is rather flexible and we believe it applies to other related problems, such as the *viability* of the set K. This weaker property says that for all initial points $x \in K$ there exists at least one trajectory of the system that remains in K for all $t > 0$ almost surely. Viability has a large literature in the case of deterministic differential inclusions, see Aubin and Cellina (1981), Aubin (1991), Ledyaev (1994) and the references therein, and was

studied recently by Aubin and Da Prato (1998) in the case of stochastic differential inclusions.
Here we test our method to prove a viability theorem for controlled systems affected by a bounded disturbance whose statistics are not known. More precisely we consider

$$\begin{cases} X'_t = f(X_t, a_t, b_t), & t > 0 \\ X_0 = x, \end{cases}$$

where a_t is an unknown measurable disturbance taking values in a compact set A, b_t is the control taking values in a compact set B, and the controller is allowed to use relaxed controls. We say that K is viable if for all $x \in K$ there exists a nonanticipating strategy for the controller that keeps the trajectory X_t forever in K for all disturbances. In this problem we consider the value function of a suitable differential game, see Sect. 2 for all the definitions. By means of the Hamilton-Jacobi-Isaacs equation of this game we prove that K is viable if and only if

$$\forall\, x \in \partial K,\ a \in A,\ p \in \mathcal{N}_K^1(x),\ \exists\, b_x \in B\ :\ f(x, a, b_x) \cdot p \geq 0.$$

Note that if B is a singleton we obtain the characterization of invariance (1.3) for system (1.2), while in the case that A is a singleton we obtain a viability theorem for a (parametrized) differential inclusion. In Sect. 4 we also reformulate these results in terms of the contingent cone $T_K(x)$, which is the usual tool in viability theory, and of other objects of nonsmooth analysis (such as proximal normals and Clarke tangent cone). Most of these statements are known, our contribution is the new proof based on viscosity solutions of the Hamilton-Jacobi equation. In particular, the main result of the section, Theorem 4.1, is a variant of a result of Cardaliaguet (1996), see also Cardaliaguet and Plaskacz (1996) for the nonautonomous case and Guseinov, Subbotin and Ushakov (1985), Aubin (1991) (1997), Cardaliaguet (1997) for related problems with different notions of strategies.

We refer the reader to Crandall, Ishii, Lions (1992) and Bardi, Crandall, Evans, Soner, Souganidis (1997) for surveys of the theory of viscosity solutions, to the books by Fleming and Soner (1993) and Bardi and Capuzzo-Dolcetta (1997) for the applications, respectively, to stochastic control and to deterministic control and games, to Fleming and Souganidis (1989) for stochastic differential games.
The paper is organized as follows. Section 2 recalls some results on discontinuous viscosity solutions and their relation with the value functions of stochastic control problems and differential games involving the exit time from a closed set. Section 3 is devoted to the invariance theorem for (SDE) and to several examples. Section 4 deals with viability and invariance for deterministic systems with and without disturbances.

2 Value functions and Dynamic Programming Equations

We consider the boundary value problem

$$(BVP)\begin{cases} F(x,u,Du,D^2u)=0 & \text{in } \Omega := \overset{\circ}{K}, \\ u=0 \text{ or } F(x,u,Du,D^2u)=0 & \text{on } \partial K, \end{cases} \tag{2.1}$$

where F is a nonlinear degenerate elliptic operator and K is an arbitrary closed set and $\overset{\circ}{K}$ is its interior. The boundary condition in (BVP) is to be understood in the pointwise viscosity sense, see the precise definition (2.6) below. In particular, we are interested in the Isaacs-Bellman operator

$$F(x,r,p,X) = \sup_{\alpha}\inf_{\beta}[-trace(A^{\alpha,\beta}(x)X) - f^{\alpha,\beta}(x)\cdot p + c^{\alpha,\beta}(x)r - l^{\alpha,\beta}(x)], \tag{2.2}$$

and we assume that for all $x \in I\!R^N$, $A^{\alpha,\beta}(x) = \frac{1}{2}\sigma^{\alpha,\beta}(x)\sigma^{\alpha,\beta}(x)^T$, where $\sigma^{\alpha,\beta}(x)$ is a $N\times M$ matrix, ^{T} denotes the transpose matrix, $\sigma^{\alpha,\beta}$, $f^{\alpha,\beta}$, $c^{\alpha,\beta}$, $l^{\alpha,\beta}$ are bounded and uniformly continuous, uniformly with respect to α,β, so that the operator is continuous. Moreover we require that $A^{\alpha,\beta}(x) \geq 0$ for all α,β and there exist $C>0$ and $c_0>0$ such that

$$||\sigma^{\alpha,\beta}(x) - \sigma^{\alpha,\beta}(y)|| \leq C|x-y|, \text{ for all } x,y \in \overline{\Omega} \text{ and all } \alpha,\beta \tag{2.3}$$

$$|f^{\alpha,\beta}(x) - f^{\alpha,\beta}(y)| \leq C|x-y|, \text{ for all } x,y \in \overline{\Omega} \text{ and all } \alpha,\beta, \tag{2.4}$$

$$c^{\alpha,\beta}(x) \geq c_0, \text{ for all } x \in \overline{\Omega}, \text{ and all } \alpha,\beta. \tag{2.5}$$

We denote by $\mathcal{S}_v$ the set of all subsolutions of (BVP), that is

$$\mathcal{S}_v = \{w : K \to I\!R \text{ bounded, upper semicontinuous, and subsolution of } (BVP)\},$$

and by $\mathcal{Z}_e$ the following set

$$\begin{aligned} \mathcal{Z}_e = \{ & W : \overline{\mathcal{O}} \to I\!R \text{ is bounded and lower semicontinuous,} \\ & \mathcal{O} \text{ is open, } K \subset \mathcal{O}, \\ & W \text{ is supersolution of } F=0 \text{ in } \mathcal{O},\ W \geq 0 \text{ on } \partial\mathcal{O}\}. \end{aligned}$$

For the definitions of viscosity sub- and supersolution we refer to Ishii, Lions (1990) or Crandall, Ishii, Lions (1992). By a subsolution of (BVP) we mean a subsolution in viscosity sense not only of the PDE in Ω but also of the boundary condition; this means that, for all $x \in \partial K$,

$$\min\{u(x), F(x,u(x),p,Y)\} \leq 0 \ \ \forall (p,Y) \in \mathcal{J}_K^{2,+}u(x), \tag{2.6}$$

where

$$\mathcal{J}_K^{2,+}u(x) := \{(p,Y) \in I\!R^N \times S(N) : \text{ for } K \ni y \to x$$
$$u(y) \le u(x) + p\cdot(y-x) + \frac{1}{2}(y-x)\cdot Y(y-x) + o(|y-x|^2)\}$$

In Bardi and Goatin (1997) we proved that for general nonlinear degenerate elliptic equations, there exists the maximal subsolution of (BVP) and

$$u(x) := \max_{w\in\mathcal{S}_v} w(x) = \inf_{W\in\mathcal{Z}_e} W(x). \tag{2.7}$$

This is the natural generalized solution of (BVP), and we call it *envelope solution* (briefly *e-solution*), or Perron-Wiener solution, of (BVP). Here is a precise statement for the Isaacs-Bellman operator (2.2).

Theorem 2.1 *Assume $A^{\alpha,\beta} = \frac{1}{2}\sigma^{\alpha,\beta}(\sigma^{\alpha,\beta})^T$ for some $N\times M$ matrix valued functions $\sigma^{\alpha,\beta}$, and $\sigma^{\alpha,\beta}, f^{\alpha,\beta}, c^{\alpha,\beta}, l^{\alpha,\beta}$ be uniformly bounded with respect to α,β and satisfy (2.3), (2.4), (2.5) with $c_0 > 0$. Suppose also $l^{\alpha,\beta}$ are uniformly continuous, uniformly in α,β. Then for any closed set K there exists the maximal subsolution u of (BVP) and it satisfies (2.7).*

For the proof see Bardi and Goatin (1997); see also Bardi, Goatin and Ishii (1998) for further properties of the e-solution.

In the sequel we will study the time-optimal stochastic control problem and a deterministic differential game. The two following theorems state that the value function of the control problem and the differential game are the e-solutions of the corresponding (BVP).

For the time-optimal stochastic control problem we consider a probability space $(\Omega', \mathcal{F}, P)$ with a right-continuous increasing filtration of complete sub-σ fields $\{\mathcal{F}_t\}$, a Brownian motion B_t in $I\!R^M$ $\mathcal{F}_t$-adapted, a compact set A, and call $\mathcal{A}$ the set of progressively measurable processes α_t taking values in A. We are given bounded maps σ from $I\!R^N \times A$ into the set of $N\times M$ matrices and $f : I\!R^N \times A \to I\!R^N$ satisfying (2.3), (2.4) and consider the controlled stochastic differential equation

$$(SDE)\begin{cases} dX_t = \sigma^{\alpha_t}(X_t)dB_t + f^{\alpha_t}(X_t)dt, & t > 0, \\ X_0 = x. \end{cases}$$

This has a pathwise unique solution X_t which is $\mathcal{F}_t$-progressively measurable and has continuous sample paths. We are given also two bounded and uniformly continuous maps $l, c : I\!R^N \times A \to I\!R$, $c^a(x) \ge c_0 > 0$ for all x, a, and consider the cost functional

$$J(x,\alpha_.) := E\left(\int_0^{\hat{t}_x(\alpha_.)} l^{\alpha_t}(X_t)e^{-\int_0^t c^{\alpha_s}(X_s)ds}dt\right),$$

where E denotes the expectation and

$$\hat{t}_x(\alpha_.) := \inf\{t \ge 0 : X_t \notin K\}$$

for a given compact set K (of course $\hat{t}_x(\alpha_.) = +\infty$ if $X_t \in K$ for all $t \geq 0$). We define the value function

$$v(x) := \inf_{\alpha_. \in \mathcal{A}} J(x, \alpha_.),$$

and consider the Bellman equation

$$F(x, u, Du, D^2u) := \max_{\alpha \in A}\{-trace(A^\alpha(x)D^2u - f^\alpha(x) \cdot Du + +c^\alpha(x)u - l^\alpha(x)\} = 0,$$

where the matrix $A^\alpha(x) := \frac{1}{2}\sigma^\alpha(x)\sigma^\alpha(x)^T$. We consider the boundary value problem (BVP) under the additional assumption

$$l^\alpha(x) \geq 0 \text{ for all } x \in I\!R^N, \alpha \in A. \tag{2.8}$$

Theorem 2.2 *Assume (2.8). Then the value function*

$$v(x) = \inf_{\alpha_. \in \mathcal{A}} E\left(\int_0^{\hat{t}_x(\alpha_.)} l^{\alpha_t}(X_t) e^{-\int_0^t c^{\alpha_s}(X_s)ds} dt\right)$$

is the unique e-solution of (BVP), i.e. $v = u$ with the property (2.7).

For the proof see Bardi and Goatin (1997).

Next we consider a two-person zero-sum deterministic differential game. We are given a controlled dynamical system

$$\begin{cases} y' = f(y(t), a(t), b(t)), & t > 0 \\ y(0) = x, \end{cases} \tag{2.9}$$

where $f : I\!R^N \times A \times B \to I\!R^N$ is continuous, A, B are compact metric spaces and $a = a(\cdot) \in \mathcal{A} :=$ {measurable functions $[0, +\infty) \to A$} is the control function of the first player. For the second player we will use relaxed controls $b = b(\cdot) \in \mathcal{B}^r :=$ {measurable functions $[0, +\infty) \to B^r$} where B^r is the set of Radon probability measures on B. For the definitions of relaxed trajectories of (2.9) we refer for instance to Warga (1972), Bardi and Capuzzo Dolcetta (1997). We will always assume that the system satisfies, for some constant L,

$$(f(x, a, b) - f(y, a, b)) \cdot (x - y) \leq L|x - y|^2 \tag{2.10}$$

for all $x, y \in I\!R^N$, $a \in A$, $b \in B$.
The cost functional, which the first player wants to minimize and the second player wants to maximize, is

$$J(x, a, b) = \int_0^{\hat{t}_x(a,b)} e^{-t} dt = 1 - e^{-\hat{t}_x(a,b)},$$

where $\hat{t}_x(a,b)$ is the first exit time from a given closed set $K \subseteq \mathbb{R}^N$, i.e. $\hat{t}_x(a,b) := \inf\{t \in [0,+\infty) : y_x(t,a,b) \notin K\}$, $y_x(t,a,b)$ being the solution of (2.9) corresponding to $a \in \mathcal{A}$, $b \in \mathcal{B}^r$. Note that $\hat{t}_x$ is the time taken by the system to reach the *open target* $\mathbb{R}^N \setminus K$, and that J is a bounded and increasing rescaling of $\hat{t}_x$.
A relaxed strategy for the second player is a map $\beta : \mathcal{A} \to \mathcal{B}^r$; it is nonanticipating if, for any $t > 0$ and $a, \tilde{a} \in \mathcal{A}$, $a(s) = \tilde{a}(s)$ for all $s \le t$ implies $\beta[a](s) = \beta[\tilde{a}](s)$ for all $s \le t$, see Elliot and Kalton (1972), Bardi and Capuzzo-Dolcetta (1997). We will denote with Δ^r the set of nonanticipating relaxed strategies for the second player.
Now we can define the upper value of this differential game, which is

$$\hat{u}(x) = \sup_{\beta \in \Delta^r} \inf_{a \in \mathcal{A}} J(x, a, \beta[a]).$$

It is well known by the Dynamic Programming Principle that the upper value is a viscosity solution of the upper Hamilton-Jacobi-Isaacs (briefly HJI) equation

$$u(x) + \tilde{H}(x, Du(x)) := u(x) + \max_{a \in A} \min_{b \in B} \{-f(x,a,b) \cdot Du(x) - 1\} = 0 \text{ in } \overset{\circ}{K}.$$

Theorem 2.3 *Assume (2.10). Then the upper value of the minimum time problem*

$$\hat{u}(x) = \sup_{\beta \in \Delta^r} \inf_{a \in \mathcal{A}} \int_0^{\hat{t}_x(a,\beta[a])} e^{-t} dt$$

is the maximal subsolution of

$$(BVP^{\sim}) \begin{cases} u + \tilde{H}(x, Du) = 0 & \text{in } \overset{\circ}{K}, \\ u = 0 \text{ or } u + \tilde{H}(x, Du) = 0 & \text{on } \partial K, \end{cases} \tag{2.11}$$

and therefore the e-solution $(BVP^{\sim})$.

The proof can be found in Bardi and Goatin (1997). Let us recall that, since the PDE in (2.11) is of 1st order, a formulation of the boundary condition equivalent to (2.6) is the following:

$$u(x) \le 0 \quad \text{or} \quad u(x) + \tilde{H}(x,p) \le 0 \ \ \forall p \in \mathcal{J}_K^{1,+}u(x), \tag{2.12}$$

for all $x \in \partial K$, where

$$\mathcal{J}_K^{1,+}u(x) := \{p \in \mathbb{R}^N : u(y) \le u(x) + p \cdot (y - x) + o(|y - x|) \text{ for } K \ni y \to x\}.$$

3 Invariant sets for diffusions

In this section we consider the stochastic controlled system (SDE) of equation (1.1), a closed set $K \subset \mathbb{R}^N$, and characterize the invariance of K

for the trajectories of (SDE). We recall that K is invariant for (SDE) if $\hat{t}_x(\alpha_.) = +\infty$ almost surely for all $x \in K$ and admissible controls $\alpha_.$, see the definitions in Section 2. Following the approach outlined in the Introduction we consider the value function v defined by (1.6) that satisfies Theorem 2.2 with $l = c \equiv 1$. The corresponding HJB equation is then

$$u + \max_{\alpha \in A} \left\{ -\frac{1}{2} trace(\sigma^\alpha (\sigma^\alpha)^T D^2 u) - f^\alpha \cdot Du \right\} - 1 = 0.$$

Theorem 3.1 *Assume σ^α, f^α be uniformly bounded with respect to α and satisfy (2.3), (2.4). Then K is invariant if and only if for any $x \in \partial K$ and any $\alpha \in A$*

$$\frac{1}{2} trace(\sigma^\alpha(x)\sigma^\alpha(x)^T Y) + f^\alpha(x) \cdot p \geq 0 \ \forall (p, Y) \in \mathcal{N}_K^2(x). \tag{3.1}$$

Proof. By the definitions, the set K is viable if and only if the value function $v \equiv 1$. To prove this we observe that $v \leq 1$ by definition and, by Theorem 2.2, v is the maximal subsolution of the boundary value problem (BVP) of equation (2.1). Then $v \equiv 1$ if and only if the constant function $u(x) := 1$ is a subsolution of (BVP). The HJB equation is trivially satisfied by u in $\overset{\circ}{K}$, while the boundary condition (2.6) holds if and only if for all $x \in \partial K$ and $(p, Y) \in \mathcal{J}_K^{2,+} u(x)$

$$\max_{\alpha \in A} \left\{ -\frac{1}{2} trace(\sigma^\alpha(x)\sigma^\alpha(x)^T Y) - f^\alpha \cdot p \right\} \leq 0.$$

Now we note that $\mathcal{J}_K^{2,+} u(x) = \mathcal{N}_K^2(x)$ and get the conclusion. □

In the sequel we will show what (3.1) becomes in some particular cases. Some of these examples are treated also by Aubin and Da Prato (1990).

Example 3.2 (K closure of a smooth open set) If $K = \overline{\Omega}$, where Ω is a C^2 N-submanifold of $I\!R^N$ with boundary, and $x \in \partial\Omega$, then the contingent cone $T_K(x)$ is a halfspace and Ω has an exterior normal $\vec{n}$ at x. In this event we have the following characterization of $\mathcal{N}_K^2(x)$ (see Crandall, Ishii and Lions (1992)): if we represent Ω near $x = 0$ in the form

$$\{(\tilde{x}, x_N) : x_N \leq \Psi(\tilde{x})\},$$

where $\tilde{x} = (x_1, ..., x_{N-1})$, and we assume that the boudary of Ω is twice differentiable at 0 and rotated so that the normal is $\vec{n} = e_N = (0, ..., 0, 1)$, it is not hard to see that

$$(p, Y) \in \mathcal{N}_K^2(0) \Longleftrightarrow \begin{cases} p = 0, \ Y \geq 0; \\ \text{or} \\ p = -\lambda \vec{n}, \ \lambda > 0, \text{and for } \tilde{x} \to 0 \\ \lambda \tilde{x} \cdot D^2\Psi(0)\tilde{x} - (\tilde{x}, 0) \cdot Y(\tilde{x}, 0) \leq o(|\tilde{x}|^2). \end{cases}$$

If in (SDE) we let σ^i be the i-th column of the $N \times M$ matrix σ, and we drop the dependence on α, we can write the equation as

$$dX_t = f(X_t)dt + \sum_{i=1}^{M} \sigma^i(X_t)dB_t^i, \tag{3.2}$$

where B_t^i are the components of B_t, so they are one-dimensional independent Brownian motions. Then we can write (3.1) as

$$\frac{1}{2}\sum_{i=1}^{M} \sigma^i(0)\cdot Y\sigma^i(0) + f(0)\cdot p \geq 0 \text{ for any } (p,Y) \in \mathcal{N}_K^2(0).$$

By the characterization of the elements of $\mathcal{N}_K^2(0)$, we have that (3.1) holds at $x = 0$ if and only if

$$\sigma_N^i(0) = 0 \ \ \forall i = 1, ..., M, \ \text{ and } \ \frac{1}{2}\sum_{i=1}^{M} \tilde{\sigma}^i(0)\cdot D^2\Psi(0)\tilde{\sigma}^i(0) \geq f(0)\cdot \vec{n}, \tag{3.3}$$

where $\tilde{\sigma}^i = (\sigma_1^i, ..., \sigma_{N-1}^i)$. Note that if K is strictly convex (near 0) and $\sigma(0) \neq 0$, the drift vector f must be directed to the interior of K, because the matrix $D^2\Psi(0)$ is negative definite. On the other hand, if $\mathbb{R}^N \setminus K$ is strictly convex and $\sigma(0) \neq 0$, the drift can have a small component directed outward K because in this case the diffusion pushes the system inside K. Returning to the original coordinates, the condition (3.1) reads

$$\vec{n}(x)^T\sigma(x) = 0 \ \text{ and } \ \frac{1}{2}\sum_{i=1}^{M} S(P\sigma(x)) \geq f(x)\cdot\vec{n}(x), \tag{3.4}$$

where P is the orthogonal projection on the tangent space to K at x and S is the second fundamental form of ∂K at x, oriented with the exterior normal $\vec{n}(x)$.

We can also rewrite (3.4) in terms of the Hessian matrix of any function ρ of class C^2 in a neighborhood of ∂K and such that $\rho(x) = 0$ and $\nabla\rho(x) = \vec{n}(x)$ for all $x \in \partial K$ (e.g., the signed distance function from ∂K). In fact $S(Pw) = -w\cdot D^2\rho(x)w$ for any tangent vector w at $x \in \partial K$, so (3.4) is equivalent to

$$\vec{n}(x)^T\sigma^\alpha(x) = 0 \ \text{ and } \ f^\alpha(x)\cdot\vec{n}(x) + trace(A^\alpha(x)D^2\rho(x)) \leq 0, \tag{3.5}$$

where $A^\alpha = \frac{1}{2}\sigma^\alpha(\sigma^\alpha)^T$. A direct proof of the sufficiency of this condition (for all $x \in \partial K$) for the invariance of K can be also found in the book of Friedman (1976) (without control α).

Example 3.3 (Balls and exterior of balls) Here we consider two special cases of the previous example. The first is the ball $K = \{x : |x| \leq R\}$

(see also Aubin and Da Prato (1990)). We obtain that K is invariant if and only if, for all $\alpha \in A$ and $|x| = R$,

$$x^T \sigma^\alpha(x) = 0 \text{ and } f^\alpha(x) \cdot x + \frac{1}{2R}\|\sigma^\alpha(x)\|^2 \leq 0, \tag{3.6}$$

where $\|\cdot\|$ denotes the Euclidean matrix norm.
The second example is the complement of a ball, $K = \{x : |x| \geq R\}$. Now the condition of invariance is, for all $\alpha \in A$ and $|x| = R$,

$$x^T \sigma^\alpha(x) = 0 \text{ and } f^\alpha(x) \cdot x + \frac{1}{2R}\|\sigma^\alpha(x)\|^2 \geq 0.$$

Both examples show clearly the role of the curvature of ∂K for the invariance property.

Example 3.4 (K hypersurface) If K is a (N-1)-dimensional manifold of class C^2 without boundary in $\mathbb{R}^N$, the contingent cone at a point $x \in K$ turns out to be the tangent hyperplane at x. We can change coordinates as in Example 3.2 and represent K near $x = 0$ as the graph of a twice differentiable function Ψ of the first $N-1$ variables, with normal vector $\vec{n}(0) = (0, \dots, 0, 1)$. Using the same notations as in Example 3.2 we compute

$$(p, Y) \in \mathcal{N}_K^2(0) \Longleftrightarrow \begin{cases} p = 0,\ (\tilde{x}, 0) \cdot Y(\tilde{x}, 0) \geq 0\ \forall \tilde{x} \in \mathbb{R}^{N-1}; \\ \text{or} \\ p = \lambda \vec{n},\ \lambda \neq 0, \text{and for } \tilde{x} \to 0 \\ \lambda \tilde{x} \cdot D^2\Psi(0)\tilde{x} + (\tilde{x}, 0) \cdot Y(\tilde{x}, 0) \geq o(|\tilde{x}|^2). \end{cases}$$

So in this case (3.3) becomes

$$\sigma_N^i(0) = 0 \ \forall i = 1, ..., M, \text{ and } \frac{1}{2}\sum_{i=1}^{M} \tilde{\sigma}^i(0) \cdot D^2\Psi(0)\tilde{\sigma}^i(0) = f(0) \cdot \vec{n}.$$

Then the invariance condition in the original coordinates is (3.4) or (3.5) with the inequality replaced by an equality; for the sphere of radius R the condition is (3.6) with $=$ instead of $\leq$.

Example 3.5 (Smooth diffusion matrix and Stratonovich equations) If the matrix σ^α is twice differentiable the conditions of the previous examples can be also written as follows. We use the notation of Example 3.2 and observe that $\nabla\rho \cdot \sigma^i \equiv 0$ on ∂K and σ^i tangent to ∂K imply

$$\sigma^i \cdot \nabla(\nabla\rho \cdot \sigma^i) \equiv 0 \text{ on } \partial K.$$

This gives

$$\sigma^i \cdot D^2\rho\,\sigma^i = -\sigma^i \cdot D\sigma^i \nabla\rho,$$

where $D\sigma^i$ is the Jacobian matrix of the vector field σ^i. Then the inequality in (3.5) becomes

$$\vec{n}(x)\cdot\phi(x) := \vec{n}(x)\cdot\left(f(x)-\frac{1}{2}\sum_{i=1}^{M} D\sigma^i(x)\,\sigma^i(x)\right)\leq 0, \qquad (3.7)$$

where we temporarily dropped α. The new vector field ϕ appearing in this inequality is the drift of the Stratonovich equation equivalent to (3.2), and we can rewrite its components as

$$\phi_j^\alpha := f_j^\alpha - \frac{1}{2}\sum_{i=1}^{M}\sum_{k=1}^{N}\frac{\partial\sigma_{ji}^\alpha}{\partial x_k}\sigma_{ki}^\alpha.$$

Now the invariance condition (3.4) or (3.5) can be written as

$$\vec{n}(x)^T\sigma^\alpha(x)=0 \text{ and } \vec{n}(x)\cdot\phi^\alpha(x)\leq 0 \qquad (3.8)$$

for all $x\in\partial K$ and $\alpha\in A$.
In the case that K is a hypersurface, as in Example 3.4, the inequalities in (3.7) and (3.8) become equalities, and the invariance condition now is

$$(\sigma^\alpha)^i(x)\in T_K(x) \;\; \forall i=1,\dots,M, \;\; \phi^\alpha(x)\in T_K(x)$$

for all $x\in\partial K$ and $\alpha\in A$, where $T_K(x)$ is the tangent vector space to K at x. This is the classical condition for the existence of a diffusion process on a manifold in Ikeda and Watanabe (1981). See also Milian (1997) for similar results for manifolds with boundary.

Example 3.6 (Polyhedra) Let K be a halfspace, i.e.

$$K=\{x\in I\!R^N : (x-a)\cdot\vec{n}\leq 0\} =: P(a,\vec{n})$$

for some point $a\in I\!R^N$ and some vector $\vec{n}\in I\!R^N$, or, up to rotation,

$$K=\{(\tilde{x},x_N) : x_N\leq a+\lambda\cdot\tilde{x}\}.$$

So in this case (3.3) reads :

$$f\cdot\vec{n}\leq 0, \;\; \sigma^i\cdot\vec{n}=0 \;\; \forall i=1,...,M.$$

In general, if we consider a polyhedron, that is any set of the form

$$K=\bigcap_{k\in I}P(a_k,\vec{n}_k),$$

where $I=\{1,...,r\}$ is a finite set, we recover the following result of Milian (1995).

Theorem 3.7 *Let $K = \bigcap_{k\in I} P(a_k, \vec{n}_k)$ be a polyhedron in $\mathbb{R}^N$. Suppose that f, σ satisfy conditions (2.3), (2.4). Then K is invariant for (3.2) if and only if the following condition holds:*

$$f(x)\cdot\vec{n}_k \le 0 \text{ and } \sigma^i(x)\cdot\vec{n}_k = 0 \ \forall i = 1, ..., M,$$
$$\text{for all } k \in I \text{ and } x \in \partial K \text{ such that } (x - a_k, \vec{n}_k) = 0.$$

This result follows from the previous one on hyperplanes, by extending f and σ out of K in a suitable way. Of course it is not hard to extend it to the case that f and σ^i in (3.2) are controlled vector fields.

Example 3.8 (Union of smooth sets) Assume $K = K_1 \cup K_2$, where K_i, $i = 1, 2$, are closures of smooth open sets in $\mathbb{R}^N$ as in Example 3.2. If $x \in \partial K_1 \cap \partial K_2$, and $\vec{n}_1(x) \neq \vec{n}_2(x)$ ($\vec{n}_i$ denotes the exterior normal to K_i), then $\mathcal{N}^2_K(x) = \{(0, Y) : Y \geq 0\}$. In fact, since $T_K(x) = T_{K_1}(x) \cup T_{K_2}(x)$, and $T_{K_i}(x)$ are halfspaces, if $(p, Y) \in \mathcal{N}^2_K(x)$ then p is parallel to both $\vec{n}_1(x)$ and $\vec{n}_2(x)$. So at the points $x \in \partial K_1 \cap \partial K_2$ we need no conditions on f and σ for (3.1) to hold, while at the other points of ∂K the condition is the same as in Example 3.2.

4 Viability for deterministic systems with disturbances

In this section we consider the deterministic controlled dynamical system (2.9). We recall the definition of 1st order normal cone to the set $K \subset \mathbb{R}^N$ at the point x

$$\mathcal{N}^1_K(x) := \left\{ p \in \mathbb{R}^N : \liminf_{K \ni y \to x} \ p \cdot \frac{y - x}{|y - x|} \geq 0 \right\};$$

we note that, up to a change of sign, it coincides with the regular normal cone in the terminology of Rockafellar and Wets (1998). We recall also the definition of the Bouligand-Severi contingent cone (see Aubin and Cellina (1981) and Aubin and Frankowska (1990)), which is simply called tangent cone by Rockafellar and Wets (1998),

$$T_K(x) := \left\{ v \in \mathbb{R}^N : \liminf_{h \to 0+} \frac{d_K(x + hv)}{h} = 0 \right\},$$

where d_K is the distance function from the set K, and of the proximal normal cone generated by the Bony generalized exterior normals to K

$$P_K(x) := \left\{ q \in \mathbb{R}^N \mid \exists \varepsilon > 0 \ : \ q \cdot (y - x) \leq \varepsilon |y - x|^2 \ \forall y \in K \right\}.$$

Theorem 4.1 *Assume (2.10). Then the following statements are equivalent:*

(i) K is viable, i.e., for all $x \in K$ there exists a relaxed strategy $\beta \in \Delta^r$ such that $\hat{t}_x(a, \beta[a]) = +\infty$ for all $a(\cdot) \in \mathcal{A}$;
(ii) for all $x \in \partial K$, $p \in \mathcal{N}^1_K(x)$, and $a \in A$, there exists $b \in B$ such that $f(x,a,b) \cdot p \geq 0$;
(iii) for all $x \in \partial K$, $q \in P_K(x)$, and $a \in A$, there exists $b \in B$ such that $f(x,a,b) \cdot q \leq 0$;
(iv) for all $x \in \partial K$ and $a \in A$, there exists $b \in B$ such that $f(x,a,b) \in \overline{co}T_K(x)$.

For the proof of the equivalence of *(iv)* we need one more concept of nonsmooth analysis, namely, the Clarke tangent cone to K at x

$$C_K(x) := \left\{ v \in \mathbb{R}^N : \liminf_{K \ni x' \to x, h \to 0+} \frac{d_K(x' + hv)}{h} = 0 \right\}.$$

For equivalent definitions and some properties of all these cones we refer to Aubin and Frankowska (1990) and Rockafellar and Wets (1998). We only recall the following inclusions

$$C_K(x) = \liminf_{K \ni y \to x} T_K(y) = \liminf_{K \ni y \to x} \overline{co}T_K(y) \subseteq T_K(x), \tag{4.1}$$

where $\liminf_{y \to x} S(x) = \{z = \lim_n z_n : z_n \in S(y_n), y_n \to x\}$ (see, e.g., Theorem 4.1.10 in Aubin and Frankowska (1990)) . We refer to Clarke and Aubin (1977) for earlier applications of the Clarke cone to invariance. We introduce also the (positive) polar cone to a subset $S \subset \mathbb{R}^N$ defined by

$$S^\circ := \{p \in \mathbb{R}^N : p \cdot q \geq 0 \text{ for any } q \in S\}.$$

Lemma 4.2 $\mathcal{N}^1_K(x) = (T_K(x))^\circ$.

Proof. $\mathcal{N}^1_K(x) \subseteq (T_K(x))^\circ$ follows from the definitions. On the contrary, assume that $p \notin \mathcal{N}^1_K(x)$. Then there exists a sequence of points $K \ni y_n \to x$ such that $\lim_{n\to\infty} p \cdot \frac{y_n - x}{|y_n - x|} < 0$. We have that $\frac{y_n - x}{|y_n - x|} \to q \in T_K(x)$. Then $p \cdot q < 0$, and $p \notin (T_K(x))^\circ$. □

Lemma 4.3 *A vector $q \in P_K(x)$ if and only if $(-q, Y) \in \mathcal{N}^2_K(x)$ for some matrix Y.*

The proof of this Lemma is straightforward.

Proof of Theorem 4.1. By the definitions,the set K is viable if and only if the value function $\hat{u} \equiv 1$, and this occurs if and only if the constant $u \equiv 1$ is a subsolution of $(BVP^\sim)$ in equation (2.11); in fact $\hat{u} \leq 1$ and it is the maximal subsolution of $(BVP^\sim)$ by Theorem 2.3. Substituting $u \equiv 1$ in the equation we see that it is a solution in $\overset{\circ}{K}$, and it is a subsolution of the boundary condition (2.12) if and only if for any $x \in \partial K$ and $p \in \mathcal{J}^{1,+}_K u(x)$

$$\max_{a \in A} \min_{b \in B} \{-f(x,a,b) \cdot p\} \leq 0. \tag{4.2}$$

Since $\mathcal{J}_K^{1,+}u(x) = \mathcal{N}_K^1(x)$ we get the equivalence between (i) and (ii). The equivalence between (ii) and (iii) is obtained in a similar way: now we use the boundary condition (2.6), we get (4.2) for all p such that $(p, Y) \in \mathcal{J}_K^{2,+}u(x) = \mathcal{N}_K^2(x)$, and then we conclude by Lemma 4.3.
The equivalence between (ii) and (iv) follows easily from Lemma 4.2, since $S^{\circ\circ} = \overline{co}S$. □

Remark 4.4 The equivalence of *(i)* and *(iii)* was proved by Cardaliaguet (1996) and extended to the nonautonomous case by Cardaliaguet and Plaskacz (1996). Guseinov, Subbotin and Ushakov (1985) analyzed a similar problem in the study of stable bridges for pursuit-evasion games in the framework of the Krasovskii-Subbotin theory of positional differential games. They found two equivalent conditions related to *(iv)*. See also Aubin (1991) (1997) and Cardaliaguet (1997) for related results.

The next two results are consequences of Theorem 4.1.

Theorem 4.5 (Invariance) *Let (2.10) hold and B be a singleton. Then the following statements are equivalent:*
(i) K is invariant for the player a;
(ii) $f(x, a) \cdot p \geq 0$ for every $x \in \partial K$, $p \in \mathcal{N}_K^1(x)$, and $a \in A$;
(iii) $f(x, a) \cdot q \leq 0$ for every $x \in \partial K$, $q \in P_K(x)$, and $a \in A$;
(iv) $f(x, A) \subseteq \overline{co}T_K(x)$ for every $x \in \partial K$;
(v) $f(x, A) \subseteq T_K(x)$ for every $x \in \partial K$;
(vi) $f(x, A) \subseteq C_K(x)$ for every $x \in \partial K$.

Proof. By Theorem 4.1 it is enough to prove the equivalence of the last three statements. By (4.1) (vi) implies (v) and (iv). Now assume (iv), fix $x \in \partial K$ and take $K \ni x_n \to x$. By assumption (iv), $f(x_n, a) \in \overline{co}T_K(x_n)$ for all $a \in A$. Then $f(x, a) \in \liminf_{K \ni y \to x} \overline{co}T_K(y)$ and we get the result by (4.1). □

In the last result of this section we assume A is a singleton; then the definition of viability simplifies because we do not have to use strategies.

Theorem 4.6 (Viability) *Let (2.10) hold and A be a singleton. Then the following statements are equivalent:*
(j) K is viable, i.e., for all $x \in K$ there exists a relaxed control $b(\cdot) \in \mathcal{B}^r$ such that $\hat{t}_x(b) = +\infty$;
(jj) for all $x \in \partial K$ and $p \in \mathcal{N}_K^1(x)$ there is $b \in B$ such that $f(x, b) \cdot p \geq 0$;
(jjj) for all $x \in \partial K$ and $q \in P_K(x)$ there is $b \in B$ such that $f(x, b) \cdot q \leq 0$;
(jv) $f(x, B) \cap \overline{co}T_K(x) \neq \emptyset$ for all $x \in \partial K$.

Remark 4.7 Under the assumptions of this theorem the solutions of (2.9) associated with relaxed controls $b(\cdot)$ correspond to the trajectories of the differential inclusion $y' \in \overline{co}f(y, B)$. Then the classical viability theorem (see Aubin and Cellina (1981)) implies that the statements *(j)* - *(jv)* are also equivalent to

(v) $\overline{co}f(x, B) \cap T_K(x) \neq \emptyset$ for all $x \in \partial K$.

We refer the reader to Aubin and Frankowska (1990) and Aubin (1991) for different proofs of Theorem 4.6 and for more results and references on viability and its applications.

Acknowledgments: Partially supported by M.U.R.S.T. project "Problemi nonlineari nell'analisi e nelle applicazioni fisiche, chimiche, biologiche." We thank J.-P. Aubin, P. Baldi, G. Colombo, G. Da Prato and W. Runggaldier for bibliographical suggestions.

References

Aubin, J.-P. (1991). *Viability Theory*, Birkäuser, Boston.

Aubin, J.-P. (1997). *Dynamic Economic Theory*, Springer, Berlin.

Aubin, J.-P. and Cellina, A. (1984). *Differential Inclusions*, Springer, Berlin.

Aubin, J.-P. and Da Prato, G. (1990). *Stochastic viability and invariance*, Ann. Sc. Norm. Sup. Pisa (IV) 17, 595-613.

Aubin, J.-P. and Da Prato, G. (1995). *Stochastic Nagumo's viability theorem*, Stochastic Anal. Appl. 13, 1-11.

Aubin, J.-P. and Da Prato, G. (1998). *The viability theorem for stochastic differential inclusions*, Stochastic Anal. Appl. 16, 1-15.

Aubin, J.-P. and Frankowska, H. (1990). *Set-valued analysis*, Birkhäuser, Boston.

Bardi, M. and Capuzzo Dolcetta, I. (1997). *Optimal control and viscosity solutions of Hamilton-Jacobi-Bellman equations*, Birkhäuser, Boston.

Bardi, M., Crandall, M.G., Evans, L.C., Soner, H.M., and Souganidis, P.E. (1997). *Viscosity solutions and applications*, I. Capuzzo Dolcetta and P.-L. Lions eds., Springer Lecture Notes in Mathematics 1660, Berlin.

Bardi, M. and Goatin, P. (1997). *A Dirichlet type problem for nonlinear degenerate elliptic equations arising in time-optimal stochastic control*, Preprint n. 50, SISSA-ISAS, Trieste.

Bardi, M., Goatin, P. and Ishii, H. (1998). *A Dirichlet type problem for nonlinear degenerate elliptic equations arising in time-optimal stochastic control*, Preprint n. 1, Dipartimento di Matematica, Università di Padova.

Cardaliaguet, P. (1996). *A differential game with two players and one target*, SIAM J. Control Optim. 34, 1441-1460.

Cardaliaguet, P. (1997). *Nonsmooth semipermeable barriers, Isaacs' equation, and application to a differential game with one target and two players*, Appl. Math. Optim. 36, 125-146.

Cardaliaguet, P., Plaskacz, S. (1996) *Viability and invariance for differential games with applications to Hamilton-Jacobi-Bellman equations*, in "Topology in Nonlinear Analysis", 149-158, Banach Center Publications 35, Polish Acad. Sci., Warsaw.

Clarke, F.H., Aubin, J.-P. (1977) *Monotone invariant solutions to differential inclusions*, J. London Math. Soc. 16, 357-366.

Crandall, M.G., Ishii, H. and Lions, P.L. (1992). *User's guide to viscosity solutions of second order partial differential equations*, Bull. Amer. Math. Soc. 27, 1-67.

Elliott, R.J. and Kalton, N.K. (1972). *The existence of value in differential games*, Mem. Amer. Math. Soc. 126.

Fleming, W.H. and Soner, H.M. (1993). *Controlled Markov Process and Viscosity Solutions*, Springer-Verlag, New York.

Fleming, W.H. and Souganidis, P.E. (1989). *On the existence of value functions of two-players, zero-sum stochastic differential games*, Indiana Univ. Math. J. 38, 293-314.

Friedman, A. (1976). *Stochastic differential equations and applications*, vol. 2, Academic Press, New York.

Gautier, S. and Thibault, L. (1993). *Viability for constrained stochastic differential equations*, Differential Integral Equations 6, 1395-1414.

Guseinov, H.G., Subbotin, A.I. and Ushakov, V.N. (1985). *Derivatives for multivalued mappings with applications to game-theoretical problems of control*, Problems Control Inform. Theory 14, 155-167.

Ishii, H. (1989). *A boundary value problem of the Dirichlet type for Hamilton-Jacobi equations*, Ann. Scuola Norm. Sup. Pisa (IV) 16, 105-135.

Ishii, H. and Lions, P.L. (1990). *Viscosity solutions of fully nonlinear second order elliptic partial differential equations*, J. Differential Equations 83, 26-78.

Ikeda, N. and Watanabe, S. (1981). *Stochastic differential equations and diffusion processes*, North Holland, Amsterdam.

Ledyaev, Yu.S. (1994). *Criteria for viability of trajectories of nonautonomous differential inclusions and their applications*, J. Math. Anal. Appl. 182, 165-188.

Lions, P.L. (1983a). *Optimal control of diffusion processes and Hamilton-Jacobi-Bellman equations. Part 1: The dynamic programming principle and applications, Part 2: Viscosity solutions and uniqueness*, Comm. Partial Differential Equations 8, 1101-1174 and 1229-1276.

Lions, P.L. (1983b). *On the Hamilton-Jacobi-Bellman equations*, Acta Appl. Math. 1, 17-41.

Milian, A. (1993). *A note on the stochastic invariance for the Ito equations*, Bull. Polish Acad. Sci. Math. 41, 139-150.

Milian, A. (1995). *Stochastic viability and a comparison theorem*, Colloq. Math. 68, 297-316.

Milian, A. (1997). *Invariance for stochastic equations with regular coefficients*, Stochastic Anal. Appl. 15, 91-101.

Nagumo, N. (1942). *Uber die lage der integralkurven gewöhnlicher differentialgleichungen*, Proc. Phys. Math. Soc. Japan 24, 551-559.

Rockafellar, R.T. and Wets, R.J-B. (1998). *Variational Analysis*, Springer-Verlag, Berlin.

Warga, J. (1972) *Optimal control of differential and functional equations*, Academic Press, New York.

12

Remarks on the Dirichlet Problem for Quasilinear Elliptic and Parabolic Equations

Guy Barles
Elisabeth Rouy
Panagiotis E. Souganidis

ABSTRACT: We prove comparison results between continuous and discontinuous viscosity sub- and super-solutions of the generalized Dirichlet problem for quasilinear elliptic and parabolic equations. The main consequence of these results is the uniqueness of continuous solutions of such problems, when they exist.

KEYWORDS: Quasilinear elliptic and parabolic equations, generalized Dirichlet problem, viscosity solutions, comparison results.

1 Introduction

Quasilinear elliptic and parabolic equations have been extensively studied in the literature with a particular focus on the solvability of the Dirichlet problem. The most interesting questions arise when the equations are not uniformly elliptic. In this case the classical Dirichlet problem may not be solvable for any boundary data unless the boundary of the domain satisfies suitable geometric-curvature conditions. Such questions were in particular motivated by the study of the minimal surface and the prescribed mean curvature equations. We refer to the books of Gilbarg and Trudinger (1983) and Ladyzhenskaya and Ural'tseva (1968) for a presentation of a large collection of known results in this direction.

The starting point of this article and our main motivation are the study of the Dirichlet problem for quasilinear elliptic and parabolic equations, when classical solutions do not exist. This article represents at least a first step in this direction. Our goal is to prove the uniqueness of continuous viscosity solutions of such equations.

The notion of viscosity solutions was introduced by Crandall and Lions (1983). For a general overview of the scope of the theory of viscosity solutions we refer to the "User's guide" of Crandall, Ishii and Lions (1992) and the book of Fleming and Soner (1993).

Viscosity solutions turn out to be the correct class of weak solutions for fully nonlinear second-order, possibly degenerate, elliptic equations, i.e., equations of the form

$$F(x,u,Du,D^2u) = 0 \quad \text{in } \Omega\,, \tag{1}$$

where Ω is an open subset of $\mathbb{R}^N$ and F is a real-valued continuous function defined on $\overline{\Omega} \times \mathbb{R} \times \mathbb{R}^N \times \mathcal{S}^N$, where $\mathcal{S}^N$ is the space of $N \times N$ symmetric matrices. The notations Du and D^2u stand respectively for the gradient and Hessian matrix of the real-valued solution u. Throughout the paper we assume that F is elliptic, i.e., that it satisfies for all $x \in \overline{\Omega}$, $u \in \mathbb{R}$, $p \in \mathbb{R}^N$, $M, N \in \mathcal{S}^N$,

$$F(x,u,p,M) \leq F(x,u,p,N) \quad \text{if } M \geq N\,,$$

where "$\geq$" denotes the usual partial ordering of symmetric matrices.

The notion of viscosity solutions not only allows to consider merely continuous functions u as solutions of (1), but also leads to the correct formulation of the boundary conditions for degenerate elliptic pdes. For example, the standard Dirichlet boundary condition

$$u = \varphi \quad \text{on } \partial\Omega\,, \tag{2}$$

with φ a continuous function, is relaxed and reads in the viscosity sense as

$$\min\left(F(x,u,Du,D^2u), u - \varphi\right) \leq 0 \quad \text{on } \partial\Omega\,, \tag{3}$$

and

$$\max\left(F(x,u,Du,D^2u), u - \varphi\right) \geq 0 \quad \text{on } \partial\Omega\,. \tag{4}$$

Roughly speaking, these relaxed conditions mean that the equation must hold up to the boundary, if the Dirichlet boundary condition does not hold in the classical sense. We refer again the reader to the above mentioned references on viscosity solutions for a more complete presentation of the formulation of boundary conditions in the viscosity sense. To justify this weak formulation, here we only remark that the generalized boundary conditions arise naturally, when passing in the limit in the vanishing viscosity method, as a by-product of the existence of boundary layers. Such boundary layers clearly exist if the Dirichlet problem does not have a (viscosity) solution which assumes continuously the boundary data (2).

We address such problems in this article concentrating on quasilinear equations of the form

$$-\mathrm{Tr}\left(A(x,Du)D^2u\right) + H(x,u,Du) = 0 \quad \text{in } \Omega\,, \tag{5}$$

where A and H are continuous functions with values in the space of $N \times N$-matrices and $\mathbb{R}$ respectively, with A satisfying for all $x \in \overline{\Omega}$, $p, q \in \mathbb{R}^N$ the ellipticity condition

$$A(x,p)q \cdot q \geq 0 \,. \tag{6}$$

In the theory of viscosity solutions, there are essentially two approaches to deal with such type of problems. The first one considers only *continuous* viscosity solutions. In this approach the uniqueness properties are usually obtained in a rather standard way using the method of Soner (1986), which is described for general fully nonlinear second-order elliptic equations in Crandall, Ishii and Lions (1992). This method works well as long as the function $F(x, u, p, M)$ is uniformly continuous with respect to p and M, uniformly in x and u, at least for x in a neighborhood of $\partial\Omega$, a property which is clearly not satisfied by quasilinear equations. This creates a real difficulty, since the classical proof uses the uniform continuity of F in several steps and, in particular, for a localization type argument. An additional difficulty of this approach is the existence of a continuous solution, at least for problems where losses of boundary data are expected. In this situation it is rather difficult to establish the continuity of the solution, because the usual approximation schemes (such as the vanishing viscosity method) lead to boundary layers which cannot be handled easily.

The second approach, which deals with *discontinuous* viscosity solutions, provides easy existence results through the Perron's method of Ishii (1987). Since, however, the aim is to obtain the existence of continuous solutions, this method also requires a so-called Strong Comparison Result (SCR), i.e. a comparison result between discontinuous solutions. Such SCR are rather difficult to obtain for the Dirichlet problems which really present losses of boundary conditions. To the best of our knowledge, no such result exists for quasilinear equations.

We refer the reader interested in SCR for first-order equations to Barles (1994), for semilinear equations to Barles and Burdeau (1995) and for Hamilton-Jacobi-Bellman equations to Barles and Rouy (1998). A comparison between the difficulties of the two above mentioned approaches can be obtained by comparing the article of Katsoulakis (1994), which uses the first one, with the article of Barles and Rouy (1998), which uses the second one to the same type of questions.

In this paper we are in the framework of the first approach. Our main result is a comparison between sub- and super-solutions of (1)-(3)-(4), one of them being assumed to be continuous. Our contribution is to overcome the difficulty that $F(x, u, p, M)$ is NOT assumed to be uniformly continuous with respect to p and M. The main point is that our proof does not use any localization argument.

In the case of (5), our result applies, for example, to the case when A is bounded, symmetric and of the form $A = \sigma\sigma^T$, with σ satisfying for some constant C and for all $x \in \overline{\Omega}$, $p \in I\!R^N$

$$|D_x\sigma(x,p)| \leq C \quad \text{and } |D_p\sigma(x,p)| \leq \frac{C}{1+|p|}\,, \tag{7}$$

and H satisfies, for some constant $C > 0$ and for any $x \in \overline{\Omega}$, $u \in I\!R$ and $p \in I\!R^N$,

$$D_u H(x,u,p) \geq 1/C > 0\,, \tag{8}$$

and

$$|D_x H(x,u,p)| \leq C(1+|p|) \quad \text{and} \quad |D_p H(x,u,p)| \leq C\,. \tag{9}$$

The paper is organized as follows: The second section is devoted to the description of a classical example (the minimal surface equation) which illustrates losses of boundary data for a quasilinear equation. In Section 3, we state and prove the main comparison result. In Section 4, we analyse the loss of boundary data and, in particular, the connection with the geometric properties of the boundary. Finally, in Section 5, we describe the extensions of the comparison result to time-dependent problems.

2 An example

We describe here a well-known example of a quasilinear elliptic equation, exhibiting loss of boundary data and we revisit it using viscosity solutions. To this end, we set $\Omega = \{x \in I\!R^N\ ;\ 1 < |x| < 2\}$, $\partial\Omega^+ = \{x \in I\!R^N\ ;\ |x| = 2\}$ and $\partial\Omega^- = \{x \in I\!R^N\ ;\ |x| = 1\}$, we fix a positive constant R, and we consider the minimal surface equation

$$-\Delta u + \frac{D^2u Du \cdot Du}{1+|Du|^2} = 0 \quad \text{in } \Omega\,,$$

and the associated Dirichlet boundary condition

$$u = 0 \quad \text{on } \partial\Omega^+ \quad \text{and} \quad u = R \quad \text{on } \partial\Omega^-\,.$$

Assuming that u is smooth and radially symmetric, a simple computation yields that u is necessarily given by

$$u(x) := \psi(C,|x|) = \int_{|x|}^{2} \frac{\sqrt{C}}{\sqrt{r^{2(N-1)} - C}}dr\,,$$

for some $0 < C \leq 1$. Since the map $C \mapsto \psi(C,1)$ is increasing in C, it is clear that a solution of the above Dirichlet problem is obtained by an appropriate choice of the constant C, *provided that* $0 = \psi(0,1) \leq R \leq \overline{R} = \psi(1,1) < +\infty$. If, however, $R > \overline{R}$, this computation does not give

a solution of the Dirichlet problem in the classical sense. It is also known that there is no smooth solution assuming continuously the boundary data both on $\partial\Omega^+$ and $\partial\Omega^-$.
In fact $U(x) = \psi(1, |x|)$ is a solution of the generalized Dirichlet problem for $R > \overline{R}$. Indeed, on $\partial\Omega^+$ the Dirichlet boundary condition has to be understood in the classical sense. But since $U(x) < R$ on $\partial\Omega^-$, the Dirichlet boundary condition reduces to a so-called State-Constraints boundary condition, i.e.

$$-\Delta u + \frac{D^2uDu \cdot Du}{1+|Du|^2} \geq 0 \quad \text{on } \partial\Omega^- .$$

It is worth mentioning that for any $x \in \partial\Omega^-$, $J^{2,-}U(x) = \emptyset$ because the normal derivative of U is equal to $-\infty$, i.e. $\frac{\partial\psi}{\partial r}(1,1) = -\infty$. Hence the boundary condition actually imposes no restriction in this case.

3 The Comparison Result

We state and prove here a comparison result for the generalized Dirichlet problem (1)-(3)-(4). To this end, we introduce the following structure assumptions on F :

(H1) *For all R there exists $\gamma_R > 0$ such that for all $x \in \overline{\Omega}$, $-R \leq v \leq u \leq R$, $p \in \mathbb{R}^N$ and $M \in \mathcal{S}^N$*

$$F(x,u,p,M) - F(x,v,p,M) \geq \gamma_R(u-v) .$$

(H2) *For all $R, K > 0$, there exists a function $m_{R,K} : \mathbb{R}^+ \to \mathbb{R}$ such that $m_{R,K}(t) \to 0$ when $t \to 0$ and , for all $\eta > 0$ and $\varepsilon > 0$*

$$F(y,u,q,Y) - F(x,u,p,X) \leq m_{R,K}(\eta + \varepsilon(1+|p|+|q|))$$

for all $x, y \in \overline{\Omega}$, $|u| \leq R$, $p, q \in \mathbb{R}^N$ and for all matrices $X, Y \in \mathcal{S}^N$ satisfying the following properties:

$$-\frac{K\eta}{\varepsilon^2}Id \leq \begin{pmatrix} X & 0 \\ 0 & -Y \end{pmatrix} \leq \frac{K\eta}{\varepsilon^2}\begin{pmatrix} Id & -Id \\ -Id & Id \end{pmatrix} + K\eta Id, \quad (10)$$

$$|p-q| \leq K\varepsilon(1+|p|+|q|) \quad (11)$$

$$|x-y| \leq K\varepsilon. \quad (12)$$

Finally we need the following regularity type assumption on $\partial\Omega$.

(H3) *There exist a continuous function n defined in a $\mathbb{R}^N$-neighborhood $\mathcal{W}$ of $\partial\Omega$ and $\eta, t_0 > 0$ such that, for all $z \in \mathcal{W} \cap \overline{\Omega}$ and for all $t \leq t_0$*

$$B(z - tn(z), \eta t) \subset \Omega \, .$$

Our result is the following

Theorem 3.1 : *Assume that* **(H1)**, **(H2)** *and* **(H3)** *hold and let u and v be respectively a bounded usc sub-solution of* (1)-(3) *and a bounded lsc super-solution of* (1)-(4). *If either*

$$u(x) = \limsup_{\substack{y \to x \\ y \in \Omega}} u(y) \quad \text{for all } x \in \partial\Omega \quad \text{and} \quad v \in C(\overline{\Omega})$$

or

$$u \in C(\overline{\Omega}) \quad \text{and} \quad v(x) = \liminf_{\substack{y \to x \\ y \in \Omega}} v(y) \quad \text{for all } x \in \partial\Omega \, ,$$

then

$$u \leq v \quad \text{on } \overline{\Omega} \, .$$

It is not clear that any usc sub-solution u (resp. lsc super-solution v) satisfies the condition of this theorem. We show, however, in the next section that, under suitable and rather natural assumptions on F, one can modify u (resp. v) on the boundary in order to fulfill this condition. In this case, the comparison between u and v will be only valid in Ω and not up to the boundary, a common restriction for comparison results for Dirichlet problems.
The assumption **(H3)** is classical. We refer to the "User's guide" of Crandall, Ishii and Lions (1992) and to Fleming and Soner (1993) for more detailed comments about it. We only remark here that it is clearly satisfied if the boundary $\partial\Omega$ is C^1 by choosing $n(z) = -Dd(z)$ where d is the signed distance function to $\partial\Omega$ ($d > 0$ in Ω and $d < 0$ in $\overline{\Omega}^c$). On the other hand, it may also be satisfied by domains with corners, like a square in $\mathbb{R}^2$.

Proof of Theorem 3.1 : We only need to prove the result in the first case, since the second one follows by changing u in $-u$ and v in $-v$.
We define $M = \max_{\overline{\Omega}}(u - v)$ and argue by contradiction assuming that $M > 0$. Since u is usc and v is continuous this maximum is achieved. Let $\mathcal{M}$ be the set of all points where it is achieved. We introduce the sets

$$\Gamma_u = \{x \in \partial\Omega \cap \mathcal{M} \text{ s.t. } u(x) \leq \varphi(x)\},$$
$$\Gamma_v = \{x \in \partial\Omega \cap \mathcal{M} \text{ s.t. } v(x) \geq \varphi(x)\}.$$

If $\Gamma_u \cup \Gamma_v = \emptyset$ or $M > \max_{\Gamma_u \cup \Gamma_v}(u - v)$, a straightforward adaptation of the standard comparison arguments yields a contradiction. Therefore we may assume that $\Gamma_u \cup \Gamma_v \neq \emptyset$ and $M = \max_{\Gamma_u \cup \Gamma_v}(u - v)$.

Our first remark is the

Lemma 3.1 : *For any bounded usc function u and any bounded lsc function v, if $M = \max_{\overline{\Omega}}(u - v) > 0$, then, for any continuous function φ, Γ_u and Γ_v are closed, disjoint, possibly empty subsets of $\partial\Omega$.*

Proof : The fact that Γ_u and Γ_v are disjoint is clear from the assumption $M > 0$. Below we prove that Γ_u is closed, the proof for Γ_v being analogous. If $(x_\varepsilon)_\varepsilon$ is a sequence of points of Γ_u converging to $\overline{x}$, then, by the upper semi-continuity of u and the lower semi-continuity of v, we have

$$u(\overline{x}) \geq \limsup_\varepsilon u(x_\varepsilon) \quad \text{and} \quad v(\overline{x}) \leq \liminf_\varepsilon v(x_\varepsilon) ,$$

and therefore

$$u(\overline{x}) - v(\overline{x}) \geq \limsup_\varepsilon (u(x_\varepsilon) - v(x_\varepsilon)) .$$

The definition of M clearly implies that $u(\overline{x}) - v(\overline{x}) = M$ and necessarily

$$u(\overline{x}) = \lim_\varepsilon u(x_\varepsilon) \quad \text{and} \quad v(\overline{x}) = \lim_\varepsilon v(x_\varepsilon) .$$

As a by-product, the last observation and the continuity of φ yield $u(\overline{x}) \leq \varphi(\overline{x})$, since $u(x_\varepsilon) \leq \varphi(x_\varepsilon)$ for all ε.

It follows from the above lemma that there exists a smooth function $\psi : \mathbb{R}^N \to \mathbb{R}$, with compact support in $\mathcal{W}$ given by **(H3)**, such that $\psi \equiv 1$ in a neighborhood of Γ_u and $\psi \equiv -1$ in a neighborhood of Γ_v. Let $\tilde{n}$ be a smooth approximation of n in $\mathcal{W}$ such that **(H3)** still holds with $\tilde{n}$ in place of n, by changing, if necessary, η in $\eta/2$. We define the smooth function $\chi : \mathbb{R}^N \to \mathbb{R}^N$ by

$$\chi(x) = \begin{cases} \psi(x)\tilde{n}(x) & \text{if } x \in \mathcal{W}, \\ 0 & \text{otherwise.} \end{cases}$$

Next we introduce the test-function $\Phi_\varepsilon : \overline{\Omega} \times \overline{\Omega} \to \mathbb{R}$ by

$$\Phi_\varepsilon(z, w) = u(z) - v(w) - |\frac{z - w}{\varepsilon} + \chi(\frac{z + w}{2})|^4 .$$

Let (x, y) be a global maximum point of Φ_ε on $\overline{\Omega} \times \overline{\Omega}$. For notational simplicity here we drop the dependence of x and y on ε.
We first prove that

$$\Phi_\varepsilon(x, y) \to M \text{ as } \varepsilon \to 0. \tag{13}$$

It is clear that $x - y = O(\varepsilon)$. It follows that along a subsequence $x, y \to \overline{x} \in \overline{\Omega}$. Then $\limsup_\varepsilon \Phi_\varepsilon(x, y) \leq M$. The converse inequality is obtained as follows. We have seen above that $\Gamma_u \cup \Gamma_v \neq \emptyset$ and let x_0 be a point of $\Gamma_u \cup \Gamma_v$.

If $x_0 \in \Gamma_v$, using the continuity of v and the fact that $M = u(x_0) - v(x_0)$, we may write for ε small enough,

$$M + O(\varepsilon) = \Phi_\varepsilon(x_0, x_0 - \varepsilon\tilde{n}(x_0)) \le \Phi_\varepsilon(x,y) ,$$

and the conclusion follows. Otherwise, $x_0 \in \Gamma_u$ and the assumption on u yields the existence of a sequence $(x_p)_{p\ge1}$ of points of Ω converging to $x_0 \in \Gamma_u$ such that $u(x_p) \to u(x_0)$, as $p \to +\infty$. Then, for p large enough and for ε small enough, we have

$$M + O(\varepsilon + |x_p - x_0|) = \Phi_\varepsilon(x_p, x_p + \varepsilon\tilde{n}(x_p)) \le \Phi_\varepsilon(x,y).$$

In this inequality, we first let $\varepsilon \to 0$ and then $p \to +\infty$. The conclusion follows using once more the continuity of v.
Standard arguments allow us to deduce from (13) that

$$\begin{aligned} &u(x) \to u(\overline{x}) \ , \ v(y) \to v(\overline{x}), \\ &\frac{x-y}{\varepsilon} + \chi(\frac{x+y}{2}) = o(1). \end{aligned} \tag{14}$$

The first consequence of these properties is that $\overline{x} \in \mathcal{M}$. The next step consists in showing that, for ε small enough, the viscosity inequalities associated to the equation $F = 0$ hold for u and v.
To this end, we have to prove that, for ε small enough, either $x \in \Omega$ or $u(x) > \varphi(x)$ and either $y \in \Omega$ or $v(y) < \varphi(y)$. Because of the first property in (14), this is obviously the case if $\overline{x} \notin \Gamma_u \cup \Gamma_v$. If $\overline{x} \in \Gamma_u$, the conclusion is true for y by the same argument. The fact that $x \in \Omega$ follows using the second property of (14) together with **(H3)**. Indeed the continuity of χ and its definition in a neighborhood of Γ_u yield

$$x - (y - \varepsilon\tilde{n}(y)) = \varepsilon o(1) .$$

The proof for the case $\overline{x} \in \Gamma_v$ is similar.

Next denote by ξ the function given by

$$\xi(z,w) = |\frac{z-w}{\varepsilon} + \chi(\frac{z+w}{2})|^4 .$$

For ε small enough and for all $\alpha > 0$, there exist $(p, X) \in \overline{J}^{2,+}u(x)$ and $(q, Y) \in \overline{J}^{2,-}v(y)$ such that

$$p = D_x\xi(x,y) ,$$

$$q = -D_y\xi(x,y) ,$$

$$-(\frac{1}{\alpha} + ||D^2\xi(x,y)||)Id \le \begin{pmatrix} X & 0 \\ 0 & -Y \end{pmatrix} \le (Id + \alpha D^2\xi(x,y))D^2\xi(x,y) ,$$

and

$$F(x,u(x),p,X) \leq 0 \quad \text{and} \quad F(y,v(y),q,Y) \geq 0\,. \tag{15}$$

It remains to check that the estimates required by **(H2)** on p, q, X, Y hold. Recall that we already checked the estimate on $|x-y|$. For some constant $C > 0$, we also have

$$\begin{aligned}
|p-q| &= 2|\frac{x-y}{\varepsilon} + \chi(\frac{x+y}{2})|^2 |D\chi(\frac{x+y}{2})(\frac{x-y}{\varepsilon} + \chi(\frac{x+y}{2}))| \\
&= \frac{1}{4}|D\chi(\frac{x+y}{2})\varepsilon(p+q)| \\
&\leq C\varepsilon(|p|+|q|)\,.
\end{aligned}$$

For X and Y analogous straightforward computations show that, for a suitable choice of α,

$$-\frac{o(1)}{\varepsilon^2}Id \leq \begin{pmatrix} X & 0 \\ 0 & -Y \end{pmatrix} \leq \frac{o(1)}{\varepsilon^2}\begin{pmatrix} Id & -Id \\ -Id & Id \end{pmatrix} + o(1)Id\,.$$

The end of the proof just consists of using the above estimates together with **(H1)** and **(H2)** in the inequality

$$F(x,u(x),p,X) - F(x,v(y),p,X) \leq F(y,v(y),q,Y) - F(x,v(y),p,X)\,,$$

which is a consequence of (15). It follows that $M \leq o(1)$, which is in contradiction with the assumption $M > 0$.

4 Discontinuities and boundary conditions

We discuss here the conditions that we imposed on the discontinuous sub-solution u and the discontinuous super-solution v in the statement of Theorem 3.1, namely that for all $x \in \partial\Omega$,

$$u(x) = \limsup_{\substack{y \to x \\ y \in \Omega}} u(y)\,,$$

or that for all $x \in \partial\Omega$,

$$v(x) = \liminf_{\substack{y \to x \\ y \in \Omega}} v(y)\,.$$

If one of these conditions does not hold, the question is whether we can we redefine u and/or v on the boundary in order for them to hold without losing the sub- and/or super-solution property.

In this section, instead of **(H3)** we assume that $\partial\Omega$ is of class C^2. This yields the existence of a neighborhood of $\partial\Omega$ in which the signed distance function to the boundary $d(.)$, normalized to be positive in Ω and negative outside, is of class C^2.

We start with the

Proposition 4.1 : *Assume that u (resp. v) is a usc sub-solution of* (1)-(3) *(resp. lsc super-solution) of* (1)-(4)*) and that, at $x_0 \in \partial\Omega$,*

$$u(x_0) > \limsup_{\substack{y \to x_0 \\ y \in \Omega}} u(y) \quad \textit{and} \quad u(x_0) > \varphi(x_0)$$

(resp.

$$v(x_0) < \liminf_{\substack{y \to x_0 \\ y \in \Omega}} v(y) \quad \textit{and} \quad v(x_0) < \varphi(x_0)\ ).$$

Then, for all $(p, M) \in J^{2,+}u(x_0)$ (resp. $\in J^{2,-}v(x_0)$) and for all λ, $\mu \in \mathbb{R}$,

$$F(x_0, u(x_0), p + \lambda Dd(x_0), M + \lambda D^2 d(x_0) + \mu Dd(x_0) \otimes Dd(x_0)) \leq 0$$

(resp. $F(x_0, u(x_0), p + \lambda Dd(x_0), M + \lambda D^2 d(x_0) + \mu Dd(x_0) \otimes Dd(x_0)) \geq 0$*).*

The main consequence of this result is the

Corollary 4.1 : *Assume that F satisfies the following property :*
(H4) *For all $x \in \partial\Omega$, $r \in \mathbb{R}$, $p \in \mathbb{R}^N$ and $M \in \mathcal{S}^N$, there exist $\lambda_1, \mu_1 \in \mathbb{R}$ (resp. λ_2, $\mu_2 \in \mathbb{R}$) such that*

$$F(x, r, p + \lambda_1 Dd(x), M + \lambda_1 D^2 d(x) + \mu_1 Dd(x) \otimes Dd(x)) > 0$$

$$(\textit{resp. } F(x, r, p + \lambda_2 Dd(x), M + \lambda_2 D^2 d(x) + \mu_2 Dd(x) \otimes Dd(x)) < 0\ ).$$

Let u (resp. v) be a bounded usc sub-solution of (1)-(3) *(resp. a bounded lsc super-solution of* (1)-(4)*) and define $\tilde{u}$ (resp. $\tilde{v}$) by*

$$\tilde{u}(x) = \begin{cases} u(x) & \textit{if } x \in \Omega \\ \limsup\limits_{\substack{y \to x \\ y \in \Omega}} u(y) & \textit{if } x \in \partial\Omega \end{cases}$$

(resp.

$$\tilde{v}(x) = \begin{cases} v(x) & \textit{if } x \in \Omega \\ \liminf\limits_{\substack{y \to x \\ y \in \Omega}} v(y) & \textit{if } x \in \partial\Omega \end{cases}),$$

then $\tilde{u}$ (resp. $\tilde{v}$) is still a sub-solution of (1)-(3) *(resp. a super-solution of* (1)-(4)*).*

The point is, of course, that now the discontinuous sub- and super-solutions $\tilde{u}$ and $\tilde{v}$ satisfy the requirements of Theorem 3.1.
Before providing the proof of these results, we examine the case of equation (5). It is clear that the first point to consider is the dependence in μ since this is a measurement of the ellipticity of the equation in the $Dd(x)$

direction. Indeed, if the condition **(H4)** is not satisfied, this implies that, for some $x \in \partial\Omega$ and $p \in \mathbb{R}^N$ and for all $\lambda \in \mathbb{R}$

$$\sigma(x, p + \lambda Dd(x))Dd(x) = 0 .$$

Therefore if we assume, for example, that $A(x,p) \geq \nu(p)Id$ with $\nu(p) > 0$ – in the example of section 2, for instance, $\nu(p) = (1+|p|^2)^{-1}$ –, then Corollary 4.1 applies to equation (5). As a consequence, Theorem 3.1 yields the comparison in Ω without the additional assumptions for the discontinuous sub or supersolution on the boundary.

We now turn to the **proof of Proposition 4.1**. We only consider the subsolution case, the super-solution case being proved by analogous arguments. If $(p, M) \in J^{2,+}u(x_0)$, there exists a smooth function ϕ such that $p = D\phi(x_0)$, $M = D^2\phi(x_0)$ and $u - \phi$ has a local maximum at x_0. Since

$$u(x_0) > \limsup_{\substack{y \to x_0 \\ y \in \Omega}} u(y) ,$$

it is clear that x_0 is still a local maximum of $x \mapsto u(x) - \phi(x) - \Psi(d(x))$, for any smooth function Ψ, the points on the boundary being the only ones playing a role in the maximization. Since $u(x_0) > \varphi(x_0)$ and u is a subsolution, the result follows from the inequality $F \leq 0$, setting $\lambda = \Psi'(0)$, $\mu = \Psi''(0)$.

The proof of Corollary 4.1 is a consequence of the following lemma.

Lemma 4.1 *Assume* **(H4)** *and let u (resp. v) be a usc bounded subsolution of* (1)-(3) *(resp. a lsc bounded super-solution of* (1)-(4)*). Then, for all $x \in \partial\Omega$,*

$$\textit{either} \quad u(x) \leq \varphi(x) \quad \textit{or} \quad u(x) = \tilde{u}(x)$$

$$(\textit{resp. either} \quad v(x) \geq \varphi(x) \quad \textit{or} \quad v(x) = \tilde{v}(x) \quad).$$

Proof : We only prove this result for u. Let $x \in \partial\Omega$ be such that $u(x) > \tilde{u}(x)$. For $\varepsilon > 0$, let $x_\varepsilon \in \overline{\Omega}$ be the global maximum of $y \mapsto u(y) - |y-x|^2/\varepsilon^2$. Classical arguments show that $x_\varepsilon \to x$, $u(x_\varepsilon) \to u(x)$ and therefore, for ε small enough, $u(x_\varepsilon) > \tilde{u}(x_\varepsilon)$. This implies that $x_\varepsilon \in \partial\Omega$ and, since $J^{2,+}u(x_\varepsilon) \neq \emptyset$, Proposition 4.1 and **(H4)** yield $u(x_\varepsilon) \leq \varphi(x_\varepsilon)$. Hence $u(x) \leq \varphi(x)$.

Proof of Corollary 4.1 : Again we just present the proof in the subsolution case, the other one being analogous. We first notice that $\tilde{u}$ is actually usc. Let ϕ be a smooth function and x be a local maximum of $\tilde{u} - \phi$. The only problem with checking the sub-solution property is on the boundary since $u = \tilde{u}$ on Ω. For $x \in \partial\Omega$, two cases have to be considered.

The first one is when $u(x) \leq \varphi(x)$. Since $\tilde{u}(x) \leq u(x)$ by definition, we have $\tilde{u}(x) \leq \varphi(x)$. The subsolution inequality then holds.
The second case is when $u(x) > \varphi(x)$. Lemma 4.1 yields that $\tilde{u}(x) = u(x)$. It is then easy to see that x is also a local maximum of $u - \phi$. Indeed if y is sufficiently close to x and such that $u(y) > \tilde{u}(y)$ then $y \in \partial\Omega$ and, again by Lemma 4.1, $u(y) \leq \varphi(y)$. Since $u(x) > \varphi(x)$ this proves that x is a local maximum of $u - \phi$. Since the viscosity inequality $F \leq 0$ holds for u at x and $\tilde{u}(x) = u(x)$, the proof is complete.

5 The time-dependent case

We consider the initial-boundary value problem

$$u_t + F(x, t, u, Du, D^2u) = 0 \quad \text{in } \Omega \times (0,T)\,, \tag{16}$$

$$u(x, 0) = u_0 \quad \text{on } \overline{\Omega}\,, \tag{17}$$

$$u(x,t) = \varphi(x,t) \quad \text{on } \partial\Omega \times (0,T)\,, \tag{18}$$

where as above Ω is a smooth bounded domain in $\mathbb{R}^N$, F is a continuous function satisfying the ellipticity condition and the initial condition u_0 and the boundary condition φ are continuous functions satisfying

$$u_0(x) = \varphi(x, 0) \quad \text{on } \partial\Omega\,. \tag{19}$$

In order to state our result, we use the following convention. We say that F satisfies **(H1')** or **(H2')** if, for any fixed $t \in (0,T)$, $F(.,t,.,.,.)$ satisfies **(H1)** or **(H2)** with γ_R and $m_{R,K}$ independent of t and with γ_R not necessarily non-negative. It is, of course, clear that the problem (16)-(17)-(18) has to be understood in the viscosity sense, although above it is written in a classical form.
Our result is the

Theorem 5.1 : *Assume that* (19) *holds and that F satisfies* **(H1')** *and* **(H2')**. *Let u, v be respectively a bounded usc sub-solution and a bounded lsc super-solution of* (16)-(17)-(18). *If either*

$$u \in C(\overline{\Omega} \times (0,T)) \text{ and } \quad v(x,t) = \liminf_{\substack{y \to x \\ y \in \Omega}} v(y,t) \text{ for all } (x,t) \in \partial\Omega \times (0,T)\,,$$

or

$$u(x,t) = \limsup_{\substack{y \to x \\ y \in \Omega}} u(y,t) \text{ for all } (x,t) \in \partial\Omega \times (0,T) \text{ and } v \in C(\overline{\Omega} \times (0,T))\,,$$

then

$$u \leq v \quad \text{on } \overline{\Omega} \times [0,T)\,.$$

This result applies in particular to the equation

$$u_t - \Delta u + \frac{D^2 u Du \cdot Du}{1 + |Du|^2} = 0 \quad \text{in } \Omega \times (0, T)\,,$$

which describes the mean curvature motion of graphs but with Dirichlet boundary conditions. Theorem 5.1 implies that, *if there exists* a continuous solution to the initial boundary value problem, then this solution is unique. The proof of such existence result, when there is actually loss of boundary data, is, to the best of our knowledge, still an open problem.

The **proof of Theorem 5.1** follows exactly the one of Theorem 3.1 provided that we can check that the maximum of $u - v$ is not achieved at $t = 0$. This is the aim of the

Lemma 5.1 : *Under the assumptions of Theorem 5.1, we have*

$$u(x, 0) \leq u_0(x) \quad \text{and} \quad v(x, 0) \geq u_0(x) \qquad \text{on } \overline{\Omega}\,.$$

Proof : Fix $x_0 \in \overline{\Omega}$. For $\varepsilon > 0$ and $C_\varepsilon > 0$, the function

$$(x, t) \mapsto u(x, t) - \frac{|x - x_0|^2}{\varepsilon^2} - C_\varepsilon t\,,$$

attains its global maximum over $\overline{\Omega} \times [0, T)$ at $(x_\varepsilon, t_\varepsilon)$. It is then immediate that, as $\varepsilon \to 0$, and for all C_ε, $(x_\varepsilon, t_\varepsilon) \to (x_0, 0)$. Choosing C_ε sufficiently large depending on ε, we find that $(x_\varepsilon, t_\varepsilon) \in (\partial\Omega \times [0, T)) \cup (\Omega \times \{0\})$. Indeed, if not, at $(x_\varepsilon, t_\varepsilon)$ we would have

$$C_\varepsilon + F(x_\varepsilon, t_\varepsilon, u(x_\varepsilon, t_\varepsilon), \frac{2(x_\varepsilon - x_0)}{\varepsilon^2}, \frac{2}{\varepsilon^2} I) \leq 0\,,$$

which is impossible if C_ε is large enough.

So we have either $t_\varepsilon = 0$ and $u(x_\varepsilon, 0) \leq u_0(x_\varepsilon)$ or $t_\varepsilon > 0$ and $x_\varepsilon \in \partial\Omega$ with $u(x_\varepsilon, t_\varepsilon) \leq \varphi(x_\varepsilon, t_\varepsilon)$. In either case, letting $\varepsilon \to 0$, we obtain the desired inequality for u.

The argument for the other inequality is similar.

Acknowledgments: This work was partially supported by the TMR Programme "Viscosity Solutions and their Applications".

References

Barles G. (1994) *Solutions de viscosité des équations de Hamilton-Jacobi.* Collection "Mathématiques et Applications" of SMAI, n°17, Springer-Verlag.

Barles G. and Burdeau J. (1995) *The Dirichlet problem for semilinear second-order degenerate elliptic equations and applications to stochastic exit time control problems,* Comm. in PDE 20 (1& 2), pp 129-178.

Barles G. and Rouy E. (1998) *A Strong Comparison Result for the Bellman Equation arising in Stochastic Exit Time Control Problems and its Applications*, Comm. in PDE, to appear.

Crandall M.G., Ishii H. and Lions P.-L. (1992) *User's guide to viscosity solutions of second order Partial differential equations*, Bull. Amer. Soc., **27**, pp 1-67.

Crandall M.G. and Lions P.-L. (1983) *Viscosity solutions of Hamilton-Jacobi Equations*, Trans. Amer. Math. Soc., **277**, pp 1-42.

W.H Fleming W.H and Soner H.M (1993) *Controlled Markov Processes and Viscosity Solutions*, Applications of Mathematics, Springer-Verlag, New-York.

Gilbarg D. and Trudinger N.S. (1983) *Elliptic Partial Differential Equations of Second-Order*. Springer, New-York.

Ishii H. (1987) *Perron' s method for Hamilton-Jacobi Equations*. Duke Math. J., **55**, pp 369-384.

Katsoulakis M. (1994) *Viscosity solutions of second-order fully nonlinear elliptic equations with state constraints*. Indiana Univ. Math. J., **43**, No 2, pp 493-519.

Ladyzhenskaya O.A and Ural'tseva N.N (1968) *Linear and Quasilinear Elliptic Equations*. Academic Press.

Lions P.-L (1985) *Neumann type boundary conditions for Hamilton-Jacobi Equations*. Duke Math. J., **52**, pp 793-820.

Soner M.H (1986) *Optimal control problems with state-space constraints*. SIAM J. on Control and Optimisation, **24**, Part I: pp 552-562, Part II: pp 1110-1122.

13

A Generalized Hamilton–Jacobi–Bellman Equation for Deterministic Optimal Control Problems

Leonard D. Berkovitz

1 Introduction

We shall be concerned with the following problem. Minimize $g(t_1, x_1)$ subject to

$$\frac{dx(t)}{dt} = f(t, x(t), u(t)) \qquad x(\tau) = \xi \tag{1}$$

$$u(t) \in Z \text{ a.e.} \tag{2}$$

$$(t_1, x_1) \in \mathcal{T}, \tag{3}$$

where t_1 is the first time that the solution $x(\cdot)$ of (1) hits a preassigned set $\mathcal{T}$ and $x_1 = x(t_1)$.

We now state the problem more precisely. Let $T > 0$, and let $\mathcal{U}$ be an open set in $\mathbf{R^n}$. Let f be a function from $[0, T] \times \mathbf{R^n} \times \mathcal{U}$ to $\mathbf{R^n}$. We shall denote points in $[0, T]$ by the letter t, points in $\mathbf{R^n}$ by x and points in $\mathbf{R^m}$ by z. Let Z be a fixed set contained in $\mathcal{U}$. Let $\mathcal{T}$ be a fixed set contained in $[0, T] \times \mathbf{R^n}$ and let g be a real valued function defined in $\mathcal{T}$. A measurable function $u(\cdot)\colon [0, T] \to \mathbf{R^n}$ that satisfies (2) will be called an *admissible control*. A function $\varphi(\cdot) = \varphi(\cdot; \tau, \xi)$ that satisfies (1) will be called a *trajectory starting at* (τ, ξ) *corresponding* to u. If the trajectory has the additional property that there exists $\tau < t_1 < T$ such that $(t_1, \varphi(t_1; \tau, \xi)) \in \mathcal{T}$ and $(t, \varphi(t; \tau, \xi)) \notin \mathcal{T}$ for $\tau \leq t < t_1$, then φ will be called an *admissible trajectory*. The pair (φ, u) will be called an *admissible process* starting at (τ, ξ). We shall denote the set of all admissible processes starting at (τ, ξ) by $\mathcal{A}(\tau, \xi)$. The control problem can be formulated precisely as:

(P) "Minimize $g(t_1, \varphi(t_1, \tau, \xi))$ over all admissible processes starting at (τ, ξ)."

Existence theorems for this problem can be found in [Ber; 1974] and in [Ces]. The assumptions that we make about the data of the problem will be given in Section 2.

Let $\mathcal{D}$ be a subset of $[0,T] \times \mathbf{R}^n$ such that for each (τ,ξ) in $\mathcal{D}$ the set of admissible processes $\mathcal{A}(\tau,\xi)$ is not empty and (P) has a solution. Let

$$W(\tau,\xi) = \min\{g(t_1, \varphi(t_1;\tau,\xi)) : (\varphi,u) \in \mathcal{A}(\tau,\xi)\}. \tag{4}$$

The function W is called the value function. It is known (See [Ber;1974], [Ces], [Fle.-Ris.]) that under appropriate hypotheses on the data of the problem that if W is C' on the interior of $\mathcal{D}$, then W satisfies the Hamilton–Jacobi–Bellman equation (HJB)

$$-W_t(t,x) + H(t,x,-W_x(t,x)) = 0 \tag{5}$$

on the interior of $\mathcal{D}$ and boundary condition

$$W(t,x) = g(t,x) \qquad (t,x) \in \mathcal{T}. \tag{6}$$

Here,

$$H(t,x,p) = \sup\{\langle p, f(t,x,z)\rangle : z \in Z\}, \tag{7}$$

where $\langle a,b\rangle$ denotes the inner product of two vectors a and b in euclidean space.

It is also known that if a function V is a C' solution of (5) and (6), then V is a verification function [Fle.-Ris.]. By this is meant that if (φ^*, u^*) is an admissible process with initial point (τ,ξ) such that

$$H(t,\varphi^*(t), V_x(t,\varphi^*(t)) = \langle -V_x(t,\varphi^*(t)),\ f(t,\varphi^*(t),u^*(t))\rangle,$$

then (φ^*, u^*) is optimal for (P). Moreover, $V(\tau,\xi) = W(\tau,\xi)$. Thus C' solutions of (5) and (6) coincide with the value function and furnish a means of checking for optimality of a trajectory.

The *optimal synthesis problem* or *optimal feedback control* problem is to find a possibly set valued function U defined on $\mathcal{D}$ such that for each (t,x) in $\mathcal{D}$, $U(t,x) \subseteq Z$, for each initial point (τ,ξ), the differential inclusion

$$x'(t) \in f(t,x,U(t,x)) \qquad x(\tau) = \xi \tag{8}$$

has solutions $\psi(\cdot;\tau,\xi)$, and each solution is a solution to (P). In principle, the function U defined by

$$U(t,x) = \arg\max\{\langle -W_x(t,x), f(t,x,z)\rangle : z \in Z\}$$

should furnish an optimal synthesis.

The analysis outlined above is sometimes referred to as the dynamic programming method or Hamilton–Jacobi theory in optimal control. Its applicability is severely limited by the requirement that the value function be C' or that a C' solution of the HJB equation exists, since in most problems of interest this is not the case. In [Fle.-Ris.] and [Ber;1970] a Hamilton–Jacobi theory was developed for problems in which it was assumed that a piecewise continuous synthesis U exists and that the resulting field of trajectories of (8) had certain structure. The value functions W for such problems are piecewise C' and satisfy (5) at points of differentiability of W.

An entirely different approach is to replace (5) by a generalized HJB equation. One method used in problems where W is assumed to be Lipschitz continuous is to replace (W_t, W_x) in (5) by the Clarke generalized derivative. This method is given in [Cla.]. Another method is to replace the partials of W by lower Dini derivates.

If L is a function defined on an open set $\mathcal{O}$ in $\mathbf{R}^n$, then the *lower Dini derivate* at x_0 in the direction v, denoted by $D^- L(x_0; v)$, is defined to be

$$D^- L(x_0; v) = \liminf_{\substack{\delta \to 0 \\ v' \to v}} \frac{L(x_0 + \delta v') - L(x_0)}{\delta}. \tag{9}$$

If L is Lipschitz continuous in a neighborhood of x_0, then (9) is equivalent to

$$D^- L(x_0; v) = \liminf_{\delta \to 0} \frac{L(x_0 + \delta v) - L(x_0)}{\delta}. \tag{10}$$

In [Ber;1989] the lower Dini derivate was used to obtain a HJB theory and optimal synthesis for control problems with compact Z, n–dimensional terminal sets and conditions on f and g that guaranteed that the value function is Lipschitz continuous. Vinter and Rowland [Row.-Vin.] considerably improved the synthesis results for problems with fixed terminal time and free end points. Frankowska [Fra;1989] [Fra;1993] and Mirica [Mir.] developed Hamilton–Jacobi theories for value functions that are not assumed to be locally Lipshitz continuous. The results in this case involve both upper and lower Dini derivates.

In this paper we shall study value functions and verification functions that are continuous on $\mathcal{D}$ but not necessarily locally Lipschitz continuous. We shall use a derivate that we call "lower tangential derivate" rather than the derivates defined in (9) or (10). The lower tangential derivate of a function L involves the difference quotients of the form $[L(\tau + s, \xi + x) - L(\tau, \xi)]/s$, where $(\tau + s, x)$ is "reachable" at time $\tau + s$. In (9) we may be considering points near (τ, ξ) that are not "reachable" from (τ, ξ) and therefore are extraneous to our considerations. In (10) we may not be considering enough points near (τ, ξ). The use of the lower tangential derivate simplifies some arguments and results in sharper results in some cases. We shall give the precise definition of lower tangential derivate in the next section.

2 Assumptions and Definitions

The norm of a vector x will be denoted by $\|x\|$. Thus, $\|x\| = \langle x, x\rangle^{1/2}$. The symbol $\mathcal{N}_\delta(t_0, x_0)$ will denote a δ–neighborhood of (t_0, x_0); i. e. $\mathcal{N}_\delta(t_0, x_0) = \{(t, x) : \|(t, x) - (t_0, x_0)\| < \delta\}$.
We shall change the formulation of (P) given in the introduction by taking the terminal time to be fixed, say at $t_1 = 1$ and taking the terminal set to be a set in $\mathbf{R}^{n+1}$ lying in the hyperplane $t_1 = 1$. That is, we assume the terminal time to be fixed at $t_1 = 1$ and the terminal states x_1 to lie in some set $\mathcal{T} \subseteq \mathbf{R}^n$. There is no loss of generality in formulating the terminal conditions this way rather than as in (P). To see this we take t to be a state variable, adjoin another scalar state variable y and take s to be the new "time" or independent variable. The new state equations are $dy/ds = 0$, $dt/ds = y$ and $dx/dt = f(t, x, z)y$. The new initial conditions are $s = 0, t(0) = \tau$ and $x(0) = \xi$; the new terminal conditions are $s = 1$, $y(1) = t(1) - \tau$, and $(t(1), x(1))$ in the original set $\mathcal{T}$.
Assumption A1. The function $f = (f_1, \ldots, f_n)$ has range in $\mathbf{R}^n$ and is defined and continuous on $[0, T] \times \mathbf{R}^n \times \mathcal{U}$, where $T > 1$ and $\mathcal{U}$ is an open set in $\mathbf{R}^n$.
Assumption A2. There exists a function $k(\cdot)$ in $L_1[0, T]$ such that for all (t, x, z) in the domain of definition of f,

$$\|f(t, x, z)\| \leq k(t)(1 + \|x\|).$$

Assumption A3. Let Z be a fixed set contained in $\mathcal{U}$. For each (t, x) in $[0, T] \times \mathbf{R}^n$ let

$$F(t, x) = \{y : y = f(t, x, z),\ z \in Z\}.$$

Then:
(i) For each (t, x) in $[0, T] \times \mathbf{R}^n$ the set $F(t, x)$ is closed and convex, and
(ii) $F(t, x) = \bigcap_{\delta > 0} clcoF(\mathcal{N}_\delta(t, x))$,
where $F(\mathcal{N}\delta(t, x)) = \bigcup_{(t', x') \in \mathcal{N}_\delta(t, x)} F(t', x')$.
Assumption A4. (i) The target set $\mathcal{T}$ is a closed subset of $\mathbf{R}^n$. (ii) The function g is continuous on $\mathcal{T}$ and is bounded below on $\mathcal{T}$.
Assumption A5. There exists a maximal set $\mathcal{D} \subset [0, 1] \times \mathbf{R}^n$ with the following properties.
(i) $\mathcal{T} \subset \partial\mathcal{D}$, where $\partial\mathcal{D}$ denotes the boundary of $\mathcal{D}$.
(ii) The set $\mathcal{D}$ has non–empty interior. There exists a sequence $\{\mathcal{D}_m\}$ of regions (open connected sets) such that: (a) $\mathcal{D}_m \subseteq \mathcal{D}_{m+1}$; (b) for each m, $\overline{\mathcal{D}}_m$ (the closure of $\mathcal{D}_m$) is contained in the interior of $\mathcal{D}$; (c) $\bigcup \mathcal{D}_m =$ interior of $\mathcal{D}$; and (d) for each m greater than some m_0, the intersection $[\{1 - 1/m\} \times \mathbf{R}^n] \cap \mathcal{D}_m \neq \phi$.
(iii) For each (τ, ξ) in $\mathcal{D}$, the set $\mathcal{A}(\tau, \xi)$ of admissible processes starting at (τ, ξ) is not empty and (P) has a solution. If (τ, ξ) is an interior point of

$\mathcal{D}$, then at least one of the optimal trajectories remains in the interior of $\mathcal{D}$ for $\tau \le t < 1$.
(iv) The value function W is continuous on $\mathcal{D} \cup \mathcal{T}$.
We note that since $W = g$ on $\mathcal{T}$, that (iv) of A5 could be stated as $\lim_{(t,x)\to(1,x_1)} W(t,x) = g(x_1)$.
Assumptions A1 and A2 guarantee that for each (τ,ξ) in $[0,1) \times \mathbf{R}^n$ and each measurable function u on $[\tau, T]$ with $u(t) \in Z$ there exists a solution to (1) and (2). There is no guarantee, however, that the solution will hit the target set $\mathcal{T}$ if $\mathcal{T}$ is not the hyperplane $t = 1$ in $\mathbf{R}^{n+1}$. Assumptions A1–A4 guarantee that if $A(\tau,\xi)$ is not empty; i.e. there exist solutions that hit $\mathcal{T}$, then (P) has a solution (See the existence theorems in [Ber;1974], [Ces.]). Thus in (iii) of A5 we could have omitted the phrase "and (P) has a solution," since it is a consequence of A1–A4 and $\mathcal{A}(\tau,\xi) \ne \phi$.
Assumption A4 is a standard condition in existence theorems in which compactness of $F(t,x)$ is not assumed. It is called the Cesari property in [Ber;1974] and Property Q in [Ces.]. The convexity of $F(t,x)$ always holds in relaxed problems:
The *reachable set* at *time* $\tau \le t \le T$ from (τ,ξ), denoted by $R(t;\tau,\xi)$ is defined by

$$R(t;\tau,\xi) = \{x : x = \varphi(t;\tau,\xi);\ \varphi \text{ a solution of (1) for some } u\}.$$

The *reachable set from* (τ,ξ), denoted by $R(\tau,\xi)$ is defined by $R(\tau,\xi) = \bigcup_{\tau\le t\le T} R(t;\tau,\xi)$.
Definition 1. Let (τ,ξ) be in $\mathcal{D}$. A vector v in $\mathbf{R}^n$ will be called a *direction vector* for $R(\tau,\xi)$ at the point (τ,ξ) if there exists a sequence $\{s_k\}, s_k > 0, s_k \to 0$ and a sequence $\{x_k\}$ of points in $\mathcal{D}$ such that $x_k \in R(\tau+s_k;\tau,\xi)$ and

$$\lim x_k = \xi \qquad \lim(x_k - \xi)/s_k = v.$$

We shall call the sequence $\{(s_k,x_k)\}$ a *secant sequence* for v. We shall denote a secant sequence for v by $\sigma(v)$.
Let $\mathcal{G}(\tau,\xi)$ denote the graph of the set valued map $s \to R(\tau+s;\tau,\xi)$, $s > 0$. It is easy to show that a vector v is a direction vector of $R(\tau,\xi)$ if and only if the vector $(1,v)$ is a tangent vector in the sense of Bouligand of the set $\mathcal{G}(\tau,\xi)$ at the point (τ,ξ).
Definition 2. Let L be a real valued function defined on $\mathcal{D}$. We define the *lower tangential derivate of* L *at* (τ,ξ) *in the direction* $(1,v)$, denoted by $D_T^-(L(\tau,\xi);(1,v))$, to be

$$D_T^-(L(\tau,\xi);(1,v)) = \inf_{\sigma(v)} [\liminf_{k\to\infty} \frac{L(\tau+s_k,x_k) - L(\tau,\xi)}{s_k} : \{(s_k,x_k)\} \in \sigma(v)].$$

That is, for each secant sequence $\sigma(v)$ we calculate the indicated lim inf and them take the infimum over all $\sigma(v)$.
We now show that the notion of direction vector is not vacuous, and hence neither is the notion of lower tangential derivate. In fact, we shall show in

Theorem 2.1 below that at an interior point (τ, ξ) of $[0, T) \times \mathbf{R^n}$, the set of direction vectors is $F(\tau, \xi)$.

In establishing these statements and elsewhere we shall utilize the following result.

Lemma 2.1 *For each (τ, ξ) in $[0, T) \times \mathbf{R^n}$ there exists an $M > 0$ such that for all (t, x) in $R(\tau, \xi)$*

$$\|F(t, x)\| \leq Mk(t). \tag{11}$$

It follows from A2 (See [Ber;1974]) that the set $R(\tau, \xi)$ is compact. The inequality (11) then again follows from A2.

We shall also need the following result.

Lemma 2.2 *Let K be a set in $\mathbf{R^n}$ of positive Lebesgue measure. Let $f: K \to \mathbf{R^m}$ be Lebesgue integrable on K. Let $\mathcal{K}$ denote the set of probability measures μ on K such that $\int_K f d\mu$ exists and is finite. Let $P = clco\{f(x): x \in K\}$ and let $\mathcal{P} = cl\{\int_K f d\mu: \mu \in \mathcal{K}\}$. Then $P = \mathcal{P}$.*

Proof: We first show that $P \subseteq \mathcal{P}$. Let x be an arbitrary point in K and let μ_x in $\mathcal{K}$ denote the δ-measure concentrated at x. Then $f(x) = \int_K f d\mu_x$ and so

$$\{f(x) : x \in \mathcal{K}\} \subseteq \{\int_{\mathcal{K}} f d\mu : \mu \in \mathcal{K}\}.$$

The set on the right is clearly convex, so the convex hull of the set on the left is contained in the set on the right. Hence, $P \subseteq \mathcal{P}$.

We now show that $P = \mathcal{P}$. If this were not true, then there would exist an element $\hat{p}$ in $\mathcal{P}$ that is not in P. We shall show that this leada to a conteadiction. Since P is closed and convex and $\hat{p} \notin P$, there exists a hyperplane Γ in $\mathbf{R^n}$ containing the point $\hat{p}$ such that P is strictly on one side of Γ. Thus, if we take ζ to be the normal to Γ that points to P, then there exists a real number $c > 0$ such that $\langle \zeta, y - \hat{p} \rangle \geq c$ for all y in P. In particular, for all x in K, we have $\langle \zeta, f(x) - \hat{p} \rangle \geq c$.

Since $\hat{p} \in \mathcal{P}$, there exists a sequence $\{\mu_n\}$ in $\mathcal{K}$ such that $\int_K f d\mu_n \to \hat{p}$. Also, $\hat{p} = \int_K \hat{p} d\mu_n$, since the μ_n are probability measures. Thus $\int_K (f(x) - \hat{p}) d\mu_n \to 0$ and so

$$\int_K \langle \zeta, f(x) - \hat{p} \rangle d\mu_n \to 0.$$

But this cannot be, since $\langle \zeta, f(x) - \hat{p} \rangle \geq c > 0$ for all x in K.

Theorem 2.1 *Let $\{s_k\}$ be a sequence of positive numbers such that* $\lim s_k = 0$. *Let $\{x_k\}$ be a sequence of points such that $x_k \in R(\tau + s_k; \tau, \xi)$. Then* $\lim x_k = \xi$ *and there exists a subsequence, which we again label as $\{(s_k, x_k)\}$, and a vector v in $F(\tau, \xi)$ such that*

$$\lim (x_k - \xi)/s_k = v. \tag{12}$$

Conversely, for every v in $F(\tau,\xi)$ there exists a sequence $\{(s_k, x_k)\}$ with $s_k > 0$, $\lim s_k = 0$, $x_k \in R(\tau+s_k;\tau,\xi)$, $\lim x_k = \xi$ and such that (12) holds.

Proof: Since $x_k \in R(\tau+s_k;\tau,\xi)$, there exists a control u and a trajectory ψ_k such that $x_k = \psi_k(\tau+s_k)$. For $0 \le s \le s_k$,

$$\|\psi_k(\tau+s)-\xi\| \le \int_0^s \|\psi_k'(\tau+\sigma)\|d\sigma \le \int_0^{s_k} k(s)M\,ds = Ms_k\|k\|_1, \quad (13)$$

where $\|\ \|_1$ denotes L_1 norm and the last inequality follows from (11). On setting $s = s_k$ in (13) we get $\lim x_k = \xi$. We also get that the sequence h_k defined by

$$h_k \equiv (\psi_k(\tau+s_k)-\xi)/s_k$$

is bounded. Hence there exists a subsequence of the h_k and a corresponding subsequence of $\{(s_k, x_k)\}$, which we again write as h_k and $\{(s_k, x_k)\}$, such that h_k converges to some vector v. We henceforth shall be working with this subsequence.

Let $\varepsilon > 0$ be given. It follows from (13) that there exists a $K(\varepsilon)$ such that for $k > K(\varepsilon)$ and $0 \le s \le s_k$, we have $(\tau+s, \psi_k(\tau+s)) \in \mathcal{N}_\varepsilon(\tau,\xi)$. Since $\psi_k'(\tau+s) \in F(\tau+s, \psi_k(\tau+s))$ a.e., it follows that for $k > K(\varepsilon)$ and $0 \le s \le s_k$,

$$\psi_k'(\tau+s) \in F(\mathcal{N}_\varepsilon(\tau,\xi)), \qquad \text{a.e.} \quad (14)$$

From (14) and from Lemma 2.2, we get that for $k > K(\varepsilon)$

$$\begin{aligned}\int_0^{s_k} \psi_k'(\tau+s)\left(\frac{ds}{s_k}\right) &\in clco\{\psi_k'(\tau+s) : 0 \le s \le s_k\}\\ &\subseteq clco\{F(\mathcal{N}_\varepsilon(\tau,\xi))\}.\end{aligned}$$

Therefore

$$h_k = \int_0^{s_k} \psi_k'(\tau+s)\left(\frac{ds}{s_k}\right) \in clco\{F(\mathcal{N}_\varepsilon(\tau,\xi))\}.$$

But $\lim h_k = v$, so $v \in clco\{F(\mathcal{N}_\varepsilon(\tau,\xi))\}$ for arbitrary $\varepsilon > 0$. Hence

$$v \in \bigcap_{\varepsilon>0} clco\{F(\mathcal{N}_\varepsilon(\tau,\xi))\} = F(\tau,\xi),$$

where the equality follows from (ii) of A3.

We now show that if $v \in F(\tau,\xi)$, then there exists a sequence $\{(s_k, x_k)\}$ such that $x_k \in R(\tau+s_k;\tau,\xi)$, $s_k \to 0$, $x_k \to \xi$ and such that (12) holds. If $v \in F(\tau,\xi)$ then there exists a z_v in Z such that $v = f(\tau,\xi,z_v)$. Let $\psi(\cdot) = \psi(\cdot;\tau,\xi)$ denote the trajectory that satisfies $\psi'(\tau+s) = f(\tau+s, \psi(\tau+s), z_v)$

and $\psi(\tau) = \xi$. Thus for $s > 0$, $\psi(\tau + s) \in R(\tau + s; \tau, \xi)$, and

$$\left\| \frac{\psi(\tau + s) - \xi}{s} - f(\tau, \xi, z_v) \right\| \leq \frac{1}{s} \int_0^s \|f(\tau + t, \psi(\tau + t), z_v) - f(\tau, \xi, z_v)\| dt. \tag{15}$$

Let $\varepsilon > 0$ be given. Since f and ψ are continuous and $\psi(\tau) = \xi$, there exists a $\delta > 0$ such that if $0 < t < \delta$ then $\|f(\tau + t, \psi(\tau + t), z_v) - f(\tau, \xi, z_v)\| < \varepsilon$. Hence for $0 < s < \delta$, the right hand side of (15) is less then ε. From this the assertion follows.

3 The Value Function

Theorem 3.1 *At each point (τ, ξ) in the interior of $\mathcal{D}$,*

$$\min_{v \in F(\tau,\xi)} D_T^-(W(\tau, \xi); (1, v)) = 0. \tag{16}$$

Proof: Let $v \in F(\tau, \xi)$. Let $\{(s_k, x_k)\}$ be any secant sequence for v. By Theorem 2.1, such sequences exist. Since (τ, ξ) is an interior point of $\mathcal{D}$, there exists a positive integer K such that for $k > K$, the points (s_k, x_k) are in $\mathcal{D}$. Also, for each such (s_k, x_k) there exists a trajectory $\varphi_k(\cdot; \tau, \xi)$ such that $x_k = \varphi_k(\tau + s_k; \tau, \xi)$. It then follows from the principle of optimality that

$$W(\tau + s_k, x_k) - W(\tau, \xi) = W(\tau + s_k, \varphi_k(\tau + s_k; \tau, \xi)) - W(\tau, \xi) \geq 0.$$

Hence $D_T^-(W(\tau, \xi); (1, v)) \geq 0$.
Now let $\varphi(\cdot; \tau, \xi)$ denote an optimal trajectory with initial point (τ, ξ). Let $\{s_k\}$ be a sequence of positive numbers converging to zero. Let $x_k = \varphi(\tau + s_k; \tau, \xi)$. Then by Theorem 2.1 there exists a subsequence of $\{(s_k, x_k)\}$, which we again denote by $\{(s_k, x_k)\}$, that is a secant sequence for some vector v_0 in $F(\tau, \xi)$. But, for this sequence,

$$W(\tau + s_k, x_k) - W(\tau, \xi) = W(\tau + s_k, \varphi(\tau + s_k; \tau, \xi)) - W(\tau, \xi) = 0.$$

Hence $D_T^-(W(\tau, \xi); (1, v_0)) = 0$, and the theorem is proved.
It is a straightforward elementary calculation to show that (16) implies that if W is differentiable at (τ, ξ), then (5) holds at (τ, ξ). More generally, we have the following.

Theorem 3.2 *The value function is a viscosity solution of*

$$-V_t + H(t, x, -V_x) = 0 \tag{17}$$

on the interior of $\mathcal{D}$, where $H(t, x, p) = \sup\{\langle p, f(t, x, z)\rangle : z \in Z\}$. Moreover, the value function satisfies the boundary condition

$$V(1, x_1) = g(x_1) \qquad x_1 \in \mathcal{T}. \tag{18}$$

Proof: That W satisfies (18) follows from the definition of W. We shall use the test–function definition of viscosity solution ([Cra.-Ish.-Lio.] or [Fle.-Son.]) to show that in the interior of $\mathcal{D}$, W is a viscosity solution.
Let (τ, ξ) be an arbitrary point in the interior of $\mathcal{D}$, and let w be a C' function on the interior of $\mathcal{D}$ such that $W - w$ has a local maximum at (τ, ξ). Then there exists a $\delta > 0$ such that for all (s, x) in $\mathbf{R}^{n+1}$ with $\|(s, x - \xi)\| < \delta$,

$$W(\tau + s, x) - w(\tau + s, x) \leq W(\tau, \xi) - w(\tau, \xi).$$

Hence

$$W(\tau + s, x) - W(\tau, \xi) \leq w(\tau + s, x) - w(\tau, \xi). \tag{19}$$

Let $v \in F(\tau, \xi)$. Then there exists a sequence $\{(s_k, x_k)\}$ in $\mathbf{R}^{n+1}$ that is a secant sequence for v. Since w is C', there exists a real number θ, such that $0 < \theta < 1$ and

$$w(\tau + s_k, x_k) - w(\tau, \xi) = w_t(\tau + \theta s_k)s_k + \langle w_x(\tau + \theta s_k, \xi + \theta(x_k - \xi)), x_k - \xi\rangle.$$

Hence if we divide through by $s_k > 0$ in (19), then take lim inf on both sides as $k \to \infty$ and use (16) we get that $w_t(\tau, \xi) + \langle w_x(\tau, \xi), v\rangle \geq 0$ for all $v \in F(\tau, \xi)$. Thus, $-w_t(\tau, \xi) + \langle -w_x(\tau, \xi), v\rangle \leq 0$ for all $v \in F(\tau, \xi) = f(\tau, \xi, Z)$. Therefore W is a subsolution of (17).
We now let w be a C' function such that $W - w$ has a local minimum at (τ, ξ). Then by reasoning similar to that in preceding paragraph we get that for arbitrary v in $F(\tau, \xi)$

$$D^-_{\mathcal{T}}(W(\tau, \xi); (1, v)) \geq w_t(\tau, \xi) + \langle w_x(\tau, \xi), v\rangle.$$

It now follows from (16) that

$$0 = \min_{v \in F(\tau, \xi)} D^-_{\mathcal{T}}(W(\tau, \xi); (1, v)) \geq \inf_{v \in F(\tau, \xi)} [w_t(\tau, \xi) + \langle w_x(\tau, \xi), v\rangle].$$

Hence

$$\sup_{v \in F(\tau, \xi)} [-w_t(\tau, \xi) + \langle -w_x(\tau, \xi), v\rangle] \geq 0.$$

Thus W is a supersolution of (17), and hence a viscosity solution.
The question now arises, "Does (16) characterize the value function in the sense that the value function is the unique continuous function on the union of $\mathcal{T}$ and the interior of $\mathcal{D}$ that satisfies (16) on the interior of $\mathcal{D}$ and (18) on $\mathcal{T}$?" One might approach this question by observing that any such V is viscosity solution of (17) and (18) and then invoke viscosity solution techniques to prove that V is the value function. We shall not follow this path, but shall show that $V = W$ using a "systems approach." Such an approach, independent of viscosity solution theory, was first carried out by R. Vintner [Vin.] for problems with Lipschitz continuous data.

Theorem 3.3 *Let V be a function that is continuous on the union of $\mathcal{T}$ and the interior of $\mathcal{D}$ such that for all (τ,ξ) in the interior of $\mathcal{D}$*

$$\min_{v\in F(\tau,\xi)} D_T^-(V(\tau,\xi);(1,v)) = 0 \tag{20}$$

and for all x_1 in $\mathcal{T}$

$$V(1,x_1) = g(x_1). \tag{21}$$

Then $V = W$ on the interior of $\mathcal{D}$ and on $\mathcal{T}$.

The first step in the proof is the establishment of the following lemma.

Lemma 3.1 *Let V be as in the statement of Theorem 3.3. Let (τ,ξ) be an interior point of $\mathcal{D}$ and let $\varphi(\ ;\tau,\xi)$ be any admissible trajectory that lies in the interior of $\mathcal{D}$ for $\tau \le t < 1$. Then the function θ defined by $\theta(t) = V(t,\varphi(t;\tau,\xi))$ for $\tau \le t \le 1$ is nondecreasing in $[\tau,1]$ and*

$$V(\tau,\xi) \le V(1,x_1) = g(x_1), \tag{22}$$

where $x_1 = \varphi(1,\tau,\xi)$.

The lower right hand Dini derivate of θ at a point t will be denoted by $d_-\theta(t)$ and is given by

$$\begin{aligned} d_-\theta(t) &= \liminf_{s\to 0+} \frac{\theta(t+s)-\theta(t)}{s} \\ &= \liminf_{s\to 0+} \frac{V(t+s,\varphi(t+s;\tau,\xi)) - V(t,\varphi(t;\tau,\xi))}{s}. \end{aligned}$$

Now let $\{s_k\}$ be a sequence such that

$$d_-\theta(t) = \lim_{s_k\to 0} \frac{V(t+s_k,\varphi(t+s_k;\tau,\xi)) - V(t,x)}{s_k},$$

where $x = \varphi(t;\tau,\xi)$. By Theorem 3.2 there exists a subsequence of $\{(s_k,\varphi(t+s_k;\tau,\xi))\}$ that is a secant sequence converging to an element v_* in $F(t,x)$. It then follows from (20) that $d_-\theta(t) \ge 0$ for all t in $[\tau,1)$. Hence ([Bru; Cor. 4.2, p. 128]) θ is non-decreasing, from which (22) follows.
Since $W(\tau,\xi) = \inf\{g(\varphi(t_1;\tau,\xi)) : \varphi(\ ;\tau,\xi) \text{ admissible}\}$ and since we are assuming that the optimal trajectory itself is interior to $\mathcal{D}$, except for the terminal point, it follows that we may restrict the infimum to such admissible trajectories. It then follows from (22) that

$$V(\tau,\xi) \le W(\tau,\xi). \tag{23}$$

We shall complete the proof of the theorem by showing that the opposite inequality holds.

Since (τ, ξ) is an interior point of $\mathcal{D}$, it follows from A5 that there exists an m_1 such that for $m > m_1$, $(\tau, \xi) \in \mathcal{D}_m$ and $[\{1 - 1/m\} \times \mathbf{R}^n] \cap \mathcal{D}_m \neq \phi$. Let $\mathcal{G}(\tau, \xi)$ denote the graph of the mapping $t \to R(t; \tau, \xi)$, $\tau \leq t \leq T$. It can be shown that our hypotheses imply that $\mathcal{G}(\tau, \xi)$ is compact ([Ber;1974]; Thm. 5.1, p. 146]. By Lemma 2.1, there exists an $M > 0$ such that $\|F(t, x)\| \equiv \|f(t, x, Z)\| \leq M$ for all (t, x) in $\mathcal{G}(\tau, \xi)$. Let $\mathcal{K}_m = \mathcal{G}(\tau, \xi) \cap \overline{\mathcal{D}}_m$. Then F is bounded on $\mathcal{K}_m$ and $\mathcal{K}_m$ is contained in the interior of $\mathcal{D}$.

If in Theorem 1, p. 296 of [Aub.-Cel.] we take $W = 0$ and $K = \mathcal{K}_m$, then our F, V and $\mathcal{K}_m$ satisfy the hypotheses of this Theorem with two exceptions. The first is that V takes on non–negative values and the second is that the sets $F(t, x)$ are compact. It is known that if the mapping F is upper semi–continuous, compact, and convex valued then F satisfies A3(ii), the Cesari property. (See e.g. [Ber;1974;p72]). An examination of the proof of Theorem 1, p. 296 of [Aub.-Cel.] shows that it is this consequence of F being compact valued that is used. A line by line imitation of the Aubin–Cellina theorem's proof shows that, with minor changes, we can ignore the requirement that V be non–negative and obtain a somewhat stronger statement than we obtain directly from the theorem. Specifically, one can prove the following.

Lemma 3.2 *There exists a positive integer m_1 such that for each $m > m_1$ there exists a trajectory $\varphi_m(\cdot; \tau, \xi)$ defined on $[\tau, 1 - 1/m]$ such that $(t, \varphi_m(t; \tau, \xi))$ is in the interior of $\mathcal{D}$ for $\tau \leq t \leq 1 - 1/m$ and*

$$V(1 - 1/m,\ \varphi_m(1 - 1/m; \tau, \xi)) - V(\tau, \xi) = 0.$$

Now let $s_m = 1 - 1/m$, and let $y_m = \varphi_m(s_m; \tau, \xi)$. Since (s_m, y_m) is interior to $\mathcal{D}$, there exists an optimal trajectory for the problem with (s_m, y_m) as initial point, say $\lambda_m(\cdot; s_m, y_m)$, defined on $[s_m, 1]$ and satisfying $\lambda_m(1; s_m, y_m) \in \mathcal{T}$. Define a function $\psi_m(\cdot; \tau, \xi)$ by

$$\begin{aligned} \psi_m(t; \tau, \xi) &= \varphi_m(t; \tau, \xi) \quad \tau \leq t \leq s_m \\ &= \lambda_m(t; s_m, y_m) \quad s_m \leq t \leq 1. \end{aligned}$$

Then for each $m > m_1$, ψ_m is an admissible trajectory. By Lemma 2.1, the derivatives ψ_m' are uniformly bounded. Hence the sequence ψ_m is uniformly bounded and equicontinuous. Hence there exists a subsequence, that we again denote as ψ_m, that converges uniformly to a function ψ. It follows from the closure theorems ([Ber;1974: Cor. 4.1, p. 59] or [Ces., p. 299]) that ψ is also a trajectory. Since $\psi_m(1; \tau, \xi) \in \mathcal{T}$ and since $\mathcal{T}$ is closed, we have $\psi(1; \tau, \xi)$ in $\mathcal{T}$. Thus $\psi(\ ; \tau, \xi)$ is admissible.

We have

$$\begin{aligned} &V(1, \psi_m(1; \tau, \xi)) - V(\tau, \xi) = V(1; \psi_m(1, \tau, \xi)) - V(s_m, y_m) \\ &+V(s_m, y_m) - V(\tau, \xi) = V(1, \psi_m(1; \tau, \xi)) - V(s_m, y_m), \end{aligned} \tag{24}$$

where the last equality follows from Lemma 3.2. Also

$$y_m = \psi_m(1-1/m;\tau,\xi) - \psi_m(1;\tau,\xi) + \psi_m(1;\tau,\xi) = \int_{1-1/m}^{1} \psi_m'(t)dt + \psi_m(1;\tau,\xi). \tag{25}$$

Now let $m \to \infty$, in (24) and (25). It then follows from Lemma 2.1, from the convergence of ψ_m to ψ, and the continuity of V that $V(1,\psi(1;\tau,\xi)) = V(\tau,\xi)$. If we now use (21) we get that $g(\psi(1,\tau,\xi)) = V(\tau,\xi)$. Hence $V(\tau,\xi) \geq W(\tau,\xi)$. Comparing this with (23) completes the proof.

References

Aubin, J.P. and Cellina, A. (1984). *Differential Inclusions*, Springer–Verlag, Berlin.

Berkovitz, L.D. (1970). *A Hamilton–Jacobi theory for a class of control problems*, Colloque sur la Théorie Mathématique du Contrôle Optimal, CBRM, Vander, Louvain, 1–23.

Berkovitz, L.D. (1974). *Optimal Control Theory*, Springer–Verlag, New York.

Berkovitz, L.D. (1989). *Optimal feedback controls*, SIAM Journal on Control and Optimization, **27**, 991–1006.

Bruckner, A. (1994). *Differentiation of Real Functions*, CRM Monograph Series, Vol. 5, American Mathematical Society, Providence, RI.

Cesari, L. (1983). *Optimization–Theory and Applications. Problems with Ordinary Differential Equations*, Springer–Verlag, New York.

Clarke, F.H. (1983). *Optimization and Nonsmooth Analysis*. John Wiley, New York.

Crandall, M.G., Ishii, H., and Lions, P-L. (1992). *User's guide to viscosity solutions of second order partial differential equations*, Bulletin of the American Mathematical Society, **27**, 1–67.

Fleming, W.H. and Rishel, R.W. (1975). *Deterministic and Stochastic Optimal Control*, Springer–Verlag, New York.

Fleming, W.H. and Soner, H.M. (1993). *Controlled Markov Processes and Viscosity Solutions*, Springer–Verlag, New York.

Frankowska, H. (1989). *Optimal trajectories associated with a solution of the contingent Hamilton–Jacobi equation*, Applied Mathematics and Optimization, **19**, 291–311.

Frankowska, H. (1993). *Lower semi continuous solutions of Hamilton–Jacobi–Bellman equation*, SIAM Journal on Control and Optimization, **31**, 257–272.

Mirica, S. (1990). *Bellman–Isaacs equation in deterministic optimal control*, Studii Si Cercetari Mathematice, **42**, 437–447.

Rowland, J.D.L. and Vinter, R.B. (1991). *Construction of optimal feedback controls*, Systems and Control Letters, **16**, 357–367.

Vinter, R. (1994). *Uniqueness of solutions to the Hamilton–Jacobi equation: a system theoretic proof*, Systems and Control Letters, **32**, 267–275.

14

Regular Solutions of Stochastic Burgers Equation

P. L. Chow

ABSTRACT: The paper is concerned with some strong regularity properties of the solutions to a stochastic conservative Burgers equation. By means of the Cole-Hopf transformation, the Burgers equation is linearized to a stochastic heat equation. Based on the Hölder regularity of solutions of the linearized equations, the existence of a unique classical or a strong solution is proved and the dependence of regularity on the smoothness of data is ascertained.

KEYWORDS: Stochastic Burgers equation, stochastic heat equation, Cole-Hopf transformation, regularity of solutions.

1 Introduction

Stochastic partial differential equations have arisen from many important applications in engineering and science, ranging from the nonlinear filtering theory [Pardoux (1979)] to the population genetic model of Fleming and Viot (1979). The Burgers equation was introduced to mimic the Navier-Stokes equations in hydrodynamics. The statistical solution of a random initial-value problem for such equation was first studied by Burgers (1974) and more recently by several authors. As a simple model of turbulence, the randomly forced Burgers equation was proposed by Jeng (1969) and has been the subject of investigation by Bertini, Canerini and Jona-Lasino (1994), Da Prato, Debussche and Temam (1994) and Handa (1996), to mention just a few. In all these papers the main analytical tool is based on a logarithmic transformation or the related Hopf-Cole transformation [Hopf (1950)]. Such a transformation, in the conservative case, linearizes the stochastic Burgers equation with dimension d $\geq$ 2. This linear equation is a parabolic equation with a random potential term, which can be analyzed more easily. This article is closely related to the work of Handa (1996), in which he showed that, for sufficiently smooth data, the random solution of the linearized equation has a certain regularity in a L_2- Sobolev space, but the associated solution to the stochastic Burgers equation is a generalized random field which satisfies the equation in a weak (distributional) sense.

In contrast the present paper will provide some stronger regularity results. It will be shown that, under suitable conditions, the solution to the random heat equation has a certain degree of Hölder regularity property depending on the smoothness of data, and the associated solution to the stochastic Burgers equation can be a strong or a classical solution in a sense to be defined later.

As preliminaries, let $H = L_2(\mathbb{R}^d)$ be a Hilbert space with the usual inner product $(.,.)$ and norm $\|\cdot\|$. Let $\{W_t, t \geq 0\}$ be a $H-$ valued Wiener process defined in a complete probability space $(\Omega, \mathcal{F}, \mu)$ with covariance operator Λ of trace class if it is a centered Gaussian process in H with $W_0 = 0$ and

$$E(W_t, f)(W_s, g) = (t \wedge s)(\Lambda f, g) \quad f, g \in H, \tag{1.1}$$

where $(t \wedge s) = \min\{t, s\}$ and the trace $tr\ \Lambda < \infty$.

Let $\{e_k\}$ be the complete orthonormal set of eigenfunctions of Λ with the corresponding set $\{\lambda_k\}$ of eigenvalues. Then W_t has the following representation [Da Prato and Zabczyk (1992)]:

$$W_t = \sum_{k=1}^{\infty} \sqrt{\lambda_k} e_k w_t^k, \tag{1.2}$$

where $\{w_t^k\}$ is a sequence of i.i.d. Brownian motions in one dimension. Let $Q \in \mathcal{L}(H)$ be a smooth linear operator on H with kernel $q(x, y)$ so that

$$(Qf)(x) = \int_{\mathbb{R}^d} q(x, y) f(y) dy.$$

Introduce a smoothed Wiener field

$$W_t^Q := QW_t = \sum_{k=1}^{\infty} \sqrt{\lambda_k}(Qe_k) w_t^k, \tag{1.3}$$

with the covariance operator

$$R = (Q^* \Lambda Q), \tag{1.4}$$

where * means the adjoint.

For Hölder regularity, let $C_{m+\alpha} := C_{m+\alpha}(\mathbb{R}^d)$ denote the set of real functions f whose partial derivatives up to the order m are Hölder continuous with modulus $\alpha \in [0, 1)$. The set $C_{m+\alpha}$ is said to be a Hölder space of order $m + \alpha$ with norm $\|f\|_{m+\alpha}$ [Friedman (1964)]. We also define $C_{m+\alpha,\gamma}$ to be the space of functions $\varphi : [0, T] \to C_{m+\alpha}$ having a finite norm

$$\|\varphi\|_{m+\alpha,\gamma} := \sup_{0 \leq t \leq T} \|\varphi(t, .)\|_{m+\alpha} + \sup_{0 \leq s \leq t \leq T} \frac{\|\varphi(t, .) - \varphi(s, .)\|_{m+\alpha}}{|t - s|^\gamma},$$

for $\alpha, \gamma \in [0, 1)$.

2 Bilinear Stochastic Heat Equation

Consider the Cauchy problem for the linear parabolic Itô equation in $\mathbb{R}^d$:

$$d\varphi_t = \frac{1}{2}\left[\Delta\varphi_t + r(x)\varphi_t\right]dt + \varphi_t dW_t^Q, \quad 0 < t \leq T, \\ \varphi_0 = \theta, \tag{2.1}$$

where Δ is the Laplacian; $r(x) := r(x,x)$ with $r(x,y)$ being the covariance function, which is the kernel of the covariance operator R for the Wiener random field W_t^Q as defined by (1.3) and (1.2), and θ is a given function on $\mathbb{R}^d$. If $r \in L_\infty\left(\mathbb{R}^d\right)$ and $\theta \in H = L_2\left(\mathbb{R}^d\right)$, the equation (2.1) has a unique solution [see, e.g. Pardoux (1979)]

$$\varphi \in C([0,T], H) \cap L_2\left((0,T), H_1\left(\mathbb{R}^d\right)\right) \text{ a.s. (almost surely.)}$$

However, if the data r and θ are smooth enough, the solution φ can be Hölder-continuous in t and higher degree of regularity in x. In fact we have the following regularity result.

Theorem 2.1. *Suppose that $\theta \in C_{m+\alpha}$ and $Q : H \to C_{m+\alpha}$ is a bounded linear operator. Then the solution of the Cauchy problem (2.1) has the regularity: $\varphi \in L_p(\Omega, C_{m+\beta,\nu})$ such that*

$$E\|\varphi\|^p_{m+\beta,\nu} \leq C\left(1 + \|\theta\|^p_{m+\alpha}\right) \tag{2.2}$$

for any $p \geq 1$, $m \geq 2$, $0 < \beta < \alpha < 1$, and $0 < \nu < \frac{(\alpha-\beta)}{2}$, where C is a constant independent of θ.

Proof. This follows from Theorem A given in the appendix. It suffices to verify that the conditions (A.1) and (A.2) are satisfied. In (2.1) we define $b_t(g)(x) := \frac{1}{2}r(x)\cdot g(x)$ and $[\sigma_t(g)h](x) := g(x)\cdot(Qh)(x)$ for $g \in C_{m+\alpha}$, $h \in H$. Now we have

$$\|b_t(g)\|_{m-2+\alpha} = \frac{1}{2}\|rg\|_{m-2+\alpha} \\ \leq \frac{1}{2}\|r\|_{m-2+\alpha}\|g\|_{m+\beta} \text{ for any } \beta \in (0,1).$$

By (1.3) and the assumption on Q,

$$\|r\|_{m-2+\alpha} \leq \sum_{k=1}^{\infty}\lambda_k\|Qe_k\|^2_{m-2+\alpha} \leq K^2 trR < \infty,$$

so that

$$\|b_t(g)\|_{m-2+\alpha} \leq (K^2 trR)\|g\|_{m+\beta}, \text{ for } g \in C_{m+\alpha} \tag{2.3}$$

with $0 < \beta < \alpha < 1$.
Next we have

$$\begin{aligned}\|\sigma(g)\,h\|_{m-1+\alpha} &= \|g\,(Qh)\|_{m-1+\alpha} \\ &\leq \|g\|_{m-1+\alpha}\,\|Qh\|_{m-1+\alpha} \leq K\,\|g\|_{m-1+\alpha}\,\|h\|,\end{aligned} \tag{2.4}$$

for $g \in C_{m+\alpha}$ and $h \in H$
In view of (2.3) and (2.4), condition (A.1) is satisfied. Due to the linearity, the Lipschitz condition (A.2) can be easily verified. □

The stochastic Feynman-Kac representation of the solution of equation (2.1) was done by Kunita (1990) and others. Here we shall use a representation as in Handa (1996) with a stronger regularity property. Let $b_t = (b_t^1, ..., b_t^d)$ be a standard Wiener process in $\mathbb{R}^d$ defined on another probability space $(\Lambda, \mathcal{G}, \nu)$, independent of the Wiener field W_t. Let $P = \mu \times \nu$ be the product measure defined on a completed product space $\left(\Omega \times \Lambda, \overline{\mathcal{F} \times \mathcal{G}}\right)$. Formally, we introduce the stochastic integral

$$Z(t,x) \;:= \int_0^t \left(\delta_{\cdot}(x + b_{t-s}), dW_s\right), \tag{2.5}$$

where $\delta_y(x) = \delta(y-x)$ is the Dirac $\delta-$ function. It is well known that, in the sense of distribution, $\delta(y-x)$ is the weak limit of the sequence of functions

$$\delta_y^n(x) = \sum_{k=1}^n e_k(x)\, e_k(y)\,.$$

In view of (1.2) and (2.6), for each Brownian path b_t, the following stochastic integral is well defined in the Itô sense

$$Z_n(t,x) \;:= \int_0^t \left(\delta_{\cdot}^n(x + b_{t-s}), dW_s\right) = \sum_{k=1}^n \sqrt{\lambda_k} \int_0^t e_k(x + b_{t-s})\, dw_s^k. \tag{2.6}$$

In fact we can show that the stochastic integral $Z(t,x)$ can be defined as the L_2- limit of Z_n with a certain Hölder regularity.

Lemma 2.1. *The stochastic integral $Z(t,x)$ in (2.5) is well defined as the $L_2(\Omega)-$ limit of the sequence $Z_n(t,x)$ given by (2.6). If $Q : H \to C_{m+\alpha}$ is continuous, the family $\{Z(t,.),\ t \geq 0\}$ is a $C_{m+\beta}$ − valued, continuous stochastic process defined on $\left(\Omega \times \Lambda, \overline{\mathcal{F} \times \mathcal{G}}, P\right)$ such that*

$$E_p Z(t,x) = 0,\;\; E_p Z^2(t,x) = \int_0^t (\Phi_{t-s} r)(x)\, dx, \tag{2.7}$$

where

$$(\Phi_t r)(x) = \int_{\mathbb{R}^d} (2\pi t)^{-\frac{d}{2}} \exp\left\{-\frac{|x-y|^2}{2t}\right\} r(y)\, dy. \tag{2.8}$$

Proof. It is easy to show that, for a fixed Brownian path b_t, $\{Z_n(t,x)\}$ is a Cauchy sequence in $L_2(\Omega)$ converging to $Z(t,x)$. Clearly we have $E_p Z_n(t,x) = 0$ and

$$E_p\left\{Z_n^2(t,x)\right\} = E_\nu\left\{\sum_{k=1}^n \lambda_k \int_0^t q_k^2(x+b_{t-s})\,ds\right\},$$

where $q_k = Qe_k$. As $n \to \infty$, it yields $E_p Z(t,x) = 0$ and the equation (2.7). For Hölder regularity, we rewrite (3.11) as

$$Z(t,x) = \sum_{k=1}^\infty \sqrt{\lambda_k}\int_0^t (Qe_k)(x+b_{t-s})\,dw_s^k := \int_0^t (\sigma(s,x), dW_s).$$

The above type of stochastic integral was studied in Chow and Jiang (1994) in which Theorem 3.3 can be applied to assert that $Z(t,.) \in C_{m+\beta}$ a.s. for any $\beta \in (0,\alpha)$, $t \geq 0$. □

The following theorem pertains to the stochastic Feynman-Kac formula as mentioned before.

Theorem 2.2. *The solution φ of the stochastic Cauchy problem (2.1) has the probabilistic representation:*

$$\varphi(t,x) = E_\nu\left\{\theta(x+b_t)\exp Z(t,x)\right\}, \tag{2.9}$$

where $Z(t,x)$ is defined by (2.5). Moreover, if θ is positive and not identically zero, then the solution is strictly positive, that is

$$\varphi(t,x) > 0\ \mu-\ \text{a.s. for any } t>0,\ x\in \mathbb{R}^d.$$

Proof. Noting (2.6), we define

$$\varphi^N(t,x) := E_\nu\left\{\theta(x+b_t)\exp Z_n(t,x)\right\}.$$

By the Itô formula, it can be shown that $\varphi_t^N := \varphi^N(t,.)$ satisfies the integral equation

$$\varphi_t^N = \Phi_t\theta + \frac{1}{2}\sum_{k=1}^N \lambda_k \int_0^t \Phi_{t-s}\left(q_k^2\varphi_s^N\right)ds + \frac{1}{2}\sum_{k=1}^N \sqrt{\lambda_k}\int_0^t \Phi_{t-s}\left(q_k^2\varphi_s^N\right)dW_s^k, \tag{2.10}$$

where Φ is defined by (2.8). Consider the difference

$$E_\mu\left\{\varphi(t,x)-\varphi^N(t,x)\right\}^2 = E_p\left\{\theta^2(x+b_t)\left[e^{Z(t,x)}-e^{Z_N(t,x)}\right]^2\right\}$$
$$\leq \theta_0^2 E_p\left\{e^{Z_n(t,x)}\,\Delta Z_N(t,x)\int_0^1 e^{\lambda\Delta Z_N(t,x)}d\lambda\right\}^2$$
$$\leq \theta_0^2\left\{E_p e^{4Z_N(t,x)}\right\}^{\frac{1}{2}}\left\{E_p\left|\Delta Z_N(t,x)\right|^8\right\}^{\frac{1}{4}}\left\{\int_0^1 E_p e^{8\lambda\Delta Z_N(t,x)}d\lambda\right\}^{\frac{1}{4}},$$

where $\Delta Z_N := (Z - Z_N)$. It is not hard to check that $E_p \exp\{\alpha Z_N(t,x)\}$ and $E_p \exp\{\alpha \Delta Z_N(t,x)\}$ are bounded for any real α and, for some $M > 0$,

$$E_\mu |\Delta Z_N(t,x)|^2 = \sum_{k=N+1}^{\infty} \lambda_k \int_0^t q_k^2(x + b_{t-s})\, ds \le MT \sum_{k=N+1}^{\infty} \lambda_k,$$

which converges to zero as $N \to \infty$. Since $Z_N(t,x)$ is a centered Gaussian random field, we deduce that

$$\lim_{N\to\infty} \sup_{0\le t\le T} \sup_{x\in\mathbb{R}^d} E_\mu \{\varphi(t,x) - \varphi^N(t,x)\}^2 = 0.$$

To show that the limit φ satisfies the equation (2.1), we take the limits in equation (2.10) termwise. To show the convergence, for example, we shall verify

$$\lim_{N\to\infty} \sum_{k=1}^{N} \lambda_k \int_0^t \Phi_{t-s}(q_k^2 \varphi_s^N)(x)\, ds = \int_0^t \Phi_{t-s}(r\varphi_s)(x)\, ds \tag{2.11}$$

in $L_2(\Omega)$ for $t \in [0,T]$, $x \in \mathbb{R}^d$. Since $r(x) = \sum_{k=1}^{\infty} \lambda_k q_k^2(x)$, we write

$$\int_0^t \Phi_{t-s}(r\varphi_s)(x)\, ds = \sum_{k=1}^{\infty} \lambda_k \int_0^t \Phi_{t-s}(q_k^2 \varphi_s)(x)\, dx.$$

Then

$$\begin{aligned} &\left| \int_0^t \Phi_{t-s}(r\varphi_s)\, ds - \sum_{k=1}^{N} \lambda_k \int_0^t \Phi_{t-s}(q_k^2 \varphi_s)\, ds \right|^2 \\ &\le 2\left\{ \left| \sum_{k=1}^{N} \lambda_k \int_0^t \Phi_{t-s}[q_k^2(\varphi_s - \varphi_s^N)]\, ds \right|^2 \right. \\ &\quad \left. + \left| \sum_{k=N+1}^{\infty} \lambda_k \int_0^t \Phi_{t-s}(q_k^2 \varphi_s)\, ds \right|^2 \right\}. \end{aligned} \tag{2.12}$$

We have

$$\begin{aligned} &E_\mu \left| \sum_{k=1}^{N} \lambda_k \int_0^t \Phi_{t-s}[q_k^2(\varphi_s - \varphi_s^N)](x)\, ds \right|^2 \\ &\le C\left(\sum_{k=1}^{\infty} \lambda_k \right)^2 \sup_{0\le t\le T, x\in\mathbb{R}^d} E_\mu |\varphi(t,x) - \varphi^N(t,x)|^2 \to 0 \end{aligned} \tag{2.13}$$

and, similarly,

$$\begin{aligned} &E_\mu \left| \sum_{k=N+1}^{\infty} \lambda_k \int_0^t \Phi_{t-s}(q_k^2 \varphi_s)(x)\, ds \right|^2 \\ &\le C\left(\sum_{k=N+1}^{\infty} \lambda_k \right)^2 \sup_{0\le t\le T, x\in\mathbb{R}^d} E_\mu \varphi^2(t,x) \to 0 \text{ as } N \to \infty, \end{aligned} \tag{2.14}$$

where

$$C = q_0^4 T \text{ and } q_0^2 = \sup_{x \in \mathbb{R}^d} \int_{\mathbb{R}^d} q^2(x,y)\, dy.$$

In view of (2.12) - (2.14), the equation (2.11) is verified. The convergence of the stochastic integral term can be shown in a similar fashion, and, hence, the representation (2.9) holds true.

As shown in Handa (1996), the strict positivity of the solution can be proved by making use of (2.9) and the Cauchy-Schwarz inequality

$$\begin{aligned}\left\{E_\nu \theta^{\frac{1}{2}}(x+b_t)\right\}^2 &= \left\{E\left[\theta(x+b_t)\, e^{Z(t,x)}\right]^{\frac{1}{2}} e^{-\frac{1}{2}Z(t,x)}\right\}^2 \\ &\le \varphi(t,x)\, E_\nu e^{-Z(t,x)}\end{aligned}$$

which implies that

$$\varphi(t,x) \ge \frac{\left\{E_\nu \theta^{\frac{1}{2}}(x+b_t)\right\}^2}{E_\nu e^{-Z(t,x)}} > 0, \text{ for any } t > 0, x \in \mathbb{R}^d \text{ a.s. } \Box$$

3 Random Conservative Burgers Equation

Consider the Burgers equation in $\mathbb{R}^d$ with a random perturbation:

$$\begin{aligned}&\frac{\partial}{\partial t}u + (u \cdot \nabla)u = \frac{1}{2}\Delta u + \dot{V}_t, \; t > 0, \; x \in \mathbb{R}^d, \\ &u(0,x) = v(x),\end{aligned} \tag{3.1}$$

where

$$u = (u_1, ..., u_d), \; (u \cdot \nabla)u = \sum_{i=1}^{d} u_i \frac{\partial u}{\partial x_i}, \; v = (v_1, ..., v_d)$$

and $V_t = \left(V_t^1, ..., V_t^d\right)$ being a Wiener random field with $\dot{V}_t = \frac{\partial}{\partial t}V_t$. The formal equation (3.1) is interpreted as an Itô equation similar to equation (2.1). The equation (3.1) is said to be conservative if the noise term V_t and the initial state v are derivable from a potential as follows

$$V_t = \nabla W_t^Q = \left(\frac{\partial W_t^Q}{\partial x_1}, \ldots, \frac{\partial W_t^Q}{\partial x_d}\right), \tag{3.2}$$

$$v = \nabla g = \left(\frac{\partial g}{\partial x_i}, \ldots, \frac{\partial g}{\partial x^d}\right), \tag{3.3}$$

where W_t^Q is the Wiener field defined by (1.3) and g is a given differentiable function. In this case the solution of (3.1) is also a conservative random field derivable from a random potential. In fact, if ψ is a smooth solution of the Cauchy problem

$$\begin{aligned} &\frac{\partial}{\partial t}\psi + \frac{1}{2}|\nabla\psi|^2 = \frac{1}{2}\Delta\psi - \dot{W}_t^Q, \ t>0, \ x\in\mathbb{R}^d, \\ &\psi(0,x) = g(x), \end{aligned} \tag{3.4}$$

then the solution of (3.1) is given by

$$u(t,x) = \nabla\psi(t,x). \tag{3.5}$$

The nonlinear equation (3.4) can be linearized to the random heat equation (2.1) by the logarithmic transformation:

$$\psi = -\log\varphi. \tag{3.6}$$

with $\varphi(0,x) = \theta(x) = \exp\{-g(x)\}$.

Lemma 3.1. *Assume that $g \in C_{m+\alpha}$ and $Q : H \to C_{m+\alpha}$ being a bounded linear operator. If φ is the solution of the random heat equation (2.1) with $\theta = \exp\{-g(x)\}$, then the random function defined by (3.6) is the unique solution of the Cauchy problem (3.4) with $\psi \in C\left([0,T], C_m\left(\mathbb{R}^d\right)\right)$* a.s.

Proof. By the assumption $g \in C_{m+\alpha}$, we have $\theta = \exp\{-g\} \in C_{m+\alpha}$ and $\theta > 0$ a.e. Hence, by Theorem 2.1, the solution φ of (2.1) belongs to $L_p(\Omega, C_{m+\beta,\nu})$ and $\varphi(t,x) > 0$ for $t \geq 0$, $x \in \mathbb{R}^d$ a.s. It follows that $\psi = -\log\varphi(t,x)$ is well defined and $\psi \in C\left([0,T], C_m\left(\mathbb{R}^d\right)\right)$ a.s. Let us introduce $\psi_t^\delta := -\log[\delta + \varphi(t,.)]$ with $\delta > 0$. It can be shown that ψ as defined satisfies the equation (3.4) by applying the Itô formula to (ψ_t^δ, h) for $H \in C\left(\mathbb{R}^d\right)$ and taking the limit as $\delta \to 0$. The uniqueness follows from that of the deterministic case. □

Now consider the Burgers equation (3.1). We say that u is *a classical solution* if $u \in \left\{C\left([0,T], C_2\left(\mathbb{R}^d\right)\right)\right\}^d$ a.s. and it satisfies the equation

$$u(t,x) = v(x) + \int_0^t \left[\frac{1}{2}\Delta u(s,x) - (u\cdot\nabla)u(s,x)\right] ds + V_t(x), \tag{3.7}$$

for any $(t,x) \in [0,T] \times \mathbb{R}^d$ a.s.
On the other hand, if $u \in C([0,T], C_1(\mathbb{R}^d))$ and, for any $w \in C_1(\mathbb{R}^d, \mathbb{R}^d)$ with compact support, it satisfies

$$(u_t, w) = (v, w) - \frac{1}{2}\int_0^t [(\nabla u_s, \nabla w) + ((u_s\cdot\nabla)u_s, w)]\, ds - (W_t, \nabla\cdot w), \tag{3.8}$$

for any $t \in [0,T]$ a.s., then u is said to be *a strong solution*, where $(\cdot,\cdot)$ denotes the inner product in H or H^d as the case may be.

Theorem 3.1. *Let the Burgers equation (4.1) satisfy the conservation conditions (3.2) and (3.3). If* $g \in C_{m+2+\alpha}$ *and* $q \in C_{m+2+\alpha}\left(\mathbb{R}^d \times \mathbb{R}^d\right)$, *then the equation (4.1) has a unique solution* $u \in C\left([0,T], C_{m+1}\left(\mathbb{R}^d\right)\right)$, *which is a classical solution for* $m \geq 1$ *but only a strong solution for* $m = 0$. *Moreover, in either case, the solution* u *is related to the solution* φ *of the random heat equation (2.1) by*

$$u(t,x) = -\frac{\nabla\varphi(t,x)}{\varphi(t,x)}, \text{ for } (t,x) \in [0,T] \times \mathbb{R}^d, \text{ a.s.} \tag{3.9}$$

Proof. Let $u(t,x) := \nabla\psi(t,x)$. By Lemma 3.1, for the assumed smoothness of data, we have $\psi \in C\left([0,T], C_{m+2}\left(\mathbb{R}^d\right)\right)$ so that

$$u \in \left[C\left([0,T], C_{m+1}\left(\mathbb{R}^d\right)\right)\right]^d \text{ a.s.}$$

For $w \in \left[C_1\left(\mathbb{R}^d\right)\right]^d$ with a compact support, in view of equation (3.4), we have

$$\begin{aligned}(\psi_t, \nabla\cdot w) &= (g, \nabla\cdot w) + \frac{1}{2}\int_0^t \left(\left[\Delta\psi_s - |\nabla\psi_s|^2\right], \nabla\cdot w\right) ds \\ &\quad + \left(W_t^Q, \nabla\cdot w\right) \text{ a.s.}\end{aligned} \tag{3.10}$$

For $m \geq 1$, after applying integration by parts to free $\nabla\cdot w$ from the divergence, the equation (3.10) yields

$$(u_t, w) = (v, w) + \int_0^t \left[\frac{1}{2}(\Delta u_s, w) - ((u_s\cdot\nabla)u_s, w)\right] ds + (V_t, w), \text{ a.s.}$$

Since $u \in C\left([0,T], C_2\left(\mathbb{R}^d\right)\right)$, the above equation implies that u is a classical solution. If $m = 0$, $u \in C\left([0,T], C_1\left(\mathbb{R}^d\right)\right)$ and the equation (3.8) clearly follows from equation (3.10) by integrating by parts once. The validity of equation (3.9) is obvious. □

4 Appendix

Consider the stochastic PDE for $u_t := u(t,\cdot)$:

$$\begin{aligned}&\frac{\partial}{\partial t}u_t = \mathcal{L}u_t + b_t(u_t) + \sigma_t(u_t)\dot{W}_t^Q \text{ in } \mathbb{R}^d,\ t > 0, \\ &u(0,x) = \varphi(x),\ x \in \mathbb{R}^d;\quad \lim_{|x|\to\infty} u(t,x) = 0, \quad t > 0,\end{aligned} \tag{4.1}$$

where

$$\mathcal{L} := \sum_{i,j=1}^{d} a_{ij}(t) \frac{\partial^2}{\partial x_i \partial x_j} + \sum_{i=1}^{d} b_i(t) \frac{\partial}{\partial x_i} + c(t), \tag{4.2}$$

is a strongly elliptic operator, $\sigma_t(u) : H \to C_{m-1+\alpha}$ is a continuous linear operator for each $t > 0$ and $u \in C_{m+\beta}$, and $b. : [0,T] \times C_{m+\beta} \to C_{m-2+\alpha}$ is a given mapping. Then the following theorem holds.

Theorem A. *Let $m \geq 2$ be an integer, $0 < \beta < \alpha < 1$, and $0 < \nu \leq \frac{(\alpha-\beta)}{2}$. Suppose that f and g satisfy the following conditions:*

(A.1) $\|b_t(u)\|_{m-2+\alpha}^2 + |||\sigma_t(u)|||_{m-1+\alpha}^2 \leq M\left(1 + \|u\|_{m+\beta}^2\right)$, for any $u \in C_{m+\beta}$ and some constant $M > 0$.
(A.2) For every $n > 0$, there exists a constant K_n such that

$$\|b_t(u) - b_t(v)\|_{m-2+\alpha} + |||\sigma_t(u) - \sigma(v)|||_{m-1+\alpha} \leq K_n \|u - v\|_{m+\beta},$$

for any $u, v \in C_{m+\beta}$ with $\max\left(\|u\|_{m+\beta}, \|v\|_{m+\beta}\right) \leq n$, where $|||\cdot|||_{m+\alpha}$ denotes the norm of an operator from H into $C_{m+\alpha}$.
(A.3) The Wiener field W_t^Q is defined by (1.3) with $Q : H \to C_{m+\alpha}$ being continuous.

Then, for any $p > 0$ and $\varphi \in C_{m+\beta+2\nu}$, the Cauchy problem (4.1) has a unique classical solution $u. \in L_p(\Omega, C_{m+\beta,\nu})$. Moreover, there is a constant $C(\alpha, \beta, \nu, p, T, M)$ independent of φ, and K_n such that

$$E\|u.\|_{m+\beta,\nu}^p \leq C\left(1 + \|\varphi\|_{m+\beta+2\nu}^p\right). \square$$

This theorem is a corrected version of Theorem 2.1 of our paper (Chow and Jiang (1994)), which holds true only for the domain $D = \mathbb{R}^d$. In this case the Green's function associated with the heat operator $(\partial_t - \mathcal{L})$ is spatially homogeneous. This property was assumed inadvertently in the proof of Theorem 2.1 in the case of a bounded domain.

References

Bertini, L., Canerini, N., Jona-Lasino, G. (1994). The stochastic Burgers equation, Communications in Mathematics and Physics, **165**, 211-232.

Burgers, J. (1974). The Nonlinear Diffusion Equations, Riedel, Dordrecht.

Chow, P.L., Jiang, J.L. (1994). Stochastic partial differential equations in Hölder spaces, Probability Theory and Related Fields, **99**, 1-27.

Da Prato, G., Debussche, A., Temam, R. (1994). Stochastic Burgers equation, Nonlinear Differential Equations and Applications, 388-402.

Da Prato, G., Zabczyk, J. (1992). Stochastic Equations in Infinite Dimensions, Encyclopedia of Mathematics and Its Applications, Cambridge University Press, Cambridge, England.

Fleming, W. H., Viot, M. (1979). Some measure-valued Markov process in population genetics theory, Indiana University Mathematics Journal, **28**, 817-843.

Friedman, A. (1964). Partial Differential Equations of Parabolic Type, Prentice Hall, Englewood Cliffs, N.J.

Handa, K. (1996). On a stochastic PDE related to Burgers' equation with noise, in Nonlinear Stochastic PDE's: Hydrodynamic Limit and Burgers Turbulence, IMA volume 77, Springer-Verlag, New York.

Hopf, E. (1950). The partial differential equation $u_t + uu_x = \mu u_{xx}$, Communications on Pure and Applied Mathematics, **3**, 201-230.

Jeng, D.T. (1969). Forced model equation for turbulence, The Physics of Fluids, **12**, 2006-2010.

Kunita, H. (1990). Stochastic Flows and Stochastic Differential Equations, Cambridge University Press, Cambridge, England.

Pardoux, E. (1979). Stochastic partial differential equations and filtering of diffusion processes, Stochastics, **3**, 127-167.

15
Piecewise-Deterministic Processes and Viscosity Solutions

Mark H. A. Davis
Mohammad Farid

ABSTRACT: This article concerns the optimal control of piecewise deterministic processes in the viscosity solutions context. Boundary conditions given in Vermes (1985) are weakened and replaced by boundary conditions in exit-time optimal control problems as given in Barles (1994). It is proved that the value function of piecewise-deterministic process optimal control is the unique viscosity solution of its associated Hamilton-Jacobi-Bellman equation.

KEYWORDS: Piecewise-deterministic processes, viscosity solutions, Markov processes, optimal control, Hamilton-Jacobi-Bellman equation.

1 Introduction

Wendell Fleming and Mete Soner, in their book (1993) demonstrate very clearly the power of viscosity solution techniques in optimal control of diffusion processes. Here we aim to convince the reader that viscosity solutions provide an equally effective treatment of controlled piecewise-deterministic processes (PDP). These processes are constructed roughly as follows (Davis (1993)). The sample path $(x_t, t \geq 0)$ takes values in a state space $E \subset \mathbb{R}^d$. It follows the integral curve of the ordinary differential system

$$\dot{x}_t = g(x_t, u_t), \tag{1}$$

up to a random time T_1 whose distribution is given by

$$P[T_1 > t] = I_{(t<t_*)} \exp\left(-\int_0^t \lambda(x_s, u_s)\, ds\right), \tag{2}$$

where $t_* = \inf\{t : x_t \notin E\}$. This means that T_1 has "hazard rate" $\lambda(x_t, u_t)$ in the interior of E, but T_1 must occur at the first boundary hitting time t_* if it has not already occurred prior to t_*. In these equations u_t is an "open loop" control function (depending in practice on the initial state x_0); see Section 2 below. At time T_1 the trajectory of x_t is left-continuous, so $x_{T_1^-}$

is just the solution of (1) calculated at $t = T_1$, while the post-jump value x_{T_1} is selected at random from a given transition function $Q(A; x, u_o)$ if $T_1 < t_*$ or $R(A, x, u_\Gamma)$ if $T_1 = t_*$ (in the latter case $x_{T_1^-} \in \partial E$ and u_Γ is a "boundary control" parameter). At time T_1 we choose a new control function $u_o(\cdot, x_{T_1})$ and the process "restarts" from x_{T_1}. In this way we construct a process whose sample path consists of controlled deterministic trajectories between jumps which occur at isolated random times $T_1, T_2, \ldots$. Conditions must be imposed to ensure that $T_n \to \infty$ a.s. We want to choose interior and boundary controls to minimise the infinite horizon discounted cost

$$J_x(u) = \mathbb{E}\left(\int_0^\infty e^{-\delta t}\ell(x_t, u_t)\, dt\right), \tag{3}$$

or the more general functional described in Section 2. The basic approach to solving this control problem is as follows. We temporarily ignore the question of boundary control, so if $x_{T_i^-} \in \partial E$ for some i then the distribution of x_{T_i} is $R(\cdot, x_{T_i^-})$ for a fixed transition measure $R(A, x)$.
Define the value function

$$V(x) = \inf_u J_x(u), \tag{4}$$

and for bounded measurable functions ψ on E,

$$\mathcal{G}\psi(x) = \inf_u \mathbb{E}\left(\int_0^{T_1} e^{-\delta t}\ell(x_t, u_t)\, dt + e^{-\delta T_1}\psi(x_{T_1})\right). \tag{5}$$

In (4) the infimum is taken over all "piecewise open loop" controls as informally described above, whereas in (5) only the first "segment" of deterministic trajectory is involved and the infimum is over time functions $t \mapsto u(t)$. Further, the only random element in (5) is the time T_1, whose distribution function is given by (2). Integrating with respect to this distribution we find that (5) corresponds to a deterministic optimal control problem which terminates at time t_*, the first time the boundary is reached. The cost function for this control problem is given by (19) below. In summary, our results are

- The operator $\mathcal{G}$ is a two-stage contraction mapping,
- its unique fixed point is V,
- V satisfies a Hamilton-Jacobi-Bellman equation ((28) and (29) below) in the viscosity sense.

The technical heart of the subject is therefore a study of deterministic control problems over intervals $[0, t_*^u]$, t_*^u is the first boundary hitting time for the trajectory corresponding to control u, and some given cost is paid on reaching the boundary, in addition to a "running cost" of the form (3).

In Chapter 4 of Davis (1993) the PDP optimal control problem was studied using methods of non-smooth analysis, following work by Dempster and Ye (1992). While this provides a viable theory, some unpleasantly strong conditions are required that severely restrict the applicability of the results. First, some kind of stability condition is required to ensure that the value function is Lipschitz continuous. An example from Davis (1993; Example (46.7)) demonstrates this point: for the scalar dynamical system $\dot{x}(t) = bx(t)$ with $b > 0$ and for $\ell(x) = 0 \vee (x \wedge 1)$ we find that

$$V(x) = \frac{1}{b-\delta}\left(\frac{b}{\delta}x^{\frac{\delta}{b}} - x\right), \quad 0 \leq x \leq 1,$$

so that V is not Lipschitz if $\delta < b$, even though ℓ is bounded. But stability conditions do not seem "relevant" to PDP control problem in, say, a compact state space. A second and more serious point concerns boundary behaviour. Vermes (1985) introduced the following condition:

$$\exists \alpha > 0 \text{ such that } g(x,u) \cdot n(x) \geq \alpha \quad \forall u \in U_o, \ \forall x \in \Gamma.$$

Here U_o is the control constraint set, $\Gamma \in \partial E$ is the "active" boundary of the state space (see Section 2) and $n(x)$ is the outward normal to Γ at $x \in \Gamma$. Under continuity assumptions this condition implies that trajectories near $x \in \Gamma$ exit through Γ at minimum speed α. This condition is violated in standard examples such as the capacity expansion problem, Davis (1993, (21.13)).

By moving to viscosity solutions we are able to dispense with stability conditions altogether (for viscosity theory the value function only needs to be continuous, which of course it is in the above example) and significantly weaken the assumptions on boundary behaviour. The boundary conditions are similar to deterministic exit-time optimal control problem given in Barles (1994) and Assumptions (3.2) in Section 3. It is finally proved that V the value function of PDP optimal control problem is the unique viscosity solution of (28)-(29).

2 Optimal feedback control of PDPs

In this section we describe our formulation for the optimal control of PDPs without going into the details of our assumptions. Later in Section 3 we make all the assumptions clear. We assume that the state of process (x_t) between jumps is given by the following ODE

$$\dot{x}_t = g(x_t, u_o(x_{T_n}, t - T_n)), \qquad T_n \leq t < T_{n+1}, \quad \text{for } n = 0, 1, 2, \ldots, \tag{6}$$

where $x_{T_n} = Y_n \in E$, with $T_0 = 0$ and $Y_0 = x$. The type of feedback control used in (6) is *piecewise open-loop control*, Vermes (1985). At time t after

the nth jump, the piecewise open-loop control *only* depends on the state of process at the last jump time (Y_n) and the time elapsed since the last jump, i.e. $t - T_n$.
In the optimal control of PDPs we are looking for an optimal policy $u = (u_0, u_\Gamma)$, which minimises the expected cost

$$J_x(u) = \mathbb{E}_x^u \left\{ \int_0^\infty \Delta(t)\, \ell(x_t, u_o(x,t))\, dt + \int_0^\infty \Delta(t)\, c(x_{t^-}, u_\Gamma(x_{t^-})\, dp^*(t) \right\}, \tag{7}$$

where

$$\Delta(t) = \exp\left\{ - \int_0^t \delta(x_s) ds \right\},$$

$$p^*(t) = \sum_{i=1}^{\infty} I_{(t \geq T_i)} I_{(x_{T_i^-} \in \Gamma)}.$$

Δ is a discounting function which under some assumptions assures the boundedness of our cost function and p^* is a counting process which counts the number of times that the process hits the boundary. Thus the second part of the cost function associates a cost to hitting the boundary.
The distribution of the first jump time is given by

$$P(T_1 \leq t) = 1 - \Lambda(t) I_{(t < t_*^{u_o}(x))}, \tag{8}$$

where

$$\Lambda(t) = \exp\left\{ - \int_0^t \lambda(x_s, u_o(x,s))\, ds \right\},$$

$$t_*^{u_o}(x) = \inf\{t \geq 0 \colon x_t \notin E, x_0 = x \in E\}.$$

Here, λ is the jump rate for the points in E, which generally depends on the control signal u_o. For the points on the boundary Γ, jump happens immediately; i.e. we always have $T_1 \leq t_*^{u_o}(x)$. Post jump locations are determined by the probability measures Q and R depending on whether the jump started from a state $x \in E$ or $x \in \Gamma$. More precisely

$$P_x(x_{t^+} \in A \mid x_{t^+} \neq x_{t^-} = x) = \begin{cases} Q(A; x, u_o) & \text{if} \quad x \in E, \\ R(A; x, u_\Gamma) & \text{if} \quad x \in \Gamma. \end{cases}$$

We define the value function of our optimal control problem as follows

$$V(x) = \inf_{u \,\in\, \mathcal{U}_o \times \mathcal{U}_\Gamma} J_x(u), \quad x \in E, \tag{9}$$

where $J_x(u)$ is defined in (7), $\mathcal{U}_o$ and $\mathcal{U}_\Gamma$ are respectively the set of control functions for the points in E and for the points on Γ. We will define these sets later in section 3. Now let us show that the stochastic optimal control problem defined in (9), can be reduced to a deterministic optimal control problem. PDPs are strong Markov processes Davis (1993), thus one can write the value function as follows

$$
\begin{aligned}
V(x) = \inf_{u \in \mathcal{U}_o \times \mathcal{U}_\Gamma} \mathbb{E}_x^u \Bigg\{ & \int_0^{T_1} \Delta(t)\, \ell(x_t, u_o(x,t))\, dt I_{(T_1 \le t_*^{u_o})} \\
& + I_{(T_1 < t_*^{u_o})} \Delta(T_1) \int_E V(y)\, Q(dy; x_{T_1^-}, u_o(x, T_1^-)) \\
& + I_{(t_*^{u_o} = T_1)} \Delta(t_*^{u_o})\, c(x_{t_*^{u_o}}, u_\Gamma(x_{t_*^{u_o}})) \\
& + I_{(t_*^{u_o} = T_1)} \Delta(t_*^{u_o}) \int_E V(y)\, R(dy; x_{t_*^{u_o}}, u_\Gamma(x_{t_*^{u_o}})) \Bigg\}.
\end{aligned}
$$

The probability density function is

$$
\frac{d}{dt} P_x(T_1 \le t) = \lambda(x_t, u_o(x,t))\, \Lambda(t)\, I_{(t < t_*^{u_o})} + \Lambda(t_*^{u_o})\, \delta(t - t_*^{u_o}),
$$

where δ is Dirac function, so we have

$$
\begin{aligned}
V(x) = \inf_{u \in \mathcal{U}_o \times \mathcal{U}_\Gamma} \Bigg\{ & \int_0^{t_*^{u_o}} \lambda(x_t, u_o)\, \Lambda(t) \left(\int_0^t \Delta(s)\, \ell(x_s, u_o)\, ds \right) dt \\
& + \Lambda(t_*^{u_o}) \int_0^{t_*^{u_o}} \Delta(t)\, \ell(x_t, u_o) dt \\
& + \int_0^{t_*^{u_o}} \lambda(x_t, u_o)\, \Lambda(t)\, \Delta(t) \int_E V(y)\, Q(dy; x_{t^-}, u_o)\, dt \\
& + \Lambda(t_*^{u_o})\, \Delta(t_*^{u_o})\, c(x_{t_*^{u_o}}, u_\Gamma(x_{t_*^{u_o}})) \\
& + \Lambda(t_*^{u_o})\, \Delta(t_*^{u_o}) \int_E V(y)\, R(dy; x_{t_*^{u_o}}, u_\Gamma(x_{t_*^{u_o}})) \Bigg\}.
\end{aligned}
$$

After some manipulations, $V(x)$ is obtained as follows

$$
\begin{aligned}
V(x) = \inf_{u \in \mathcal{U}_o \times \mathcal{U}_\Gamma} \Bigg\{ & \int_0^{t_*^{u_o}} \Lambda(t)\, \Delta(t)\, \ell(x_t, u_o(x,t))\, dt \\
& + \int_0^{t_*^{u_o}} \Lambda(t)\, \Delta(t) \int_E V(y)\, Q(dy; x_{t^-}, u_o(x, t^-))\, \lambda(x_t, u_o(x,t))\, dt \\
& + \Lambda(t_*^{u_o})\, \Delta(t_*^{u_o})\, c(x_{t_*}, u_\Gamma(x_{t_*})) \\
& + \Lambda(t_*^{u_o})\, \Delta(t_*^{u_o}) \int_E V(y)\, R(dy; x_{t_*}, u_\Gamma(x_{t_*})) \Bigg\}.
\end{aligned}
$$

Now define

$$
f_0^V(z, u_o) = \ell(z, u_o) + \int_E V(y)\, Q(dy; z, u_o)\, \lambda(z, u_o),
$$

$$F_0^V(z, u_\Gamma) = c(z, u_\Gamma) + \int_E V(y)\ R(dy; z, u_\Gamma),$$
$$\chi(t) = \Lambda(t)\ \Delta(t),$$

then

$$V(x) = \inf_{u\ \in\ \mathcal{U}_o \times \mathcal{U}_\Gamma} \left\{ \int_0^{t_*^{u_o}} \chi(t)\ f_0^V(x_{t-}, u_o(x, t^-))\ dt + \chi(t_*^{u_o})\ F_0^V(x_{t_*^{u_o}}, u_\Gamma(x_{t_*^{u_o}})) \right\}.$$

Later we will show that there exists a Borel measurable function $\hat{u}_\Gamma \colon \partial E \mapsto U_\Gamma$ such that for all $x \in \partial E$ we have

$$c(x, \hat{u}_\Gamma(x)) + \int_E V(y)\ R(dy; x, \hat{u}_\Gamma(x)) = \min_{u\ \in\ U_\Gamma} \left\{ c(x, u) + \int_E V(y)\ R(dy; x, u) \right\},$$

where U_Γ is a compact metrizable set. Now let us define

$$F^V(x) = c(x, \hat{u}_\Gamma(x)) + \int_E V(y)\ R(dy; x, \hat{u}_\Gamma(x)),$$

then one can easily see that the value function can be written as follows

$$V(x) = \inf_{u_o\ \in\ \mathcal{U}_o} \left\{ \int_0^{t_*^{u_o}} \chi(t)\ f_0^V(x_{t-}, u_o(x, t^-))\ dt + \chi(t_*^{u_o})\ F^V(x_{t_*^{u_o}}) \right\}. \tag{10}$$

The value function in (10) is very similar to the exit-time optimal control problems as discussed in Barles (1994). The major difference here is that, as a result of the stochastic nature of our problem, the value function itself appears in the terminal and running costs. In this chapter our aim is to exploit the similarities of (10) with exit-time optimal control problems and prove the existence and uniqueness results for the viscosity solutions of the associated Hamilton-Jacobi-Bellman (HJB) equation.
In the following sections, for the sake of notational simplicity, with slight abuse of notation we will use the following form for the value function

$$V(x) = \inf_{u_o\ \in\ \mathcal{U}_o} \left\{ \int_0^{t_*} \chi(t)\ f_0^V(x_t, u_o(t))\ dt + \chi(t_*)\ F^V(x_{t_*}) \right\}, \quad x \in E,$$

but it must be understood that, (10) is what we actually mean. In particular we use the more compact forms; t_* and $u_o(t)$, instead of $t_*^{u_o}$ and $u_o(x, t)$, as long as no confusion may arise.

3 Assumptions

In this section we state all the assumptions needed in our proofs. First we start by giving the hypothesis on the PDPs state space, transition measures, system dynamics and all other functions involved in the definition of cost function.

Assumptions 3.1

(A1) The state space E is a bounded open subset of $\mathbb{R}^d$ and of class $W^{2,\infty}$. This regularity, in particular, implies that d_∂, the distance to ∂E, is of class $W^{2,\infty}$; thus the normal vector to the boundary is given by $n(x) = -Dd_\partial(x)$. Note that E is not necessarily connected.

(A2) The control parameter sets U_o and U_Γ are closed subsets of compact metrizable spaces.

(A3) $g: \bar{E} \times U_o \mapsto \mathbb{R}^d$ is a bounded uniformly continuous function satisfying

$$\begin{cases} \forall x, y \in \bar{E}, \quad | g(x,u) - g(y,u) | \leq L \mid x - y \mid, \\ \forall x \in \bar{E} \text{ and } \forall u \in U_o, \quad | g(x,u) | \leq K, \end{cases}$$

where L and K are constants independent of $u \in U_o$.

(A4) $\ell: \bar{E} \times U_o \mapsto \mathbb{R}_+$, $c: \partial E \times U_\Gamma \mapsto \mathbb{R}_+$, and $\lambda: \bar{E} \times U_o \mapsto \mathbb{R}_+$ are non-negative bounded uniformly continuous functions.

(A5) $\delta: \bar{E} \mapsto [\delta_0, \infty[, \quad \delta_0 > 0$; is a bounded uniformly continuous function.

(A6) $Q: \bar{E} \times U_o \mapsto \mathcal{P}(E)$ and $R: \partial E \times U_\Gamma \mapsto \mathcal{P}(E)$ are continuous stochastic kernels (see proposition 7.30 in Bertsekas and Shreve (1978)) such that for all $f \in C_b(E)$, the maps $(x, u_o) \mapsto \int_E f(y)\, Q(dy; x, u_o)$ $(x \in \bar{E},\ u_o \in U_o)$ and $(x, u_\Gamma) \mapsto \int_E f(y)\, R(dy; x, u_\Gamma)$ $(x \in \partial E,\ u_\Gamma \in U_\Gamma)$ are bounded uniformly continuous. Here $C_b(E)$ and $\mathcal{P}(E)$ respectively denote, the set of all bounded continuous functions on E and the set of probability measures on E.

(A7) $\mathcal{U}_o$ and $\mathcal{U}_\Gamma$ are respectively the space of all Borel measurable functions $u_o: E \times \mathbb{R}_+ \mapsto U_o$ and $u_\Gamma: \partial E \mapsto U_\Gamma$.

(A8) There exists $\epsilon > 0$ such that $R(E_\epsilon; z, u_\Gamma) = 1$ for all $z \in \partial E$ and $u_\Gamma \in U_\Gamma$, where

$$E_\epsilon = \{x \in E: \inf_{u_o \in \mathcal{U}_o} t_*^{u_o}(x) \geq \epsilon\}.$$

Informally, (A8) says that jumps from the boundary are always to points whose distances from the boundary (as measured by the boundary hitting time) are uniformly bounded away from zero.
Now let us give the hypothesis on the behaviour of state trajectory near the boundary of the state space.

Assumptions 3.2

(A9) For all $x \in \partial E$ if there exists $u \in U_o$ such that $g(x,u) \cdot n(x) \leq 0$, then there exists $u' \in U_o$ such that $g(x,u') \cdot n(x) < 0$.

(A10) For all $x \in \partial E$ if $g(x,u)\cdot n(x) \leq 0$ for all $u \in U_o$, then $g(x,u)\cdot n(x) < 0$ for all $u \in U_o$.

Assumptions (3.2) are the hypothesis of non-degeneracy defined to remove difficulties arising from trajectories which are tangent to the boundary, Barles (1994). One can translate (A9) in a more picturesque way by "if for $x \in \partial E$ there exists a trajectory which re-enters E, then there is one re-entering strictly; which enters without being tangent at the boundary". Assumption (A10) can be expressed in the contrapositive sense: "if at $x \in \partial E$ there is a trajectory which exits from the open set E, then there exists one which exits strictly, i.e. which exits without being tangential to the boundary".

4 Discontinuous viscosity solutions

In this section we explain the definition of discontinuous viscosity solutions as given in Barles and Perthame (1988), Barles and Perthame (1990) and Barles (1994). Let us consider equations posed on closed sets

$$G(x, V, DV) = 0 \quad \text{on } \bar{E}, \tag{11}$$

where E is an open set in $\mathbb{R}^d$; further the real valued function, G, is not supposed to be continuous but is simply defined at all points and is locally bounded on $\bar{E}\times\mathbb{R}\times\mathbb{R}^d$. The idea in this approach is to consider simultaneously the equation formulated in the interior and the boundary condition, let us for example consider a Dirichlet problem,

$$\begin{array}{ll} H(x, V(x), DV(x)) = 0, & x \in E, \\ V(x) = \varphi(x), & x \in \partial E, \end{array} \tag{12}$$

where, H is a continuous function on $\bar{E} \times \mathbb{R} \times \mathbb{R}^d$ and φ is continuous on ∂E. We thus define G by,

$$G(x, v, p) = \begin{cases} H(x, v, p) & \text{in } E, \\ v - \varphi(x) & \text{on } \partial E. \end{cases}$$

More generally for an arbitrary boundary condition we have

$$F_\partial(x, V, DV) = 0 \quad \text{on } \partial E,$$

where F_∂ is continuous function on $\partial E \times \mathbb{R} \times \mathbb{R}^d$, one has

$$G(x, v, p) = \begin{cases} H(x, v, p) & \text{in } E, \\ F_\partial(x, v, p) & \text{on } \partial E. \end{cases}$$

z_* and z^* denote the lower semi-continuous (l.s.c.) and upper semi-continuous (u.s.c.) envelopes of the locally bounded function z, respectively. These envelopes can be defined by

$$z_*(x) = \liminf_{y \to x} z(y),$$

and

$$z^*(x) = \limsup_{y \to x} z(y).$$

If the function z depends on several variables (as in the case of Hamiltonian G) the notations z_* and z^* will always denote the l.s.c. and u.s.c. envelopes of z with respect to *all variables.*
The definition of discontinuous viscosity solution of a discontinuous equation is as follows.

Definition 4.1 : Discontinuous viscosity solution
A locally bounded u.s.c. function, v, on $\bar{E}$, is a viscosity subsolution of (11) if and only if,
$\forall \phi \in C^1(\bar{E})$, *if $x_0 \in \bar{E}$ is a local maximum of $v - \phi$ one has*

$$G_*(x_0, v(x_0), D\phi(x_0)) \leq 0. \tag{13}$$

A locally bounded l.s.c. function, w, on $\bar{E}$ is a viscosity supersolution of (11) if and only if,
$\forall \phi \in C^1(\bar{E})$, *if $x_0 \in \bar{E}$ is a local minimum of $w - \phi$ one has*

$$G^*(x_0, w(x_0), D\phi(x_0)) \geq 0. \tag{14}$$

A solution V *is either a continuous function which satisfies both (13) and (14) or else a locally bounded function whose u.s.c. and l.s.c. envelopes satisfy (13) and (14) respectively.*

Let us show that how this definition applies to boundary value problems, for example the Dirichlet problem. From the definition of G we can calculate easily the l.s.c. and u.s.c envelopes, as follows

$$\begin{cases} G_*(x,v,p) = G^*(x,v,p) = H(x,v,p), & \forall x \in E, \\ G_*(x,v,p) = \min\{H(x,v,p), v - \varphi(x)\}, & \forall x \in \partial E, \\ G^*(x,v,p) = \max\{H(x,v,p), v - \varphi(x)\}, & \forall x \in \partial E. \end{cases}$$

Recall that the functions H and φ are supposed to be continuous respectively on $\bar{E} \times \mathbb{R} \times \mathbb{R}^d$ and on ∂E. Thus the boundary conditions are expressed in the viscosity sense by,

$$\begin{cases} \min\{H(x,V,DV), V - \varphi\} \leq 0 & \text{on } \partial E, \\ \max\{H(x,V,DV), V - \varphi\} \geq 0 & \text{on } \partial E. \end{cases}$$

So, essentially one must understand that at boundary points where $V \neq \varphi$ the equation holds up to the boundary in a certain sense made precise by the two viscosity inequalities.
More generally, a general boundary condition is given by

$$F_\partial(x, V, DV) = 0 \quad \text{on } \partial E, \tag{15}$$

where F_∂ is a continuous function, and it must be understood in the viscosity sense as follows

$$\begin{cases} \min\{H(x,V,DV), F_\partial(x,V,DV)\} \leq 0 & \text{on } \partial E, \\ \max\{H(x,V,DV), F_\partial(x,V,DV)\} \geq 0 & \text{on } \partial E. \end{cases} \tag{16}$$

5 PDP optimal control: revisited

In Section 2 we assumed that there exists a Borel-measurable function $\hat{u}_\Gamma: \partial E \mapsto U_\Gamma$ such that for all $x \in \partial E$ we have

$$F_0^V(x, \hat{u}_\Gamma(x)) = \min_{u \in U_\Gamma} F_0^V(x,u).$$

Here we will show how one can utilise Borel-measurable selection theorems to prove this. Moreover we show that there exists a contraction mapping $\mathcal{G}$, such that V is the *unique* fixed point of $\mathcal{G}$. We will also use the fixed point property to show that V is bounded and continuous in E. Now let us start with the following theorem.

Theorem 5.1 *Assume (A1)–(A2), (A4), (A6) and $V: E \mapsto \mathbb{R}_+$ is bounded and continuous in E, then there exists a Borel-measurable function $\hat{u}_\Gamma: \partial E \mapsto U_\Gamma$ such that for all $x \in \partial E$*

$$c(x, \hat{u}_\Gamma(x)) + \int_E V(y)\ R(dy; x, \hat{u}_\Gamma(x)) = \min_{u\ \in\ U_\Gamma}\left\{c(x,u) + \int_E V(y)\ R(dy; x, u)\right\},$$

moreover $F^V(x) := c(x, \hat{u}_\Gamma(x)) + \int_E V(y)\ R(dy; x, \hat{u}_\Gamma(x))$ is a bounded uniformly continuous function on ∂E.

This is a consequence of the following Lemmas which can be proved using Bertsekas and Shreve (1978; Propositions 7.31, 7.32, 7.33).

Lemma 5.2 *$RV: \partial E \times U_\Gamma \mapsto \mathbb{R}_+$ defined by*

$$RV(x,u) = \int_E V(y) R(dy; x, u),$$

is bounded and continuous.

Lemma 5.3 *Let*

$$F^V(x) = \inf_{u \in U_\Gamma} F_0^V(x,u), \tag{17}$$

then $F^V: \partial E \mapsto \mathbb{R}_+$ is a bounded continuous function and the infimum in (17) is attained by some $u \in U_\Gamma$.

Lemma 5.4 *There exists a Borel-measurable function $\hat{u}_\Gamma: \partial E \mapsto U_\Gamma$ such that $\{(x, \hat{u}_\Gamma(x)): x \in \partial E\} \subset \partial E \times U_\Gamma$ and*

$$F^V(x) = F_0^V(x, \hat{u}_\Gamma(x)), \quad \forall x \in \partial E.$$

Now let us define $\mathcal{G}: B(E)_+ \mapsto B(E)_+$ as follows

$$\mathcal{G}\psi(x) = \inf_{u \in \mathcal{U}_o \times \mathcal{U}_\Gamma} \mathbb{E}_x^u \left\{ \int_0^{T_1} \Delta(t)\ell(x_t, u_o(x,t))\, dt \right.$$
$$\left. + \int_0^{T_1} \Delta(t)c(x_{t-}, u_\Gamma(x_{t-}))\, dp^*(t) + \Delta(T_1)\psi(x_{T_1}) \right\} \quad (18)$$

where, $\psi: E \mapsto \mathbb{R}_+$ is a positive Borel measurable function, i.e., $\psi \in B(E)_+$. Under our hypothesis in Assumptions 3.1, it is possible to show that $\mathcal{G}$ is a two-stage contraction, i.e., $\mathcal{G}^2$ is a contraction, Davis (1993; p. 178). The following theorem gives this result.

Theorem 5.5 *If Assumptions (A1)-(A5), (A7) and (A8) hold, then $\mathcal{G}$ is a two-stage contraction mapping.*

Proof. It suffices to show that there exists $\rho \in]0,1[$ such that for all $\psi_1, \psi_2 \in B(E)$, we have

$$\| \mathcal{G}^2\psi_1 - \mathcal{G}^2\psi_2 \| \leq \rho \| \psi_1 - \psi_2 \| .$$

In general, $\mathcal{G}$ is not a contraction if $t_*(x) < \infty$ for some $x \in E$. This is mainly because the accumulation of jumps may occur. One can show that

$$\| \mathcal{G}^2\psi_1 - \mathcal{G}^2\psi_2 \| \leq \sup_{u \in \mathcal{U}_o \times \mathcal{U}_\Gamma} \mathbb{E}_x^u \Delta(T_2) \| \psi_1 - \psi_2 \|,$$

and Assumption (A8) implies that,

$$\sup_{u \in \mathcal{U}_o \times \mathcal{U}_\Gamma} \mathbb{E}_x^u \Delta(T_2) \leq \rho < 1.$$

For a complete proof refer to Davis (1993) and Farid (1997). □
By strong Markov property in Davis (1993), it is easy to verify that

$$V(x) = \mathcal{G}V(x),$$

i.e. V is a fixed point of $\mathcal{G}$. We know that $\mathcal{G}^2$ is a contraction mapping, thus one can state the following corollary.

Corollary 5.6 *The value function of PDPs optimal control problem, i.e. V, is the unique fixed point of $\mathcal{G}$.*

This is a direct corollary of Apostol (1974; Theorem 4.48). It is easily envisaged that the definition of $\mathcal{G}$-operator in (18) is equivalent to

$$\mathcal{G}\psi(x) = \inf_{u_o \in \mathcal{U}_o} \left\{ \int_0^{t_*} \chi(t)\, f_0^\psi(x_t, u_o(t))\, dt + \chi(t_*)\, F^\psi(x_{t_*}) \right\}. \quad (19)$$

Now we want to show that the value function of PDPs optimal control problem is bounded and continuous in E. Assumptions 3.1 guarantee the boundedness of the cost function $J_x(u)$, as defined in Equation (7). In particular (A8) says that there exists some $t_\epsilon > 0$ such that it is equal to the minimum time between two jumps from the boundary.

Theorem 5.7 *Under Assumptions 3.1, the cost function of PDPs optimal control problem is bounded.*

Proof. One can show that

$$0 \leq J_x(u) \leq J_m,$$

where

$$J_m := \frac{K_1}{\delta_0} + \frac{K_2}{1 - e^{-\delta_0 t_e}},$$

and K_1, K_2 are respectively the upper bounds for ℓ and c. □

To prove the continuity of the value function we need to define a sequence of Borel measurable functions $(v_n, n \in \mathbb{Z}_+)$, generated as follows

$$\begin{aligned} v_{n+1}(x) &= \mathcal{G}v_n(x), \qquad (20) \\ &= \inf_{u_o \in \mathcal{U}_o} \left\{ \int_0^{t_*} \chi(t)\, f_0^{v_n}(x_t, u_o(t))\, dt + \chi(t_*)\, F^{v_n}(x_{t_*}) \right\}. \end{aligned}$$

According to Corollary 5.6, $V(x) = \lim_{n \to \infty} v_n$ because V is the unique fixed point of $\mathcal{G}$. In the following theorem we use this fact to prove the existence of two monotone sequences of continuous functions converging to V, one from above and one from below and thus concluding the continuity of V in E.

Theorem 5.8 *Under Assumptions 3.1 and 3.2, the value function of PDPs optimal control problem, i.e. $V: E \mapsto \mathbb{R}_+$, is bounded and continuous in E.*

This is a consequence of the following two lemmas.

Lemma 5.9 *Let $v_0 \equiv 0$ then $(v_n, n \in \mathbb{Z}_+)$ is an increasing sequence of functions and $v_n \uparrow V$ as $n \to \infty$, thus V is bounded from below and l.s.c. in E.*

Proof. It is an easy extension of Theorem 5.3 in Barles (1994) that all functions in the sequence $(v_n, n \in \mathbb{Z}_+)$ are continuous in E, i.e. $v_n \in C(E)$ for all $n \in \mathbb{Z}_+$. After the first iteration we have

$$v_1(x) = \inf_{u \in \mathcal{U}_o} \left\{ \int_0^{t_*} \chi(t)\, \ell(x_t, u_o(t))\, dt + \chi(t_*)\, c(x_{t_*}, \hat{u}_\Gamma(x_{t_*})) \right\},$$

thus $v_0 \leq v_1$. Now assume $v_{n-2} \leq v_{n-1}$ then it is easy to see $\mathcal{G}v_{n-2} \leq \mathcal{G}v_{n-1}$ and hence $v_{n-1} \leq v_n$. So $(v_n, n \in \mathbb{Z}_+)$ with $v_0 \equiv 0$ is an increasing sequence and we know that $v_n \uparrow V$ as $n \to \infty$. Using Lemma 7.14, p. 147 of Bertsekas and Shreve (1978) one can observe that V is lower semi-continuous (l.s.c.) in E and bounded below. □

For the next Lemma we need a supplementary condition, which can be removed later:

(A11) For some $\epsilon_0 > 0$ process never jumps into $\tilde{E}_{\epsilon_0} = E - E_{\epsilon_0}$, where E_{ϵ_0} is defined in (A8).

Under this condition E_ϵ is an invariant set for the Markov chain $\{x, x_{T_1}, x_{T_2}, \ldots\}$, for any $\epsilon < \epsilon_0$ and $x \in E_\epsilon$.

Lemma 5.10 *For $0 < \epsilon < \epsilon_0$ (ϵ_0 is as in (A11)), there exists a constant K_ϵ such that if $w_0 \equiv K_\epsilon$, then $(w_n, n \in \mathbb{Z}_+)$ is a decreasing sequence of functions and $w_n \downarrow V$ in E_ϵ as $n \to \infty$, thus V is bounded from above and u.s.c. in E_ϵ.*

Proof. For all $x \in E_\epsilon$ and $u \in \mathcal{U}_o \times \mathcal{U}_\Gamma$ we have $\mathbb{E}_x^u \Delta(T_1) \leq \alpha_\epsilon$, where

$$\alpha_\epsilon = \left(\delta_0 e^{-(\delta_0+\lambda_0)\epsilon} + \lambda_0\right) / (\delta_0 + \lambda_0),$$

in view of (A4) and (A5). Here λ_0 is an upper bound for $\lambda(x, u)$. Hence if $w_0 \equiv \kappa$

$$\begin{aligned} w_1(x) &= \inf_{u \in \mathcal{U}_o \times \mathcal{U}_\Gamma} \mathbb{E}_x^u \left\{ \int_0^{T_1} \Delta(t)\, \ell(x_t, u_o(t))\, dt \right. \\ &\qquad \left. + \Delta(T_1)\, c(x_{T_1}, u_\Gamma(x_{T_1}))\, I_{\{T_1 = t_*\}} + \Delta(T_1)\, w_0(x_{T_1}) \right\}, \\ &\leq K_1/\delta_0 + K_2 e^{-\delta_0 \epsilon} + \alpha_\epsilon \kappa, \end{aligned}$$

where K_1, K_2 are respectively upper bounds for ℓ, and c. Since $\alpha_\epsilon < 1$ we see that $w_1(x) \leq \kappa$ if

$$\kappa \geq K_1/(1-\alpha_\epsilon)\delta_0 + K_2 e^{-\delta_0 \epsilon}/(1-\alpha_\epsilon).$$

Hence for each $\epsilon > 0$ there exists a K_ϵ such that if $w_0(x) \equiv K_\epsilon$ then $w_1(x) \leq w_0(x)$ for all $x \in E_\epsilon$. Now assume $w_{n-1} \leq w_{n-2}$ which implies $w_n \leq w_{n-1}$, hence $(w_n, n \in \mathbb{Z}_+)$ is a decreasing sequence of continuous functions in E_ϵ and according to Bertsekas and Shreve (1978; Lemma 7.14) one can conclude that V is upper semi-continuous in E_ϵ and bounded above.
□

Under the continuity conditions imposed in Assumptions 3.1, (A11) is not needed. We can start by modifying the transition measures so that (A11) is satisfied and then use a limiting argument, Farid and Davis (1998b).

6 Principle of dynamic programming

In this section we will prove the principle of dynamic programming for the value function V of the PDPs optimal control problem. This is an essential step in proving that V is a viscosity solution of its associated HJB equation Barles (1994), Ishii (1989) and Lions and Souganidis (1985).

Let us consider the following exit-time optimal control problem

$$W(x_0) = \inf_{u \in \mathcal{U}_o} \left\{ \int_0^{t_*^u(x_0)} \chi(t)\ \ell(x(t), u(t))\ dt \right. \\ \left. + \chi(t_*^u(x_0))\ \varphi(x(t_*^u(x_0))) \right\},\ x_0 \in E, \quad (21)$$

where χ is the same function defined in Section 2.

Theorem 6.1 *Under hypothesis (A3) and (A4), if φ is continuous on ∂E, the value function W of the exit-time optimal control problem satisfies the dynamic programming principle:*

$$W(x_0) = \inf_{u \in \mathcal{U}_o} \left\{ \int_0^{t_*^u(x_0)\wedge T} \chi(t)\ \ell(x(t), u(t))\ dt \right. \\ + I_{(T<t_*^u(x_0))}\ \chi(T)\ W(x(T)) \\ \left. + I_{(t_*^u(x_0)\leq T)}\ \chi(t_*^u(x_0))\ \varphi(x(t_*^u(x_0))) \right\}, \quad (22)$$

for all $T > 0$.

Proof. Refer to Ishii (1989), Barles (1994) and Farid (1997). □
Now we use this result to prove the dynamic programming principle for the value function of the PDPs optimal control problem.

Theorem 6.2 *Under hypothesis in Assumptions (3.1), the value function of PDPs optimal control problem satisfies the dynamic programming principle:*

$$V(x_0) = \inf_{u_o(\cdot) \in \mathcal{U}_o} \left\{ \int_0^{t_*^{u_o}(x_0)\wedge T} \chi(t)\ f_0^V(x(t), u_o(t))\ dt \right. \\ + I_{(T<t_*^{u_o}(x_0))}\chi(T)\ V(x(T)) \\ \left. + I_{(t_*^{u_o}(x_0)\leq T)}\chi(t_*^{u_o}(x_0))F^V(x(t_*^{u_o}(x_0))) \right\},$$

for all $T > 0$.

Proof. Using Theorem 6.1 and Theorem 5.1, for all $v \in C_b(E)$ one can write

$$\mathcal{G}v(x_0) = \inf_{u_o(\cdot) \in \mathcal{U}_o} \left\{ \int_0^{t_*^{u_o}(x_0)\wedge T} \chi(t)\ f_0^v(x(t), u_o(t))\ dt \right. \\ + I_{(T<t_*^{u_o}(x_0))}\chi(T)\ \mathcal{G}v(x(T)) \quad (23) \\ \left. + I_{(t_*^{u_o}(x_0)\leq T)}\chi(t_*^{u_o}(x_0))\ F^v(x(t_*^{u_o}(x_0))) \right\},$$

and it is also known that

$$V(x_0) = \mathcal{G}V(x_0).$$

So replacing v with V in (23), yields the desired result. □

7 Viscosity solutions

In this section we will utilise the same method as in Chapter 4 of Barles (1994) and prove that V is the unique viscosity solution of its associated HJB equation. This result is very useful because there are numerical methods for solving HJB equations which converge to viscosity solutions (e.g. see Fleming and Soner (1993) and Kushner and Dupuis (1992)).
First let us re-state some of the theorems in Chapter 5 of Barles (1994), this will help us to obtain results for the PDPs optimal control case.
Consider the value function of an exit-time optimal control problem:

$$W(x_0) = \begin{cases} \inf_{u_o \in \mathcal{U}_o} \left\{ \int_0^{t_*} \chi(t)\, \ell(x(t), u_o(t))\, dt + \chi(t_*)\, \varphi(x(t_*)) \right\} & \text{in } E, \\ \varphi(x_0) & \text{on } \partial E. \end{cases}$$

First we want to prove that W is a viscosity solution of its associated HJB equation with Dirichlet type boundary condition given by

$$\sup_{u \in U_o} \{(\lambda(x,u) + \delta(x))W - DW \cdot g(x,u) - \ell(x,u)\} = 0 \text{ in } E, \tag{24}$$

$$W = \varphi \text{ on } \partial E. \tag{25}$$

Theorem 7.1 *Under hypothesis in Assumptions 3.1, if φ is continuous on ∂E, then W is a viscosity solution (discontinuous) of (24)–(25).*

Proof. For a complete proof refer to Chapter 5 of Barles (1994). □
As one may have noticed, under hypothesis in Assumptions 3.1, W is not necessarily a continuous function in E. To make this happen we need some further assumptions about the behaviour of the state trajectory near the boundary of the state space. The following theorem gives the required result.

Theorem 7.2 *Under hypothesis in Assumptions 3.1 and 3.2, if φ is continuous on ∂E, the value function W of the exit-time optimal control problem is continuous in E and is the unique viscosity solution of (24)–(25).*

Proof. For a complete proof refer to Chapter 5 of Barles (1994). Assumptions 3.2 imply that $\partial E = \partial E_1 \cup \partial E_2 \cup \partial E_3$, where ∂E_1, ∂E_2, ∂E_3 are each union of connected components of ∂E and are described as follows:

∂E_1: The set of all points $x \in \partial E$ for which there exists $u, u' \in U_o$ such that

$$g(x,u) \cdot n(x) > 0, \quad g(x,u') \cdot n(x) < 0.$$

∂E_2: The set of all points $x \in \partial E$ for which for all $u \in U_o$

$$g(x,u) \cdot n(x) > 0.$$

∂E_3: The set of all points $x \in \partial E$ for which for all $u \in U_o$

$$g(x,u) \cdot n(x) < 0.$$

□

Now we would like to prove similar results for PDPs optimal control problem, but first let us define the following HJB equation with Dirichlet type boundary condition

$$H_\psi(x, V, DV) = 0 \quad \text{in } E, \tag{26}$$
$$V(x) = F^\psi(x) \quad \text{on } \partial E, \tag{27}$$

where

$$H_\psi(x,v,p) = \sup_{u \in U_o} \left\{ (\lambda(x,u) + \delta(x))v - g(x,u) \cdot p - f_0^\psi(x,u) \right\},$$

and $\psi: \bar{E} \to \mathbb{R}_+$ is a Borel measurable function. In the special case of $\psi = V$, we need to clearly define what do we mean by a viscosity solution of

$$H_V(x, V, DV) = 0 \quad \text{in } E, \tag{28}$$
$$V(x) = F^V(x) \quad \text{on } \partial E. \tag{29}$$

Definition 7.3 *A bounded u.s.c. function, v, on $\bar{E}$, is a viscosity subsolution of (28)–(29) if and only if,*
$\forall \phi \in C^1(\bar{E})$, if $x_0 \in \bar{E}$ is a local maximum of $v - \phi$ one has

$$\begin{aligned} &H_v(x_0, v(x_0), D\phi(x_0)) \le 0 && \text{if } x_0 \in E,\\ &\min\{H_v(x_0, v(x_0), D\phi(x_0)), v(x_0) - F^v(x_0)\} \le 0 && \text{if } x_0 \in \partial E. \end{aligned}$$

A bounded l.s.c. function, w, on $\bar{E}$ is a viscosity supersolution of (28)–(29) if and only if,
$\forall \phi \in C^1(\bar{E})$, if $x_0 \in \bar{E}$ is a local minimum of $w - \phi$ one has

$$\begin{aligned} &H_w(x_0, w(x_0), D\phi(x_0)) \ge 0 && \text{if } x_0 \in E,\\ &\max\{H_w(x_0, w(x_0), D\phi(x_0)), w(x_0) - F^w(x_0)\} \ge 0 && \text{if } x_0 \in \partial E. \end{aligned}$$

A solution, V, is either a continuous function which is both subsolution and supersolution of (28)–(29); or else a bounded function whose u.s.c. and l.s.c. envelopes are respectively subsolution and supersolution of (28)–(29).

For viscosity solutions of (26)–(27) we have the following theorem.

Theorem 7.4 *Under Assumptions 3.1 and 3.2, for all positive functions $\psi \in C_b(E)$;*

$$v(x) = \begin{cases} \mathcal{G}\psi(x) & \text{in } E,\\ F^\psi(x) & \text{on } \partial E, \end{cases}$$

is bounded and continuous in E and is the unique viscosity solution of (26)–(27).

Proof. This is exactly similar to Theorem 7.2 and all we have to show is that $f_0^{\psi}: \bar{E} \times U_o \to \mathbb{R}_+$ and $F^{\psi}: \partial E \to \mathbb{R}_+$ are bounded continuous functions; using hypothesis (A6) and Theorem 5.1, one can easily conclude this. □
Now using Theorem 7.4 we want to prove that the value function V of PDPs optimal control problem, is the unique viscosity solution of (28)–(29). The proof follows from Lemmas 7.6 and 7.7 below.

Theorem 7.5 *Under Assumptions 3.1 and 3.2, V, the value function of PDPs optimal control problem; is continuous in E and is the unique viscosity solution of (28)–(29).*

According to Theorem 5.8, V is bounded and continuous in E; now in order to prove that V is the unique viscosity solution of (28)–(29), we need the following lemmas.

Lemma 7.6 *V is a viscosity solution of (28)–(29).*

Proof. Using Theorem 6.2, we know that V satisfies the dynamic programming principle; thus using exactly the same steps as in Theorem 7.1 one can show that V is a viscosity solution of (28)–(29). □

Lemma 7.7 *Any viscosity solution of (28)–(29) is a fixed point of $\mathcal{G}$.*

Proof. Assume V' is a viscosity solution of (28)–(29), then it is easy to see that it is also a viscosity solution of

$$H_{V'}(x, \bar{V}, D\bar{V}) = 0 \quad \text{in } E, \tag{30}$$
$$\bar{V}(x) = F^{V'}(x) \quad \text{on } \partial E, \tag{31}$$

which is regarded as an equation for unknown $\bar{V}$ with V' "frozen". According to Theorem 7.4, $\mathcal{G}V'$ is the unique viscosity solution of (30)–(31) in E, thus we must have $V' = \mathcal{G}V'$ in E. On the other hand we know that (Corollary 5.6) V is the unique fixed point of $\mathcal{G}$, hence $V' = V$ in E and $F^{V'} = F^V$ on ∂E, which means $V' = V$ in $\bar{E}$. □
Now we want to prove that there is only one viscosity solution of (28)–(29) which is also continuous in $\bar{E}$.

Theorem 7.8 *Let*

$$\tilde{V}(x) = \begin{cases} V(x) & \text{in } E, \\ \lim\limits_{y \to x,\ y \in E} V(y) & \text{on } \partial E. \end{cases}$$

then $\tilde{V}$ is the only viscosity solution of (28)–(29) which is continuous in $\bar{E}$.

Proof. We know that V is continuous in E, thus $V(x) = V^*(x) = V_*(x)$ for all $x \in E$ and it is clear that V is a subsolution and supersolution in E. Since $\tilde{V}(x) = V(x)$ for all $x \in E$, the above argument implies that $\tilde{V}$ is a viscosity solution of (28)–(29) in E. In the following paragraphs one should

note that $F^V = F^{V_*} = F^{V^*} = F^{\tilde{V}}$ and $f_0^V = f_0^{V_*} = f_0^{V^*} = f_0^{\tilde{V}}$ because $V = V_* = V^* = \tilde{V}$ for all $x \in E$.
Now we want to prove that $\tilde{V}$ is also a viscosity solution for the points on the boundary. Assume $V_*(x_0) \geq F^V(x_0)$ for some $x_0 \in \partial E$, then we have $\tilde{V}(x_0) - F^V(x_0) \geq 0$ since $\tilde{V}(x) \geq V_*(x)$ for all $x \in \bar{E}$. This implies that $\tilde{V}$ satisfies the supersolution inequality at x_0. Now assume $V_*(x_0) < F^V(x_0)$ for some $x_0 \in \partial E$. Since V_* is a supersolution, for all $\phi \in C^1(\bar{E})$ if x_0 is a local minimum of $V_* - \phi$ we have

$$H_V(x_0, V_*(x_0), D\phi(x_0)) \geq 0.$$

Because $V_*(x_0) < F^V(x_0)$, there must exist $(x_n, n \in \mathbb{N}) \in E$ such that $x_n \to x_0$ as $n \to +\infty$ and

$$V_*(x_0) = \lim_{n\to\infty} V(x_n),$$

thus $V_*(x_0) = \tilde{V}(x_0)$. Moreover there exists $\delta > 0$ such that

$$V_*(x_0) - \phi(x_0) \leq V_*(x) - \phi(x),$$

for all $x \in B_\delta(x_0)$. Using $\tilde{V}(x) \geq V_*(x)$ for all $x \in \bar{E}$ and $V_*(x_0) = \tilde{V}(x_0)$, for all $x \in B_\delta(x_0)$ one gets

$$\tilde{V}(x_0) - \phi(x_0) \leq \tilde{V}(x) - \phi(x),$$

thus, x_0 is a local minimum of $\tilde{V} - \phi$ and

$$H_V(x_0, \tilde{V}(x_0), D\phi(x_0)) \geq 0.$$

Finally one can conclude that $\tilde{V}$ is a viscosity supersolution in $\bar{E}$. Similarly one can prove that $\tilde{V}$ is a viscosity subsolution of (28)–(29).
The uniqueness part can be argued as follows. Assume v and w are viscosity solutions of (28)–(29). We know that if V_1 is a viscosity subsolution and V_2 is a viscosity supersolution then

$$V_1 \leq V_2 \quad \text{in } E.$$

Thus we have

$$v^* \leq w_* \quad \text{in } E,$$

and

$$w^* \leq v_* \quad \text{in } E,$$

now let $v \equiv V$ and this implies $w^* \leq V \leq w_*$ in E. Thus $w^* = V = w_*$ in E and hence there is only one continuous viscosity solution in E. Since this viscosity solution has got a unique continuous extension to $\bar{E}$, i.e. $\tilde{V}$ (see Proposition 11, p. 149 in Royden (1988)), and its extension is a viscosity solution in $\bar{E}$; one can conclude that the equations in (28)–(29) have at most one continuous viscosity solution in $\bar{E}$. □

Now let us denote $E_{PDP} = E \cup \partial E_3$ and $\Gamma = \partial E_1 \cup \partial E_2$ and define

$$V(x)=\begin{cases} \inf\limits_{u_o \in \mathcal{U}_o}\left\{\int_0^{t_*^{u_o}} \chi(t) f_0^V(x_t, u_o(t))\, dt + \chi(t_*^{u_o}) F^V(x(t_*^{u_o}))\right\}, & x \in E_{PDP}, \\ F^V(x), & x \in \Gamma, \end{cases}$$

where

$$t_*^{u_o}(x) = \inf\{t \geq 0\colon x_t \notin E_{PDP},\ x \in E_{PDP}, u_o \in \mathcal{U}_o\}.$$

Thus the discontinuity of the value function, as defined above, only occurs on ∂E_1 and on other parts of the boundary, i.e. $\partial E_2 \cup \partial E_3$, the above value function is continuous. In fact Theorem 7.8 suggests that by defining $V(x) = \lim\limits_{y \to x,\ y \in E} V(y)$ for all $x \in \partial E$ one can achieve continuity in $\bar{E}$. According to the above definition for V, we have $V(x) = \lim\limits_{y \to x,\ y \in E} V(y)$ for all $x \in \partial E_2 \cup \partial E_3$.

As a final point one can conclude that the boundary conditions given in Barles (1994) and stated in Assumptions 3.2 are more generalised than the previous boundary conditions considered in the same context of exit-time optimal control problems. Obviously the cost of this generalisation is losing the continuity of the value function on the boundary, which is more realistic in many applications since the optimal policy may not include hitting some parts of the boundary.

Acknowledgments: This article summarises the results of a chapter in the second author's PhD thesis (Davis and Farid (1996), Farid (1997), Farid and Davis (1998a), Farid and Davis (1998b)) written while both authors were at the Center for Process Systems Engineering, Imperial College, London, UK, funded by an EPSRC grant. The second author would also like to thank City University, London, UK, for supporting this work.

References

Apostol, Tom M. (1974). Mathematical analysis. Addison-Wesley Publishing Company.

Barles, G. (1994). Solutions de viscosité des équations de Hamilton- Jacobi. Mathematiques & Applications 17. Paris, Springer-Verlag.

Barles, G. and Perthame, B. (1988). Exit time problems in optimal control and vanishing viscosity method. SIAM J. Control and Optimization, 26(5):1133-1148.

Barles, G. and Perthame, B. (1990) Comparison principle for Dirichlet type Hamilton-Jacobi equations and singular perturbations of degenerated elliptic equations. Applied Mathematics and Optimization 21:21-44.

Bertsekas, Dimitri P. and Shreve, Steven E. (1978). Stochastic optimal control: the discrete time case. Mathematics in Science and Engineering 139. Academic Press.

Davis, Mark H. A. (1993). Markov models and optimization. Monographs on Statistics and Applied Probability 49. Chapman & Hall.

Davis, Mark H. A. and Farid, Mohammad (1996). A target recognition problem: sequential analysis and optimal control. SIAM J. Control and Optimization, 34(6):2116-2132.

Dempster , M. A. H. and Ye, J. J. (1992). Necessary and sufficient optimality conditions for control of piecewise-deterministic Markov processes, Stochastics and Stochastic Reports, 40:125-145.

Farid, M. (1997). Optimal control, piecewise-deterministic processes and viscosity solutions. PhD thesis, Imperial College of Science, Technology and Medicine, University of London, UK.

Farid, M. and Davis, Mark H. A. (1998a). Optimal consumption and exploration: a case study in piecewise-deterministic Markov modelling, to be published in Annals of Operations Research, Baltzer Science Publishers.

Farid, M. and Davis, Mark H. A. (1998b). Exit-time optimal control problems and piecewise-deterministic Markov processes, to be submitted to SIAM Journal on Control and Optimization.

Fleming W. H. and Soner, M. H. (1993). Controlled Markov processes and viscosity solutions. Applications of Mathematics 25. Springer-Verlag.

Ishii, Hitoshi (1989). A boundary value problem of the Dirichlet type for Hamilton-Jacobi equations. Ann. Scuola. Norm. Pisa Cl. Sci. Series 4, 16(1):105-135.

Kushner, H. J. and Dupuis, Paul G. (1992). Numerical methods for stochastic control problems in continuous time. Applications of Mathematics 24. Springer-Verlag.

Lions, P. L. and Souganidis, P. E. (1985). Differential games, optimal control and directional derivatives of viscosity solutions of Bellman's and Isaacs equations. SIAM J. Control and Optimization, 23(4):566-583.

Royden, H. L. (1988). Real Analysis. Collier Macmillan Publishing.

Vermes, D. (1985). Optimal control of piecewise deterministic Markov processes. Stochastics, 14:165-207.

16
Mathematical Approaches to the Problem of Noise-Induced Exit

Martin V. Day

ABSTRACT: We provide an overview of some of the methods that have been used to study sub-large deviations phenomena in the small noise exit problem. Both formal asymptotic methods and rigorous probabilistic methods are considered and compared.

KEYWORDS: Exit Problem; Small Noise.

1 Introduction

We consider here the basic problem of small noise induced exit of a finite dimensional system from an asymptotically stable equilibrium point. In $\mathbb{R}^d$ let $\omega(t)$ be a standard Wiener process and consider the diffusion process

$$dx^\epsilon(t) = b(x^\epsilon(t))\,dt + \epsilon^{1/2}\sigma(x^\epsilon(t))\,d\omega(t), \tag{1}$$

starting at $x^\epsilon(0) \in D$, where D is a basin of attraction for an asymptotically stable critical point at 0 for the deterministic system

$$\dot{x}(t) = b(x(t)). \tag{2}$$

For an initial $x = x^\epsilon(0) \in D$ we follow $x^\epsilon(t)$ up to its first contact with ∂D:

$$\tau_D^\epsilon = \inf\{t > 0 : \; x^\epsilon(t) \in \partial D\}.$$

Our interest is in the asymptotic behavior ($\epsilon \downarrow 0$) of the *exit point distribution*

$$\mu_x^\epsilon(dy) = P_x[x^\epsilon(\tau_D^\epsilon) \in dy] \tag{3}$$

and the *mean exit time*

$$v_D^\epsilon(x) = E_x[\tau_D^\epsilon]. \tag{4}$$

The main offering of this paper is the overview in Section 3 of some of the approaches which have been used to obtain sub-large deviations results. By this we mean results that describe properties of (3) and (4) that cannot be determined from the usual large deviations results [17] describing exponential dependence on ϵ. It is the mathematical concepts behind the methods that we discuss, rather than specific results which apply under diverse and various hypotheses on ∂D. Some of the approaches we describe have been given rigorous treatments in the literature, others are based only on formal asymptotic calculations. The latter have played an important role in the discovery and explanation of significant features, though they do not by themselves provide "proof" in the mathematical sense. Beyond theoretical and formal asymptotic approaches, analogue simulations have also been used to observe features of the exit problem firsthand; see [11] and references. More sophisticated digital simulations based on the importance sampling technique were conducted in [10].

A long-standing interest of Wendell Fleming's has been small noise ($\epsilon \downarrow 0$) limits in stochastic control problems, particularly as regards the Hamilton-Jacobi-Bellman equation which describes the associated value function ϕ^ϵ, and its asymptotic relation to that of its deterministic ($\epsilon = 0$) counterpart. A logarithmic transformation $\phi^\epsilon = -\epsilon \log(u^\epsilon)$ typically converts solutions u^ϵ of linear equations equations (such as those associated with $\mathcal{L}^\epsilon$ below) into Hamilton-Jacobi-Bellman equations of this small noise type. This allows the small noise asymptotics of appropriate control problems to be used to obtain large deviations results for the exit problem. Fleming's paper [14] broke new ground in this area. The advent of viscosity solution techniques has made these analyses possible without explicit recourse to control arguments; see [13] for instance. The 1990 paper of Perthame [29] is a development of these ideas specifically in the context of the exit problem as we are considering it here. We will comment more on that work in Section 3.2 below. This general approach is not limited to Wentzel-Freidlin type large deviations but can also be used for results on occupation measures, principal eigenvalues and other issues; see [15] and [16]. In fact the logarithmic connection to stochastic optimization problems has been developed by Dupuis and Ellis [9] into a general approach to large deviations analysis of broad scope.

2 Basic Results

We review in this section a number of results to which our discussion in Section 3 refers. Two special cases of the exit problem will be mentioned frequently. These depend on whether (2) crosses ∂D with a positive angle or is tangent to it. Let n denote the (unit) outward normal to ∂D. The *nontangential case* is succinctly described by $n \cdot b < 0$ at all points $y \in \partial D$. This was the case originally treated by the large deviations analysis in [17]. In many applications D is taken to be the maximal domain of attraction of the stable point at $x = 0$. This makes the *characteristic boundary*, $n \cdot b = 0$, a more natural hypothesis, though more difficult mathematically.

2.1 PDE Formulations

The generator of the diffusion process (1) is the differential operator

$$\mathcal{L}^\epsilon u(x) = \frac{\epsilon}{2}\sum_{i,j} a_{ij}(x)\frac{\partial^2}{\partial x_i \partial x_j}u(x) + \sum_i b_i(x)\frac{\partial}{\partial x_i}u(x). \tag{5}$$

The diffusion matrix is $a(x) = \sigma(x)\sigma(x)^T$. Typical regularity assumptions would require $a(\cdot), b(\cdot)$ to be C^3, with a positive definite hypothesis on $a(\cdot)$. Similarly D is assumed to be bounded with C^3 boundary. However, we will not concern ourselves with precise hypotheses, since our goal is to provide only a heuristic overview.

The quantities of interest in the exit problem can be characterized in terms of appropriate boundary value problems associated with $\mathcal{L}^\epsilon$. The exit point distribution (3) is intimately connected with the Dirichlet problem in D: given a continuous function $f(y)$ defined on ∂D, find $u(x) = u_f^\epsilon(x)$ solving

$$\mathcal{L}^\epsilon u(x) = 0,\ x \in D; \quad u(y) = f(y),\ y \in \partial D. \tag{6}$$

Indeed as is well-known,

$$u_f^\epsilon(x) = E_x[f(x^\epsilon(\tau_D^\epsilon))] = \int_{\partial D} f(y)\,\mu_x^\epsilon(dy).$$

With regard to the mean exit time (4), its distribution is linked to $\mathcal{L}^\epsilon$ through the Poisson equation

$$\mathcal{L}^\epsilon v(x) + \phi(x) = 0,\ x \in D; \quad v(y) = 0,\ y \in \partial D \tag{7}$$

by the solution formula

$$v(x) = E_x[\int_0^{\tau_D^\epsilon} \phi(x^\epsilon(t))\,dt].$$

In particular, for $\phi \equiv 1$ we get the mean exit time, $v_D^\epsilon(x) = E_x[\tau_D^\epsilon]$.

2.2 Green's Identity

A principal tool in exploiting these PDE characterizations is Green's identity, which provides a fundamental relation between $\mathcal{L}^\epsilon$ and its formal adjoint:

$$\mathcal{L}^{\epsilon *} g(x) = \frac{\epsilon}{2}\sum_{i,j}\frac{\partial^2}{\partial x_i \partial x_j}[a_{ij}(x)f(x)] - \sum_i \frac{\partial}{\partial x_i}[b_i(x)f(x)]. \tag{8}$$

The unit outward normal and *conormal* at $y \in \partial D$ will be

$$n(y) = (n_i(y)), \quad \eta(y) = a(y)n(y).$$

Two operators defined on ∂D appear in (11) below. One is a scaled conormal derivative:

$$\mathcal{N}^\epsilon f(y) = \frac{\epsilon}{2}\frac{\partial}{\partial \eta} f(y) = \frac{\epsilon}{2}\sum_{i,j} n_i a_{ij}\frac{\partial}{\partial x_j} f(y). \tag{9}$$

The second is a sort of "boundary adjoint" to $\mathcal{N}^\epsilon$:

$$\mathcal{N}^{\epsilon *} g(y) = \frac{\epsilon}{2}\sum_{i,j} n_i \frac{\partial}{\partial x_j}[a_{ij} g(y)] - \langle n, b\rangle \cdot g(y). \tag{10}$$

Green's (second) identity, equation (6.5) of [27], is a fundamental relation among these operators:

$$\int_D g(x)\mathcal{L}^\epsilon f(x) - f(x)\mathcal{L}^{\epsilon *} g(x)\, dx = \int_{\partial D} g(y)\mathcal{N}^\epsilon f(y) - f(x)\mathcal{N}^{\epsilon *} g(y)\, dy. \tag{11}$$

2.3 Leveling Results

The exit point distributions μ_x^ϵ need not in general converge as $\epsilon \downarrow 0$. However the dependence on the initial point x vanishes in the limit. This is manifested in the leveling property of the solution u_f^ϵ to the Dirichlet problem (6): for any compact $K \subseteq D$,

$$\sup_{x,y\in K} |u_f^\epsilon(x) - u_f^\epsilon(y)| \to 0 \text{ as } \epsilon \downarrow 0.$$

This was proven in [8] for the nontangential case, but the concluding remarks show that it holds for characteristic boundaries as well. See [12] also. If in fact μ_x^ϵ does converge as $\epsilon \downarrow 0$ then the leveling property implies that the limit $\mu_x^\epsilon \Rightarrow \mu^0$ is independent of $x \in D$ and that for a given f there is a constant

$$C_f^0 = \int_{\partial D} f\, d\mu^0 \tag{12}$$

so that $u_f^\epsilon \to C_f^0$ uniformly on compacts. In this case finding μ^0 is equivalent to identifying a formula for C_f^0.

A relative leveling result for the mean exit time is provided by Corollary 1 of [7]:

$$\lambda^\epsilon E_x[\tau_D^\epsilon] \to 1, \tag{13}$$

holding uniformly on any compact $K \subseteq D$. Here $\lambda^\epsilon > 0$ is the principal eigenvalue for $\mathcal{L}^\epsilon$ with Dirichlet boundary data on ∂D: $\mathcal{L}^\epsilon\phi^\epsilon + \lambda^\epsilon\phi^\epsilon = 0$ for $\phi^\epsilon > 0$ in D with $\phi^\epsilon = 0$ on ∂D. Because of (13) analysis of the principal eigenvalue is often considered equivalent to analysis of the mean exit time. The treatment in [7] only considered the nontangential case. To our knowledge (13) has not been worked out in the characteristic boundary case, although it may implied by some of the sophisticated work on exponential loss of memory and long-time asymptotics: [23], [24] and [30] for instance. As we will see in Section 3.2, (13) would be useful in the characteristic boundary case.

2.4 Hasminksii's Formula

Many of the approaches to sub-large deviations phenomena can be viewed as expressions involving an invariant density for some version of $x^\epsilon(t)$. We summarize here a general formula for the invariant distribution in terms of an embedded Markov chain. Although this formula goes back much further in the context of Markov chains, it was developed in the context of diffusion processes by Hasminskii in 1960. An exposition can be found in [18]; see Ch. IV, §4. We will use this formula below for several variants of the original $x^\epsilon(t)$ of (1), to produce rather explicit relations among the invariant distribution, exit point distribution and exit time. We describe it here in general for a process $x(t)$, which will refer to any of the variants actually used below.
The idea is to construct a sequence of stopping times

$$0 < \tau_1 < \tau_2 < \dots$$

associated with contact times of a (compact) subset Γ. This is typically done using a second disjoint subset Γ': with $\tau_0 = 0$ define

$$\begin{aligned} \tau_i' &= \inf\{t > \tau_i : x(t) \in \Gamma'\} \\ \tau_{i+1} &= \inf\{t > \tau_i' : x(t) \in \Gamma\}. \end{aligned}$$

The essential technical features are that $E_x[\tau_1] < \infty$, uniformly over $x \in \Gamma$, and that there is some sort of separation between Γ and Γ' which insures $\tau_i < \tau_{i+1}$ with positive probability. The idea is to view the history of $x(t)$ as a sequence of cycles, $\tau_i < t \leq \tau_{i+1}$, punctuated by the Markov chain on Γ determined by

$$X_i = x(\tau_i).$$

Generally X_i will have a (unique) invariant measure ν on Γ, from which the (unique) invariant measure π for $x(t)$ is constructed by means of *Hasminskii's formula*:

$$\int f(x)\,\pi(dx) = \frac{1}{E_\nu[\tau_1]} \int_\Gamma E_z[\int_0^{\tau_1} f(x(s))\,ds]\,\nu(dz). \tag{14}$$

3 Methods for Sub-Large Deviations Asymptotics

Large deviations results provide definitive descriptions of the exponential ϵ-dependence of the mean exit time and the exit point distribution. However, in some situations there are interesting phenomena which can not be seen through this kind of exponential asymptotics. This is true particularly in cases of characteristic boundary (see [3]). For instance large deviations analysis will identify a special subset $\Gamma^* \subseteq \partial D$ on which the exit point distribution μ_x^ϵ must concentrate as $\epsilon \downarrow 0$. However the distribution of μ_x^ϵ within Γ^* is not discernible through large deviations results. In particular if ∂D is a limit cycle of (2) then Γ^* will necessarily be all of ∂D, so something other than large deviations is needed to study the exit point distribution. If Γ^* consists of isolated points there is the question of the relative weighings of these points in μ_x^ϵ as $\epsilon \downarrow 0$. Even if Γ is a single point, large deviations does not tell us anything about the asymptotic "shape" of μ_x^ϵ. The issue of skewing of the exit point distribution is a prime example. One may be even more interested in the effect of these features on the mean exit time, and less than exponential aspects of its ϵ-dependence. These are all examples of what we call "sub-large deviations" phenomena. It is the mathematical ideas that have been used to address these kind of questions that we overview in this section.

A variety of methods have been used to study sub-large deviations features. Some are probabilistic in nature, while others are based on PDE techniques. Our purpose in this section is to summarize a number of these methods, and indicate their relations to each other using Green's identity and Hasminkskii's formula. We have tried to use a consistent notation to do this. As a result the notation below is sometimes different that that of the original sources.

3.1 The Original Matkowsky-Schuss Method

The 1977 paper [25] of Matkowsky and Schuss introduced a fruitful formal approach to sub-large deviations results for the exit problem. Initially developed in the nontangential case $n \cdot b < 0$, it was subsequently applied to cases with characteristic boundary; see [26]. The basic idea is to use a solution z^ϵ of the adjoint equation $\mathcal{L}^{\epsilon *} z^\epsilon = 0$ together with u_f^ϵ of (6) in Green's identity, resulting in

$$0 = \int_{\partial D} z^\epsilon \mathcal{N}^\epsilon u_f^\epsilon - f \mathcal{N}^{\epsilon *} z^\epsilon. \tag{15}$$

This is exploited by postulating suitable asymptotic forms for z^ϵ and u_f^ϵ:

$$z^\epsilon(x) \sim e^{\phi(x)/\epsilon} \sum \epsilon^{n/2} w_n(x). \tag{16}$$

and

$$u_f^\epsilon(x) \sim C_f^0 + (f(x) - C_f^0) e^{\zeta(x)/\epsilon} \tag{17}$$

The function $\zeta(x)$ should be 0 on ∂D but negative inside D, so that (17) produces a function which adjusts from the interior constant C_f^0 to the boundary data f in a layer near ∂D. Assuming that the exit point distributions do converge, (17) agrees nicely with the leveling property (12).

Careful asymptotic calculations proceed from (17) and (16). In the course of these calculations one finds that the term $f\mathcal{N}^{\epsilon *} z^\epsilon$ is negligible with respect to the first term in (15), and is therefore dropped from the calculations. The determination of C_f^0 is thus based on using (16) and (17) in

$$0 \sim \int_{\partial D} z^\epsilon \mathcal{N}^\epsilon u_f^\epsilon. \tag{18}$$

The resulting formulas involve terms that describe the distribution of μ^0 within the support set Γ^*, thus giving sub-large deviations information. For the mean exit time (4) we know that $\mathcal{L}^{\epsilon *} v_D^\epsilon = -1$ with $v_D^\epsilon(y) = 0$ on ∂D. Using these facts and z^ϵ in Green's identity implies

$$\int_D -z^\epsilon = \int_{\partial D} z^\epsilon \mathcal{N}^\epsilon v_D^\epsilon.$$

by postulating the asymptotic expression

$$v_D^\epsilon(x) \sim c^\epsilon e^{K/\epsilon}(1 - e^{\zeta/\epsilon}).$$

Expressions for c^ϵ are derived, which provide sub-large deviations information on the mean exit time.

3.2 Boundary Local Time Formulation

The success of [25] prompted several efforts to put those results on a rigorous basis: [19], [6], [29]. These developments all assumed the nontangential case, $n \cdot b < 0$. In [5] the probabilistic approach of [6] was refined for use in the context of characteristic boundaries. This is the method we describe in this section. It provides a probabilistic redevelopment of the original Matkowsky-Schuss method by interpreting $z^\epsilon(x)$ as the equilibrium density $p^\epsilon(x)$ of a reflected version of the diffusion $x^\epsilon(t)$. Shortly after [5], the exit conditioning approach described in Section 3.5 below emerged and provided a way to obtain qualitative results without needing precise, rigorous asymptotics of an equilibrium density. As a consequence these boundary local time ideas have not been exploited to the point of producing explicit sub-large deviations results. Even so, the method remains interesting as a probabilistic interpretation of the original Matkowsky-Schuss method.
Consider the diffusion process $x^r(t)$ in $\bar{D}$ which obeys (1) in D but with conormal reflection from ∂D. This can be characterized as the solution to the following stochastic differential equation with reflection term (see [1]):

$$dx^r(t) = b(x^r(t))\,dt + \epsilon^{1/2}\sigma(x^r(t))\,dw(t) - \frac{\epsilon}{2}\eta(x^r(t))\,d\ell^\epsilon(t). \tag{19}$$

Here $\ell^\epsilon(t)$ is the *boundary local time* process, a continuous monotone process which increases only when $x^r(t) \in \partial D$: $d\ell^\epsilon(t) = 1_{\partial D}(x^r(t))\,d\ell^\epsilon(t)$.

We will not go into the existence, uniqueness theory of (19). Suffice it to say that unique solutions exist with appropriate regularity properties, with $x^r(t) \in \bar{D}$ for all $t \geq 0$, satisfying a version of Itô's lemma: for $u \in C^2(\bar{D})$:

$$\begin{aligned} du(x^r(t)) &= \mathcal{L}^\epsilon u(x^r(t))\,dt + \epsilon^{1/2}\nabla u(x^r(t)) \cdot \sigma(x^r(t))\,d\omega(t) \\ &\quad -\mathcal{N}^\epsilon u(x^r(t))\,d\ell^\epsilon(t). \end{aligned} \tag{20}$$

From this it is clear that, for smooth functions u, the generator of $x^r(t)$ is given by $\mathcal{L}^\epsilon u$ restricted to those functions with $\mathcal{N}^\epsilon u = 0$ on ∂D.
The reflected process $x^r(t)$ has a unique invariant density $p^\epsilon(x)$ (see [5]). We want to consider p^ϵ as the function z^ϵ in the Matkowsky-Schuss approach above. The invariance implies $0 = \int_D p^\epsilon \mathcal{L}^\epsilon u$ for all smooth u with $\mathcal{N}^\epsilon u = 0$ on ∂D. Applying this in Green's identity for u having compact support in D leads to $\mathcal{L}^{\epsilon *} p^\epsilon = 0$. Now Green's identity implies $0 = \int_{\partial D} u \mathcal{N}^{\epsilon *} p^\epsilon$ for all $\mathcal{N}^\epsilon u = 0$. From this follows the necessary boundary conditions for p^ϵ on ∂D:

$$\mathcal{N}^{\epsilon *} p^\epsilon = 0. \tag{21}$$

Moreover, applying Green's identity yet again to u_f^ϵ and p^ϵ tells us that

$$0 = \int_{\partial D} p^\epsilon \mathcal{N}^\epsilon u_f^\epsilon, \tag{22}$$

for all nice f defined on ∂D. Thus with p^ϵ as z^ϵ in the Matkowsky-Schuss development, (21) makes the negligibility of $f\mathcal{N}^{\epsilon *} z^\epsilon$ precise, and (22) is an exact version of (18).
Further probabilistic reasoning reveals a mechanism behind the asymptotic formulas resulting from this method. Let $C \subseteq D$ be a neighborhood of 0 inside D. Suppose we start $x^r(t)$ at $x^r(0) = y \in \partial D$ and follow it until its first contact with ∂C and then its subsequent return τ_1 to ∂D:

$$\begin{aligned} \tau_C^\epsilon &= \inf\{t > 0 : \ x^r(t) \in C\} \\ \tau_1 &= \inf\{t > \tau_C^\epsilon : \ x^r(t) \in \partial D\}. \end{aligned}$$

For $y \in \partial D$ and $g(x) \in C(\bar{D})$, define the *mean excursion operator*,

$$T^\epsilon g(y) = E_y[g(x^r(\tau_1))].$$

For smooth $f(y)$ on ∂D, define the *boundary generator*

$$\mathcal{G}^\epsilon f(y) = -\mathcal{N}^\epsilon u_f^\epsilon(y)$$

and the *boundary local time operator*

$$B^\epsilon f(y) = E_y[\int_0^{\tau_1} f(x^r(t))\,d\ell^\epsilon],$$

both for $y \in \partial D$. Applying Itô's formula (20) to u_f^ϵ, integrating over $[0, \tau_1]$ and taking the expectation produces the relationship among these operators:

$$T^\epsilon f - f = B^\epsilon[\mathcal{G}^\epsilon f] \quad \text{on } \partial D. \tag{23}$$

The asymptotic behavior of the exit location distribution is given by the behavior of T^ϵ as $\epsilon \downarrow 0$. Suppose for instance that a limiting exit measure $\mu_x^\epsilon \Rightarrow \mu^0$ exists. Then we have

$$T^\epsilon f(y) \to \int_{\partial D} f \, d\mu^0,$$

with leveling results implying that the convergence is uniform in $y \in \partial D$. As we have described it, the Matkowsky-Schuss approach is essentially the application in (22) (or (18)) of appropriate asymptotic formulas for $\mathcal{G}^\epsilon f = -\mathcal{N}^\epsilon u_f^\epsilon$ and p^ϵ on ∂D. Now the primary goal of [5] was to establish an appropriate scaled limit for B^ϵ: with $c^\epsilon = \epsilon^{1/2}$:

$$c^\epsilon B \to B, \tag{24}$$

along with an inversion formula, B^{-1}. (The scaling factor $c^\epsilon = \epsilon^{1/2}$ is for non-degenerate characteristic boundary. For the nontangential case, $c^\epsilon = 1$ would be appropriate.) So instead of arguing from (22), we can use a scaled limit of (23) to find the relationship between μ^0 and limiting forms of the various operators. Proceeding purely speculatively, suppose that with some scaling factor k^ϵ the invariant density converges to a probability measure on ∂D:

$$k^\epsilon p^\epsilon(y)\, dy \Rightarrow \nu(dy). \tag{25}$$

Convergence of T^ϵ in (23) suggests $1/c^\epsilon$ as a scaling constant for $\mathcal{G}^\epsilon$: $\frac{1}{c^\epsilon}\mathcal{G}^\epsilon \to \mathcal{G}$. The scaled limits of (23) and (22) would then be

$$\begin{aligned} \textstyle\int_{\partial D} f \, d\mu^0 - f &= B[\mathcal{G} f], \\ \textstyle\int_{\partial D} \mathcal{G} f \, d\nu &= 0. \end{aligned}$$

Taking B^{-1} of the first line and then integrating with respect to ν produces the formula

$$\int_{\partial D} f \, d\mu^0 = \frac{\int B^{-1}[f]\, d\nu}{\int B^{-1}[1]\, d\nu}. \tag{26}$$

This is of course purely formal. However it makes the important point that the asymptotic analysis of normal derivatives ($\mathcal{G}^\epsilon$) ought to be replaceable by asymptotic analysis of the boundary local time operator, whose asymptotics have been worked out rigorously in [5].

The equation (22) at the heart of the Matkowsky-Schuss approach is an expression of the invariance of the density p^ϵ. Hasminskii's formula provides an alternate expression of this invariance, which links it directly to the boundary local time operator. This provides a rigorous approach to the relationship (26) among the exit time, exit location and invariant density which our hypothetical calculation suggested. Moreover such an approach avoids the conormal derivative $\mathcal{G}^\epsilon$ and other differential operators altogether, using only mean path integral expressions. This reduces the exit problem completely to the study of (25), or its counterpart in nonconvergent situations. The idea is to use the excursions from ∂D to ∂C and back again as the basic cycles in Hasminskii's formula, and then form a sort of conormal derivative of the resulting version of (14). The basic formula that results (see [5] for details) is the following, holding for any $\epsilon > 0$:

$$\int_{\partial D} g(y) p^\epsilon(y)\, dy = \frac{\int_{\partial D} B^\epsilon g(y)\, \mu_D^\epsilon(dy)}{E_{\mu_D^\epsilon}[\tau_1]}, \quad \text{all } g \in C(\partial D). \tag{27}$$

Here μ_D^ϵ is the (unique) T^ϵ-invariant probability measure on ∂D. The leveling results tell us that $\mu_D^\epsilon \sim \mu_x^\epsilon$ for all $x \in D$. Note that (22) follows from this by taking $g = \mathcal{G}^\epsilon f$ and using (23): $\int_{\partial D} p^\epsilon \mathcal{G}^\epsilon f\, dy = c \int_{\partial D} (T^\epsilon f - f)\, d\mu_D^\epsilon = 0$. Thus (27) is an expression of the same relationship between p^ϵ and u_f^ϵ as before, but in which the three basic quantities p^ϵ, μ_D^ϵ and $E[\tau_1]$ appear explicitly.

To illustrate the use of (27) assume again the convergent case (25). We have a scaled limit (24) of B^ϵ. Then using $g \equiv 1$ in (27) we obtain the appropriate scaling of the mean excursion time:

$$\frac{c^\epsilon}{k^\epsilon} E_{\mu_D^\epsilon}[\tau_1] \to c = \int B[1]\, d\mu^0.$$

Using this in (27) gives the general limiting relationship:

$$c \int_{\partial D} g\, d\nu = \int_{\partial D} B[g]\, d\mu^0.$$

Replacing g with $B^{-1}[f]$ we find

$$\int f\, d\mu^0 = c \cdot \int B^{-1}[f]\, d\nu, \tag{28}$$

holding for all $f \in C(\partial D)$. Using $f \equiv 1$ gives the alternate formula $c = 1/\int_{\partial D} B^{-1}[1]\, d\nu$, so that (28) confirms the speculative formula (26), when (25) holds. If $k^\epsilon p^\epsilon(y)\, dy$ does not converge, a similar analysis with the known asymptotics of B^ϵ will connect the possible limit points of μ^0 with those of $k^\epsilon p^\epsilon(y)\, dy$ on ∂D.

One could also obtain asymptotic expressions for the mean exit time. With a version of (13) for the characteristic boundary case one could justify writing

$$\begin{aligned} E_{\mu_D^\epsilon}[\tau_1] &= E_{\mu_D^\epsilon}[\tau_C^\epsilon] + E_{\mu_D^\epsilon}[E_{x^\epsilon(\tau_C^\epsilon)}[\tau_D^\epsilon]] \\ &= (o(1) + 1) E_0[\tau_D^\epsilon], \end{aligned}$$

for some $\delta > 0$. Therefore $E_0[\tau_D^\epsilon] \sim E_{\mu_D^\epsilon}[\tau_1] \sim \frac{k^\epsilon}{c^\epsilon} c$.

Perthame [29] gives a rigorous development of the Matkowsky-Schuss approach based on PDE-viscosity methods. Although it uses no probabilistic reasoning, there are a number of close parallels with the boundary local time ideas above that we wish to point out. First note that in place of Matkowsky and Schuss' $z^\epsilon(x)$ he uses $v_\epsilon > 0$ solving $\mathcal{L}^{\epsilon *} v_\epsilon = 0$ with boundary conditions $\mathcal{N}^{\epsilon *} v_\epsilon = 0$. Thus his v_ϵ corresponds exactly with the equilibrium density p^ϵ for the conormally reflected process $x^r(t)$. Perthame's hypothesis (H3) rules out fully characteristic ∂D. In the nontangential case $n \cdot b < 0$ it turns out that the scaled limit in (24) with $c^\epsilon = 1$ is just a multiplicative operator:

$$B[f] = \frac{-1}{n \cdot b} f; \quad B^{-1]}[f] = -(n \cdot b) f.$$

(This was not derived in [5], but follows by applying the same methods in the simpler nontangential case.) In light of this we recognize Perthame's equation (7) as an approximate version of our (23): $\mathcal{G}^\epsilon f \sim B^{-1}[T^\epsilon f - f]$. (He attributes this to Kamin [19] originally.) His Theorem 3 is an exponential leveling result, with the limiting formula for the nontangential case.

3.3 Green's Function Approach

A new approach to formal asymptotic calculations was presented by Naeh, Kłosek, Matkowsky and Schuss in [28]. In contrast to the method of Section 3.1 which involved asymptotic representations of solutions to both forward and backward equations, as well as some matching conditions, the newer method seemed more direct. It involves the asymptotic representation of a single quantity: the classical Green's function.

The Green's function $G(x, y) = G^\epsilon(x, y)$ provides an integral representation of the solution to the Poisson equation (7):

$$v(x) = \int_D G(x, y)\phi(y)\, dy. \tag{29}$$

The construction of G is a classical approach to the study of elliptic PDEs; see [27]. G is characterized as a solution of

$$\mathcal{L}^\epsilon{}_x G(x, y) = 0, \ \mathcal{L}^{\epsilon *}{}_y G(x, y) = 0 \text{ for } x \neq y; \quad G(x, y) = 0 \text{ for } y \in \partial D,$$

with a singularity of appropriate type at $x = y$. (The subscripts on $\mathcal{L}^\epsilon{}_x$ and $\mathcal{L}^{\epsilon *}{}_y$ indicate which of the two variables of $G(x, y)$ the operators act on.) A classical formula extends (29) to solutions of the Poisson equation (7) with nonzero Dirichlet boundary data, $v(y) = f(y)$, $y \in \partial D$:

$$v(x) = \int_D G(x, y)\phi(y)\, dy - \int_{\partial D} [\mathcal{N}^{\epsilon *}{}_y G(x, y)] f(y)\, dy.$$

This is [27] equation (10.4) (after conversion to our notation). The function

$$K(x, y) = -\mathcal{N}^{\epsilon *}{}_y G(x, y) \tag{30}$$

is sometimes called the "Poisson kernel" (at least in the case of $\mathcal{L}^\epsilon = \frac{1}{2}\Delta$). It yields an integral representation of the solution to the Dirichlet problem (6)

$$u_f^\epsilon(x) = \int_{\partial D} K(x,y) f(y)\, dy,$$

and is therefore the density of the exit point distribution with respect to surface measure on ∂D:

$$\mu_x^\epsilon(dy) = K(x,y)\, dy.$$

In other words (30) is the density of $x^\epsilon(\tau_D^\epsilon)$ for $x^\epsilon(0) = x$.
The notation of [28] uses $p(x,y)$ in place of $G(x,y)$ and expresses (30) as

$$-\mathcal{N}^{\epsilon *}{}_y G(x,y) = J^G(x,y) \cdot n(y), \tag{31}$$

where $J^G = (J_i)$ is the *probability current density*:

$$J_i(x,y) = b_i(y) G(x,y) - \frac{\epsilon}{2} \sum_j \frac{\partial}{\partial x_j} [a_{ij}(y) G(x,y)], \tag{32}$$

So the point is that if we can produce the Green's function $G(x,y)$, or a suitable asymptotic representation of it as $\epsilon \downarrow 0$, then the following formulas give the asymptotic behavior of the mean exit time and exit point distribution:

$$E_x[\tau_D^\epsilon] = \int_D G(x,y)\, dy \tag{33}$$

$$E_x[f(x^\epsilon(\tau_D^\epsilon))] = \int_{\partial D} J^G(x,y) \cdot n(y) f(y)\, dy \tag{34}$$

These are equations (2.32) and (2.41) of [28] respectively. They were developed in [28] with the case of singular diffusion in mind, but for nonsingular diffusion they follow from the classical theory of PDEs.
Formal asymptotic calculations proceed by supposing an asymptotic approximation to the Green's function of the form

$$G(x,y) = p(y) q(x,y) \tag{35}$$

The factor $p(y)$ (called a *quasi-stationary density*) is taken to be a solution of $\mathcal{L}^{\epsilon *} p = 0$ of the form

$$p(y) = K^\epsilon(y) e^{-\psi(y)/\epsilon},$$

with appropriate normalization. The factor $q(x,y)$ is a boundary layer function with $q(x,y) = 0$ for $y \in \partial D$ but should be essentially constant for y in the interior of D (away from an asymptotically thin boundary layer).

In general (35) is only formal, unable to capture all the detailed mathematical structure of the true Green's function $G(x,y)$. For instance, the boundary layer function q will not capture the essential singularity of G at $x = y$. In fact as the asymptotic calculations in [28] proceed, no x-dependence of q is included. However expressions of the form (35) can be worked out explicitly in some special cases. Based on these, there is reasonable confidence that the asymptotic calculations stemming from (35) will lead to appropriate conclusions. However the conclusions of such an approach are limited by the presumptions and "ansatzes" which feed the calculations. They failed in particular to predict the cycling phenomena of exit to limit cycles. Once that type of possible behavior is recognized and built into the presumed asymptotic forms, formal methods become effective is studying its detailed structure; see [21].
We have explained (33) and (34) apart from any probabilistic interpretation. The development in [28] discusses them in terms of a process $x^+(t)$, which obeys the same stochastic dynamics (1) as $x^\epsilon(t)$ in D, but upon contact with ∂D instantaneously jumps to a specified $x_0 \in D$ and then resumes evolution in accord with (1). It is intuitively reasonable to expect that $x^+(t)$ has a unique stationary distribution π^+ in D. We merely wish to to point out the relation of π^+ to G using Hasminskii's formula. The evolution of $x^+(t)$ can be decomposed into the sequence of independent cycles between successive jumps from ∂D to x_0 by using $\Gamma = \partial D$ and $\Gamma' = \{x_0\}$ in Section 2.4. The mean cycle length is $E_{x_0}[\tau_D^\epsilon]$. Hasminskii's formula identifies the stationary distribution according to

$$\int f(y)\,\pi^+(dy) = \frac{1}{E_{x_0}[\tau_D^\epsilon]} E_{x_0}[\int_0^{\tau_D^\epsilon} f(x^+(s))\,ds].$$

Now we know that

$$E_{x_0}[\int_0^{\tau_D^\epsilon} f(x^+(s))\,ds] = E_{x_0}[\int_0^{\tau_D^\epsilon} f(x^\epsilon(s))\,ds] = \int_D G(x_0,y)f(y)\,dy.$$

Thus we see that

$$\pi^+(dy) = \frac{1}{E_{x_0}[\tau_D^\epsilon]} G(x_0,y)\,dy,$$

identifying the Green's function $G(x_0,y)$ as the (unnormalized) stationary density for $x^+(t)$ in D. This interpretation is described in [28].

3.4 Maier and Stein Principal Eigenfunction Method

R. Maier and D. Stein have written a number of papers exploring particular phenomena of the exit problem by means of formal asymptotic calculations. Their real contribution is not so much the mathematical foundation of their method (described below) but how they have effectively used it to study some of the interesting sub-large deviations phenomena in the exit problem, particularly in cases for which the quasipotential function of large deviations analysis is nonsmooth. For instance in [22] and [20] they look at the situation in which the exit location concentrates at a saddle point on ∂D. They point out that singularities in the quasipotential are quite possible precisely at this saddle point, and that they effect the asymptotic description of the mean exit time. Their calculations offer a more detailed description of the "skewing" phenomena of the exit location distribution than in other treatments, such as [3].

The calculations of Maier and Stein are based on an approach closely related to the Green's function approach of Section 3.3. The central object is the principal adjoint eigenfunction ψ^ϵ for $\mathcal{L}^{\epsilon *}$ with Dirichlet conditions on ∂D:

$$\mathcal{L}^{\epsilon *}\psi^\epsilon + \lambda^\epsilon \psi^\epsilon = 0 \text{ in } D; \quad \psi^\epsilon(y) = 0 \text{ on } \partial D. \tag{36}$$

Here λ^ϵ is the principal eigenvalue mentioned in Section 2. General results imply that $\psi^\epsilon(x) > 0$ everywhere in D. We are free to specify a normalization of ψ^ϵ. The natural choice is

$$1 = \int_D \psi^\epsilon(x)\, dx, \tag{37}$$

allowing us to interpret $\psi^\epsilon(x)$ as a probability density in D, called the *quasi-stationary density* in [20]. They compute the probability current associated with ψ^ϵ:

$$J^\psi(y) \cdot n(y) = -\mathcal{N}^{\epsilon *}\psi^\epsilon(y)$$

and use this as the (unnormalized) density of exit points. The "reaction rate" $\lambda^\epsilon \sim 1/E_x[\tau_D^\epsilon]$ describes the mean exit time and is given the formula

$$\lambda^\epsilon = \int_{\partial D} J^\psi(y) \cdot n(y)\, dy \Big/ \int_D \psi^\epsilon(x)\, dx. \tag{38}$$

We will see that (38) follows directly from Green's formula. Consider any smooth u in $\bar{D}$. Since $\psi^\epsilon(y) = 0$ on ∂D, Green's formula implies,

$$\lambda^\epsilon \int_D \psi^\epsilon u = -\int_D (\mathcal{L}^{\epsilon *}\psi^\epsilon)u = \int \psi^\epsilon \mathcal{L}^\epsilon u + \int_{\partial D} (-\mathcal{N}^{\epsilon *}\psi^\epsilon)u.$$

Taking $u \equiv 1$ and our normalization (37) yields (38) above. Using the solution $u = u_f^\epsilon$ of the Dirichlet problem we find

$$\int_D \psi^\epsilon u_f^\epsilon = \frac{1}{\lambda^\epsilon} \int_{\partial D} (-\mathcal{N}^{\epsilon *}\psi^\epsilon) f. \tag{39}$$

Equation (39) shows us that $-\frac{1}{\lambda^\epsilon}\mathcal{N}^{\epsilon *}\psi^\epsilon = \frac{1}{\lambda^\epsilon}J^\psi \cdot n$ is the density of the exit position of $x^\epsilon(\tau_D^\epsilon)$ if $x^\epsilon(0)$ is distributed within D according to density $\psi^\epsilon(x)$:

$$\int_D \psi^\epsilon u_f^\epsilon = \int_D E_x[f(x^\epsilon(\tau_D^\epsilon))]\psi^\epsilon(x)\, dx.$$

Recognizing $\psi^\epsilon(x)$ as the stationary density of the process $x^*(t)$ below, it seems clear that $\psi^\epsilon(x)\,dx \Rightarrow \delta_0(dx)$. More detailed large deviations results on ψ^ϵ should be possible. The leveling properties of u_f^ϵ show that $\int_D \psi^\epsilon u_f^\epsilon \to u_f^\epsilon$ as $\epsilon \downarrow 0$. Thus the accuracy of the approximation

$$E_0[f(x^\epsilon(\tau_D^\epsilon))] \approx \frac{1}{\lambda^\epsilon}\int_D J^\psi(y)\cdot n(y)f(y)\,dy$$

can be studied directly in terms of leveling and large deviations results. The probability current densities associated with $G(x,y)$ and $\psi^\epsilon(y)$ respectively should produce the same asymptotic results as $\epsilon \downarrow 0$. However, they are not identical for $\epsilon > 0$; they produce slightly different exit point distributions:

$$\begin{aligned} E_x[f(x^\epsilon(\tau_D^\epsilon))] &= \textstyle\int_{\partial D} J^G(x,y)\cdot n(y)f(y)\,dy \\ \textstyle\int_D E_x[f(x^\epsilon(\tau_D^\epsilon))]\psi^\epsilon(x)\,dy &= \tfrac{1}{\lambda^\epsilon}\textstyle\int_{\partial D} J^\psi(y)\cdot n(y)f(y)\,dy. \end{aligned}$$

The second of these is an averaged version of the first. As the formal asymptotic calculations in [20] proceed, an asymptotic expression for ψ^ϵ is developed by setting λ^ϵ to 0 in (36) and using $\mathcal{L}^{\epsilon *}\psi^\epsilon = 0$ as an approximate equation. This is the same equation used for $p(y)$ of (35). Thus, as the asymptotic calculations are actually carried out, any distinction between the Green's function interpretation and the principal eigenfunction interpretation is lost.

We observe that the formulas arising from this eigenfunction approach can once again be interpreted in terms of Hasminskii's formula for an appropriately defined process. Let $x^*(t)$ be the stochastic process which follows (1) between times of contact with ∂D, but upon contact with ∂D is instantaneously restarted at a random point in D distributed with density ψ^ϵ. The successive times of contact with ∂D form the sequence of cycle times. The expected cycle time is given by

$$E[\tau_1] = \int_D E_x[\tau_D^\epsilon]\psi^\epsilon(x)\,dx = \int_D \psi^\epsilon v_D^\epsilon\,dx.$$

Notice that (using Green's formula)

$$\lambda^\epsilon \int_D \psi^\epsilon v_D^\epsilon = -\int_D (\mathcal{L}^{\epsilon *}\psi^\epsilon)v_D^\epsilon = -\int_D \psi^\epsilon \mathcal{L}^\epsilon v_D^\epsilon = \int_D \psi^\epsilon = 1.$$

Thus the mean cycle time for $x^*(t)$ is given exactly by

$$E[\tau_1] = 1/\lambda^\epsilon.$$

The solution $v(x)$ of the Poisson equation (7) provides the mean path integral over one cycle: $E[\int_0^{\tau_D^\epsilon} \phi(x^\epsilon(s))\,ds] = \int_D \psi^\epsilon v$. Hasminskii's formula would identify the stationary distribution π^* for $x^*(t)$ as

$$\begin{aligned}\int \phi(x)\,\pi^*(dx) &= \tfrac{1}{E^*[\tau]}E^*[\int_0^\tau \phi(x^*(s))\,ds]\\ &= \lambda^\epsilon \int_D \psi^\epsilon v\\ &= \int_D -(\mathcal{L}^{\epsilon *}\psi^\epsilon)v\\ &= \int_D -\psi^\epsilon \mathcal{L}^\epsilon v\\ &= \int_D \psi^\epsilon \phi.\end{aligned}$$

Thus ψ^ϵ is the stationary density of $x^*(t)$. We can view (38) and (39) as a method for determining the mean exit time and exit point distribution from the invariant density $\psi^\epsilon(x)$.

3.5 Exit Conditioning

We conclude with a brief mention of the method of exit conditioning: [4]. The approaches described so far relate strongly to Green's identity and quantities which are natural from a PDE approach to the problem. Exit conditioning is more exclusively probabilistic. It was developed expressly to treat cases of characteristic boundary and has been successful in producing qualitative results on sub-large deviations phenomena without using formal asymptotic expansions. The idea is to consider a subregion $G \subseteq D$ with $0 \in G$ for which the exit problem (τ_G^ϵ) is nontangential: $n \cdot b < 0$ on ∂G. Large deviations results are used to describe the exit point distribution from G, i.e. $\nu_x^\epsilon(dz) = P_x[x^\epsilon(\tau_G^\epsilon) \in dz]$. The exit point distribution $\mu_x^\epsilon(dy)$ from D is then represented in terms of ν_x^ϵ and a *conditional exit kernel* $Q^\epsilon(z, dy)$, $z \in \partial G$:

$$\mu^\epsilon(dy) = \int_{\partial G} c^\epsilon g(z) Q^\epsilon(z, dy)\,\nu^\epsilon(dz) + \mathrm{o}(1). \tag{40}$$

Here g is a specific positive continuous function on ∂D, and the o(1) accounts for dependence on the initial point, which we know vanishes as $\epsilon \downarrow 0$. To define Q^ϵ we use another subdomain C: $0 \in C \subseteq G \subseteq D$. One might think of $D \setminus C$ as a boundary strip. Define τ_C^ϵ to be the first time of contact with ∂C for $x^\epsilon(t)$, started at $x^\epsilon(0) = z \in \partial G$. Then

$$Q^\epsilon(z, A) = P_z[x^\epsilon(\tau_D^\epsilon) \in A | \tau_D^\epsilon < \tau_C^\epsilon]. \tag{41}$$

What makes (40) effective is the possibility of describing the asymptotic behavior of Q^ϵ in terms of qualitative features of (2). The details are more involved than we can summarize here. See [3] and [2] for a fuller description, and some results of this approach.

Acknowledgments: This article is an expansion and update of the author's presentation at the Workshop on Asymptotic Methods for Stochastic Systems - Theory and Practice, Univ. Maryland and U. S. Army, Oct. 25–27,

1987. Sections 3.1 and 3.2 in particular were originally presented there. We wish to thank Wendell Fleming for his suggestion in 1979 that we consider these problems, and his continued support and interest in our work on them.

References

[1] R. F. Anderson and S. Orey (1976). *Small random perturbations of dynamical systems with reflecting boundary*, Nagoya Mathematical Journal , **60**, 189–216.

[2] M. V. Day (1996). *Exit cycling for the van der Pol oscillator and quasi-potential calculations*, Journal of Dynamics and Differential Equations, **8**, 573–601.

[3] M. V. Day (1994). *Cycling and skewing of exit measures for planar systems*, Stochastics and Stochastic Reports **48**, 227–247.

[4] M. V. Day (1992). *Conditional exits for small noise diffusions*, The Annals of Probability **20**, 1385–1419.

[5] M. V. Day (1989). *Boundary local time and small parameter exit problems with characteristic boundaries*, SIAM Journal on Mathematical Analysis **20**, 222–248.

[6] M. V. Day (1987). *Recent progress on the small parameter exit problem*, Stochastics **20**, 121–150.

[7] M. V. Day (1983). *On the exponential exit law in the small parameter exit problem*, Stochastics **8**, 297–323.

[8] M. V. Day (1982). *Exponential leveling for stochastically perturbed dynamical systems*, SIAM Journal on Mathematical Analysis **13**, 532–540.

[9] Paul Dupuis and Richard S. Ellis (1997). A WEAK CONVERGENCE APPROACH TO THE THEORY OF LARGE DEVIATIONS, J. Wiley, New York.

[10] P. Dupuis and H. J. Kushner (1987). *Stochastic systems with small noise, analysis and simulation: a phase locked loop example*, SIAM Journal on Applied Mathematics **47**, 643–660.

[11] M. I. Dykman, P. V. E. McClintock, V. N. Smelyanski, N. D. Stein, and N. G. Stocks (1992). *Optimal paths and the prehistory problem for large fluctuations in noise driven systems*, Physical Review Letters **68**, 2718–2721.

[12] A. Eizenberg (1990). *The exponential leveling in elliptic singular perturbation problems with complicated attractors*, Journal d'Analyse Mathmatique **55**, 229–249.

[13] L. C. Evans and H. Ishii (1985). *A PDE approach to some asymptotic problems concerning random differential equations with small noise intensities*, Annales de l'Institut Henri Poincaré. Analyse Non Linearé **2**, 1–20.

[14] Wendell H. Fleming (1977/78). *Exit probabilities and optimal stochastic control*, Applied Mathematics and Optimization **4**, 329–346.

[15] Wendell H. Fleming (1992). *Stochastic control and large deviations*, Future Tendencies in Computer Science, Control and Applied Mathematics, 291–300, Lecture Notes in Computer Science **653** Springer, Berlin.

[16] Wendell H. Fleming and Shuenn-Jyi Sheu, *Asymptotics for the principal eigenvalue and eigenfunction of a nearly first order operator with large potential,* The Annals of Probability (to appear).

[17] M. I. Freidlin and A. D. Wentzell (1984). RANDOM PERTURBATIONS OF DYNAMICAL SYSTEMS, Springer-Verlag, Berlin, New York.

[18] R. Z. Hasminskii (1980). STOCHASTIC STABILITY OF DIFFERENTIAL EQUATIONS, Sijthoff and Noordhoff, Alphen aan den Rijn, the Netherlands.

[19] S. Kamin (1978). *Elliptic perturbation of a first order operator with a singular point of attracting type*, Indiana University Mathematics Journal **27**, 935–952.

[20] R. S. Maier and D. L. Stein (1997). *Limiting exit location distribution in the stochastic exit problem*, SIAM Journal on Applied Mathematics **57**, 752–790.

[21] R. S. Maier and D. L. Stein (1996). *Oscillatory behavior of the rate of escape through an unstable limit cycle*, Physical Review Letters **77**, 4860–4863.

[22] R. S. Maier and D. L. Stein (1996). *A scaling theory of bifurcations in the symmetric weak-noise escape problem*, Journal of Statistical Physics **83**, 291–357.

[23] F. Martinelli, and E. Scopploa (1988). *Small random perturbations of dynamical systems: exponential loss of memory of the initial condition*, Communications in Mathematical Physics **120**, 25–69.

[24] P. Mathieu (1995). *Spectra, exit times and long time asymptotics in the zero-white-noise limit*, Stochastics and Stochastics Reports **55**, 1–20.

[25] B. J. Matkowsky and Z. Schuss (1977). *The exit problem for randomly perturbed dynamical systems*, SIAM Journal on Applied Mathematics **33**, 365–382.

[26] B. J. Matkowsky, Z. Schuss and C. Tier (1983). *Diffusion across characteristic boundaries with critical points*, SIAM Journal on Applied Mathematics **43**, 673–695.

[27] C. Miranda (1970). PARTIAL DIFFERENTIAL EQUATIONS OF ELLIPTIC TYPE, Springer-Verlag, Heidelberg.

[28] T. Naeh, M. M. Kłosek, B. J. Matkowsky and Z. Schuss (1990). *A direct approach to the exit problem*, SIAM Journal on Applied Mathematics **50** 595–627.

[29] B. Perthame (1990). *Perturbed dynamical systems with an attracting singularity and weak viscosity limits in Hamilton-Jacobi equations*, Transactions of the American Mathematical Society **317** 723–748.

[30] M. Sigura (1995). *Metastable behaviors of diffusion processes with small parameter*, Journal of the Mathematical Society of Japan **47** 755–788.

17

An Approximation Scheme for Evolutive Hamilton-Jacobi Equations

Maurizio Falcone
Tiziana Giorgi

ABSTRACT: An approximation scheme for the evolutive Hamilton-Jacobi equation associated with the finite horizon deterministic problem is presented. The scheme allows a variable time step and can be applied to structured and unstructured grids. Depending on the regularity of the value function different orders of convergence are proved. In particular, under Lipschitz assumptions the convergence behavior of conservative monotone hyperbolic schemes is retrieved. Namely, if $\frac{k}{h} \leq$ *constant* then we have convergence of order $\frac{1}{2}$, where k and h are the space and the time step, respectively.

KEYWORDS: Numerical schemes, viscosity solutions, evolutive Hamilton–Jacobi–Bellman equations.

1 Introduction

In this paper, we propose an algorithm for the approximation of the value function associated with the non-autonomous finite horizon deterministic problem.

Consider a system of controlled ordinary differential equations

$$\begin{aligned} \dot{y}(t) &= b(y(t), t, \alpha(t)) \qquad t \in (s, T), \\ y(s) &= x \end{aligned} \tag{1}$$

where $\alpha(\cdot) \in \mathcal{A} \equiv L^\infty([0,T]; A)$ is the *control*, A is a compact subset of $\mathbb{R}^m$, and the *state* $y \in \mathbb{R}^n$. The *cost functional* related to the finite horizon problem is

$$J(x, s, \alpha(\cdot)) \equiv \int_s^T f(y(t), t, \alpha(t)) e^{-\lambda(t-s)} dt + \psi(y(T)) e^{-\lambda(T-s)} \tag{2}$$

where $\lambda > 0$ is a parameter, f is the *running cost* and ψ is the *final cost*. We are interested in the numerical computation of the *value function* associated to this problem, namely

$$u(x,s) \equiv \inf_{\alpha(\cdot)\in\mathcal{A}} J(x,s,\alpha(\cdot)). \tag{3}$$

Capuzzo Dolcetta (1983) proposed a convergent time approximation of the value function for the autonomous infinite horizon case. Subsequentially, Capuzzo Dolcetta and Ishii (1984) proved convergence of order $\frac{\gamma}{2}$, where γ happens to be the Hölder exponent of the value function. Falcone in (1987) considered a time-space approximation based on the one proposed by Dolcetta and, in the case $\gamma = 1$, he proved convergence of the kind $\frac{k}{h}$, where k and h represent respectively the (fixed) space and the time steps. Other approximation schemes for first order Hamilton–Jacobi equations have been proposed by several authors using finite differences as in Crandall and Lions (1984) or finite elements as in Gonzales and Rofman (1985) (see also the book Kushner and Dupuis (1992) and the recent survey paper Falcone (1997) for a more extensive and up-dated list of references).
In this work, following the idea of Capuzzo Dolcetta (1983), we use the fact that u can be characterized (see Lions (1982)) as the unique viscosity solution of the Hamilton-Jacobi equation

$$\begin{aligned} &-\frac{\partial u}{\partial s} + \lambda u + \max_{a\in A}\{-b(x,s,a)\cdot\nabla_x u - f(x,s,a)\} = 0 \text{ in } \mathbb{R}^n \times [0,T) \\ &u(x,T) = \psi(x) \text{ in } \mathbb{R}^n, \end{aligned} \tag{4}$$

and as a first step we consider an approximation in time of (4), recursively defined. Our contribution here is to extend the scheme allowing a *variable* time step and deriving a-priori estimates on the rate of convergence in the L^∞ norm. A variable time step is mainly motivated by the necessity to have an accurate numerical resolution around the singularities of the solution (to be understood in this framework as the curves where the ∇u has jumps) and by the use of domain decomposition techniques without overlapping where handling the boundary conditions on the internal interfaces is crucial (see Camilli, Falcone, Lanucara and Seghini (1994) for more informations).
We refer the reader interested in the analysis of the set of singularities of viscosity solutions to Hamilton-Jacobi-Bellman equations to the papers Fleming (1969) and Cannarsa and Soner (1987).
It should be noted that, in the Lipschitz case and after a suitable change of variable, our scheme belongs to a class of schemes introduced by Souganidis (1985) (see Remark 2.2). As a consequence the approximate solution u_h converges to u as $h^{1/2}$. If u is only Hölder continuous of order γ a combination of the arguments of Capuzzo Dolcetta and Ishii (1984), and Souganidis (1985) gives convergence of order $\gamma/2$. As second step, in order to get a complete discretized approximation of u, we consider a linear approximation in space of u_h denoting by u_h^k the solution of the fully discrete scheme. We then prove that u_h^k converges to u as $\frac{k^\gamma}{h^{\gamma/2}} + h^{\gamma/2}$. Moreover, our proof can be easly modified to the infinite horizon case to obtain that the scheme proposed by Falcone in (1987) converges as $\frac{k^\gamma}{h^{\gamma/2}}$.

We assume in the sequel that for any (x,s), $(y,t) \in \mathbb{R}^n \times [0,T]$ and $a \in A$, there exists $\gamma \in (0,1]$ and positive constants such that

$$\begin{aligned} &|b(x,s,a) - b(y,t,a)| \le L_b(|x-y| + |s-t|), \\ &|b(x,s,a)| \le M_b \quad \text{and} \quad b(x,s,\cdot) \in C(A), \end{aligned} \tag{5}$$

$$\begin{aligned} &|f(x,s,a) - f(y,t,a)| \le L_f(|x-y|^\gamma + |s-t|^\gamma), \\ &|f(x,s,a)| \le M_f \quad \text{and} \quad f(x,s,\cdot) \in C(A), \end{aligned} \tag{6}$$

$$\begin{aligned} &|\psi(x) - \psi(y)| \le L_\psi |x-y|^\gamma, \\ &|\psi(x)| \le M_\psi. \end{aligned} \tag{7}$$

Note that under our assumptions (1) has a unique solution, see Coddington (1955).

Let us now recall the definition of viscosity solution and some properties of the value function that we will need afterwards.

Definition 1.1 *A function $u \in C(\mathbb{R}^n \times [0,T))$ is a viscosity solution* of (4) if for every $\phi \in C^1(\mathbb{R}^n \times [0,T))$ the following holds:
(i) if $u - \phi$ attains a local maximum at $(x_0, s_0) \in \mathbb{R}^n \times [0,T)$ then

$$\begin{aligned} -\frac{\partial \phi(x_0,s_0)}{\partial s} &+ \lambda u(x_0,s_0) + \\ &+ \max_{a \in A}\{-b(x_0,s_0,a) \cdot \nabla_x \phi(x_0,s_0) - f(x_0,s_0,a)\} \le 0; \end{aligned}$$

(ii) if $u - \phi$ attains a local minimum at $(x_0, s_0) \in \mathbb{R}^n \times [0,T)$ then

$$\begin{aligned} -\frac{\partial \phi(x_0,s_0)}{\partial s} &+ \lambda u(x_0,s_0) + \\ &+ \max_{a \in A}\{-b(x_0,s_0,a) \cdot \nabla_x \phi(x_0,s_0) - f(x_0,s_0,a)\} \ge 0. \end{aligned}$$

Moreover, if $u \in C(\mathbb{R}^n \times [0,T])$ and $u(x,T) = \psi(x)$ in $\mathbb{R}^n$ we say that u is a viscosity solution of (4).

Theorem 1.1 *Under the assumptions (5)-(7) the value function u is the viscosity solution in $BUC(\mathbb{R}^n \times [0,T])$ of (4), and for any (x,s), $(y,t) \in \mathbb{R}^n \times [0,T]$ satisfies*

$$|u(x,s) - u(y,t)| \le L_u(|x-y|^\gamma + |s-t|^\gamma).$$

For a proof of this theorem we refer to Crandall, Ishii and Lions (1992).

The next theorem tells us that the value function u verifies a so called dynamic programming principle.

Theorem 1.2 *Suppose (5)-(7), let $(x,s) \in \mathbb{R}^n \times [0,T]$, for each control α choose $\tau \in [s,T]$ (τ may depend on α), then we have:*

$$u(x,s) = \inf_{\alpha \in \mathcal{A}} \left\{ \int_s^\tau f(y(t),t,\alpha(t))e^{-\lambda(t-s)}dt + u(y(\tau),\tau)e^{-\lambda(\tau-s)} \right\}.$$

2 Time Approximation

As in the case of the autonomous infinite horizon problem, see Capuzzo Dolcetta (1983), the time approximation we consider is based on the fact that a direct discretization of the state and of the cost functional gives a discrete version of the dynamic programming principle in Theorem 1.2, which is nothing else but an approximation in time of the equation (4).
For $(x,s) \in \mathbb{R}^n \times [0,T]$ fixed, given any $N \in \mathbb{N}$, set $h^N = \frac{T-s}{N}$ and for any $\alpha(\cdot) \in \mathcal{A}$ define

$$\begin{aligned} y_{j+1} &= y_j + h^N b(y_j, s_j, \alpha(s_j)) \qquad j = 0..(N-1), \\ y_0 &= x, \end{aligned} \tag{8}$$

where $s_j = s + jh^N$ with $j = 0..N$.
Moreover, consider

$$J_N(x,s,\alpha(\cdot)) \equiv \sum_{j=0}^{N-1} h^N f(y_j, s_j, \alpha(s_j))e^{-\lambda j h^N} + \psi(y_N)e^{-\lambda(T-s)}, \tag{9}$$

$$u_N(x,s) \equiv \inf_{\alpha(\cdot)\in\mathcal{A}} J_N(x,s,\alpha(\cdot)). \tag{10}$$

Lemma 2.1 *Suppose (5)-(7), then for $n = 1..(N-1)$ we have:*

$$\begin{aligned} u_N(x,s) = \quad & \inf_{\alpha(\cdot)\in\mathcal{A}} \left\{ \sum_{j=0}^{n-1} h^N f(y_j, s_j, \alpha(s_j))e^{-\lambda j h^N} + \right. \\ & \left. + u_{N-n}(y_n, s_n)e^{-\lambda n h^N} \right\}. \end{aligned} \tag{11}$$

Proof. Let $\bar{u}_N(x,s)$ be the right hand side of (11). By definition, for any $\epsilon > 0$ there exist $\alpha_0(\cdot)$, and $\alpha_1(\cdot) \in \mathcal{A}$ such that

$$\begin{aligned} \bar{u}_N(x,s) \quad & \le \sum_{j=0}^{n-1} h^N f(y_j^0, s_j, \alpha_0(s_j))e^{-\lambda j h^N} + \\ & + u_{N-n}(y_n^0, s_n)e^{-\lambda n h^N} \le \bar{u}_N(x,s) + \epsilon, \end{aligned} \tag{12}$$

and, since $\frac{T-s_n}{N-n} = h^N$,

$$u_{N-n}(y_n^0, s_n)e^{-\lambda n h^N} \leq$$
$$\leq \left\{ \sum_{j=0}^{(N-n)-1} h^N f(y_j^1, s_{n+j}, \alpha_1(s_{n+j}))e^{-\lambda j h^N} + \right. \tag{13}$$
$$\left. +\psi(y_{N-n}^1)e^{-\lambda(T-s_n)} \right\} e^{-\lambda n h^N} \leq u_{N-n}(y_n^0, s_n)e^{-\lambda n h^N} + \epsilon,$$

here y_j^0 and y_j^1 are the solutions of (8) for $\alpha_0(\cdot)$, x and $\alpha_1(\cdot)$, y_n^0, respectively. Set

$$\bar{\alpha}(t) = \begin{cases} \alpha_0(t) & t \in [s, s_n] \\ \alpha_1(t) & t \in (s_n, T], \end{cases}$$

then $\bar{\alpha}(\cdot) \in \mathcal{A}$, and

$$\bar{y}_j = \begin{cases} y_j^0 & j = 0..n \\ y_{j-n}^1 & j = (n+1)..N \end{cases}$$

is the solution of (8) for $\alpha(\cdot) = \bar{\alpha}(\cdot)$.
As a consequence, we have

$$u_N(x, s) \leq J_N(x, s, \bar{\alpha}(\cdot)) = \sum_{j=0}^{n-1} h^N f(y_j^0, s_j, \alpha_0(s_j))e^{-\lambda j h^N} +$$
$$+\textstyle\sum_{j=n}^{N-1} h^N f(y_{j-n}^1, s_j, \alpha_1(s_j))e^{-\lambda j h^N} + \psi(y_{N-n}^1)e^{-\lambda(T-s)},$$

therefore, using (13) we get

$$u_N(x, s) \leq \sum_{j=0}^{n-1} h^N f(y_j^0, s_j, \alpha_0(s_j))e^{-\lambda j h^N} + u_{N-n}(y_n^0, s_n)e^{-\lambda n h^N} + \epsilon,$$

and, due to (12), $u_N(x, s) \leq \bar{u}_N(x, s) + 2\epsilon$ for any $\epsilon > 0$. We obtain the other inequality in a similar way, thus the theorem follows. □

Remark 2.1 *Equation (11) for n=1 yields a discrete version of the dynamic programming principle, that is*

$$u_N(x, s) = \inf_{\alpha(\cdot) \in \mathcal{A}} \{h^N f(x, s, \alpha(s)) +$$
$$+ e^{-\lambda h^N} u_{N-1}(x + h^N b(x, s, \alpha(s)), s + h^N)\} \tag{14}$$

Moreover, $s = T$ implies $h^N = 0$ for any $N > 0$, hence (10) leads $u_N(x, T) = \psi(T)$.

The previous remark suggests an approximation in time of the value function $u(x,s)$. Equation (14) will be modified in order to allow a variable time step. This makes the method more flexible and efficient.

Consider a partition $P \equiv \{0 = s_0 < s_1 < .. < s_N = T\}$ of the interval $[0,T]$, and set $h = \max_j h_j$, where $h_{j+1} = s_{j+1} - s_j$ $(j = 0..(N-1))$, then for any $(x,s) \in \mathbb{R}^n \times [0,T]$ define

$$\begin{aligned} u_h(x,s) = \quad & \inf_{a\in A}\{(s_{j+1}-s)f(x,s,a) + e^{-\lambda(s_{j+1}-s)} \\ & u_h(x+(s_{j+1}-s)\,b(x,s,a), s_{j+1})\} \quad \text{if } s \in [s_j, s_{j+1}) \\ u_h(x,T) \quad & = \psi(x), \end{aligned} \tag{15}$$

with $j = 0..(N-1)$.

Remark 2.2 *If we perform the change of variable $t = T - s$, it is straightforward to see that, when $\gamma = 1$, (15) is a particular case of the schemes consider by Souganidis (1985).*

Lemma 2.2 *Assume (5)-(7), for any $x, y \in \mathbb{R}^n$ and $j = 0..(N-1)$ we have*

1. $|u_h(x,s_j) - u_h(y,s_j)| \le L_1 e^{L_b^\gamma (N-j)h^\gamma} |x-y|^\gamma$
2. $|u_h(x,s_j) - u_h(x,T)| \le L_2((T-s_j) + (N-j)h^\gamma)$,
3. *let $j \neq 0$, $s = s_j$ and $t \in [s_{j-1}, s_j)$ then*
$$|u_h(x,s_j) - u_h(x,t)| \le L_3(|s_j - t| + e^{L_b^\gamma (N-j)h^\gamma}|s_j - t|^\gamma),$$
4. *if $s,t \in [s_j, s_{j+1})$*
$$|u_h(x,s) - u_h(x,t)| \le L_4(|s-t| + e^{L_b^\gamma (N-j)h^\gamma}|s-t|^\gamma),$$
5. *$u_h(x,s)$ is bounded and uniformly continuous in $\mathbb{R}^n \times [0,T]$,*

where L_1, L_2, L_3 and L_4 are positive constants that do not depend on P.

Proof. We start by noticing that since $\psi(x)$ is continuous, then the inf in (15) is actually a min. In fact, if $u_h(x,s_j)$ is continuous in x, then the inf in the definition of $u_h(x,s_{j-1})$ is a min, so *1.* can be derived for $j-1$ which gives $u_h(x,s_{j-1})$ continuous in x and so on down to $j = 0$.
For $s = T$, we defined u_h to be ψ, thus by (7) we get

$$\begin{aligned} & |u_h(x,s_{N-1})| \le M_f h_N + e^{-\lambda h_N} M_\psi, \\ & |u_h(x,s_{N-1}) - u_h(y,s_{N-1})| \le \\ & \qquad \le (h_N L_f + L_\psi e^{-\lambda h_N}(1 + L_b^\gamma h_N^\gamma))\,|x-y|^\gamma, \\ & |u_h(x,s_{N-1}) - u_h(x,s_N)| \le \\ & \qquad \le h_N(M_f + \lambda M_\psi) + L_\psi e^{-\lambda h_N} M_b^\gamma h_N^\gamma. \end{aligned} \tag{16}$$

It is then easy to prove by induction the inequalities below

$$|u_h(x,s_j)| \leq M_f\left(h_{j+1} + \sum_{n=j+2}^{N} h_n\, e^{-\lambda \sum_{m=j+1}^{n-1} h_m}\right) + M_\psi\, e^{-\lambda \sum_{m=j+1}^{N} h_m}, \quad (17)$$

$$|u_h(x,s_j) - u_h(y,s_j)| \leq$$

$$\leq \left[L_f\left(h_{j+1} + \sum_{n=j+2}^{N} h_n\, e^{-\lambda \sum_{m=j+1}^{n-1} h_m} \prod_{m=j+1}^{n-1}(1 + L_b^\gamma h_m^\gamma)\right) + \right.$$

$$\left. + e^{-\lambda \sum_{m=j+1}^{N} h_m} L_\psi \prod_{m=j+1}^{N}(1 + L_b^\gamma h_m^\gamma)\right] |x-y|^\gamma, \quad (18)$$

$$|u_h(x,s_j) - u_h(x,T)| \leq (M_f + \lambda M_\psi)(h_{j+1} + \sum_{n=j+2}^{N} h_n\, e^{-\lambda \sum_{m=j+1}^{n-1} h_m}) +$$

$$+ L_\psi M_b^\gamma \left(\sum_{n=j+1}^{N} h_n^\gamma\, e^{-\lambda \sum_{m=j+1}^{n} h_m}\right), \quad (19)$$

where $j = 0..(N-2)$.
The inequalities *1.* and *2.* directly follow from (18), and (19) with

$$L_1 = (L_f T + L_\psi) \quad \text{and} \quad L_2 = M_f + \lambda M_\psi + L_\psi M_b^\gamma.$$

Now, (17) implies $|u_h(x,s)| \leq 2TM_f + M_\psi$, which together with *1.* yield *3.*, and *4.* if we notice that for $s \in [s_{N-1}, s_N)$

$$|u_h(x,s) - u_h(x,T)| \leq |T-s|M_f + e^{-\lambda(T-s)} L_\psi M_b^\gamma |T-s|^\gamma + \lambda M_\psi |T-s|,$$

$$|u_h(x,s) - u_h(x,t)| \leq (M_f + L_f)|s-t| + L_\psi(L_b^\gamma + M_b^\gamma)|s-t|^\gamma.$$

Moreover,

$$L_3 = M_f + \lambda(2TM_f + M_\psi) + L_1 M_b^\gamma \quad \text{and} \quad L_4 = M_f + L_f + L_1(L_b^\gamma + M_b^\gamma).$$

Finally, the uniform continuity is a consequence of *1.* ,*3.* , and *4.* □

Remark 2.3 *In the case $\gamma = 1$, starting from the inequality (8), it is possible to prove the Lipschitz continuity of $u_h(x,s)$.*

A combination of the arguments used by Capuzzo Dolcetta and Ishii (1984) and by Souganidis (1985) gives convergence of order $\frac{\gamma}{2}$ of u_h to u as $h \to 0$.

Theorem 2.1 *Let u_h be defined as in (15), and u the value function defined in (3), then if the assumptions (5)-(7) are verified we have for any $(x,s) \in \mathbb{R}^n \times [0,T]$*

$$|u_h(x,s) - u(x,s)| \leq Ch^{\frac{\gamma}{2}}, \tag{20}$$

where C depends on T but not on the partion P.

Proof. Given $\epsilon, \alpha > 0$ we define

$$\varphi(y,t,x,s) = u_h(y,t) - u(x,s) + \beta_\epsilon(y-x) + \eta_\alpha(t-s),$$

where $\beta_\epsilon(x) = -\frac{|x|^2}{\epsilon^2}$, and $\eta_\alpha(s) = -\frac{|s|^2}{\alpha^2}$. By Theorem 1.1 and Lemma 2.1, for any $\delta > 0$ there exist $(y_1,t_1),(x_1,s_1) \in \mathbb{R}^n \times [0,T]$ such that

$$\varphi(y_1,t_1,x_1,s_1) > \sup \varphi - \delta.$$

We choose $\theta \in C_0^\infty(\mathbb{R}^n \times \mathbb{R}^n)$ with $\theta(y_1,x_1) = 1$ and $0 \leq \theta(y,x) \leq 1$, $|D\theta| < 1$, so that for any $\delta \in (0,1)$

$$\zeta(y,t,x,s) = \varphi(y,t,x,s) + \delta\theta(y,x)$$

has a max at a point (y_0,t_0,x_0,s_0), with $y_0, x_0 \in \text{supp}\theta$ and $t_0, s_0 \in [0,T]$. Therefore, if we set

$$\phi(x,s) = u_h(y_0,t_0) + \beta_\epsilon(y_0 - x) + \eta_\alpha(t_0 - s) + \delta\theta(y_0,x),$$

$u(x_0,s_0) - \phi(x_0,s_0) = -\zeta(y_0,t_0,x_0,s_0)$ is a local min for $u(x,s) - \phi(x,s)$. Recalling the definition of ζ, for any $(y,t),(x,s) \in \mathbb{R}^n \times [0,T]$ we have

$$\begin{aligned} &u_h(y_0,t_0) - u(x_0,s_0) + \beta_\epsilon(y_0 - x_0) + \eta_\alpha(t_0 - s_0) + \delta\theta(y_0,x_0) \geq \\ &\geq u_h(y,t) - u(x,s) + \beta_\epsilon(y-x) + \eta_\alpha(t-s) + \delta\theta(y,x). \end{aligned} \tag{21}$$

Now, taking in (21) $x = y = y_0$ and $s = s_0$, $t = t_0$ we obtain, if $\delta < \frac{1}{\epsilon^2}$

$$|y_0 - x_0| \leq C_{\epsilon,\delta}\, \epsilon^{\frac{2}{2-\gamma}}, \tag{22}$$

where $C_{\epsilon,\delta} = \max\left\{ \left(\frac{L_u}{(1-\delta\epsilon^2)^2}\right)^{\frac{1}{2-\gamma}}, (L_u + \delta)^{\frac{1}{2-\gamma}} \right\}$. Similarly, we find

$$|s_0 - t_0| \leq \alpha^{\frac{2}{2-\gamma}} L_u^{\frac{1}{2-\gamma}}. \tag{23}$$

Always from (21), using (22) and (23), for $x = y$ and $s = t$ we derive

$$\begin{aligned} &u_h(x,s) - u(x,s) \leq \\ &\leq u_h(y_0,t_0) - u(x_0,s_0) + C_{\epsilon,\delta}^2\, \epsilon^{\frac{2\gamma}{2-\gamma}} + \alpha^{\frac{2\gamma}{2-\gamma}} L_u^{\frac{2}{2-\gamma}} + 2\delta. \end{aligned} \tag{24}$$

We need to consider three different cases:
(case1) $t_0 = T$. In this case $u_h(y_0, T) = \psi(y_0) = u(y_0, T)$, so Theorem 1.1 with (22)-(24) yields

$$u_h(x,s) - u(x,s) \leq (C^{\gamma}_{\epsilon,\delta} + C^{2}_{\epsilon,\delta})\epsilon^{\frac{2\gamma}{2-\gamma}} + L_u^{\frac{2}{2-\gamma}} \alpha^{\frac{2\gamma}{2-\gamma}} + 2\delta,$$

and since if $\delta \to 0$ then $C_{\epsilon,\delta} \to$ constant, taking $\epsilon = \alpha = h^{\frac{2-\gamma}{4}}$ we have (20).
(case2) $t_0 \neq T$, $s_0 = T$. We have $u(x_0, T) = \psi(x_0) = u_h(x_0, T)$, moreover we know $t_0 \in [s_{j-1}, s_j)$ for some j, so that *2.* and *3.* of Lemma 2.2 lead to

$$u_h(y_0, t_0) - u(y_0, T) \leq C((T - s_j) + (N-j)h^{\gamma} + h + e^{L_b^{\gamma}(N-j)h^{\gamma}}|s_j - t_0|^{\gamma})$$

On the other hand, (23) yields

$$|s_j - T| \leq \alpha^{\frac{2}{2-\gamma}} L_u^{\frac{1}{2-\gamma}}$$
$$(N-j) \leq \frac{\alpha^{\frac{2}{2-\gamma}}}{h_{min}} L_u^{\frac{1}{2-\gamma}},$$

where $h_{min} = \min_{1 \leq j \leq N} h_j$. Combining these with (24), we derive

$$u_h(x,s) - u(x,s) \leq C\Big[h + \alpha^{\frac{2\gamma}{2-\gamma}} + \alpha^{\frac{2}{2-\gamma}} +$$
$$+ \left(\frac{\alpha^{\frac{2}{2-\gamma}}}{h_{min}} + e^{L_b^{\gamma} L_u^{\frac{1}{2-\gamma}} h^{\frac{\gamma}{2}}}\right) h^{\gamma} + (C^{\gamma}_{\epsilon,\delta} + C^{2}_{\epsilon,\delta})\,\epsilon^{\frac{2\gamma}{2-\gamma}}\Big] + 2\delta.$$

Hence, the choices $\epsilon = h^{\frac{2-\gamma}{4}}$ and $\alpha = (h_{min} h^{-\frac{\gamma}{2}})^{\frac{2-\gamma}{2}}$ if $h_{min} < 1$, and $\alpha = h^{-\frac{\gamma(2-\gamma)}{4}}$ otherwise, give us (20). If $h > 1$ we use $h \leq T^{1-\frac{\gamma}{2}} h^{\frac{\gamma}{2}}$, $h^{-\frac{\gamma}{2}} \leq 1 \leq h^{\frac{\gamma}{2}}$ and so on.
(case3) $s_0, t_0 \neq T$. Since u is a viscosity solution of (4) there exists $a^* \in A$ such that

$$-\frac{\partial \phi(x_0, s_0)}{\partial s} + \lambda u(x_0, s_0) +$$
$$-b(x_0, s_0, a^*) \cdot \nabla_x \phi(x_0, s_0) - f(x_0, s_0, a^*) \geq 0, \tag{25}$$

and by definition of u_h if $t_0 \in [s_{j-1}, s_j)$, then

$$u_h(y_0, t_0) - (s_j - t_0) f(y_0, t_0, a^*) +$$
$$-e^{-\lambda(s_j - t_0)} u_h(y_0 + (s_j - t_0) b(y_0, t_0, a^*), s_j) \leq 0. \tag{26}$$

Consider (21) for $y = y_0 + (s_j - t_0) b(y_0, t_0, a^*)$, $x = x_0$ and $t = s_j$, $s = s_0$, then (26) becomes

$$(1 - e^{-\lambda(s_j - t_0)}) u_h(y_0, t_0) - (s_j - t_0) f(y_0, t_0, a^*) +$$

$$+e^{-\lambda(s_j-t_0)}\Big[(s_j-t_0)\nabla_y\beta_\epsilon(y_0-x_0)\cdot b(y_0,t_0,a^*)-\frac{1}{\epsilon^2}(s_j-t_0)^2$$
$$|b(y_0,t_0,a^*)|^2-\frac{2}{\alpha^2}(s_j-t_0)(t_0-s_0)-\frac{1}{\alpha^2}(s_j-t_0)^2\Big]+$$
$$+e^{-\lambda(s_j-t_0)}(s_j-t_0)[\delta\, b(y_0,t_0,a^*)\cdot\nabla_y\theta(y',x_0)]\le 0, \tag{27}$$

for some y'.
Dividing (27) by $(s_j-t_0)>0$, and subtracting (25) from it, we have

$$\frac{(1-e^{-\lambda(s_j-t_0)})}{(s_j-t_0)}u_h(y_0,t_0)-\lambda u(x_0,s_0)+f(x_0,s_0,a^*)-f(y_0,t_0,a^*)+$$
$$+\nabla_y\beta_\epsilon(y_0-x_0)\cdot\left(e^{-\lambda(s_j-t_0)}b(y_0,t_0,a^*)-b(x_0,s_0,a^*)\right)-\delta M_b-$$
$$-\frac{1}{\alpha^2}(s_j-t_0)+\delta\nabla_x\theta(y_0,x_0)\cdot b(x_0,s_0,a^*)-\frac{1}{\epsilon^2}(s_j-t_0)M_b^2+$$
$$+\left(1-e^{-\lambda(s_j-t_0)}\right)\frac{2}{\alpha^2}(t_0-s_0)\le 0,$$

which implies

$$\frac{(1-e^{-\lambda(s_j-t_0)})}{(s_j-t_0)}u_h(y_0,t_0)-\lambda u(x_0,s_0)\le L_f(|y_0-x_0|^\gamma+|t_0-s_0|^\gamma)+$$
$$+2\delta M_b+(s_j-t_0)(\frac{1}{\alpha^2}+\frac{M_b^2}{\epsilon^2})+\frac{2}{\alpha^2}\lambda(t_0-s_0)(s_j-t_0)+$$
$$+\frac{2}{\epsilon^2}|y_0-x_0|\Big[\lambda(s_j-t_0)M_b+L_b(|y_0-x_0|+|t_0-s_0|)\Big].$$

Once again, setting $\epsilon=\alpha=h^{\frac{2-\gamma}{4}}$, we obtain (20). Note that if $h>1$ we use $h\le T^{1-\frac{\gamma}{2}}h^{\frac{\gamma}{2}}$ and $h^{\frac{\gamma}{2}-\frac{1}{2}}\le 1\le T^{\frac{1}{2}}h^{\frac{\gamma}{2}}$ and so on.
Using the fact that there exists a^* for which the min in (15) is obtained, we can prove in a similar way the other inequality and the theorem follows. □

3 Space approximation

We will consider a space approximation of $u_h(x,s)$. As we said the algorithm can work on structured or unstructured grids, we will give the convergence proof for a general unstructured grid.
For simplicity, we assume $\Omega\subset\mathbb{R}^n$ to be a polyhedron and consider a finite family T_i of simplices which give a regular triangulation of $\Omega=\cup_i T_i$.
Let M be the number of vertices, x_i, of the triangulation. Set

$$k\equiv\max_i\{\text{diam } T_i\},\ G\equiv\{x_i,i=1..M\}\text{ and } S\equiv\{s_j,j=0..N\}.$$

We suppose, for any $(x_i, s_j) \in G \times S$, and $a \in A$

$$x_i + h_j b(x_i, s_j, a) \in \Omega, \tag{28}$$

then for $x \in \Omega$, and $t \in [0,T]$ we define

$$\begin{aligned} u_h^k(x,s) &= \sum_{i=1}^{M} \lambda_i(x) u_h^k(x_i, s_{j+1}) \qquad \text{if } s \in [s_j, s_{j+1}), \\ u_h^k(x,T) &= \psi(x) \qquad \text{if } s = T, \end{aligned} \tag{29}$$

with $x = \sum_{i=1}^{M} \lambda_i(x) x_i$, where $\sum_{i=1}^{M} \lambda_i(x) = 1$, $\lambda_i(x) > 0$, and

$$\begin{aligned} u_h^k(x_i, s_j) &= \min_{a \in A}\{h_{j+1} f(x_i, s_j, a) + e^{-\lambda h_{j+1}} \sum_{l=1}^{M} \lambda_l^i(a, s_j) u_h^k(x_l, s_{j+1})\}, \\ u_h^k(x,T) &= \psi(x), \end{aligned} \tag{30}$$

$j = 0..(N-1)$, here $x_i + h_{j+1} b(x_i, s_j, a) = \sum_{l=1}^{M} \lambda_l^i(a, s_j) x_l$.

Lemma 3.1 *Suppose (5)-(7) and let $u_h^k(x,s)$ be defined as in (28)-(30), for any $x_i \in I$ and $j = 0..(N-2)$ we have*

$$|u_h^k(x_i, s_j) - u_h^k(x_i, T)| \leq (M_f + \lambda M_\psi)(h_{j+1} + \sum_{n=j+2}^{N} h_n e^{-\lambda \sum_{m=j+1}^{n-1} h_m})$$

$$+ L_\psi M_b^\gamma (\sum_{n=j+1}^{N} h_n^\gamma e^{-\lambda \sum_{m=j+1}^{n} h_m}) + L_\psi k^\gamma (\sum_{n=j+1}^{N} e^{-\lambda \sum_{m=j+1}^{n} h_m}). \tag{31}$$

Moreover,

$$|u_h^k(x_i, s_{N-1}) - u_h^k(x_i, T)| \leq (M_f + \lambda M_\psi) h_N + L_\psi e^{-\lambda h_N}(k^\gamma + h_N^\gamma M_b^\gamma)$$

Proof. The result follows by induction using the definition of $u_h^k(x,s)$.

Remark 3.1 *From Lemma 3.1 we infer, for j=0..(N-1),*

$$\begin{aligned} &|u_h^k(x_i, s_j) - u_h^k(x_i, T)| \leq \\ &\leq (M_f + \lambda M_\psi)(T - s_j) + L_\psi (M_b^\gamma h^\gamma + k^\gamma)(N - j). \end{aligned} \tag{32}$$

Theorem 3.1 *Let $u(x,s)$ be the value function defined in (3), and suppose the hypotheses of Lemma 3.1 be verified, then for any $x \in \Omega$ and $s \in [0,T]$*

$$|u_h^k(x,s) - u(x,s)| \leq C\left(\frac{k^\gamma}{h^{\gamma/2}} + h^{\gamma/2}\right), \tag{33}$$

if there exist positive constants such that

$$\text{for any } j = 1..N \qquad h \leq C_1 h_j, \qquad \text{and} \qquad \frac{k}{h} \leq C_2.$$

Proof. Following Theorem 2.1, define for $(x,s) \in \mathbb{R}^n \times [0,T]$ and $(x_i, s_j) \in G \times S$

$$\varphi(i,j,x,s) = u_h^k(x_i,s_j) - u(x,s) + \beta_\epsilon(x_i - x) + \eta_\alpha(s_j - s).$$

It is easy to verify that

$$\varphi_M \equiv \sup_{(x_i,s_j)\in G\times S,\,(x,s)\in\mathbb{R}^n\times[0,T]} \varphi(i,j,x,s) =$$
$$= \max_{(x_i,s_j)\in G\times S}\left[\sup_{(x,s)\in\mathbb{R}^n\times[0,T]} \varphi(i,j,x,s)\right] < \infty.$$

Fix $\delta \in (0,1)$, then there exists $(x_{i_1}, s_{j_1}, x_1, s_1) \in G \times S \times \mathbb{R}^n \times [0,T]$ such that $\varphi(i_1,j_1,x_1,s_1) > \varphi_M - \delta$. Consider $\theta \in C_0^\infty(\mathbb{R}^n \times \mathbb{R}^n)$ with $\theta(x_{i_1}, x_1) = 1$, $0 \leq \theta(y,x) \leq 1$, $|D\theta| < 1$ and define

$$\zeta(i,j,x,s) = \varphi(i,j,x,s) + \delta\theta(x_{i_1}, x).$$

Since $(\text{supp}\theta)_i \equiv \{x \text{ s.t. } (x_i,x) \in \text{supp}\theta\} \subset \mathbb{R}^n$ is compact, we claim

$$\zeta_M \equiv \sup_{(x_i,s_j)\in G\times S,\,(x,s)\in\mathbb{R}^n\times[0,T]} \zeta(i,j,x,s) =$$
$$= \max_{(x_i,s_j)\in G\times S}\left[\max_{(x,s)\in(\text{supp}\theta)_i\times[0,T]} \zeta(i,j,x,s)\right].$$

Thus, $\zeta_M = \zeta(i_0,j_0,x_0,s_0)$ for some $(x_{i_0}, s_{j_0}, x_0, s_0) \in G \times S \times (\text{supp}\theta)_{i_0} \times [0,T]$. Set $(y_0,t_0) = (x_{i_0}, s_{j_0})$, and define

$$\phi(x,s) = u_h^k(y_0,t_0) + \beta_\epsilon(y_0 - x) + \eta_\alpha(t_0 - s) + \delta\theta(y_0,x),$$

then $u(x_0,s_0) - \phi(x_0,s_0) = -\zeta(i_0,j_0,x_0,s_0)$ is a local min for $u(x,s) - \phi(x,s)$. So, for any $(x_i,s_j,x,s) \in G \times S \times \mathbb{R}^n \times [0,T]$

$$u_h^k(y_0,t_0) - u(x_0,s_0) + \beta_\epsilon(y_0 - x_0) + \eta_\alpha(t_0 - s_0) + \delta\theta(y_0,x_0) \geq$$
$$\geq u_h^k(x_i,s_j) - u(x,s) + \beta_\epsilon(x_i - x) + \eta_\alpha(s_j - s) + \delta\theta(x_i,x). \tag{34}$$

At this point, as in Theorem 2.1, we can derive formulas (22), (23) and

$$u_h^k(x_i,s_j) - u(x_i,s_j) \leq$$
$$\leq u_h^k(y_0,t_0) - u(x_0,s_0) + C_{\epsilon,\delta}^2\,\epsilon^{\frac{2\gamma}{2-\gamma}} + \alpha^{\frac{2\gamma}{2-\gamma}} L_u^{\frac{2}{2-\gamma}} + 2\delta. \tag{35}$$

(case1) $t_0 = T$. Since by (30) $u_h^k(y_0, T) = \psi(y_0) = u(y_0, T)$, as in Theorem 2.1 (case1), we have $u_h^k(x_i, s_j) - u(x_i, s_j) \le Ch^{\frac{\gamma}{2}}$.
(case2) $t_0 \ne T$, $s_0 = T$. In this case $u(x_0, T) = \psi(x_0) = u_h^k(x_0, T)$, so (35) reads

$$u_h^k(x_i, s_j) - u(x_i, s_j) \le |u_h^k(y_0, t_0) - u_h^k(y_0, T)| + \\ + L_\psi C_{\epsilon,\delta}^\gamma \epsilon^{\frac{2\gamma}{2-\gamma}} + C_{\epsilon,\delta}^2 \epsilon^{\frac{2\gamma}{2-\gamma}} + \alpha^{\frac{2\gamma}{2-\gamma}} L_u^{\frac{2}{2-\gamma}} + 2\delta.$$

But, by Remark 3.1 since $t_0 = s_{j_0}$,

$$|u_h^k(y_0, t_0) - u_h^k(y_0, T)| \le (M_f + \lambda M_\psi)(T - s_{j_0}) + L_\psi (M_b^\gamma h^\gamma + k^\gamma)(N - j_0).$$

Hence, proceeding as in Theorem 2.1 (case2), we can prove that (33) holds true.
(case3) $s_0, t_0 \ne T$. First note that there exists $a^* \in A$ for which (25) is still true, and the analogous of (26) is

$$u_h^k(y_0, t_0) - h_{j_0+1} f(y_0, t_0, a^*) - e^{-\lambda h_{j_0+1}} \sum_{l=1}^{M} \lambda_l(a^*, s_{j_0}) u_h^k(x_l, s_{j_0+1}) \le 0, \tag{36}$$

where $y_0 + h_{j_0+1} b(y_0, t_0, a^*) = \sum_{l=1}^{M} \lambda_l(a^*, s_{j_0}) x_l$.
Similarly to what was done in Theorem 2.1, we consider (34) for $x = x_0$, $x_i = x_l$, $s = s_0$, $s_j = s_{j_0+1}$, and we substitute it in (36) to obtain

$$(1 - e^{-\lambda h_{j_0+1}}) u_h^k(y_0, t_0) - h_{j_0+1} f(y_0, t_0, a^*) + e^{-\lambda h_{j_0+1}} \sum_{l=1}^{M} \lambda_l(a^*, s_{j_0}) \\ \Big[\nabla_y \beta_\epsilon(y_0 - x_0) \cdot (x_l - y_0) - \frac{2}{\alpha^2}(t_0 - s_0) h_{j_0+1} - \frac{|x_l - y_0|^2}{\epsilon^2} - \\ - \frac{h_{j_0+1}^2}{\alpha^2} + \delta \nabla_x \theta(y_l, x_0) \cdot (x_l - y_0) \Big] \le 0, \tag{37}$$

again (25), and (37) give

$$\frac{(1 - e^{-\lambda h_{j_0+1}})}{h_{j_0+1}} u_h^k(y_0, t_0) - \lambda u(x_0, s_0) \le \\ \le L_f(|y_0 - x_0|^\gamma + |t_0 - s_0|^\gamma) + \nabla_y \beta_\epsilon(y_0 - x_0) \cdot \Big[b(x_0, s_0, a^*) - \\ - e^{-\lambda h_{j_0+1}} \Big(\sum_{l=1}^{M} \lambda_l(a^*, s_{j_0}) \frac{(x_l - y_0)}{h_{j_0+1}} \Big) \Big] + \\ + \frac{1}{\epsilon^2} \sum_{l=1}^{M} \lambda_l(a^*, s_{j_0}) \frac{|x_l - y_0|^2}{h_{j_0+1}} + \delta \sum_{l=1}^{M} \lambda_l(a^*, s_{j_0}) \frac{|x_l - y_0|}{h_{j_0+1}} + \\ + \frac{h_{j_0+1}}{\alpha^2} + \delta M_b + \frac{2}{\alpha^2} \lambda (t_0 - s_0) h_{j_0+1}. \tag{38}$$

On the other hand, $\lambda_l(a^*, s_{j_0}) \neq 0$ implies $|y_0 + h_{j_0+1}b(y_0, t_0, a^*) - x_l| \leq k$ thus, for any l, we know either $|y_0 - x_l| \leq k + h_{j_0+1}M_b$ or $\lambda_l(a^*, s_{j_0}) = 0$. Moreover, $\sum_{l=1}^{M} \lambda_l(a^*, s_{j_0})(x_l - y_0) = h_{j_0+1}b(y_0, t_0, a^*)$ hence, if we use (22) and (23), and we let δ tend to 0, (38) becomes

$$u_h^k(y_0, t_0) \quad -u(x_0, s_0) \leq C\Bigg(h_{j_0+1} + \epsilon^{\frac{2\gamma}{2-\gamma}} + \alpha^{\frac{2\gamma}{2-\gamma}} + h_{j_0+1}\epsilon^{\frac{2\gamma-2}{2-\gamma}} + \\ +\epsilon^{\frac{2\gamma-2}{2-\gamma}}\alpha^{\frac{2}{2-\gamma}} + \frac{1}{\epsilon^2}\left(\frac{k}{h_{j_0+1}^{1/2}} + h_{j_0+1}^{1/2}\right)^2 + \frac{h_{j_0+1}}{\alpha^2}\Bigg). \tag{39}$$

If we choose $\alpha = h_{j_0+1}^{\frac{2-\gamma}{4}}$, and $\epsilon^2 = \frac{1}{(1+C_2)^{2-\gamma}}(\frac{k}{h_{j_0+1}^{1/2}} + h_{j_0+1}^{1/2})^{2-\gamma}$ by (35), and (39) we may conclude

$$u_h^k(x_i, s_j) - u(x_i, s_j) \leq C\left(\frac{k}{h^{1/2}} + h^{1/2}\right)^\gamma$$

and (33) follows. If $h > 1$ we use $h_{j_0+1} \leq T^{1-\frac{\gamma}{2}}h^{\frac{\gamma}{2}}$, and $h_{j_0+1} \leq T^{\frac{1}{2}}h^{\frac{1}{2}}$.

□

References

Camilli, F., Falcone, M., Lanucara, P. and Seghini, A. (1994). A domain decomposition method for Bellman equations, in D.E. Keyes and J.Xu (eds), *Domain Decomposition methods in Scientific and Engineering Computing*, Contemporary Mathematics n.180, AMS, 477-483.

Cannarsa, P. and Soner, H.M. (1987). On the singularities of the viscosity solutions to Hamilton–Jacobi–Bellman equations, *Ind. Univ. Math. J.* **36**, 501-524.

Capuzzo Dolcetta, I. (1983). On a discrete approximation of the Hamilton–Jacobi equation of dynamic programming, *Appl. Math. Optim.* **10**, 367-377.

Capuzzo Dolcetta, I. and Ishii, H. (1984). Approximate solutions of the Bellman equation of deterministic control theory, *Appl. Math. Optim.* **11**, 161-181.

Coddington, E. A. (1955). *Theory of Ordinary Differential Equations*, Mc Graw-Hill, New York, Toronto, London.

Crandall, M. G. and Lions, P. L. (1983). Viscosity solutions of Hamilton–Jacobi equations, *Tran. Amer. Math. Soc.* **277:1**, 1-42.

Crandall, M. G. and Lions, P. L. (1984). Two approximations of solutions of Hamilton–Jacobi equations, *Math. Comp.* **43**, 1-19.

Crandall, M. G., Ishii, H. and Lions, P. L. (1992). User's guide to viscosity solutions of second order partial differential equations, *Bull. Amer. Math. Soc.* **27:1**, 1-67.

Crandall, M. G., Evans, L. C. and Lions, P. L. (1984). Some properties of viscosity solutions of Hamilton–Jacobi equations, *Tran. Amer. Math. Soc.* **282:1**, 487-502.

Falcone, M. (1997). Numerical solution of dynamic programming equations, Appendix in the book by M. Bardi and I. Capuzzo Dolcetta, *Optimal control and viscosity solutions of Hamilton-Jacobi-Bellman equations*, Birkhäuser, Boston.

Falcone, M. (1987). A numerical approach to the infinite horizon problem of deterministic control theory, *Appl. Math. Optim.* **15**, 1-13 and (1991) **23**, 213-214.

Fleming, W. H. (1969). The Cauchy problem for a nonlinear first-order partial differential equation, *J. Diff. Eq.* **5**, 515-530.

Fleming, W. H. and Soner, H. M. (1993). *Controlled Markov Processes and Viscosity Solutions*, Springer-Verlag, New York.

Gonzalez, R.L.V. and Rofman, E. (1985). On deterministic control problems: an approximation procedure for the optimal cost, part 1 and 2, *SIAM J. Control Optim.* **23**, 242-285.

Kushner, H. and Dupuis, P. (1992). *Numerical methods for stochastic control problems in continuous time*, Springer Verlag.

Lions, P. L. (1982). *Generalized Solutions of Hamilton–Jacobi Equations*, Pitman, London.

Souganidis, P. E. (1985). Approximation schemes for viscosity solutions of Hamilton–Jacobi equations, *J. Differential Equations* **59**, 1-43.

18

Homogenization of the Cauchy Problem for Hamilton-Jacobi Equations

Hitoshi Ishii

ABSTRACT: We study the asymptotic behavior of solutions of the Cauchy problem for Hamilton-Jacobi equations with periodic coefficients as the frequency of periodicity tends to infinity. The limit functions are characterized as unique solutions of Hamilton-Jacobi equations with the Hamiltonians determined by the corresponding cell problems. Our result applies to the case where the initial data oscillate periodically and so does the Hamiltonian both in the spatial and time variables.

KEYWORDS: Homogenization, Hamilton-Jacobi equations, Cauchy problem.

1 Introduction and main results

In this paper we shall be concerned with the Cauchy problem

$$u_t^\varepsilon(x,t) + H(x,\, t,\, x/\varepsilon,\, t/\varepsilon,\, Du^\varepsilon(x,t)) = 0 \quad \text{in} \;\; \mathbf{R}^N \times (0,\infty) \tag{1.1}$$

with the initial condition

$$u^\varepsilon(x,0) = u_0(x,\, x/\varepsilon) \quad \text{for} \;\; x \in \mathbf{R}^N, \tag{1.2}$$

where $\varepsilon > 0$ is a parameter, and study the asymptotic behavior of the solution u^ε as $\varepsilon \searrow 0$.

Here H is a real-valued function on $\Gamma \equiv \mathbf{R}^N \times [0,\infty) \times \mathbf{R}^N \times \mathbf{R} \times \mathbf{R}^N$, u_0 is a given real-valued function on $\mathbf{R}^N \times \mathbf{R}^N$, u^ε is the real-valued unknown function on $\mathbf{R}^N \times [0,\infty)$, and u_t^ε and Du^ε denote the time and spatial derivatives of u^ε, i.e., $u_t^\varepsilon = \partial u^\varepsilon/\partial t$ and $Du^\varepsilon = (\partial u^\varepsilon/\partial x_1, ..., \partial u^\varepsilon/\partial x_N)$.

Our question concerning the asymptotic behavior of solutions of (1.1) – (1.2) is a sort of homogenization of partial differential equations, and the studies in [LPV] and [E] are closely related to ours. Indeed, P.-L. Lions, G. Papanicolaou, and S. R. S. Varadhan [LPV] studied (1.1) – (1.2) and obtained results similar to Theorem 1.1 below in the case when $H(x, t, \xi, \tau, p)$

and $u_0(x,\xi)$ are independent, respectively, of τ and ξ. W. E generalized their results so that the case when $u_0(x,\xi)$ depends on ξ is covered.

Our purpose here is to study the asymptotic problem for (1.1) – (1.2) in its full generality in the sense that we do not impose any of the restrictions on $H(x,t,\xi,\tau,p)$ and $u_0(x,\xi)$ described above. Our result, Theorem 1.1 below, is, in its statement, a straightforward generalization of those of [LPV] and [E], but those estimates used in [LPV] to control the solutions of (1.1) – (1.2) which allowed their limiting arguments are not available in our generality and we need here a new estimate (see, e.g.,Theorem 3.1) on solutions of (1.1) to control the limit function of the solutions of (1.1) – (1.2) and to establish the result. The new estimate is somehow related to the fact that the underlying controlled (or differential game) system is exactly controllable on the torus $\mathbf{R}^{N+1}/\mathbf{Z}^{N+1}$ and, in this connection, we refer the reader to [Ar].

Our proof of the main result is based on the so-called perturbed test function method developed by Evans [Ev1, Ev2] together with standard techniques in the theory of viscosity solutions [CIL2].

We refer the interested reader for related work on homogenization of Hamilton-Jacobi equations to [Al, Co1, Co2, Ev2, H, HI].

The referee of this paper kindly pointed out that the asymptotic problem (1.1) – (1.2) with $H(x,t,\xi,\tau,p)$ and $u_0(x,\xi)$ independent of ξ is referred as "an averaging problem" rather than "a homogenization problem" and the averaging problem for (1.1) – (1.2) has already been studied in [Ch] and [Br].

Now, let us give the assumptions on H and u_0.

(A1) $H \in C(\Gamma)$ and moreover $H \in BUC(\Gamma_R)$ for all $R > 0$, where

$$\Gamma_R = \mathbf{R}^N \times [0,\infty) \times \mathbf{R}^N \times \mathbf{R} \times B(0,R),$$

Here $B(0,R)$ denotes the closed ball with radius R and center at 0, and $BUC(X)$ denotes the space of real-valued functions on a metric space X which are bounded and uniformly continuous on X.

(A2) $\lim_{R\to\infty} \inf\{H(x,\, t,\, \xi,\, \tau,\, p) \mid (x,\, t,\, \xi,\, \tau,\, p) \in \Gamma,\ |p| \geq R\} = \infty.$

(A3) For each $(x,\, t,\, p) \in \mathbf{R}^{2N+1}$ the function $H(x,\, t,\, \xi,\, \tau,\, p)$ of $(\xi,\, \tau)$ is periodic with period $\mathbf{Z}^{N+1}$.

(A4) There is a function $\omega \in C([0,\, \infty))$, with $\omega(0) = 0$, such that

$$|H(x,\, t,\, \xi,\, \tau,\, p) - H(x,\, s,\, \xi,\, \sigma,\, p)| \leq \omega(|t-s| + |\tau - \sigma|)$$

for all $(x,\, t,\, \xi,\, \tau,\, p),\, (x,\, s,\, \xi,\, \sigma,\, p) \in \Gamma$.

(A5) $u_0 \in BUC(\mathbf{R}^N \times \mathbf{R}^N)$ and for each $x \in \mathbf{R}^N$ the function $u_0(x, \xi)$ of ξ is periodic with period $\mathbf{Z}^N$.

Concerning the solvability of the Cauchy problem (1.1) – (1.2) we have

Proposition 1.1 *Under the assumptions* (A1), (A2), (A4), *and* (A5) *there is a unique viscosity solution* $u^\varepsilon \in \cap_{T>0} BUC(\mathbf{R}^N \times [0, T])$ *of* (1.1) – (1.2).

We refer the reader to [Bl, I2] for the proof of the above proposition.

In order to describe the asymptotic behavior of the solution u^ε of (1.1) – (1.2) as $\varepsilon \searrow 0$, we introduce the cell problems associated with (1.1) – (1.2). For each $(\hat{x}, \hat{t}, \hat{p}) \in \mathbf{R}^N \times [0, \infty) \times \mathbf{R}^N$ consider the problem of finding a pair $(\lambda, v) \in \mathbf{R} \times B(\mathbf{R}^N \times \mathbf{R})$ such that v is a solution of

$$v_\tau(\xi, \tau) + H(\hat{x}, \hat{t}, \xi, \tau, \hat{p} + Dv(\xi, \tau)) = \lambda \quad \text{in } \mathbf{R}^N \times \mathbf{R}, \tag{1.3}$$

where $B(\mathbf{R}^N \times \mathbf{R})$ denotes the space of all bounded functions on $\mathbf{R}^N \times \mathbf{R}$.

Proposition 1.2 *Under the assumptions* (A1) – (A4), *there is a unique* $\lambda \in \mathbf{R}$ *for which problem* (1.3) *has a viscosity subsolution and a viscosity supersolution which are bounded and periodic with period* $\mathbf{Z}^{N+1}$.

The above proposition allows us to define the function $\overline{H}$ on $\mathbf{R}^N \times [0, \infty) \times \mathbf{R}^N$ by the formula $\overline{H}(\hat{x}, \hat{t}, \hat{p}) = \lambda$, where λ is the real number for which (1.3) has a viscosity subsolution and a viscosity supersolution which are bounded and periodic with period $\mathbf{Z}^{N+1}$.

Proposition 1.3 *The function* G *on* Γ *defined by* $G(x, t, \xi, \tau, p) = \overline{H}(x, t, p)$ *satisfies conditions* (A1), (A2), *and* (A4) *with* $H = G$.

In view of Propositions 1.1 and 1.3, under the assumptions (A1) – (A5) there is a unique viscosity solution $u \in \cap_{T>0} BUC(\mathbf{R}^N \times [0, T])$ of the Cauchy problem

$$u_t(x, t) + \overline{H}(x, t, Du(x, t)) = 0 \quad \text{in } \mathbf{R}^N \times (0, \infty) \tag{1.4}$$

with the initial condition

$$u(x, 0) = \underline{u}_0(x) \quad \text{for } x \in \mathbf{R}^N, \tag{1.5}$$

where

$$\underline{u}_0(x) = \min_{\xi \in \mathbf{R}^N} u_0(x, \xi). \tag{1.6}$$

The main result in this paper is this theorem.

Theorem 1.1 *Assume that* (A1) – (A5) *hold. For* $\varepsilon > 0$ *let* $u^\varepsilon \in \cap_{T>0} BUC(\mathbf{R}^N \times [0, T])$ *be the viscosity solution of* (1.1) – (1.2) *and* $u \in \cap_{T>0} BUC(\mathbf{R}^N \times [0, T])$ *the viscosity solution of* (1.4) – (1.5). *Then, as* $\varepsilon \searrow 0$,

$$u^\varepsilon(x, t) \to u(x, t) \text{ uniformly on } \mathbf{R}^N \times [\delta, T]$$

for all $0 < \delta < T$.

Our plan in this paper is as follows. In Section 2 we present a few observations on solutions of a simple Hamilton-Jacobi-Bellman-Isaacs equation by studying a corresponding differential game. Using the observations in Section 2, in Section 3 we establish some estimates of solutions of Hamilton-Jacobi equations with coercive Hamiltonian. We establish Proposition 1.2 in Section 4 and Proposition 1.3 and Thereom 1.1 in the final section, Section 5.

2 A simple Hamilton-Jacobi equation and a differential game

In this section we consider the Hamilton-Jacobi equation

$$v_t(x,t) + g(|Dv(x,t)|) = 0 \quad \text{in } \mathbf{R}^N \times (0,\infty) \tag{2.1}$$

and a differential game associated with it.

Here the function g on $[0,\infty)$ is given by

$$g(r) = \inf_{a\in I}(ar + h(a)), \tag{2.2}$$

where $I = (0,b]$, with $b > 0$, and $h \in C(I)$ is assumed to be nonincreasing and to satisfy

$$\inf_I h = 0, \tag{2.3}$$

and

$$\lim_{a\searrow 0} h(a) = \infty. \tag{2.4}$$

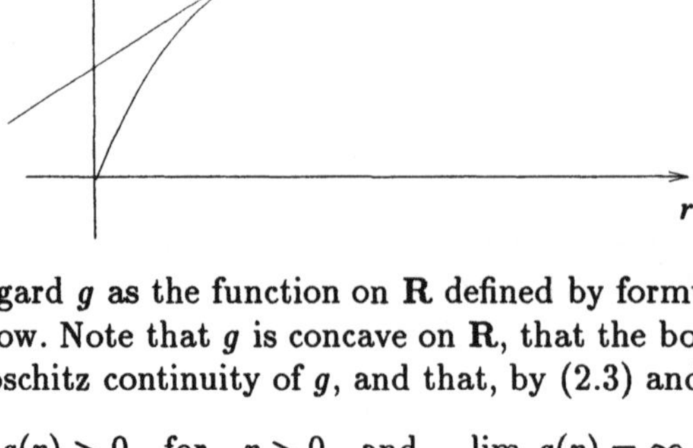

We shall regard g as the function on $\mathbf{R}$ defined by formula (2.2) in our discussions below. Note that g is concave on $\mathbf{R}$, that the boundedness of I implies the Lipschitz continuity of g, and that, by (2.3) and (2.4),

$$g(r) > 0 \quad \text{for} \quad r > 0 \quad \text{and} \quad \lim_{r\to\infty} g(r) = \infty. \tag{2.5}$$

Observe that for all $p \in \mathbf{R}^N$,

$$g(|p|) = \max_{\xi \in S^{N-1}} \inf_{a \in I} (-a\langle \xi, p\rangle + h(a)).$$

We introduce a differential game associated with (2.1) as follows. Let $\mathcal{A}$ and $\mathcal{B}$ be the sets

$$\mathcal{A} = \left\{ \alpha : [0,\infty) \to (0,b] \;\middle|\; \alpha \text{ is measurable, } \inf_{s\geq 0} \alpha(s) > 0 \right\}$$

and

$$\mathcal{B} = \{\beta : [0,\infty) \to B(0,\,1) \mid \beta \text{ is measurable}\},$$

and define Λ as the set of mappings $\lambda : \mathcal{A} \to \mathcal{B}$ such that if $\alpha, \beta \in \mathcal{A}$ and $t > 0$ satisfy $\alpha = \beta$ on $[0,\,t)$ then $\lambda[\alpha] = \lambda[\beta]$ on $[0,\,t)$.

Fix $v_0 \in BUC(\mathbf{R}^N)$, and define the function V on $\mathbf{R}^N \times [0,\infty)$ by

$$V(x,t) = \inf_{\lambda \in \Lambda} \sup_{\alpha \in \mathcal{A}} \left\{ -\int_0^t h(\alpha(s))ds + v_0\left(x + \int_0^t \alpha(s)\lambda[\alpha](s)ds\right) \right\}.$$

Define also the functional J on $\mathbf{R}^N \times [0,\infty) \times \mathcal{A} \times \mathcal{B}$ by

$$J(x,t,\alpha,\beta) = -\int_0^t h(\alpha(s))ds + v_0\left(x + \int_0^t \alpha(s)\beta(s)ds\right).$$

In the terminology of differential games this function V is called the value function. Also, the mappings $\lambda \in \Lambda$ are referred as strategies, while $\alpha \in \mathcal{A}$ and $\beta \in \mathcal{B}$ are referred as controls. See, e.g., [EK, ES, I1].

Proposition 2.1 *We have*

$$\sup_{\mathbf{R}^N \times [0,\infty)} |V| \leq \|u_0\|_{L^\infty(\mathbf{R}^N)}. \tag{2.6}$$

Proof. Fix $(x,t) \in \mathbf{R}^N \times (0,\infty)$. Let $\alpha_0(s) \equiv b$. Since $h(b) = 0$, we have

$$\begin{aligned} V(x,t) &\geq \inf_{\lambda\in\Lambda} J(x,t,\alpha_0,\lambda[\alpha_0]) \\ &= \inf_{\lambda\in\Lambda} v_0\left(x + \int_0^t b\lambda[\alpha_0](s)ds\right) \geq -\|v_0\|_{L^\infty(\mathbf{R}^N)}. \end{aligned}$$

On the other hand, since $h(a) \geq 0$ for all $a \in I$, we have

$$V(x,t) \leq \sup_{\alpha\in\mathcal{A}} J(x,t,\alpha,\lambda[\alpha]) \leq \|v_0\|_{L^\infty(\mathbf{R}^N)}$$

for any $\lambda \in \Lambda$. $\square$

Proposition 2.2 (Dynamic programming principle) *Let $x \in \mathbf{R}^N$ and $0 \le \tau < t$. Then*

$$V(x, t) = \inf_{\lambda\in\Lambda} \sup_{\alpha\in\mathcal{A}} \left\{ -\int_0^{t-\tau} h(\alpha(s))ds + V\left(x + \int_0^{t-\tau} \alpha(s)\lambda[\alpha](s)ds,\ \tau\right)\right\}. \tag{2.7}$$

See [ES, I1] for a proof.

Proposition 2.3 $V \in BUC(\mathbf{R}^N \times [0, \infty))$. *Moreover,*

$$|V(x,t) - V(y,s)| \le \omega_0(|x-y|) + \omega_0(b|t-s|). \tag{2.8}$$

for all (x,t), $(y,s) \in \mathbf{R}^N \times [0,\infty)$, where ω_0 denotes the modulus of continuity of the function v_0, i.e., $\omega_0(r) = \sup_{|x-y|\le r} |v_0(x) - v_0(y)|$ for $r \ge 0$.

Proof. Fix $x, y \in \mathbf{R}^N$ and $0 \le s < t$. Let $\lambda \in \Lambda$ and $\alpha \in \mathcal{A}$, and observe that

$$|J(x,t,\alpha,\lambda[\alpha]) - J(y,t,\alpha,\lambda[\alpha])| \le \omega_0(|x-y|).$$

Hence

$$|V(x,t) - V(y,t)| \le \omega_0(|x-y|).$$

Define $\alpha_0 \in \mathcal{A}$ by $\alpha_0(s) = b$. By Proposition 2.2, we have

$$\begin{aligned} V(x,t) &\ge \inf_{\lambda\in\Lambda} V\left(x + \int_0^{t-s} b\lambda[\alpha_0](\sigma)d\sigma,\ s\right) \\ &\ge V(x,s) - \omega_0(b|t-s|). \end{aligned}$$

Next, fix $\varepsilon > 0$ and $\lambda \in \Lambda$ and choose $\alpha \in \mathcal{A}$ in view of Proposition 2.2 so that

$$V(x,t) < \varepsilon + V\left(x + \int_0^{t-s} \alpha(\sigma)\lambda[\alpha](\sigma)d\sigma,\ s\right).$$

From this we see that $V(x,t) \le V(x,s) + \omega_0(b|t-s|)$, and conclude that

$$|V(x,t) - V(x,s)| \le \omega_0(b|t-s|).$$

The proof is now complete. □

Proposition 2.4 *The value function V is a viscosity solution of* (2.1).

Since the proof of the above proposition is standard, we refer for a proof to [ES, I1].

Theorem 2.1 *For each $\varepsilon > 0$ there is a positive number $T(\varepsilon) > 0$ such that for all $(x,t) \in \mathbf{R}^N \times (0,\infty)$, if $t \ge T(\varepsilon)$ then*

$$V(x,t) \le \min_{B(x,\ \varepsilon)} v_0. \tag{2.9}$$

Proof. The first step is to define the function $T : (0,\infty) \to (0,\infty)$.
Setting

$$\mathcal{A}_t = \left\{\alpha \in \mathcal{A} \;\middle|\; \int_0^t h(\alpha(s))ds \le 2M_1\right\} \quad (t > 0),$$

where $M_1 = \|v_0\|_{L^\infty(\mathbf{R}^N)} + 1$, we observe that for $(x,t) \in \mathbf{R}^N \times (0,\infty)$,

$$V(x,t) = \inf_{\lambda\in\Lambda} \sup_{\alpha\in\mathcal{A}_t} J(x,\, t,\, \alpha,\, \lambda[\alpha]). \tag{2.10}$$

Indeed, for any $\lambda \in \Lambda$, if $\alpha \in \mathcal{A} \setminus \mathcal{A}_t$, then

$$J(x,t,\alpha,\lambda[\alpha]) < -2M_1 + v_0\Big(x + \int_0^t \alpha(s)\lambda[\alpha](s)ds\Big) < -M_1 < V(x,t).$$

This shows that (2.10) is valid.

Now, by (2.2) we have

$$g(r) \le \alpha(s)r + h(\alpha(s)) \quad (r \ge 0,\ \alpha \in \mathcal{A})$$

and integrating this over $(0,\, t)$, we get

$$tg(r) \le r\int_0^t \alpha(s)ds + \int_0^t h(\alpha(s))ds \le r\int_0^t \alpha(s)ds + 2M_1$$

for all $r \ge 0$ and $\alpha \in \mathcal{A}_t$. Plugging $r = \Big(\int_0^t \alpha(s)ds\Big)^{-1}$ into the above, we get

$$t \le \frac{2M_1 + 1}{g\Big(1 \big/ \int_0^t \alpha(s)ds\Big)} \tag{2.11}$$

for $t > 0$ and $\alpha \in \mathcal{A}_t$.

We define $T : (0,\infty) \to (0,\infty)$ by $T(r) = (2M_1 + 1)\, g\,(1/r)^{-1}$. The function T is continuous and strictly increasing and satisfies

$$\lim_{r\searrow 0} T(r) = 0.$$

Moreover (2.11) now reads

$$t \le T\Big(\int_0^t \alpha(s)ds\Big) \quad \text{for all } t > 0,\ \alpha \in \mathcal{A}_t. \tag{2.12}$$

The second step is to define an appropriate strategy $\mu \in \Lambda$.

Fix $x \in \mathbf{R}^N$ and $\varepsilon > 0$. Choose a $\xi \in B(0,\, 1)$ so that $\min_{B(x,\varepsilon)} v_0 = v_0(x + \varepsilon\xi)$. For $\alpha \in \mathcal{A}$ set

$$\tau[\alpha] = \inf\Big\{t \ge 0 \;\Big|\; \int_0^t \alpha(s)ds \ge \varepsilon\Big\}.$$

It is clear that $0 < \tau[\alpha] < \infty$ for $\alpha \in \mathcal{A}$.

Now define $\mu : \mathcal{A} \to \mathcal{B}$ by

$$\mu[\alpha](s) = \begin{cases} \xi & (0 \leq s < \tau[\alpha]), \\ 0 & (\tau[\alpha] \leq s). \end{cases}$$

It is not hard to check that this μ is a strategy, i.e., $\mu \in \Lambda$. (We leave it to the reader to check this.)

To finish the proof, we first observe that, in view of (2.12) and the monotonicity of T, if $\alpha \in \mathcal{A}_t$ and $t \geq T(\varepsilon)$ then

$$\int_0^t \alpha(s)ds \geq \varepsilon$$

which means that $\tau[\alpha] \leq t$. Hence we see that if $t \geq T(\varepsilon)$, then

$$\begin{aligned} V(x,t) &\leq \sup_{\alpha \in \mathcal{A}_t} J(x,t,\alpha,\mu[\alpha]) \leq \sup_{\alpha \in \mathcal{A}_t} v_0\left(x + \int_0^{\tau[\alpha]} \alpha(s)ds \cdot \xi\right) \\ &= v_0(x + \varepsilon\xi) = \min_{B(x,\varepsilon)} v_0. \qquad \square \end{aligned}$$

3 Hamilton-Jacobi equations with coercive Hamiltonian

In this section we consider the Hamilton-Jacobi equation

$$v_t(x,t) + G(x,t,Dv(x,t)) = 0 \quad \text{in } \mathbf{R}^N \times (0,\infty). \tag{3.1}$$

Here we make the following assumptions:

(B1) $G \in C(\mathbf{R}^N \times [0,\infty) \times \mathbf{R}^N)$ and moreover $G \in BUC(\mathbf{R}^N \times [0,\infty) \times B(0,R))$ for all $R > 0$.

(B2) $\lim_{R\to\infty} \inf\{G(x,t,p) \mid (x,t,p) \in \mathbf{R}^N \times [0,\infty) \times \mathbf{R}^N,\ |p| \geq R\} = \infty.$

It should be remarked that these assumptions for G are the same as the conditions (A1) and (A2) for $H(x,t,\xi,\tau,p) = G(x,t,p)$.

We begin with a simple lemma.

Lemma 3.1 *Under the assumptions* (B1) – (B2) *there are a function* $g \in C([0,\infty))$ *and constants* $C \in \mathbf{R}$ *and* $b > 0$ *such that for all* $(x,t,p) \in \mathbf{R}^N \times [0,\infty) \times \mathbf{R}^N$,

$$G(x,t,p) \geq g(|p|) + C, \tag{3.2}$$

and such that

$$g(r) = \inf_{a\in I}(ar + h(a)) \quad \textit{for all } r \geq 0, \tag{3.3}$$

where $I = (0,\, b]$ *and* h *is a continuous nonincreasing function on* I *satisfying*

$$h(b) \left(= \inf_I h\right) = 0 \quad \textit{and} \quad \lim_{a \searrow 0} h(a) = \infty. \tag{3.4}$$

Proof. It is clear from (B1) – (B2) that there is a nondecreasing function $f \in C[0,\, \infty))$ such that $G(x,t,p) \geq f(|p|)$ for all $(x,t,p) \in \mathbf{R}^N \times [0,\, \infty) \times \mathbf{R}^N$ and such that $\lim_{r\to\infty} f(r) = \infty$. For instance, the function f on $[0,\, \infty)$ defined by

$$f(r) = \inf\{G(x,t,p) \mid (x,\, t,\, p) \in \mathbf{R}^N \times [0,\, \infty) \times \mathbf{R}^N,\ |p| \geq r\}$$

has the required properties.

The rest of the proof is basically to replace the function f by a concave function which is smaller than or equal to f pointwise.

To this end, we define the sequences $\{r_n\}_{n=0}^{\infty}$ and $\{e_n\}_{n=0}^{\infty}$ by

$$r_n = \inf\{r \geq 0 \mid f(r) = f(0) + n\} \quad \text{and} \quad e_n = f(0) + n - 1.$$

It is then clear that

$$f(r) \geq e_n + 1 \quad \text{for } r \in [r_n,\, \infty),\ n = 0,\, 1,\, 2,\, \ldots$$

and that $r_0 = 0$, the sequence $\{r_n\}_{n=0}^{\infty}$ is increasing, and $\lim_{n\to\infty} r_n = \infty$.

We next define the sequence $\{s_n\}_{n=0}^{\infty}$ by $s_n = \sum_{k=0}^{n} r_k$. Since $r_n \leq s_n$ for all $n \geq 0$, it follows that

$$f(r) \geq e_n + 1 \quad \text{for all } r \in [s_n,\, \infty),\ n = 0,\, 1,\, 2,\, \ldots$$

Note also that $s_0 = 0$ and $s_{n+1} - s_n > s_n - s_{n-1}$ for all $n \geq 1$.

We set $C = e_0$ and define $g \in C([0,\, \infty))$ by

$$g(r) = \frac{1}{s_{n+1} - s_n}(r - s_n) + e_n - e_0 \quad \text{for } r \in [s_n,\, s_{n+1}],\ n = 0,\, 1,\, 2,\, \ldots$$

and $h \in C((0,\, b])$, with $b = 1/s_1$, by

$$h(a) = e_n - as_n - e_0 \quad \text{for } a \in \left[\frac{1}{s_{n+1} - s_n},\, \frac{1}{s_n - s_{n-1}}\right],\ n = 1,\, 2,\, \ldots$$

It is not difficult to see that these g and h have the required properties. We leave it to the reader to check the details. □

Note that under the assumptions (B1) and (B2), as we have seen in the above proof, there is a function $f \in C([0,\, \infty))$ such that

$$\lim_{r\to\infty} f(r) = \infty, \tag{3.5}$$

and

$$G(x, t, p) \geq f(|p|) \quad \text{for } (x, t, p) \in \mathbf{R}^N \times [0,\infty) \times \mathbf{R}^N, \tag{3.6}$$

and also, as is easily seen, there is a function $m \in C([0,\infty))$ such that for all $(x, t, p) \in \mathbf{R}^N \times [0, \infty) \times \mathbf{R}^N$,

$$G(x, t, p) \leq m(|p|). \tag{3.7}$$

Theorem 3.1 *Assume that* (B1) – (B2) *hold. Let* $f \in C([0, \infty))$ *be a function satisfying* (3.5) – (3.6). *Then there is a constant* $C = C(f) \in \mathbf{R}$ *depending only on* f *and for each* $\gamma > 0$ *a constant* $T = T(\gamma, f) > 0$ *depending only on* γ, f *such that*

$$\lim_{\gamma \searrow 0} T(\gamma, f) = 0$$

and such that if $u \in \cap_{S>0} BUC(\mathbf{R}^N \times [0, S])$ *is a viscosity subsolution of* (3.1) *and if* $t \geq T$ *then*

$$u(x,t) \leq \min_{y \in B(x,\gamma)} u(y, 0) - Ct \quad \textit{for } x \in \mathbf{R}^N.$$

Proof. Choose a function $g \in C([0, \infty))$ and a constant C so that (3.2) – (3.3) hold for some function $h \in C(I)$, where I is an interval of the form $(0, b]$. Here g and C can be chosen so that they depend only on f.

With this function g and $v_0(x) = u(x,0)$, we define $V \in BUC(\mathbf{R}^N \times [0, \infty))$ as in Section 2. Define $v \in \cap_{S>0} BUC(\mathbf{R}^N \times [0, S])$ by $v(x, t) = u(x, t) + Ct$. Observe that v is a viscosity subsolution of

$$v_t(x, t) + g(|Dv(x, t)|) = 0 \quad \text{in } \mathbf{R}^N \times (0, \infty).$$

Since V is a viscosity solution of (2.1), by the standard comparison result for solutions of Hamilton-Jacobi equations, we see that $v(x, t) \leq V(x, t)$ for all $(x, t) \in \mathbf{R}^N \times [0, \infty)$. Now Theorem 2.1 guarantees that for each $\gamma > 0$ there is a positive constant $T = T(\gamma, f)$ depending only on γ and f such that if $(x,t) \in \mathbf{R}^N \times [T, \infty)$, then

$$\begin{aligned} u(x, t) &\leq V(x, t) - Ct \leq \min_{y \in B(x,\gamma)} V(y, 0) - Ct \\ &= \min_{y \in B(0,\gamma)} u(y, 0) - Ct. \end{aligned}$$

This completes the proof. □

Proposition 3.1 *Assume that* (B1) – (B2) *hold. Let* $u \in \cap_{T>0} BUC(\mathbf{R}^N \times [0, T])$ *be a viscosity supersolution of* (3.1). *Let* $m \in C([0, \infty))$ *be a function satisfying* (3.7) *and* $v_0 \in BUC(\mathbf{R}^N)$ *be a function such that* $u(x,0) \geq v_0(x)$ *for* $x \in \mathbf{R}^N$. *Then there is a function* $\omega \in C([0, \infty))$ *satisfying* $\omega(0) = 0$ *and depending only on* m *and* v_0 *such that*

$$u(x,t) \geq v_0(x) - \omega(t) \quad \textit{for } (x,t) \in \mathbf{R}^N \times [0, \infty).$$

This is a rather standard observation in the theory of viscosity solutions.

Proof. Fix $\varepsilon > 0$ and choose a constant $A_\varepsilon > 0$ so that for all $x, y \in \mathbf{R}^N$,

$$|v_0(x) - v_0(y)| \le \left(\varepsilon^2 + A_\varepsilon^2|x-y|^2\right)^{1/2}.$$

Next choose $B_\varepsilon > 0$ so that

$$B_\varepsilon \ge m(A_\varepsilon).$$

Now, fix $y \in \mathbf{R}^N$ and set

$$v(x,t) = v_0(y) - \left(\varepsilon^2 + A_\varepsilon^2|x-y|^2\right)^{1/2} - B_\varepsilon t \quad \text{for } (x,\, t) \in \mathbf{R}^N \times [0,\, \infty).$$

Then it is easy to see that v is a viscosity subsolution of

$$v_t(x,t) + m(|Dv(x,\, t)|) = 0 \quad \text{in } \mathbf{R}^N \times (0,\, \infty). \tag{3.8}$$

Since u is a viscosity supersolution of (3.8), by the standard comparison result, we see that for all $t \ge 0$,

$$u(y,t) \ge v(y,t) = v_0(y) - \varepsilon - B_\varepsilon t.$$

This implies the existence of ω with the required properties. □

4 The cell problems

Fix $(\hat{x},\, \hat{t},\, \hat{p}) \in \mathbf{R}^N \times [0,\, \infty) \times \mathbf{R}^N$ and set

$$F(\xi,\, \tau,\, q) = H(\hat{x},\, \hat{t},\, \xi,\, \tau,\, \hat{p} + q) \quad \text{for } (\xi,\, \tau,\, q) \in \mathbf{R}^N \times \mathbf{R} \times \mathbf{R}^N.$$

Here we study the cell problem

$$v_\tau(\xi,\, \tau) + F(\xi,\, \tau,\, Dv(\xi,\, \tau)) = \lambda \quad \text{in } \mathbf{R}^N \times \mathbf{R}, \tag{4.1}$$

where the unknowns are a function v and a constant λ.

Note that, under the assumptions (A1) – (A4), the function F satisfies:

(C1) $F \in BUC(\mathbf{R}^N \times \mathbf{R} \times B(0,\, R))$ for all $R > 0$.

(C2) $\lim_{R\to\infty} \inf\{F(\xi,\, \tau,\, p) \mid (\xi,\, \tau,\, p) \in \mathbf{R}^N \times \mathbf{R} \times \mathbf{R}^N,\ |p| \ge R\} = \infty.$

(C3) For each $p \in \mathbf{R}^N$ the function $F(\xi,\, \tau,\, p)$ of $(\xi,\, \tau)$ is periodic with period $\mathbf{Z}^{N+1}$.

(C4) There is a function $\omega \in C([0, \infty))$, with $\omega(0) = 0$, such that

$$|F(\xi, \tau, p) - F(\xi, \sigma, p)| \le \omega(|\tau - \sigma|)$$

for all $(\xi, \tau, p), (\xi, \sigma, p) \in \mathbf{R}^{2N+1}$.

In order to solve (4.1), we approximate it by a "regular" problem, which is a quite standard technique in the study of cell problems.

We thus consider for each $\varepsilon > 0$ the Hamilton-Jacobi equation

$$\varepsilon v(\xi, \tau) + v_\tau(\xi, \tau) + F(\xi, \tau, Dv(\xi, \tau)) = 0 \quad \text{in } \mathbf{R}^N \times \mathbf{R}. \tag{4.2}$$

Due to a result [Bl] by G. Barles, under the assumptions (C1), (C2), and (C4), there is a unique viscosity solution $v \in BUC(\mathbf{R}^{N+1})$ of (4.2). It is clear by the uniqueness that the viscosity solution is periodic with period $\mathbf{Z}^{N+1}$.

Proposition 4.1 *Let v^ε be the unique viscosity solution of* (4.2) *in the class $BUC(\mathbf{R}^{N+1})$. Then :* (1) *It holds that for all $(y, s) \in \mathbf{R}^{N+1}$,*

$$\inf_{(\xi,\tau)\in\mathbf{R}^{N+1}} F(\xi, \tau, 0) \le -\varepsilon v^\varepsilon(y, s) \le \sup_{(\xi,\tau)\in\mathbf{R}^{N+1}} F(\xi, \tau, 0). \tag{4.3}$$

(2) *There is a constant $M > 0$ depending only on F such that for any $\varepsilon > 0$,*

$$\sup_{\mathbf{R}^{N+1}} v^\varepsilon - \inf_{\mathbf{R}^{N+1}} v^\varepsilon \le M. \tag{4.4}$$

Proof. The proof of (1) is standard and we just need to look at a maximum and a minimum points of the function v^ε and apply the definition of viscosity solution. For instance, if (ξ, τ) is a maximum point of v^ε, then we have

$$\varepsilon v^\varepsilon(\xi, \tau) + F(\xi, \tau, 0) \le 0,$$

and hence

$$\min_{(\xi,\tau)\in\mathbf{R}^{N+1}} F(\xi,\tau,0) \le -\varepsilon v^\varepsilon(y,s) \quad \text{for all } (y,s) \in \mathbf{R}^{N+1}.$$

The other inequality of (4.3) follows similarly.

To check that (4.4) holds, fix $\varepsilon > 0$ and let $(y, s) \in \mathbf{R}^{N+1}$ be a minimum point of v^ε. Set $\Omega = y + [0, 1)^N$ and choose $r > 0$ so that if $x \in \Omega$ then $y \in B(x, r)$. (For instance, $r = \sqrt{N}$ does this job.) Noting that, according to (4.3), v^ε is a viscosity subsolution of

$$v_\tau(\xi, \tau) + F(\xi, \tau, Dv(\xi, \tau)) - \sup_{(\xi,\tau)\in\mathbf{R}^{N+1}} F(\xi, \tau, 0) = 0 \quad \text{in } \mathbf{R}^{N+1},$$

we apply Theorem 3.1 to conclude that there are numbers $T > 0$ and $C > 0$ depending only on r and F (or more precisely, a function f on

$[0, \infty)$ satisfying (3.5) and (3.6) with F in place of G) such that for all $(x,t) \in \mathbf{R}^{N+1}$, if $t \geq s+T$, then

$$v^\varepsilon(x,t) \leq \min_{B(x,r)} v^\varepsilon(\cdot, s) + C(t-s).$$

Hence we have

$$\sup_{\mathbf{R}^{N+1}} v^\varepsilon = \sup_{\Omega\times[T,T+1)} v^\varepsilon \leq v^\varepsilon(y,s) + C(T+1).$$

The proof is now complete. □

Let us restate Proposition 1.2 as a theorem here for convenience and prove it.

Theorem 4.1 *Under the assumptions* (C1) - (C4), *there is a unique* $\lambda \in \mathbf{R}$ *for which problem* (4.1) *has a viscosity subsolution and a viscosity supersolution which are bounded and periodic with period* $\mathbf{Z}^N$.

Proof. (Uniqueness) Let μ and ν be real numbers such that (4.1) with $\lambda = \mu$ and $\lambda = \nu$ both have a bounded viscosity subsolution and a bounded viscosity supersolution.

We want to show that $\mu = \nu$. To do this, we suppose that $\mu > \nu$. Let v and w be a bounded viscosity subsolution and a bounded viscosity supersolution of (4.1), respectively, with $\lambda = \nu$ and $\lambda = \mu$. We may assume by adding a constant if necessary that $v > w$ in $\mathbf{R}^{N+1}$.

Fix $\lambda \in \mathbf{R}$ be such that $\nu < \lambda < \mu$. Choose $\varepsilon > 0$ so small that v and w are a viscosity subsolution and a viscosity supersolution of (4.1) with λ just fixed above. Now, a comparison result [Bl] due to Barles yields that $v \leq w$ in $\mathbf{R}^{N+1}$. This is a contradiction, which implies that $\mu = \nu$.

(Existence) For $\varepsilon > 0$ let $v^\varepsilon \in BUC(\mathbf{R}^{N+1})$ be the unique viscosity solution of (4.2). Set

$$w^\varepsilon(\xi, \tau) = v^\varepsilon(\xi, \tau) - \inf v^\varepsilon \quad \text{for } (\xi, \tau) \in \mathbf{R}^{N+1},$$

and $\lambda^\varepsilon = -\varepsilon \inf v^\varepsilon$.

In view of Proposition 4.1, there is a constant $M > 0$ independent of ε such that $|\lambda^\varepsilon| \leq M$ and $0 \leq w^\varepsilon \leq M$ in $\mathbf{R}^{N+1}$.

Since $-\lambda^\varepsilon \leq \varepsilon v^\varepsilon(\xi, \tau) \leq -\lambda^\varepsilon + \varepsilon M$ for all $(\xi, \tau) \in \mathbf{R}^{N+1}$, we see that w^ε satisfies

$$w^\varepsilon_\tau(\xi, \tau) + F(\xi, \tau, Dw^\varepsilon(\xi, \tau)) \leq \lambda^\varepsilon \quad \text{in } \mathbf{R}^{N+1},$$

and

$$w^\varepsilon_\tau(\xi, \tau) + F(\xi, \tau, Dw^\varepsilon(\xi, \tau)) \geq \lambda^\varepsilon - \varepsilon M \quad \text{in } \mathbf{R}^{N+1}$$

in the viscosity sense.

Since the set $\{\lambda^\varepsilon\}_{\varepsilon>0}$ is bounded, we can choose a sequence $\{\varepsilon_k\}_{k\in\mathbf{N}}$ of positive numbers, with $\lim_{k\to\infty}\varepsilon_k = 0$, for which the sequence $\{\lambda^{\varepsilon_k}\}$ is convergent.

Set $\lambda = \lim_{k\to\infty}\lambda^{\varepsilon_k}$ and, noting that the family $\{w^\varepsilon\}_{\varepsilon>0}$ is uniformly bounded, define functions $\overline{w}$ and $\underline{w}$ by

$$\overline{w}(\xi,\,\tau) = \lim_{r\searrow 0}\sup\{w^\varepsilon(y,\,s) \mid \varepsilon = \varepsilon_k,\; k\geq 1/r,\; (y,\,s)\in B(\xi,\,\tau;\,r)\}$$

and

$$\underline{w}(\xi,\,\tau) = \lim_{r\searrow 0}\inf\{w^\varepsilon(y,\,s) \mid \varepsilon = \varepsilon_k,\; k\geq 1/r,\; (y,\,s)\in B(\xi,\,\tau;\,r)\}.$$

It is clear that $\overline{w}$ and $\underline{w}$ are bounded on $\mathbf{R}^{N+1}$ and periodic with period $\mathbf{Z}^{N+1}$. By a result [BP] due to Barles and Perthame, we have that $\overline{w}$ and $\underline{w}$ are, respectively, a viscosity subsolution and a viscosity supersolution of (4.1) with λ defined as above. □

Remark. By (4.3), the constant λ above satisfies

$$\inf_{(\xi,\,\tau)\in\mathbf{R}^{N+1}} F(\xi,\,\tau,\,0) \leq \lambda \leq \sup_{(\xi,\,\tau)\in\mathbf{R}^{N+1}} F(\xi,\,\tau,\,0).$$

5 Proof of Proposition 1.3 and Theorem 1.1

Proof of Proposition 1.3. The remark just made above yields

$$\inf_{(\xi,\tau)\in\mathbf{R}^{N+1}} H(x,\,t,\,\xi,\,\tau,\,p) \leq \overline{H}(x,\,t,\,p) \leq \sup_{(\xi,\,\tau)\in\mathbf{R}^{N+1}} H(x,\,t,\,\xi,\,\tau,\,p). \tag{5.1}$$

The first inequality above guarantees that

$$\lim_{R\to\infty}\inf\{\overline{H}(x,\,t,\,p) \mid (x,t,p)\in\mathbf{R}^N\times[0,\,\infty)\times\mathbf{R}^N,\; |p|\geq R\} = \infty.$$

The second inequality shows that $\overline{H}$ is bounded on the set $\mathbf{R}^N\times[0,\,\infty)\times B(0,\,R)$ for any $R>0$.

Next, we show that $\overline{H}$ is uniformly continuous on the set $\mathbf{R}^N\times[0,\,\infty)\times B(0,\,R)$ for any $R>0$.

To this end, we argue by contradiction, and thus we suppose that $\overline{H}$ were not uniformly continuous on $E\equiv\mathbf{R}^N\times[0,\,\infty)\times B(0,\,R)$ for some $R>0$. This means that there are a positive constant ε and sequences $\{(x_n,\,t_n,\,p_n)\}_{n\in\mathbf{N}},\ \{(y_n,\,s_n,\,q_n)\}_{n\in\mathbf{N}}\subset E$ such that

$$|x_n-y_n|+|t_n-s_n|+|p_n-q_n|\to 0 \quad\text{as}\quad n\to\infty \tag{5.2}$$

and

$$\overline{H}(x_n,\,t_n,\,p_n)\geq\overline{H}(y_n,\,s_n,\,q_n)+\varepsilon \quad\text{for all } n\in\mathbf{N}.$$

Set $\lambda_n = \overline{H}(x_n, t_n, p_n)$ and $\mu_n = \overline{H}(y_n, s_n, q_n)$. The problem

$$v_\tau(\xi, \tau) + H(x_n, t_n, \xi, \tau, p + Dv(\xi, \tau)) = \lambda_n \quad \text{in } \mathbf{R}^{N+1}$$

has a viscosity subsolution v_n and a viscosity supersolution w_n which are bounded and periodic with period $\mathbf{Z}^{N+1}$.

Reviewing carefully the proof of (4.4) and that of the existence part of Theorem 4.1, we see that we may assume that the families $\{v_n\}$ and $\{w_n\}$ are uniformly bounded on $\mathbf{R}^{N+1}$.

We can choose an increasing sequence $\{n_k\} \subset \mathbf{N}$ such that $\{\lambda_{n_k}\}$ and $\{\mu_{n_k}\}$ are convergent. Set $\lambda = \lim_{k\to\infty} \lambda_{n_k}$ and $\mu = \lim_{k\to\infty} \mu_{n_k}$. Clearly, we have that $\lambda \geq \mu + \varepsilon$.

By assumption (A1) the family of functions $H(x_n, t_n, \xi, \tau, p_n + q)$ of (ξ, τ, q) on $\mathbf{R}^N \times \mathbf{R} \times B(0, L)$ is uniformly bounded and equi-continuous for each $L > 0$. Thanks to the Ascoli-Arzela theorem, we may assume that as $k \to \infty$,

$$H(x_{n_k}, t_{n_k}, \xi, \tau, p_{n_k} + q) \to F(\xi, \tau, q)$$

uniformly on $\mathbf{R}^N \times \mathbf{R} \times B(0, L)$ for each $L > 0$ and for some function $F \in C(\mathbf{R}^{2N+1})$. From (5.2) we see that as $k \to \infty$,

$$H(y_{n_k}, s_{n_k}, \xi, \tau, q_{n_k} + q) \to F(\xi, \tau, q)$$

uniformly on $\mathbf{R}^N \times \mathbf{R} \times B(0, L)$ for each $L > 0$. From (A1) – (A4) it follows that F satisfies (C1) – (C4).

As in the last part of the proof of Theorem 4.1, we deduce that the problem

$$v_\tau(\xi, \tau) + F(\xi, \tau, Dv(\xi, \tau)) = \lambda \quad \text{in } \mathbf{R}^{N+1} \tag{5.3}$$

has a viscosity subsolution and a viscosity supersolution which are bounded and periodic with period $\mathbf{Z}^{N+1}$. Similarly, the same holds true for (5.3) with λ replaced by μ. This contradicts the uniqueness assertion of Theorem 4.1, which proves the required uniform continuity of $\overline{H}$.

It remains to show that the function $G(x, t, \xi, \tau, p) = \overline{H}(x, t, p)$ satisfies (A4). This assertion can be proved by an argument parallel to the above, the details of which we leave to the reader. □

Proof of Theorem 1.1. Let $u^\varepsilon \in \cap_{T>0} BUC(\mathbf{R}^N \times [0, T])$ be the viscosity solution of (1.1) – (1.2). Define functions $\overline{u}$ and $\underline{u}$ on $\mathbf{R}^N \times (0, \infty)$ by

$$\overline{u}(x, t) = \lim_{r \searrow 0} \sup\{u^\varepsilon(y, s) \mid 0 < \varepsilon < r,\ (y, s) \in B(x, t; r)\}$$

and

$$\underline{u}(x, t) = \lim_{r \searrow 0} \inf\{u^\varepsilon(y, s) \mid 0 < \varepsilon < r,\ (y, s) \in B(x, t; r)\}.$$

Recall that $\underline{u}_0(x) = \inf_{\xi \in \mathbf{R}^N} u_0(x, \xi)$.

The main part of the proof is divided into two propositions below.

Proposition 5.1 *There is a function $\omega \in C([0, \infty))$ satisfying $\omega(0) = 0$ such that for all $(x, t) \in \mathbf{R}^N \times (0, \infty)$,*

$$\underline{u}_0(x) - \omega(t) \le \underline{u}(x,t) \le \overline{u}(x,t) \le \underline{u}_0(x) + \omega(t).$$

Proof. It is clear by definition that $\underline{u} \le \overline{u}$ in $\mathbf{R}^N \times (0, \infty)$.

According to (A1), there is a function $m \in C([0, \infty))$ such that $H(x, t, \xi, \tau, p) \le m(|p|)$ for all $(x, t, \xi, \tau, p) \in \Gamma$.

It follows that, by (A5), $\underline{u}_0 \in BUC(\mathbf{R}^N)$ and that $u_0(x, \xi) \ge \underline{u}_0(x)$ for all $(x, \xi) \in \mathbf{R}^{2N}$.

Proposition 3.1 now ensures that there is a function $\omega \in C([0, \infty))$, with $\omega(0) = 0$, such that $u^\varepsilon(x,t) \ge \underline{u}_0(x) - \omega(t)$ for all $(x, t) \in \mathbf{R}^N \times (0, \infty)$ and $\varepsilon > 0$, whence

$$\underline{u}(x, t) \ge \underline{u}_0(x) - \omega(t) \quad \text{for all } (x, t) \in \mathbf{R}^N \times (0, \infty).$$

Next, by (A2) there is a function $f \in C([0, \infty))$ satisfying

$$\lim_{r \to \infty} f(r) = \infty$$

such that for all $(x, t, \xi, \tau, p) \in \Gamma$,

$$H(x, t, \xi, \tau, p) \ge f(|p|).$$

According to Theorem 3.1, there are a function T on $(0, \infty)$ and a constant $C > 0$, both of which depend only on f, such that

$$\lim_{r \searrow 0} T(r) = 0,$$

and such that for all $x \in \mathbf{R}^N$, $r > 0$, $t \ge T(r)$, and $\varepsilon > 0$,

$$u^\varepsilon(x,t) \le \inf_{y \in B(x,r)} u^\varepsilon(y, 0) + Ct.$$

Setting $r = \sqrt{N}\varepsilon$, we have

$$u^\varepsilon(x,t) \le \underline{u}_0(x) + Ct \quad \text{for all } x \in \mathbf{R}^N,\ t \ge T(\sqrt{N}\varepsilon), \text{ and } \varepsilon > 0.$$

This implies that $\overline{u}(x,t) \le \underline{u}_0(x) + Ct$. for all $(x, t) \in \mathbf{R}^N \times (0, \infty)$, completing the proof. □

Proposition 5.2 *The functions $\overline{u}$ and $\underline{u}$ are a viscosity subsolution and a viscosity supersolution of* (1.4), *respectively.*

Proof. We only give the proof that $\overline{u}$ is a viscosity subsolution of (1.4), and leave it to the reader to fill the rest of the proof.

Let $\varphi \in C(\mathbf{R}^N \times (0, \infty))$ and $(\hat{x}, \hat{t}) \in \mathbf{R}^N \times (0, \infty)$, and assume that $\overline{u} - \varphi$ attains a maximum at $(\hat{x}, \hat{t})$. We may moreover assume that

$$\overline{u} = \varphi \quad \text{at } (\hat{x}, \hat{t}) \quad \text{and} \quad \overline{u}(x, t) < \varphi(x, t) \quad \text{if } (x, t) \ne (\hat{x}, \hat{t}).$$

Set $\hat{p} = D\varphi(\hat{x}, \hat{t})$. Let v be a bounded viscosity supersolution of

$$v_\tau(\xi, \tau) + H(\hat{x}, \hat{t}, \xi, \tau, \hat{p} + Dv(\xi, \tau)) = \overline{H}(\hat{x}, \hat{t}, \hat{p}) \quad \text{in } \mathbf{R}^{N+1},$$

the existence of which is guaranteed by Theorem 4.1 (or equivalently, Proposition 1.2).

Fix $\gamma > 0$. Consider the inf-convolution of v in the τ variable, i.e.,

$$v_\delta(\xi, \tau) = \inf_{s \in \mathbf{R}} \left(v(\xi, s) + \frac{1}{2\delta}(\tau - s)^2 \right),$$

where $\delta > 0$ is a constant to be chosen. Indeed, we choose $\delta > 0$ small enough so that

$$\frac{\partial}{\partial \tau} v_\delta + H(\hat{x}, \hat{t}, \xi, \tau, \hat{p} + Dv_\delta) \geq \overline{H}(\hat{x}, \hat{t}, \hat{p}) - \gamma \quad \text{in } \mathbf{R}^{N+1} \tag{5.4}$$

in the viscosity sense. This is possible due to assumption (A4) and the well-known property that inf-convolutions approximate well supersolutions. Another nice property of inf-convolutions is that they make functions Lipschitz continuous, and hence we have

$$\left\| \frac{\partial}{\partial \tau} v_\delta \right\|_{L^\infty(\mathbf{R}^{N+1})} \leq C_\gamma \tag{5.5}$$

for some constant $C_\gamma > 0$. (C_γ can be chosen so as to depend only on δ and $\|v\|_{L^\infty}$, δ depends on γ, and after all, C_γ depends only on γ and $\|v\|_{L^\infty}$.) See [CIL2] for these basic properties of inf-convolutions.

Fix $\varepsilon > 0$ and $\alpha > 0$ we consider the function

$$\Phi(x, t, y, s) = u^\varepsilon(x, t) - \varphi(x, t) - \varepsilon v_\delta(y/\varepsilon, s/\varepsilon) - \alpha|x - y|^2 - \alpha(t - s)^2$$

defined for $(x, t) \in B(\hat{x}, \hat{t}; r)$ and $(y, s) \in \mathbf{R}^{N+1}$, where $r > 0$ is chosen so that $B(\hat{x}, \hat{t}; r) \subset \mathbf{R}^N \times (0, \infty)$.

The function Φ clearly attains a maximum. Let $(x_\alpha, t_\alpha, y_\alpha, s_\alpha)$ be a maximum point of Φ.

Using (5.5) and that

$$\Phi(x_\alpha, t_\alpha, y_\alpha, s_\alpha) \geq \Phi(x_\alpha, t_\alpha, y_\alpha, t_\alpha),$$

we have

$$\alpha\,(t_\alpha - s_\alpha)^2 \leq \varepsilon v_\delta(y_\alpha/\varepsilon, t_\alpha/\varepsilon) - \varepsilon v_\delta(y_\alpha/\varepsilon, s_\alpha/\varepsilon) \leq C_\gamma |t_\alpha - s_\alpha|,$$

i.e.,

$$\alpha |t_\alpha - s_\alpha| \leq C_\gamma. \tag{5.6}$$

We can choose a sequence $\alpha_j \to \infty$ so that as $\alpha = \alpha_j \to \infty$,

$$(x_\alpha, t_\alpha, y_\alpha, s_\alpha) \to (x_\varepsilon, t_\varepsilon, x_\varepsilon, t_\varepsilon) \quad \text{and} \quad q_\alpha \equiv 2\alpha(t_\alpha - s_\alpha) \to q_\varepsilon$$

for some $(x_\varepsilon, t_\varepsilon) \in B(\hat{x}, \hat{t}; r)$ and $q_\varepsilon \in \mathbf{R}$, with $|q_\varepsilon| \le 2C_\gamma$.

It is easy to see that $(x_\varepsilon, t_\varepsilon)$ is a maximum point of the function

$$\Psi(x,t) \equiv u^\varepsilon(x,t) - \varphi(x,t) - \varepsilon v_\delta(x/\varepsilon, t/\varepsilon)$$

on $B(\hat{x}, \hat{t}; r)$. According to the definition of the function $\overline{u}$, there is a sequence $\varepsilon_k \searrow 0$ such that as $\varepsilon = \varepsilon_k \to 0$,

$$(x_\varepsilon, t_\varepsilon) \to (\hat{x}, \hat{t}) \quad \text{and} \quad q_\varepsilon \to \hat{q}$$

for some $\hat{q} \in \mathbf{R}$.

Fix such $\varepsilon = \varepsilon_k$ and $\alpha = \alpha_j$. (Notice that the choice of α_j depends on ε as well.) We may assume that $(x_\alpha, t_\alpha) \in \operatorname{int} B(\hat{x}, \hat{t}; r)$.

Now, since u^ε is a viscosity solution of (1.1), we have

$$\varphi_t(x_\alpha, t_\alpha) + q_\alpha + H(x_\alpha, t_\alpha, x_\alpha/\varepsilon, t_\alpha/\varepsilon, p_\alpha + r_\alpha) \le 0, \tag{5.7}$$

where

$$p_\alpha = D\varphi(x_\alpha, t_\alpha) \quad \text{and} \quad r_\alpha = 2\alpha(x_\alpha - y_\alpha).$$

Since v_δ is a viscosity supersolution of (5.4), we have

$$q_\alpha + H(\hat{x}, \hat{t}, y_\alpha/\varepsilon, s_\alpha/\varepsilon, \hat{p} + r_\alpha) \ge \overline{H}(\hat{x}, \hat{t}, \hat{p}) - \gamma. \tag{5.8}$$

By (5.6), (5.7) and (A2), we find a constant $M_\gamma > 0$ independent of ε and α such that

$$|r_\alpha| \le M_\gamma.$$

Hence, we may assume that as $\alpha = \alpha_j \to \infty$, $r_\alpha \to r_\varepsilon$ for some $r_\varepsilon \in \mathbf{R}^N$, with $|r_\varepsilon| \le M_\gamma$. Sending $\alpha = \alpha_j \to \infty$, we get

$$\varphi_t(x_\varepsilon, t_\varepsilon) + q_\varepsilon + H(x_\varepsilon, t_\varepsilon, x_\varepsilon/\varepsilon, t_\varepsilon/\varepsilon, p_\varepsilon + r_\varepsilon) \le 0,$$

where $p_\varepsilon = D\varphi(x_\varepsilon, t_\varepsilon)$, and

$$q_\varepsilon + H(\hat{x}, \hat{t}, x_\varepsilon/\varepsilon, t_\varepsilon/\varepsilon, \hat{p} + r_\varepsilon) \ge \overline{H}(\hat{x}, \hat{t}, \hat{p}) - \gamma.$$

Subtracting the latter inequality from the former and using (A1), if $\varepsilon = \varepsilon_k$ is small enough then we have

$$\varphi_t(x_\varepsilon, t_\varepsilon) \le -\overline{H}(\hat{x}, \hat{t}, \hat{p}) + 2\gamma.$$

Sending $\varepsilon = \varepsilon_k \to 0$, we have

$$\varphi_t(\hat{x}, \hat{t}) + \overline{H}(\hat{x}, \hat{t}, \hat{p}) \le 2\gamma.$$

Because of the arbitrariness of $\gamma > 0$, we conclude that $\varphi_t(\hat{x}, \hat{t}) + \overline{H}(\hat{x}, \hat{t}, \hat{p}) \le 0$ and that $\overline{u}$ is a viscosity subsolution of (1.1). □

Completion of the proof. Let $u \in \cap_{T>0} BUC(\mathbf{R}^N \times [0, T])$ be the viscosity solution of (1.4) – (1.5). Using a comparison result (see, e.g., [I2, Bl]), we see from Propositions 5.1 and 5.2 that $u(x, t) = \overline{u}(x, t) = \underline{u}(x, t)$ for all $(x, t) \in \mathbf{R}^N \times (0, \infty)$. This means that as $\varepsilon \searrow 0$, $u^\varepsilon(x, t) \to u(x, t)$ uniformly on $\mathbf{R}^N \times [\delta, T]$ for each $0 < \delta < T$, which completes the proof. □

Acknowledgments: This research was supported in part by Grant-in-Aid for Scientific Research (No. 0944067), the Ministry of Education, Science, Sports and Culture, Japan.

References

[Al] Alvarez, O. Homogenization of Hamilton-Jacobi equations in perforated sets. preprint.

[Ar] Arisawa, M. Ergodic problem for the Hamilton-Jacobi-Bellman equation. II, to appear in Ann. Inst. H. Poincaré, Anal. Non Linéaire.

[Bl] Barles, G. (1990) Uniqueness and regularity results for first-order Hamilton-Jacobi equations. Indiana Univ. Math. J., **39**, 443–466.

[Br] Barron, E. N. (1993) Averaging in Lagrange and minimax problems of optimal control. SIAM J. Control Optim. **31**, 1630–1652.

[BP] Barles, G. and Perthame, B. (1987) Discontinuous solutions of deterministic optimal stopping-time problems, Modèl. Math. Anal. Num., **21**, 557–579.

[BLP] Bensoussan, A. Lions, J.-L., and Papanicolaou, G. (1978) *Asymptotic Analysis for Periodic Structures*, North Holland, Amsterdam-New York-Oxford.

[Ch] Chaplais, F. (1987) Averaging and deterministic optimal control. SIAM Control Optim., **25**, 767–780.

[Co1] Concordel, M. C. (1996) Periodic homogenization of Hamilton-Jacobi equations: 1. Additive eigenvalues and variational formula, Indiana Univ. Math. J., **45**, 1095–1117.

[Co2] Concordel, M. C. (1997) Periodic homogenization of Hamilton-Jacobi equations: 2. Eikonal equations, Proc. Roy. Soc. Edinburgh Sect. A, **127**, 665 – 689.

[CIL1] Crandall, M. G., Ishii, H. and Lions, P.-L. (1987) Uniqueness of viscosity solutions of Hamilton-Jacobi equations revisited, J. Math. Soc. Japan, **39**, 581–596.

[CIL2] Crandall, M. G., Ishii, H. and Lions, P.-L. (1992) User's guide to viscosity solutions of second order partial differential equations, Bull. Amer. Math. Soc. (N.S.), **27**, 1–67.

[E] E, W. (1991) A class of homogenization problems in the calculus of variations, Comm. Pure Appl. Math., **44**, 733 – 759.

[EK] Elliott, R. J. and Kalton, N. K. (1972) *The existence of value in differential games*, Memoirs of Amer. Math. Soc., No. 126, Amer. Math. Soc., Providence, Rhode Island.

[Ev1] Evans, L. C. (1989) The perturbed test function technique for viscosity solutions of partial differential equations, Proc. Roy. Soc. Edinburgh Sect. A, **111**, 359–375.

[Ev2] Evans, L. C. (1992) Periodic homogenisation of fully nonlinear partial differential equations, Proc. Roy. Soc. Edinburgh Sect. A, **120**, 245–265.

[ES] Evans, L. C. and Souganidis, P. E. (1984) Differential games and representation formulas for solutions of Hamilton-Jacobi equations, Indiana Univ. Math. J., **33**, 773–797.

[H] Horie, K. (1997) Homogenization of Hamilton-Jacobi equations on perforated domains, Master's thesis.

[HI] Horie, K. and Ishii, H. Homogenization of Hamilton-Jacobi equations on domains with small scale periodic structure, to appear in Indiana Univ. Math. J.

[I1] Ishii, H. (1988) Representation of solutions of Hamilton-Jacobi equations. Nonlinear Anal., **12**, 121–146.

[I2] Ishii, H. Comparison results for Hamilton-Jacobi equations without growth condition on solutions from above, to appear in Appl. Anal.

[LPV] Lions, P.-L., Papanicolaou, G. and Varadhan, S. R. S. Homogenization of Hamilton-Jacobi equations, preprint.

19

The Critical Exponent for a Stochastic PDE to Hit Zero

Carl Mueller
Etienne Pardoux

ABSTRACT: Earlier, the first author had studied the stochastic partial differential equation $u_t = u_{xx} + u^{-\alpha} + g(u)\dot{W}(t,x)$ with the initial function $u(0,x) = u_0(x)$ continuous and strictly positive. Here, x lies on a circle and $\dot{W}(t,x)$ is 2-parameter white noise. $g(x)$ satisfies a growth condition. It was shown that if $\alpha > 3$, then $\inf_x u(t,x)$ never reaches 0. In this paper, we show that for $0 < \alpha < 3$, $\inf_x u(t,x)$ has a positive probability of reaching 0 in finite time. Thus $\alpha = 3$ is the critical parameter for u to hit 0.

KEYWORDS: stochastic partial differential equations; white noise; singular drift.

1 Introduction

Let $\mathbf{S}$ be the interval $[0, J]$ with endpoints identified. In other words, $\mathbf{S} = \mathbf{R}/(J\mathbf{Z})$, where $\mathbf{R}$ denotes the real line, and $\mathbf{Z}$ the set of positive and negative integers. The following stochastic partial differential equation (SPDE) was considered in Mueller (1998).

$$\begin{aligned} u_t &= u_{xx} + u^{-\alpha} + g(u)\dot{W}(t,x) \qquad t \geq 0, x \in \mathbf{S}; \\ u(0,x) &= u_0(x). \end{aligned} \tag{1.1}$$

Here $\dot{W}(t,x)$ stands for the "space–time white noise". A rigorous formulation of our SPDE, together with a precise definition of the noise term will be given below. The main point of Mueller (1998) was to show that, if $\tilde{u}(t,x)$ satisfies

$$\begin{aligned} \tilde{u}_t &= \tilde{u}_{xx} + g(\tilde{u})\dot{W}(t,x) \qquad t \geq 0, x \in \mathbf{S}; \\ \tilde{u}(0,x) &= u_0(x). \end{aligned}$$

and

$$|g(r)| \leq C(1 + |r|^{\gamma}) \tag{1.2}$$

holds for some constant $C > 0$ and for some $\gamma \in (0, 3/2)$, then with probability 1, $|\tilde{u}(t,x)|$ does not blow up to ∞ in finite time. As a byproduct, the following Proposition was also shown.

Proposition 1 *Let $u(t,x)$ satisfy (1.1). Suppose that $u_0(x)$ is a strictly positive continuous function on $\mathbf{S}$, and suppose that $\alpha > 3$. Assume that (1.2) holds for some constant C and some $\gamma \in (0, 3/2)$. Then with probability 1, $u(t,x) > 0$ for all $t \geq 0$, $x \in \mathbf{S}$.*

Proposition 1 raises the question of whether $\alpha = 3$ is the critical exponent for $u(t,x)$ to reach 0. Our main theorem gives a positive answer to this question, in the case where the diffusion coefficient g is bounded away from zero.
More precisely, we assume that for some $c, C > 0$, $\gamma \in (0, 3/2)$,

$$c \leq |g(r)| \leq C(1 + |r|^\gamma), \tag{1.3}$$

and that the mapping $r \to g(r)$ is locally Lipschitz.
It follows from the results in Mueller (1998) that equation (1.1) has a unique solution on the random time interval $[0, \tau]$, where

$$\tau = \inf\{t > 0 \; ; \; \inf_{x \in \mathbf{S}} u(t,x) = 0\},$$

provided the initial condition u_0 satisfies $\inf_{x \in S'} u_0(x) > 0$. The aim of this paper is to show the

Theorem 1 *Suppose $u_0 \in C(\mathbf{S})$ and $\inf_{x \in S} u_0(x) > 0$, and let $0 < \alpha < 3$. If u solves equation (1.1), then $P(\tau < \infty) > 0$. In other words, the solution of equation (1.1) hits 0 in finite time with positive probability.*

We now introduce some notation. We let

$$\bar{G}(t,x) = (4\pi t)^{-1/2} \exp\left(-\frac{x^2}{4t}\right).$$

Recall that $\bar{G}(t,x)$ is the fundamental solution of the heat equation on $\mathbf{R}$. Next, let $G(t,x,y)$ be the fundamental solution of the heat equation on $\mathbf{S}$, and recall that

$$G(t,x,y) = \sum_{m \in \mathbf{Z}} \bar{G}(t, \rho(x,y) + mJ) \tag{1.4}$$

where $\rho(x,y)$ is the shortest distance from x to y on $\mathbf{S}$.
A rigorous framework for equations such as (1.3) appears in Walsh's notes Walsh (1986), chapter 3. We regard (1.1) as a shorthand for the following integral equation.

$$\begin{aligned} u(t,x) &= \int_{\mathbf{S}} G(t,x,y) u_0(y) dy + \int_0^t \int_{\mathbf{S}} G(t-s,x,y) u(s,y)^{-\alpha} dy\, ds \\ &\quad + \int_0^t \int_{\mathbf{S}} G(t-s,x,y) g(u(s,y)) W(dy,ds) \end{aligned} \tag{1.5}$$

Since we will often refer to the final integral in (1.5), we give it a name.

$$N(t,x) = \int_0^t \int_{\mathbf{S}} G(t-s,x,y)g(u(s,y))W(dy,ds),$$

where $\{W(A),\ A \in \mathcal{B}(\mathbf{R}_+ \times \mathbf{S})\}$ is a generalized zero mean Gaussian random field, such that $E[W(A)W(B)] = m(A \cap B)$, where m denotes Lebesgue's measure, and $\mathcal{B}(\mathbf{R}_+ \times \mathbf{S})$ stands for the class of Borel subsets of $\mathbf{R}_+ \times \mathbf{S}$. In fact, $N(t,x)$ is a white noise integral in the sense of Walsh (1986), Chapter 2, which means that it is a kind of Itô's integral.

Existence and uniqueness of solutions to equations such as (1.1) and (1.5) are proved in Walsh (1986), Exercise 3.4. Actually, Walsh's results deal with $x \in \mathbf{R}$, but it is easy to extend his proofs to the case $x \in \mathbf{S}$. However, since neither $u^{-\alpha}$ nor $g(u)$ is a Lipschitz function of u, we cannot use Walsh's result directly. Instead, we truncate and take limits to obtain existence and uniqueness for (1.3) up to the first time τ that u hits zero, and we use Mueller's result in Mueller (1998) in order to show that u does not explode. Let τ_n be the first time t that $\inf_{x \in \mathbf{S}} u(t,x) \le 1/n$,and let

$$f_n(u) = [\max(u, 1/n)]^{-\alpha}.$$

Consider the following equation.

$$\begin{aligned} u^{(n)}(t,x) &= \int_{\mathbf{S}} G(t,x,y)u_0(y)dy + \int_0^t \int_{\mathbf{S}} G(t-s,x,y)f_n(u^{(n)}(s,y))dy\,ds \\ &\quad + \int_0^t \int_{\mathbf{S}} G(t-s,x,y)g(u^{(n)}(s,y))W(dy,ds) \end{aligned} \tag{1.6}$$

Then, as shown in Mueller (1998), (1.6) has a unique solution $u^{(n)}(t,x)$. Since $f_n(u) = u^{-\alpha}$ for $u \ge 1/n$, it follows that $u(t,x) = u^{(n)}(t,x)$ for $t \le \tau_n$, where $u(t,x)$ is any solution to (1.1). Now $\lim_{n\to\infty} \tau_n = \tau$. Then we get existence and uniqueness for (1.1) for $t < \tau$.

Alternatively, we could get existence and uniqueness in law for (1.1) from Girsanov's theorem for space–time white noise, which appears as Lemma 2 in the next section. In fact, our proof of theorem 1 is based on Girsanov's theorem. First we use a Girsanov transformation to remove the drift term

$$\int_0^t \int_{\mathbf{S}} G(t-s,x,y)f(u(s,y))dy\,ds$$

from (1.5). The solution of the remaining SPDE is then shown to hit zero in finite time a.s.

The heart of the proof is to show that this Girsanov transformation leads to an absolutely continuous change of measure, at least up to time τ.

2 Proof of Theorem 1

In this section, we will refer to solutions of several different SPDE. Each of these SPDE will be given in differential form, as in (1.1). The rigorous meaning is always given by an integral equation like (1.5). For the SPDE we consider, existence and uniqueness is established as for (1.1). Let $\mathcal{F}_t$ denote the σ-field generated by $\dot{W}(s,y)$ for $0 \leq s \leq t$ and $y \in \mathbf{S}$. More precisely, $\mathcal{F}_t$ is the σ-field generated by the random variables $\int_0^t \int_{\mathbf{S}} \varphi(s,y) W(dy,ds)$ for all continuous functions φ. Finally, we will often use c, C to denote strictly positive constants which are small and large, respectively. Their values may change from line to line, but they will never depend on the parameter n, which we define later. As a final piece of notation, θ_s will denote the time shift, so that $\theta_s W(dx,dt) = W(dx, d(s+t))$ (with s fixed). Finally, if X is a random variable and A is an event, then $E[X;A]$ means $E[X\mathbf{1}_A]$.
Our strategy is to consider the equation with no drift:

$$\begin{aligned} v_t &= v_{xx} + g(v)\dot{W}(t,x) \qquad t \geq 0, x \in \mathbf{S}; \\ v(0,x) &= u_0(x). \end{aligned} \tag{2.1}$$

We can solve for $v(t,x)$ as follows

$$v(t,x) = \int_{\mathbf{S}} G(t,x,y) u_0(y) dy + N(t,x),$$

where

$$N(t,x) = \int_0^t \int_{\mathbf{S}} G(t-s,x,y) g(v(s,y)) W(dy,ds).$$

We note that uniqueness for our SPDE implies that $h(t,x) = v(s+t,x)$ satisfies the following SPDE :

$$\begin{aligned} h_t &= h_{xx} + g(h)\theta_s \dot{W}(t,x), \qquad t \geq 0, x \in \mathbf{S}\,; \\ h(0,x) &= v(s,x). \end{aligned}$$

It is easy to see that

Lemma 1 *Suppose that $v(t,x)$ satisfies (2.2). Then*

$$P\left(v(t,x) = 0 \textit{ for some } t > 0, x \in \mathbf{S}\right) = 1.$$

Proof. Let

$$\bar{v}_t = \int_{\mathbf{S}} v(t,x)\, dx.$$

It suffices to show that

$$\inf_{t \geq 0} \bar{v}_t < 0.$$

But, since $\int_{\mathbf{S}} G(t,x,y)\,dx = J$,

$$\bar{v}_t = \int_{\mathbf{S}} u_0(x)\,dx + J\int_0^t\int_{\mathbf{S}} g(v(s,y))\,W(dy,ds),$$

and the second term in the above right hand side is a continuous martingale, with increasing process given by

$$\langle \int_0^{\cdot}\int_{\mathbf{S}} g(v(s,y))\,W(dy,ds)\rangle_t = \int_0^t\int_{\mathbf{S}} g^2(v(s,y))\,dyds \geq Jc^2 t.$$

It then follows from a well-known result –see theorem V.1.6 in Revuz–Yor (1991)– that there exists a one dimensional standard Brownian motion $\{B_t,\, t \geq 0\}$ such that the above martingale is a time–changed of B, hence

$$\inf_{t\geq 0}\int_0^t\int_{\mathbf{S}} g(v(s,y))\,W(dy,ds) = \inf_{t\geq 0} B_t = -\infty.$$

◇

Let

$$\alpha_n = \inf\{t > 0;\int_0^t\int_{\mathbf{S}} v^{-2\alpha}(s,x)dxds > n\},$$

and for $T > 0, T_n \triangleq \tau_v \wedge \alpha_n \wedge T$, where

$$\tau_v = \inf\{t > 0; \inf_{x\in\mathbf{S}} v(t,x) \leq 0\}.$$

Given a function $h : [0,\infty)\times\mathbf{S}$, let h^{T_n} denote the truncated function

$$h^{T_n}(t,x) = h(t,x)\mathbf{1}_{\{t\leq T_n\}}.$$

Let $P_u^{T_n}$, $P_v^{T_n}$ denote the measures on the space of continuous functions $f : [0,\infty)\times\mathbf{S}\to\mathbf{R}$ induced by $u^{T_n}(t,x)$, $v^{T_n}(t,x)$ respectively. The following Girsanov theorem is proved as the "usual one" (for a similar statement, see e.g. Nualart, Pardoux (1994). Let $h(r) = g^{-1}(r)r^{-\alpha}$.

Lemma 2 *For each $n\in N$, $P_u^{T_n}$ is absolutely continuous with respect to $P_v^{T_n}$, and*

$$\frac{dP_u^{T_n}}{dP_v^{T_n}} = \exp\left(\int_0^{T_n}\int_{\mathbf{S}} h(v(t,x))W(dx,dt) - \frac{1}{2}\int_0^{T_n}\int_{\mathbf{S}} h(v(t,x))^2dxdt\right).$$

We claim that the following lemma is enough to prove Theorem 1.

Lemma 3 *For any $T > 0$*

$$\int_0^{T_v}\int_{\mathbf{S}} h(v(t,x))^2dxdt < \infty \quad a.s.$$

In view of condition (1.2), Lemma 3 is equivalent to the fact that for all $T > 0$,

$$\int_0^{\tau_v \wedge T} \int_{\mathbf{S}} v(t,x)^{-2\alpha} dx dt < \infty \ a.s.$$

If this last condition holds, then $P\left(\bigcup_{n\in N, T>0} \Omega_{n,T}\right) = 1$, where $\Omega_{n,T} = \{\alpha_n \wedge T \geq \tau_v\}$. But on each $\Omega_{n,T}$, $\inf_{x\in S} u(T_n, x) = 0 \ P_u^{T_n} a.s.$ It thus remains to proceed to the

Proof of Lemma 3.

Since $0 < \alpha < 3$, we can choose a constant $\varepsilon > 0$ so small that

$$0 < \varepsilon < \frac{3-\alpha}{2}. \tag{2.2}$$

For each $K > 0$, define

$$A(K) \triangleq \left\{ \sup_{(t,x)\in[0,T]\times\mathbf{S}} v(t,x) \leq K \right\}.$$

Since $\{v(t,x), (t,x) \in \mathbf{R}_+ \times \mathbf{S}\}$ is a.s continuous, $\sup_{(t,x)\in[0,T]\times\mathbf{S}} v(t,x)$ is a.s. finite, and

$$\lim_{K\to\infty} P(A(K)) = 1.$$

On the set $A(K)$,

$$\begin{aligned} &\int_0^{\tau_v \wedge T} \int_{\mathbf{S}} v(t,x)^{-2\alpha} dx dt \\ \leq\ & \sum_{n=0}^{\infty} 2^{2\alpha(n+1)} \\ &\times K^{-2\alpha} m\left(\{(t,x) \in [0, \tau_v \wedge T] \times S; 2^{-n-1}K < v(t,x) \leq 2^{-n}K\}\right), \end{aligned} \tag{2.3}$$

where m denotes Lebesgue's measure.

We need to estimate the right hand side of (2.3). For that sake, we introduce some notation. For each $n \in \mathbf{N}^*$, let

$$D_n = \{(t,x); 0 \leq t \leq 2^{-(4-2\varepsilon)n}, 0 \leq x \leq 2^{-(2-2\varepsilon)n}\},$$

and for each $(t,x) \in D_n$, we define the grids

$$\begin{aligned} \Gamma_n(t,x) &= \Big[[0,T] \times \mathbf{S}\Big] \bigcap \Big[\bigcup_{k,l\in\mathbf{N}} (t + k2^{-(4-2\varepsilon)n}, x + l2^{-(2-2\varepsilon)n})\Big], \\ \bar{\Gamma}_n(t,x) &= \Big[[0,\tau_v] \times S\Big] \cap \Gamma_n(t,x). \end{aligned}$$

Then we have that

$$m(\{(t,x)\in[0,\tau_v\wedge T]\times\mathbf{S};\ 2^{-n-1}K<v(t,x)\le 2^{-n}K\})$$
$$\le\int\int_{D_n}\#\{(s,y)\in\bar\Gamma_n(t,x)\ ;\ v(s,y)\le 2^{-n}K\}dxdt \tag{2.4}$$

where # denotes the number of points in a set.
Let (t,x) be an arbitrary point in D_n. For $(s,y)\in\Gamma_n(t,x)$, we have

$$v(s,y)=\int_{\mathbf{S}}G(2^{-(4-2\varepsilon)n},y,z)v(s_n^-,z)dz$$
$$+\int_0^{2^{-(4-2\varepsilon)n}}\int_{\mathbf{S}}G(2^{-(4-2\varepsilon)n}-r,y,z)g(v(s_n^-+r,z))\theta_{s_n^-}W(dz,dr),$$

with $s_n^-=s-2^{-(4-2\varepsilon)n}$. Set $\lambda_n=2^{-(4-2\varepsilon)n}$, $\pi_n=2^{-(2-2\varepsilon)n-1}$.

$$\begin{aligned}
v(s,y)&=V_n(s,y)+N_n(s,y)+E_n(s,y),\\
&\text{where}\\
V_n(s,y)&=\int_{\mathbf{S}}G(\lambda_n,y,z)v(s_n^-,z)dz\\
N_n(s,y)&=\int_0^{\lambda_n}\int_{\{|z-y|\le\pi_n\}}G(\lambda_n-r,y,z)g(v(s_n^-,y))\theta_{s_n^-}W(dz,dr)\\
E_n(s,y)&=\int_0^{\lambda_n}\int_{\mathbf{S}}G(\lambda_n-r,y,z)g(v(r,z))\theta_{s_n^-}W(dz,dr)-N_n(s,y)
\end{aligned}$$

Let $A_n=A(K)\bigcap B_n\bigcap C_n$, where

$$B_n=\left\{\sup_{(s,y)\in\Gamma_n(t,x)}|E_n(s,y)|\le 2^{-n}\right\},$$

$$C_n=\left\{\sup_{(s,y)\in\Gamma_n(t,x)}|N_n(s,y)|\le 2^{-(1-\varepsilon)n}\right\}.$$

Let us admit for a moment the following two lemmas.

Lemma 4 *The following holds :*

$$\sum_{n=1}^{\infty}P(B_n^c)<\infty,\qquad \sum P(C_n^c)<\infty.$$

Lemma 5 *For each $K>0$, there exists a constant c_K such that for each $n\in\mathbf{N},(t,x)\in D_n$,*

$$E\left[\#\{(s,y)\in\bar\Gamma_n(t,x);v(s,y)\le 2^{-n}K\};A_n\right]\le c_K.$$

We note that Lemma 3 is a consequence of the fact for each $K > 0$,

$$\mathbf{1}_{A(K)} \int_0^{\tau_v \wedge T} \int_{\mathbf{S}} v(t,x)^{-2\alpha} dx dt < \infty \ a.s.$$

Hence the result follows from (2.3), (2.4), Lemma 4, Lemma 5, the facts

$$\int_{D_n} \int dx dt = 2^{-(6-4\varepsilon)n}, 6 - 4\varepsilon > 2\alpha,$$

and the following

Lemma 6 *Let $\{X_n, n \in \mathbf{N}\}$ be a sequence of $\mathbf{R}_+$–valued random variables, and $\{F_n, n \in N\}$ a sequence of events, such that*

(i) $\sum_{n=0}^{\infty} P(F_n^c) < \infty$,

(ii) $\sum_{n=0}^{\infty} E(X_n; F_n) < \infty$.

then $\sum_{n=0}^{\infty} X_n < \infty \ a.s.$

Proof : Let $F = \{\sum_{n=0}^{\infty} X_n = +\infty\}$. On the set $F \bigcap \{\liminf F_n\}$,

$$\sum_{n=0}^{\infty} X_n \mathbf{1}_{F_n} = +\infty.$$

Then (ii) implies that $P(F \bigcap \{\liminf F_n\}) = 0$.
But (i) and Borel Cantelli tell us that

$$P(\limsup F_n^c) = 0,$$
hence
$$P(\liminf F_n) = 1.$$

Consequently $P(F) = 0$, which is our result. ◇

It remains to prove Lemma 4 and Lemma 5.
Proof of Lemma 4 : We first estimate $P(B_n^c)$.

$$E_n = E_n' + E_n'', \text{ where}$$

$$E_n' = \int_0^{\lambda_n} \int_{\{|z-y| \leq \pi_n\}} G(\lambda_n - r, y, z)[g(v(r,z)) - g(v(s_n^-, y))]\theta_{s_n^-} W(dz, dr)$$

$$E_n'' = \int_0^{\lambda_n} \int_{\{|z-y|>\pi_n\}} G(\lambda_n - r, y, z) g(v(r,z)) \theta_{s_n^-} W(dz, dr)$$

$$\Delta_n(s,y) = \{r \in I\!R_+, z \in S; |r-s| \le \lambda_n, |z-y| \le \pi_n\}.$$

$$E(E_n'^{2p}) \le C \sup_{(r,z)\in\Delta_n(s,y)} E\left(|v(r,z) - v(s,y)|^2\right)^p \left(\int_0^{\lambda_n} \int_{\mathbf{S}} G^2(r,y,z) dz dr\right)^p.$$

From the computation in Walsh (1986), page 320,

$$\sup_{(r,z)\in\Delta_n(s,y)} E|v(r,z) - v(s,y)|^{2p} \le c\ 2^{-2(1-\varepsilon)n(p-4)}$$

and

$$\begin{aligned}
\int_0^{\lambda_n} \int_{\mathbf{S}} G^2(r,y,z) dz dr &= \int_0^{\lambda_n} G(2r,y,y) dr \\
&= \sum_{m\in\mathbf{Z}} \int_0^{\lambda_n} \bar{G}(2r, mJ) dr \\
&\le \int_0^{\lambda_n} \bar{G}(2r,0) dr + \frac{1}{J} \int_0^{\lambda_n} \int_{\mathbf{R}} \bar{G}(2r,x) dx dr \\
&= c 2^{-(2-\varepsilon)n} + \frac{1}{J} 2^{-(4-2\varepsilon)n}.
\end{aligned}$$

Finally

$$\begin{aligned}
E\left(E_n'^{2p}\right) &\le c\, 2^{-4(1-\varepsilon)n(p-2)} \\
E\left(E_n''^{2p}\right) &\le c \left(\int_0^{\lambda_n} \int_{\{|z-y|>\pi_n\}} G^2(r,y,z) dz dr\right)^p \\
&\le c(K) \left(\int_0^{\lambda_n} \int_{\{|x|>\pi_n\}} t^{-1} e^{-x^2/2t} dx dt\right)^p \\
&\le c \left(\int_0^{\lambda_n} \int_{\{|x|>\pi_n\}} 2^{(2-2\varepsilon)n+1} \frac{x}{t} e^{-x^2/2t} dx dt\right)^p \\
&= 4c 2^{2(1-\varepsilon)pn} \left(\int_0^{\lambda_n} \exp\left(-\frac{2^{-(4-4\varepsilon)pn}}{8t}\right) dt\right)^p \\
&\le 4c 2^{2(1-\varepsilon)pn} \left(\int_0^{\lambda_n} \frac{e^{-(8-4\varepsilon)n}}{t^2} \exp\left(-\frac{2^{-(4-4\varepsilon)n}}{8t}\right) dt\right)^p \\
&\le 2^5 c 2^{-2(1+\varepsilon)pn} \left(\int_0^{\lambda_n} \frac{e^{-(4-4\varepsilon)n}}{8t^2} \exp\left(-\frac{2^{-(4-4\varepsilon)n}}{8t}\right) dt\right)^p \\
&\le 2^5 c 2^{-2(1+\varepsilon)pn} e^{-2^{2\varepsilon pn/8}}.
\end{aligned}$$

ε being fixed, for n large enough

$$E\left(E_n^{2p}\right) \le c2^{-4(1-\varepsilon)n}(p-2).$$

Now

$$B_n^c = \bigcup_{(s,y)\in\Gamma_n(t,x)} \{|E_n(s,y)| > 2^{-n}\},$$

and consequently for each $p \ge 1$,

$$P(B_n^c) \le c2^{(6-4\varepsilon)n}P\left(|E_n(s,y)| > 2^{-n}\right) \le c2^{-[2-4\varepsilon]np+[14-12\varepsilon]n},$$

and the first statement of Lemma 4 follows upon choosing $p > (7-6\varepsilon)/(1-2\varepsilon)$.
Similarly,

$$\begin{aligned} P(C_n^c) &\le c2^{(6-4\varepsilon)n}P(|N_n(s,y)| > 2^{-(1-\varepsilon)n}) \\ &\le c2^{(6-4\varepsilon)n}P\left(|X| > 2^{\frac{\varepsilon}{2}n}\right) \\ &\le c2^{(6-4\varepsilon)n}e^{-2^{\varepsilon n-1}}, \end{aligned}$$

where X stands for a $N(0,1)$ random variable. The second statement of Lemma 4 follows.

Proof of Lemma 5

It follows from the definition of A_n that

$$\begin{aligned} \xi_n &\triangleq E\Big[\#\{(s,y) \in \bar{\Gamma}_n(t,x); v(s,y) \le 2^{-n}K\}; A_n\Big] \\ &\le E\Big[\#\{(s,y) \in \bar{\Gamma}_n(t,x); 0 < v(s,y) \le 2^{-n}K; \\ &\qquad |E_n(s,y)| \le 2^{-n}; |N_n(s,y)| \le 2^{-(1-\varepsilon)n}\}\Big] \\ &\le E\Big[\#\{(s,y) \in \bar{\Gamma}_n(t,x); -2^{-n} < \bar{v}_n(s,y) \le 2^{-n}(K+1); \\ &\qquad |N_n(s,y)| \le 2^{-(1-\varepsilon)n}\}\Big] \\ &\le E\Big[\#\{(s,y) \in \bar{\Gamma}_n(t,x); \\ &\qquad -2^{-n} < \bar{v}_n(s,y) \le 2^{-n}(K+1); 0 \le V_n(s,y) \le \delta 2^{-(1-\varepsilon)n}\}\Big], \end{aligned}$$

where $\bar{v}_n(s,y) = V_n(s,y) + N_n(s,y)$, for some $\delta > 1$.
We order the points in $\Gamma_n(t,x)$ lexicographically, calling the i-th point (s_i, x_i). This means that if $i < j$ then $s_i \le s_j$, and if $s_i = s_j, x \le x_i < x_j$ (mod J).

Let $\mathcal{F}_i^n$ denote the σ–algebra generated by the white noise $\dot{W}$ in the set

$$\Delta_i^n = \Delta^n(s_i, y_i),$$

where

$$\Delta^n(s,x) = \mathbf{S} \times [0, s_n^-],$$

and for any $y \neq x$ such that $(s,y) \in \Gamma_n(t,x)$,

$$\Delta^n(s,y) = \mathbf{S} \times [0, s_n^-] \bigcup [x - \pi_n, y - \pi_n] \times [s_n^-, s].$$

Note that $[x - \pi_n, y - \pi_n]$ is the usual notation whenever $x - \pi_n < y - \pi_n$, and denotes $[x - \pi_n, J] \bigcup [0, y - \pi_n]$ whenever $y - \pi_n \leq x - \pi_n$.
Now for each $i = 1, 2, \ldots$, $V_n(s_i, y_i)$ is $\mathcal{F}_i^n$ measurable, and

$$N_n(s_i, y_i) = g(v((s_i)_n^-, y_i))c_n U_i,$$

where

$$c_n^2 = \int_0^{\theta n} \int_{\{|z-y| \leq \pi_n\}} G^2(r, y, z) dy dr,$$

and U_i is a Gaussian random variable with zero mean and unit variance, independent of $\mathcal{F}_i^n$. It follows from some computations in the proof of Lemma 4 and the fact $\lambda_n^{-1/2}\pi_n \to \infty$, that

$$c_n = \bar{c}_n 2^{-(1-\varepsilon/2)n},$$

where $\bar{c}_n \to \bar{c} > 0$ as $n \to \infty$.
Let $\sigma_n(k)$ denote the k–th integer i such that both $\bar{v}_n(s_i, y_i) \leq 2^{-n}(K+1)$ and $0 \leq V_n(s_i, y_i) \leq \delta 2^{-(1-\varepsilon)n}$, $k = 1, 2, \ldots$. Define

$$\bar{\sigma}_n = \inf\{k; \bar{v}_n(s_{\sigma_n(k)}, x_{\sigma_n(k)}) \leq -2^{-n}\}.$$

Clearly

$$\xi_n \leq E\bar{\sigma}_n.$$

Let us admit for a moment that

Lemma 7 *There exists $d_K > 0$ such that for each i, a.s. on the set $\{0 \leq V_n(s_i, y_i) \leq \delta 2^{-(1-\varepsilon)n}\}$*

$$P^{\mathcal{F}_i^n}(\bar{v}_n(s_i, y_i) \leq -2^{-n} / \bar{v}_n(s_i, y_i) \leq 2^{-n}(K+1)) \geq d_K.$$

Define for $k = 1, 2, \cdots$ the random variable

$$X_k = \mathbf{1}_{\{\bar{v}_n(s_{\sigma_n(k)}, x_{\sigma_n(k)}) \leq -2^{-n}\}}$$

The X_k's are dependent random variables taking values in $\{0, 1\}$, and satisfying

$$P(X_1 = 1) \geq d_K,$$

and for $k = 2, 3, \cdots$,

$$P(X_k = 1/X_1, \cdots, X_{k-1}) \geq d_K$$

Moreover $\bar{\sigma}_n = \inf\{k; X_k = 1\}$.
Suppose now that we are given a sequence of mutually independent r.v.'s $\{Z_k; k \geq 1\}$, which are globally independent of the X_k's, and whose probability law is uniform on the interval $[0, 1]$.
Let $\{Y_k, k \geq 1\}$ be $\{0, 1\}$–valued r.v.'s, defined as follows :

$$Y_k = \begin{cases} 0, & \text{if } X_k = 0 \text{ or } Z_k > d_K/f_k(X_1, \cdots, X_{k-1}); \\ 1, & \text{if } X_k = 1 \text{ and } Z_k \leq d_K/f_k(X_1, \cdots, X_{k-1}); \end{cases}$$

where $f_k(X_1, \cdots, X_{k-1}) \triangleq P(X_k = 1/X_1, \cdots, X_{k-1})$.
It is easily seen that

$$P(Y_k = 1/Y_1, \cdots, Y_{k-1}) = d_K, k = 2, 3, \cdots$$

Consequently $\{Y_k\}$ is a sequence of independent and identically distributed random variables. Let

$$\tilde{\sigma} = \inf\{k; Y_k = 1\}.$$

Clearly

$$\bar{\sigma}_n \leq \tilde{\sigma},$$

hence

$$E\bar{\sigma}_n \leq E\tilde{\sigma} = d_K^{-1}.$$

Lemma 5 is proved, provided we prove Lemma 7. Let us first establish

Lemma 8 *Let X be a $N(0, 1)$ random variable, and $a, b > 0$. Then*

$$1+(a+b)b \leq \frac{P(X > a)}{P(X > a + b)} \leq \frac{1}{2P(X > 1)} \vee \left(1 + \frac{\sqrt{e}}{1 - e^{-1}}(a + b)be^{ab+\frac{b^2}{2}}\right)$$

Proof
The lower bound (which we shall not use) follows from

$$\begin{aligned} \frac{P(a < X \leq a + b)}{P(X > a + b)} &\geq \frac{be^{\frac{-(a+b)^2}{2}}}{\sqrt{2\pi}P(X > a + b)} \\ &= b\left(\int_0^\infty e^{\frac{-y^2}{2}-(a+b)y}dy\right)^{-1} \\ &> b\left(\int_0^\infty e^{-(a+b)y}dy\right)^{-1} = b(a + b). \end{aligned}$$

In case $a + b \leq 1$, the upper bound follows from $P(X > a) \leq \frac{1}{2}$, $P(X > a + b) \geq P(X > 1)$.

We finally establish the upper bound in the case $a+b>1$.

$$
\begin{aligned}
\frac{P(a<X\le a+b)}{P(X>a+b)} &< \frac{be^{-\frac{a^2}{2}}}{\sqrt{2\pi}P(X>a+b)} \\
&= \frac{be^{ab+\frac{b^2}{2}}}{\int_0^\infty e^{-\frac{y^2}{2}-(a+b)y}\,dy} \\
&< \frac{\sqrt{e}be^{ab+\frac{b^2}{2}}}{\int_0^1 e^{-(a+b)y}\,dy} \\
&= \frac{\sqrt{e}(a+b)be^{ab+\frac{b^2}{2}}}{1-e^{-(a+b)}} \\
&< \frac{\sqrt{e}}{1-e^{-1}}(a+b)be^{ab+\frac{b^2}{2}}
\end{aligned}
$$

$\diamond$

We can now finally proceed to the
Proof of Lemma 7

$$
\begin{aligned}
P^{\mathcal{F}_i^n} & (\bar{v}_n(s_i,y_i)\le -2^{-n}/\bar{v}_n(s_i,y_i)\le 2^{-n}(K+1)) \\
= & \frac{P^{\mathcal{F}_i^n}(N_n(s_i,y_i)\le -2^{-n}-V_n(s_i,y_i))}{P^{\mathcal{F}_i^n}(N_n(s_i,y_i)\le 2^{-n}(K+1)-V_n(s_i,y_i))} \\
= & \frac{P^{\mathcal{F}_n^i}(U_i\le -\rho_{n,i}(2^{-\frac{\varepsilon}{2}n}+2^{(1-\frac{\varepsilon}{2})n}V_n(s_i,y_i)))}{P^{\mathcal{F}_i^n}(U_i\le -\rho_{n,i}(-2^{-\frac{\varepsilon}{2}n}(K+1)+2^{(1-\frac{\varepsilon}{2})n}V_n(s_i,y_i)))},
\end{aligned}
$$

where $\rho_{n,i}=(g(v((s_i)_n^-,y_i))\bar{c}_n)^{-1}$ is bounded and bounded away from zero, uniformly in i and n.
On the set

$$-2^{-\frac{\varepsilon}{2}n}(K+1)+2^{(1-\frac{\varepsilon}{2})n}V_n(s_i,y_i)\le 0,$$

the above is greater than or equal to

$$P(U_i\le -\rho_{n,i}(K+2)2^{-\frac{\varepsilon}{2}n})\ge 1/3,$$

for n large enough.
On the set

$$-2^{-\frac{\varepsilon}{2}n}(K+1)+2^{(1-\frac{\varepsilon}{2})n}V_n(s_i,y_i)>0,$$

we use Lemma 8, with $a=-2^{-\frac{\varepsilon}{2}n}(K+1)+2^{(1-\frac{\varepsilon}{2})n}V_n(s_i,y_i)$, $b=2^{-\frac{\varepsilon}{2}n}(K+2)$.
Hence, since moreover $V_n(s_i,y_i)\le \delta 2^{-(1-\varepsilon)n}$,

$$P^{\mathcal{F}_i^n}(\bar{v}_n(s_i,y_i)\le -2^{-n}|\bar{v}_n(s_i,y_i)\le 2^{-n}(K+1))$$

$$\ge \left\{c+\frac{\sqrt{e}}{1-e^{-1}}(\delta+1)(k+2)\exp\left[(\delta+\frac{1}{2})(K+2)\right]\right\}^{-1}.$$

Lemma 8 is proved.

Acknowledgments: The first author was supported by an NSA grant. Part of this research was done while both authors were visiting the MSRI at Berkeley, whose support is gratefully aknowledged.

References

C. Mueller (1998) Long-time existence for signed solutions to the heat equation with a noise term, *Probability Theory and Related Fields* **110**, 51–68.

D. Nualart, E. Pardoux (1994) Markov field properties of solutions of white noise driven quasi–linear parabolic PDEs, *Stochastics* **48**, 17–44.

D. Revuz, M. Yor (1991). *Continuous martingales and Brownian motion*, Springer.

J.B. Walsh (1986) An introduction to stochastic partial differential equations, In P. L. Hennequin editor, *Ecole d'Eté de Probabilités de Saint Flour XIV-1984, Lecture Notes in Mathematics* **1180**, Berlin, Heidelberg, New York, Springer-Verlag.

PART III

STOCHASTIC CONTROL, FILTERING AND PARAMETER ESTIMATION

20

Robustness of Zakai's Equation via Feynman-Kac Representations

Rami Atar
Frederi Viens
Ofer Zeitouni

ABSTRACT: We propose to study the sensitivity of the optimal filter to its initialization, by looking at the distance between two differently initialized filtering processes in terms of the ratio between two simple Feynman-Kac integrals in the product space. We illustrate, by considering two simple examples, how this approach may be employed to study the asymptotic decay rate, as the difference between the growth rates of the two integrals. We apply asymptotic methods, such as large deviations, to estimate these growth rates. The examples we consider are the linear case, where we recover known results, and a case where the drift term in the state process is nonlinear. In both cases, only the small noise regime and only one-dimensional diffusions are studied.

KEYWORDS: Non-linear filtering; Feynman-Kac; Large deviations.

1 Introduction

Zakai's equation of nonlinear filtering theory represents the solution of the filtering problem consisting of evaluating the conditional law of a Markov process observed in white Gaussian noise. In this paper we study the stability of Zakai's equation with respect to perturbations in its initial conditions. It is known since the work of Kunita (1971) that under mild conditions, the conditional law, viewed as a random process taking values in the space of probability measures, is stationary when appropriately initialized. Stettner (1989) shows that whenever the state process is a Feller Markov process converging in law to its unique invariant measure, so is its conditional law. Actually, c.f. Stettner (1991), the joint law of the state and its filtering process is Markovian even if the filter is wrongly initialized. It thus seems natural to investigate the rate of convergence and to study the sensitivity of the optimal filter to its initialization with the wrong initial measure. This issue is also highly relevant for numerical and practical computation of the optimal filter or its approximations, for almost never does one have access to the true initial distribution.

Several approaches exist to analyze this exponential sensitivity, taking full advantage of the linear structure of Zakai's equation. In a recent article Ocone and Pardoux (1996) have studied L^p type of convergence, and showed that the nonlinear filter initialized at the wrong initial condition converges (in an L^p sense) to the nonlinear filter initialized at the correct initial condition. In particular cases (most notably, the Kalman filter), this convergence is exponential (see also Ocone (1997a) for a study of the Beneš case from a different point of view). In general, however, no rates of convergence are given by this approach.

Another approach, which does yield exponential rate of convergence, extends earlier Lyapunov exponent techniques suitable for the finite state space case as in Atar and Zeitouni (1997a). It is based on evaluating the rate of contraction of solutions of Zakai's equation in the projective Hilbert metric, using the Birkhoff coefficient associated with the kernel of the solution, see Atar and Zeitouni (1997b) for a development of this idea, and more recent work Budhiraja and Ocone (1997), Borkar, Mitter and Tatikonda (1997). This technique can be applied rather well in the case of compact state space, and yields results which are usually not tight when the noise level in the observation is weak. A different procedure, which as a by-product yields contraction in the Hilbert projective metric by controlling logarithmic derivatives of Zakai's kernel, is announced in Da Prato, Fuhrman and Malliavin (1995). An approach based on studying the relative entropy is reported in Clark, Ocone and Coumarbatch (1997) and in Ocone (1997b). Model robustness over the infinite time interval, and the relation of this problem to the sensitivity to initial conditions is dealt with in Budhiraja and Kushner (1997a), Budhiraja and Kushner (1997b).

The results based on Hilbert Projective metric are restricted to the compact state space case, and are usually not tight in the limit of low observation noise, for reasons described in Atar and Zeitouni (1997b). Some exceptional cases where contraction results for one dimensional, $\mathbb{R}$-valued diffusions exist and are tight are described in Atar (1997). Our goal in this paper is to suggest a different point of view, looking at the contraction in the space of positive measures as a ratio of two expectations, for which a simple Feynman-Kac representation can be achieved. Asymptotic methods, such as large deviations, can be then applied to the estimation of the growth rate of the latter Feynman-Kac integrals. We content ourselves here with presenting the idea and analyzing a simple one dimensional Gaussian diffusions, for which results are available in more generality by different methods (see, e.g., Ocone and Pardoux (1996)). We also present some immediate consequences for a class of one dimensional nonlinear diffusions. The case of general diffusion processes, even in one dimension, requires additional work and ideas and at present is not resolved.

2 A Feynman-Kac representation for the decay rate

For any measurable space $(\Omega, \mathcal{F})$, let $M(\Omega, \mathcal{F})$ and $M_1(\Omega, \mathcal{F})$ denote the spaces of finite signed measures on $(\Omega, \mathcal{F})$ and of probability measures on $(\Omega, \mathcal{F})$, respectively. For $p \in M(\Omega, \mathcal{F})$ and f measurable on $(\Omega, \mathcal{F})$ let $p(f) := \int f dp$. Define on $M(\Omega, \mathcal{F})$ the norm $\|\cdot\|$ compatible with the variation distance i.e.,

$$\|p\| = \sup \{p(f) : f \text{ is measurable on } (\Omega, \mathcal{F}), |f| \leq 1\}, \quad p \in M(\Omega, \mathcal{F}).$$

One fixed notation, namely $\|\cdot\|$, will be used to denote the above norm for measures on different measurable spaces. Next, let $\mathcal{F} \otimes \mathcal{F}$ denote the product σ-field, and for $p, q \in M(\Omega, \mathcal{F})$, define the exterior product $p \wedge q \in M(\Omega \times \Omega, \mathcal{F} \otimes \mathcal{F})$ by

$$p \wedge q = \frac{1}{2} (p \times q - q \times p).$$

We then have

Lemma 1 *Let $(\Omega, \mathcal{F})$ be a measurable space and let $p, q \in M_1(\Omega, \mathcal{F})$. Then*

$$\|p \wedge q\| \leq \|p - q\| \leq 2\|p \wedge q\|. \tag{1}$$

Proof: Note that for $r, s \in M(\Omega, \mathcal{F})$ one has $\|r \times s\| = \|r\|\|s\|$. Indeed, let f be measurable and such that $|f| \leq 1$, then by Jordan's decomposition for a signed measure and Fubini's theorem,

$$(r \times s)(f) = \int \left[\int f(x, y) r(dx)\right] s(dy) =: \int g(y) s(dy).$$

But $|g(y)| \leq \|r\|$, so $(r \times s)(f) \leq \|r\|\|s\|$. The reverse inequality is trivial. The leftmost inequality in (1) follows from

$$2(p \wedge q) = (p - q) \times q - q \times (p - q),$$

implying that

$$2\|p \wedge q\| \leq \|(p - q) \times q\| + \|q \times (p - q)\| = 2\|p - q\|.$$

The rightmost inequality in (1) follows from

$$p(f) - q(f) = (p \times q)(f \times \mathbf{1}) - (q \times p)(f \times \mathbf{1}) = 2(p \wedge q)(f \times 1).$$

□

Let $(S, \mathcal{S})$ be a Polish space, and let Ω denote the set of RCLL paths ω_t, $t \geq 0$ on S. Consider a strong Markov process, $x_t, t \geq 0$, with paths in Ω, possessing the Feller property. Let $g : S \to \mathbb{R}^d$ be measurable and define

$$y_t = \int_0^t g(x_s)ds + \sigma d\nu_t,$$

where $\nu_t, t \geq 0$ is a standard Brownian motion on $\mathbb{R}^d$, independent of x_t. In the context of filtering theory, x_t is referred to as the state process, and y_t as the observation process.

Let b_t, b'_t, $t \geq 0$ be two processes on $(S, \mathcal{S})$ with the same transition law as x_t, but possibly different initial laws $p, p' \in M_1(S, \mathcal{S})$, respectively, and which are independent of x_t, ν_t and of each other. Let E_b ($E_{b'}$, $E_{b,b'}$) denote expectation w.r.t. b (resp. b', (b, b')) alone.

Let P be the measure induced by (x_t, y_t), $t \geq 0$, and denote by $\mathcal{Y}_t$ the sigma-field generated by $\{y_s, 0 \leq s \leq t\}$. We assume that

$$E \int_0^t |g(x_s)|^2 ds < \infty, \quad t \geq 0. \tag{2}$$

Now let

$$\Lambda_t = \exp\left(\frac{1}{\sigma^2}\int_0^t (g(b_s), dy_s) - \frac{1}{2\sigma^2}\int_0^t |g(b_s)|^2 ds\right), \quad t \geq 0,$$

and define the measure valued processes ρ_t and p_t by

$$\rho_t(\phi) = E_b\left[\phi(b_t)\Lambda_t\right], \quad t \geq 0, \phi \in C_b(\mathbb{R}),$$

$$p_t(\phi) = \rho_t(\phi)/\rho_t(\mathbf{1}). \tag{3}$$

Let also Λ'_t, ρ'_t and p'_t be the processes defined as above, with b replaced by b'. Then it is well known that in case where p equals the initial law of x_t, one has P-a.s. that for all $t \geq 0$, p_t equals the conditional law of x_t given $\mathcal{Y}_t$ under P. In this work however, we are interested mainly in the case where p is arbitrary. p_t may then be interpreted as the filtering process with perturbed initial condition p. In particular, we shall look at the decay rate of the distance between differently perturbed filtering processes:

$$\limsup_{t\to\infty} \frac{1}{t} \log \|p_t - p'_t\|.$$

In view of Lemma 1 and equation (3), the decay rate above may be studied in terms of $\|\rho_t \wedge \rho'_t\|$ and $\|\rho_t \times \rho'_t\|$, namely one has

$$\limsup_{t\to\infty} \frac{1}{t} \log \|p_t - p'_t\| = \limsup_{t\to\infty} \frac{1}{t} \log \frac{\|\rho_t \wedge \rho'_t\|}{\|\rho_t \times \rho'_t\|},$$

and a similar statement for the lim inf. While by definition the denominator can be written as:

$$\|\rho_t \times \rho'_t\| = E_{b,b'}\Lambda_t\Lambda'_t,$$

we also have the following representation for the numerator.

Lemma 2 *Let V_t denote the event*

$$b_s \neq b'_s, \quad all\ s \in [0,t].$$

Then

$$(\rho_t \wedge \rho'_t)(A) = \frac{1}{2} E_{b,b'} \Lambda_t \Lambda'_t \mathbf{1}_{V_t} \left[\mathbf{1}_{(b_t,b'_t)\in A} - \mathbf{1}_{(b'_t,b_t)\in A} \right], \quad A \in \mathcal{S} \otimes \mathcal{S},$$

and hence

$$\|\rho_t \wedge \rho'_t\| \leq E_{b,b'} \Lambda_t \Lambda'_t \mathbf{1}_{V_t}, \tag{4}$$

implying that

$$\limsup_{t\to\infty} \frac{1}{t} \log \|p_t - p'_t\| \leq \limsup_{t\to\infty} \frac{1}{t} \log \left(\frac{E_{b,b'} \Lambda_t \Lambda'_t \mathbf{1}_{V_t}}{E_{b,b'} \Lambda_t \Lambda'_t} \right). \tag{5}$$

Proof: By definition we have

$$(\rho_t \wedge \rho'_t)(A) = \frac{1}{2} E_{b,b'} \Lambda_t \Lambda'_t \left[\mathbf{1}_{V_t} + \mathbf{1}_{V_t^c} \right] \left[\mathbf{1}_{(b_t,b'_t)\in A} - \mathbf{1}_{(b'_t,b_t)\in A} \right].$$

Using strong Markovity of the process (b_t, b'_t), a standard argument leads to

$$E_{b,b'} \Lambda_t \Lambda'_t \mathbf{1}_{V_t^c} \left[\mathbf{1}_{(b_t,b'_t)\in A} - \mathbf{1}_{(b'_t,b_t)\in A} \right] = 0, \tag{6}$$

and the result follows. Indeed, if we let $\{\mathcal{F}_t\}$ denote the filtration generated by $\{(b_t, b'_t)\}$, and

$$\tau = \inf\{s \geq 0 : b_s = b'_s\},$$

with $\inf \emptyset = \infty$, then τ is a stopping time on $\{\mathcal{F}_t\}$. We let θ_s, $s \geq 0$ denote the family of shift operators $\theta_s : \Omega \times \Omega \to \Omega \times \Omega$. Namely, we have

$$(b_{s+t}, b'_{s+t})(\omega) = (b_t, b'_t)(\theta_s \omega), \quad \omega \in \Omega, \quad s,t \geq 0.$$

The strong Markov property then states that for any bounded measurable H,

$$E_{b,b'} \left[H_\tau \circ \theta_\tau \,|\, \mathcal{F}_\tau \right] = \phi(b_\tau, b'_\tau, \tau) \quad \text{on } \{\tau < \infty\}$$

where $\phi(x, x', u) = E_{b,b'} \left[H_u \,|\, b_0 = x, b'_0 = x' \right]$. Let us fix $t > 0$ and $A \in \mathcal{S} \otimes \mathcal{S}$ and for $n > 0$ define

$$H^n_s(b, b') =$$
$$\min\left\{n,\ \Lambda_{t-s} \Lambda'_{t-s}\right\} \mathbf{1}_{V^c_{t-s}} \left[\mathbf{1}_{(b(t-s),b'(t-s))\in A} - \mathbf{1}_{(b'(t-s),b(t-s))\in A} \right].$$

Note that on $\{\tau < \infty\}$ we have $H^n_\tau \circ \theta_\tau = H^n_0$. Moreover, for $x = x'$ and any u one has $\phi(x, x', u) = 0$. This proves that on $\{\tau < \infty\}$, $E_{b,b'} [H^n_0 | \mathcal{F}_\tau] = 0$, while on $\{\tau = \infty\}$ this fact is trivial. Hence $E_{b,b'} H^n_0 = 0$, and since assumption (2) and the independence of $\{x_t\}$ and $\{\nu_t\}$ imply that $\Lambda_t \Lambda'_t$ is $E_{b,b'}$-integrable, we obtain (6). □

In many cases the bound (4) is useless, in particular if $S = \mathbb{R}^d$, $d \geq 3$, and $\{x_t\}$ a standard Brownian motion, where V_t occurs almost surely for every t (given $P_{b,b'}(V_0) = 1$). However it is sharp in some other situations. For example, it holds with equality in the case $S = \mathbb{R}$ and $\{x_t\}$ a diffusion process, provided that $P_{b,b'}(b_0 > b'_0) = 1$. In high dimension, one needs to replace the coupling time τ by a more general coupling time, at which time the joint law of (b_t, b'_t) is exchangeable. For example, if $S = \mathbb{R}^d$ and the components of $b.$ are independent, one may take as coupling time the maximum of the collision times for each coordinate. In more generality, one may take as coupling time τ any stopping time at which the law of (b_τ, b'_τ) is the stationary law. For some examples where coupling times are explicitly constructed, see Lindvall (1992).

3 A Gaussian Example

In this section we present a particularly simple example of a filtering problem where computations using Section 2 can be carried out rather explicitly. Let x_t denote a stationary Orenstein-Uhlenbeck process, i.e.

$$dx_t = -\frac{1}{2} x_t dt + dw_t \,, \tag{7}$$

observed linearly in Gaussian white noise of intensity σ:

$$dy_t = x_t dt + \sigma d\nu_t \,. \tag{8}$$

Here, w_t, ν_t denote independent standard Brownian motions. We let b, b' denote independent stationary solutions to (7), and use p_s to denote the density of the standard normal law. We let

$$\bar{\Lambda}_t = \exp\left(\frac{1}{\sigma}\int_0^t (b_s - x_s)\, d\nu_s - \frac{1}{2\sigma^2}\int_0^t (b_s - x_s)^2 ds\right), \quad t \geq 0,$$

and

$$\bar{\Lambda}'_t = \exp\left(\frac{1}{\sigma}\int_0^t (b'_s - x_s)\, d\nu_s - \frac{1}{2\sigma^2}\int_0^t (b'_s - x_s)^2 ds\right), \quad t \geq 0 \,.$$

It is straightforward to check that now, (5) reads

$$\limsup_{t\to\infty} \frac{1}{t} \log \|p_t - p'_t\| \leq \limsup_{t\to\infty} \frac{1}{t} \log \left(\frac{E_{b,b'}\left(\frac{p(b_0)p'(b_0)}{p_s(b_0)p_s(b'_0)} \bar{\Lambda}_t \bar{\Lambda}'_t \mathbf{1}_{V_t}\right)}{E_{b,b'}\left(\frac{p(b_0)p'(b'_0)}{p_s(b_0)p'_s(b'_0)} \bar{\Lambda}_t \bar{\Lambda}'_t\right)} \right) . \tag{9}$$

In order not to be burdened by (inessential) technicalities, we make the following:

Assumption A There exists a constant $C > 0$ such that

$$C^{-1} \le \inf_{x\in\mathbb{R}} \frac{p(x)}{p_s(x)} \le \sup_{x\in\mathbb{R}} \frac{p(x)}{p_s(x)} \le C\,, \; C^{-1} \le \inf_{x\in\mathbb{R}} \frac{p'(x)}{p_s(x)} \le \sup_{x\in\mathbb{R}} \frac{p'(x)}{p_s(x)} \le C\,.$$

Our goal is to prove the:

Theorem 1 *Assume p, p' satisfy assumption A. Then,*

$$\limsup_{\sigma\to 0} \limsup_{n\to\infty} \frac{\sigma}{n} \log \|p_n - p'_n\| < 0\,, \; a.s.. \tag{10}$$

In fact, there exist $\alpha, \beta > 0$ such that for all t large enough,

$$P(\|p_t - p'_t\| > \exp(-\alpha t/\sigma)) \le \exp(-\beta t/\sigma)\,. \tag{11}$$

A control of the right hand side of (10) is possible, however since our bounds are not expected to be particularly tight we do not try to make it explicit.
Proof of Theorem 1
Throughout this proof, we use t to denote the time index, but except when computing expectations we will always think of $t = 0, 1, \ldots$. Thus, the statement

$$\limsup_{t\to\infty} a_t = 0, \text{a.s.},$$

is taken to say that $\limsup_{n\to\infty} a_n = 0$, a.s. We also note that (11) follows readily from our proof of (10) by a Chebycheff inequality. Thus, we concentrate here on proving the later.
Obviously, under assumption A, the right hand side of (9) reads

$$\begin{aligned} \limsup_{t\to\infty} \frac{1}{t} \log \|p_t - p'_t\| &\le \limsup_{t\to\infty} \frac{1}{t} \log \left(\frac{E_{b,b'}\left(\Lambda_t \Lambda'_t 1_{V_t}\right)}{E_{b,b'}\left(\bar{\Lambda}_t \bar{\Lambda}'_t\right)} \right) \\ &:= \limsup_{t\to\infty} \frac{1}{t} \log \frac{B_1^t}{B_2^t}\,. \end{aligned} \tag{12}$$

For $t < \infty$, let $(\lambda_i^t)^2$ denote the Karhunen-Löeve eigenvalues and let $\varphi_i^t(\cdot)$ denote the associated eigenfunctions. Let $\{a_i\}, \{a'_i\}$ denote independent sequences of independent standard normal random variables, and let $\phi_i^t = \int_0^t \varphi_i^t(s) x_s ds$, $\nu_i^t = \int_0^t \varphi_i^t(s) d\nu(s)$. Then, ϕ_i^t are independent normal variables of zero mean and variance $1/(\lambda_i^t)^2$, ν_i^t are independent standard normal variables, and one has the identities

$$\begin{aligned} I_1^t := \int_0^t (b_s - x_s)^2 ds &= \sum_{i=1}^{\infty} \left(\frac{a_i}{\lambda_i^t} - \phi_i^t \right)^2, \\ I_1'^t := \int_0^t (b'_s - x_s)^2 ds &= \sum_{i=1}^{\infty} \left(\frac{a'_i}{\lambda_i^t} - \phi_i^t \right)^2, \end{aligned}$$

$$I_2^t := \int_0^t (b_s - x_s) d\nu_s = \sum_{i=1}^{\infty} \left(\frac{a_i}{\lambda_i^t} - \phi_i^t \right) \nu_i^t ,$$

$$I_2'^t := \int_0^t (b_s' - x_s) d\nu_s = \sum_{i=1}^{\infty} \left(\frac{a_i'}{\lambda_i^t} - \phi_i^t \right) \nu_i^t .$$

Further, one knows that $\lambda_i^t \sim \pi i/t$. The definitions above are used in the proof of the following lemma, which is presented later.

Lemma 3 *For* $p \in [1, \infty)$,

$$\limsup_{\sigma \to 0} \limsup_{t \to \infty} \frac{\sigma}{t} \log E_b(\bar{\Lambda}_t^p) = \frac{1}{4} \left(p^{3/2} - 3\sqrt{p} \right) , \; a.s.. \tag{13}$$

In particular,

$$\limsup_{\sigma \to 0} \limsup_{t \to \infty} \frac{\sigma}{t} \log B_2^t = -1 , \; a.s.. \tag{14}$$

The key to the evaluation of a bound on the right hand side of (12) lies in restricting the domain of integration in B_1^t. For K a fixed constant (later taken as $K = 1 + \eta$ with $\eta > 0$ arbitrary small), define the events

$$A_t := \{\{a_i\} : I_1^t \leq K\sigma t\} , \quad A_t' := \{\{a_i'\} : I_1'^t \leq K\sigma t\} .$$

Then,

$$\begin{aligned} \frac{B_1^t}{B_2^t} &= \frac{E_{b,b'}(\bar{\Lambda}_t \bar{\Lambda}_t' 1_{V_t})}{E_b(\bar{\Lambda}_t)^2} \\ &\leq \frac{E_{b,b'}(\bar{\Lambda}_t \bar{\Lambda}_t' 1_{A_t} 1_{A_t'} 1_{V_t})}{E_b(\bar{\Lambda}_t)^2} + 2 \frac{E_b(\bar{\Lambda}_t 1_{A_t^c})}{E_b(\bar{\Lambda}_t)} := \frac{B_{11}^t}{B_2^t} + 2C_1^t . \end{aligned} \tag{15}$$

The proof of the following lemma is postponed to the end of this section:

Lemma 4 *With* K *as above, and any* $\eta > 0$,

$$\limsup_{\sigma \to 0} \limsup_{t \to \infty} \frac{\sigma}{t} \log C_1^t < 0 , \; a.s.. \tag{16}$$

Equipped with Lemmas 3 and 4, our next task is to achieve a control on B_{11}^t, which is contained in the following.

$$B_{11}^t := E_{b,b'}(\bar{\Lambda}_t \bar{\Lambda}_t' 1_{A_t} 1_{A_t'} 1_{V_t}) \leq E_b^{2/p}(\bar{\Lambda}_t^p) E_{b,b'}^{1/q}(1_{A_t} 1_{A_t'} 1_{V_t}) . \tag{17}$$

In view of (12), (15) and Lemma 4, Lemma 3), (17) and the following large deviations computation complete the proof of Theorem 1:

Lemma 5

$$\limsup_{\sigma \to 0} \limsup_{t \to \infty} \frac{\sigma}{t} \log E_{b,b'}(1_{A_t} 1_{A_t'} 1_{V_t}) \leq -K^{-1} \left(\frac{3 + \sqrt{3}}{4} \right)^2 , \; a.s.$$

Indeed, taking $p = 1 + \delta$ with δ small enough and $\eta > 0$ small enough one obtains that

$$\limsup_{\sigma\to 0}\limsup_{t\to\infty}\frac{\sigma}{t}\log\left(\frac{B_{11}^t}{B_2^t}\right) \le \delta\left(1-\left(\frac{3+\sqrt{3}}{4}\right)^2+g_\eta\right)+o(\delta)<0\,,\ \text{a.s.},$$

where $g_\eta \to 0$ as $\eta \to 0$. The proof of Lemma 5 is presented below. □

Proof of Lemma 5: It is enough to show that

$$\limsup_{n\to\infty}\frac{\sigma}{n}\log E\left(E_{b,b'}(\mathbf{1}_{A_n}\mathbf{1}_{A_n'}\mathbf{1}_{V_n})\right) \le -K^{-1}\left(\frac{3+\sqrt{3}}{4}\right)^2+g(\sigma)\,, \tag{18}$$

where $g(\sigma) \to 0$ as $\sigma \to 0$. Indeed, assume (18). Then, for any $0 < \beta < 1$,

$$P\left(E_{b,b'}(\mathbf{1}_{A_n}\mathbf{1}_{A_n'}\mathbf{1}_{V_n}) > \exp\left(-\frac{\beta n}{\sigma}\left(K^{-1}\left(\frac{3+\sqrt{3}}{4}\right)^2+g(\sigma)\right)\right)\right)$$
$$\le \exp\left(-\frac{(1-\beta)n}{\sigma}\left(K^{-1}\left(\frac{3+\sqrt{3}}{4}\right)^2+g(\sigma)+\bar{g}(n)\right)\right),$$

where $\bar{g}(n) \to 0$ as $n \to \infty$, implying by the Borel-Cantelli lemma that

$$\limsup_{\sigma\to 0}\limsup_{n\to\infty}\frac{\sigma}{n}\log E_{b,b'}(\mathbf{1}_{A_n}\mathbf{1}_{A_n'}\mathbf{1}_{V_n}) \le -K^{-1}\left(\frac{3+\sqrt{3}}{4}\right)^2\,,\ \text{a.s.}$$

For $t \in \mathbb{R}_+$ let $\lfloor t\rfloor$ ($\lceil t\rceil$, respectively) denote the greatest (smallest) integer smaller than or equal to t (greater than or equal to t). Then the conclusion of the lemma follows by noting that, from monotonicity,

$$\frac{\sigma}{t}\log E_{b,b'}(\mathbf{1}_{A_t}\mathbf{1}_{A_t'}\mathbf{1}_{V_t}) \le \frac{\sigma}{\lceil t\rceil}\log E_{b,b'}(\mathbf{1}_{A_{\lfloor t\rfloor}}\mathbf{1}_{A'_{\lfloor t\rfloor}}\mathbf{1}_{V_{\lfloor t\rfloor}})\,.$$

Turning to the proof of the key estimate (18), the following transformation reduces the three-dimensional problem into two-dimensional one:

$$\begin{pmatrix} b_\cdot - x_\cdot \\ b'_\cdot - x_\cdot \end{pmatrix} = \begin{pmatrix} \sqrt{3/2}u_\cdot + v_\cdot/\sqrt{2} \\ \sqrt{3/2}u_\cdot - v_\cdot/\sqrt{2} \end{pmatrix}, \tag{19}$$

where $u_\cdot, v_\cdot$ are two independent Orenstein-Uhlenbeck processes satisfying the SDE (7). Rewrite now (18) in terms of $u_\cdot, v_\cdot$ to obtain that

$$\limsup_{n\to\infty}\frac{\sigma}{n}\log E\left(E_{b,b'}(\mathbf{1}_{A_n}\mathbf{1}_{A_n'}\mathbf{1}_{V_n})\right)$$
$$\le \limsup_{n\to\infty}\frac{\sigma}{n}\log P\Big(\frac{1}{t}\int_0^t(\sqrt{3}u_s - v_s)^2ds \le 2K\sigma,$$

$$\frac{1}{t}\int_0^t (\sqrt{3}u_s + v_s)^2 ds \le 2K\sigma, \frac{1}{t}\int_0^t \mathbf{1}_{v_s<0} ds = 0\Big)$$

$$\le -\inf\Big\{\frac{1}{8}\iint \frac{f_x^2+f_y^2}{f}(x,y)m(dx,dy):$$

$$f\ge 0, \iint f(x,y)m(dx,dy)=1,$$

$$\iint (\sqrt{3}x+y)^2 f(x,y)m(dx,dy) \le 2K\sigma,$$

$$\iint (\sqrt{3}x-y)^2 f(x,y)m(dx,dy) \le 2K\sigma,$$

$$f(x,y)=f(x,y)\mathbf{1}_{y\ge 0}\Big\}, \tag{20}$$

where $m(dx,dy) = \exp(-(x^2+y^2)/2)dxdy/2\pi$, f_x, f_y denote the partial derivatives of $f(x,y)$ w.r.t. x, y, respectively, and the last inequality in (20) is a consequence of the Donsker-Varadhan theorem for occupation measures, c.f. Deuschel and Stroock (1989) pg. 241 for this version. The conclusion now follows by solving the calculus of variations problem (20), as outlined below.

The cost function in the constrained problem is

$$J(f) = \iint \left(\frac{1}{8}\frac{f_x^2+f_y^2}{f} + r(x,y)f\right) m(dx,dy)$$

where

$$r(x,y) = \lambda_1(\sqrt{3}x+y)^2 + \lambda_2(\sqrt{3}x-y)^2 + \mu.$$

The Euler-Lagrange equation then has the form

$$\frac{1}{8}\|\nabla \log f\|^2 + \frac{1}{4}(\Delta \log f - x(\log f)'_x - y(\log f)'_y) = r(x,y)f.$$

The solution to the above, which is supported on $\{y \ge 0\}$ and satisfies

$$\iint f(x,y)m(dx,dy) = 1, \quad \iint (\sqrt{3}x \pm y)^2 f(x,y)m(dx,dy) = 2K\sigma$$

is given by

$$f(x,y) = c_1 y^2 e^{-(\gamma x^2+y^2)/2\alpha^2}\mathbf{1}_{y\ge 0}$$

where c_1 is a normalization constant, $\gamma = -\alpha^2 + \sqrt{\alpha^4 + 3(1+2\alpha^2)}$ and for small values of σ,

$$\alpha^2 = \frac{2K\sigma}{3+\sqrt{3}} + o(\sigma).$$

Substituting into the rate function one obtains

$$\frac{1}{8}\iint \frac{f_x^2+f_y^2}{f}(x,y)m(dx,dy) = \left(\frac{3+\sqrt{3}}{4}\right)^2 (K\sigma)^{-1} + o(\sigma^{-1})$$

and (18) follows. $\square$

Proof of Lemma 3 Write

$$\bar{\Lambda}_t^p = \exp\left(\frac{p}{\sigma}\sum_{i=1}^{\infty}(\frac{a_i}{\lambda_i^t} - \phi_i^t)\nu_i^t - \frac{p}{2\sigma^2}\sum_{i=1}^{\infty}(\frac{a_i}{\lambda_i^t} - \phi_i^t)^2\right).$$

Noting that the exponent is nothing but a quadratic form in the random variables a_i, taking the expectation reveals that

$$E_b\bar{\Lambda}_t^p = \prod_{i=1}^{\infty}\frac{1}{\sqrt{1+p(\sigma\lambda_i^t)^{-2}}}\exp\left(\sum_{i=1}^{\infty}\frac{p(\nu_i^t)^2 - 2\phi_i^t\nu_i^t\sigma(\lambda_i^t)^2 - (\phi_i^t\lambda_i^t)^2}{2(\sigma^2\lambda_i^2/p+1)}\right).$$

An analysis of the first term reveals that

$$\begin{aligned}\log\prod_{i=1}^{\infty}\frac{1}{\sqrt{1+p(\sigma\lambda_i^t)^{-2}}} &\sim -\frac{1}{2}\sum_{i=1}^{\infty}\log(1+pt^2/\pi^2\sigma^2 i^2)\\ &\sim -\frac{t\sqrt{p}\int_0^{\infty}\log(1+x^{-2})dx}{2\sigma\pi} + o(t)) = -\frac{t\sqrt{p}}{2\sigma}.\end{aligned}$$

while for the second term one notes that

$$E_t := \sum_{i=1}^{\infty}\frac{p(\nu_i^t)^2 - 2\phi_i^t\nu_i^t\sigma(\lambda_i^t)^2 - (\phi_i^t\lambda_i^t)^2}{2(\sigma^2\lambda_i^2/p+1)}$$

is a sum of independent, random variables, whose variance is of order t, and

$$\lim_{t\to\infty} E(E_t)/t = \frac{(p-1)\sqrt{p}\int_0^{\infty}(1+x^2)^{-1}dx}{2\pi\sigma} = \frac{(p-1)\sqrt{p}}{4\sigma}.$$

Hence, $E_t/t \to (p-1)\sqrt{p}/4\sigma$, in probability, and a.s. on the sequence $t_n = n$ because $E(E_t - E(E_t))^4 = O(t^2)$ (for fixed σ). $\square$

Proof of Lemma 4 The key to the proof is Chebycheff's inequality: Fix $\beta > 0$, then

$$E_b(\bar{\Lambda}_t \mathbf{1}_{A_t^c}) \le e^{\frac{-\beta K t}{2\sigma}} E_b(\bar{\Lambda}_t \exp(\beta I_1^t/2\sigma^2)).$$

One now repeats the analysis of Lemma 3 to conclude that

$$\begin{aligned}\limsup_{t\to\infty}\frac{2\sigma}{t}\log E_b(\bar{\Lambda}_t\mathbf{1}_{A_t^c}) &\le -\beta K - \frac{\sqrt{1-\beta}\int_0^{\infty}\log(1+x^{-2})dx}{\pi}\\ &\quad + \frac{\beta}{\sqrt{1-\beta}\pi}\int_0^{\infty}(1+x^2)^{-1}dx\\ &= -\beta K - \sqrt{1-\beta} + \frac{\beta}{2\sqrt{1-\beta}}.\end{aligned}$$

Optimization over $\beta \ge 0$ and an application of Lemma 3 with $p = 1$ yield the result, as soon as $K > 1$. $\square$

4 A non-Gaussian Example

Consider next the stochastic process x_t solution of the SDE

$$dx_t = c(x_t)dt + dw_t\,, \tag{21}$$

and let y_t be as in (8). Define $F(x) = \int_0^x (c(\theta)+\theta/2)d\theta$. To avoid unessential technical difficulties, we use the following assumption:
Assumption B The functions $|F(x)|$, $|c'(x)|$ and $|c^2(x)-x^2/2|$ are bounded by a global constant C.
We now have the

Corollary 1 *Assume assumptions A and B. Then,*

$$\limsup_{\sigma\to 0}\limsup_{n\to\infty}\frac{\sigma}{n}\log\|p_n - p'_n\| < 0\,,\ a.s.. \tag{22}$$

Proof: Using assumptions A and B, one knows that the solution to (21) is absolutely continuous w.r.t. to the Orenstein-Uhlenbeck $b_\cdot$, with Radon-Nikodym derivative equal to

$$\begin{aligned}&\frac{p(b_0)}{p_s(b_0)}\exp\left(\int_0^t (c(b_s)+b_s/2)db_s - \frac{1}{2}\int_0^t (c^2(b_s)-(b_s/2)^2)ds\right)\\ &\leq Ce^{2C}\exp 2Ct\,,\end{aligned} \tag{23}$$

where the inequality above uses Ito's formula:

$$F(b_t) = F(b_0) + \int_0^t (c(b_s)+b_s/2)db_s + \frac{1}{2}\int_0^t (c'(b_s)+1/2)ds\,.$$

Hence, exactly as in the argument leading to (9),

$$\|p_t - p'_t\| \leq C^3 e^{4C} e^{4Ct}\frac{E_{b,b'}(\bar\Lambda_t\bar\Lambda'_t \mathbf{1}_{V_t})}{(E_b\bar\Lambda_t)^2}\,, \tag{24}$$

with the difference from the Gaussian case lying in the fact that in the R.H.S. of (24), the random variables ϕ_i^t are neither independent nor Gaussian or uncorrelated. However, due to the proof of (11), there exist $\alpha,\beta>0$ such that

$$P_G\left(\frac{E_{b,b'}(\bar\Lambda_t\bar\Lambda'_t \mathbf{1}_{V_t})}{(E_b\bar\Lambda_t)^2} > \exp(-\alpha t/\sigma)\right) \leq \exp(-\beta t/\sigma)\,, \tag{25}$$

where P_G denotes the measure under which ϕ_i^t are normal independent of variance $(\lambda_i^t)^2$. Applying now again a change of measure as in (23), we see that for t large enough,

$$\begin{aligned}&P\left(\frac{E_{b,b'}(\bar\Lambda_t\bar\Lambda'_t \mathbf{1}_{V_t})}{(E_b\bar\Lambda_t)^2} > \exp(-\alpha t/\sigma)\right)\\ &\leq e^{3Ct}P_G\left(\frac{E_{b,b'}(\bar\Lambda_t\bar\Lambda'_t \mathbf{1}_{V_t})}{(E_b\bar\Lambda_t)^2} > \exp(-\alpha t/\sigma)\right)\,,\end{aligned}$$

and the conclusion follows from (25). □

Remark Obviously, assumption B is far from optimal, and one may relax it by using appropriate Hölder inequalities. However, even at best, the technique of this section is quite limited and does not seem to allow one to go beyond the case of linear observation functions $h(\cdot)$ or, more generally, beyond the case of non-constant diffusion coefficients in the state process. What is needed in order to better employ the bound (5) is a direct way of controlling $E(\bar{\Lambda}_t^p)$ which does not involve the Karhunen-Löeve expansion. Rather, a conditional large deviation principle for the joint occupation measure of (b, b', x, ν) is needed. This will be reported elsewehere.

Acknowledgments: The work of R. Atar was supported in part by the Office of Naval Research (ONR-N00014-96-1-0276). The work of O. Zeitouni was partially supported by a grant from the basic research fund administered by the Israeli Academy of Sciences.

References

R. Atar (1997) *Exponential Stability for Nonlinear Filtering of Diffusion Processes in Non-Compact Domain,* submitted.

R. Atar and O. Zeitouni (1997a) *Lyapunov Exponents for Finite State Nonlinear Filtering,* Siam J. Contr. Optim., 35, pp. 36–55.

R. Atar and O. Zeitouni (1997b) *Exponential Stability for Nonlinear Filtering,* Ann. Inst. Hen. Poincare, 33, pp. 697–725.

V. S. Borkar, S. K. Mitter and S. Tatikonda (1997) *Optimal Sequential Vector Quantization of Markov Sources,* preprint.

A. Budhiraja and H. J. Kushner (1997a) *Robustness of Nonlinear Filters Over the Infinite Time Interval,* to appear in Siam J. Contr. Optim.

A. Budhiraja and H. J. Kushner (1997b) *Approximation and Limit Results for Nonlinear Filters Over an Infinite Time Interval,* preprint.

A. Budhiraja and D. L. Ocone (1997) *Exponential Stability of Discrete Time Filters for Bounded Observation Noise,* Systems and Control Letters, 30, pp. 185–193.

J. M. C. Clark, D. L. Ocone and C. Coumarbatch (1997) *Relative Entropy and Error Bounds for Filtering of Markov Process,* preprint.

J. D. Deuschel and D. W. Stroock (1989) Large Deviations. Academic Press, Boston.

G. Da Prato, M. Fuhrman, P. Malliavin (1995) *Asymptotic Ergodicity for the Zakai Filtering Equation*, C. R. Acad. Sci. Paris, t. 321, Série I, pp. 613–616.

H. Kunita (1971) *Asymptotic Behavior of the Nonlinear Filtering Errors of Markov Processes,* J. Multivariate Anal., 1, pp. 365–393.

T. Lindvall (1992) Lectures on the coupling method, Wiley, New York.

D. L. Ocone (1997a) *Asymptotic Stability of Beneš Filters*, preprint.

D. L. Ocone (1997b) this volume.

D. L. Ocone and E. Pardoux (1996) *Asymptotic Stability of the Optimal Filter with respect to its Initial Condition*, Siam J. Contr. Optim., 34, pp. 226–243.

L. Stettner (1989) *On Invariant Measures of Filtering Processes,* Stochastic Differential Systems, Proc. 4th Bad Honnef Conf., 1988, Leture Notes in Control and Inform. Sci. 126, edited by Christopeit, N., Helmes, K. and Kohlmann, M., Springer, pp. 279–292.

L. Stettner (1991) *Invariant Measures of the Pair: State, Approximate Filtering Process,* Colloq. Math., LXII, pp. 347–351.

21

Estimation of Probability Distributions for Individual Parameters Using Aggregate Population Data

H.T. Banks
B.G. Fitzpatrick
Laura K. Potter
Yue Zhang

ABSTRACT: In this paper we discuss a general methodology for estimating the distribution of individual growth rates in a size-structured population using aggregate population data. The method, for which rigorous theoretical formulations have been developed, is presented in the context of an inverse problem methodology and its use is illustrated with application to mosquitofish, *Gambusia affinis*, populations in rice fields.

KEYWORDS: size-structured populations, distributed growth rates, inverse problems, mosquitofish

1 Introduction

In this paper, we present results using inverse problem techniques for estimation of growth distribution in size-structured population models using aggregate population data. The models employed here are based on ideas initially discussed in Banks, Botsford, Kappel and Wang (1988), which entail models wherein growth rates may vary across individuals of the population as well as with size and time.

These models are in contrast to the usual stochastic partial differential equation models as described for example in Fleming (1974), (1975), Fleming and Su (1974) and in the next section. Although they are not stochastic in the usual sense, they are probabilistic in that one models growth, mortality, etc. via probability distributions across the population.

Simulation studies were presented in Banks, Botsford, Kappel and Wang (1988) to demonstrate that such ideas could lead to population densities that exhibit dispersion and bimodality. Rigorous theoretical developments

of the associated inverse problem technique and initial illustrations with computational examples were given in Banks and Fitzpatrick (1991), Fitzpatrick (1993) and (1994). A survey of results and other references can be found in Banks (1994). The general philosophy
underlying our approach of using aggregate population data to estimate individual rates is given (along with an application to susceptibility and vaccination efficiency in populations) in Banks, Fitzpatrick and Zhang (1996). As detailed in earlier references, our efforts on such problems were initiated in collaboration with marine biologists (specifically Lou Botsford and his colleagues) in studies related to the introduction of mosquitofish into rice fields, in attempts to control mosquito populations without chemicals. To establish viable control strategies, it is very important to have quantitative models which predict accurately the evolution of the populations. In the paper Botsford, Vandracek, et al. (1987), the mosquitofish population was modeled using the Sinko-Streifer (also called the McKendrick-Von Foerster) model for size-structured population density evolution. As we shall discuss below, this leads to a number of conceptual difficulties.

2 Modeling Philosophy

There is a huge literature (we won't even attempt to cite these here) on the modeling of populations in the quantitatively oriented biological research literature. The efforts range from modeling population growth, death, mortality in insects, marine life, plants and animals to gene frequency mutation and drift, as well as treatments of susceptibility to disease in vaccinated populations of humans. In such a diverse scientific literature it is often difficult to discern an underlying commonality. However, there are some usual features in much of the literature. The modeling often relies on data about certain observables, while it is knowledge about other nonobservable parameters that is of interest to investigators. For example, one usually has total population (aggregate densities) counts in models where growth, mortality and migration rates of a typical individual are of most interest. Or one may have aggregate data on percent of a population vaccinated, and numbers of those who fall ill and recover (or not) in studies of disease prophylaxis. In the case of genetic studies one has data on phenotypes, where it is gene frequencies and their changes that one wishes to understand.
A second feature of most modeling attempts is the presence of uncertainty. This may arise in the model itself, in parameters in the model, in distribution over the population of unobservable traits or characteristics. A conceptually important question is how to treat these uncertainties when attempting to extract the maximum information about the population from the data. The standard approach to treat probabilistic or stochastic aspects of populations is through the use of stochastic differential equations. A brief review of examples early on in this approach was given in Fleming (1974),

where he discussed distributed parameter or partial differential equation models for geographically-structured (including size or age-structured) populations. He also discussed population genetics models (equations for the mean and the covariance of gene frequency in spatially migrating populations — see Fleming (1975), Fleming and Su (1974)). Our focus here will be on size-structured population models and our use of rate distribution models as an alternative to stochastic differential equations. To put this in context of the more standard approach, we introduce and discuss briefly the Fokker-Planck size-structured model.

The Fokker-Planck equation is the basis of a stochastic size-structured model (see Banks (1988), Banks, Tran and Woodward (1993)) which is based on the assumption that movement from one size class to another can be described by a Markov process. The "physiological age" interpretation of the Fokker-Planck equation was first suggested by Weiss (1968); a careful derivation based on the paradigm of Brownian motion of particles which is applicable to growth processes is given in Okubo (1980). The Fokker-Planck equation, under the assumption of a Markov transition for the growth process, is

$$\frac{\partial u}{\partial t}(t,x) + \frac{\partial}{\partial x}(\mathcal{M}_1(t,x)u(t,x)) = \frac{1}{2}\frac{\partial^2}{\partial x^2}(\mathcal{M}_2(t,x)u(t,x)), \tag{1}$$

where $u(t,x)$ is the population density at time t and size x and the moments are given by

$$\mathcal{M}_j(t,x) \equiv \lim_{\Delta t\to 0}\frac{1}{\Delta t}\int_{-\infty}^{\infty}(y-x)^j p(t,x;t+\Delta t,y)dy. \tag{2}$$

The function $p(t,x;t+\Delta t,y)$ is the probability density for the transition from size x at time t to size y at time $t+\Delta t$; i.e., $p(t,x;t+\Delta t,y)\Delta x$ is the probability that members in the size interval $[x,x+\Delta x]$ at time t will move to size y at time $t+\Delta t$. The moments $\mathcal{M}_1$, $\mathcal{M}_2$ have probabilistic interpretations: $\mathcal{M}_1$ is the mean (or first moment) of the rate of increase in size

$$\mathcal{M}_1(t,x) = \lim_{\Delta t\to 0}\mathcal{E}\left\{\frac{x(t+\Delta t)-x(t)}{\Delta t}\right\},$$

where $\mathcal{E}[X]$ denotes the expected value of a random variable X, while $\mathcal{M}_2$ is the second moment of the rate of increase in size

$$\mathcal{M}_2(t,x) = \lim_{\Delta t\to 0}\mathcal{E}\left\{\frac{(x(t+\Delta t)-x(t))^2}{\Delta t}\right\}.$$

Appropriate boundary conditions must be formulated for (1). Since $\mathcal{M}_1 u - \frac{\partial}{\partial x}\left(\frac{\mathcal{M}_2}{2}u\right)$ represents the population flux, we have

$$\left[\mathcal{M}_1(t,x)u(t,x) - \frac{\partial}{\partial x}(\mathcal{M}_2(t,x)u(t,x))\right]^{x=x_0}$$

$$= \int_{x_o}^{x_m} K(t,x)u(t,x)dx \tag{3}$$

$$\left[\mathcal{M}_1(t,x)u(t,x) - \frac{\partial}{\partial x}(\mathcal{M}_2(t,x)u(t,x))\right]^{x=x_m} = 0 \tag{4}$$

while the initial conditions are given by

$$u(t_0,x) = \Phi(x). \tag{5}$$

Here x_m is maximum size, x_0 is the minimum size, and K is the fecundity rate as explained below in our discussion of the Sinko-Streifer model.

The system (1), (3)-(5) comprises an initial-boundary value problem for a transport-dominated diffusion process (the $\mathcal{M}_1$ term is typically dominant over the $\mathcal{M}_2$ term) that offers significant computational challenges. First, the moments $\mathcal{M}_1, \mathcal{M}_2$ almost always (for the applications to populations) depend on both t and x and must be estimated; the expressions(2) is not helpful since, in general, p is unknown. Moreover, even simulation with (1), (3)-(5) is nontrivial. Traditional finite difference and finite element methods produce erroneous oscillatory solutions. Thus the spline-based, fixed node methods first proposed in Banks (1988) are of very limited interest. However, high promise can be found (Banks, Tran and Woodward (1993)) in a moving node finite element technique first suggested in Harrison (1988). Even so, estimation or inverse problems built on Fokker-Planck models are extremely different computationally and have enjoyed limited use in the literature.

Our efforts to treat both unobservable individual parameters as well as a certain amount of stochasticity is built upon the Sinko-Streifer model and thus we briefly outline its features.

The classical Sinko-Streifer model, henceforth referred to simply as (SS), is given by

$$\begin{aligned} \tfrac{\partial}{\partial t}v(t,x) + \tfrac{\partial}{\partial x}(g(t,x)v(t,x)) &= -\mu(t,x)v(t,x) & t > t_0 \ , \\ v(0,x) &= \Phi(x) & x_0 < x < x_m \ , \\ g(t,x_0)v(t,x_0) &= \textstyle\int_{x_0}^{x_m} K(t,\xi)v(t,\xi)d\xi & \\ g(t,x_m) &= 0 & \end{aligned} \tag{6}$$

and simulates the time evolution of a population with respect to the size x of the individuals. In our use of this model, the parameter $x \in [x_0, x_m]$ denotes the length of the fish, and the function v represents the size density function so that

$$N(t) = \int_a^b v(t,x)dx \tag{7}$$

is the number of fish in the population at time t whose size is between a and b. The function g represents the growth rate of the individuals,

$$\frac{dx(t)}{dt} = g(t,x) \ ,$$

so that in this simple model all individuals have the same growth rate. The last condition in (6), $g(t, x_m) = 0$, indicates that growth ceases when individuals reach maximum size x_m. The initial condition Φ describes the initial size density for the population. The function K is a fecundity kernel, and is used to express the recruitment rate $R(t, v) = \int_{x_0}^{x_m} K(t, \xi)v(t, \xi)d\xi$. The mortality rate is given by μ.

Difficulties arise, however, in applying the model (SS) to observed data since the data often exhibit features that the model cannot simulate or predict. One such phenomenon is a dispersion in size as time progresses. Another is that the population begins with a unimodal density and in time develops into one with a bimodal density. Both of these features are present in the mosquitofish data (see Banks, Botsford, Kappel and Wang (1988), Botsford, Vandracek, et al. (1987), and the data plot in Figure 1).

This dispersion in size and the transition from a unimodal to a bimodal density is qualitatively inconsistent with solutions of the Sinko-Streifer model under biologically feasible assumptions on the model. We are able to capture both phenomena with growth rate distribution (GRD) models.

2.1 GRD Models with Identical Initial Size Densities

Here we use a modification of the model (SS) (in actuality, a continuum of (SS) across families of growth rates) which exhibits features present in the data. We use as the aggregate population density (APD) the function

$$u(t, x) = \int_G v(t, x; g)dP(g), \tag{8}$$

where G is a collection of growth rates and P is a probability measure on G. This approach was first suggested in Banks, Botsford, Kappel and Wang (1988). As detailed in Banks and Fitzpatrick (1991), the resulting model is rich enough to exhibit the phenomena of interest, namely, dispersion and development of two modes from one. As also explained in Banks and Fitzpatrick (1996), measure theoretic results allow us to approximate the continuum measure in (8) by a discrete measure corresponding to a finite dimensional set of growth rates.

Briefly, we assume that we have a family $G = \{g_{jk}\}$, $j = 1, \cdots, M_1$, $k = 1, \cdots, M_2$ of individual growth rates; individuals are grouped into the same subpopulation if they have the same growth rate. The subpopulations grow according to the (SS) with $g = g_{jk}$ in (6). Here we take $g_{jk}(x) = r_j(\gamma_k - x)$, $\mu = 0$ and $K = 0$. Note, however, that a generalization to allow distribution of mortality and fecundity over subpopulations is readily achieved.

The size density for the subpopulation jk is given by $v(t, x; g_{jk})$, and the aggregate population density is given by

$$u(t, x) = \sum_{j,k} v(t, x; g_{jk})p_{jk}, \tag{9}$$

where p_{jk} is the probability of an individual being in the jk subpopulation. Note that this formulation embodies the assumption that each subpopulation has the same initial size density, i.e., $v(0,x;g_{jk}) = \Phi(x)$ for all j,k. We used the above outlined formulation in a least squares inverse problem involving fitting several sets of field data $\{\hat{u}(t,x)\}$. The probabilities $\{p_{jk}\}$ for $1 \le j \le M_1$ and $1 \le k \le M_2$ in (9) were estimated by solving the following inverse problem:

$$\begin{aligned}\min_{\mathbf{p}\in P^M(G)} J(\mathbf{p}) &= \sum_{n=1}^{N}\int_{x_0}^{x_1}[\sum_{j,k} v(t_n,x;g_{jk})p_{jk} - \hat{u}(t_n,x)]^2 dx \qquad (10)\\ &= \sum_{n=1}^{N}\int_{x_0}^{x_1}([\sum_{j,k} v(t_n,x;g_{jk})p_{jk}]^2 \\ &\qquad -2\hat{u}(t_n,x)\sum_{j,k} v(t_n,x;g_{jk})p_{jk} + [\hat{u}(t_n,x)]^2)dx.\end{aligned}$$

Here $P^M(G)$ is a finite dimensional approximation to the probability measure space $P(G)$ defined by

$$P^M(G) = \{P \in P(G) | P = \sum_{j,k} p_{jk}\delta_{g_{jk}},\ \sum_{j,k} p_{jk} = 1\}$$

where $M = M_1 \times M_2$ and $\delta_{g_{jk}}$ is the Dirac measure with an atom at g_{jk}. We denote by $\mathbf{p}$ the array that contains p_{jk}, $1 \le j \le M_1$, $1 \le k \le M_2$, and set

$$a_{jk\ell m} = \sum_{n=1}^{N}\int_{x_0}^{x_1} v(t_n,x;g_{jk})v(t_n,x;g_{\ell m})dx$$

$$b_{jk} = -\sum_{n=1}^{N}\int_{x_0}^{x_1} \hat{u}(t_n,x)v(t_n,x;g_{jk})dx$$

$$c = \sum_{n=1}^{N}\int_{x_0}^{x_1} [\hat{u}(t_n,x)]^2 dx\ .$$

Moreover, we define $\mathbf{A}$ and $\mathbf{b}$ to be the arrays that contain $a_{jk\ell m}$ and b_{jk} respectively, for $1 \le j,\ell \le M_1$ and $1 \le k,m \le M_2$.
The inverse problem (10) now becomes a quadratic programming problem in which one minimizes $\mathbf{p}^T\mathbf{A}\mathbf{p} + 2\mathbf{p}^T\mathbf{b} + c$ over $P^M(G)$. To include the equality constraint, $\sum p_{jk} = 1$, we introduce the Lagrange multiplier λ and solve the unconstrained problem $\min F(\mathbf{p},\lambda)$, where

$$F(\mathbf{p},\lambda) \equiv \mathbf{p}^T\mathbf{A}\mathbf{p} + 2\mathbf{p}^T\mathbf{b} + c + \lambda[\sum_{j,k} p_{jk} - 1]\ . \qquad (11)$$

We used the "method of characteristics" based techniques discussed in Ito, Kappel and Peichl (1991) to solve (6) for the densities $v(t,x;g_{jk})$, which were calculated in parallel using PVM software on eight IBM RISC 6000 workstations. These densities were then used with the IMSL subroutine *DQPROG* to solve for the optimal solution $(\mathbf{p}^*, \lambda^*)$. Finally, the aggregate population density $u(t, x)$ was determined by substituting p^*_{jk} into (9).

In the above parameter estimation problem, for each subgroup jk, a proportioned initial size density $\Phi(x)p_{jk}$ was assumed. The Day 195 data (see Figure 1) was chosen so that

$$\sum_{j,k} v(t_0, x; g_{jk})p_{jk} = \sum_{j,k} \Phi(x)p_{jk}.$$

Note that since the first data set is used here as the initial size density, it cannot be used in solving the inverse problem. This leaves one less data set available for use in estimating the optimal parameter set.

Example 1: We present in the following an example which is typical of the field-data fitting results. The fish were stocked on June 28, 1982 into four rice paddies with parallel water flow. Each paddy had an outflow trap to measure emigration. Measurements were taken weekly (two paddies one day and two the next) with the use of fifteen traps per paddy. The total number of fish caught was greater than or equal to the number actually measured.

The size distribution frequency for size class i is defined as $f_i = n_{m,i}/N_m$, where $n_{m,i}$ is the number of fish measured in size class i, and N_m is the total number of fish measured.

The total population was divided into five hundred and twelve subpopulations with

$$r_j = 0.2 + 1/31 * 4.8 * (j-1) \qquad j = 1, 2, \ldots, 32$$

$$\gamma_1 = 16/38, \ \gamma_2 = 22/38, \ \gamma_3 = 24/38,$$

$$\gamma_k = 16/38 + 1/15 * 22/38 * (k-1) \qquad k = 4, \ldots, 16.$$

The range for r_j was chosen arbitrarily, while the γ_k were based on the field data. The choice for $\gamma_1 = 16/38$ was made by inspecting the data from Day 195. At $x = 16$mm, a significant decrease in the number of fish occurs, which we interpreted as the smallest possible maximum length for some subpopulation. We obtained $\gamma_1 = 16/38$ and $\gamma_{16} = 1.0$ after normalizing against the largest possible length, $x_1 = 38$mm, for the total population. The initial size density $\Phi(x)$ (for all subpopulations) was approximated by interpolating the Day 195 data.

The computed results manifest dispersion and bimodality as we expected (see Figure 1), with a residual of $J_1 = 3.7626 \times 10^{-4}$. Figures 2 and 3 illustrate the probability densities and distribution respectively as a function of r and γ.

Although the computed results do provide dispersion and bimodality, the mismatches between the computed solutions and the field data are still significant. Furthermore, the discrepancies did not diminish as fast as hoped when we increased the number of subpopulations, $M = M_1 \times M_2$. The idea of varying the initial size density among the subgroups became appealing.

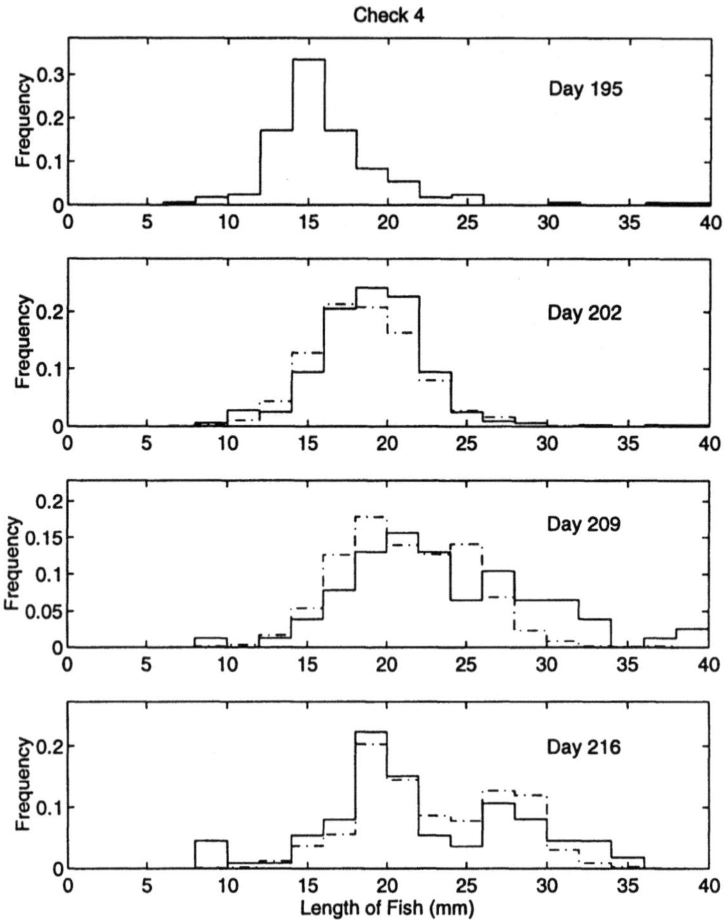

FIGURE 1. Computed results (- .) vs. field data (—) for Example 1.

2.2 GRD Models with Subpopulation Dependent Initial Size Densities

A further refinement (and one which is appropriate for the mosquitofish population presented here) entails a parameterization of the initial density $\Phi(x) = \sum_{j,k} \phi_{jk}(x)$ with $\phi_{jk}(x) = \sum_{\ell} \chi_{\ell}(x)\alpha_{\ell;j,k}$. Here $\chi_{\ell}(x)$ is the characteristic function of the interval $[x_{\ell}, x_{\ell+1})$ corresponding to a parti-

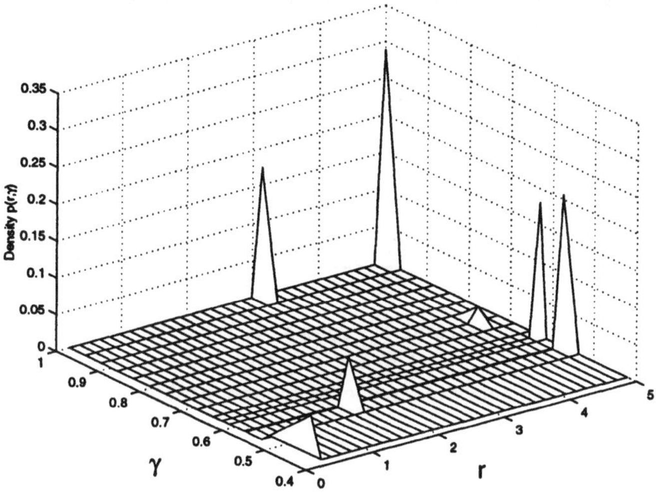

FIGURE 2. Probability densities p_{jk} for Example 1.

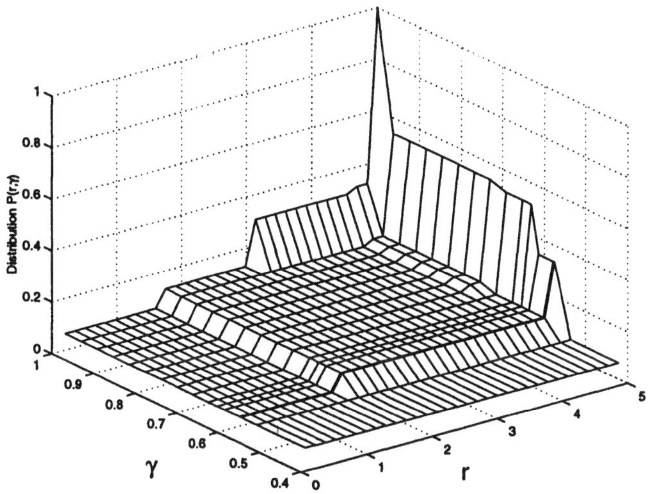

FIGURE 3. Probability distribution P_{jk} for Example 1.

tion $0 = x_0 = x_1 < x_2 < \cdots < x_{M_3+1} = 1$, and we have normalized the size range $[x_0, x_m]$ to [0,1]. If one solves (6) with $g = g_{jk}, \Phi = \chi_\ell$, then $v(t, x; g_{jk}, \chi_\ell)$ represents the density for individuals with growth rate g_{jk} that initially have size structure density given by χ_ℓ.
The subpopulation jk density is given by

$$v(t, x; g_{jk}) = \sum_\ell v(t, x; g_{jk}, \chi_\ell)\alpha_{\ell;j,k}$$

so that the aggregate population density is then given by

$$u(t, x) = \sum_{j,k} v(t, x; g_{jk})p_{jk} = \sum_\ell \sum_{j,k} v(t, x; g_{jk}, \chi_\ell)p_{jk}\alpha_{\ell;j,k}. \tag{12}$$

As before, we require the constraint $\sum_{j,k} p_{jk} = 1$. We denote α as the array containing $\alpha_{\ell;j,k}$ for $1 \le j \le M_1$, $1 \le k \le M_2$ and $1 \le \ell \le M_3$.
This formulation leads to a new inverse least squares problem with objective function

$$\begin{aligned} J(\mathbf{p}, \alpha) &= \sum_{n=0}^{N} \int_0^1 |u(t_n, x) - \hat{u}(t_n, x)|^2 \, dx \qquad (13) \\ &= \sum_{n=0}^{N} \int_0^1 \left| \sum_{j,k} \sum_{\ell=1}^{M_3} v(t_n, x; g_{jk}, \chi_\ell)p_{jk}\alpha_{\ell;j,k} - \hat{u}(t_n, x) \right|^2 dx \end{aligned}$$

subject to $\sum_{j,k} p_{jk} = 1$, and $v(t, x; g_{jk}, \chi_\ell)$ being the solution of (6) corresponding to $g = g_{jk}$ and $\Phi = \chi_\ell$. Minimizing (13) over both $\mathbf{p}$ and α requires solving a nonlinear programming problem.
An alternative and simpler approach to directly solving (13) is to reduce the nonlinear programming problem back to a quadratic programming problem. This can be accomplished by eliminating either $\mathbf{p}$ or α from the minimization process.

2.2.1 Minimizing Over α Only

One method for reducing (13) to a quadratic programming problem is to remove $\mathbf{p}$ from the parameter estimation problem and simply minimize over α. That is, if we can pre-determine a good choice for $\mathbf{p}$, we can fix the probabilities p_{jk} in (13) and reformulate it as a quadratic programming problem with objective function

$$F(\alpha) \equiv \alpha^{\mathrm{T}}\mathbf{A}\alpha + 2\alpha^{\mathrm{T}}\mathbf{b} + c. \tag{14}$$

Here $\mathbf{A}$ is the $M_1M_2M_3 \times M_1M_2M_3$ array containing the elements

$$a_{jk\ell,qst} = \sum_{n=0}^{N} \int_0^1 [v(t_n, x; g_{jk}, \chi_\ell)p_{jk}][v(t_n, x; g_{qs}, \chi_t)p_{qs}]dx, \tag{15}$$

b is the vector with elements

$$b_{jk\ell} = -\sum_{n=0}^{N} \int_0^1 \hat{u}(t_n, x) v(t_n, x; g_{jk}, \chi_\ell) p_{jk} dx, \tag{16}$$

and

$$c = \sum_{n=0}^{N} \int_0^1 [\hat{u}(t_n, x)]^2 dx. \tag{17}$$

A good way to pre-determine the probabilities p_{jk} is to first solve the problem for (11) with identical initial size densities. We can then use the solutions $\mathbf{p}^*$ as the fixed probabilities in the arrays **A** and **b**.

A major advantage of this approach is that the formulation allows for different initial size densities across subpopulations. Moreover, since the initial size densities are not estimated directly with the first data set, all of the data sets may be used in solving the inverse problem.

As before, we used "method of characteristics" based techniques, parallel computation, and the IMSL subroutine *DQPROG* to solve for the optimal solution α^*. The resulting $\alpha^*_{\ell;j,k}$ and the pre-determined p^*_{jk} were then substituted into (12) to solve for $u(t, x)$ corresponding to $(\mathbf{p}^*, \alpha^*)$.

Example 2: In the following example, we use the 1982 field data as described in Example 1. First the probability measures p_{jk} were obtained by optimizing (11), where the total population was divided into thirty-two subpopulations according to the following:

$$\begin{aligned} r_j &= 0.2 + 1/7 * 4.8 * (j - 1) \qquad j = 1, 2, \ldots, 8 \\ \gamma_1 &= 16/38,\ \gamma_2 = 22/38,\ \gamma_3 = 24/38,\ \gamma_4 = 1. \end{aligned} \tag{18}$$

As in Example 1, the Day 195 data was used as the initial size distribution $\Phi(x)$ for all subgroups.

The computed probabilities p^*_{jk} were then used in solving the reformulated quadratic programming problem for (14) with subpopulation-dependent initial size densities. The entire population was further divided into a total of six hundred and eight subpopulations, with r_j and γ_k as in (18) for $j = 1, 2, \ldots, 8$, $k = 1, 2, 3, 4$, and

$$\chi_\ell = [x_\ell, x_{\ell+1}) = [1/19 * (\ell - 1), 1/19 * \ell) \qquad \ell = 1, \ldots, 19.$$

The intervals χ_ℓ each correspond to a 2mm size class as seen in the field data.

The computed weights $\alpha^*_{\ell;j,k}$ were then substituted into (12) along with the precomputed p^*_{jk} to obtain the aggregate population density $u(t, x)$ (see Figure 4). This calculated density gave a better fit to the field data than the fit obtained in Example 1. Figures 5 and 6 illustrate the probability densities and distribution respectively as a function of r and γ.

The overall residual was $J_2 = 3.08780 \times 10^{-4}$. Since the parameter estimation carried out in Example 1 did not utilize the Day 195 data, we define a second residual $\tilde{J}_2$ which is taken only over Days 202, 209 and 216 of the data. This provides a means of comparing the accuracy of Example 1 versus Example 2.
In this manner we obtained $\tilde{J}_2 = 2.9305 \times 10^{-5}$, which is quite an improvement over $J_1 = 3.7626 \times 10^{-4}$.

2.2.2 Iterative Quadratic Programming

An alternative approach for simplifying the nonlinear programming problem for (13) is an iterative process involving two separate quadratic programming problems. The first of these quadratic programming problems is formulated by minimizing (13) over p_{jk} while $\alpha_{\ell;j,k}$ is held constant, and the second minimizes over $\alpha_{\ell;j,k}$ while p_{jk} is held constant. These quadratic programming problems can be solved in succession, creating an iterative process that progressively improves upon $\mathbf{p}$ and α.
Since the variables $\alpha_{\ell;j,k}$ are associated with the initial size densities ϕ_{jk}, we can approximate $\alpha_{\ell;j,k}$ using the initial time data. That is, we let

$$\chi_\ell = [x_\ell, x_{\ell+1}) \qquad \ell = 1, \ldots, M_3,$$

subdividing the size classes in the same manner as the data. For each ℓ, we then set

$$\alpha_{\ell;j,k} = \hat{u}(0, x_\ell)$$

for every j, k.
These values of $\alpha_{\ell;j,k}$ can then be inserted as fixed quantities in (13) to arrive at a quadratic programming problem with objective function

$$F(\mathbf{p}, \lambda) \equiv \mathbf{p}^T \mathbf{A} \mathbf{p} + 2\mathbf{p}^T \mathbf{b} + c + \lambda[\sum_{j,k} p_{jk} - 1]. \tag{19}$$

Here $\mathbf{A}$ is the $M_1 M_2 \times M_1 M_2$ array with elements

$$a_{jk\ell m} = \sum_{n=0}^{N} \int_0^1 [\sum_{i=1}^{M_3} v(t_n, x; g_{jk}, \chi_i)\alpha_{i;j,k}][\sum_{i=1}^{M_3} v(t_n, x; g_{\ell m}, \chi_i)\alpha_{i;\ell,m}]dx,$$

the vector $\mathbf{b}$ contains elements

$$b_{jk} = -\sum_{n=0}^{N} \int_0^1 \hat{u}(t_n, x)[\sum_{i=1}^{M_3} v(t_n, x; g_{jk}, \chi_i)\alpha_{i;j,k}]dx,$$

and

$$c = \sum_{n=0}^{N} \int_0^1 [\hat{u}(t_n, x)]^2 dx.$$

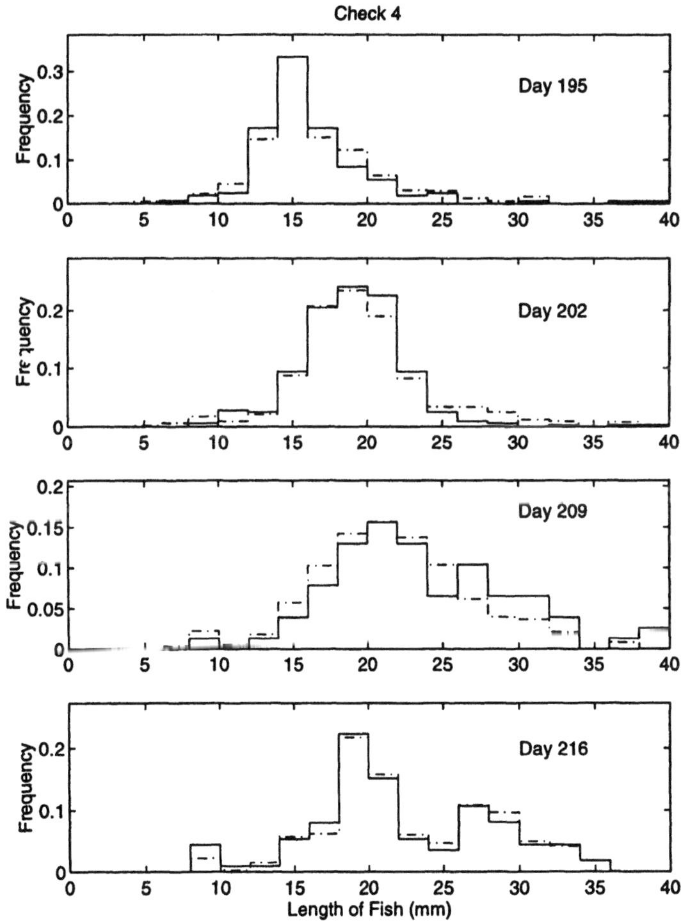

FIGURE 4. Computed results (- .) vs. field data (—) for Example 2.

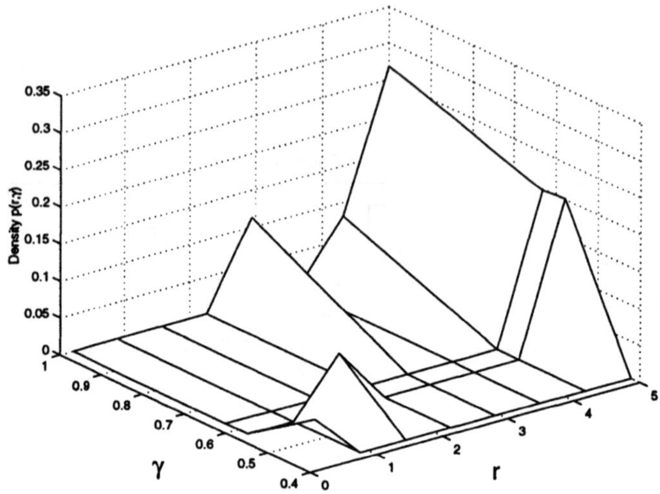

FIGURE 5. Probability densities p_{jk} for Example 2.

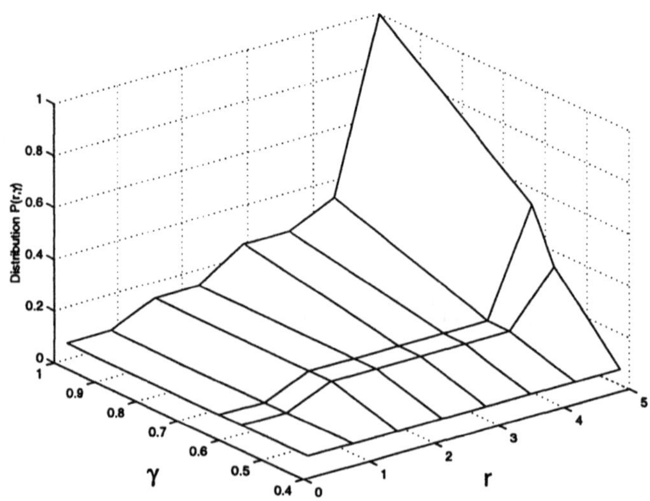

FIGURE 6. Probability distribution P_{jk} for Example 2.

Again we used the "method of characteristics" based techniques, parallel computation and the IMSL subroutine *DQPROG* to obtain the optimal $(\mathbf{p}^*, \lambda^*)$.
These probability measures p^*_{jk} may now in turn be used to solve a quadratic programming problem over $\alpha_{\ell;j,k}$. In fact, by holding $\mathbf{p}$ constant in the original problem for (13), we have reduced the nonlinear programming problem to the quadratic programming problem involving (14)-(17). From this we obtain the optimal α^*.
This process can be repeated, leading to the following iterative method:

1. Partition the size interval into subintervals $\{x_\ell, x_{\ell+1}\}, \ell = 1, 2, \ldots, M_3$, so that the subintervals correspond to the subdivisions of size classes in the data. For each ℓ, let $\alpha^0_{\ell;j,k} = \hat{u}(0, x_\ell)$.

2. (a) For $i = 0, 1, \ldots$, let $\alpha_{\ell;j,k} = \alpha^i_{\ell;j,k}$ and solve the quadratic programming problem for (19) to obtain the optimal p^i_{jk}.

   (b) With $p_{jk} = p^i_{jk}$ for all j, k, solve the quadratic programming problem for (14) to obtain the optimal $\alpha^{i+1}_{\ell;j,k}$.

3. Stop iterating when $|\mathbf{p}^{i+1} - \mathbf{p}^i| < TOL_1$ and $|u^{i+1}(0, x) - u^i(0, x)| < TOL_2$, for some user-specified tolerances TOL_1 and TOL_2.

Our choice of the above stopping criteria was motivated by the structure of our model. The p_{jk} are probabilities associated with the subpopulations g_{jk}, and enter the model through (9). We would like $|\mathbf{p}^i - \mathbf{p}^{i+1}|$ to be within a specified tolerance. An additional stopping criterion should involve the variables $\alpha_{\ell;j,k}$, which are weights associated with the parameterization of the initial size density $\Phi(x) = \sum_{j,k} \sum_\ell \chi_\ell(x) \alpha_{\ell;j,k}$. The individual $\alpha_{\ell;j,k}$ themselves need not converge upon iteration, but we would like the initial size densities to converge. That is, we would like
$|u^i(0, x) - u^{i+1}(0, x)| < TOL_2$, for some specified tolerance TOL_2.
Example 3: In this example we continue to use the 1982 field data as detailed in Example 1. To initialize the iterative process, we approximated the weights $\alpha^0_{\ell;j,k}$ by the initial data from Day 195. That is, we let

$$\chi_\ell = [x_\ell, x_{\ell+1}) = [1/19 * (\ell - 1), 1/19 * \ell) \qquad \ell = 1, \ldots, 19,$$

and for each ℓ we set $\alpha^0_{\ell;j,k} = \hat{u}(0, x_\ell)$ for all j, k.
We divided the total population into six hundred and eight subpopulations with

$$r_j = 0.3 + 1/3 * 4.7 * (j - 1) \qquad j = 1, 2, 3, 4$$

$$\gamma_1 = 16/38, \ \gamma_2 = 22/38, \ \gamma_3 = 24/38, \tag{20}$$

$$\gamma_k = 16/38 + 1/7 * 22/38 * (k - 1) \qquad k = 4, \ldots, 8.$$

The variables $\alpha^0_{\ell;j,k}$ were then inserted into (19) and the quadratic programming problem was solved for the probability measures p^0_{jk}. The corresponding aggregate population density $u^0(t,x)$ was formed by substituting $\alpha^0_{\ell;j,k}$ and p^0_{jk} into (12).
These computed solutions $u^0(t,x)$ did not fit the field data as well as any of the previous computed solutions (see Figure 7), returning a residual of $J^0_3 = 0.0020$, and a residual over Days 202, 209 and 216 of $\tilde{J}^0_3 = 0.0018$.
Continuing the iterative process, we used p^0_{jk} to optimize (14) for $\alpha^1_{\ell;j,k}$. This led to a new set of computed solutions $u^1(t,x)$, which fit the field data better than the previous solution set (see Figure 7), and returned the improved residuals $J^1_3 = 5.6266 \times 10^{-4}$ and $\tilde{J}^1_3 = 2.1284 \times 10^{-4}$. We also found that $|u^0(0,x) - u^1(0,x)|_2 = 0.0600$.
Using $\alpha^1_{\ell;j,k}$ we then optimized (19) for p^1_{jk}. The resulting $u^2(t,x)$ did not appear to improve the fit to the field data, as illustrated by the residuals $J^2_3 = 5.6266 \times 10^{-4}$ and $\tilde{J}^2_3 = 2.1284 \times 10^{-4}$. In addition, we noted that $|\mathbf{p}^0 - \mathbf{p}^1|_2 = 1.9451 \times 10^{-6}$, which also suggested that the latest iteration did not bring much improvement. The computed solutions $u^2(t,x)$ are not shown here since the difference between $u^1(t,x)$ and $u^2(t,x)$ are not perceptible to the eye at this resolution.
Finally, we used p^1_{jk} to optimize (14) for $\alpha^2_{\ell;j,k}$. The resulting fit to the field data did improve slightly over the previous fit, returning the residuals $J^3_3 = 4.6647 \times 10^{-4}$ and $\tilde{J}^3_3 = 1.4301 \times 10^{-4}$. We also found that $|u^2(0,x) - u^3(0,x)|_2 = 0.0043$. Even though there was an improvement in the residuals, there was no perceptible improvement visually in the graph of $u^3(t,x)$ compared to $u^2(t,x)$.
The probability densities $p^0(r,\gamma)$ and distribution $P^0(r,\gamma)$ are shown in Figures 8 and 9 respectively. The densities and distributions for $\mathbf{p}^1$ are not shown here because the difference between $\mathbf{p}^0$ and $\mathbf{p}^1$ is not perceptible to the eye at this resolution. Note that for fixed r, the probability distribution $P(\gamma)$ suggests a bimodal population density, as obtained in Banks and Fitzpatrick (1991). The probability densities $p(r,\gamma)$ do not contribute to this conclusion, and in fact limiting densities themselves may not exist for measures in the set $\{P(g) : g \in G\}$ for general infinite dimensional families G. (see Banks and Fitzpatrick (1991); the *distributions, not the densities,* can be approximated as we have done in these computations.)

Acknowledgments: This research was supported in part by the U.S. Air Force Office of Scientific Research under grants AFOSR F49620-95-1-0236, AFOSR F49620-93-1-0355, AFOSR F49620-95-1-0375, and in part by the U.S. Department of Education through a GAANN Fellowship to Y.Z. under grant P200A40730 and by an NSF-GRT fellowship to L.K. Potter under grant GER-9454175.

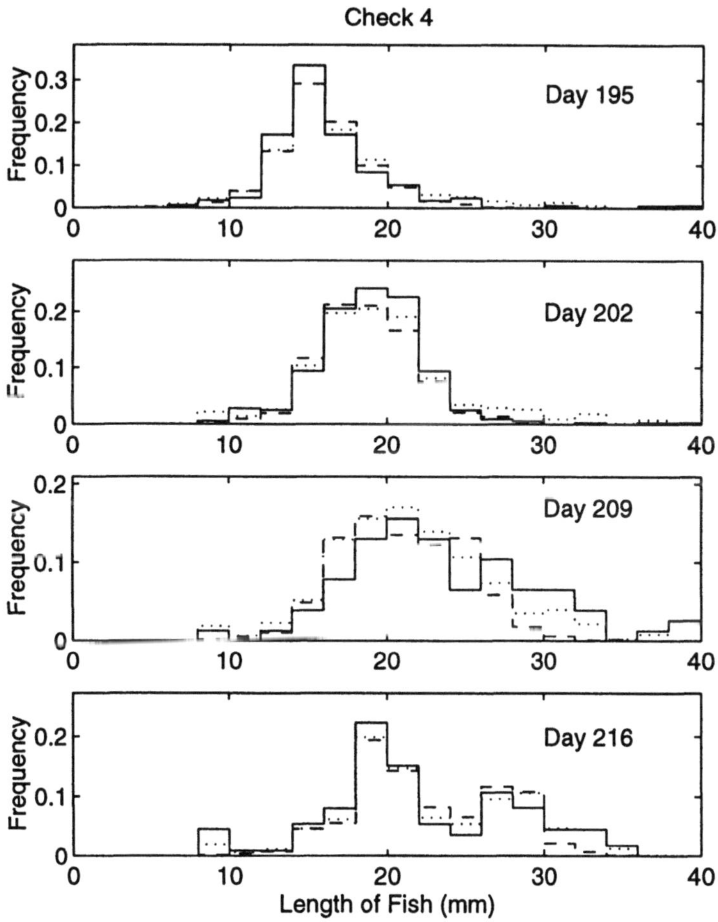

FIGURE 7. Computed results $u^0(t,x)$ (- -) and $u^1(t,x)$ (· ·) vs. field data (—) for Example 3.

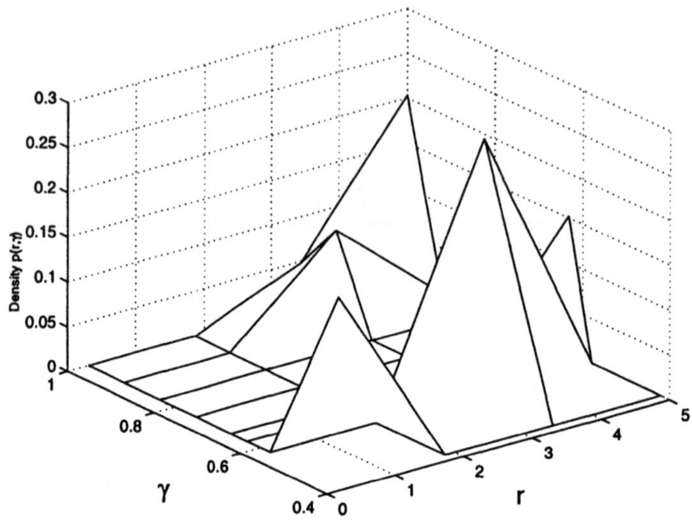

FIGURE 8. Probability densities p^0_{jk} for Example 3.

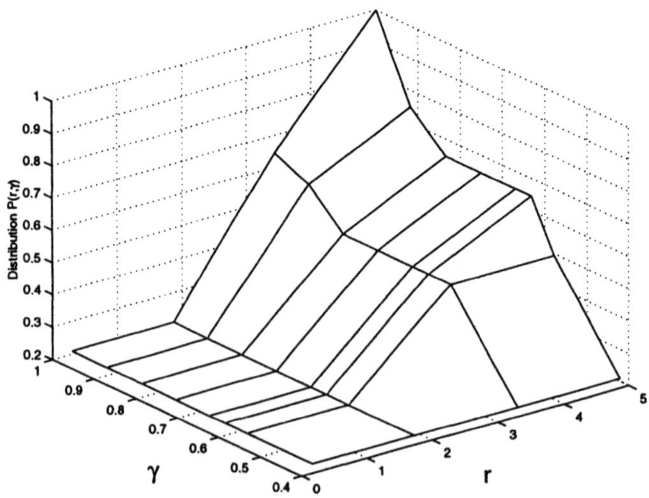

FIGURE 9. Probability distribution P^0_{jk} for Example 3.

References

Banks, H.T. (1988). In *Proceedings IFIP Conference on Optimal Control of Systems Governed by PDE, (Santiago de Compostela, July, 1987) Springer LN in Control and Info. Science,* Springer **114**, 3-10.

Banks, H.T. (1994). In *Frontiers of Theoretical Biology, Lecture Notes in Biomathematics,* Springer-Verlag, Berlin, **100**, 609-623.

Banks, H.T., Botsford, L.W., Kappel, F. and Wang, C. (1988). In *Proceedings 2nd Course on Mathematical Ecology, (Trieste, December 8-12, 1986),* World Press, Singapore, 521-541.

Banks, H.T. and Fitzpatrick, B.G. (1991). Quart. Appl. Math **49**, 215-235.

Banks, H.T., Fitzpatrick, B.G. and Zhang, Y. (1996). In *Differential Equations and Applications to Biology and to Industry,* World Science Press, 13-22.

Banks, H.T., Tran, H.T. and Woodward, D.E. (1993). SIAM J. Num. Anal. **30**, 1574-1602.

Botsford, L.W., Vandracek, B., Wainwright, T., Linden, A., Kope, R., Reed, D. and Cech, J.J. (1987). Journal of Mathematical Analysis and Applications, **172**, 73-91.

Fitzpatrick, B.G. (1993). Journal of Mathematical Analysis and Applications, **172**, 73-91.

Fitzpatrick, B.G. (1994). In *Control and Estimated Parameter Systems: Nonlinear Phenomena, Intl. Series in Num. Math.,* Birkhäuser **118**, 131-141.

Fleming, W.H. (1974). In *Control Theory, Numerical Methods and Computer Systems Modelling, Lecture Notes in Economics and Mathematical Systems,* Springer-Verlag, Berlin, **107**, 180-191.

Fleming, W.H. (1975). J. Math Biology **2**, 219-233.

Fleming, W.H. and Su, C.H. (1974). Theor. Population Biology **5**, 431-449.

Harrison, G.W. (1988). Num. Meth. for PDE, **4**, 219-232.

Ito, K., Kappel, F. and Peichl, G. (1991). Siam J. Numerical Analysis **28**, 923-954.

Okubo, A. (1980). *Diffusion and Ecological Problems: Mathematical Models,* Springer-Verlag, New York.

Weiss, G.H. (1968). Bull. Math. Biophys., **30**, 427-435.

22

Solvable Infinite Time Horizon Stochastic Control Problems in Noncompact Symmetric Spaces

T. E. Duncan

ABSTRACT: A family of infinite time horizon, discounted cost and ergodic cost control problems are formulated and explicitly solved in noncompact symmetric spaces that are symmetric tubes or Siegel domains. Similar results can be verified for symmetric cones so that all classical symmetric spaces can be included as well as four exceptional symmetric spaces. For each spherical polynomial, a solvable control problem on each of these symmetric spaces is given. The problems are to control a Brownian motion in the symmetric space by a drift vector field so that the resulting process remains close to the origin. The control problem has an invariance under the maximal compact subgroup K for the symmetric space that is inherited from the K-invariance of the spherical polynomials.

KEYWORDS: infinite time horizon stochastic control, solvable stochastic control, discounted cost stochastic control, ergodic cost stochastic control.

1 Introduction

While many results are available for the existence of optimal controls and the existence and the uniqueness of the solution of the Hamilton-Jacobi-Bellman equation for a stochastic control problem, there are relatively few examples where an optimal control is expressed explicitly as a function of the state of the system. This phenomenon seems to be especially true for the infinite time horizon stochastic control problems. The most well known example is the linear quadratic Gaussian problem (Fleming and Rishel (1975)). Some other examples occur in the mathematics of finance for portfolio and portfolio/consumption problems (e.g., Davis and Norman (1990), Merton (1971), Taksar, Klass, and Assaf (1988)). In this paper some examples are given for a discounted cost and an ergodic cost (average cost per unit time) where the controlled diffusion evolves in a noncompact symmetric space.

These stochastic control problems were motivated by some finite time horizon solvable stochastic control problems (Duncan, 1987; Duncan 1991; Duncan and Upmeier, 1995). These problems are specific examples of stochastic control in Riemannian manifolds (Duncan (1977), Duncan (1979)). For real hyperbolic three space, an infinite time horizon discounted cost and an ergodic cost control problems are solved explicitly in Duncan (1997).

The noncompact symmetric spaces of arbitrary rank are a large family of manifolds that are useful models for many physical phenomena and have good mathematical properties especially from their relation with Lie groups. The stochastic control formulation and explicit solution can include a symmetric space that is associated with any classical Lie group and some exceptional Lie groups. Two general families of noncompact symmetric spaces are i) symmetric cones and ii) symmetric tubes and Siegel domains.

A well known example of a symmetric cone is a family of positive definite matrices of fixed rank and an example of a symmetric tube is a family of nonsquare matrices.

Since the Killing form on the Lie algebra induces a (natural) Riemannian structure on the symmetric space (Helgason (1984)), there are a Laplacian and a standard Brownian motion on the symmetric space. Since a symmetric space is a homogeneous space, there is a natural origin for the space. The stochastic control problem is to keep the Brownian motion near the origin using a drift vector field with a criterion that is an infinite horizon discounted cost or an ergodic cost. To solve explicitly a stochastic control problem, the relation between the cost functional and the infinitesimal generator of the controlled process is important. For the control problem in a symmetric space it is necessary to express the Laplacian in a "good" coordinate system. The cost functional for the control problem depends on a spherical function which induces a symmetry on the control problem that reduces the analysis to a maximal abelian subspace.

2 Preliminaries

Let Γ be a bounded symmetric domain. A transitive (reductive) Lie group of isometries of Γ has a Lie algebra $\mathfrak{g}$ with the Cartan decomposition $\mathfrak{g} = \mathfrak{k} \oplus \mathfrak{p}$. It is more convenient here to use the unbounded realization Π of the symmetric domain as a Siegel domain. The Cayley transform maps Γ onto Π.

There is a Peirce decomposition of the tangent space Z at an arbitrary base point as

$$Z = X \oplus U \oplus V$$

where

$$X := \{x \in Z : \{ee^*x\} = x = \{ex^*e\}\}$$

is a formally real Jordan algebra with unit element e and product $x \circ y =$

$\{xe^*y\}$,

$$U := \{\xi \in Z : \{ee^*\xi\} = \xi = -\{e\xi^*e\}\}$$

and

$$V := \{z \in Z : \{ee^*z\} = z/2\}.$$

The interior of the family of squares in X is a cone Λ and the unbounded model of the symmetric R-space is the Siegel domain

$$\Pi := \{x + \xi + v \in Z : 2x - \Phi(v,v) \in \Lambda\}$$

where $\Phi(v_1, v_2) := 2\{v_1 v_2^* e\}$.
The unbounded domain Π is separated into three cases i) $Z = X$, ii) $Z = X \oplus U$ and iii) $Z = X \oplus U \oplus V$. For the case i), Π is the cone Λ, for ii), Π is a symmetric tube domain $\Lambda \times U$ and for iii), Π is a Siegel domain. If these cases are irreducible then the resulting symmetric spaces correspond to the root systems i) A_{r-1}, ii) C_r or D_r and iii) B_r or BC_r. Four exceptional spaces of dimensions 16, 27, 32, and 54 are also included in the analysis.
A Riemannian symmetric space can be algebraically characterized by a ternary product structure (Lie triple system) on its tangent space Z at an arbitrary base point (e.g., Loos (1969)). Another ternary product (Jordan triple system) on Z that is more important here is easily related to the Lie ternary product (Loos (1971)). The symmetric spaces admitting such a Jordan triple structure include all of the classical spaces and four of the exceptional ones.
Each (formally real) Jordan algebra X with a product induces a (positive definite) Jordan triple product. The Cayley transform maps the bounded domain description of the Jordan triple system onto the symmetric cone

$$\Lambda = \{x^2 : x \in X \text{ invertible}\} \tag{1}$$

associated with X. The cone is homogeneous under its (reductive) linear automorphism group

$$G^0 = GL(\Lambda) := \{P \in GL(X) : P\Lambda = \Lambda\}$$

and

$$K^0 = O(\Lambda) := \{P \in GL(\Lambda) : Pe = e\}$$

is a maximal compact subgroup whose elements are the Jordan algebra automorphisms of X.
A well known example of a symmetric cone is a matrix cone. For $\mathbb{K} = \mathbb{R}$, $\mathbb{C}$ or $\mathbb{H}$ (quaternions), the self-adjoint $r \times r$ matrices

$$X_{\mathbb{K}} = \{x \in \mathbb{K}^{r \times r} : x = x^*\}$$

form a Jordan algebra with unit $e = 1$ (the identity matrix) under the anticommutator product

$$x \circ y := \frac{1}{2}(xy + yx).$$

The corresponding Jordan triple product is

$$\{xy^*z\} = \frac{1}{2}(xy^*z + zy^*x)$$

and

$$\Lambda_{\mathbb{K}} = \{x = x^* \in \mathbb{K}^{r \times r} : x \text{ positive definite}\}$$

is a matrix cone. If $P_g x := gxg^*$ then

$$GL(\Lambda_{\mathbb{K}}) = \{P_g : g \in GL(r, \mathbb{K})\}$$

and

$$O(\Lambda_{\mathbb{K}}) = \{P_g : g \in U(r, \mathbb{K}) \text{ (unitary)}\}.$$

Endow Λ with the unique G^0-invariant Riemannian structure such that the basis $(t_1, \ldots, t_r)$ of $\mathfrak{a} \subset \mathfrak{p}^0 \equiv T_e\Lambda$ is orthonormal where $\mathfrak{a}$ ia a maximal abelian subspace of $\mathfrak{p}^0$ and $(t_1, \ldots, t_r)$ is a natural basis (Duncan and Upmeier (1995)). Define the map $\varphi : \mathfrak{a} \to \Lambda$ given by

$$\varphi(t_1, \ldots, t_r) := \exp\left(\sum_i t_i M_{e_i}\right)(e) = \sum_i e^{t_i} e_i. \; eqno(2)$$

Now consider the maximal compact subgroup $K^0 = O(\Lambda)$ of $GL(\Lambda)$. Let

$$\mathfrak{a}_+^0 := \left\{\sum_{i=1}^{r} t_i M_{e_i} : t_1 < \cdots < t_r\right\} \tag{3}$$

denote the "positive Weyl chamber." It follows (Helgason (1984)) that

$$\Lambda = K^0 \cdot \varphi(\mathfrak{a}_+^0).$$

Let $D(\Lambda)$ be the algebra of all (scalar) differential operators on Λ that are G invariant. Each $D \in \mathcal{D}(\Lambda)$ has a K^0-radial part D_K which is a differential operator on $\mathfrak{a}_+^0$ satisfying

$$(Df) \circ \varphi = D_K(f \circ \varphi)$$

for all K^0-invariant functions $f \in C^\infty(\Lambda)$.
The K^0-radial part of the Laplace-Beltrami operator Δ^0 is

$$\Delta_K^0 = \sum_{i=1}^{r} \left(\frac{\partial^2}{\partial t_i^2} - \frac{a}{2}(r-1)\frac{\partial}{\partial t_i}\right) + a \sum_{i \neq j} \frac{1}{\exp(t_j - t_i) - 1} \frac{\partial}{\partial t_i} \tag{4}$$

where a is a nonnegative integer from the Lie structure (Duncan and Upmeier (1995)).
To construct the spherical functions, the *minors* of an irreducible Jordan algebra X with respect to a frame $\{e_1, \ldots, e_r\}$ of idempotents are introduced. The determinant of X is the unique irreducible polynomial N_r of degree r on X that satisfies

$$N_r(Px) = N_r(x)N_r(Pe) \tag{5}$$

for all $P \in GL(\Lambda)$ and $x \in X$ and is normalized so that $N_r(e) = 1$. For each $k \geq 1$ the Jordan subalgebra

$$X_k := \sum_{1 \leq i \leq j \leq k} \oplus X_{ij}$$

with unit element $e_1 + \cdots + e_k$ has a determinant $N_k : X_k \to \mathbb{R}$ of degree k. The determinant N_k can be extended to X by the orthogonal projection $X \mapsto X_k$. The polynomials $(N_1, \ldots, N_r)$ are called the minors of X.
For a real or complex matrix, the kth minor is

$$N_k \begin{pmatrix} x_{11} & \cdots & x_{1r} \\ \vdots & & \vdots \\ x_{r1} & & x_{rr} \end{pmatrix} := \det \begin{pmatrix} x_{11} & \cdots & x_{1k} \\ \vdots & & \vdots \\ x_{k1} & & x_{kk} \end{pmatrix} \tag{6}$$

By integrating over the maximal compact subgroup K^0, a smooth K^0-invariant function

$$K_\alpha^0(x) := \int_{K^0} N_\alpha(sx)ds$$

on Λ is obtained that is called the spherical function of type $\alpha \in \mathbb{C}^r$. If $\alpha \in \vec{\mathbb{N}}^r$, then K_α^0 is a polynomial on X that is called the spherical polynomial of type α. A spherical function is an eigenfunction of Δ^0, that is,

$$\Delta^0 K_\alpha^0 = \chi_\alpha^0 K_\alpha^0$$

for each $\alpha \in \mathbb{C}^r$. For $\alpha \in \vec{\mathbb{N}}^r$, the spherical polynomial K_α^0 has a finite expansion

$$K_\alpha^0 \left(\sum_{i=1}^r x_i e_i \right) = \sum_{\beta \in \vec{\mathbb{N}}^r} c_\beta q_\beta(x_1, \ldots, x_r) \tag{7}$$

where $c_\beta \geq 0$ and q_β is a monomial symmetric function (Proposition 4.21 in Duncan and Upmeier (1995)).
These spherical polynomials for a symmetric cone of positive definite matrices over $\mathbb{R}$ or $\mathbb{C}$ were constructed in statistics by taking the expectation of a matrix with entries that are independent Gaussian random variables (e.g., Farrell (1985), Muirhead (1982), and Takemura (1984)).

Every Jordan *algebra with unit and involution $*$ induces a Jordan triple product. The Cayley transform (Korányi and Wolf (1965)) maps this space onto a symmetric tube domain

$$\Pi = \{z \in Z : z + z^* \in \Lambda\} = \Lambda \times U \tag{8}$$

where Λ is a symmetric cone of the formally real Jordan algebra

$$X := \{x \in Z : x^* = x\}$$

and

$$U := \{\xi \in Z : \xi^* = -\xi\}.$$

If $U = iX$, then Z is the complexification of the formally real Jordan algebra X and $\Pi = \Lambda \times iX$ is a complex Hermitian symmetric space. A basic example of this type is Siegel's (right) half-space that corresponds to the cone $\Lambda_{\mathbb{R}}$ of positive definite matrices over $\mathbb{R}$.

For the more general symmetric Siegel domain, the Jordan triple Z has the Peirce decomposition $Z = X \oplus U \oplus V$. The Cayley transform maps the bounded domain realization onto the symmetric Siegel domain

$$\Pi = \{x + \xi + v : 2x - \Phi(v, v) \in \Lambda\}$$

where Λ is the cone associated with X and Φ is an Hermitian form. If $U = iX$ and V is complex, then Π is an Hermitian symmetric space. An example of Z is the matrix space

$$Z = {}_r\mathbb{K}^R = {}_r\mathbb{K}^r \oplus {}_r\mathbb{K}^{(R-r)}$$

where $r < R$ are positive integers and $\mathbb{K} = \mathbb{R}, \mathbb{C}$ or $\mathbb{H}$ (the quaternions).

Associated with a bounded domain induced from Z is a transitive (reductive) Lie group of isometries G. Let K be a maximal compact subgroup of G. Let $\mathcal{D}(\Pi)$ be the algebra of all (scalar) differential operators on Π that are G-invariant, i.e., $(g_* D)f = D(f \circ g) \circ g^{-1} = Df$ for all $f \in C^\infty(\Pi)$ and $g \in G$. The Laplace-Beltrami operator Δ is an element of $\mathcal{D}(\Pi)$.

Each $D \in \mathcal{D}(\Pi)$ has a K-radial part D_K that is a differential operator on $\mathfrak{a}_+$, the "positive Weyl chamber," that is K-invariant. For the Laplace-Beltrami operator Δ the K radial part is

$$\begin{aligned}\Delta_K = &\sum_{i=1}^{r} \left(\frac{\partial^2}{\partial t_i^2} + \left(c\coth(t_i) + b\coth\left(\frac{t_i}{2}\right)\right) \frac{\partial}{\partial t_i} \right) \\ &+ a \sum_{i \neq j} \frac{\sinh(t_i)}{\cosh(t_j) - \cosh(t_i)} \frac{\partial}{\partial t_i}\end{aligned} \tag{9}$$

where a, b, c are nonnegative integers from the Lie algebra decomposition (Duncan and Upmeier (1995)).

Define the map $\psi : \mathfrak{a}_+ \to \Lambda \subset \Pi$ given by

$$\psi(t_1, \ldots, t_r) = \sum_{i=1}^{r} \sinh^2\left(\frac{t_i}{2}\right) e_i. \tag{10}$$

A spherical polynomial on a symmetric tube or Siegel domain Π is "induced" from the spherical polynomials on the associated symmetric cone. Specifically, for each $\alpha \in \vec{\mathbb{N}}^r$ there is a spherical polynomial K_α on Π with eigenvalue

$$\chi_\alpha = \sum_i \alpha_i \left(\alpha_i + a(r-1) + c + \frac{b}{2}\right) \tag{11}$$

which satisfies

$$K_\alpha \circ \varphi \circ \psi^{-1} = \sum_{\beta \leq \alpha} c_\alpha^\beta K_\beta^0 \tag{12}$$

where the sum is over all partitions β that are dominated by α and the coefficients c_α^β are nonnegative (Theorem 8.16 in Duncan and Upmeier (1995)). Using (7) the spherical function K_α on Π has a finite expansion

$$(K_\alpha \circ \varphi \circ \psi^{-1})(x) = \sum_{\beta \leq \alpha} a_\alpha^\beta q_\beta(x) \tag{13}$$

in terms of monomial symmetric functions q_β with nonnegative coefficients a_α^β.

3 Main Results

The infinite time horizon, discounted cost and ergodic cost stochastic control problems are formulated and explicitly solved where the controlled Brownian motion evolves in a symmetric tube or Siegel domain. The cost functional arises from a spherical polynomial so that there is an invariance under the maximal compact subgroup K and the cost increases as the process moves farther away from the origin.
The controlled diffusion in Π has the infinitesimal generator

$$\frac{1}{2}\Delta + \langle u(\tau), \nabla_\tau \rangle$$

where $\tau = (\tau_1, \ldots, \tau_k)$ are the coordinates for the positive Weyl chamber and Δ is the Laplace-Beltrami operator. Since the K-invariance of the cost functional and the control are exploited, it suffices to consider the controlled diffusion projected to the positive Weyl chamber $\mathfrak{a}_+$ that has the infinitesimal generator

$$\frac{1}{2}\Delta_K + \langle u(\tau), \nabla_\tau \rangle. \tag{14}$$

Let $(X(t), t \geq 0)$ be the controlled diffusion with the infinitesimal generator (14) where the dependence on the control u has been suppressed. Fix $\alpha \in \vec{\mathbb{N}}^r$ and let K_α be the associated spherical function given by (12). The discounted cost is

$$J_d(U) = E_{X(0)} \int_0^\infty e^{-\rho t}[(K_\alpha \circ \psi)(X(t)) + \langle A(X(t))U(X(t)), U(X(t))\rangle]dt \tag{15}$$

and the ergodic cost is

$$J_e(U) = \limsup_{T\to\infty} \frac{1}{T} E_{X(0)} \int_0^\infty [(K_\alpha \circ \psi)(X(t)) + \langle A(X(t))U(X(t)), U(X(t))\rangle]dt \tag{16}$$

where $\rho > 0$, $A(\tau) := \operatorname{diag}(f_1(\tau), \dots, f_r(\tau))$

$$f_i(\tau) = \frac{1}{({}_iK_\alpha \circ \psi)(\tau)} \left[\frac{\partial}{\partial \tau_i}(K_\alpha \circ \psi)(\tau)\right]^2 \tag{17}$$

and

$${}_iK_\alpha\left(\sum_{j=1}^r x_j e_j\right) = x_i \frac{\partial}{\partial x_i} \sum_\mu \frac{1}{|\mu|} c_\alpha^\mu q_\mu(x_1, \dots, x_r)$$

for $i = 1, 2, \dots, r$. It is clear that

$$K_\alpha \circ \psi = \sum_{i=1}^r {}_iK_\alpha \circ \psi + {}_0K_\alpha \tag{18}$$

where ${}_0K_\alpha$ is a constant.

Since the infinitesimal generator of the controlled diffusion $(X(t), t \geq 0)$ is (14), the family of stochastic differential equations that $(X(t), t \geq 0)$ satisfies is

$$dX_k(t) = \left[\frac{c}{2}\coth(X_k(t)) + \frac{b}{2}\coth\left(\frac{X_k(t)}{2}\right) + \frac{a}{2}\sum_{j\neq k} \frac{\sinh(X_k(t))}{\cosh(X_j(t)) - \cosh(X_k(t))} + U_k(X(t))\right]dt + dB_k(t) \tag{19}$$

for $k = 1, \dots, r$, $X(0) = (X_1(0), \dots, X_r(0))$, $X_1(0) > X_2(0) > \cdots > X_r(0) > 0$, $X(t) \in \mathfrak{a}_+$, and $((B_1(t), \dots, B_r(t)), t \geq 0)$ is a standard r-dimensional Brownian motion.

The family of admissible (Markovian) controls, $\mathcal{U}$, is all $\mathbb{R}^r$-valued Borel measurable functions of the controlled diffusion $X(\cdot)$ such that the associated family of stochastic differential equations (19) has one and only one strong solution and the associated costs are finite.

Initially it is necessary to verify that the stochastic control problems (15, 19) and (16, 19) are well posed, that is, there is an admissible control that gives a finite value to (15) and (16).

Lemma 1 *Let* $(X(t), t \geq 0)$ *be the* $\mathfrak{a}_+$*-valued diffusion that is the strong solution of the family* (19) *with the control* $U^1(X(t)) = -b_1 X(t)$ *for* $t \geq 0$ *where* $b_1 > 0$. *Then for* $\rho > 0$

$$E_{X(0)} \int_0^\infty e^{-\rho t}[(K_\alpha \circ \psi)(X(t)) + \langle A(X(t))U^1(X(t)), U^1(X(t))\rangle]dt < \infty \tag{20}$$

and

$$\limsup_{T\to\infty} \frac{1}{T} E_{X(0)} \int_0^\infty (K_\alpha \circ \psi)(X(t)) + \langle A(X(t))U^1(X(t)), U^1(X(t))\rangle]dt < \infty. \tag{21}$$

Proof. Initially assume that the Lie algebra is rank one so that $(X(t), t \geq 0)$ is an $\mathbb{R}_+$-valued process that satisfies the stochastic differential equation

$$dX(t) = \left[\frac{c}{2}\coth(X(t)) + \frac{b}{2}\coth\left(\frac{X(t)}{2}\right) + U(t)\right]dt + dB(t) \tag{22}$$

for suitable nonnegative integers b and c. A control $U^1(X(t)) = -b_1 X(t)$ for $t \geq 0$ is used where $b_1 > 0$. By a comparison theorem for the stochastic differential equation (p. 437, Ikeda and Watanabe (1989)) it suffices for the verification of (20,21) to consider the stochastic differential equation

$$dY(t) = \left(\frac{\alpha}{Y(t)} - \beta Y(t)\right)dt + dB(t) \tag{23}$$

where $\alpha > 0$ and $\beta > 0$. It can be shown (Duncan and Upmeier (1995)) that a solution of (22) or (23) with a positive initial condition does not hit the origin with probability 1 so that both of these equations have unique strong (pathwise) solutions.
It is shown that

$$\sup_{t\geq 0} E[\exp(\gamma Y(t))] < \infty$$

for each $\gamma > 0$ by computing the moments of $(Y(t), t \geq 0)$. Fix $p \in \mathbb{N}$. A differential equation and a differential inequality are obtained for $EY^p(t)$. The differential equation can be obtained by applying Itô's formula to $(Y^p(t), t > 0)$ and taking the expectation. The stochastic integral term is a martingale by the integrability from the Picard iteration construction of the solution or using a family of stopping times. Alternatively, Dynkin's formula can be obtained directly from the (local) integrability of the generator acting on the function $f(x) = x^p$. The differential equation is

$$\frac{d}{dt}EY^p(t) = \left(\alpha p + \frac{1}{2}p(p-1)\right)EY^{p-2}(t) - \beta p EY^p(t).$$

Using Hölder's inequality the following differential inequality is obtained

$$\frac{d}{dt}EY^p(t) \le \left(\alpha p + \frac{1}{2}p(p-1)\right)(EY^p(t))^{(p-2)/p} - \beta p EY^p(t). \tag{24}$$

Let $v(t) = EY^p(t)$ so the inequality (24) can be written as

$$\frac{dv}{dt} \le \left(\alpha p + \frac{1}{2}p(p-1)\right) v^{(p-2)/p} - \beta p v. \tag{25}$$

For a suitable $K_1 > 0$ on the set $\Gamma = \{x : K_1 p^{p/2} < x\}$ the right hand side of the inequality (25) is negative so for $t \ge T$ where T does not depend on p it follows that

$$v(t) \le K_2 p^{p/2}$$

or equivalently

$$EY^p(t) \le K_2 p^{p/2}.$$

where K_2 does not depend on p. Thus,

$$\begin{aligned} E\exp[\gamma Y(t)] &\le \sum_{k=0}^{\infty} \frac{\gamma^k}{k!} EY^k(t) \\ &\le \sum_{k=0}^{\infty} \frac{\gamma^k}{k!} K_2 k^{k/2} \le K < \infty \end{aligned} \tag{26}$$

where K is a constant that does not depend on t.
By definition of ψ in (10) $K_\alpha \circ \psi$ is a finite sum of exponentials. Given the inequality (26) it follows that for $\rho > 0$

$$E_{X(0)} \int_0^\infty e^{-\rho t}[(K_\alpha \circ \psi)(X(t)) + \langle A(X(t))U^1(X(t)), U^1(X(t))\rangle]dt < \infty$$

and

$$\limsup_{T\to\infty} \frac{1}{T} E_{X(0)} \int_0^\infty (K_\alpha \circ \psi)(X(t)) + \langle A(X(t))U^1(X(t)), U^1(X(t))\rangle]dt$$
$$< \infty.$$

If $\operatorname{rk} \mathfrak{g} > 1$ then a reduction method from Riemannian geometry to a scalar-valued process (p. 463 Ikeda and Waranabe (1989)) is used. Consider the point 0 that is the origin for the (Riemannian) symmetric space M. Consider the geodesic polar coordinates on T_0M, $(r, \theta^1, \ldots, \theta^{d-1})$ where $\dim M = d$ and $(\theta^1, \ldots, \theta^{d-1})$ are coordinates on the unit sphere in T_0M. Use this (global) coordinate system to express the Laplace-Beltrami operator Δ as

$$\begin{aligned} \Delta = \frac{\partial^2}{\partial r^2} + \frac{1}{\sqrt{\det g}} \\ \cdot \left(\frac{\partial}{\partial r}(\sqrt{\det g}) \frac{\partial}{\partial r} + \frac{1}{\sqrt{\det g}} \sum_{i,j=1}^{d-1} \frac{\partial}{\partial \theta^i} \left(g^{ij} \sqrt{\det g} \frac{\partial}{\partial \theta^j} \right) \right) \end{aligned} \tag{27}$$

where $g = (g_{ij})$ is the Riemannian metric in these coordinates and $g^{-1} = (g^{ij})$.
Let h be the Jacobian of $\exp_0$, the exponential map at 0. It follows that

$$\frac{\partial}{\partial r}\log h + \frac{d-1}{r} = \frac{\partial}{\partial r}\log\sqrt{\det g}.$$

Since M is a noncompact symmetric space, it has nonpositive sectional curvature (p. 241, Helgason (1978)) and the Ricci curvature on unit vectors is bounded below. It follows (p. 464, Ikeda and Watanabe (1989)) that for $r > 0$

$$\frac{\partial}{\partial r}\log h \leq \frac{\partial}{\partial r}A(r,b) - \frac{d-1}{r}$$

and

$$\frac{\partial}{\partial r}\log h \geq 0$$

where $b < 0$ and

$$A(r,b) = \left(\frac{1}{\sqrt{-b}}\sinh\sqrt{-b}\,r\right)^{d-1}.$$

by a suitable choice of control the stochastic differential equation in the direction r that is induced from (27) and a control vector field can be compared to (23) so an upper bound for $E\exp[\gamma\sum_{i=1}^{r} X_i(t)]$ can be obtained from the previous result for rank 1 □

The solution of the infinite time horizon, discounted expected cost, control problem is given now.

Theorem 1 *The stochastic control problem given by* (15, 19) *has an admissible optimal control* U^* *given by*

$$U^*(x) = -\frac{1}{2}gA^{-1}(x)\nabla_x(K_\alpha \circ \psi)(x) \tag{28}$$

where g *is the positive real root of*

$$\frac{1}{2}g^2 + g\left(\rho - \frac{1}{2}\chi_\alpha\right) - 1 = 0 \tag{29}$$

and the optimal cost is

$$J_d(U^*) = g(K_\alpha \circ \psi)(x) + h \tag{30}$$

where

$$h = \frac{g^2\,{}_0K_\alpha}{4\rho} \tag{31}$$

$$X(0) = x$$

${}_0K_\alpha$ *is given in* (18) *and* χ_α *is given by* (11).

Proof. The Hamilton-Jacobi-Bellman (HJB) equation for an infinite time horizon, discounted expected cost, control problem (Fleming and Soner (1993), Kushner and Dupuis (1992)) is

$$0 = \min_{v \in U}[A^v W(x) - \rho W(x) + L(x, v)] \tag{32}$$

where A^v is the infinitesimal generator of the controlled diffusion with the control v and the cost functional is

$$J(u) = E_x \int_0^\infty e^{-\rho t} L(x(t), u(t)) dt.$$

It is assumed that

$$E_x \int_0^\infty e^{-\rho t} |L(x(t), u(t))| dt < \infty.$$

Applying (32) to the control problem (15,19) it is clear by the K-invariance of the control problem that the HJB equation (32) can be simplified to an equation on $\mathfrak{a}_+$, the positive Weyl chamber, using the radial part of the infinitesimal generator (14) as

$$0 = \min_{v \in \mathbb{R}^r} \left[\frac{1}{2} \Delta_K W + \sum_{j=1}^r v_j \frac{\partial W}{\partial \tau_j} - \rho W + (K_\alpha \circ \psi)(\tau) + \sum_{j=1}^r f_j v_j^2 \right]. \tag{33}$$

It is clear that the family of controls obtained by the minimization in (33) is

$$U_j^*(\tau) = -\frac{1}{2 f_j(\tau)} \frac{\partial W}{\partial \tau_j} \tag{34}$$

for $j = 1, \ldots, r$.
Assume a solution of the form

$$W(\tau) = g(K_\alpha \circ \psi)(\tau) + h. \tag{35}$$

To apply a verification theorem (p. 147, Fleming and Soner (1993)) it is necessary to verify that for each admissible control

$$\liminf_{T \to \infty} W(X(T)) \leq 0. \tag{36}$$

For all functions β whose growth is bounded by an affine function of the spherical polynomial $K_\alpha \circ \psi$, the Dynkin formula can be applied. Since W is an affine function of one term in the cost functional (15) it follows from the finiteness of the cost functional for each admissible control to be considered that (36) is satisfied.

Substitute (34) and (35) in (33) to obtain

$$
\begin{aligned}
0 &= \frac{1}{2}\chi_\alpha g(K_\alpha \circ \psi) + \sum_j -\frac{1}{2f_j} g^2 \left[\frac{\partial}{\partial \tau_j}(K_\alpha \circ \psi)\right]^2 \\
&\quad - \rho(g(K_\alpha \circ \psi) + h) + (K_\alpha \circ \psi) + \frac{1}{4}\sum_j \frac{1}{f_j} g^2 \left[\frac{\partial}{\partial \tau_j}(K_\alpha \circ \psi)\right]^2 \\
&= (K_\alpha \circ \psi)\left[\frac{1}{2}\chi_\alpha g - \frac{g^2}{4} - \rho g + 1\right] - \rho h + \frac{1}{4} g^2 \, {}_0K_\alpha
\end{aligned}
\tag{37}
$$

Let g be the positive real root of

$$
\frac{1}{4}g^2 + g\left(\rho - \frac{1}{2}\chi_\alpha\right) - 1 = 0 \tag{38}
$$

and

$$
h = \frac{g^2 \, {}_0K_\alpha}{4\rho}. \tag{39}
$$

Then (35) is a classical (smooth) solution of (33). Since the solution of the stochastic differential equation (19) with U^* does not hit the boundaries of the positive Weyl chamber then the local smoothness of the coefficients of the family of stochastic differential equations implies the existence of a unique strong global solution (e.g., Duncan and Upmeier (1995)) □
The explicit solution of the ergodic cost control problem (16,19) is given now.

Theorem 2 *The stochastic optimal control problem* (16, 19) *has an admissible optimal control* U^* *given by*

$$
U^*(x) = -\frac{1}{2} g A^{-1}(x) \nabla_x (K_\alpha \circ \psi)(x) \tag{40}
$$

where g *is the positive real root of*

$$
\frac{g^2}{4} - \frac{1}{2}\chi_\alpha g - 1 = 0 \tag{41}
$$

and the optimal cost is

$$
J_e(U^*) = \frac{g^2 \, {}_0K_\alpha}{4} \tag{42}
$$

where ${}_0K_\alpha$ *is given in* (18).

Proof. Since the inequality (21) is satisfied, the ergodic control problem is well posed. The HJB equation for an ergodic control problem (Bensoussan and Frehse (1992), Kushner and Dupuis (1992)) is

$$
0 = \min_{v \in U}[A^v V(x) - \gamma + L(x, v)] \tag{43}
$$

where the cost, C, is

$$C(U) = \limsup_{T\to\infty} \frac{1}{T} E_{X(0)} \int_0^T L(X(s), U(s))ds. \tag{44}$$

A^v is the infinitesimal generator of the controlled diffusion and γ is the optimal cost.

Since there is an inherent degeneracy or ambiguity in (43) because it is necessary to solve for both V and γ, a direct construction of an optimal control and the optimal cost is made. The construction uses the controls and the optimal costs from a family of finite time horizon control problems. This family of finite time horizon control problems arises essentially from the cost function (16) by fixing T. More precisely, for $T > 0$ consider the finite time horizon control problem with the system (19) and the cost function

$$\begin{aligned} C_T(U) = E_{X(0)} \int_0^T & [(K_\alpha \circ \psi)(X(t)) \\ & + \langle A(X(t))U(X(t)), U(X(t))\rangle]dt \end{aligned} \tag{45}$$

The HJB equation for the control problem (19,45) is

$$\begin{aligned} 0 = \frac{\partial W_T}{\partial s} + \min_{v\in\mathbb{R}^r} \Bigg[& \frac{1}{2}\Delta_K W_T + \sum_{j=1}^r v_j \frac{\partial W_T}{\partial \tau_j} \\ & + (K_\alpha \circ \psi)(\tau) + \sum_{j=1}^r f_j v_j^2 \Bigg]. \end{aligned} \tag{46}$$

Performing the minimization in (46) the family of controls is

$$U_j(s, x; T) = -\frac{1}{2f_j(x)} \frac{\partial W_T}{\partial x_j}(s, x). \tag{47}$$

Assume a solution W_T of (47) as

$$\begin{aligned} W_T(s, \tau) &= g_T(s)(K_\alpha \circ \psi)(\tau) + h_T(s) \\ W_T(T, \tau) &\equiv 0. \end{aligned} \tag{48}$$

The validity of a verification theorem (p. 141, Fleming and Soner (1993)) follows as in the proof of Theorem 1.

Substitute (47, 48) in (46) to obtain

$$
\begin{aligned}
0 &= g_T'(K_\alpha \circ \psi) + h_T' + \frac{1}{2}\chi_\alpha g_T(K_\alpha \circ \psi) \\
&+ g_T^2 \sum_j \frac{-1}{2f_j}\left[\frac{\partial}{\partial \tau_j}(K_\alpha \circ \psi)\right]^2 + (K_\alpha \circ \psi) \\
&+ \frac{1}{4} g_T^2 \sum_j \frac{1}{f_j}\left[\frac{\partial}{\partial \tau_j}(K_\alpha \circ \psi)\right]^2 \\
&= (K_\alpha \circ \psi)\left[g_T' + \frac{1}{2}\chi_\alpha g_T - \frac{g_T^2}{4} + 1\right] \\
&+ h_T' + \frac{g_T^2}{4}\, {}_0K_\alpha . \qquad (49)
\end{aligned}
$$

Let

$$
\begin{aligned}
g_T' + \frac{1}{2}\chi_\alpha g_T - \frac{g_T^2}{4} + 1 &= 0 \qquad (50) \\
g_T(T) &= 0
\end{aligned}
$$

and

$$
\begin{aligned}
h_T'(s) + \frac{g_T^2}{4}\, {}_0K_\alpha &= 0 \qquad (51) \\
h_T(T) &= 0.
\end{aligned}
$$

Then W_T given by (48) satisfies (46).
It is clear that the family of optimal controls $(U_T^*, T > 0)$ converges (at least pointwise) to

$$
U^*(x) = -\frac{g}{2} A^{-1}(x) \nabla_x (K_\alpha \circ \psi)(x) \qquad (52)
$$

because $g_T \to g$ as $T \to \infty$ by the stability of (50) or from the explicit solution of g_T.
It is simple to verify by contradiction that U^* is an optimal control for (16, 19).
The optimal cost using the cost function

$$
\begin{aligned}
\tilde{C}_T(U) = \frac{1}{T} E_{X(0)} \int_0^T & [(K_\alpha \circ \psi)(X(t)) \\
&+ \langle A(X(t))U(t), U(t)\rangle] dt \qquad (53)
\end{aligned}
$$

is

$$
\begin{aligned}
\tilde{W}_T(0,x) &= \frac{1}{T} W_T(0,x) \\
&= \frac{1}{T}\left(g_T(0)(K_\alpha \circ \psi)(x) + \int_0^T \frac{g_T^2}{4}\, {}_0K_\alpha dt\right). \qquad (54)
\end{aligned}
$$

It easily follows from (54) that

$$\lim_{T\to\infty} \tilde{W}_T(0,x) = \frac{g^2\, {}_0K_\alpha}{4} \tag{55}$$

is the optimal cost. □

Acknowledgments: The author thanks Y. Z. Hu for some useful discussions. This research was supported in part by NSF Grant DMS 9623439.

References

Bensoussan, A. and Frehse, J. (1992). On Bellman equations of ergodic control in $\mathbb{R}^N$. *J. Reine Angew. Math.* **429**, 125–160.

Davis, M. H. A. and Norman, A. R. (1990). Portfolio selection with transaction costs. *Math. Operations Res.* **15**, 676–713.

Duncan, T. E. (1977). Dynamic programming optimality criteria for stochastic systems in Riemannian manifolds. *Appl. Math. Optim.* **3**, 191–208.

Duncan, T. E. (1979). Stochastic systems in Riemannian manifolds. *J. Optim. Theory Appl.* **27**, 399-421.

Duncan, T. E. (1987). A solvable stochastic control problem in hyperbolic three space. *Systems Control Lett.* **8**, 435–439.

Duncan, T. E. (1991). Some solvable stochastic control problems in noncompact symmetric spaces of rank one. *Stochastics Stochastics Rep.* **35**, 129–142.

Duncan, T. E. (1997). Some solvable infinite time horizon stochastic control problems in hyperbolic three space. Proc. 1997 Summer Res. Inst. on Differential Geometry and Control, Amer. Math. Soc. (to appear).

Duncan, T. E. and Upmeier, H. (1995). Stochastic control problems and spherical functions on symmetric spaces. *Trans. Amer. Math. Soc.* **347**, 1083–1130.

Farrell, R. H. (1985). *Multivariate Calculation*, Springer-Verlag, New York, Berlin.

Fleming, W. H. and Rishel, R. W. (1975). *Deterministic and Stochastic Optimal Control*, Springer-Verlag.

Fleming, W. H. and Soner, H. M. (1993). *Controlled Markov Processes and Viscosity Solutions*, Springer-Verlag.

Helgason, S. (1978). *Differential Geometry, Lie Groups, and Symmetric Spaces*, Academic Press.

Helgason, S. (1984). *Groups and Geometric Analysis*, Academic Press.

Ikeda, N. and Watanabe S. (1989). *Stochastic Differential Equations and Diffusion Processes*, North-Holland.

Korányi, A. and Wolf, J. A. (1965). Generalized Cayley transformations of bounded symmetric domains. *Amer. J. Math.* **87**, 899–939.

Kushner, H. J. and Dupuis, P. G. (1992), *Numerical Methods for Stochastic Control Problems in Continuous Time*, Springer-Verlag.

Loos, O. (1969). *Symmetric Spaces, I, II*, Benjamin.

Loos, O. (1971). Jordan triple systems, *R*-spaces and bounded symmetric domains. *Bull. Amer. Math. Soc.* **77**, 558–561.

Merton, R. C. (1971). Optimal consumption rules in a continuous-time model. *J. Economic Theory* **3**, 373–413.

Muirhead, R. J. (1982). *Aspects of Multivariate Statistical Theory*, Wiley, New York.

Takemura, A. (1984). *Zonal Polynomials*, Inst. Math. Stat. Lecture Notes—Monographs Series, 4.

Taksar, M., Klass, M. J., and Assaf, D. (1988). A diffusion model for optimal portfolio selection in the presence of brokerage fees. *Math. Operations Res.* **13**, 277–294.

23

Exact Finite Dimensional Filters for Exponential Functionals of the State

Robert J. Elliott
Vikram Krishnamurthy

ABSTRACT: In this paper, we derive finite dimensional filters for certain exponential functionals of the state of a continuous-time linear Gaussian process. Apart from being of mathematical interest, these new filters have applications in the state reconstruction of doubly-stochastic auto-regressive processes. We also derive similar filters for exponential functionals of the state of nonlinear Benes systems.

KEYWORDS: Stochastic systems, Finite dimensional filters, Kalman filter, Doubly stochastic AR model, Zakai equation, Benes filter

1 Introduction

The Kalman filter is widely used in engineering, economics and other fields. It considers linear Gaussian dynamics for the state x, and observation y processes of the form

$$\begin{aligned} dx_t &= A_t\, x_t\, dt + B_t\, dw_t, \quad x_0 \in \mathbb{R}^m \\ dy_t &= C_t\, x_t\, dt + D_t\, dv_t, \quad y_0 = 0 \in \mathbb{R}^n\ . \end{aligned}$$

where v and w are independent vector Brownian motions.
The Kalman filter determines the conditional density of the unobserved signal x_t given the observations to time t, $\mathcal{Y}_t = \sigma\{y_s : s \leq t\}$, as a Gaussian random variable with mean m_t and variance Σ_t. Here m_t is the conditional mean of x_t given $\mathcal{Y}_t$ and Σ_t is the conditional variance of x_t, though Σ_t turns out to be deterministic and given by the Riccati equation. Thus the Kalman filter is a finite dimensional filter with two sufficient statistics.
In this paper our aim is to derive finite dimensional filters for certain nonlinear functionals η_t of the state x_t, i.e., our aim is to compute $\mathbf{E}\{\eta_t|\mathcal{Y}_t\}$. Regarding the existence of finite dimensional filters for functionals of the state of a linear Gaussian system, the following results are known in the literature:

- As shown in [8], [14], finite dimensional filters exist when

$\eta_t = \int_0^t p(x_s)\, ds$ and $p(x_t)$ is any polynomial function of the state x_t. Note that finite dimensional filters exist for any function $p(x)$ because the smoothed conditional density of x_s given $\mathcal{Y}_t$, $s \leq t$, is finite dimensional.

- In [8] we have recently derived finite dimensional filters for $\eta_t = \int_0^t x_s\, dx_s$ and $\eta_t = \int_0^t x_s\, dy_s$. These filters were then used to estimate the parameters of the linear systems (A, B, C and D) via the Expectation Maximization (EM) algorithm.

The main result of this paper, Theorem 4.4 provides a finite dimensional filter for exponential functionals of the form $\eta_t = \exp(\int_0^t f(x_s)\, ds)$ where $f(x_s)$ is a quadratic polynomial function of the state x_s. Apart from the fact that there are very few finite dimensional filters, the filter presented in this paper is interesting for three reasons:
1. Whereas finite dimensional filters can be derived for any polynomial functional of the state [14], [8], in our case $f(x_s)$ has to be a second order polynomial. For higher order polynomials, we have not been able to find finite dimensional filters.
2. The motivation for our filtering problem lies in the context of state estimation of the *doubly stochastic* auto-regressive time series z_t defined as

$$dz_t = f(x_t)\, z_t\, dt + du_t, \quad z_0 = 1$$

Here z_t can be viewed as continuous-time auto-regressive process with random coefficient $f(x_t)$ where x_t itself evolves according to another auto-regressive process. In [10] we derived finite dimensional filters for such problems in discrete time. This paper can be viewed as a continuous time version of [10]. However, there are two significant differences between the continuous-time results in this paper and the discrete-time results in [10]. First, the discrete-time results hold for any order polynomial $f(x)$, whereas the continuous time result only holds for quadratic $f(x)$. Second, unlike the discrete-time case where at discrete-time k, the filtered density is characterized by k sufficient statistics, in continuous-time the filtered density is characterized by a fixed number of sufficient statistics (this number is independent of the time). This is intriguing in that a discretization of the continuous-time filters presented in this paper would be computationally much more efficient than the optimal discrete-time filters derived in [10].
3. Finally, we derive similar finite dimensional filters for *nonlinear* systems of the Benes type. The Benes filter [1], [2] considers continuous-time nonlinear dynamics for a diffusion state x, and a linear observation process y of the form

$$\begin{aligned} dx_t &= \psi_t(x_t)\, dt + B_t\, dw_t, \quad x_0 \in \mathbb{R}^m \\ dy_t &= C_t\, x_t\, dt + D_t\, dv_t, \quad y_0 = 0 \in \mathbb{R}^n\,. \end{aligned} \tag{1}$$

where v and w are independent vector Brownian motions. As long as the nonlinearity $\psi_t(\cdot)$ is of the Benes type (specified in Sec.5), the Benes filter is a finite dimensional filter that yields the conditional density of the state x_t given the observations to time t. In Sec.5 of this paper we derive a finite dimensional filter for exponential functionals $\eta_t = \exp(\int_0^t f(x_s)\,ds)$ where $f(x_s)$ is a quadratic polynomial functions of the state x_s of the above Benes system.

Caveat: If it was possible to explicitly compute the conditional density of $\int_0^t f(x_s)\,ds$ given $\mathcal{Y}_t$, then computing $\mathbf{E}\left\{\exp(\int_0^t f(x_s)ds)\Big|\mathcal{Y}_t\right\}$ is straightforward. Unfortunately, there exists no explicit expression in the literature for this filtered density in terms of a finite dimensional sufficient statistic. Hence the motivation for this paper. Note that in [8], although a finite dimensional filter for the conditional mean $\mathbf{E}\{\int_0^t f(x_s)\,ds|\mathcal{Y}_t\}$ is given, this conditional mean is not computed from the filtered density for $\int_0^t f(x_s)\,ds$. Thus the results of this paper cannot be obtained from [8]. (See also the remark following Definition 4.1).

In other words, the results in this paper are a special case of non universal finite dimensional filtering. We are not interested in finite dimensional computability of $\mathbf{E}\{g(z_t)|\mathcal{Y}_t\}$ for every test function g with compact support, but for the single identity function $g(z) = z$.

Applications: Doubly stochastic models such as (4) are used in image enhanced tracking of the co-ordinates of maneuvering targets [5], [15]. In image enhanced tracking (see [5], [15] and references therein), the aim is to estimate the coordinates of the target z_t given noisy measurements y_t from an imager which at each time t receives orientation information x_t of the target. In [5], [15], the underlying process x_t was assumed to be a finite state Markov chain (instead of a Gaussian process). Our signal model (2), (4), (3) can be considered as a natural progression of the "jump Markov" models considered in [5] and [15]. Instead of x_t being a finite state Markov chain, in our case x_t is a continuous valued Gaussian process.

Another possible application is in queueing networks. It was shown in [11] that when the driving process is a finite-state Markov chain, the doubly stochastic AR process can be used to model the buffer occupancy of a queuing system with a state dependent service rate. In [11] the data source was modelled as a Markov modulated fluid source. However, it may also be sensible to model the traffic generation process as a positive function of the state of a linear Gaussian process. In this case, filters similar to those derived in this paper would lead to optimal estimates of buffer levels based on noisy observations of the state of the traffic sources feeding the buffer. Such estimates allow traffic control policies to be implemented without the overhead of sending back buffer level information to the source.

Finally, we mention that the conditional expectations computed by these filters are required for the risk sensitive stochastic control of partially observed linear and Benes systems. These conditional expectations correspond

to the information states for such problems. We refer the reader to [4] for details including an alternative derivation of the filters via gauge transformations.
The paper is organized as follows: In Sec.2 we define the signal model and estimation objectives. In Sec.3 we introduce a measure change that facilitates easy derivation of the filters. The new finite dimensional filter is presented in Sec.4. In Sec 5, a similar filter is derived for non-linear Benes type systems.

2 Problem Formulation and Aim

Consider the classical linear Gaussian model for the signal model and observation processes. That is, the signal $\{x_t\}$ and observations y_t, $t \geq 0$ are modelled as

$$dx_t = A_t\, x_t\, dt + B_t\, dw_t, \qquad x_0 \in \mathbb{R}^m \tag{2}$$

$$dy_t = C_t\, x_t\, dt + D_t\, dv_t, \qquad y_0 = 0 \in \mathbb{R}^n\ . \tag{3}$$

Here w and v are independent r dimensional and n dimensional Brownian motions, respectively, defined on a probability space $(\Omega, \mathcal{F}, P)$. Further, w and v are independent of x_0. We assume that x_0 a random variable with normal density $\pi_0(x)$ which is $N(\hat{x}_0, P_0)$.
The matrix functions $A_t \in \mathbb{R}^{m\times m}$, $B_t \in \mathbb{R}^{m\times r}$, $C_t \in \mathbb{R}^{n\times m}$ and $D_t \in \mathbb{R}^{n\times n}$ are measurable functions of t. We assume D_t is a positive definite matrix. We assume that the Gaussian process x_t drives the scalar doubly stochastic AR process $z_t \in \mathbb{R}$ as follows:

$$dz_t = f(x_t)\, z_t\, dt + du_t, \quad z_0 = 1\ . \tag{4}$$

Here u is a Brownian process independent of v and w and $f : \mathbb{R}^m \to \mathbb{R}$ is a scalar valued function.
Remarks:
(1) In Sec. 4, we will consider examples of f for which finite dimensional filters exist.
(2) The results in this paper also hold for time-varying functions f. However, for notational convenience we will use $f(x)$ instead of $f_t(x)$. Moreover, it is straightforward to extend the filters to models which include the state noise in their dynamics, e.g.

$$dz_t = (f(x_t)dt + p(x_t)dw_t)z_t + du_t$$

where $p(x)$ is a linear in x. This extension is discussed further in Sec.4.
Notation: Consider the complete right continuous filtrations

$$\mathcal{G}_t = \sigma\{x_s, y_s : s \leq t\}, \quad \mathcal{Y}_t = \sigma\{y_s : s \leq t\}\ .$$

Vectors x, y will be considered as column vectors. Write

$$\nabla = \left(\frac{\partial}{\partial x_1}, \frac{\partial}{\partial x_2}, \ldots, \frac{\partial}{\partial x_n}\right)' .$$

For any function $g : \mathbb{R}^m \to \mathbb{R}$, write

$$\nabla^2 g = \begin{bmatrix} \frac{\partial^2 g}{\partial x_1^2} & \cdots & \frac{\partial^2 g}{\partial x_1 \partial x_m} \\ \vdots & \vdots & \vdots \\ \frac{\partial^2 g}{\partial x_m \partial x_1} & \cdots & \frac{\partial^2 g}{\partial x_m^2} \end{bmatrix} .$$

For a vector field $g(x) = [g_1(x)\ g_2(x)\ \cdots\ g_m(x)]'$ defined on $\mathbb{R}^m$, define

$$\operatorname{div}(g) = \frac{\partial g_1}{\partial x_1} + \frac{\partial g_2}{\partial x_2} + \cdots + \frac{\partial g_m}{\partial x_m} .$$

Aim: The aim is to derive a finite dimensional filter for computing $\mathbf{E}\{z_t|\mathcal{Y}_t\}$ where $\mathbf{E}$ denotes expectation under the measure P.
Note that we can solve (4) explicitly for z_t; this yields

$$z_t = \exp\left(\int_0^t f(x_s)\, ds\right)\left[1 + \int_0^t \exp\left(-\int_0^s f(x_r) dr\right) du_s\right] . \quad (5)$$

Since u is independent of x and y, our aim can be re-expressed as follows: Compute the filtered estimate

$$\mathbf{E}\{z_t|\mathcal{Y}_t\} = \mathbf{E}\left\{\exp\left(\int_0^t f(x_s)\, ds\right)|\mathcal{Y}_t\right\} .$$

Thus, as mentioned in Sec.1, our aim is to derive a finite dimensional filter for $\eta_t = \exp\left(\int_0^t f(x_s)\, ds\right)$.

3 Dynamics

In this section we present a measure change which will facilitate easy derivation of the filters. The content of this section is fairly standard (e.g. see [7] or [8]).
We model the above dynamics by supposing that initially we have a probability space $(\Omega, \mathcal{F}, \bar{P})$ such that under $\bar{P}$

1. w is r-dimensional Brownian motion and $\{x_t\}$ is defined by (2).

2. $\{y_t\}$ is n-dimensional Brownian motion, independent of w and x_0, and having quadratic variation $\langle y\rangle_t = D_t > 0$, i.e., D_t is a positive definite matrix.

Consider the exponential

$$\Lambda_t = \exp\left(\int_0^t (C_s\, x_s)'\,(D_s^{-1})'\, D_s^{-1}\, dy_s - \frac{1}{2}\int_0^t x_s'\, C_s'\,(D_s^{-1})'\, D_s^{-1}\, C_s\, x_s\, ds\right). \tag{6}$$

Then

$$d\Lambda_t = \Lambda_t\, x_t'\, C_t'\,(D_s^{-1})'\, D_s^{-1}\, dy_t \tag{7}$$

and $\bar{\mathbf{E}}\{\Lambda_t\} = 1$, where $\bar{\mathbf{E}}$ denotes expectation under $\bar{P}$.
Define a measure P in terms of $\bar{P}$ by setting

$$\left.\frac{dP}{d\bar{P}}\right|_{\mathcal{G}_t} = \Lambda_t\ .$$

Girsanov's theorem implies that under P, v_t is a standard n-dimensional Brownian motion if we define $dv_t = D_t^{-1}\,(dy_t - C_t\, x_t\, dt)$, $v_0 = 0$. That is, under P, $dy_t = C_t\, x_t\, dt + D_t\, dv_t$. Note that under P, the process $\{x_t\}$ still satisfies (2). Consequently, under P the processes $\{x_t\}$ and $\{y_t\}$ satisfy the real world dynamics (2) and (3). However, $\bar{P}$ is a more convenient measure with which to work.
In the sequel we assume that $g : \mathbb{R}^m \to \mathbb{R}$ is an arbitrary "test" function in $\mathbb{C}^2$ with compact support.
Filtering is concerned with estimates of the form $\mathbf{E}\{g(x_t)|\mathcal{Y}_t\}$. Using Bayes' theorem, [7], we have

$$\mathbf{E}\{g(x_t)|\mathcal{Y}_t\} = \frac{\bar{\mathbf{E}}\{\Lambda_t\, g(x_t)|\mathcal{Y}_t\}}{\bar{\mathbf{E}}\{\Lambda_t|\mathcal{Y}_t\}}\ . \tag{8}$$

Note that $\bar{\mathbf{E}}\{\Lambda_t\, g(x_t)|\mathcal{Y}_t\}$ is a measure valued process; it is an unnormalized conditional expectation of $g(x_t)$ given $\mathcal{Y}_t$.
Suppose $\bar{\mathbf{E}}\{\Lambda_t\, g(x_t)|\mathcal{Y}_t\}$ has a density $\alpha_t(x)$; then

$$\bar{\mathbf{E}}\{\Lambda_t\, g(x_t)|\mathcal{Y}_t\} = \int_{\mathbb{R}^m} g(x)\,\alpha_t(x)\, dx\ . \tag{9}$$

We summarize the well-known Zakai equation for $\alpha_t(x)$ and its explicit solution given by the Kalman filter. See [2] for a proof.

Theorem 3.1 *(Kalman Filter) The un-normalized density $\alpha_t(x)$ satisfies the Zakai equation*

$$\alpha_t(x) = \alpha_0(x) - \int_0^t \operatorname{div}\left(A_s\, x\, \alpha_s(x)\right)\, ds + \frac{1}{2}\int_0^t \operatorname{Tr}\left(\nabla^2\alpha_s(x)\, B_s\, B_s'\right)\, ds + \int_0^t \alpha_s(x)\, x'\, C_s'\,(D_s^{-1})'\, D_s^{-1}\, dy_s \tag{10}$$

where $\alpha_0(x) = \pi_0(x)$, *the density of* x_0. *Furthermore the explicit solution of (10) is given by*

$$\alpha_t(x) = \frac{\nu_t}{(2\,\pi)^{m/2}\,|\Sigma_t|^{1/2}} \exp\left(-\frac{1}{2}(x - m_t)'\,\Sigma_t^{-1}\,(x - m_t)\right) . \tag{11}$$

where $m_t = \mathbf{E}\{x_t|\mathcal{Y}_t\}$, $m_0 = \hat{x}_0$, $\Sigma_t = \mathbf{E}\{(x_t - m_t)(x_t - m_t)'|\mathcal{Y}_t\}$ *and* $\Sigma_0 = P_0$. *The sufficient statistics* m_t *and* Σ_t *are given by the Kalman filter equations*

$$dm_t = A_t\,m_t\,dt + \Sigma_t\,C_t'\,(D_t^{-1})'\,D_t^{-1}\,(dy_t - C_t\,m_t\,dt) \tag{12}$$

$$\dot{\Sigma}_t = \Sigma_t\,A_t' + A_t\,\Sigma_t + B_t\,B_t' - \Sigma_t\,C_t'\,(D_t^{-1})'\,D_t^{-1}\,C_t\,\Sigma_t\,. \tag{13}$$

Also, in (11) ν_t *is a normalizing factor:*

$$\nu_t = (2\,\pi)^{m/2} \exp\left(\int_0^t m_s'\,C_s'(D_s^{-1})'\,D_s^{-1}\,dy_s - \frac{1}{2}\int_0^t m_s'\,C_s'\,(D_s^{-1})'\,D_s^{-1}\,c_s\,m_s\,ds\right) . \tag{14}$$

4 Finite Dimensional Filters for Exponential Functionals

Motivated by the techniques in [7], we define a measure associated with z_t as follows:

Definition 4.1 *For any test function* $g : \mathbb{R}^m \to \mathbb{R}$, *define a measure valued process* $\bar{\mathbf{E}}\{\Lambda_t\,z_t\,g(x_t)|\mathcal{Y}_t\}$. *Suppose this has a density* $\beta_t(x)$ *so that*

$$\bar{\mathbf{E}}\{\Lambda_t\,z_t\,g(x_t)|\mathcal{Y}_t\} = \int_{\mathbb{R}^m} \beta_t(x)\,g(x)\,dx\,. \tag{15}$$

Remarks:
(1) It is important to note that $\beta_t(x)$ is not the un-normalized conditional density of z_t given $\mathcal{Y}_t$. In fact, $\beta_t(x)$ is used for determining the conditional mean estimate of z_t given $\mathcal{Y}_t$ via Bayes theorem (8) as follows:

$$\mathbf{E}\{z_t|\mathcal{Y}_t\} = \frac{\bar{\mathbf{E}}\{\Lambda_t\,z_t|\mathcal{Y}_t\}}{\bar{\mathbf{E}}\{\Lambda_t|\mathcal{Y}_t\}} = \frac{\int_{\mathbb{R}^m}\beta_t(x)\,dx}{\int_{\mathbb{R}^m}\alpha_t(x)\,dx}\,.$$

(2) The existence of the density $\beta_t(x)$ follows from the existence and uniqueness of solutions of stochastic partial differential equations such as (16) below. This is established in Sec.4.2 of [2].

Lemma 4.2 *Formally, $\beta_t(x)$ satisfies the Zakai equation*

$$\begin{aligned}\beta_t(x) &= \beta_0(x) - \int_0^t \operatorname{div}(\beta_s(x)\, A_s\, x)\, ds + \frac{1}{2}\int_0^t \operatorname{Tr}\left(\nabla^2 \beta_s(x)\, B_s\, B_s'\right) ds \\ &\quad + \int_0^t \beta_s(x)\, x'\, C_s'\, (D_s^{-1})'\, D_s^{-1}\, dy_s + \int_0^t f(x)\, \beta_s(x)\, ds\,. \qquad (16)\end{aligned}$$

PROOF. From Ito's rule, for any test function g

$$\begin{aligned}g(x_t) = g(x_0) &+ \int_0^t (\nabla g(x_s))'\, A_s\, x_s\, ds + \int_0^t (\nabla g(x_s))'\, B_s\, dw_s \\ &+ \frac{1}{2}\int_0^t \operatorname{Tr}\left(\nabla^2 g(x_s)\, B_s\, B_s'\right) ds\,. \qquad (17)\end{aligned}$$

From (7) and (17)

$$\begin{aligned}\Lambda_t\, g(x_t) = g(x_0) &+ \int_0^t \Lambda_s\, (\nabla g(x_s))'\, A_s\, x_s\, ds + \int_0^t \Lambda_s\, (\nabla g(x_s))'\, B_s\, dw_s \\ &+ \frac{1}{2}\int_0^t \Lambda_s \operatorname{Tr}\left(\nabla^2 g(x_s) B_s\, B_s'\right) ds + \int_0^t \Lambda_s\, g(x_s)\, x_s'\, C_s'\, (D_s^{-1})'\, D_s^{-1}\, dy_s\,. \qquad (18)\end{aligned}$$

$$\begin{aligned}\Lambda_t\, z_t\, g(x_t) = g(x_0) &+ \int_0^t \Lambda_s\, z_s\, (\nabla g(x_s))'\, A_s\, x_s\, ds + \int_0^t \Lambda_s\, z_s\, (\nabla g(x_s))'\, B_s\, dw_s \\ &+ \frac{1}{2}\int_0^t \Lambda_s\, z_s \operatorname{Tr}\left(\nabla^2 g(x_s)\, B_s\, B_s'\right) ds + \int_0^t \Lambda_s\, z_s\, g(x_s)\, x_s'\, C_s'\, (D_s^{-1})'\, D_s^{-1}\, dy_s \\ &+ \int_0^t \Lambda_s\, g(x_s)\, f(x_s)\, z_s\, ds + \int_0^t \Lambda_s\, g(x_s)\, du_s.\end{aligned}$$

Conditioning on $\mathcal{Y}_t$ under $\bar{P}$ (see Lemma 3.2, pg 262 in [16]) we have

$$\begin{aligned}\bar{\mathbf{E}}\{\Lambda_t\, z_t\, g(x_t)|\mathcal{Y}_t\} &= \bar{\mathbf{E}}\{g(x_0)|\mathcal{Y}_0\} + \int_0^t \bar{\mathbf{E}}\{\Lambda_s\, z_s\, (\nabla g(x_s)'\, A_s\, x_s)\, |\mathcal{Y}_s\}\, ds \\ &+ \frac{1}{2}\int_0^t \bar{\mathbf{E}}\{\Lambda_s\, z_s \operatorname{Tr}\left(\nabla^2 g(x_s) B_s\, B_s'\right) |\mathcal{Y}_s\} \\ &+ \int_0^t \bar{\mathbf{E}}\{\Lambda_s\, z_s\, g(x_s)\, x_s'\, C_s'\, (D_s^{-1})'\, D_s^{-1}|\mathcal{Y}_s\}\, dy_s + \int_0^t \bar{\mathbf{E}}\{\Lambda_s\, g(x_s)\, f(x_s)\, z_s|\mathcal{Y}_s\}\, ds\,.\end{aligned}$$

That is, in terms of the densities $\beta_t(x)$

$$\int_{\mathbb{R}^m} \beta_t(x)\, g(x)\, dx = \int_{\mathbb{R}^m} g(x)\, \beta_0(x)\, dx + \int_0^t \int_{\mathbb{R}^m} \beta_s(x)\, (\nabla g(x))'\, A_s\, x\, dx\, ds$$
$$+ \frac{1}{2} \int_0^t \int_{\mathbb{R}^m} \beta_s(x) \operatorname{Tr}\left(\nabla^2 g(x)\, B_s\, B_s'\right)\, dx\, ds$$
$$+ \int_0^t \int_{\mathbb{R}^m} \beta_s(x)\, g(x)\, x'\, C_s'\, (D_s^{-1})'\, D_s^{-1}\, dx\, dy_s + \int_0^t \int_{\mathbb{R}^m} \beta_s(x)\, g(x)\, f(x)\, dx\, ds\,.$$

Integrating each term by parts (as on page 277 of [6]) and because the resulting equation holds for all test functions g, (16) follows. □
Our aim now is to find explicit solutions of the Zakai equation (16). We will look for solutions of (16) of the form

$$\bar{\beta}_t(x) = G_t(x)\, \alpha_t(x) \tag{19}$$

where the "gauge transformation" [13] $G_t(x)$ is a deterministic function of A_t, B_t, C_t, D_t, Σ_t and m_t, and $\alpha_t(x)$ is the un-normalized Kalman state density (10). As noted below Defn.4.1, if such a solution exists it is unique.

Theorem 4.3 *Suppose $\beta_t(x) = G_t(x)\, \alpha_t(x)$ is an explicit solution of (16). Then $G_t(x)$ satisfies the parabolic linear partial differential equation*

$$\frac{\partial G_t(x)}{\partial t} = -(\nabla G_t(x))'\, A_t\, x + \frac{1}{2} \operatorname{Tr}\left(\nabla^2 G_t(x)\, B_t\, B_t'\right)$$
$$- \operatorname{Tr}\left(\nabla G_t(x)\, (x - m_t)'\, \Sigma_t^{-1}\, B_t\, B_t'\right) + f(x)\, G_t(x) \tag{20}$$

with initial condition $G_0(x) = 1$.

Remark: (20) is a linear parabolic partial differential equation (pde). For an arbitrary $f(x)$ in (4), the above theorem reduces the construction of a finite dimensional filter for $\mathbf{E}\{z_t|\mathcal{Y}_t\}$, to finding an explicit solution $G_t(x)$ of the parabolic pde (20). The parametrix method is often used to construct fundamental solutions for such pdes [9]. Conditions for the existence of unique solutions for such pdes are given in Chapter 1 of [9].
PROOF. First notice that from (11)

$$\nabla \alpha_s(x) = -\Sigma_s^{-1}(x - m_s)\, \alpha_s(x)\,.$$

Let us evaluate the various terms on the right hand side of (16):

$$\begin{aligned}
\operatorname{div}\left(\bar{\beta}_s(x)\, A_s\, x\right) &= (\nabla G_s(x))'\, A_s\, x\, \alpha_s(x) + G_s(x) \operatorname{div}\left(\alpha_s(x)\, A_s\, x\right) \\
\nabla \bar{\beta}_s(x) &= \nabla G_t(x)\, \alpha_t(x) + G_t(x)\, \nabla \alpha_t(x) \\
&= \nabla G_t(x)\, \alpha_t(x) - G_t(x) \Sigma_t^{-1}\, (x - m_t)\, \alpha_t(x)
\end{aligned}$$

$$\begin{aligned}
\nabla^2\bar{\beta}_s(x) &= \nabla^2 G_s(x)\,\alpha_s(x) + 2\,\nabla G_s(x)\,(\nabla\alpha_s(x))' \\
&\quad + G_s(x)\,\nabla^2\alpha_s(x) \\
\operatorname{Tr}\left(\nabla^2\bar{\beta}_s(x)\,B_s\,B_s'\right) &= \alpha_s(x)\operatorname{Tr}\left(\nabla^2 G_s(x)\,B_s\,B_s'\right) \\
&\quad + 2\operatorname{Tr}\left(\nabla G_s(x)\,\nabla\alpha_s(x) B_t B_t'\right) \\
&\quad + G_t(x)\operatorname{Tr}\left(\nabla^2\alpha_s(x)\,B_s\,B_s'\right) .
\end{aligned}$$

Substituting $\bar{\beta}_t(x)$, given by (19) in the differential form of the right hand side of (16) we obtain

$$\begin{aligned}
d\bar{\beta}_t(x) &= -\left[\nabla G_t(x)\right]'\,A_t\,x\,\alpha_t(x)\,dt - G_t(x)\operatorname{div}\left[\alpha_t(x)\,A_t\,x\right]\,dt \\
&+ \frac{1}{2}\alpha_t(x)\operatorname{Tr}\left(\nabla^2 G_t(x) B_t\,B_t'\right)\,dt - \alpha_t(x)\operatorname{Tr}\left(\nabla G_t(x)(x-m_t)'\,\Sigma_t^{-1}\,B_t\,B_t'\right)\,dt \\
&\qquad + \frac{1}{2}\operatorname{Tr}\left(G_t(x)\,\nabla^2\alpha_t(x)\,B_t\,B_t'\right)\,dt + f(x)\,G_t(x)\,\alpha_t(x)\,dt \\
&\qquad\qquad + G_t(x)\,\alpha_t(x)\,x'\,C_t'\,\left(D_t^{-1}\right)'\,D_t^{-1}\,dy_t\,. \qquad (21)
\end{aligned}$$

Also

$$\begin{aligned}
d\bar{\beta}_s(x) &= dG_t(x)\,\alpha_t(x) + d\alpha_t(x)\,G_t(x) \\
&= \frac{\partial G_t(x)}{dt}\,\alpha_t(x)\,dt + G_t(x)\Big[-\operatorname{div}\left(\alpha_t(x)\,A_t\,x\right)\,dt \\
&\qquad + \alpha_t(x)\,x'\,C_t'\,\left(D_t^{-1}\right)'\,D_t^{-1}\,dy_t + \frac{1}{2}\operatorname{Tr}\left(\nabla^2\alpha_t(x)\,B_t\,B_t'\right)\,dt\Big]\,. \qquad (22)
\end{aligned}$$

Consequently $\bar{\beta}_s(x)$ given by (19) is a solution of (16) if (21) equals (22). Equating (21) and (22) yields (20). □

Let us now consider a specific example admitting a finite dimensional filter.

Theorem 4.4 *Consider the linear system (2), (3). Suppose the non-linearity f in (4) is the quadratic*

$$f(x) = f^{(0)} + x'\,f^{(1)} + x'\,f^{(2)}\,x \qquad (23)$$

where $f^{(0)} \in \mathbb{R}$, $f^{(1)} \in \mathbb{R}^m$ and $f^{(2)} \in \mathbb{R}^{m\times m}$ are known and $f^{(2)}$ is a symmetric negative semi-definite matrix. Then at time t, the un-normalized density $\beta_t(x)$ is completely characterized by the 5 statistics ϕ_t, h_t, l_t, Σ_t and m_t as follows:

$$\beta_t(x) = G_t(x)\,\alpha_t(x) \quad \textit{where} \quad G_t(x) = \exp\left(\phi_t + x'\,h_t + x'\,l_t\,x\right)\,. \qquad (24)$$

In (24), $\phi_t \in \mathbb{R}$, $h_t \in \mathbb{R}^m$ and $l_t \in \mathbb{R}^{m\times m}$ is a symmetric matrix. They satisfy the following ordinary differential equations:

$$\frac{d\phi_t}{dt} = \frac{1}{2} h_t'\, B_t\, B_t'\, h_t + \mathrm{Tr}\,[l_t\, B_t\, B_t'] + m_t'\, \Sigma_t^{-1}\, B_t\, B_t'\, h_t + f^{(0)} \tag{25}$$

$$\frac{dh_t}{dt} = -A_t'\, h_t + 2\, l_t\, B_t\, B_t'\, h_t - \Sigma_t^{-1}\, B_t\, B_t'\, h_t + 2\, l_t\, B_t\, B_t'\, \Sigma_t^{-1}\, m_t + f^{(1)}, \tag{26}$$

$$\frac{dl_t}{dt} = -l_t\, A_t - A_t'\, l_t + 2\, l_t\, B_t\, B_t'\, l_t - \Sigma_t^{-1}\, B_t\, B_t'\, l_t - l_t\, B_t\, B_t'\, \Sigma_t^{-1} + f^{(2)} \tag{27}$$

with initial conditions $\phi_0 = 0$, $h_0 = 0$ and $l_0 = 0$. Finally, providing $\Sigma_t^{-1} - 2\, l_t$ is positive definite, a finite dimensional filter for z_t is given by

$$\mathbf{E}\{z_t|\mathcal{Y}_t\} = |\Sigma_t|^{-1/2}\, \left|\Sigma_t^{-1} - 2\, l_t\right|^{-1/2} \exp\Big[\phi_t - \frac{m_t'\Sigma_t^{-1} m_t}{2} - \frac{1}{4}(h_t + \Sigma_t m_t)'\, \left(\Sigma_t^{-1} + l_t\right)^{-1} (h_t + \Sigma_t m_t)\Big] \tag{28}$$

PROOF. Let us evaluate the various terms in (20):

$$\frac{\partial G_t(x)}{dt} = \left[\frac{d\phi_t}{dt} + x'\,\frac{dh_t}{dt} + x'\,\frac{dl_t}{dt}\,x\right] G_t(x), \quad \nabla G_t(x) = (h_t + 2\, l_t\, x)\, G_t(x)$$

$$\mathrm{Tr}\left[\nabla^2 G_t(x)\, B_t\, B_t'\right] = \Big(h_t'\, B_t\, B_t'\, h_t + 2\, x'\, l_t\, B_t\, B_t'\, h_t + 2\, h_t'\, B_t\, B_t'\, l_t\, x + 4x'\, l_t\, B_t\, B_t'\, l_t\, x + 2\mathrm{Tr}[l_t\, B_t\, B_t']\Big)\, G_t(x)$$

$$\mathrm{Tr}\left[\nabla G_t(x)(x - m_t)'\, \Sigma_t^{-1}\, B_t\, B_t'\right] = \Big(x'\Sigma_t^{-1} B_t B_t' h_t - m_t'\Sigma_t^{-1} B_t B_t' h_t + 2x'\Sigma_t^{-1} B_t B_t' l_t x - 2m_t'\Sigma_t^{-1} B_t B_t' l_t x\Big)\, G_t(x)$$

Substituting for the above expressions in (20) and equating coefficients of x, $x'\,x$ and the constants, yields (25), (26) and (27).
To prove (28), note that by Bayes theorem (8) we have

$$\mathbf{E}\{z_t|\mathcal{Y}_t\} = \frac{\bar{\mathbf{E}}\{\Lambda_t\, z_t|\mathcal{Y}_t\}}{\bar{\mathbf{E}}\{\Lambda_t|\mathcal{Y}_t\}} = \frac{\int_{\mathbb{R}^m} \beta_t(x)\, dx}{\int_{\mathbb{R}^m} \alpha_t(x)\, dx} \tag{29}$$

Note that (with ν_t defined in (14)) $\int_{\mathbb{R}^m} \alpha_t(x)\, dx = \nu_t$. Let us now evaluate $\int_{\mathbb{R}^m} \beta_x(x)\, dx$:

$$\beta_t(x) = \alpha_t(x)\, G_t(x)$$

$$= \frac{\nu_t}{(2\pi)^{m/2}\, |\Sigma_t|^{1/2}} \exp\left[\phi_t + x'h_t + x'l_t x - \frac{1}{2}(x - m_t)'\Sigma_t^{-1}(x - m_t)\right] \tag{30}$$

By completing the square, we can re-express the term within the exponential as

$$\left[x+\frac{1}{2}(-\Sigma_t^{-1}/2+l_t)(h_t+\Sigma_t^{-1}m_t)\right]'(-\Sigma_t^{-1}/2+l_t)$$
$$\left[x+\frac{1}{2}(-\Sigma_t^{-1}/2+l_t)(h_t+\Sigma_t^{-1}m_t)\right]$$
$$-\frac{1}{4}(h_t+\Sigma_t m_t)'(-\Sigma_t^{-1}/2+l_t)(h_t+\Sigma_t m_t)+(\phi_t-\frac{1}{2}m_t'\Sigma_t^{-1}m_t) \quad (31)$$

Therefore as long as $\Sigma_t^{-1}-2\,l_t$ is positive definite,

$$\int_{\mathbb{R}^m}\beta_t(x)\,dx=\nu_t|\Sigma_t|^{-1/2}\left|\Sigma_t^{-1}-2\,l_t\right|^{-1/2}\exp\left[\phi_t-\frac{1}{2}m_t'\Sigma_t^{-1}m_t\right.$$
$$\left.-\frac{1}{4}(h_t+\Sigma_t m_t)'\left(-\Sigma_t^{-1}/2+l_t\right)^{-1}(h_t+\Sigma_t m_t)\right]$$

Finally substituting for $\int_{\mathbb{R}^m}\beta_t(x)dx$ and $\int_{\mathbb{R}^m}\alpha_t(x)dx$ in (29) yields (28). □
Remark: Note that the ordinary differential equation (27) is a Riccati equation for l. Also (26) can be solved by variation of constants and (25) can be integrated directly for ϕ.

Discussion

We discuss several issues relating to Theorem 4.3 and Theorem 4.4.
(1) The solution $G_t(x)$ in (24) of the linear parabolic pde (20) is identical to the fundamental solution of the heat equation. This is not surprising; the parametrix method for constructing fundamental solutions of linear parabolic pdes (see Chapter 1, [9]) first constructs the principle part of the solution as the solution to the following heat equation (see pg.4, [9]) $\partial G_t(x)/\partial t=(1/2)\mathrm{Tr}\left(\nabla^2 G_t(x)\,B_t\,B_t'\right)$.
(2) The above exact solution (24) to Zakai equation (16) only holds when $f(x)$ is quadratic (or linear or constant) in x. We have not been able to find finite dimensional filters when $f(x)$ is a cubic or higher order polynomial. To see why this is so, consider for simplicity the scalar case ($m=1$) and suppose $f(x)=f^{(0)}+f^{(1)}\,x+f^{(2)}\,x^2+f^{(3)}\,x^3$. Then it is easy to verify that $\bar{\beta}(x)=G_t(x)\,\alpha_t(x)$ where $G_t(x)=\exp(\phi_t+h_t\,x+l_t\,x^2+\mu_t\,x^3)$ is *not* a solution of (16). In other words $G_t(x)=\exp(\phi_t+h_t\,x+l_t\,x^2+\mu_t\,x^3)$ is not a solution of (20). In particular, the LHS of (20) involves up to cubic powers of x, whereas the RHS involves up to 4th powers of x. By a similar reasoning, it can be shown that for any polynomial $f(x)$ of order greater than 2, the solution $G_t(x)$ of (20) cannot be an exponential of a polynomial in x.
In contrast, as shown in [8], finite dimensional filters can be derived for any polynomial functional of the state. This is because deriving the filters in [8]

involves solving a linear parabolic partial differential equation of the form (20) with the key difference that the last term is $f(x)$ instead of $f(x)G_t(x)$.
(3) In the case when A, B, C and D are constants (rather than time-varying), and A, B are scalar (hence $f^{(1)}$, $f^{(2)}$, $f^{(3)}$ are scalars), it is easy to relate the condition $\Sigma^{-1} - 2\,l_t > 0$ (28) to $f^{(2)}$:
If $A > 0$, then $\Sigma^{-1} - 2\,l_t > 0$ is satisfied for all t as long as $f^{(2)}$ is negative.
If $A < 0$, then $\Sigma^{-1} - 2\,l_t > 0$ if $\Sigma^{-2}B^2/2 + \Sigma^{-1}A < f^{(2)} < 0$.
These can be shown by solving the algebraic versions of the Riccati equations (27) and (13) for l and Σ.
(4) Theorem 4.4 can be straightforwardly extended to the following doubly stochastic model which includes the process noise in its dynamics. We summarize this in the following Corollary.

Corollary 4.5 *Consider the signal model (2), (3) and*

$$dz_t = (f(x_t)dt + p'(x_t)dw_t)z_t + du_t, \qquad z_0 = 1 \tag{32}$$

where $p(x) = p^{(0)} + p^{(1)}x$ *where* $p^{(0)} \in \mathbb{R}^r$ *and* $p^{(1)} \in \mathbb{R}^{r\times m}$, *and* $f(x)$ *is given by (23). Then Theorem 4.4 holds with the following extra terms on the right hand sides of (25), (26) and (27)*
Eq. (25): $-(m_t'\,\Sigma_t^{-1}B_t + h_t'B_t)p^{(0)} - \mathrm{Tr}\left[B_t p^{(1)}\right]$
Eq. (26): $\Sigma_t^{-1}B_t p^{(0)} - p^{(1)'}B_t'\Sigma_t^{-1}m_t - 2l_t B_t p^{(0)} - p^{(1)'}B_t'h_t$
Eq. (27): $\Sigma_t^{-1}B_t p^{(1)} - 2l_t B_t p^{(1)}$

5 Finite Dimensional Filters for Benes Systems

This section presents finite dimensional filters for exponential functionals of the state of a Benes type nonlinear system. The derivation is similar to the linear case developed in Sec.3 and 4. For related results see [12].
Dynamics: Consider the following continuous-time nonlinear signal model: The state $\{x_t\}$, $t \geq 0$ as

$$dx_t = \psi_t(x_t)\,dt + B_t\,dw_t, \qquad x_0 \in \mathbb{R}^m \tag{33}$$

where $\psi_t : \mathbb{R}^m \to \mathbb{R}^m$. The observation process $\{y_t\}$, $t \geq 0$ is given by (3) repeated below for convenience:

$$dy_t = C_t\,x_t\,dt + D_t\,dv_t, \qquad y_0 = 0 \in \mathbb{R}^n\,. \tag{3}$$

Apart from the nonlinear dynamics in (33), we make the same assumptions and definitions as in Sec.2. Also everything in Sec.3 before Theorem 3.1 holds, including of the definition of $\alpha_t(x)$ in (9). Moreover, $\beta_t(x)$ is defined as in Definition 4.1.
Assumption: Assume that $\psi_t(x)$ in (2) satisfies the following condition [2]: Define $\Psi_t(x) : \mathbb{R}^m \to \mathbb{R}$ such that

$$(B_t\,B_t')\,\nabla\Psi_t(x) = \psi_t(x), \quad x \in \mathbb{R}^m\,. \tag{34}$$

Then Ψ_t satisfies the following *Benes non-linearity* condition (page 198, [2]):

$$\frac{\partial \Psi}{\partial t} + \frac{1}{2}\mathrm{Tr}\left(B_t\, B_t'\, \nabla^2 \Psi\right) + \frac{1}{2}(\nabla \Psi)'\, (B_t\, B_t')\, \nabla \Psi = \frac{1}{2} x'\, \Lambda_t\, x + x'\, \sigma_t + \delta_t \tag{35}$$

where $\Lambda_t \in \mathbb{R}^{m \times m}$ is an arbitrary symmetric matrix satisfying

$$\Lambda_t + C_t'\, (D_t\, D_t')^{-1} \quad C_t \geq 0$$

$\sigma_t \in \mathbb{R}^m$ is an arbitrary vector and $\delta_t \in \mathbb{R}$ is an arbitrary scalar. Several examples of Benes nonlinearities are given in pp.199-200 of [2].
Aim: Derive a filter for z_t in (4) given observations y_t in (3) where x_t is given by (33).
Benes Filter: The following theorem summarizes the Benes filter which is a finite dimensional filter for reconstructing the state of (33) given the observations (3). For a proof please see [2], pg 197-203.

Theorem 5.1 *(Benes Filter) The Zakai equation for the un-normalized density $\alpha_t(x)$ is given by*

$$\alpha_t(x) = \alpha_0(x) - \int_0^t \mathrm{div}\left(\psi_s(x)\, \alpha_s(x)\right)\, ds + \frac{1}{2}\int_0^t \mathrm{Tr}\left(\nabla^2 \alpha_s(x)\, B_s\, B_s'\right)\, ds + \int_0^t \alpha_s(x)\, x'\, C_s'\, (D_s^{-1})'\, D_s^{-1}\, dy_s \tag{36}$$

where $\alpha_0(x) = \pi_0(x)$, the density of x_0. When the nonlinearity $\psi_t(x)$ satisfies (35), then the explicit solution of the above Zakai equation is given as follows

$$\alpha_t(x) = \exp\left(\Psi_t(x) - \frac{1}{2} x' M_t x + r_t' x + \rho_t\right) \tag{37}$$

where $\Psi_t(x)$ is defined in (34), $M_t \in \mathbb{R}^{m \times m}$ is a symmetric positive definite matrix, $r_t \in \mathbb{R}^m$ and $\rho_t \in \mathbb{R}$. Moreover, M_t, r_t and ρ_t are given by the Benes filter equations:

$$\begin{aligned}
\frac{dM_t}{dt} &= -M_t\, B_t\, B_t'\, M_t + \Lambda_t + C_t'\, (D_t\, D_t')'\, C_t \\
dr_t &= -\left(M_t\, B_t\, B_t'\, r_t + \sigma_t\right)\, dt + C_t'\, (D_t\, D_t')^{-1}\, dy_t \\
\frac{d\rho_t}{dt} &= \frac{1}{2} r_t'\, (B_t\, B_t')^{-1}\, r_t - \delta_t - \frac{1}{2}\mathrm{Tr}\left(B_t B_t' M_t\right)
\end{aligned} \tag{38}$$

□

The following theorem for the un-normalized filtered density β_t associated with z_t is the Benes system equivalent of Lemma 4.2 and Theorem 4.3.

Theorem 5.2 *The un-normalized density $\beta_t(x)$ satisfies the Zakai equation*

$$\beta_t(x) = \beta_0(x) - \int_0^t \operatorname{div}(\beta_s(x)\,\psi_s(x))\,ds + \frac{1}{2}\int_0^t \operatorname{Tr}\left(\nabla^2\beta_s(x)\,B_s\,B_s'\right)\,ds$$
$$+\int_0^t \beta_s(x)\,x'\,C_s'\,(D_s^{-1})'\,D_s^{-1}\,dy_s + \int_0^t f(x)\,\alpha_s(x)\,ds\,. \qquad (39)$$

Suppose $\bar{\beta}_t(x) = G_t(x)\,\alpha_t(x)$ is an explicit solution of (39). Then $G_t(x)$ satisfies the parabolic pde

$$\frac{\partial G_t(x)}{\partial t} = \frac{1}{2}\operatorname{Tr}\left(\nabla^2 G_t(x) B_t B_t'\right) + \operatorname{Tr}\left(\nabla G_t(x)(r_t - M_t\,x)' B_t B_t'\right) + f(x)\,G_t(x) \qquad (40)$$

PROOF. The proof of (39) is identical to Lemma 4.2 with $\psi_s(x)$ replacing $A_s\,x$.
The proof of (40) is similar to Theorem 4.3. The LHS of (39) is

$$\begin{aligned} d\bar{\beta}_t(x) &= \frac{\partial G_t(x)}{\partial t}\,\alpha_t(x)\,dt + G_t(x)\,d\alpha_t(x)\\ &= \frac{\partial G_t(x)}{\partial t}\,\alpha_t(x)\,dt + G_t(x)\Big[-\operatorname{div}(\alpha_t(x)\,\psi_t(x))\,dt \\ &+\frac{1}{2}\operatorname{Tr}\left(\nabla^2\alpha_t(x)\,B_t\,B_t'\right)\,dt + \alpha_t(x)\,x'\,C_t'\,\left(D_t^{-1}\right)'\,D_t^{-1}\,dy_t\Big] \end{aligned} \qquad (41)$$

Let us evaluate the RHS of (39).

$$\begin{aligned} \nabla\alpha_t(x) &= \alpha_t(x)\left[(B_t\,B_t')^{-1}\,\psi_t(x) - M_t\,x + r_t\right]\\ \operatorname{div}\left[\bar{\beta}_t(x)\,\psi_t(x)\right] &= (\nabla G_t(x))'\,\psi_t(x)\,\alpha_t(x) + G_t(x)\operatorname{div}(\alpha_t(x)\,\psi_t(x))\\ \nabla\bar{\beta}_t(x) &= \nabla G_t(x)\,\alpha_t(x) + G_t(x)\,\alpha_t(x) \qquad (42)\\ &\times\left[(B_t'\,B_t')^{-1}\,\psi_t(x) - M_t\,x + r_t\right]\\ \nabla^2\,\bar{\beta}_t(x) &= \nabla^2 G_t(x)\,\alpha_t(x) + 2\,\nabla G_t(x)\,\nabla\alpha_t'(x) + G_t(x)\,\nabla^2\alpha_t(x) \end{aligned} \qquad (43)$$

Substituting for these terms in the RHS of (39) and equating with the (41) yields (40). □
The following theorem presents an example of a finite dimensional filter and is the Benes system equivalent of Theorem 4.4. The proof is similar to Theorem 4.4 and hence omitted.

Theorem 5.3 *Consider the nonlinear Benes system (33), (3). Suppose the non-linearity f in (4) is the quadratic $f(x) = f^{(0)} + x'\,f^{(1)} + x'\,f^{(2)}\,x$ where $f^{(0)} \in \mathbb{R}$, $f^{(1)} \in \mathbb{R}^m$ and $f^{(2)} \in \mathbb{R}^{m\times m}$ are known constants and*

$f^{(2)}$ is a symmetric negative semi-definite matrix. Then at time t, the unnormalized density $\beta_t(x)$ is completely characterized by $\beta_t(x) = G_t(x)\,\alpha_t(x)$ where $G_t(x)$ is given by

$$\bar{\beta}_t(x) = G_t(x)\,\alpha_t(x) \quad \text{where} \quad G_t(x) = \exp\left(\phi_t + x'\,h_t + x'\,l_t\,x\right) \,. \tag{44}$$

In (44), $\phi_t \in \mathbb{R}$, $h_t \in \mathbb{R}^m$ and $l_t \in \mathbb{R}^{m\times m}$ is a symmetric matrix. These sufficient statistics satisfy the following ordinary differential equations:

$$\frac{d\phi_t}{dt} = \frac{1}{2}\,h_t'\,B_t\,B_t'\,h_t + \mathrm{Tr}\,[l_t\,B_t\,B_t'] + r_t'\,B_t\,B_t'\,h_t + f^{(0)}, \quad \phi_0 = 0$$
$$\frac{dh_t}{dt} = 2\,l_t\,B_t\,B_t'\,h_t + M_t\,B_t\,B_t'\,h_t + 2\,l_t\,B_t\,B_t'\,r_t + f^{(1)}, \quad h_0 = 0$$
$$\frac{dl_t}{dt} = 2\,l_t\,B_t\,B_t'\,l_t + M_t\,B_t\,B_t'\,l_t + l_t\,B_t\,B_t'\,M_t + f^{(2)}, \quad l_0 = 0$$

Remark: Unlike the linear case (Corollary 4.5), we have not been able to extend the above theorem to the case when the dynamics include the state noise, e.g. $dz_t = (f(x_t)dt + p'(x_t)dw_t)z_t + du_t$. This is because (40) now has the following extra terms on the RHS

$$G_t(x)\left[(B_tB_t')^{-1}\,\psi_t(x) - M_tx + r_t\right]B_t\,p(x)$$
$$-\left(\nabla G_t(x)\right)'\,B_t\,p(x) - G_t(x)\mathrm{div}\,[B_t\,p(x)]$$

Unfortunately, the presence of the $\psi_t(x)$ term above means that the solution $G_t(x)$ of (40) is no longer an exponential of a quadratic in x (apart from the trivial case when $\psi_t(x)$ is linear).

6 An Extension

The results in this paper can easily be generalized to the following observation model:

$$dy_t = C_tx_tdt + D_t\,dv_t + E_tdx_t \tag{45}$$

Compared to the standard observation equation (3), the above equation has the extra term $E_t\,dx_t$. Note that for Benes systems, (45) allows for nonlinear observations. The conditions which yield finite dimensional filters are exactly the same as (35). The reader is referred to [4] for further details, including applications in optimal stochastic control.

Acknowledgments: Vikram Krishnamurthy acknowledges the support of an Australian Research Council grant and the Co-operative Research Center for Sensor Signal and Information Processing.
Robert Elliott wishes to thank NSERC for its support under grant A7964. The authors also thank an anonymous referee for useful comments.

Bibliography

[1] V.E. Benes, *Exact Finite Dimensional Filters for Certain Diffusions with Non-linear Drift*, Stochastics, Vol.5, pp.65-92, 1981.

[2] A. Bensoussan, *Stochastic Control of Partially Observable Systems*, Cambridge University Press, Cambridge, 1992.

[3] A. Bensoussan and R.J. Elliott, *General Finite Dimensional Risk Sensitive Problems and Small Noise Limits*, to appear.

[4] C.D. Charalambous and R.J. Elliott, *Certain Classes of Nonlinear Partially Observable Stochastic Optimal Control Problems with Explicit Optimal Control Laws Equivalent to LEQG/LQG Problems*, IEEE Transactions on Automatic Control, vol.42, no.4, pp.482-497, 1997.

[5] F. Dufour and P. Bertrand, *An Image-based Filter for Discrete-time Markovian Jump Linear Systems*, Automatica, Vol.32, No.2, pp.241-247, 1996.

[6] R.J. Elliott, *Stochastic Calculus and Applications*, Applications of Mathematics, Vol.18, Springer Verlag, 1982.

[7] R.J. Elliott, L. Aggoun and J.B. Moore, *Hidden Markov Models: Estimation and Control*, Springer Verlag, 1995.

[8] R.J. Elliott, V. Krishnamurthy, *Exact Finite Dimensional Filters for Maximum Likelihood Parameter Estimation of Continuous-time Linear-Gaussian Systems*, SIAM Journal on Control and Optimization, Vol.35, No.6, pp 1908-1923, November 1997,

[9] A Friedman, *Partial Differential Equations of Parabolic Type*, Prentice Hall, N.J., 1964

[10] V. Krishnamurthy, R.J. Elliott, *Exact Finite Dimensional Filters for Doubly Stochastic Auto-regressive Processes*, IEEE Trans. Automatic Control, Vol.42. No.9, September 1997, pp 1289-1293.

[11] V. Krishnamurthy and J.S. Evans, *Covert Tracking of Markov Jump Linear Systems*, Automatica, Vol 34, No.5, May 1998 (to appear).

[12] J. Levine, *Finite Dimensional Realizations of stochastic pde's and the application to filtering*, Stochastics and Stochastics Reports, Vol.37, pp.75-103, 1991.

[13] S. I. Marcus, *Algebraic and geometric methods in nonlinear filtering*, SIAM Journal on Control and Optimization, Vol.22, No.7, pp.817-844, 1984.

[14] D.L. Ocone, J.S. Baras and S.I. Marcus, *Explicit Filters for Diffusions with Certain Nonlinear Drifts*, Stochastics, Vol8, pp 1-16, 1982.

[15] D.D. Sworder and R.G. Hutchins, *Image-Enchanced Tracking*, IEEE Trans. Aerospace and Electronic Systems, Vol.25, No.5, pp.701-709, 1989.

[16] E. Wong and B. Hajek, *Stochastic Processes in Engineering Systems*, Springer Verlag, NY 1985.

24

A Lyapunov Theory of Nonlinear Observers

A. J. Krener

ABSTRACT: We will develop a Lyapunov and a partial converse Lyapunov theory of nonlinear observers.

1 Convergent Observers

The problem of estimating the state of a dynamical system from partial and possibly noisy measurements has a long history. In its nonlinear state space form, one assumes that the dynamics satisfies a known nonlinear differential equation with unkown initial condition and the measurement is a known nonlinear function of the state

$$\begin{aligned} \dot{x} &= f(x) \\ x(0) &= x^0 \\ y &= h(x) \end{aligned} \tag{1}$$

The linear form of the problem is

$$\begin{aligned} \dot{x} &= Ax \\ x(0) &= x^0 \\ y &= Cx \end{aligned} \tag{2}$$

One is given an estimate $\hat{x}^0$ of x^0 and the observations $y(s),\ 0 \leq s \leq t$ up to time t. The problem is generate an estimate $\hat{x}(t)$ of $x(t)$ in real time, as the process evolves. The estimate should converge to the true state as $t \to \infty$. Ideally the estimation process should be robust to noise both in the dynamics and in the observations, to the initial state error and also to modeling errors in the functions $f,\ h$.

One way of approaching this problem is to assume that the dynamics, initial condition and observations are corrupted by noise with a known distribution and then to find the conditional density of the state given the past observations. When the dynamics and observations are linear functions of the state and the noises and initial condition are independent and jointly Gaussian then the conditional density is Gaussian and explicitly computable.

Wiener [We] solved this problem for stationary Gaussian processes using the method of spectral factorization. Kalman [Ka],[KB] extended this to nonstationary Gaussian processes and reduced the problem to solving off-line a Riccati equation and on-line a linear differential equation driven by the observations.

When the dynamics and/or observations are nonlinear then the unnormalized conditional density satisfies the Zakai equation, a parabolic PDE driven by the observations [DM]. Accurately computing its solution in real time for all but the smallest state dimensions is a very difficult task.

The Extended Kalman Filter [Ge] is a widely used alternative method for estimating the state of a nonlinear system. It is obtained by linearizing the nonlinear dynamics and the observation along the trajectory of the estimate. It requires the on-line solution of a Riccati differential equation and a linear differential equation driven by the observations. The Extended Kalman Filter is globally defined but it is only a local method, one expects the estimate to converge to the true state if the initial estimation error is not too large but there is no general proof of this.

There are several nonstochastic approaches to state estimation. For linear systems with linear observations, Luenberger [Lu] developed the concept of an observer. This is another linear system that is driven by the observations in such a way that the error dynamics is asymptotically stable.

Several nonstochastic methods have been proposed for nonlinear estimation. Some of these are surveyed by Misawa and Hendrick [MH]. Other methods include linearization [KI],[BZ],[KR], high gain observers [GHO] and H_∞ methods [Kr2].

Our approach to the Lyapunov theory of nonlinear observers follows that sketched out by Krener and Duarte [KD]. It parallels the excellent paper of Lin, Sontag and Wang [LSW], in which a Lyapunov and converse Lyapunov theory for the stability of nonlinear dynamical systems with respect to sets is presented. Many of our proofs are straight forward adaptations of theirs.

The systems that we are considering are of the form

$$\dot{x} = f(x, u) \tag{3}$$

$$y = h(x) \tag{4}$$

$$x(0) = x^0 \tag{5}$$

with state x, control u and observation y. The basic problem is to estimate the current state $x(t)$ from knowledge of the system and the past values of the control $u(\tau)$ and observation $y(\tau)$, $0 \le \tau \le t$ and some information about the initial state. For simplicity we assume that the information available about the initial state is an initial state estimate $\hat{x}^0$.

Definition 1 *An observer is a causal functional*

$$\begin{bmatrix} \hat{x}^0 \\ u(\tau) \\ y(\tau) \end{bmatrix} \longmapsto \hat{x}(t), \qquad 0 \le \tau \le t \tag{6}$$

from the initial state estimate, $\hat{x}^0$ and the past control and observation, $u(\tau)$, $y(\tau)$, $0 \le \tau \le t$, to the current state estimate, $\hat{x}(t)$ such that the estimate $\hat{x}(t)$ as a function of t is continuous from the left and limits exist from the right.

Recall a functional $\xi(\cdot) \longmapsto \zeta(\cdot)$ is causal if $\zeta(t)$ only depends on $\xi(\tau)$ for $\tau \le t$. We will need some terminology. A function $\alpha(r)$ is of class K if it is continuous, strictly increasing and $\alpha(0) = 0$. A function $\alpha(r)$ is of class K_∞ if it is of class K and in addition $\alpha(r)$ goes to ∞ as r goes to ∞. A function $\gamma(r, s)$ is of class K^2 if it is continuous in both variables, strictly increasing in each variable when the other is fixed and $\gamma(0,0) = 0$. A function $\gamma(r, s)$ is of class K^2_∞ if it is of class K^2 and goes to ∞ as either r or s goes to ∞. We introduce noises into the model to reflect our uncertainty about the dynamics $w(\cdot)$, the measurments $v(\cdot)$ and the error of the initial state estimate $\tilde{x}^0 = x^0 - \hat{x}^0$.

$$\dot{x} = f(x,u) + g(x)w \tag{7}$$
$$y = h(x) + v \tag{8}$$
$$x(0) = x^0 = \hat{x}^0 + \tilde{x}^0 \tag{9}$$

We will measure the size of the three noises using L_2 norms but other norms or ways of measuring size are possible. For example, the size of $\tilde{x}^0$ could be measured by a nonegative function $Q^0(x)$ which has a global minimum at $x = \hat{x}^0$. We denote the L_2 norm of the driving and observation noises on $[\tau, t]$ by

$$||w, v||_{[\tau,t]} = \left(\int_\tau^t |w(\sigma)|^2 + |v(\sigma)|^2 \, d\sigma \right)^{\frac{1}{2}}$$

Definition 2 *An observer is said to be convergent if there exist class K_∞ functions $\delta_i(\epsilon)$, $i = 1, 2$, and a function $T(r, s, \epsilon)$ with the following properties. For any times $0 \le \tau_1 \le \tau_2 \le t$, any initial initial state, any initial state estimate $\hat{x}^0$, any control $u(\cdot)$ and any noises $w(\cdot)$, $v(\cdot)$ with support in $[\tau_1, \tau_2]$, the state trajectory and state estimate satisfy*
(1) for any $\epsilon > 0$ if

$$|x(\tau_1) - \hat{x}(\tau_1)| \le \delta_1(\epsilon)$$
$$||w, v||_{[\tau_1,\tau_2]} \le \delta_2(\epsilon)$$

then

$$|x(t) - \hat{x}(t)| \le \epsilon,$$

(2) for any r, s, $\epsilon > 0$, if

$$|x(\tau_1) - \hat{x}(\tau_1)| < r$$
$$||w, v||_{[\tau_1,\tau_2]} < s$$
$$T(r, s, \epsilon) \le t - \tau_2$$

then

$$|x(t) - \hat{x}(t)| < \epsilon$$

Condition (1) is the stability of the estimation error. Loosely speaking it states that the estimation error at a later time is a uniformly continuous function of the estimation error at a prior time and the driving and observation noises between the times. Condition (2) is the asymptotic stability of the noise free estimation error. It states that if the noises stop then the estimation error asymptotically converges to zero uniformly with respect to time, prior state, prior state estimate, control, prior driving noise and prior observation noise. Of course there are obvious local versions of the definition, e. g., (1) and (2) hold only if $|x(\tau_1) - \hat{x}(\tau_1)|$ and $\|w, v\|_{[\tau_1,\tau_2]}$ are sufficiently small.
Condition (1) implies that if an observer is convergent, if the estimate is exact at some time, $x(\tau) = \hat{x}(\tau)$ and if the noise is zero at subsequent times $w(\sigma) = 0$, $v(\sigma) = 0$, $\tau \le \sigma \le t$ then the estimate is exact at subsequent times $x(t) = \hat{x}(t)$ for all $t \ge \tau$.
A function $\beta(r,t)$ is of class KL if for fixed t, it is of class K and for fixed r, it is strictly decreasing to 0 as t goes to ∞. A function $\gamma(r,s,t)$ is of class K^2L if for fixed t, it is of class K^2 and for fixed r, s, it is strictly decreasing to 0 as t goes to ∞.

Theorem 1 *An observer is convergent iff there exists a function $\gamma(r,s,t)$ of class K^2L such that for any $0 \le \tau_1 \le \tau_2 \le t$, any initial state x^0, control history $u(\cdot)$, and noise triple $\tilde{x}^0$, $w(\cdot)$, $v(\cdot)$, with support in $[\tau_1, \tau_2]$, the state trajectory and estimate satisfy*

$$|x(t) - \hat{x}(t)| \le \gamma(|x(\tau_1) - \hat{x}(\tau_1)|, \|w, v\|_{[\tau_1,\tau_2]}, t - \tau_2) \tag{10}$$

Proof: Suppose $\gamma(r,s,t)$ exists. First we show (1) of Definition 2 holds. Define a class K_∞ function $\bar{\gamma}(r) = \gamma(r,r,0) + r$ and let $\delta(\epsilon) = \bar{\gamma}^{-1}(\epsilon)$. If

$$\begin{aligned} |x(\tau_1) - \hat{x}(\tau_1)| &\le \delta(\epsilon) \\ \|w, v\|_{[\tau_1,\tau_2]} &\le \delta(\epsilon) \end{aligned}$$

then

$$\begin{aligned} |x(t) - \hat{x}(t)| &\le \gamma(\delta(\epsilon), \delta(\epsilon), t - \tau_2) \\ &\le \gamma(\delta(\epsilon), \delta(\epsilon), 0) \\ &\le \bar{\gamma}(\delta(\epsilon)) = \epsilon \end{aligned}$$

for all $0 \le \tau_1 \le \tau_2$.
Given any r, s, ϵ choose $T = T(r,s,\epsilon)$ so that

$$\gamma(r,s,T) < \epsilon.$$

This is always possible because γ is K^2L. Then (2) holds.

To show the converse, we first show that the $T(r,s,\epsilon)$ can be chosen with the following properties;
(i) for fixed $r,\ s>0$, the mapping $\epsilon \mapsto T(r,s,\epsilon)$ is a continuous, strictly decreasing map from $(0,\infty)$ to $(0,\infty)$,
(ii) for fixed $r,\ \epsilon>0$, the mapping $s \mapsto T(r,s,\epsilon)$ is a strictly increasing map which goes to ∞ as $s\rightarrow\infty$,
(iii) for fixed $s,\ \epsilon>0$, the mapping $r \mapsto T(r,s,\epsilon)$ is a strictly increasing map which goes to ∞ as $r\rightarrow\infty$.
Define

$$A(r,s,\epsilon)=\{T\geq 0: \text{Condition (2) is satified for } r,s,\epsilon\}.$$

This set is not empty for any $r,\ s,\ \epsilon>0$ and is decreasing in $r,\ s$ and increasing in ϵ. Define $\bar{T}(r,s,\epsilon)=\inf A(r,s,\epsilon)$ then $\bar{T}$ is increasing in $r,\ s$ and decreasing in ϵ. Because they are class K_∞, the functions $\delta_i(\epsilon)\rightarrow\infty$ as $\epsilon\rightarrow\infty$ so $\bar{T}(r,s,\epsilon)=0$ for ϵ sufficiently large.
Define

$$T(r,s,\epsilon)=\frac{2}{\epsilon}\int_{\frac{\epsilon}{2}}^{\epsilon} T(r,s,\nu)\,d\nu+\frac{r+s}{\epsilon}$$

then T is strictly increasing in $r,\ s$, continuous and strictly decreasing in ϵ and

$$\begin{aligned} T(r,s,\epsilon) &> \bar{T}(r,s,\epsilon) \\ \lim_{\epsilon\rightarrow\infty} T(r,s,\epsilon) &= 0 \\ \lim_{\epsilon\rightarrow 0+} T(r,s,\epsilon) &= \infty \end{aligned}$$

Define a function $\phi(r,s)=\max\{\delta_1^{-1}(r),\ \delta_2^{-1}(s)\}$ then by Condition (1) for any noises with support in $[\tau_1,\tau_2]$ for any $t\geq\tau_2$

$$|x(t)-\hat{x}(t)|\leq\phi(|x(\tau_1)-\hat{x}(\tau_1)|,\|w,v\|_{[\tau_1,\tau_2]}) \tag{11}$$

Define a function $\psi(r,s,\cdot)=T^{-1}(r,s,\cdot)$ where $T(r,s,\epsilon)$ is as above. Then by Condition (2) for any noises with support in $[\tau_1,\tau_2]$ and any $t\geq\tau_2$

$$|x(t)-\hat{x}(t)|<\psi(r,s,t-\tau_2)$$

for any $r>|x(\tau_1)-\hat{x}(\tau_1)|$ and $s>\|w,v\|_{[\tau_1,\tau_2]}$.
Define

$$\bar{\psi}(r,s,t)=\min\left\{\phi(r,s),\inf_{\rho>r,\sigma>s}\psi(\rho,\sigma,t)\right\}$$

then by the above for any noises with support in $[\tau_1,\tau_2]$ and any $t\geq\tau_2$

$$|x(t)-\hat{x}(t)|\leq\bar{\psi}(|x(\tau_1)-\hat{x}(\tau_1)|,\|w,v\|_{[\tau_1,\tau_2]},t-\tau_2) \tag{12}$$

If $\bar{\psi}$ was K^2L we would be done but it may not be. However we can construct a K^2L function which majorizes it. By its definition, $\bar{\psi}(r,s,t)$ is increasing in $r,\ s$ and decreases to 0 as $t\rightarrow\infty$.

Define

$$\hat{\psi}(r,s,t) = \int_r^{r+1} \int_s^{s+1} \bar{\psi}(\rho,\sigma,t)\, d\sigma\, d\rho$$

then $\hat{\psi}(r,s,t)$ is continuously increasing in r, s and $\hat{\psi}(r,s,t) \geq \bar{\psi}(r,s,t)$. Note that since $\psi(r,s,t) \to \infty$ as $t \to 0$

$$\bar{\psi}(r,s,t) \leq \bar{\psi}(r,s,0) = \phi(r,s)$$

which is a continuous function. By the Lebesgue dominated convergence theorem, for any fixed $r, s \geq 0$

$$\lim_{t\to\infty} \hat{\psi}(r,s,t) = \int_r^{r+1} \int_s^{s+1} \lim_{t\to\infty} \bar{\psi}(\rho,\sigma,t)\, d\sigma\, d\rho$$

Since $\bar{\psi}(r,s,t)$ decreases to 0 as $t \to \infty$, so does $\hat{\psi}(r,s,t)$.
Define

$$\tilde{\psi}(r,s,t) = \hat{\psi}(r,s,t) + \frac{r+s}{(r+s+1)(t+1)}$$

then this function is continuous and strictly increasing in r, s for fixed t and for fixed r, s it strictly decreases to 0 as $t \to \infty$. Moreover from (12) for any noises with support in $[\tau_1, \tau_2]$ and any $t \geq \tau_2$

$$\begin{aligned} |x(t) - \hat{x}(t)| &\leq \bar{\psi}(|x(\tau_1) - \hat{x}(\tau_1)|, \|w,v\|_{[\tau_1,\tau_2]}, t-\tau_2) \\ &\leq \hat{\psi}(|x(\tau_1) - \hat{x}(\tau_1)|, \|w,v\|_{[\tau_1,\tau_2]}, t-\tau_2) \\ &\leq \tilde{\psi}(|x(\tau_1) - \hat{x}(\tau_1)|, \|w,v\|_{[\tau_1,\tau_2]}, t-\tau_2) \end{aligned} \quad (13)$$

The only thing that is lacking for it to be K^2L is that $\tilde{\psi}(0,0,t)$ might not be 0. This can remedied that by defining

$$\gamma(r,s,t) = \sqrt{\phi(r,s)}\sqrt{\tilde{\psi}(r,s,t)}.$$

then (11,13)

$$|x(t) - \hat{x}(t)| \leq \gamma(|x(\tau_1) - \hat{x}(\tau_1)|, \|w,v\|_{[\tau_1,\tau_2]}, t-\tau_2)$$

□

Lemma 1 *Suppose α is a class K function then there exists a K^2L function $\gamma_\alpha(r,s,t)$ with the following property.*
If $z(\cdot)$, $\zeta(\cdot)$ are any absolutely continuous functions satisfying the differential inequality

$$\begin{aligned} \dot{z}(\tau) &\leq -\alpha(z(\tau)) + \dot{\zeta}(\tau) \\ z(0) &\geq 0 \\ \zeta(0) &= 0 \\ \dot{\zeta}(t) &\geq 0 \end{aligned}$$

where the support of $\dot{\zeta}(t)$ *is* $[0, \tau_2] \subseteq [0, t]$ *then*

$$z(t) \leq \gamma_\alpha(z(0), \zeta(\tau_2), t - \tau_2) \tag{14}$$

Proof: Define $\bar{\alpha}$ by $\bar{\alpha}(0) = 0$ and for $r > 0$

$$\bar{\alpha}(r) = \frac{1}{r}\int_0^r \min\{s^2, \alpha(s)\}\, ds$$

Then $\bar{\alpha}$ is a smooth class K function and for $r > 0$

$$\begin{aligned}
\bar{\alpha}(r) &< \alpha(r) \\
\bar{\alpha}(r) &< r^2 \\
\frac{d}{dr}\bar{\alpha}(r) &= \frac{1}{r}\left(\min\{r^2, \alpha(r)\} - \bar{\alpha}(r)\right) > 0
\end{aligned}$$

Define for any $r > 0$

$$\eta(r) = -\int_1^r \frac{ds}{\bar{\alpha}(s)}$$

then

$$\lim_{r \to 0+} \eta(r) = \infty$$

It is not hard to see that η is a strictly decreasing differentiable function mapping $(0, \infty)$ to some interval $(a, +\infty)$. Then η^{-1} is a strictly decreasing function from (a, ∞) to $(0, \infty)$.
Define

$$\beta_\alpha(r, t) = \begin{cases} 0, & r = 0 \\ \eta^{-1}(\eta(r) + t), & r > 0 \end{cases}$$

Note that $\beta_\alpha(r, t)$ is a class KL function which satisfies

$$\begin{aligned}
\frac{d}{dt}\beta_\alpha(r, t) &= -\bar{\alpha}(\beta_\alpha(r, t)) \\
\beta_\alpha(r, 0) &= r
\end{aligned}$$

Trivially

$$\dot{z}(\tau) \leq -\bar{\alpha}(z(\tau)) + \dot{\zeta}(\tau)$$

so $z(t) \leq \bar{z}(t)$ where

$$\begin{aligned}
\frac{d}{d\tau}\bar{z}(\tau) &= -\bar{\alpha}(\bar{z}(\tau)) + \dot{\zeta}(\tau) \\
\bar{z}(0) &= z(0)
\end{aligned}$$

Now consider the mapping

$$\theta(\tau) = \beta_\alpha(\bar{z}(\tau), \tau_2 - \tau)$$

then

$$\begin{aligned}\theta(0) &= \beta_\alpha(z(0),\tau_2)\\ \theta(\tau_2) &= \bar{z}(\tau_2)\\ \frac{d}{d\tau}\theta(\tau) &= \frac{d}{dz}\beta_\alpha(\bar{z}(\tau),\tau_2-\tau)\dot{\zeta}(\tau)\end{aligned}$$

so

$$\bar{z}(\tau_2) = \beta_\alpha(z(0),\tau_2) + \int_0^{\tau_2} \frac{d}{dz}\beta_\alpha(\bar{z}(\tau),\tau_2-\tau)\dot{\zeta}(\tau)d\tau$$

But

$$\frac{d}{dz}\beta_\alpha(\bar{z}(\tau),\tau_2-\tau) \le 1$$

because it satisfies the linear differential equation

$$\begin{aligned}\frac{d}{dt}\frac{d}{dz}\beta_\alpha(\bar{z}(\tau),\tau_2-\tau) &= -\frac{d}{dz}\bar{\alpha}(\bar{z}(\tau))\frac{d}{dz}\beta_\alpha(\bar{z}(\tau),\tau_2-\tau)\\ \frac{d}{dz}\beta_\alpha(r,0) &= 1\end{aligned}$$

with

$$\frac{d}{dz}\bar{\alpha}(\bar{z}(\tau)) \ge 0$$

Hence

$$\bar{z}(\tau_2) \le \beta_\alpha(z(0),\tau_2) + \zeta(\tau_2)$$

and

$$\begin{aligned}\bar{z}(t) &\le \beta_\alpha(\beta_\alpha(z(0),\tau_2)+\zeta(\tau_2),t-\tau_2)\\ &\le \beta_\alpha(z(0)+\zeta(\tau_2),t-\tau_2)\\ &= \gamma_\alpha(z(0),\zeta(\tau_2),t-\tau_2)\end{aligned}$$

where γ_α is defined by

$$\gamma_\alpha(r,s,t) = \beta_\alpha(r+s,t)$$

□

Definition 3 *A Lyapunov function for the observer $\hat{x}(t)$ is a function $Q(x,t)$ with the following properties*
(1) $Q(x,t)$ is a causal functional of the initial state estimate $\hat{x}^0$ and the past control and observation, $u(\tau)$, $y(\tau)$, $0 \le \tau \le t$ such that $Q(x,t)$ has a minimum at $x = \hat{x}(t)$ and is smooth for all $x \ne \hat{x}(t)$,
(2) there exist class K_∞ functions $\alpha_i(r)$, $i = 1,2$ which satisfy

$$\alpha_1(|x-\hat{x}(t)|) \le Q(x,t) \le \alpha_2(|x-\hat{x}(t)|), \tag{15}$$

(3) there exist a class K funtion $\alpha_3(r)$ which satisfies

$$\frac{d}{dt}Q(x(t),t) \leq -\alpha_3(|x(t)-\hat{x}(t)|) + |w(t)|^2 + |v(t)|^2 \tag{16}$$

for any state trajectory $x(t)$ and noises $w(t)$, $v(t)$ consistent with the observation $y(t) = h(x(t)) + v(t)$.

Theorem 2 *The observer $\hat{x}(t)$ is convergent if there exist a Lyapunov function $Q(x,t)$.*

Proof: Suppose a Lyapunov function $Q(x,t)$ exists then

$$\begin{aligned}
\frac{d}{dt}Q(x(t),t) &\leq -\alpha_3(|x(t)-\hat{x}(t)|) + |w(t)|^2 + |v(t)|^2 \\
&= -\alpha_3(\alpha_2^{-1}(\alpha_2(|x(t)-\hat{x}(t)|))) + |w(t)|^2 + |v(t)|^2 \\
&\leq -\alpha(Q(x(t),t)) + |w(t)|^2 + |v(t)|^2
\end{aligned} \tag{17}$$

where α is the class K function defined by

$$\alpha(r) = \alpha_3(\alpha_2^{-1}(r)) \tag{18}$$

By the above lemma, there exists a class K^2L function $\gamma_\alpha(r,s,t)$ such that

$$Q(x(t),t) \leq \gamma_\alpha(Q(x(\tau_1),\tau_1), \|w,v\|^2_{[\tau_1,\tau_2]}, t-\tau_2)$$

Hence the observer is convergent because

$$\begin{aligned}
\alpha_1(|x(t)-\hat{x}(t)|) &\leq Q(x(t),t) \\
&\leq \gamma_\alpha(Q(x(\tau_1),\tau_1), \|w,v\|^2_{[\tau_1,\tau_2]}, t-\tau_2) \\
&\leq \gamma_\alpha(\alpha_2(|x(\tau_1)-\hat{x}(\tau_1)|), \|w,v\|^2_{[\tau_1,\tau_2]}, t-\tau_2)
\end{aligned}$$

or as desired

$$|x(t)-\hat{x}(t)| \leq \gamma(|x(\tau)-\hat{x}(\tau)|, \|w,v\|_{[\tau_1,\tau_2]}, t-\tau_2)$$

where $\gamma(r,s,t)$ is the K^2L function defined by

$$\gamma(r,s,t) = \alpha_1^{-1}\left(\gamma_\alpha(\alpha_2(r), s^2, t)\right).$$

□

A partial converse to this is the following.

Theorem 3 *Suppose an observer $\hat{x}(t)$ is convergent then there exist a function $Q(x,t)$ such that*
(1) $Q(x,t)$ is a causal functional of the initial state estimate $\hat{x}^0$ and the past control and observation, $u(\tau)$, $y(\tau)$, $0 \leq \tau \leq t$ such that $Q(x,t)$ has

a minimum at $x = \hat{x}(t)$
(2) there exist class K_∞ *functions* $\alpha_i(r),\ i = 1,2$ *which satisfy*

$$\alpha_1(|x - \hat{x}(t)|) \leq Q(x,t) \leq \alpha_2(|x - \hat{x}(t)|), \tag{19}$$

(3') Along any state trajectory $x(t)$, *input* $u(t)$, *observation* $y(t)$ *and noises* $w(t),\ v(t)$ *consistent with the system (3,4)*

$$Q(x(t_2),t_2) \leq Q(x(t_1),t_1) + \|w,v\|^2_{[t_1,t_2]}$$

Proof: Suppose an observer $\hat{x}(t)$ is convergent, define $Q(x,t)$ as

$$Q(x,t) = \inf\{|z(\tau_1) - \hat{x}(\tau_1)|^2 + \|w,v\|^2_{[\tau_1,t]} : 0 \leq \tau_1 \leq t\} \tag{20}$$

where the infimum is over all $z(\tau),\ w(\tau),\ v(\tau)$ satisfying

$$\frac{d}{d\tau}z(\tau) = f(z(\tau),u(\tau)) + g(z(\tau))w(\tau) \tag{21}$$

$$y(\tau) = h(z(\tau)) + v(\tau) \tag{22}$$

$$z(t) = x. \tag{23}$$

and $u(\tau),\ y(\tau)$ are the control and observation. Clearly $Q(x,t)$ is a causal functional of the the control, observation and estimate $\hat{x}(\tau)$. The latter is a causal functional of the control, observation and initial state estimate $\hat{x}^0$. We claim that

$$\alpha_1(|x - \hat{x}(t)|) \leq Q(x,t) \leq \alpha_2(|x - \hat{x}(t)|) \tag{24}$$

where α_i are the class K_∞ functions

$$\begin{aligned} \alpha_1(|x - \hat{x}(t)|) &= (\min\{\delta_1(|x - \hat{x}(t)|), \delta_2(|x - \hat{x}(t)|)\})^2 \\ \alpha_2(|x - \hat{x}(t)|) &= |x - \hat{x}(t)|^2 \end{aligned}$$

and δ_i are from the Definition 2 of a convergent observer. The second inequality is obvious, the first can be seen as follows. Suppose it is not true, i.e., there exists an x, τ_1 and $z(\tau),\ w(\tau),\ v(\tau)$ satisfying (21-23) such that

$$|z(\tau_1) - \hat{x}(\tau_1)|^2 + \|w,v\|^2_{[\tau_1,t]} < (\min\{\delta_1(|x - \hat{x}(t)|), \delta_2(|x - \hat{x}(t)|)\})^2$$

which implies that

$$\begin{aligned} |z(\tau_1) - \hat{x}(\tau_1)| &< \delta_1(|x - \hat{x}(t)|) \\ \|w,v\|_{[\tau_1,t]} &< \delta_2(|x - \hat{x}(t)|) \end{aligned}$$

By property (1) of a convergent observer, this implies the contradiction

$$|x - \hat{x}(t)| < |x - \hat{x}(t)|.$$

Condition (3') follows immediately from the definition of $Q(x,t)$, (20-23).
□

2 Conclusions

We have presented a Lyapunov theorem for convergent observers and a partial converse. The Lyapunov function is of a novel type, it is a causal functional of the past control and observation. Moreover it does not explicitly depend on the current estimate $\hat{x}(t)$ but it determines the estimate as it arg min. In some sense the function $x \mapsto Q(x,t)$ is the state at time t of an infinite dimensional observer. Further refinement of this work is needed to obtain a Lyapunov and converse Lyapunov theory of observers.

Acknowledgments: Research supported in part by AFOSR-49620-95-1-0409.

References

[BZ] D. Bestle and M. Zeitz, *Canonical form observer design for non-linear time-variable systems*, Internat. J. Control **38** (1983), 419-431.

[DM] M. H. A. Davis and S. I. Marcus, *An introduction to nonlinear filtering*, in Stochastic Systems: The Mathematics of Filtering and Identification and Applications, M. Hazewinkel and J. C. Willems, (eds.), D. Reidel Publishing, Dordrecht, (1981), 53-76.

[GHO] J. P. Gauthier, H. Hammouri, and S. Othman, *A simple observer for nonlinear systems with applications to bioreactors*, IEEE Trans, Autom. Contr., **37**, (1992), 875-880.

[Ge] A. Gelb, Applied Optimal Estimation, MIT Press, Cambridge, (1974).

[Ka] R. E. Kalman, *A new approach to linear filtering and prediction problems*, Trans. of ASME, Part D, J. of Basic Engineering, **82**, (1960), 35-45.

[KB] R. E. Kalman and R. S. Bucy, *New results in linear filtering and prediction theory*, Trans. of ASME, Part D, J. of Basic Engineering, **83**, (1961), 95-108.

[Kr1] A. J. Krener, *Nonlinear stabilizability and detectability*, in Systems and Networks: Mathematical Theory and Applications, U. Helmke, R. Mennicken and J. Saurer, (eds.), Akademie Verlag, Berlin, (1994), 231-250.

[Kr2] A. J. Krener, *Necessary and sufficient conditions for nonlinear worst case (H-infinity) control and estimation.* Summary appeared in Journal of Mathematical Systems, Estimation, and Control,**4**, (1994), 485-488, full manuscript appeared in Journal of Mathematical Systems, Estimation, and Control, **7**, (1997), 81-106.

[KD] A. J. Krener and A. Duarte, A hybrid approach to nonlinear filtering. Proc. of 35th Conference on Decision and Control, Kobe (1996), 1815-1819.

[KI] A. J. Krener and A. Isidori, *Linearization by output injection and nonlinear observers*, Systems and Control Letters , (1983), 47-52.

[KR] A. J. Krener and W. Respondek, *Nonlinear observers with linearizable error dynamics*, SIAM J. Control and Optimization, **23**, (1985), 197-216.

[LSW] Y. Lin, E. D. Sontag and Y. Wang, A smooth converse Lyapunov Theorem for robust stability. SIAM J. Control and Optimization, to appear.

[Lu] D. G. Luenberger, *Observing the state of a linear system*, IEEE Trans. on Military Electronics, **8** (1964), 74-80.

[MH] E. A. Misawa, J. K. Hedrick, *Nonlinear observers a state of the art survey*, Trans. of ASME, J. of Dynamic Systems, Measurement and Control, **111** (1989), 344-352.

[We] N. Weiner, Extrapolation, Interpolation and Smoothing of Stationary Time Series, with Engineering Applications, Technology Press of MIT, Cambridge, (1949).

25

Existence of Optimal Controls for Variance Control

Harold J. Kushner

ABSTRACT: The question of the existence of optimal controls is fundamental to many further developments in control theory, and there is a huge literature on the subject. For stochastic control, variance control problems have received much attention recently owing to their importance in financial mathematics. The existence question is answered in much the same way as it is when only the drift is controlled. In order to establish the intuition, classical results are surveyed and the necessary extensions for the variance control problem are outlined.

KEYWORDS: variance control, existence of optimal controls, stochastic optimal control, martingale measures

1 Introduction

The question of the existence of an optimal control in some appropriate sense is fundamental throughout control theory, and has been the subject of a large literature. For example, among many others, Young (1969),Warga (1962), McShane and Warfield (1967), Benes (1970, 1971), Bismut (1976), Fleming (1977), Fleming and Nisio (1884), Kushner (1975, 1978, 1990b), Krylov (1980), Borkar (1989), Haussmann (1986), Kushner and Dupuis (1992).

We are concerned with the stochastic control problem where the variance of the noise is controlled, as well as the drift. The basic ideas for this are in Kushner (Section 8, 1990a). Although that section was concerned with the convergence of numerical methods for the variance control problem, the structures and methods introduced there can be readily adapted to the question of existence. The motivation for this paper arises from the numerous questions which have arisen recently concerning variance control owing to its importance in models in financial mathematics.

There are many ways to deal with the question of existence. One might be interested in a feedback control with particular properties. We are concerned with the question of the existence of an optimal control in the class of properly defined and appropriately measurable functions of (ω, t). This is the form in which the question arises when one is concerned with the convergence of numerical methods. To keep the development simple, we focus on one type of cost function.

Since the basic ideas and techniques exist in various forms in the literature, our aim is to survey the ideas and develop the intuition behind them. The problem of existence often reduces to the proper definition of the problem. One seeks to define the problem in such a way that the set of possible solutions or the set of possible (solution, control) pairs is compact in an appropriate sense, and where the cost function and state are appropriately continuous in the control. The basic issues are similar to those which arose in deterministic control and the calculus of variations a long time ago. In order to motivate the structures which are used, we start by surveying the deterministic problem, then move on to the stochastic problem where only the drift is controlled, and finally make the changes necessary to deal with the variance control problem.

We provide a concise summary of the essential background for the controlled variance problem for those not well acquainted with the literature on existence. The proofs rely heavily on weak convergence methods (Ethier and Kurtz (1986), Kushner and Dupuis (1992)). The essential concept from control theory is that of relaxed control (Young (1969), Warga (1962), Fleming (1977), Borkar (1989), Kushner and Dupuis (1992)) which is extended to the variance control problem via the concept of martingale measure.

2 The Existence of a Deterministic Optimal Control

This section surveys some standard ideas in deterministic control, and is for motivation only. It introduces some of the issues and ideas which are also fundamental to the stochastic problem. The ODE of concern is

$$\dot{x}(t) = b(x(t), t, u(t)), \quad x \in I\!R^r - \text{Euclidean } r \text{ space}, \tag{2.1}$$

where $b(\cdot)$ is continuous in its arguments, piecewise in t, and $u(t) \in \mathcal{U}$, a compact set in some Euclidean space. We assume either that $b(\cdot)$ is bounded or that it satisfies a Lipschitz condition in x, uniformly in the other variables. A uniqueness condition on the solution will be added below.

The cost function of interest over the fixed time interval $[0, T]$ is

$$W(x, T, u, \tau) = \int_0^{T\wedge\tau} k(x(s), s, u(s))ds + g(x(T\wedge\tau), T\wedge\tau) \tag{2.2}$$

where $x(0) = x$. The τ is a stopping time, which can be chosen by the controller. The functions $k(\cdot), g(\cdot)$ are continuous (piecewise in t for $k(\cdot)$ is enough) and have at most a polynomial growth in x. For simplicity, it is always assumed that the processes are defined on all of $[0, T]$, even if $\tau < T$. If the stopping time τ is not part of the problem formulation, ignore it. Define

$$V(x,T) = \inf_{\text{admissible } u,\tau} W(x,T,u,\tau).$$

Remarks on optimal controls and limits of sequences of controls. The following remarks and examples are intended to illustrate the definitions and structures which need to be introduced to properly handle the question of existence of an optimal control.

There need not exist an optimal control in the sense that there are admissible $\bar{u}(\cdot), \bar{\tau}$ such that $V(x,T) = W(x,T,\bar{u},\bar{\tau})$. Consider the following example, where $\mathcal{U} = [-1,1]$, and $\dot{x}(t) = u(t)$, $x(0) = 0$, and $W(x,T,u) = \int_0^T [x^2(t) + (u^2(t)-1)^2]dt$.

The infimum of the costs is obviously zero. But there is no optimal control in the usual sense. A minimizing sequence alternates increasingly rapidly between the values ± 1, trying to maintain $x(t)$ as close to zero as possible.

We note that even if there are admissible $u(\cdot), \tau$ which minimize the cost function, they need not be useful in a practical sense, since the control might not be constructible by practical or convenient means.

These considerations suggest that the question of existence is essentially mathematical, since in an application one could try to select a suitable member of a minimizing sequence, although we do not usually know such a sequence a priori. The question of existence is important in applications since it is necessary for the dynamic programming and variational equations to be meaningful. These equations often serve as the basis of analysis of qualitative properties and of numerical methods. Existence is important in convergence proofs for numerical methods.

The next example will give another hint of what one must do to establish a useful existence result. For integer n, let $\dot{x}^n = b(x^n, u^n)$, $\mathcal{U} = [0,1]$, where $b(\cdot)$ is bounded and continuous and $u^n(\cdot) = \alpha_1$ on $[2i/n, 2i/n + 1/n), i = 1, 2, \ldots$, and equals α_2 on the complementary intervals. Then $\{x^n(\cdot)\}$ is equicontinuous, and the limit $x(\cdot)$ of any convergent subsequence satisfies

$$\dot{x} = \left[b(x,\alpha_1) + b(x,\alpha_2)\right]/2. \tag{2.3}$$

Thus the limit problem has taken us out of the class of solutions of (2.1) corresponding to ordinary controls. But, the limit can be arbitrarily well approximated by the solution to (2.1) with a simply piecewise constant ordinary control.

The problem of existence of an optimal control is one of "closure." One needs to enlarge the sense of solution and control so that each sequence of (solution, control) in this enlarged class has a convergent subsequence, and the limit solution satisfies the ODE in some appropriate extended sense and with the control being the limit of the control sequence. Loosely speaking, we wish to imbed the problem in a larger class which is compact in an appropriate sense and in which the classical solutions are dense. In extending the definition of solution, we need to be certain that we have not altered the inf of the costs, and that any problem (solution, control) in the extension can be well approximated by a classical one. The limits of the paths constructed in the above examples should be included in this extended class of solutions. These considerations of closure an approximation lead to the concept of relaxed control (Young (1969), Warga (1962), Fleming (1977)).

The representation of a control as a measure. Let $u(\cdot)$ be an admissible ordinary control and let $\mathcal{B}(\mathcal{U})$ and $\mathcal{B}(\mathcal{U} \times [0,T])$ denote the σ-algebras over the Borel sets in $\mathcal{U}$ and $\mathcal{U} \times [0,T]$, resp. Define the Dirac measures $m_t(\cdot)$ on $\mathcal{B}(\mathcal{U})$ and $m(\cdot)$ on $\mathcal{B}(\mathcal{U} \times [0,T])$ by

$$m_t(A) = I_A(u(t)), \quad m(A \times [0,t]) = \int_0^t m_s(A)ds \quad A \in \mathcal{B}(\mathcal{U}),$$

where $I_A(u)$ is the indicator function of the set A. We can now write (2.1) in the equivalent form

$$\dot{x}(t) = \int_{\mathcal{U}} b(x(t), t, \alpha) m_t(d\alpha), \tag{2.4a}$$

or, alternatively, as (2.4b) with cost (2.5):

$$x(t) = x + \int_0^t \int_{\mathcal{U}} b(x(s), s, \alpha) m_s(d\alpha) ds = x + \int_0^t \int_{\mathcal{U}} b(x(s), s, \alpha) m(d\alpha ds). \tag{2.4b}$$

$$W(x, T, m, \tau) = \int_0^{T \wedge \tau} \int_{\mathcal{U}} k(x(s), s, \alpha) m(d\alpha ds). \tag{2.5}$$

Relaxed Control. The constructions to follow have been standard in deterministic control for a long time (e.g., Young (1969)). An *admissible relaxed control* or simply a *relaxed control* $m(\cdot)$ is a measure on $\mathcal{B}(\mathcal{U} \times [0,T])$ such that $m(\mathcal{U} \times [0,t]) = t$ for all $t \le T$. Given a relaxed control $m(\cdot)$, there is a derivative $m_t(\cdot)$ such that $m(d\alpha dt) = m_t(d\alpha)dt$; $m_t(A)$ is measurable for each $A \in \mathcal{B}(\mathcal{U})$, and $m_t(\cdot)$ is a measure on $\mathcal{B}(\mathcal{U})$ for each t. In fact, we can define the derivative (for almost all t) by

$$m_t(A) = \lim_{\delta \to 0} m(A \times [t - \delta, t])/\delta. \tag{2.6}$$

We make the unrestrictive assumption that (2.4) has a unique solution for each relaxed control.

Example. Return to the example which led to (2.3). Let $m_t(\cdot)$ be the measure which takes the value 1/2 at $\alpha_i, i = 1, 2$. Then the limit can be written as:

$$\dot{x} = \int_{\mathcal{U}} b(x, \alpha) m_t(d\alpha).$$

With the use of a relaxed control, the set of possible velocities and cost rates $(b(x,t,\mathcal{U}), k(x,t,\mathcal{U}))$ is replaced by its convex hull. Indeed, if this set is convex for each x, t, and satisfies a upper semicontinuity property in (x, t) for almost all t, then one need not introduce relaxed controls and suitable implicit function theorems can be used to establish the existence of an optimal control. See McShane and Warfield (1967), Kushner (1977).

The topology on the space of relaxed controls. The weak topology is used. I.e., a sequence of relaxed controls $m^n(\cdot)$ converges to a relaxed control $m(\cdot)$ if and only if for any bounded and continuous function $\phi(\cdot)$

$$\int_0^T \int_{\mathcal{U}} \phi(s,\alpha) m^n(d\alpha ds) \to \int_0^T \int_{\mathcal{U}} \phi(s,\alpha) m(d\alpha ds). \tag{2.7}$$

Thus any sequence $\{m^n(\cdot), x^n(\cdot)\}$ has a convergent subsequence and the limit $(x(\cdot), m(\cdot))$ satisfies (2.4). Thus, the space of (solution, control) for (2.4) is compact.

The cost values. We need to be sure that the extended definitions of control and solution have not altered the infimum of the cost values: I.e., we need to assure that

$$\inf_{u,\tau} W(x, T, u, \tau) = \inf_{m,\tau} W(x, T, m, \tau). \tag{2.8}$$

The following classical result (Young (1969)) says that any admissible relaxed control can be well approximated by a "nice" ordinary admissible control and that (2.8) holds.

Approximation theorem: The "Chattering" Theorem. *Assume the uniqueness condition, and the given conditions on $b(\cdot), k(\cdot), g(\cdot)$. Ignore τ if it is not part of the problem statement. For any $\epsilon > 0$ and $\tau \leq T$ there is a $\delta > 0$, a finite set $\mathcal{U}_\epsilon \subset \mathcal{U}$, and a $\mathcal{U}_\epsilon$-valued ordinary control $u^\epsilon(\cdot)$ which is constant on the intervals $[i\delta, i\delta + \delta)$ and is such that*

$$\sup_{t \leq T} |x(t,m) - x(t,u^\epsilon)| \leq \epsilon, \quad |W(x,T,m,\tau) - W(x,T,u^\epsilon,\tau)| \leq \epsilon. \tag{2.9}$$

Thus (2.8) *holds.* Here $x(t, m)$ denotes the solution under the control $m(\cdot)$. Thus, the class of relaxed controls is a natural extension of the class of ordinary controls. It can be arbitrarily well approximated by a classical control. It guarantees that the set of (solution, control) is compact, and does not change the infimum of the costs.

3 Existence of a Stochastic Optimal Control

We now extend the ideas in Section 2 to the stochastic case, where only the drift is being controlled. Stochastic control problems occur in very many forms, and we are concerned with only the general ideas and not the most comprehensive results. The "closure" questions in Section 2 also occur here. The basic SDE is

$$dx = b(x,t,u)dt + \sigma(x,t)dw + dy, \tag{3.1}$$

where $x(t)$ is confined to a set $G \subset \mathbb{R}^r$, and $y(\cdot)$ is the reflection term which arises due to the reflecting part of the boundary of G, if any. The main difficulties in treating questions of existence of optimal controls and uniqueness of solutions arise owing to the behavior on and near the boundary of the allowed state space. Part of the boundary might be absorbing (the dynamics being zero there), part might be reflecting, and we might have the option of stopping when the path first reaches a target set, which is part of the boundary of G. The nature of the reflecting parts of the boundary are crucial for uniqueness. In order that the cost function be suitably continuous in the control, one needs to treat the problem of stopping on first reaching a target set with care. Since we are concerned only with the main ideas, and do not wish to encumber them with too much detail, we suppose that the set G is a simplex, perhaps infinite. Also, suppose that the faces are reflecting inward, with a constant direction on each face. The directions on the edges and corners are in the convex hull of those on the adjoining faces. Furthermore, assume that at the edges and corners, the convex hull of the directions on the adjoining faces contains a vector which points strictly interior (the so-called strictly-S condition). This is the form of many problems arising in the control of queueing type networks, and it also arises when state spaces are truncated for numerical reasons. If a face is absorbing in that the dynamical terms $b(\cdot), \sigma(\cdot)$ go to zero as that face is approached, the reflection selected for that face is irrelevant. The general theoretical background for the existence and uniqueness of solutions to such equations via the very useful Skorohod problem techniques is in Dupuis and Ishii (1990, 1991a, 1991b, 1992).

For simplicity, we suppose that there is no target set, since that involves some subtle issues of continuity if the process is degenerate on part of the exit set (although the basic ideas continue to hold, under broad conditions on the boundary of the target set and the process). See the discussion in Kushner and Dupuis [1992, pages 279-280], also Kushner (1977) and Stroock and Varadhan (1972).

Assume that $b(\cdot), \sigma(\cdot)$ are continuous (only piecewise in t needed) and bounded or else unbounded, with such continuity, but Lipschitz continuous in x, uniform in t, u. When writing equations such as (3.1), or the relaxed control version below, it is assumed implicitly that there is a filtration $\mathcal{F}_t$ such that $w(\cdot)$ is a standard vector-valued $\mathcal{F}_t$-Wiener process and that $u(\cdot), x(\cdot)$ are progressively measurable with $u(t)$ being $\mathcal{U}$-valued. For $x(\cdot)$, this entails only that it be non-anticipative. We can add a (uncontrolled) jump term to (3.1) with little additional difficulty.

Stochastic relaxed controls. Stochastic relaxed controls play the same role in the stochastic case as they do in the deterministic case. A stochastic relaxed control is a measure-valued process that is a deterministic relaxed control for each element ω of the underlying probability space, and which satisfies the nonanticipativeness condition that is also required of ordinary stochastic controls. We say $m(\cdot)$ is an admissible relaxed control for $w(\cdot)$, or that the pair $(m(\cdot), w(\cdot))$ is admissible, if $m(\cdot, \omega)$ is a deterministic relaxed control for almost all ω, and if $m(A \times [0,t])$ is $\mathcal{F}_t$-adapted (it is continuous) for all $A \in \mathcal{B}(\mathcal{U})$. There exists a progressively measurable derivative $m_t(A)$ for all $A \in \mathcal{B}(\mathcal{U})$. We continue to use the weak topology. The space of relaxed controls is compact, since $\mathcal{U}$ is compact. Thus, the set of relaxed controls in this space is tight. Let $m(\cdot)$ be an admissible relaxed control. We will also use the following simplified notation. For $t \geq 0$, let $m(t, \cdot)$ be the random measure with values $m(t, A) = m(A \times [0,t])$ for $A \in \mathcal{B}(\mathcal{U})$. Now, rewrite (3.1) in terms of the relaxed control:

$$dx = \int_{\mathcal{U}} b(x,t,\alpha) m_s(d\alpha) dt + \sigma(x,t) dw + dy$$

or, equivalently,

$$x(t) = x + \int_0^t \int_{\mathcal{U}} b(x(s), s, \alpha) m(d\alpha ds) + \int_0^t \sigma(x(s), s) dw(s) + y(t). \quad (3.2)$$

We say that (3.2) has a unique weak sense solution if the distribution of the pair $(m(\cdot), w(\cdot))$ determines the distribution of the triple $(x(\cdot), m(\cdot), w(\cdot))$. Under the Lipschitz condition, there is a unique strong sense solution. The references contain a general discussion of strong sense existence and uniqueness for the reflected problem, under the Lipschitz condition. Girsanov measure transformation techniques can often be used to deal with weak sense uniqueness. We assume the weak sense uniqueness.

For a complete and separable metric space H, let $D[H; 0.T]$ denote the space of H-valued functions which are right continuous, continuous at T, and with left hand limits on the interval $[0,T]$, with the Skorohod topology (Ethier and Kurtz (1986)) used. Thus, $x(\cdot), y(\cdot)$ take values in the (complete and separable) Skorohod space $D[\mathbb{R}^r; 0, T]$. Considering $m(\cdot)$ as a measure-valued process with values $m(t, \cdot)$ at time t, and letting $M(\mathcal{U})$ denote the space of measures on the Borel sets of $\mathcal{U}$, with the metrized weak topology used, the process $m(\cdot)$ takes values in the Skorohod space $D[M(\mathcal{U}); 0.T]$

The cost function. The control interval is $[0,T]$, unless it is stopped earlier at $\tau \leq T$ by choice of the controller. In terms of the relaxed control,

$$W(x,T,m,\tau) = E \int_0^{T\wedge\tau} \int_{\mathcal{U}} k(x(s), s, \alpha) m_s(d\alpha) ds + Eg(x(T\wedge\tau), T\wedge\tau)) + c'y(T\wedge\tau). \quad (3.3)$$

Relaxed controls are introduced for the stochastic problem for the same reasons that they were introduced for the deterministic problem; i.e., the compactification of the space of pairs (solution, control) and its approximation.

Existence of an optimal control. We need to prove the existence of an optimal relaxed control and that

$$V(x) = \inf_{m(\cdot),\tau} W(x,T,m,\tau) = \inf_{u(\cdot),\tau} W(x,T,u,\tau). \quad (3.4)$$

If there is a Lipschitz condition so that the solution to (3.2) can be defined for all relaxed controls with respect to a single Wiener process, then the minimization is over $m(\cdot), \tau$ as the notation in (3.4) indicates. However, if the solutions are defined indirectly, say by a Girsanov transformation, then the Wiener process varies with the control, and the minimization should actually read $\inf_{m(\cdot),w(\cdot),\tau}$, where the $m(\cdot), w(\cdot)$ are an admissible pair and τ is a stopping time with respect to the associated filtration.

An approximation theorem for relaxed controls. The proof of the following theorem is in Kushner [1990b, Theorem 3.5.2] See also Fleming (1977). The reference does not deal with the reflected problem, but that is easily added.

Theorem 3.1. *Let $x(0) = x$. Let the pair $(m(\cdot),\ w(\cdot))$ be admissible with filtration $\mathcal{F}_t$, and let τ be an $\mathcal{F}_t$-stopping time. Given $\epsilon > 0$, there is a finite set $\{\alpha_1^\epsilon, \ldots, \alpha_{p_\epsilon}^\epsilon\} = \mathcal{U}_\epsilon \subset \mathcal{U}$, and a $\delta > 0$ with the following properties. There is a probability space on which are defined processes $(x^\epsilon(\cdot), u^\epsilon(\cdot), w^\epsilon(\cdot), y^\epsilon(\cdot), \tau^\epsilon)$, with associated filtration $\mathcal{F}_t^\epsilon$, where $w^\epsilon(\cdot)$ is a standard $\mathcal{F}_t^\epsilon$-Wiener process, τ^ϵ is an $\mathcal{F}_t^\epsilon$-stopping time, $y^\epsilon(\cdot)$ the reflection term, and $u^\epsilon(\cdot)$ is an admissible $\mathcal{U}_\epsilon$-valued ordinary control which is constant on the intervals $[i\delta, i\delta+\delta)$. Furthermore, the processes satisfy (3.2) and $|W(x,T,m,\tau) - W(x,T,u^\epsilon,\tau^\epsilon)| \le \epsilon$. (3.4) holds.*

Theorem 3.2. *Let $\{x^n(\cdot), m^n(\cdot), w^n(\cdot), y^n(\cdot), \tau^n\}$ be a sequence of solutions (filtration $\mathcal{F}_t^n$), with τ^n an $\mathcal{F}_t^n$-stopping time, with $\{x^n(0)\}$ tight. Then $\{x^n(\cdot), m^n(\cdot), w^n(\cdot), y^n(\cdot), \tau^n\}$ is tight. Let $(x(\cdot), m(\cdot),\ w(\cdot), y(\cdot), \tau)$ denote the limit of a weakly convergent subsequence. Define*

$$\mathcal{F}_t = \mathcal{F}(x(s), m(s), w(s), y(s), \tau \wedge s, s \le t).$$

Then $w(\cdot)$ is a standard $\mathcal{F}_t$-Wiener process and τ is an $\mathcal{F}_t$-stopping time, $m(\cdot)$ is admissible with respect to $w(\cdot)$, and $x(\cdot)$ satisfies (3.2). Furthermore $W(x,T,m^n,\tau^n) \to W(x,T,m,\tau)$. This implies the existence of an optimal control.

Comments on the proof. The proof is in Kushner and Dupuis [1992, Chapter 10]. We make only a few comments. The tightness of the set $(x^n(\cdot), m^n(\cdot), w^n(\cdot), y^n(\cdot), \tau^n(\cdot))$ is simple to establish, as is the continuity of the weak sense limit processes. Also, $w^n(\cdot)$ obviously converges weakly to a standard Wiener process. Let us index the convergent subsequence by n also, and denote the weak sense limit by $(x(\cdot), m(\cdot), w(\cdot), y(\cdot), \tau)$. We need to establish the non anticipativity properties and to show that (3.2) is solved.

We have

$$x^n(t) = x^n(0) + \int_0^t \int_{\mathcal{U}} b(x^n(s), s, \alpha) m^n(d\alpha ds) + \int_0^t \sigma(x^n(s), s) dw^n(s) + y^n(t). \tag{3.5}$$

Let $q(\cdot)$ be a bounded and real-valued function of its arguments. Let $\{\phi_j(\cdot)\}$ be an arbitrary finite set of bounded and continuous functions of (t,α), and define the inner product notation $(\phi, m)_t = \int_0^t \int_{\mathcal{U}} \phi(\alpha, s)m(d\alpha ds)$. Let $t_i \leq t \leq t+s$ be given such that $P\{\tau = t_i\} = 0$ for all i. We have

$$Eq\left(x^n(t_i), w^n(t_i), y^n(t_i), \tau^n \wedge t_i, (\phi_j, m^n)_{t_i}; j, i\right)\left[w^n(t+s) - w^n(t)\right] = 0.$$

The weak convergence implies that

$$Eq\left(x(t_i), w(t_i), y(t_i), \tau \wedge t_i, (\phi_j, m)_{t_i}; j, i\right)\left[w(t+s) - w(t)\right] = 0. \tag{3.6}$$

It can also be established that

$$\begin{aligned} &Eq\left(x(t_i), w(t_i), y(t_i), \tau \wedge t_i, (\phi_j, m)_{t_i}; j, i\right) \\ &\quad \times \left[\left[w(t+s) - w(t)\right]\left[w(t+s) - w(t)\right]' - Is\right] = 0. \end{aligned} \tag{3.7}$$

The last two expressions and the arbitrariness of the $q(\cdot), t_i, t, s, \phi_j(\cdot)$, imply the non-anticipativity property, and the $\mathcal{F}_t$-Wiener property, since they imply that the Wiener process $w(t)$ and $w(t)w(t)' - It$ are $\mathcal{F}_t$-martingales. We need only identify the limit process. To do this, for $\delta < 0$, define the piecewise constant process $x_\delta(\cdot)$ by $x_\delta(t) = x(i\delta)$ for $t \in [i\delta, i\delta + \delta)$. Then rewrite (3.5) as

$$\begin{aligned} x^n(t) = x^n(0) + \int_0^t \int_{\mathcal{U}} b(x_\delta^n(s), s, \alpha)m^n(d\alpha ds) &+ \int_0^t \sigma(x_\delta^n(s), s)dw^n(s) \\ &+ y^n(t) + \rho_\delta^n(t), \end{aligned} \tag{3.8}$$

where

$$\lim_{\delta \to 0} \sup_n E \sup_{t \leq T} |\rho_\delta^n(t)| = 0. \tag{3.9}$$

We can use the weak convergence by taking limits ($n \to \infty$, fixed δ) in (3.8) to get that

$$\begin{aligned} x(t) = x(0) + \int_0^t \int_{\mathcal{U}} b(x_\delta(s), s, \alpha)m(d\alpha ds) &+ \int_0^t \sigma(x_\delta(s), s)dw(s) \\ &+ y^n(t) + \rho_\delta(t), \end{aligned} \tag{3.10}$$

where

$$\lim_{\delta \to 0} E \sup_{t \leq T} |\rho_\delta(t)| = 0. \tag{3.11}$$

Now, (3.10), (3.11) and the non-anticipativity properties imply that we can let $\delta \to 0$ and get that (3.2) holds. We omit the few details of the proof that $y(\cdot)$ is the reflection process.

4 Controlled Variance

Now, consider the case where the control also appears in the variance term, so that (3.1) is replaced by

$$dx = b(x,t,u)dt + \sigma(x,t,u)dw + dy. \tag{4.1}$$

We continue to use the cost function, the set G, and its boundary conditions, and the conditions on $b(\cdot)$ and $\sigma(\cdot)$ of the previous section. Of course, now the continuity of $\sigma(\cdot)$ includes the control as well.

The issues concerning the existence of an optimal control which arose in the previous sections also arise here, but in a more complicated form. In the previous sections, we found that the concept of relaxed control led to an enlargement of the concept of (solution, control) pairs, and with the appropriate topology and an assumption on uniqueness guaranteed that the space of such pairs is compact, and that the set of pairs corresponding to ordinary controls is dense in them. We will follow an analogous path for the present case. But, first it is helpful to see what sorts of processes arise when we take limits of sequences of solutions corresponding to ordinary controls.

Example. Let us return to the previous example, where the control varied between α_1 and α_2 on successive intervals of width $1/n$, and drop the reflection term. Write the SDE as

$$dx^n = b(x^n,t,u^n)dt + \sigma(x^n,t,u^n)dw^n, \tag{4.2}$$

Define $a(x,t,\alpha) = \sigma(x,t,\alpha)\sigma(x,t,\alpha)'$.

Let $m^n(\cdot)$ denote the relaxed control representation of $u^n(\cdot)$. Then, it is not hard to show that the set $(x^n(\cdot), m^n(\cdot), w^n(\cdot))$ is tight. We need only characterize the weak sense limit process. The differential operator of the solution process under $u^n(\cdot)$ is

$$\mathcal{L}^n f(x) = f_x'(x)\int_{\mathcal{U}} b(x,t,\alpha)m_t^n(\alpha) + \frac{1}{2}f_{xx}(x)\text{o}\left[\int_{\mathcal{U}} a(x,t,\alpha)m_t(d\alpha)\right],$$

where $f_{xx}\text{o}a$ denotes the form $\sum_{i,j} f_{x_i,x_j}a_{ij}$. Take a weakly convergent subsequence of $(x^n(\cdot), m^n(\cdot))$, with weak sense limit denoted by $(x(\cdot), m(\cdot))$. Then, by standard arguments, one can show that the limit pair solves the martingale problem with operator

$$\begin{aligned}\mathcal{L}f(x) &= f_x'(x)\left[\frac{b(x,t,\alpha_1)}{2} + \frac{b(x,t,\alpha_2)}{2}\right] \\ &\quad + \frac{1}{2}f_{xx}(x)\text{o}\left[\frac{a(x,1,\alpha_1)}{2} + \frac{a(x,t,\alpha_2)}{2}\right].\end{aligned}$$

There are two ways of representing the limit as an SDE, depending on how the limit operator is factored. We can represent the pair $(x(\cdot), m(\cdot))$

in terms of a single Wiener process in the sense that (possibly having to augment the probability space by the addition of a Wiener process which is independent of the other processes) there is a standard Wiener process $w(\cdot)$, with respect to which the other processes are non anticipative and such that

$$x(t) = x(0) + \frac{1}{2}\int_0^t [b(x(s), s, \alpha_1) + b(x(s), s, \alpha_2)]\, ds \\ + \frac{1}{\sqrt{2}}\int_0^t [a(x(s), s, \alpha_1) + a(x(s), s, \alpha_2)]^{1/2}\, dw(s),$$

where any "measurable" square root function can be used. Owing to the fact that the second order part of the limit operator can be split into two parts, each having a different control value, we can also represent the limit (again, possibly having to augment the probability space by adding a pair of mutually independent Wiener processes which are independent of the other processes) in terms of a pair of mutually independent Wiener processes $w_i(\cdot), i = 1, 2$, in the sense that

$$x(t) = x(0) + \frac{1}{2}\int_0^t [b(x(s), s, \alpha_1) + b(x(s), s, \alpha_2)]\, ds \\ + \frac{1}{\sqrt{2}}\int_0^t \sigma(x(s), s, \alpha_1) dw_1(s) + \frac{1}{\sqrt{2}}\int_0^t \sigma(x(s), s, \alpha_2) dw_2(s),$$

and where $(x(\cdot), m(\cdot))$ is non anticipative with respect to the $w_i(\cdot), i = 1, 2$.

A more general case. Now, suppose that $u^n(\cdot)$ alternatively take values $\alpha_i, i = 1, \ldots, k$, on time intervals which go to zero as $n \to \infty$, and that there are measurable functions $p_i(\cdot)$ such that the total amount of time on the interval $[0, t]$ that the control $u^n(\cdot)$ takes the value α_i converges to $\int_0^t p_i(s) ds$ as $n \to \infty$. Then the limit operator has the form

$$\mathcal{L}f(x) = f_x'(x) \sum_i p_i(t) b(x, t, \alpha_i) + \frac{1}{2} f_{xx}(x) \circ \left[\sum_i p_i(t) a(x, t, \alpha_i)\right]. \tag{4.3}$$

As for the case when $k = 2$, the pair $(x(\cdot), m(\cdot))$ can be represented in terms of a single Wiener process as

$$dx = \sum_i p_i(t) b(x, t, \alpha_i) dt + \left[\sum_i p_i(t) a(x, t, \alpha_i)\right]^{1/2} dw, \tag{4.4}$$

with any "measurable" choice of the square root process. Again, owing to the way that the second order part of the limit operator splits, each part having a different control value, we can also represent the solution in terms of a set of mutually independent vector-valued Wiener processes $w_i(\cdot), i = 1, \ldots, k$, as

$$dx = \sum_i p_i(t) b(x, t, \alpha_i) dt + \sum_i \sqrt{p_i(t)}\, \sigma(x, t, \alpha_i) dw_i. \tag{4.5}$$

Discussion. The representation (4.4) might seem simpler than the representation (4.5) in that there is a single Wiener process. But, the square root of the sum creates serious problems in analysis. Even if $\sigma(x,t,\alpha)$ satisfies a Lipschitz condition in x, the square root of the integral will not in general be Lipschitz continuous. It is difficult to approximate with the representation (4.4). Also, the role of the control is somewhat vague, being buried in both $m(\cdot)$ and in the square root. The representation (4.5) does not suffer from any of these disadvantages.

The question concerns the representation of the family of possible solutions, analogously to what was done with relaxed controls in the previous sections. Keep in mind that we are concerned with convenient representations. The limit processes are the same, no matter what the representation. To prepare ourselves for what follows, we rewrite (4.5). Define the relaxed control $m(\cdot)$ for the last example by its derivative

$$m_t(A) = \sum_i p_i(t) I_{\{\alpha_i \in A\}}, \quad A \in \mathcal{B}(\mathcal{U}), \tag{4.6}$$

and define a (vector) measure-valued martingale process $M(\cdot)$ by its values on the sets $A \in \mathcal{B}(\mathcal{U})$:

$$M(A,dt) = \sum_i \sqrt{p_i(t)}\, dw_i(t) I_{\{\alpha_i \in A\}}. \tag{4.7}$$

Recall that the $w_i(\cdot)$ are mutually independent standard vector-valued Wiener processes. Note that the quadratic variation process of $M(\cdot)$ is $m(\cdot)I$. Rewrite (4.5) in terms of $M(\cdot)$ as:

$$x(t) = x + \int_0^t \int_{\mathcal{U}} b(x(s),s,\alpha) m(d\alpha ds) + \int_0^t \int_{\mathcal{U}} \sigma(x(s),s,\alpha) M(d\alpha ds). \tag{4.8}$$

It is seen that the number of Wiener processes that are needed for the representation depends on the number of values that the control takes. But the form (4.8), which writes the stochastic integral in terms of a measure-valued martingale obtained from those Wiener processes suggests a convenient representation, no matter how many values the control takes. This leads to the notion of a martingale measure.

Martingale measures. Let $\mathcal{F}_t$ be a filtration on our probability space and $m(\cdot)$ a relaxed control, adapted to the filtration. First, let $M(\cdot)$ be a (real) measure-valued continuous random process defined on the space and with the following properties. For each $B \in \mathcal{B}(\mathcal{U})$, $M(B,\cdot)$ is a continuous $\mathcal{F}_t$-martingale. Let $M(\mathcal{U},0) = 0$, and $EM(\mathcal{U},t)^2 < \infty$, each t. If $B_i \in \mathcal{B}(\mathcal{U})$ are disjoint, then $M(B_1 \cup B_2, \cdot) = M(B_1,\cdot) + M(B_2,\cdot)$ and the two components are orthogonal. In general, we will work with vector-valued $M(\cdot) = (M_i(\cdot), i = 1,\ldots)$, where the number of components is finite and equals the number of columns in $\sigma(\cdot)$, and whose components are orthogonal: I.e., the martingales $M_i(A,\cdot), M_j(B,\cdot)$ are orthogonal for all $A, B \in \mathcal{B}(\mathcal{U})$ and all pairs $i \neq j$. The quadratic variation process is $\langle M_i(A,\cdot), M_j(B,\cdot)\rangle(t) = \delta_{ij} m(A \cap B, t)$, where $m(\cdot)$ is the relaxed control. Then $M(\cdot)$ is a measure valued $\mathcal{F}_t$-martingale. The process defined by (4.7) is a simple example. In fact, it is the canonical example, if the measure is supported by a finite number of points.

Tightness. The weak topology is used on the space of martingale measures. I.e., $M^n(\cdot)$ converges weakly to $M(\cdot)$ if and only if

$$\left\{\int_0^t \int_{\mathcal{U}} f_i(s,\alpha) M^n(d\alpha ds), i\right\} \to \left\{\int_0^t \int_{\mathcal{U}} f_i(s,\alpha) M(d\alpha ds), i\right\}$$

in distribution for each finite, bounded and continuous set $\{f_i(\cdot)\}$. By the compactness of $\mathcal{U}$, any sequence $\{M^n(\cdot), m^n(\cdot)\}$ is tight. If $M(\cdot), m(\cdot)$ is the limit of a weakly convergent subsequence, then $M(\cdot)$ is a martingale measure, and $m(\cdot)I$ is the associated quadratic variation process.

Stochastic integrals and SDE's. For full information on martingale measure, see El Karoui and Mélèard (1990). Martingale measures were used in Kushner (1990a) for the representation of solutions as done here, and to help prove the convergence of numerical approximations. Stochastic integrals are defined with respect to martingale measures in the same way that they are defined with respect to continuous martingales. Let $f(\omega, s, \alpha)$ be measurable on the product of the predictable σ-algebra and $\mathcal{B}(\mathcal{U})$. Suppose that

$$\int_0^t \int_{\mathcal{U}} f^2(s,\alpha) m_s(d\alpha) ds < \infty \tag{4.9}$$

with probability one for each t. In fact, we need only be concerned with the case where the expectation of the integral is bounded. Then define the stochastic integral

$$\int_0^t \int_{\mathcal{U}} f(s,\alpha) M(d\alpha ds),$$

by starting with simple functions which depend on only a finite number of values of α (which yields the classical stochastic integral) and working up, using (4.9) to suitably approximate $f(\cdot)$ by a sequence of simple functions, analogously to what is done for the real valued martingale case.

If $b(\cdot), \sigma(\cdot)$ satisfy a Lipschitz and linear growth condition in x, uniformly in the other variables, then one can define a unique strong sense solution to the SDE

$$x(t) = x + \int_0^t \int_{\mathcal{U}} b(x(s), s, \alpha) m_s(d\alpha) ds + \int_0^t \int_{\mathcal{U}} \sigma(x(s), s, \alpha) M(d\alpha ds) + y(s). \tag{4.10}$$

Without the reflection term, the method follows the classical Picard iteration method. The reflection is treated by a straightforward adaptation of the methods in Dupuis and Ishii (1990, 1991a, 1991b, 1992).

We say that (4.10) has a *unique weak sense solution* if the distribution of the pair $(m(\cdot), M(\cdot))$ implies the distribution of $(x(\cdot), m(\cdot), M(\cdot))$. We assume that (4.10) has a unique weak sense solution for each pair $(m(\cdot), M(\cdot))$.
The process defined by (4.10) is the natural representation for the control problem where the variance is controlled. Suppose that $(x(\cdot), m(\cdot))$ solves the martingale problem with operator (where we drop the reflection for simplicity)

$$\mathcal{L}^m f(x) = \int_{\mathcal{U}} \mathcal{L}^\alpha f(x) m_t(d\alpha), \tag{4.11}$$

where $\mathcal{L}^\alpha$ is the differential generator with control fixed at α. Then, (El Karoui and Mélèard (1990)) by augmenting the space if necessary, there always is a martingale measure such that (4.10) holds. Analogous results hold for the problem with reflection.
The cost function. The cost function is assumed to take the form (3.3). We need to keep in mind that it depends on the pair $(m(\cdot), M(\cdot))$ and not just $m(\cdot)$. The inf over $(m(\cdot), \tau)$ is shorthand for the inf over $(m(\cdot), M(\cdot), \tau)$ and analogously for the inf over $(u(\cdot), \tau)$.
Approximations and existence of an optimal control. Analogously to what was done in the last section, we need to know that

$$\inf_{m(\cdot),\tau} W(x, T, m, \tau) = \inf_{u,\tau} W(x, T, u, \tau), \tag{4.12}$$

and that any process of the type (4.10) and its associated cost can be arbitrarily well approximated by a classical controlled diffusion process and its associated cost. The following results are taken from Kushner (1990a).
Theorem 4.1. *Given the Lipschitz condition and a solution to* (4.10) *and* $\delta > 0$, *there is an ordinary control* $u^\delta(\cdot)$ *taking a finite number of values and an associated solution* $x^\delta(\cdot)$ *to* (4.1) *defined on the same probability space and such that*

$$E \sup_{s \le T} \left| x^\delta(s) - x(s) \right|^2 \le \delta. \tag{4.13}$$

Whether or not the Lipschitz condition holds, under the weak sense uniqueness condition there is a probability space on which $u^\delta(\cdot), x^\delta(\cdot)$ *are defined, together with a stopping time* τ^δ *such that*

$$\left| W(x, T, m, \tau) - W(x, T, u^\delta, \tau^\delta) \right| \le \delta. \tag{4.14}$$

Comments. The basic ideas are in Kushner (1990a), and only a few comments will be made here. We can approximate a given $(x(\cdot), m(\cdot), M(\cdot))$ by concentrating the measure on a finite number of points. Given $\delta > 0$, the approximation will be of the form

$$\begin{aligned} x^\delta(t) = x &+ \int_0^t \int_{\mathcal{U}} b(x^\delta(s), s, \alpha) m_s^\delta(d\alpha) ds \\ &+ \int_0^t \int_{\mathcal{U}} \sigma(x^\delta(s), s, \alpha) M^\delta(d\alpha ds) + y^\delta(t), \end{aligned} \tag{4.15}$$

where $m_s^\delta(\cdot)$ is concentrated on a finite number of points and, hence, $M^\delta(\cdot)$ can be written in terms of a finite number of Wiener processes and control values. The approximation is constructed as follows.

Let $C_i^\delta, i \leq k_\delta < \infty$, with $c_i^\delta \in C_i^\delta$, be a disjoint covering of $\mathcal{U}$, where the diameter of the sets goes to zero as $\delta \to 0$. Then $M(C_i^\delta, \cdot)$ are orthogonal martingales, with quadratic variation processes $m(C_i^\delta, \cdot)I$. There are mutually independent standard vector-valued Wiener processes (possibly requiring an augmentation of the probability space by Wiener processes that are independent of the other processes) $w_i^\delta(\cdot)$ such that we have the representation

$$M^\delta(C_i^\delta, t) = \int_0^t \left[m_s(C_i^\delta)\right]^{1/2} dw_i^\delta(s). \tag{4.16}$$

Define $m^\delta(\cdot), M^\delta(\cdot)$ as the restrictions of $m(\cdot), M(\cdot)$ to the sets $\{C_i^\delta\}$. Then rewrite (4.15) as

$$\begin{aligned} x^\delta(t) = x &+ \sum_i \int_0^t b(x^\delta, s, c_i^\delta) m_s(C_i^\delta) ds \\ &+ \sum_i \int_0^t \sigma(x^\delta(s), s, c_i^\delta) \left[m_s(C_i^\delta)\right]^{1/2} dw_i^\delta(s) + y^\delta(t). \end{aligned} \tag{4.17}$$

The sequence

$$(x^\delta(\cdot), M^\delta(\cdot), m^\delta(\cdot), y^\delta(\cdot))$$

converges weakly to

$$(x(\cdot), M(\cdot), m(\cdot), y(\cdot)).$$

Given the approximation (4.17), we can further approximate the controls by ordinary controls which are piecewise constant.

The existence of an optimal control. The last theorem asserts that the infimum of the costs in the class of solutions to (4.10) is the same as the infimum over solutions to (4.1). One gets a minimizing sequence $(x^n(\cdot), m^n(\cdot), M^n(\cdot), y^n(\cdot), \tau^n)$, with weak sense limit $(x(\cdot), m(\cdot), M(\cdot), y(\cdot), \tau)$ and shows that

$$W(x, T, m^n, \tau^n) \downarrow W(x, T, m, \tau)$$

which must equal the minimum.

The limit process can be identified either through the martingale problem form or via a direct approximation. The latter approach might be more convenient for the problem with reflection. Given $\Delta > 0$, write

$$\begin{aligned} x^n(t) = x &+ \int_0^t \int_{\mathcal{U}} b(x^n(s), s, \alpha) m^n(d\alpha ds) \\ &+ \sum_{i\Delta < t} \int_{i\Delta}^{i\Delta+\Delta} \int_{\mathcal{U}} \sigma(x^n(i\Delta), s, \alpha) M^n(d\alpha ds) + y^n(t) + \rho_\Delta^n(t) \end{aligned} \tag{4.19}$$

where

$$\lim_{\Delta \to 0} \sup_n E|\sup_{s \le T} \rho_\Delta^n(s)| = 0.$$

Then apply the weak convergence to the functions in (4.19) to identify the limit process directly.

Acknowledgments: The work was partially supported by Contracts DAAH 04-96-1-0075 from the Army Research Office and NSF Grant 9703895. The author also gratefully acknowledges the support and encouragement of Dr. Denis Talay of INRIA-Sophia Antipolis, which led to this work.

References

Benes, V. E. (1970) *SIAM J. Control Optim.* **8**, 79-188.

Benes, V.E. (1971) *SIAM J. Control Optim.* **9** 446-475.

Bismut, J. M. (1976) *Mem. Amer. Math. Soc. no. 167*, Providence.

Borkar,V.S. (1989) *Optimal Control of Diffusion Processes*, Longman Scientific and Technical, Harlow, Essex, UK.

Dupuis, P. and Ishii, H (1990) *Nonlinear Anal.* **15** 1123-1138.

Dupuis, P. and Ishii, H. (1991a) *Stochastics Stochastics Rep.* **35** 31-62.

Dupuis, P. and Ishii, H (1991b) *Hokkaido Math J.* **20** 135-164.

Dupuis, P. and Ishii, H (1993) *Ann. Probab.* **21** 554-580.

El Karoui, N. and Mélèard, S. (1990) *Probab. Theory Related Fields* **84** 83-101.

Ethier, S.N. and Kurtz, T.G. (1986) *Markov Processes: Characterization and Convergence*, Wiley, New York.

Fleming,W.F. (1977) In *Differential Games and Control Theory: III*, P.T. Liu E. Roxin and R. Sternberg, editors, 147-165, Marcel Dekker.

Fleming, W.H. and Nisio, M. (1984) *Nagoya Math. J.* **93** 71-108.

Haussmann, U.G. (1986) In *Stochastic Differential Systems: Lecture Notes on Control and Information Sciences*, K. Helmes N. Christopeit and M. Kohlmann, editors. **78** 171-186, Springer-Verlag, Berlin and New York.

Krylov, N.V. (1980) *Controlled Diffusion Processes*, Springer-Verlag, New York.

Kushner, H.J. (1975) *J. Optim. Theory Appl.* **15** 347-359.

Kushner, H.J. (1977) *Probability Methods for Approximations in Stochastic Control and for Elliptic Equations*, Academic Press, New York.

Kushner, H.J. (1978) *SIAM J. Control Optim.* **16** 330-346.

Kushner, H.J. (1990a) *SIAM J. Control Optim.* **28** 999-1048.

Kushner, H.J. (1990b) *Weak Convergence Methods and Singularly Perturbed Stochastic Control and Filtering Problems*, Birkhäuser, Boston.

Kushner, H.J. and Dupuis, P. (1992) *Numerical Methods for Stochastic Control Problems in Continuous Time*, Springer-Verlag, Berlin and New York.

McShane, E.J. and Warfield, R.B. (1967) *Proc. Amer. Math. Soc.* **18** 41-47.

Stroock, D.W. and Varadhan, S.R.S. (1972) *Comm. Pure Appl. Math.* **25** 651-713.

Warga, J. (1962) *J. Math. Anal. Appl.* **4** 111-128.

Young, L.C. (1969) *Lectures on the Calculus of Variations and Optimal Control Theory*, Saunders, Philadelphia.

26

On Optimal Ergodic Control of Diffusions with Jumps

Jose-Luis Menaldi
Maurice Robin

ABSTRACT: **Our purpose is to study an optimal ergodic control problem where the state of the system is given by a diffusion process with jumps in the whole space. The corresponding dynamic programming (or Hamilton-Jacobi-Bellman) equation is a quasi-linear integro-differential equation of second order. A key result is to prove the existence and uniqueness of an invariant density function for a jump diffusion, whose lower order coefficients are only locally bounded and Borel measurable. Based on this invariant probability, existence and uniqueness (up to an additive constant) of solutions to the ergodic HJB equation is established.**

Introduction

We are interested in controlling a diffusion process with jumps in the whole space. The controller can access only the drift term of the state equation, so that the state of the system is defined via the Girsanov transformation. The programming equation is an ergodic quasi-linear integro-differential equation. Solving this equation an optimal feedback can be obtained.
Most of the ergodic properties of diffusion processes and their relation with partial differential equations are well know in the classic literature. However, similar questions for diffusion processes with jumps are not so popular, only recently some attention was given, cf. [18], Garroni and Menaldi [8] and reference therein.
Due to applications in stochastic control (in particular the action of a feedback function), we have to be able to treat diffusions with jumps with only locally bounded and Borel measurable lower order coefficients (where the control is applied). Moreover, since we are interested in the whole space, an assumption related to the existence of a Liapunov's function is needed. We assume the drift is of linear growth at infinity, but the existence and regularity of the Green function or transition density function (even in the purely partial differential equations case) as proved in Garroni and Menaldi [8] does not applied.

To show the existence of an invariant measure, many arguments are based on the so-called *Doeblin's condition*, which in turn is based on the strict positivity of the Green functions deduced from the strong maximum principle. The technique, to solve an elliptic second (order linear) equation (without zero order term) in the whole space, is based on the construction of an ergodic operator. For instance, we refer to the books of Bensoussan [1] and Khasminskii [12]. Related discussions can be found in the books of Borkar [4] and Ethier and Kurtz [6].
Now we are going to describe, without all the technical assumptions, the ergodic problem we will consider. Let $v(x)$ be a Borel measurable function from $\mathbb{R}^d$ into a compact metric space V (i.e., a measurable feedback). The dynamic of the system (for a given feedback) follows a diffusion with jumps in $\mathbb{R}^d$, i.e. a (strong) Markov process $(\Omega, P, X_t, t \geq 0)$ with semigroup $(\Phi_v(t), t \geq 0)$ and infinitesimal generator A_v, as discussed in the next section. A long run average cost is associated to the controlled system by

$$J(v) = \liminf_{T\to\infty} \frac{1}{T} \int_0^T \Phi_v(t) f dt,$$

where $f = f(x, v(x))$ is the running cost of the system under the feedback v. Our purpose is to give a characterization of the optimal cost

$$\lambda = \inf\{J(v) : v(\cdot)\}$$

and to construct an optimal feedback control $\hat{v}$. Notice that the dynamic equation is given later by (4.1) and that we expect λ to be a constant.
A formal application of the dynamic programming principle (as described in Fleming and Soner [7]) yields the following Hamilton-Jacobi-Bellman equation

$$\inf_v \{A_v u(x)\} = \lambda \text{ in } \mathbb{R}^d,$$

where the infimum is calculated for each fixed x, and v in V. An optimal feedback control is obtained as the minimizer $\hat{v}(x)$ in the HJB equation.
In order to study this Hamilton-Jacobi-Bellman equation we need some previous discussion. In Section 1, we give some details on the construction of the diffusion with jumps in the whole space $\mathbb{R}^d$, under convenient assumptions. Next we recall some results (cf. [19]) related to the invariant probability measure μ_v, for any measurable feedback v. Finally, we consider the above HJB equation under appropriate assumptions

1. Ergodic Control Problem

Before setting the optimal ergodic control problem, we need to recall some facts about diffusions with jumps in the whole space. To that effect, we consider an integro-differential operator of the form

$$I_0\varphi(x) = \int_{\mathbb{R}^d_*} [\varphi(x+z) - \varphi(x) - z \cdot \nabla\varphi(x)] M_0(x, dz), \tag{1.1}$$

where ∇ is the gradient operator in x, the Levy kernel $M_0(x, dz)$ is a Radon measure on $\mathbb{R}^d_* = \mathbb{R}^d \setminus \{0\}$ for any fixed x, and satisfies

$$\int_{|z|<1} |z|^2 M_0(x, dz) + \int_{|z|\geq 1} |z| M_0(x, dz) < \infty, \quad \forall x \in \mathbb{R}^d. \tag{1.2}$$

It is clear that this operator is associated with a jump process in $\mathbb{R}^d$, e.g., Gikhman and Skorokhod [9].
Similarly, let L_0 be a second order uniformly elliptic operator associated with a diffusion process in the whole space, i.e.

$$L_0 = \sum_{i,j=1}^{d} a_{ij}(x)\partial_{ij} + \sum_{i=1}^{d} b_i(x)\partial_i, \tag{1.3}$$

where the coefficients (a_{ij}) are bounded and Lipschitz continuous, i.e. for some $c_0, M > 0$

$$\begin{cases} c_0|\xi|^2 \leq \sum_{i,j=1}^{d} a_{ij}(x)\xi_i\xi_j \leq c_0^{-1}|\xi|^2 & \forall x, \xi \in \mathbb{R}^d, \\ |a_{ij}(x) - a_{ij}(x')| \leq M|x - x'|, & \forall x, x' \in \mathbb{R}^d, \end{cases} \tag{1.4}$$

$a_{ij} = a_{ji}$, and the first order coefficients (b_i) are Lipschitz continuous, i.e. for some $M > 0$,

$$\begin{cases} |b_i(x) - b_i(x')| \leq M|x - x'|, & \forall x, x' \in \mathbb{R}^d \\ b_i(0) = 0, \quad i = 1, \ldots, d. \end{cases} \tag{1.5}$$

The fact that $b = (b_i)$ vanishes on the origin and the additional assumption

$$-\sum_{i=1}^{d} b_i(x)x_i \geq c_1|x|^2, \quad \forall x \in \mathbb{R}^d, |x| \geq r_1, \tag{1.6}$$

for some constants $c_1, r_1 > 0$, will allow us to show some "stability" on the system.
The Levy kernel $M_0(x, dz)$ is assumed to have a particular structure, namely

$$M_0(x, A) = \int_{\{\zeta: j(x,\zeta)\in A\}} m_0(x, \zeta)\pi(d\zeta), \tag{1.7}$$

where $\pi(\cdot)$ is a σ-finite measure on the measurable space $(F, \mathcal{F})$, the functions $j(x, \zeta)$ and $m_0(x, \zeta)$ are measurable for (x, ζ) in $\mathbb{R}^d \times F$, and there exist a measurable and positive function $j_0(\zeta)$ and constants $C_0 > 0$, $1 \leq \gamma \leq 2$ [γ is referred to as the order of M_0 or I_0] such that for every x, ζ we have

$$\begin{cases} |j(x,\zeta)| \leq j_0(\zeta), \qquad 0 \leq m_0(x,\zeta) \leq 1, \\ \int_F |j_0(\zeta)|^p (1 + j_0(\zeta))^{-1} \pi(d\zeta) \leq C_0, \quad \forall p \in [\gamma, 2], \end{cases} \tag{1.8}$$

the function $j(x,\zeta)$ is continuously differentiable in x for any fixed ζ and there exists a constant $c_0 > 0$ such that for any (x,ζ) we have

$$c_0 \leq \det(\mathbf{1} + \theta \nabla j(x,\zeta)) \leq c_0^{-1}, \quad \forall \theta \in [0,1], \tag{1.9}$$

where $\mathbf{1}$ denotes the identity matrix in $\mathbb{R}^d$, ∇ is the gradient operator in x, and $\det(\cdot)$ denotes the determinant of a matrix.

Depending on the assumptions on the coefficients of the operators L_0, I_0 and on the domain $\mathcal{O}$ of $\mathbb{R}^d$, we can construct the corresponding Markov-Feller process. The reader is referred to the books Bensoussan and Lions [3], Gikhman and Skorokhod [9] (among others) and references therein. Usually, more regularity on the coefficients $j(x,\zeta)$ and $m_0(x,\zeta)$ is needed, e.g.

$$\begin{cases} |m_0(x,\zeta) - m_0(x',\zeta)| \leq M|x - x'|, \quad \forall x, x' \in \mathbb{R}^d, \\ |j(x,\zeta) - j(x',\zeta)| \leq j_0(\zeta)|x - x'|, \quad \forall x, x' \in \mathbb{R}^d, \end{cases} \tag{1.10}$$

for some constant $M > 0$ and the same function $j_0(\zeta)$ as in assumption (1.8). Thus the integro-differential operator I_0 has the form

$$I_0\varphi(x) = \int_F [\varphi(x + j(x,\zeta)) - \varphi(x) - j(x,\zeta) \cdot \nabla\varphi(x)] m_0(x,\zeta) \pi(d\zeta). \tag{1.11}$$

It is possible to show that the Markov-Feller process associated with the infinitesimal generator $L_0 + I_0$ (which is referred to as the "diffusion with jumps") has a transition probability density function $G_0(x,t,y)$, which is smooth in some sense (cf. Garroni and Menaldi [8]).

Since our purpose is to treat control problems, we remark that (in general) the optimal feedback is not smooth. This forces us to consider some coefficients (e.g. those of first order) which are only measurable. To that effect, we will use the so-called Girsanov transformation.

Let $\Omega = D([0,+\infty), \mathbb{R}^d)$ be the canonical space of right continuous functions with left-hand limits ω from $[0,+\infty)$ into $\mathbb{R}^d$ endowed with the Skorokhod topology. Denote by either X_t or $X(t)$ the canonical process and by F_t the filtration generated by $\{X_s : s \leq t\}$ (universally completed and right-continuous). Now let $(\Omega, P^0, F_t, X_t, t \geq 0)$ be the (homogeneous) Markov-Feller process with transition density function $G_0(x,t,y)$ associated with the integro-differential operator $L_0 + I_0$, i.e. the density w.r.t. the Lebesgue measure of $P^0\{X(t) \in dy \mid X(s) = x\}$ is equal to $G_0(x, t-s, x)$. For the sake of simplicity, we will refer to $(P_x^0, X(t), t \geq 0)$ as the above

Markov-Feller process, where P_x^0 denote the conditional probability w.r.t. $\{X(0) = x\}$.
Hence, for any smooth function $\varphi(x)$ the process

$$Y_\varphi(t) = \varphi(X(t)) - \int_0^t (L_0 + I_0)\varphi(X(s))ds \tag{1.12}$$

is a P_x-martingale. This follows immediately from the representation

$$\left\{ \begin{array}{rcl} E_x\{\varphi(X(t))\} & = & \int_{\mathbb{R}^d} G_0(x,t,y)\varphi(y)dy + \\ & & + \int_0^t ds \int_{\mathbb{R}^d} G_0(x,t-s,y)(L_0 + I_0)\varphi(y)dy, \end{array} \right. \tag{1.13}$$

and the Markov property. Moreover, it is also possible to express the process X_t as follows

$$dX(t) = a^{1/2}(X(t))dw(t) + \int_{\mathbb{R}_*^d} z\mu_X(dt,dz) + b(X(t))dt, \tag{1.14}$$

where $(w(t), t \geq 0)$ is a standard Wiener process in $\mathbb{R}^d$, $a^{1/2}(x)$ is the positive square root of the matrix $(a_{ij}(x))$ and $b(x)$ is the vector $(b_i(x))$. The process μ_X is the martingale measure associate with the process $(X(t), t \geq 0)$, i.e. if $\eta_X(t, A)$ denotes the integer random measure defined as the number of jumps the process $X(\cdot)$ on $(0, t]$ with values in $A \subset \mathbb{R}_*^d$ (recall that $\mathbb{R}_*^d = \mathbb{R}^d \setminus \{0\}$) then

$$\mu_X(dt, A) + \pi_X(dt, A) = \eta_X(dt, A), \tag{1.15}$$

where $\mu_X(t, A)$ is a square integral (local) martingale quasi-left continuous and $\pi_X(t, A)$ is a predictable increasing process obtained via the Doob-Meyer decomposition, and

$$\pi_X(dt, dz) = M_0(X(t-), dz)dt, \tag{1.16}$$

where $M_0(x, dz)$ is the Levy kernel used to define the integro-differential operator I_0 given by (1.1), see e.g., Bensoussan and Lions [3].
Let $g(x) = (g_1(x), \ldots, g_d(x)$ and $c(x, z)$ be functions defined for x in $\mathbb{R}^d$, $z \in \mathbb{R}_*^d$ such that

$$\left\{ \begin{array}{l} g_i, c \text{ are bounded, measurable and,} \\ 0 \leq c(x,z) \leq C_0(1 \wedge |z|), \quad \forall x, z, \end{array} \right. \tag{1.17}$$

where C_0 is a constant.
Consider the exponential martingale $(e(t), t \geq 0)$ as the solution of the stochastic differential equation

$$\left\{ \begin{array}{rcl} de(t) & = & e(t)[r_X(t)dw(t) + \int_{\mathbb{R}_*^d} \gamma_X(t,z)\mu_X(dt,dz)], \\ e(0) & = & 1, \end{array} \right. \tag{1.18}$$

where

$$\begin{cases} r_X(t) &= a^{-1/2}(X(t))g(X(t)), \\ \gamma_X(t,z) &= zc(X(t),z), \end{cases} \tag{1.19}$$

i.e.,

$$\begin{cases} e(t) &= \exp\{\int_0^t r_X(s)dw(s) - \int_0^t |r_X(s)|^2 ds + \\ & + \int_0^t \int_{\mathbb{R}_*^d} \gamma_X(s,z)\mu_X(ds,dz) - \\ & - \int_0^t \int_{\mathbb{R}_*^d} [\gamma_X(s,z) - \ln(1+\gamma_X(s,z))]\pi_X(ds,dz)\}. \end{cases} \tag{1.20}$$

If we denote by

$$L = L_0 + \sum_{i=1}^d g_i(x)\partial_i \tag{1.21}$$

and

$$I\varphi(x) = I_0\varphi(x) + \int_{\mathbb{R}_*^d} [\varphi(x+z) - \varphi(x)]c(x,z)M_0(x,dz), \tag{1.22}$$

then, by means of Itô's formula one can prove that for any smooth function φ, the process

$$Z_\varphi(t) = \varphi(X(t)) - \int_0^t (L+I)\varphi(X(s))ds \tag{1.23}$$

is a P_x-martingale, where the new probability measure P_x on Ω is defined as

$$dP_x = e(t)dP_x^0 \text{ on } F_t. \tag{1.24}$$

Notice that the probability measures P_x^0 and P_x are absolutely continuous, one with respect to the other. Also, a representation of the form (1.14) is valid under the new probability measure P_x, i.e.

$$\begin{cases} dX(t) &= a^{1/2}(X(t))dw(t) + [b(X(t)) + g(X(t))]dt + \\ & + \int_{\mathbb{R}_*^d} z\mu(dt,dz), \end{cases} \tag{1.25}$$

where $(w(t), t \geq 0)$ is again a standard Wiener process and μ is the martingale measure associate with the (canonical) process $X(t)$ under the new measure P_x.

Remark 1.1 *Due to the linear growth of the coefficients $b_i(x), i = 1, \dots, d$, we can not use directly the construction in Garroni and Menaldi [8] of the Green function (or transition density).* □

Now we are ready to formulate our optimal ergodic control problem. Let $f(x,v)$, $g(x,v) = (g_1(x,v), \dots, g_d(x,v))$, and $c(x,v,z)$ be functions defined for (x,v) in $\mathbb{R}^d \times V$, z in $\mathbb{R}^d_\star$ such that

$$\left\{ \begin{array}{l} f, g_i, c \text{ are bounded, measurable,} \\ \text{and continuous in the control variable } v, \\ 0 \le c(x,v,z) \le C_0(1 \wedge |z|), \quad \forall x, v, z, \end{array} \right. \tag{1.26}$$

where C_0 is a constant and V is a compact metric space.
We consider the Markov-Feller process $(P_x, X(t), t \ge 0)$ defined on the canonical space $D([0,\infty[, \mathbb{R}^d)$ described above, corresponding to the integro-differential operator $L_0 + I_0$ in the whole space.
An admissible control is a stochastic process $(v(t), t \ge 0)$ with values in V, adapted to the filtration F_t. For any admissible control $(v(t), t \ge 0)$ we can use the Girsanov transformation (1.18),(1.24) to define an exponential martingale $e(t) = e_v(t)$ and a new probability measure denoted by $P_x = P_x^v$ such that $(P_x^v, X(t), t \ge 0)$ represents the state of the system. Notice that in this case, $e_v(t)$ is given by (1.20) with

$$\left\{ \begin{array}{lcl} r_X(t) & = & a^{-1/2}(X(t)) g(X(t), v(t)), \\ \gamma_X(t,z) & = & z\, c(X(t), v(t), z). \end{array} \right. \tag{1.27}$$

A cost is associated with the controlled system $(P_x^v, X(t), t \ge 0)$ by

$$J_x(v) = \lim_{T \to \infty} E_x^v \{ \frac{1}{T} \int_0^T f(X(t), v(t)) dt \}. \tag{1.28}$$

Our purpose is to give a characterization of the optimal cost

$$\lambda = \inf\{ J_x(v) : v(\cdot) \} \tag{1.29}$$

and to construct an optimal control $\hat{v}(t)$.
It is useful to remark that we expect to obtain an optimal Markovian control, i.e.

$$\hat{v}(t) = \hat{v}(X(t)), \quad \forall t \ge 0, \tag{1.30}$$

for some feedback function $\hat{v}(x)$ and to prove that the optimal cost λ is constant, i.e., independent of the initial condition $X(0) = x$.
For a given feedback control $v = v(x)$, the controlled state of the system $(P_x^v, X(t), t \ge 0)$ is a Markov-Feller process with infinitesimal generator of the form $L + I$, as in (1.21) and (1.22), with

$$\left\{ \begin{array}{lcl} g(x) & = & g(x, v(x)), \\ c(x,z) & = & c(x, v(x), z). \end{array} \right. \tag{1.31}$$

2. Dirichlet Problem

Denote by $v(x)$ any measurable feedback. Let $L = L_v$ and $I = I_v$ be the second order differential operator (1.21) and the integro-differential operator (1.22) as before. Also we set $f(x) = f(x, v(x))$ given by (1.26). For a given bounded and smooth domain $\mathcal{O}$, we consider first the interior Dirichlet problem

$$\begin{cases} -(L+I)u &= f \text{ in } \mathcal{O}, \\ u &= h \text{ in } \mathbb{R}^d \setminus \mathcal{O}, \end{cases} \tag{2.1}$$

and next the exterior Dirichlet problem

$$\begin{cases} -(L+I)u &= f \text{ in } \mathbb{R}^d \setminus \overline{\mathcal{O}}, \\ u &= h \text{ in } \overline{\mathcal{O}}, \end{cases} \tag{2.2}$$

where f and h are given measurable and bounded functions.

Notice the fact that there is not zero-order coefficient and the non-local character of the integro-differential operator I. So that for the interior problem [exterior problem] we need the solution u to be defined in a neighborhood of the closure $\overline{\mathcal{O}}$ [$\mathbb{R}^d \setminus \mathcal{O}$, respectively]. Thus, we seek the solution as defined in the whole space $\mathbb{R}^d$.

A natural way to handle the non-homogeneous boundary conditions is to pursue the following two-steps. First we suitably extend the boundary (or exterior) data h to the whole space, for instance if h is defined in $\mathbb{R}^d \setminus \mathcal{O}$ then we extend h to the whole $\mathbb{R}^d$ preserving its regularity properties. Next, we solve an homogeneous problem (interior Dirichlet problem with $h = 0$) for $u - h$, where we use the zero-extension to define the non-local operator I. With this in mind, we can re-consider the interior Dirichlet problem [or the exterior Dirichlet problem] as

$$\begin{cases} -(L+I)u &= f \text{ in } \mathcal{O}, \\ u &= h \text{ in } \partial\mathcal{O}. \end{cases} \tag{2.3}$$

Notice that if we modify the function h outside of $\overline{\mathcal{O}}$ then the value Iu may change (since it uses the values of u outside of $\overline{\mathcal{O}}$). Then, we understand the solution u of the above equation as $u = v + w$ where v solves a non-homogeneous Dirichlet boundary conditions second-order differential equation

$$\begin{cases} -Lv &= 0 \text{ in } \mathcal{O}, \\ v &= h \text{ in } \partial\mathcal{O}, \end{cases} \tag{2.4}$$

and w solves an homogeneous (interior) Dirichlet problem

$$\begin{cases} -(L+I)w &= f + Iv \text{ in } \mathcal{O}, \\ w &= 0 \text{ in } \mathbb{R}^d \setminus \mathcal{O}, \end{cases} \tag{2.5}$$

for the whole integro-differential operator $L + I$. Sufficient conditions to solve the PDE (2.4) are well known (cf. Gilbarg and Trudinger [10], Ladyzhenskaya and Uraltseva [14]) so we will state results concerning the existence, uniqueness and regularity for the solutions of the homogeneous interior Dirichlet problem (i.e., with $h = 0$) with an integro-differential operator of the above form. Actually, we will need a weak formulation of this problem. An excellent treatment of the interior Dirichlet problem for this class of integro-differential operators can be found in Gimbert and Lions [11] and, based on Garroni and Menaldi [8], some natural extensions to the exterior Dirichlet problem are detailed in [19].
Notice that if u denotes the solution of the non-homogeneous interior Dirichlet problem (2.1) with $f(x) = f(x, v(x))$ then we expect to have the following stochastic representation

$$u(x) = E_x^v\{\int_0^\tau f(X(t), v(X(t)))dt + h(X(\tau))\}, \tag{2.6}$$

where $\tau = \tau_v$ is the first exit time of the process $X(t)$ from the closed set $\overline{\mathcal{O}}$, i.e.

$$\tau = \inf\{t \geq 0 \ : \ X(t) \notin \overline{\mathcal{O}}\}, \tag{2.7}$$

$E_x^v\{\cdot\}$ is the mathematical expectation with respect to the measure P_x^v, and $(P_x^v, X(t), t \geq 0)$ is the diffusion with jumps corresponding to $L_v + I_v$. Sometimes it is convenient to call a *probabilistic solution* of the interior (exterior) Dirichlet problem a measurable and bounded (locally bounded, for the exterior) function u satisfying:

$$\begin{cases} u(X(t))\mathbf{1}_{(t<\tau)} + \int_0^{\tau \wedge t} f(X(s))ds + h(X(\tau))\mathbf{1}_{(t \geq \tau)} \\ \qquad \text{is a } F_t - \text{(local) martingale,} \end{cases} \tag{2.8}$$

with the exit time $\tau = \tau(\overline{\mathcal{O}})$ for the interior problem and $\tau = \tau(\mathbb{R}^d \setminus \mathcal{O})$, where

$$\tau(D) = \inf\{t \geq 0 : X(t) \notin D\}, \tag{2.9}$$

where $\tau = \infty$ if $X(t) \in D$, $\forall t \geq 0$. Certainly, for a probability solution, the above stochastic representation is valid.
Let us turn our attention to the variational formulation of the (homogeneous) interior Dirichlet problem (2.1), i.e. a solution in $W_0^{1,p}(\mathcal{O})$. The key point here is to establish that $L + I$ preserves Sobolev spaces, i.e. it maps $W_0^{1,p}(\mathcal{O})$ into $W^{-1,p}(\mathcal{O})$, and that a weak version of the maximum principle holds. We need to assume that $j(x, \zeta)$ has a bounded second derivative in x, i.e. there exist $\delta > 0$ such that for some constant $C > 0$, and with $F_\delta = \{\zeta \in F \ : \ j_0(\zeta) < \delta\}$ we have

$$||\nabla_x^2 j(\cdot, \zeta)||_{L^\infty(\mathbb{R}^d)} \leq C, \quad \forall \zeta \in F_\delta \tag{2.10}$$

as well as assumption (1.10) to make sense of $I_0\varphi$ for a test function φ not necessary smooth. Next, a version of the Maximum Principle is necessary to show that the bilinear form

$$a(\varphi,\psi) = -\langle L\varphi,\psi\rangle - \langle I\varphi,\psi\rangle \tag{2.11}$$

is continuous and coercive in $H_0^1(\mathcal{O})$. We state the main results in this direction.

Theorem 2.1 (Interior) *Let the assumptions* (1.4), (1.5), (1.8), (1.9), (1.10), (1.26) *and* (2.10) *hold. Then there exists a unique probability solution u of the (homogeneous, i.e. $h=0$) interior Dirichlet problem* (2.1) *in $W_0^{1,p}(\mathcal{O})$, for any $1<p<\infty$. Moreover, u belongs to $W_{loc}^{2,p}(\mathcal{O})$ and for some constant C (independent of the feedback) we have*

$$\|u\|_{L^\infty} \le C\|f\|_{L^\infty}\,. \quad \square \tag{2.12}$$

The above results for $j(x,\zeta)=j(\zeta)$, independent of x, have been proved in Bensoussan and Lions [3] and extended later to W_0^p in Gimbert and Lions [11]. We refer to [19] for details. The general case with $h\neq 0$, can be treated by means of the PDE problem (2.4).
The exterior Dirichlet problem presents some extra difficulties and it is not easily found in the literature. Notice that in our setting, the first order coefficient $b(x)$ has a linear growth, so that standard arguments do not apply and the meaning of the boundary conditions becomes an issue. Here we adopt the probability solution sense and the variational formulation with a weight (Liapunov's type) function of the form

$$\psi_q(x) = (2+|x|^2)^{q/2}, \quad q>0. \tag{2.13}$$

Under the assumption (1.6), for any $q>0$ there exist α_q, $c_q>0$ and a ball B_q such that

$$\begin{cases} L\psi_q(x) + I\psi_q(x) \le -\alpha_q\psi_q(x), & \forall x\in \mathbb{R}^d\setminus B_q, \\ |L\psi_q(x)| + |I\psi_q(x)| \le c_q\psi_q(x), & \forall x\in \mathbb{R}^d\setminus B_q, \end{cases} \tag{2.14}$$

provided we suppose that the function $j_0(\zeta)$ of (1.8) satisfies

$$\int_F [j_0(\zeta)]^q(1+j_0(\zeta))^{-1}\pi(d\zeta) < \infty, \tag{2.15}$$

which is always true if $q\le 1$.

Theorem 2.2 (Exterior) *Let the assumptions* (1.4), ..., (1.10), (1.26) *and* (2.10) *hold. Suppose $\mathcal{O}$ a smooth bounded domain containing the ball B_q given by* (2.14). *Then there exists one and only one probability solution u of the (homogeneous, i.e. $h=0$) exterior Dirichlet problem* (2.2) *in*

$W^{1,p}_{loc}(\mathbb{R}^d) \cap L^\infty(\mathbb{R}^d)$, *for any* $1 < p < \infty$. *Moreover,* u *belongs to* $W^{2,p}_{loc}(\mathbb{R}^d \setminus \mathcal{O})$ *and*

$$\|u\psi_{-q}\|_{L^\infty} \le \frac{1}{\alpha_q}\|f\psi_{-q}\|_{L^\infty}, \tag{2.16}$$

where α_q *is given by* (2.14). □

3. Invariant Measure

As in the previous section, we assume given a Borel measurable feedback control and we denote by $L+I$ the corresponding integro-differential operator. Now, let $\mathcal{O}$ be a sufficiently large smooth and bounded domain (e.g. a ball) so that the following non-homogeneous exterior Dirichlet problem

$$\begin{cases} (L+I)u &= 0 \;\text{ in } \mathbb{R}^d \setminus \overline{\mathcal{O}}, \\ u &= \varphi \;\text{ in } \overline{\mathcal{O}}, \end{cases} \tag{3.1}$$

can be solved in $W^{2,p}_{loc}(\mathbb{R}^d \setminus \overline{\mathcal{O}}) \cap W^{1,p}_{loc}(\mathbb{R}^d) \cap L^\infty(\mathbb{R}^d)$ for non-negative φ in $W^{1,p}(\mathcal{O}) \cap L^\infty(\mathbb{R}^d)$. Now, consider the non-homogeneous interior Dirichlet problem in a larger domain (ball) $B \supset \overline{\mathcal{O}}$,

$$\begin{cases} (L+I)v &= 0 \;\text{ in } B, \\ v &= u \;\text{ in } \mathbb{R}^d \setminus B, \end{cases} \tag{3.2}$$

which can be solved in $W^{2,p}_{loc}(B) \cap W^{1,p}(B) \cap L^\infty(\mathbb{R}^d)$, for any v in $W^{1,p}_{loc}(\mathbb{R}^d \setminus \overline{\mathcal{O}}) \cap L^\infty(\mathbb{R}^d)$. Therefore we can define the linear operator

$$\begin{cases} P : W^{1,p}(\mathcal{O}) \cap L^\infty(\mathbb{R}^d) \to W^{1,p}(\mathcal{O}) \cap L^\infty(\mathbb{R}^d), \\ P\varphi = v, \end{cases} \tag{3.3}$$

where the solution u of (3.2) has been restricted to the domain $\overline{\mathcal{O}}$. Notice that we are using *weak solutions* of problems (3.1) and (3.12). Strong solutions, i.e. u in $W^{2,p}(\mathbb{R}^d \setminus \overline{\mathcal{O}})$ and v in $W^{2,p}(B)$, require some extra assumptions on the integro-differential opearator I, c.f. Gimbert and Lions [11] and [19].

By means of the weak maximum principle, we can prove that

$$\varphi \ge 0 \;\text{ implies }\; P\varphi \ge 0. \tag{3.4}$$

Since $P\varphi = 1$ for $\varphi = 1$, the operator P can be identified with a (one-step) transition probability measure on $(\overline{\mathcal{O}}, \mathcal{B})$, so that

$$\begin{cases} P : B(\overline{\mathcal{O}}) \to B(\overline{\mathcal{O}}), \\ P\varphi(x) = \int_{\overline{\mathcal{O}}} \varphi(y) P(x, dy), \end{cases} \tag{3.5}$$

where $B(\overline{\mathcal{O}})$ is the space of bounded Borel measurable functions on $\overline{\mathcal{O}}$. Moreover, it can be proved (cf. [19]) that P is an ergodic operator, i.e. defining

$$\lambda(x,y,F) = P\mathbf{1}_F(x) - P\mathbf{1}_F(y), \tag{3.6}$$

for any x, y in X and any Borel subset F of $\overline{\mathcal{O}}$, where $\mathbf{1}_F$ is the characteristic function of the set B, we have

$$\exists \delta > 0 \;/\; \lambda(x,y,F) \leq 1-\delta, \quad \forall x,y \in \overline{\mathcal{O}},\ \forall F \in \mathcal{B}. \tag{3.7}$$

Hence, based on Doob's ergodicity theorem, there exists a unique probability measure on $(\overline{\mathcal{O}}, \mathcal{B})$, denoted by ν, such that

$$|P^n\varphi(x) - \int_{\overline{\mathcal{O}}} \varphi d\nu| \leq K e^{-\rho n} \|\varphi\|, \qquad \forall n = 1,2,\ldots \tag{3.8}$$

where $\rho = -\ln(1-\delta)$, $K = 2/(1-\delta)$. The measure ν is the unique invariant probability for $(P, \overline{\mathcal{O}}, \mathcal{B})$, i.e. the unique probability on $\overline{\mathcal{O}}$ such that

$$\int_{\overline{\mathcal{O}}} \varphi d\nu = \int_{\overline{\mathcal{O}}} P\varphi d\nu, \quad \forall \varphi \in B(\overline{\mathcal{O}}). \tag{3.9}$$

At this point we consider the interior and exterior Dirichlet problems

$$\begin{cases} -(L+I)u_0 &= f \text{ in } \mathbb{R}^d \setminus \overline{\mathcal{O}}, \\ u_0 &= 0 \text{ in } \overline{\mathcal{O}}, \end{cases} \tag{3.10}$$

and

$$\begin{cases} -(L+I)v &= f \text{ in } B, \\ v &= u_0 \text{ in } \mathbb{R}^d \setminus B. \end{cases} \tag{3.11}$$

Based on the results of the previous section, we can define the operator

$$T : L_q^\infty(\mathbb{R}^d) \longrightarrow W_{loc}^{1,p}(\mathbb{R}^d) \cap L_q^\infty(\mathbb{R}^d), \qquad Tf = v_0 \tag{3.12}$$

with possesses the property (3.4). Thus we can define a positive measure $\tilde{\mu}$ on $\mathbb{R}^d$ (un-normalized) by

$$\int_{\mathbb{R}^d} f(x) d\tilde{\mu}(dx) = \int_{\mathcal{O}} Tf(x)\nu(dx) \tag{3.13}$$

Next define the probability measure μ by

$$\mu(F) = \frac{\tilde{\mu}(F)}{\tilde{\mu}(\mathbb{R}^d)}, \quad \forall F \in \mathcal{B}(\mathbb{R}^d). \tag{3.14}$$

Theorem 3.1 (Invariant Measure) *Let the assumptions* (1.4)—(1.10), (1.26) *and* (2.10) *hold. Then* μ*, given by* (3.14)*, is an invariant probability measure for the diffusion with jumps in* $\mathbb{R}^d$*, i.e. for any bounded and Borel measurable function* f *we have*

$$\int_{\mathbb{R}^d} E_x\{f(X(t))\}\mu(dx) = \int_{\mathbb{R}^d} f(x)\mu(dx). \tag{3.15}$$

Moreover, the invariant probability measure μ *is unique and*

$$\lim_{T\to\infty} \frac{1}{T}\int_0^T E_x\{f(X(t))\}dt = \int_{\mathbb{R}^d} f(x)\mu(dx), \tag{3.16}$$

for any bounded and Borel measurable function f*. Furthermore, the measure* μ *is absolutely continuous w.r.t. the Lebesgue measure, i.e. we can write*

$$\int_{\mathbb{R}^d} f(x)\mu(dx) = \int_{\mathbb{R}^d} f(x)m(x)dx, \tag{3.17}$$

where the invariant density $m(x)$ *satisfies*

$$m \geq 0, \qquad \int_{\mathbb{R}^d} m(x)dx = 1. \quad \square \tag{3.18}$$

Now we can discuss the ergodic linear equation. Consider the space

$$L_q^\infty(\mathbb{R}^d) = \{\varphi \,:\, \varphi\psi_{-q} \in L^\infty(\mathbb{R}^d)\}, \tag{3.19}$$

for $q > 0$ and $\psi_{-q}(x) = (2+|x|^2)^{-q/2}$. The linear equation is then

$$\begin{cases} u \in W_{loc}^{2,p}(\mathbb{R}^d) \cap L_q^\infty(\mathbb{R}^d), \quad p \geq d,\ q > 0, \\ -(L+I)u = f \ \text{ a.e. in } \mathbb{R}^d. \end{cases} \tag{3.20}$$

Theorem 3.2 (Linear Equation) *Let assumptions* (1.4)—(1.10), (1.26) *and* (2.10) *hold. Then the linear integro-differential equation* (3.20) *has a solution* u *(unique up to an additive constant) if and only* f *has a zero-mean, i.e.*

$$\mu(f) \doteq \int_{\mathbb{R}^d} f(x)\mu(dx) = 0, \tag{3.21}$$

where $\mu(dx)$ *is the unique invariant probability measure defined by* (3.14)*. Moreover, under the above zero-mean conditon, there exists a solution of* (3.20) *for which we have the a priori estimate*

$$\|u\psi_{-q}\|_{L^\infty(\mathbb{R}^d)} \leq C_q\|f\psi_{-q}\|_{L^\infty(\mathbb{R}^d)} \tag{3.22}$$

for some positive constant C_q *depending only on* q, d *and the bounds imposed by the assumptions on the coefficients of the operators* L *and* I. $\square$

Remark 3.3 *Notice that the discounted linear equation*

$$\begin{cases} u \in W^{2,p}_{loc}(\mathbb{R}^d) \cap L^\infty_q(\mathbb{R}^d), \quad p \geq d,\ q > 0, \\ -(L+I)u_\alpha + \alpha u_\alpha = f \ \text{ a.e. in } \mathbb{R}^d. \end{cases} \tag{3.23}$$

with $\alpha > 0$ has a unique solution under the assumption of the previous theorem. Moreover, based the estimate (3.22), one can prove that ∇u_α and $\nabla^2 u_\alpha$ remain uniformly bounded in $L^p_{loc}(\mathbb{R}^d)$ as $\alpha \to 0$, for any finite p. □

4. Programming Equation

Let v be a given Borel measurable function from $\mathbb{R}^d$ into V, referred to as a measurable feedback. The dynamic of the system follows the stochastic integro-differential equation

$$\begin{cases} dX(t) &= a^{1/2}(X(t))dw(t) + [b(X(t)) + g(X(t), v(X(t)))]dt + \\ &+ \int_{\mathbb{R}^d_*} z\mu(dt, dz), \end{cases} \tag{4.1}$$

on the canonical probability space $\Omega = D([0,\infty), \mathbb{R}^d)$, with the probability $P = P^v_x$ satistying

$$P\{X(0) = x\} = 1, \tag{4.2}$$

and where $w(t) = w_v(t)$ is a standard Wiener process in $\mathbb{R}^d$ and $\mu(dt, dx) = \mu_v(dt, dx)$ is an integer random (martingale) measure associated with a Poisson measure with characteristic Levy kernels

$$M_0(x, dz),\ c(x, v(x), z)M_0(x, dz). \tag{4.3}$$

Then (Ω, P^x, X^v) defines a Markov-Feller process on $\mathbb{R}^d$ (so-called diffusion with jumps) with infinitesimal generator A_v, which is an extension of the integro-differential operator $L_0 + I_0 + \mathcal{L}_v$, cf. (1.3), (1.11), and where $\mathcal{L}_v$ is given by

$$\begin{cases} \mathcal{L}_v\varphi(x) &= \int_F [\varphi(x + j(x,\zeta)) - \varphi(x)]c(x, v(x), j(x,\zeta)) \times \\ &\times m_0(x,\zeta)\pi(d\zeta) + \sum_{i=1}^{d} g_i(x, v(x))\partial_i\varphi(x) \end{cases} \tag{4.4}$$

At this point, we can re-formulate our optimal ergodic problem as in the introduction i.e., for a given measurable feedback control $v(x)$ there exists a unique invariant probability measure $\mu_v(dx)$ in $\mathbb{R}^d$ of the Markov-Feller process (Ω, P^x, X^v) as above. The long run average cost associated with the controlled system is given by

$$J(v) = \int_{\mathbb{R}^d} f(x, v(x))\mu_v(dx). \tag{4.5}$$

Recall our assumptions on the data (1.4)—(1.10), (1.26) and (2.10). Our purpose is to give a characterization of the optimal cost

$$\lambda = \inf\{J(v) : v(\cdot)\} \tag{4.6}$$

and to construct an optimal feedback control $\hat{v}$.
Denote by $H(x, \varphi(x))$ the Hamiltonian

$$H(x, \varphi(x)) = \inf\{\mathcal{L}_v \varphi(x) + f(x, v) : v \in V\}, \tag{4.7}$$

where the operator $\mathcal{L}_v$ is given by (4.4), f satisfies (1.26) and φ belongs to the Sobolev space $W^{1,p}_{loc}(\mathbb{R}^d) \cap L^\infty_q(\mathbb{R}^d)$, cf. (2.13) and (3.17). It can be proved that $H(x, \varphi(x)$ belongs to $L^\infty_q(\mathbb{R}^d)$ for every φ in $W^{1,p}_{loc}(\mathbb{R}^d) \cap L^\infty_q(\mathbb{R}^d)$.
As mentioned in Remark 3.3, we can adapt the techniques used in Theorem 3.2 (cf. [19]) to show that the discounted nonlinear equation

$$\begin{cases} u \in W^{2,p}_{loc}(\mathbb{R}^d) \cap L^\infty_q(\mathbb{R}^d), \quad p \geq d,\ q > 0, \\ (L_0 + I_0)u_\alpha + H(\cdot, u_\alpha) = \alpha u_\alpha \ \text{ a.e. in } \mathbb{R}^d. \end{cases} \tag{4.8}$$

with $\alpha > 0$ possesses a unique solution and that ∇u_α and $\nabla^2 u_\alpha$ remain uniformly bounded in $L^p_{loc}(\mathbb{R}^d)$ as $\alpha \to 0$, for any finite p.
The limiting Hamilton-Jacobi-Bellman equation can be expressed as

$$\begin{cases} u \in W^{2,p}_{loc}(\mathbb{R}^d) \cap L^\infty_q(\mathbb{R}^d), \quad p \geq d,\ q > 0, \\ (L_0 + I_0)u + H(\cdot, u) = \lambda \ \text{ a.e. in } \mathbb{R}^d. \end{cases} \tag{4.9}$$

where the unknowns are the function u and the constant λ.
Essentially based on results for the linear equation, as in Bensoussan [1], we select a discounted optimal feedback, i.e.

$$v_\alpha(x) \in \operatorname{Argmin} H(x, u_\alpha(x)), \quad \forall x \in \mathbb{R}^d, \tag{4.10}$$

where u_α is the solution of the nonlinear equation (4.8).

Theorem 4.1 (HJB equation) *Let the assumptions* (1.4)—(1.10), (1.26) *and* (2.10) *hold. Define*

$$\tilde{u}_\alpha = u_\alpha - \mu_\alpha(u_\alpha), \quad \alpha > 0, \tag{4.11}$$

where μ_α is the invariant probability measure corresponding to the integro-differential operator $L_0 + I_0 + \mathcal{L}_v$, with $v = v_\alpha$ as given by (4.11). *Then there exist a constant λ and a function u in $W^{2,p}_{loc}(\mathbb{R}^d) \cap L^\infty_q(\mathbb{R}^d)$ such that*

$$\alpha u_\alpha \to \lambda, \qquad \tilde{u}_\alpha \rightharpoonup u \ \textit{weakly*} \tag{4.12}$$

as α goes to zero. The couple (λ, u) is a solution of the HJB equation (4.9). *Moreover, λ is equal to the optimal cost* (4.6) *and any stationary feedback $\hat{v}$ satisfying*

$$\hat{v} \in \operatorname{Argmin} H(x, u(x)), \quad \forall x \in \mathbb{R}^d, \tag{4.13}$$

produces the optimal cost, i.e. $\lambda = J(\hat{v})$. □

Notice that if the couple (λ, u) is a solution of the HJB equation (4.9) then λ is the optimal cost (4.6), and if v is a stationary optimal feedback (i.e. (4.13) holds for v), then u solves the linear equation

$$-(L_0 + I_0 + \mathcal{L}_v)u + \lambda = f_v \text{ a.e. in } \mathbb{R}^d, \tag{4.14}$$

where $f_v(x) = f(x, v(x))$. By means of the Itô's formula we get for every $x \in \mathbb{R}^d$, $T > 0$

$$u(x) = E_x^v\{\int_0^T [f(X(t), v(X(t))) - \lambda]dt\} + E_x^v\{u(X(T))\}. \tag{4.15}$$

Now, if the feedback and its invariant measure are such that

$$E_x^v\{u(X(T))\} \to \mu_v(u) \quad \text{as} \quad T \to \infty \tag{4.16}$$

then we deduce for every $x \in \mathbb{R}^d$, $T > 0$,

$$u(x) = E_x^v\{\int_0^\infty [f(X(t), v(X(t))) - \lambda]dt\} + \mu_v(u). \tag{4.17}$$

This is a representation of u, and gives uniqueness (up to a constant) for the potential function u. Actually, we conjecture that the transition density is strictly positive (for any stationary feedback), and therefore the strong mixing property holds, which in turn implies the convergence (4.16). To the best of our knowledge this has not been proved so far.

Bibliography

[1] A. Bensoussan, *Perturbation methods in optimal control*, Wiley, New York, 1988.

[2] A. Bensoussan and J.L. Lions, *Applications of variationals inequalities in stochastic control*, North-Holland, Amsterdam, 1982.

[3] A. Bensoussan and J.L. Lions, *Impulse control and quasi-variational inequalities*, Gauthier-Villars, Paris, 1984.

[4] V.S. Borkar, *Topics in Controlled Markov Chains*, Pitman Research Notes in Mathematics Series No 240, Longman, Essex, 1991.

[5] M.G. Crandall, H. Ishii and P.L. Lions, User's guide to viscosity solutions of second order partial differential equations, *Bull. Am. Math. Soc.*, **27** (1992), 1–67.

[6] S.N. Ethier and T.G. Kurtz, *Markov processes*, Wiley, New York, 1986.

[7] W.H. Fleming and H.M. Soner, *Controlled Markov processes and viscosity solutions*, Springer-Verlag, New York, 1992.

[8] M.G. Garroni and J.L. Menaldi *Green functions for second order integral-differential problems*, Pitman Research Notes in Mathematics Series No 275, Longman, Essex, 1992.

[9] I.I. Gikhman and A.V. Skorokhod, *Stochastic differential equations*, Springer-Verlag, Berlin, 1972.

[10] D. Gilbarg and N.S. Trudinger, *Elliptic partial differential equations of second order*, Second Edition, Springer-Verlag, New York, 1983.

[11] F. Gimbert and P.L. Lions, Existence and regularity results for solutions of second order elliptic integro-differential operators, *Ricerche di Matematica*, **33** (1984), 315–358,

[12] R.Z. Khasminskii, (Hasminskii) *Stochastic Stability of Differential Equations*, Sijthoff and Noordhoff, The Netherlands, 1980.

[13] N.V. Krylov, *Nonlinear elliptic and parabolic equations of second order*, Reidel, Dordrecht, 1987

[14] O.A. Ladyzhenskaya and N.N. Uraltseva, *Linear and quasilinear elliptic equations*, Academinc Press, New York, 1968.

[15] P.L. Lions, A remark on Bony Maximum principle, *Procedings Am. Math. Soc.*, **88** (1982), 503–508.

[16] J.L. Menaldi, On the stopping time problem for degenerate diffusions, *SIAM J. Control Optim.*, **18** (1980), 697–721.

[17] J.L. Menaldi, Optimal impulse control problems for degenerate diffusions with jumps, *Acta Appl. Math.*, **8** (1987), 165–198.

[18] J.L. Menaldi and M. Robin, Ergodic control of reflected diffusions with jumps, *Appl. Math. Optim.*, **35** (1997), 117–137.

[19] J.L. Menaldi and M. Robin, Invariant Measure for Diffusions with Jumps, *Appl. Math. Optim.*, to appear.

[20] M.H. Protter and H.F. Weinberger, *Maximum principles in differential equations*, Second edition, Springer-Verlag, New York, 1984.

[21] M. Robin, Long-term average cost control problems for continuous time Markov processes: A survey, *Acta Appl. Math.*, **1** (1983), 281–299.

[22] D.W. Stroock and S.R. Varadhan, *Multidimensional diffusion process*, Springer-Verlag, Berlin, 1979.

27

Markov Marginal Problems and Their Applications to Markov Optimal Control

Toshio Mikami

ABSTRACT: In this paper we discuss a class of Markov marginal problems (MMP). By MMP, we mean the problem to construct a Markov diffusion process with given (marginal) constraints on the path space. As an application we consider Markov optimal control problems.

KEYWORDS: Markov marginal problem, Markov optimal control, copula, covariance kernel.

1 Introduction

Let $\mathbf{M}$ and $\mathbf{X}$ be a topological space, and $\{\mu_t\}_{t\in\mathbf{M}}$ be a family of Borel probability measures on $(\mathbf{X}, \mathbf{B}(\mathbf{X}))$. As a classical marginal problem, the following is known.
(MP). Find a probability measure $\mathbf{Q}$ on $(\mathbf{X}^{\mathbf{M}}, \mathbf{B}(\mathbf{X}^{\mathbf{M}}))$ such that

$$\mathbf{Q}(f \in \mathbf{X}^{\mathbf{M}}; f(t) \in dx) = \mu_t(dx) \quad \text{for all } t \in \mathbf{M}. \tag{1.1}$$

One would like to find a solution Q, to MP, with addtional nice properties.

Let $\mathbf{P}(\mathbf{X})$ denote the space of Borel probability measures on $(\mathbf{X}, \mathbf{B}(\mathbf{X}))$, equipped with the weak topology, and let $C(\mathbf{M}, \mathbf{X})$ denote the space of continuous functions from $\mathbf{M}$ to $\mathbf{X}$, equipped with the topology of uniform convergence. With respect to MP, the following is known (see Theorem 2.1 in Blumenthal and Corson (1972)).

Theorem 1.1 *Let* $\mathbf{M}$ *and* $\mathbf{X}$ *be compact metric spaces and let* $\mathbf{X}$ *be connected and locally connected. Suppose that* $\mu : \mathbf{M} \mapsto \mathbf{P}(\mathbf{X})$ *is continuous and that* $supp(\mu_t) = \mathbf{X}$ *for all* $t \in \mathbf{M}$. *Then there exists* $\mathbf{Q} \in P(C(\mathbf{M}, \mathbf{X}))$ *such that*

$$\mathbf{Q}(f \in C(\mathbf{M}, \mathbf{X}); f(t) \in dx) = \mu_t(dx) \quad \textit{for all } t \in \mathbf{M}. \tag{1.2}$$

Instead of giving the topological structure to **M** and **X**, giving the order structure to them, similar problem has been considered in Edwards (1978), Kamae and Krengel (1978), and Kellerer (1987, 1997).

In this paper we consider the case **M** is an interval on R, such as $[0,1]$ and $[0,\infty)$, and we would like to construct a Markov diffusion process with given constraints. By a Markov diffusion process, we mean a continuous Markov process which satisfies a stochastic differential equation of Ito-type.

As a typical problem in this direction, one can mention the problem in Nelson's stochastic quantization. Let us briefly introduce it. Let $\rho(t,x)dx$ $(t \geq 0)$ be a faimly of Borel probability measures such that the following holds: for any infinitely differentiable function $f : R^d \mapsto R$ with a compact support and $t \geq 0$

$$\begin{aligned} &\int_{R^d} f(x)\rho(t,x)dx - \int_{R^d} f(x)\rho(0,x)dx \\ = &\int_0^t ds \int_{R^d} [2^{-1} \sum_{i,j=1}^{d} a(s,x)^{ij} \partial^2 f(x)/\partial x_i \partial x_j \\ &+ < b(s,x), \nabla_x f(x) >]\rho(s,x)dx, \end{aligned} \tag{1.3}$$

where $a(t,x) = (a(t,x)^{ij})_{i,j=1}^d$ is a measurable, symmetric, nonnegative definite $d\times d$-matrix, and $b(t,x) = (b(t,x)^i)_{i=1}^d$ is a measurable vector function. Here we put $\nabla_x \equiv (\partial/\partial x_i)_{i=1}^d$, and $< \cdot, \cdot >$ denotes the inner product in R^d. One of basic problems in Nelson's stochastic quantization (see Nelson (1967, 1984)) is to construct a Markov diffusion process $\{X(t)\}_{t\geq 0}$ on a probability space $(\Omega, \mathbf{B}, P)$ such that

$$P(X(t) \in dx) = \rho(t,x)dx \quad \text{for all } t \geq 0. \tag{1.4}$$

As an answer to the above problem, the following is known (see Carlen (1984) and also Carmona (1987), Nagasawa (1989) and Zheng (1985)).

Theorem 1.2 *Suppose that $a(t,x)$ is an identity matrix and that the following condition holds: for all $t > 0$,*

$$\int_0^t ds \int_{R^d} |b(s,x)|^2 \rho(s,x)dx < \infty. \tag{1.5}$$

Then there exists a Markov process $\{X(t)\}_{t\geq 0}$ such that (1.4) holds and that

$$dX(t) = b(t,X(t))dt + dW(t) \quad (t > 0), \tag{1.6}$$

where $\{W(t)\}_{t\geq 0}$ is a d-dimensional Wiener process (see e.g. Stroock and Varadhan (1979)).

When $a(t,x)$ is not an identity matrix, the problem was considered in Mikami (1990), and Quastel and Varadhan (1997).

As we discussed in Mikami (1989), this problem is related to Markov optimal control problem. Let us briefly introduce it. Let $(\Omega, \mathbf{B}, P)$ be a probability space and $\{\mathbf{B}_t\}_{t\geq 0}$ be a right continuous, increasing family of sub σ-fields of $\mathbf{B}$, and let $\{W(t)\}_{t\geq 0}$ denote a d-dimensional $(\mathbf{B}_t)$-Wiener process on $(\Omega, \mathbf{B}, P)$, and $\sigma(t,x)$ be a Borel measurable $d\times d$-matrix ($t \geq 0$, $x \in R^d$), and $u(t)$ be $(\mathbf{B}_t)$-progressively measurable ($t \geq 0$). Consider a semimartingale $\{X^u(t)\}_{t\geq 0}$ which satisfies the follows: for $t > 0$,

$$dX^u(t) = u(t)dt + \sigma(t, X^u(t))dW(t). \tag{1.7}$$

For Borel measurable functions $L(t,x;u) : [0,\infty) \times R^d \times R^d \mapsto R$ and $\Phi(x) : R^d \mapsto R$, study the following problem (Markov optimal control problem).
(MOCP). For $T > 0$, consider if the following is true:

$$\begin{aligned} & \inf\{\int_0^T E[L(t, X^u(t); u(t))]dt + \Phi(X^u(T)) : \{u(t)\}_{t\in[0,T]}\} \qquad (1.8) \\ = & \inf\{\int_0^T E[L(t, X^u(t); u(t))]dt + \Phi(X^u(T)) : u(t) = b(t, X^u(t)) \\ & \quad \text{for some } b(t,x)((t,x) \in [0,T] \times R^d)\}. \end{aligned}$$

(About Markov optimal control, we refer the reader to Fleming and Soner (1993).) (1.8) is true, if we assume that σ is an identity matrix or $d = 1$ and that $L(t,x;\cdot)$ is convex for all $(t,x) \in [0,T] \times R^d$ and that $\int_0^T E[|u(t)|^2]dt < \infty$. This can be proved from Theorem 1.2, by Jensen's inequality (see Mikami (1989)).

Remark 1.1. As a similar problem to MOCP, one can mention the Monge problem that is still open (see Rüschendorf and Uckelmann (1997) and the references therein).

In this paper we give a new approach to MOCP, by studying the marginal problem that we call Markov marginal problem (MMP). Roughly speaking, our approach is as follows: for $(X^u(t), u(t))_{t\in[0,T]}$ in (1.7), find $b(t,x)((t,x) \in [0,T] \times R^d)$ for which the following has a weak solution

$$X(t) = X^u(0) + \int_0^t b(s, X(s))ds + \int_0^t \sigma(s, X(s))dW(s), \tag{1.9}$$

($t \in [0,T]$), and for a.e. $t \in [0,T]$,

$$P((X^u(t), u(t)) \in dx) = P((X(t), b(t, X(t))) \in dx). \tag{1.10}$$

If this can be done for the minimizing X^u, then (1.8) is true.

In section 2, we give a partial answer to the above problem and give the application to MOCP in section 3. The reader will notice that our approach was inspired by the idea of covariance kernels on central limit theorems (see Remark 2.2) and that we used the result on 2-copulas (see Remark 2.3).

2 Marginal problems

In this section we state and prove theorems on Markov marginal problems (MMP) introduced in section 1. We fix $T > 0$ from now on.

The following result which should be compared with Theorems 1.1 and 1.2 is known (see Theorem 2.1 in Mikami (1995) and the proof).

Theorem 2.1 *For any family of probability density functions* $\{p(t,\cdot)\}_{0\le t\le T}$ *on* R^d, *there exists a measurable* $\phi(t,\cdot) : R^d \mapsto R^d$ $(t \in [0,T])$ *such that* $\{\phi(t, W(t+1))\}_{0\le t\le T}$ *is a Markov process and that*

$$P(\phi(t, W(t+1)) \in dx) = p(t,x)dx \quad \textit{for all } t \in [0,T], \tag{2.1}$$

where $\{W(t)\}_{t\ge 0}$ *is a d-dimensional Wiener process. Moreover if the conditional distribution functions* $F^k_{(t,x_1,\cdots,x_{k-1})}(x_k) \equiv \int_{-\infty}^{x_k} dy_k \int_{R^{d-k}} p(t, x_1, \cdots, x_{k-1}, y_k, z)dz / \int_{R^{d-k+1}} p(t, x_1, \cdots, x_{k-1}, z)dz \in C^{1,2}([0,T] \times R^k; [0,1])$ *and is strictly increasing in* x_k *for all* $(t, x_1, \cdots, x_{k-1}) \in [0,T] \times R^{k-1}$ *for* $k = 1, \cdots, d$, *then* $\phi \in C^{1,2}([0,T] \times R^d; R^d)$ *and* $\{\phi(t, W(t+1))\}_{0\le t\le T}$ *is a Markov diffusion process.*

Put $P_0(\mathbf{X}) \equiv \{\mu \in P(\mathbf{X}); supp(\mu) = \mathbf{X}\}$. The next theorem can be proved from Theorem 2.3 below in the same way as in Theorem 2.2 of Blumenthal and Corson (1972).

Theorem 2.2 *Suppose that* $\mathbf{X}$ *is a connected and locally connected compact metric space, and that* $\mu_t \in P_0(\mathbf{X})$ *for all* $t \in [0,T]$. *Then there exists a measurable* $\overline{\phi}(t,\cdot) : R \mapsto \mathbf{X}$ $(t \in [0,T])$ *such that*

$$P(\overline{\phi}(t, W(t+1)) \in dx) = \mu_t(dx) \quad \textit{for all } t \in [0,T], \tag{2.2}$$

where $\{W(t)\}_{t\ge 0}$ *is a one-dimensional Wiener process. Moreover if* $\mu : [0,T] \mapsto \mathbf{P}(\mathbf{X})$ *is continuous, then* $\overline{\phi} : [0,T] \times R \mapsto \mathbf{X}$ *is continuous, and in particular* $\{\overline{\phi}(t, W(t+1))\}_{t\in[0,T]}$ *is continuous.*

The next theorem played a crucial role in Theorem 2.2 of Blumenthal and Corson (1972) and will be used in the proof of Theorem 2.2 above.

Theorem 2.3 *Suppose that* $\mathbf{X}$ *is a connected and locally connected compact metric space. Then there exists a continuous function* $\tilde{\varphi} : P_0(\mathbf{X}) \mapsto P_0([0,1])$ *and a continuous function* $\varphi : [0,1] \mapsto \mathbf{X}$ *such that*

$$(\tilde{\varphi}(\mu))^{\varphi^{-1}} = \mu \quad \textit{for all } \mu \in P_0(\mathbf{X}). \tag{2.3}$$

Let us prove Theorem 2.2 for the sake of completeness.
(Proof of Theorem 2.2). Put for $x \in R$ and $t \in [0,T]$,

$$F_{1,t}(x) = P(W(t+1) \leq x), \tag{2.4}$$
$$F_{2,t}(x) = \tilde{\varphi}(\mu_t)([0, \max(\min(1,x),0)]), \tag{2.5}$$

where $\tilde{\varphi}$ is a function stated in Theorem 2.3.

For a distribution function F on R, put

$$F^{-1}(u) \equiv \begin{cases} \sup\{x \in R; F(x) < u\} & \text{for } u \in (0,1], \\ \sup\{x \in R; F(x) = 0\} & \text{for } u = 0. \end{cases} \tag{2.6}$$

If the set where the supremum is taken is empty, then we put the sup $= -\infty$.

In the same way as in Mikami (1995),

$$P(F_{2,t}^{-1}(F_{1,t}(W(t+1))) \in dx) = \tilde{\varphi}(\mu_t)(dx) \quad \text{for all } 0 \leq t \leq T, \tag{2.7}$$

since $F_{1,t}(W(t+1))$ is uniformly distributed on $[0,1]$, and since for a distribution function F on R and $u \in (0,1]$,

$$F^{-1}(u) \leq x \quad \text{iff} \quad u \leq F(x). \tag{2.8}$$

Putting $\overline{\phi}(t,x) = \varphi(F_{2,t}^{-1}(F_{1,t}(x)))$ $(x \in R,\ t \in [0,T])$, one get, from Theorem 2.3 and (2.7),

$$P(\overline{\phi}(t, W(t+1)) \in dx) = \tilde{\varphi}(\mu_t)^{\varphi^{-1}}(dx) = \mu_t(dx) \tag{2.9}$$

for all $0 \leq t \leq T$. The rest part of the theorem can be proved easily and we omit the proof. **Q.E.D.**

Let X_o be a R-valued, $(\mathbf{B}_0)$-adapted random variable on $(\Omega, \mathbf{B}, P)$. To give theorems on MMP discussed after Remark 1.1, let us state the assumption.

(A.1). d=1. $\sigma(t,x) : [0,T] \times R \mapsto R$ is bounded and is continuous in x, uniformly in t. There exists $\nu > 0$ such that

$$\sigma(t,x) \geq \nu^{1/2} \quad ((t,x) \in [0,T] \times R). \tag{2.10}$$

(A.2). (1.7) has a weak solution such that $X^u(0) = X_o$ (in law).
(A.3).

$$E[\int_0^T |u(t)|dt] < \infty, \quad P(\int_0^T |u(t)|^2 dt < \infty) = 1. \tag{2.11}$$

(A.4).

$$\limsup_{r\to\infty} P(|X_o| \geq r)r^4 < \infty. \tag{2.12}$$

Remark 2.1. Under (A.1), (1.7) has a unique weak solution X^0 which is strong Markov if $u \equiv 0$ (see Stroock and Varadhan (1979)). Under (A.1)-(A.3), there exist a jointly measurable $u'(t,x) : [0,T] \times C([0,T];R) \mapsto R$ for which $u'(t,\cdot)$ is $\mathbf{B}(C([0,t];R))_+$-measurable for $t \in [0,T]$, and a $\sigma[X^u(s); 0 \leq s \leq t]$-adapted Wiener process $\overline{W}(t)$ such that for $t \in [0,T]$,

$$X^u(t) = X_o + \int_0^t u'(s, X^u)ds + \int_0^t \sigma(s, X^u(s))d\overline{W}(s). \tag{2.13}$$

Moreover $P^{(X^u)^{-1}}$ is absolutely continuous with respect to $P^{(X^0)^{-1}}$ (see section 7.6 in Liptser and Shiryaev (1977)), and $P(X^0(t) \in dx)$ is absolutely continuous with respect to dx for $t \in (0,T]$ (see Stroock and Varadhan (1979)).

Let us state a technical result which can be obtained exactly in the same way as in Theorem 3.1 in Freidlin (1985), p. 371 and the proof is omitted.

Theorem 2.4 *Suppose that (A.1) holds and that $B(t,v) : [0,T] \times R \mapsto R$ is bounded, measurable, and globally Lipschitz continuous in v, uniformly in $t \in [0,T]$. Then the following integral equation has a unique solution: for $(t,x) \in [0,T] \times R$,*

$$\begin{aligned} v(t,x) &= E[\exp(\int_0^t \sigma(s, X^0(s))^{-1} B(s, v(s, X^0(s)))dW(s) \\ &\quad - \int_0^t \sigma(s, X^0(s))^{-2} B(s, v(s, X^0(s)))^2 ds/2); X^0(t) \leq x], \end{aligned} \tag{2.14}$$

where X^0 is a weak solution to (1.7) with $u(t) \equiv 0$ and with $X^0(0) = X_o$.

Let X^B be the solution of the following: for $t \in [0,T]$,

$$X^B(t) = X_o + \int_0^t B(s, v(s, X^B(s)))ds + \int_0^t \sigma(s, X^B(s))dW(s), \tag{2.15}$$

which has a unique weak solution under the assumption in Theorem 2.4 (see Stroock and Varadhan (1979)). Then the following result can be obtained and be considered as MMP. It will be used in the proof of Theorem 3.1.

Theorem 2.5 *Under the same assumption as in Theorem 2.4, for $v(t,x)$ constructed in Theorem 2.4, (2.15) has a unique weak solution, and the following holds:*

$$P(X^B(t) \le x) = v(t,x) \quad ((t,x) \in [0,T] \times R). \tag{2.16}$$

In particular, for any $\varphi \in C_o^\infty(R;R)$ *and* $t \in [0,T]$,

$$\int_R \varphi(x)v(t,dx) - \int_R \varphi(x)v(0,dx) \tag{2.17}$$

$$= \int_0^t ds \int_R (2^{-1}\sigma(s,x)^2\partial^2\varphi(x)/\partial x^2 + B(s,v(s,x))\partial\varphi(x)/\partial x)v(s,dx).$$

(Proof). The first part of the theorem can be easily proved by Cameron-Martin-Maruyama-Girsanov formula (see Stroock and Varadhan (1979)). The second part is an easy consequence of the Ito formula. **Q. E. D.**

In Theorem 2.5, we fixed a diffusion coefficient. If we allow that it can be changed, then we get the following result which will be used in the proof of Theorem 3.2.

Theorem 2.6 *Suppose that (A.1)-(A.2) and (A.4) hold, and that there exists a positive constant* C_u *which dominates* $\{|u(t)|\}_{0\le t\le T}$ *a.s.. Then there exist measurable* $a^n(\cdot,\cdot):[0,T]\times R \mapsto [\nu,\infty)$ *and* $b^n(\cdot,\cdot):[0,T]\times R \mapsto R$ *such that the following has a unique weak solution: for* $t \in [0,T]$

$$X^n(t) = X_o + \int_0^t b^n(s,X^n(s))ds + \int_0^t a^n(s,X^n(s))^{1/2}dW(s) \tag{2.18}$$

$(n \ge 1)$, *and the following holds: in the sense of weak convergence,*

$$\lim_{n\to\infty} P(X^n(t) \in dx) = P(X^u(t) \in dx) \quad (t \in [0,T]), \tag{2.19}$$

$$\lim_{n\to\infty} P(b^n(t,X^n(t)) \in dx) = P(u(t) \in dx) \quad (t \in (0,T]). \tag{2.20}$$

Let us prove Theorem 2.6 with the aid of lemmas given later.
(Proof). For $n \ge 1$, $t \in [0,T]$ and $x \in R$, put

$$F_t(x) = P(X^u(t) \le x), \tag{2.21}$$

$$C = (\int_R (1+x^2)^{-3/2}dx)^{-1}, \qquad c^n(x) = Cn(1+(nx)^2)^{-3/2},$$

$$F_t^n(x) = \int_R (\int_{-\infty}^{x-y} c^n(z)dz)F_t(dy), \qquad f_t^n(x) = \int_R c^n(x-y)F_t(dy).$$

Then the following can be shown easily by the Ito formula: for $n \ge 1$ and $x \in R$,

$$\partial f_t^n(x)/\partial t = 2^{-1}\partial^2(a^n(t,x)f_t^n(x))/\partial x^2 - \partial(b^n(t,x)f_t^n(x))/\partial x \quad (2.22)$$

$dt - a.e.$ on $(0,T)$, where we put for $n \geq 1$, $x \in R$ and $t \in [0,T]$,

$$G_t(x) = P(u(t) \leq x), \quad b(t,x) = G_t^{-1}(1 - F_t(x)) \text{ (see (2.6))}, \quad (2.23)$$

$$b^n(t,x) = \int_R b(t,y)c^n(x-y)F_t(dy)/f_t^n(x),$$

$$\tilde{a}^n(t,x) = \int_R \sigma(t,y)^2 c^n(x-y)F_t(dy)/f_t^n(x),$$

$$\tilde{b}(t,x) = E[u(t)|X^u(t) = x], \tilde{b}^n(t,x) = \int_R \tilde{b}(t,y)c^n(x-y)F_t(dy)/f_t^n(x),$$

$$a^n(t,x) = \tilde{a}^n(t,x) + 1_{(0,T]}(t)2f_t^n(x)^{-1}\int_{-\infty}^x [b^n(t,y) - \tilde{b}^n(t,y)]f_t^n(y)dy.$$

By the assumption, one can show that there exists a unique weak solution to (2.18), and that (2.19)-(2.20) hold for a^n and b^n defined in (2.23). To prove the uniqueness and the existence part, one only has to show that the following holds (see Stroock and Varadhan (1979)):

$$a^n(t,x) \geq \nu \quad (0 \leq t \leq T, x \in R), \quad (2.24)$$

$$\sup_{0\leq t\leq T, x\in R} (|\partial a^n(t,x)/\partial x| + |\partial b^n(t,x)/\partial x| + |a^n(t,0)| + |b^n(t,0)|) < \infty. \quad (2.25)$$

(2.24)-(2.25) and (2.19)-(2.20) will be proved in Lemmas 2.1-2.2 and in Lemmas 2.3-2.5 which will be given later. **Q. E. D.**

Remark 2.2. In (2.22), we determined $a^n(t,x)$ from a given function $b^n(t,x)$ to derive PDE for $f_t^n(x)$. This idea is that of covariance kernel (see Mikami (1997) and Cacoullos, Papathanasiou and Utev (1994), and the references therein).

In the rest part of this section, we prove (2.19)-(2.20) and (2.24)-(2.25) as Lemmas 2.1-2.5. Let us first prove (2.24).

Lemma 2.1 *Under the assumption in Theorem 2.6,*

$$\int_{-\infty}^x [b^n(t,y) - \tilde{b}^n(t,y)]f_t^n(y)dy \geq 0 \quad ((t,x) \in (0,T] \times R). \quad (2.26)$$

(Proof). For $x \in R$ and $t \in (0,T]$,

$$\int_{-\infty}^x [b^n(t,y) - \tilde{b}^n(t,y)]f_t^n(y)dy \quad (2.27)$$

$$= \int_R c^n(y)dy \int_{-\infty}^{x-y} [b(t,z) - \tilde{b}(t,z)]F_t(dz),$$

and

$$\int_{-\infty}^{x} b(t,z)F_t(dz) = E[G_t^{-1}(1-F_t(X^u(t))); X^u(t) \le x], \tag{2.28}$$

$$\int_{-\infty}^{x} \tilde{b}(t,z)F_t(dz) = E[u(t); X^u(t) \le x]. \tag{2.29}$$

For any distribution function Φ on R^2 such that $\int_{R^2} |y|\Phi(dxdy) < \infty$,

$$\int_{(-\infty,x]\times R} y\Phi(dzdy) = \int_0^{\infty} [\Phi(x,\infty) - \Phi(x,y)]dy - \int_{-\infty}^{0} \Phi(x,y)dy. \tag{2.30}$$

This and (2.27)-(2.29) together with the following completes the proof: for any distribution function Φ on R^2,

$$\Phi(x,y) \ge max(\Phi(x,\infty) + \Phi(\infty,y) - 1, 0) \quad (x, y \in R) \tag{2.31}$$

(see Schweizer (1991) or Schweizer and Sklar (1983)). In fact, from (2.8),

$$\begin{aligned} & P(X^u(t) \le x, b(t, X^u(t)) \le y) \qquad (2.32) \\ = \; & P(1 - G_t(y) \le F_t(X^u(t)) \le F_t(x)) \\ = \; & max(F_t(x) + G_t(y) - 1, 0) \le P(X^u(t) \le x, u(t) \le y) \text{ (from (2.31))}. \end{aligned}$$

Here we used the fact that $F_t(X^u(t))$ is uniformly distributed on $[0,1]$ for $t \in (0,T]$. **Q. E. D.**

Remark 2.3. (2.32) implies that our approach is effective only in case $d = 1$. In fact, for any distribution function $\Phi(x_1, \cdots,, x_d)$ on R^d, there exists a distribution function C_Φ on $[0,1]^d$ which is uniquely determined on $Range(F_1) \times \cdots Range(F_d)$ (see Schweizer (1991), Schweizer and Sklar (1983)) such that the following holds: for $(x_i)_{i=1}^d \in R^d$,

$$\begin{aligned} & \min(F_1(x_1), \cdots, F_d(x_d)) \qquad (2.33) \\ \ge \; & \Phi(x_1, \cdots,, x_d) = C_\Phi(F_1(x_1), \cdots,, F_d(x_d)) \\ \ge \; & max(F_1(x_1) + \cdots + F_d(x_d) + d - 1, 0). \end{aligned}$$

Here we denote by $F_i(x)$ the value of Φ when $x_k = \infty$ $(k \ne i)$ and $x_i = x$. t-norm $W(u_1, \cdots, u_d) = max(u_1 + \cdots + u_d + d - 1, 0)$ is a distribution function on $[0,1]^d$ iff $d \le 2$.

Let us prove (2.25) which will also be used in the proof of Lemma 2.3.

Lemma 2.2 *Under the assumption in Theorem 2.6, for any $n \ge 1$, there exists $C_n > 0$ such that*

$$\sup_{0\le t\le T, x\in R}(|\partial a^n(t,x)/\partial x| + |\partial b^n(t,x)/\partial x| \tag{2.34}$$
$$+|a^n(t,0)| + |b^n(t,0)|) \le C_n.$$

(Proof). We only have to show that the following holds:

$$\sup_{x\in R} c^n(x)^{-1}|\partial c^n(x)/\partial x| \le (3/2)n, \tag{2.35}$$

$$\sup_{0\le t\le T, x\in R} f_t^n(x)^{-1}(1+|x|)^{-3} < \infty, \tag{2.36}$$

$$\sup_{0\le t\le T, x\in R} |\partial f_t^n(x)/\partial x|(1+|x|)^4 < \infty, \tag{2.37}$$

$$\sup_{0\le t\le T}\{\sup_{x<0} F_t^n(x)(1+|x|)^2 + \sup_{x>0}(1-F_t^n(x))(1+|x|)^2\} < \infty. \tag{2.38}$$

This is true, since

$$\begin{aligned} &\partial a^n(t,x)/\partial x \\ = &\int_R \sigma(t,y)^2(\partial c^n(x-y)/\partial x)F_t(dy)/f_t^n(x) \\ &-a^n(t,x)[\partial f_t^n(x)/\partial x]/f_t^n(x) + 1_{(0,T]}(t)2[b^n(t,x) - \tilde{b}^n(t,x)], \end{aligned}$$

$$\begin{aligned} \partial b^n(t,x)/\partial x &= \int_R b(t,y)(\partial c^n(x-y)/\partial x)F_t(dy)/f_t^n(x) \\ &\quad -b^n(t,x)[\partial f_t^n(x)/\partial x]/f_t^n(x), \end{aligned}$$

and since

$$|b^n(t,x)| + |\tilde{b}^n(t,x)| \le \sup_{0\le s\le T, y\in R}\{|b(s,y)| + |\tilde{b}(s,y)|\} \le 2C_u,$$

$$\begin{aligned} |a^n(t,x)| \le &\sup_{0\le s\le T, y\in R}|\sigma(s,y)|^2 \\ &+1_{(0,T]}(t)4C_u\min(F_t^n(x), 1-F_t^n(x))/f_t^n(x). \end{aligned}$$

Here we used the fact that $u(t)$ and $G_t^{-1}(1-F_t(X^u(t)))$ has the same probability law for $t\in(0,T]$ and that

$$\int_R (b^n(t,x) - \tilde{b}^n(t,x))f_t^n(x)dx = E[u(t) - G_t^{-1}(1-F_t(X^u(t)))] = 0$$

(see (2.27)-(2.29) and (2.32)).

(2.35) can be shown easily and we omit the proof.
(2.36) can be shown by the follows: for $r > 0$,

$$\begin{aligned} f_t^n(x) &\geq \int_{|y|\leq r} nC(1+2n^2(x^2+y^2))^{-3/2}F_t(dy) \\ &\geq nCP(|X^u(t)| \leq r)(1+2n^2(x^2+r^2))^{-3/2}, \end{aligned} \tag{2.39}$$

and

$$\begin{aligned} & P(|X^u(t)| \geq r) \\ \leq\ & P(|X^u(0)| \geq r/3) + P(|\int_0^t u(s)ds| \geq r/3) \\ & + P(|\int_0^t \sigma(s, X^u(s))dW(s)| \geq r/3) \\ \leq\ & P(|X^u(0)| \geq r/3) + (3tC_u/r)^4 + 3t^2 \sup_{0\leq s\leq T, y\in R} |\sigma(s,y)|^4(3/r)^4 \end{aligned} \tag{2.40}$$

by Chebychev's inequality.

Next we prove (2.37). For $x \in R$,

$$\begin{aligned} & |\partial f_t^n(x)/\partial x| \\ \leq\ & \int_R (3/2)n^2C \times 2n|x-y|/(1+n^2(x-y)^2)^{5/2}F_t(dy) \leq (3/2)n^2C. \end{aligned} \tag{2.41}$$

For x for which $|x| > 2$,

$$\begin{aligned} & |\partial f_t^n(x)/\partial x| \\ \leq\ & \left(\int_{|x-y|<|x|/2} + \int_{|x-y|\geq|x|/2}\right)(3/2)n^2C \\ & \times 2n|x-y|/(1+n^2(x-y)^2)^{5/2}F_t(dy) \\ \leq\ & (3/2)n^2CP(|X^u(t)| \geq |x|/2) + (3/2)n^2C|x|/(1+(|x|/2)^2)^{5/2}, \end{aligned} \tag{2.42}$$

since $2r/(1+r^2)^{5/2}$ is decreasing on $[1,\infty)$. This together with (2.40) completes the proof from (A.4).

(2.38) can be proved easily from (2.37). **Q. E. D.**

(2.19) can be shown by the following lemma.

Lemma 2.3 *Under the assumption of Theorem 2.6,*

$$lim_{n\to\infty} P(X^n(t) \leq x) = F_t(x) \quad ((t,x) \in [0,T] \times R). \tag{2.43}$$

(Proof). Fix $n \geq 1$ and put, for $k \geq 1$, $t \in [0,T]$ and $x \in R$,

$$g^k(x) = k/(2\pi)^{1/2}\exp(-k^2x^2/2), \quad G^k(x) = \int_{-\infty}^{x} g^k(y)dy, \tag{2.44}$$
$$h_t(x) = P(X^n(t) \leq x) - F_t^n(x), \quad u^k(t,x) = \int_R G^k(x-y)h_t(dy).$$

Then one can show that the following holds: for $t \in [0,T]$

$$\int_R |h_t(x)|^2 dx \leq \int_R |h_0(x)|^2 dx \exp(tC_u^2/\nu), \tag{2.45}$$

which completes the proof.

Since the following holds by Lemma 2.2 and (2.37)

$$\sup_{0\leq t\leq T}\{E[|X^n(t)|] + \int_R |y| f_t^n(y)dy\} < \infty, \tag{2.46}$$

letting $r \to \infty$ and then $k \to \infty$ in (2.47) below, we get (2.45), from Lemma 2.2, (A.4) and (2.38): for $r > 0$, by (2.22) and the Ito formula,

$$\begin{aligned}
&\int_{-r}^{r} |u^k(t,x)|^2 dx \qquad (2.47)\\
\leq\ & \Big(\int_0^t ds \big|\big[\int_R a^n(s,y)g^k(x-y)h_s(dy)u^k(s,x)\big]_{x=-r}^{r}\big| + \int_R |u^k(0,x)|^2 dx\\
&+ \int_0^t ds \int_{R^3} C_n |y_3/k| g^k(y_1 + y_3/k - y_2)\\
&\times |\partial h_s(y_1)/\partial y_1||\partial h_s(y_2)/\partial y_2| g^1(y_3)dy_1dy_2dy_3\\
&+2\int_0^t ds \int_{R^2} C_n |y_1/k| g^1(y_1)|\partial h_s(y_2)/\partial y_2| dy_1dy_2\Big)\exp(tC_u^2/\nu).
\end{aligned}$$

(2.47) can be shown in the same way as in Chap. 3, section 2 in Ladyzhenskaya, O. A., Solonnikov, V.A. and Ural'tseva, N. N. (1968). In fact,

$$\begin{aligned}
&\int_{-r}^{r} |u^k(t,x)|^2/2dx - \int_{-r}^{r} |u^k(0,x)|^2 dx/2 \qquad (2.48)\\
=\ & \int_0^t ds \int_{-r}^{r} \{\int_R [2^{-1}a^n(s,y)(dg^k(x-y)/dx)\\
&-b^n(s,y)g^k(x-y)]h_s(dy)\}u^k(s,x)dx\\
=\ & \int_0^t ds\big[\int_R 2^{-1}a^n(s,y)g^k(x-y)h_s(dy)u^k(s,x)\big]_{x=-r}^{r}
\end{aligned}$$

$$-\int_0^t ds \int_{-r}^{r} 2^{-1}a^n(s,x)|\partial u^k(s,x)/\partial x|^2 dx$$
$$+\int_0^t ds \int_{-r}^{r} \{\int_R 2^{-1}[a^n(s,x)$$
$$-a^n(s,y)]g^k(x-y)h_s(dy)\}[\partial u^k(s,x)/\partial x]dx$$
$$-\int_0^t ds \int_{-r}^{r} b^n(s,x)[\partial u^k(s,x)/\partial x]u^k(s,x)dx$$
$$+\int_0^t ds \int_{-r}^{r} \{\int_R [b^n(s,x)-b^n(s,y)]g^k(x-y)h_s(dy)\}u^k(s,x)dx.$$

Q. E. D.

Finally we prove (2.20).

Lemma 2.4 *Under the assumption of Theorem 2.6, for $t \in (0,T]$ and the continuity point x of $G_t(\cdot)$,*

$$P(b^n(t,X^n(t)) \le x) \to G_t(x) \quad \text{as } n \to \infty. \tag{2.49}$$

(Proof). For $t \in (0,T]$, $\delta > 0$ and the continuity point x of $G_t(\cdot)$,

$$\begin{aligned} & P(G_t^{-1}(1-F_t(X^n(t)-\delta)) + 2C_uC(n^2\delta^3)^{-1}f_t^n(X^n(t))^{-1} \le x) \\ \le\ & P(b^n(t,X^n(t)) \le x) \qquad (2.50) \\ \le\ & P(G_t^{-1}(1-F_t(X^n(t)+\delta)) - 2C_uC(n^2\delta^3)^{-1}f_t^n(X^n(t))^{-1} \le x). \end{aligned}$$

This is true, since $b(t,\cdot) = G_t^{-1}(1-F_t(\cdot))$ is nonincreasing, and since

$$\begin{aligned} & \int_{|x-y|\ge\delta} c^n(x-y)F_t(dy)/f_t^n(x) \\ \le\ & nC(1+n^2\delta^2)^{-3/2}/f_t^n(x) < C(n^2\delta^3)^{-1}/f_t^n(x). \end{aligned}$$

Let us show that the probability in both ends of (2.50) converge to $G_t(x)$ as $n \to \infty$ and then $\delta \to 0$. The left end can be considered as follows:

$$\begin{aligned} & P(G_t^{-1}(1-F_t(X^n(t)-\delta)) \qquad (2.51) \\ & \quad +2C_uC(n^2\delta^3)^{-1}f_t^n(X^n(t))^{-1} \le x) \\ \ge\ & P(G_t^{-1}(1-F_t(X^n(t)-\delta)) \le x-\delta) \\ & \quad -P(2C_uC(n^2\delta^3)^{-1}f_t^n(X^n(t))^{-1} \ge \delta). \end{aligned}$$

By (2.8) and Lemma 2.3,

$$\begin{aligned} & P(G_t^{-1}(1-F_t(X^n(t)-\delta)) \le x-\delta) && (2.52)\\ \to\ & P(1-G_t(x-\delta) \le F_t(X^u(t)-\delta)) \quad \text{as } n\to\infty \\ \to\ & P(1-G_t(x) \le F_t(X^u(t))) = G_t(x) \quad \text{as } \delta \to 0. \end{aligned}$$

Take $k \ge 1$ and let $n \ge k$. Then by Lemma 2.3,

$$\begin{aligned} & P(2C_uC(n^2\delta^3)^{-1}f_t^n(X^n(t))^{-1} \ge \delta) && (2.53)\\ \le\ & P(2C_u\delta^{-4}(\int_R (k^{-2}+(X^n(t)-y)^2)^{-3/2}F_t(dy))^{-1} \ge 1). \end{aligned}$$

The limsup of the last probability in (2.53), as $n \to \infty$, is dominated by

$$P(2C_u\delta^{-4}(\int_R (k^{-2}+(X^u(t)-y)^2)^{-3/2}F_t(dy))^{-1} \ge 1) \tag{2.54}$$

which converges, as $k \to \infty$, to

$$P(2C_u\delta^{-4}(\int_R |X^u(t)-y|^{-3}F_t(dy))^{-1} \ge 1) = 0 \tag{2.55}$$

(see Lemma 2.5 given below).

The right end of (2.50) can be shown to converge to $G_t(x)$, as $n \to \infty$ and then $\delta \to 0$, in the same way as above. **Q. E. D.**

Let us prove the last equality of (2.55).

Lemma 2.5 *Suppose that (A.1)-(A.3) hold. Then for* $t \in (0, T]$,

$$P(\int_R |X^u(t)-y|^{-3}F_t(dy) < \infty) = 0. \tag{2.56}$$

(Proof). Put

$$S_t \equiv \{x \in R; \lim_{h\to 0}(F_t(x+h)-F_t(x))/h > 0\}. \tag{2.57}$$

Then we only have to show that for $x \in S_t$,

$$\int_R |x-y|^{-3}F_t(dy) = \infty, \tag{2.58}$$

since $P(X^u(t) \in S_t) = 1$ for $t \in (0, T]$. This is true, since $P(X^u(t) \in dx)$ is absolutely continuous with respect to dx for $t \in (0, T]$ (see Remark 2.1).

Let us prove (2.58). For $x \in S_t$ and $\delta > 0$,

$$\begin{aligned} & \int_R |x-y|^{-3}F_t(dy) > \int_{|x-y|<\delta} |x-y|^{-3}F_t(dy) \\ >\ & \delta^{-3}(F_t(x+\delta)-F_t(x-\delta)) \to \infty \quad \text{as } \delta \to 0. \end{aligned}$$

Q. E. D.

3 Applications to Markov optimal control

In this section we consider the applications of Markov marginal problems discussed in section 2 to Markov optimal control problems.

Let $L_1(t,x)$, $L_2(t,x) : [0,T] \times R^d \mapsto R$ and $\Phi(x) : R^d \mapsto R$ be bounded and continuous. For a R^d-valued stochastic process $\{X(t)\}_{0 \le t \le T}$ on a probability space $(\Omega, \mathbf{B}, P)$, put

$$J_1(X) \equiv E[\int_0^T L_1(t, X(t))dt + \Phi(X(T))]. \tag{3.1}$$

For a weak solution X^u to (1.7), put

$$J_2(X^u) \equiv E[\int_0^T L_2(t, u(t))dt], \tag{3.2}$$

$$J_3(X^u) \equiv E[\int_0^T [L_1(t, X^u(t)) + L_2(t, u(t))]dt + \Phi(X^u(T))]. \tag{3.3}$$

The following theorem can be proved from Theorem 2.1 easily and we omit the proof.

Corollary 3.1 *For a Borel probability measure P_o on R^d, the infimum of $J_1(X)$ over $\{X(t)\}_{t\in[0,T]}$ for which $P(X(0) \in dx) = P_o(dx)$ and for which $\{P(X(t) \in dx)/dx\}_{t\in[0,T]}$ satisfies the assumption in Theorem 2.1 is equal to that over Markov diffusion processes $\{X(t)\}_{t\in[0,T]}$ for which $P(X(0) \in dx) = P_o(dx)$ and for which $\{P(X(t) \in dx)/dx\}_{t\in[0,T]}$ satisfies the assumption in Theorem 2.1.*

The following theorem can be proved from Theorem 2.5 and will be proved later.

Theorem 3.1 *Suppose that (A.1)-(A.2) hold. Then for any $\varepsilon > 0$, there exist bounded measurable $b^\varepsilon(\cdot,\cdot) : [0,T] \times R \mapsto R$ and a unique weak solution to the following : for $t \in [0,T]$*

$$X^\varepsilon(t) = X_o + \int_0^t b^\varepsilon(s, X^\varepsilon(s))ds + \int_0^t \sigma(s, X^\varepsilon(s))dW(s) \tag{3.4}$$

such that the following holds: for $t \in (0,T]$

$$\lim_{\varepsilon \to 0} P(b^\varepsilon(t, X^\varepsilon(t)) \in dx) = P(u(t) \in dx) \quad \textit{weakly}. \tag{3.5}$$

As an application of Theorem 3.1, the following can be proved easily and the proof is omitted.

Corollary 3.2 *Suppose that (A.1) holds. Then the infimum of $J_2(X^u)$ over $\{u(t)\}_{t\in[0,T]}$ for which (A.2) holds is equal to that over $\{u(t)\}_{t\in[0,T]}$ for which (A.2) with $u(t) = b(t, X^u(t))$ holds for some $b(t,x)$.*

The following theorem can be proved from Theorem 2.6 and will be proved later.

Theorem 3.2 *Suppose that (A.1)-(A.4) hold. Then for any* $\varepsilon > 0$ *and* $n \geq 1$, *there exist bounded, measurable* $\tilde{b}^{\varepsilon,n}(\cdot,\cdot) : [0,T] \times R \mapsto R$, *and measurable* $\tilde{a}^{\varepsilon,n}(\cdot,\cdot) : [0,T] \times R \mapsto [\nu,\infty)$ *such that there exists a unique weak solution to the following: for* $t \in [0,T]$

$$\tilde{X}^{\varepsilon,n}(t) = X_o + \int_0^t \tilde{b}^{\varepsilon,n}(s,\tilde{X}^{\varepsilon,n}(s))ds + \int_0^t \tilde{a}^{\varepsilon,n}(s,\tilde{X}^{\varepsilon,n}(s))^{1/2}dW(s), \quad (3.6)$$

and such that the following holds: in the sense of weak convergence,

$$\lim_{\varepsilon\to 0}\lim_{n\to\infty} P(\tilde{X}^{\varepsilon,n}(t) \in dx) = P(X^u(t) \in dx) \quad (0 \leq t \leq T), \quad (3.7)$$

$$\lim_{\varepsilon\to 0}\lim_{n\to\infty} P(\tilde{b}^{\varepsilon,n}(t,\tilde{X}^{\varepsilon,n}(t)) \in dx) = P(u'(t,X^u) \in dx) \quad (0 < t \leq T), \quad (3.8)$$

where u' *is given in (2.13).*

Denote by $X^{u,\sigma'}$ a weak solution to (1.7) when σ is replaced by some function σ'. As an application of Theorem 3.2, the following can be proved easily and the proof is omitted.

Corollary 3.3 *Suppose that (A.4) holds. Then the infimum of* $J_3(X^{u,\sigma'})$ *over* $\{u(t)\}_{t\in[0,T]}$ *and* σ' *for which (A.1)-(A.3) with* $\sigma = \sigma'$ *hold is equal to that over* $\{u(t)\}_{t\in[0,T]}$ *and* σ' *for which (A.1)-(A.3) with* $\sigma = \sigma'$ *and* $u(t) = b(t, X^{u,\sigma'}(t))$ *hold for some* $b(t,x)$.

Let us prove Theorem 3.1 first.
(Proof of Theorem 3.1.) For any $\varepsilon > 0$, put

$$G_{t,\varepsilon}^{-1}(u) \quad (3.9)$$
$$= \begin{cases} \max(G_t^{-1}(G_t(-1/\varepsilon)), -1/\varepsilon) & \text{if } u \leq G_t(-1/\varepsilon), \\ G_t^{-1}(u) & \text{if } G_t(-1/\varepsilon) < u < G_t(1/\varepsilon), \\ G_t^{-1}(G_t(1/\varepsilon)) & \text{if } u \geq G_t(1/\varepsilon) \end{cases}$$

(see (2.6) for the notation of G_t^{-1}).
For $\varphi \in C_o^\infty(R;[0,\infty))$ which is not identically zero, put

$$B^\varepsilon(t,u) = \int_R \varphi((u-y)/\varepsilon)G_{t,\varepsilon}^{-1}(y)dy/C_\varepsilon, \quad (3.10)$$

where we put $C_\varepsilon = \varepsilon\int_R \varphi(y)dy$. Then for $\varepsilon > 0$, $B^\varepsilon(t,u)$ is bounded, measurable, and globally Lipschitz continuous in u, uniformly in $t \in [0,T]$, since

$$|G_{t,\varepsilon}^{-1}(u)| \leq 1/\varepsilon. \tag{3.11}$$

Let $v^\varepsilon(t,x)$ be the solution to (2.14) with $B = B^\varepsilon$. Then the following has a unique weak solution from Theorem 2.5: for $t \in [0,T]$

$$X^\varepsilon(t) = X_o + \int_0^t B^\varepsilon(s, v^\varepsilon(s, X^\varepsilon(s)))ds + \int_0^t \sigma(s, X^\varepsilon(s))dW(s), \tag{3.12}$$

and letting $\varepsilon \to 0$, for $t \in (0,T]$ and the continuity point x of G_t,

$$P(B^\varepsilon(t, u^\varepsilon(t, X^\varepsilon(t))) \leq x) \to G_t(x). \tag{3.13}$$

Let us show that (3.13) is true to complete the proof. Take $r > 0$ so that $supp(\varphi) \subset [-r, r]$. For $x \in R$,

$$\begin{aligned} P(G_{t,\varepsilon}^{-1}(v^\varepsilon(t, X^\varepsilon(t)) + r\varepsilon) \leq x) &\leq P(B^\varepsilon(t, v^\varepsilon(t, X^\varepsilon(t))) \leq x) \qquad (3.14)\\ &\leq P(G_{t,\varepsilon}^{-1}(v^\varepsilon(t, X^\varepsilon(t)) - r\varepsilon) \leq x), \end{aligned}$$

since $G_{t,\varepsilon}^{-1}$ is nondecreasing.

Suppose that $G_t(x) \in (0,1]$. Then for sufficiently small ε, $G_t(-1/\varepsilon) < G_t(x) \leq G_t(1/\varepsilon)$, and from (3.14),

$$\begin{aligned} &P(B^\varepsilon(t, v^\varepsilon(t, X^\varepsilon(t))) \leq x) \geq P(G_{t,\varepsilon}^{-1}(v^\varepsilon(t, X^\varepsilon(t)) + r\varepsilon) \leq x) \qquad (3.15)\\ \geq\ & P(G_{t,\varepsilon}^{-1}(v^\varepsilon(t, X^\varepsilon(t)) + r\varepsilon) \leq G_t^{-1}(G_t(x))) \quad (\text{since } G_t^{-1}(G_t(x)) \leq x)\\ =\ & P(G_{t,\varepsilon}^{-1}(v^\varepsilon(t, X^\varepsilon(t)) + r\varepsilon) \leq G_{t,\varepsilon}^{-1}(G_t(x))) \quad (\text{see } (3.9))\\ \geq\ & P(v^\varepsilon(t, X^\varepsilon(t)) + r\varepsilon \leq G_t(x)) = \max(G_t(x) - r\varepsilon, 0) \to G_t(x), \end{aligned}$$

as $\varepsilon \to 0$, since $G_{t,\varepsilon}^{-1}$ is nondecreasing (see (3.9)) and since $v^\varepsilon(t, X^\varepsilon(t))$ is uniformly distributed on $[0,1]$ for $t \in (0,T]$ (see (2.16)).

Suppose that $G_t(x) \in [0,1)$. Then for sufficiently small ε, $G_t(-1/\varepsilon) \leq G_t(x) < G_t(1/\varepsilon)$, and from (3.14),

$$\begin{aligned} &P(B^\varepsilon(t, v^\varepsilon(t, X^\varepsilon(t))) \leq x) \qquad (3.16)\\ \leq\ & P(G_{t,\varepsilon}^{-1}(v^\varepsilon(t, X^\varepsilon(t)) - r\varepsilon) \leq x)\\ \leq\ & P(v^\varepsilon(t, X^\varepsilon(t)) - r\varepsilon \leq G_t(x)) = \min(G_t(x) + r\varepsilon, 1) \to G_t(x), \end{aligned}$$

as $\varepsilon \to 0$, in the same way as in (3.15). Here we used the follows: if $u > G_t(x)$, then

$$G_{t,\varepsilon}^{-1}(u) = \begin{cases} G_t^{-1}(u) > x & \text{if } G_t(1/\varepsilon) > u > G_t(x),\\ G_t^{-1}(G_t(1/\varepsilon)) > x & \text{if } u \geq G_t(1/\varepsilon). \end{cases}$$

Q. E. D.

Finally we prove Theorem 3.2.
(Proof of Theorem 3.2). For $\varepsilon > 0$, let $\{X^{u,\varepsilon}(t)\}_{0\leq t\leq T}$ satisfy the following; for $t \in [0,T]$,

$$\begin{aligned} X^{u,\varepsilon}(t) &= X_o + \int_0^t \min(\max(u'(s,X^{u,\varepsilon}),-1/\varepsilon),1/\varepsilon)ds \\ &\quad + \int_0^t \sigma(s,X^{u,\varepsilon}(s))dW(s). \end{aligned} \tag{3.17}$$

The existence and uniqueness of a weak solution to (3.17) can be easily shown by Cameron-Martin-Maruyama-Girsanov formula (see Remark 2.1 and Liptser and Shiryaev (1977)).

From Theorem 2.6, there exist measurable $\tilde{a}^{\varepsilon,n}(\cdot,\cdot) : [0,T]\times R \mapsto [\nu,\infty)$ and $\tilde{b}^{\varepsilon,n}(\cdot,\cdot) : [0,T]\times R \mapsto R$ such that the following has a unique weak solution: for $t \in [0,T]$,

$$\tilde{X}^{\varepsilon,n}(t) = X_o + \int_0^t \tilde{b}^{\varepsilon,n}(s,\tilde{X}^{\varepsilon,n}(s))ds + \int_0^t \tilde{a}^{\varepsilon,n}(s,\tilde{X}^{\varepsilon,n}(s))^{1/2}dW(s), \tag{3.18}$$

and

$$\begin{aligned} P(\tilde{X}^{\varepsilon,n}(t)\in dx) &\to P(X^{u,\varepsilon}(t)\in dx) \\ &\to P(X^{u}(t)\in dx) \quad (0\leq t\leq T), \end{aligned} \tag{3.19}$$

$$\begin{aligned} & P(\tilde{b}^{\varepsilon,n}(t,\tilde{X}^{\varepsilon,n}(t))\in dx) \\ \to\ & P(\min(\max(u'(t,X^{u,\varepsilon}),-1/\varepsilon),1/\varepsilon)\in dx) \\ \to\ & P(u'(t,X^{u})\in dx) \quad (0< t\leq T), \end{aligned} \tag{3.20}$$

weakly as $n\to\infty$ and then $\varepsilon\to 0$. **Q. E. D.**

Acknowledgments: MOCP was introduced to me when I was a student of Professor Wendell H. Fleming's during 1988-1989. It is a great pleasure for me to submit a paper to this volume. I would like to take this oppotunity to thank him for his constant encouragement and useful suggestions since I was a student. I would also like to thank Prof. Kellerer for a useful comments on the first version of this paper and for pointing out his paper in 1987.

References

Blumenthal, R. M. and Corson, H. H. (1972). *On continuous collections of measures,* In *Proc. 6th Berkeley Sympos. Math. Statist. Probab.* **2**, Berkeley 1970/1971, Univ. California press, Berkeley, pp. 33-40.

Cacoullos, T., Papathanasiou, V. and Utev, S. A. (1994). *Variational inequalities with examples and an application to the central limit theorem, Ann. Probab.* **22**, 1607-1618.

Carlen, E. A. (1984). *Conservative diffusions, Commun. Math. Phys.* **94**, 293-315.

Carmona, R. (1987). *Probabilistic construction of Nelson processes,* in: Itô, K., Ikeda, N. (eds.) *Proc. Probabilistic Methods in Mathematical Physics,* Katata 1985, Kinokuniya, Tokyo, pp. 55-81.

Darsow, W. F., Nguyen, B. and Olsen, E. T. (1992). *Copulas and Markov processes, Illinois. J. Math.* **36**, 600-642.

Edwards, D. A. (1978). *On the existence of probability measures with given marginals, Ann. Inst. Fourier.* **28**, 53-78.

Fleming, W. H. and Soner, H. M. (1993). *Controlled Markov processes and Viscosity solutions,* Springer-Verlag, Berlin.

Freidlin, M. I. (1985). *Functional integration and partial differential equations,* Princeton University Press, Princeton.

Kamae, T. and Krengel, U. (1978). *Stochastic partial ordering, Ann. Probab.* **6**, 1044-1049.

Kellerer, H. G. (1987). *Markov property of point processes, Probab. Th. Rel. Fields* **76**, 71-80.

Kellerer, H. G. (1997). *Representation of Markov kernels by random mappings under order conditions,* in: Benes, V., Stepan, J.(eds.) *Proc. Distributions with given Marginals and Moment Problems,* Prague 1996, Kluwer Academic Publishers, Dordrecht, pp. 143-160.

Ladyzhenskaya, O. A., Solonnikov, V.A. and Ural'tseva, N. N. (1968). *Linear and quasilinear equations of parabolic type,* AMS, Providence, RI.

Liptser, R. S. and Shiryaev, A. N. (1977). *Statistics of random processes I,* Springer-Verlag, Berlin.

Mikami, T. (1989). *Small random perturbations of dynamical systems and Markov processes from marginal distributions, Thesis,* Brown University, Providence.

Mikami, T. (1990). *Variational processes from the weak forward equation, Commun. Math. Phys.* **135**, 19-40.

Mikami, T. (1995). *Copula fields and their applications, Proc. Japan Acad. Ser. A* **71**, 221-224.

Mikami, T. (1997). *Equivalent conditions on the central limit theorem for a sequence of probability measures on R,* to appear in *Statist. Probab. Lett.*.

Nagasawa, M. (1989). *Transformations of diffusion and Schrödinger process, Probab. Th. Rel. Fields* **82**, 109-136.

Nelson, E. (1967). *Dynamical theories of Brownian motion,* Princeton University Press, Princeton.

Nelson, E. (1984). *Quantum fluctuations,* Princeton University Press, Princeton.

Quastel, J. and Varadhan, S. R. S. (1997). *Diffusion semigroups and diffusion processes corresponding to degenerate divergence form operators, Comm. Pure Appl. Math.* **50**, 667-706.

Rüschendorf, L. and Uckelmann, L. (1997). *On optimal multivariate couplings,* in: Benes, V., Stepan, J.(eds.) *Proc. Distributions with given Marginals and Moment Problems,* Prague 1996, Kluwer Academic Publishers, Dordrecht, pp. 143-160.

Schweizer, B. (1991). *Thirty years of copulas,* in: Dall'Aglio, G., Kotz, S., Salinetti, G. (eds.) *Proc. Advances in Probability Distributions with Given Marginals,* Rome 1990, Kluwer Academic Publishers, Dordrecht, pp. 13-50.

Schweizer, B. and Sklar, A. (1983). *Probabilistic metric space,* North - Holland, Amsterdam.

Stroock, D. W. and Varadhan, S. R. S. (1979). *Multidimensional diffusion processes,* Springer - Verlag, Berlin.

Zheng, W. A. (1985). *Tightness results for laws of diffusion processes application to stochastic mechanics, Ann. Inst. Henri Poincaré, Sect. B* **21**, 103-124.

28

Entropy Inequalities and Entropy Dynamics in Nonlinear Filtering of Diffusion Processes

Daniel Ocone

ABSTRACT: Suppose that the nonlinear filtering equations are solved using an incorrect initial condition. It is known that the relative entropy of the actual conditional distribution with respect to this incorrectly initialized filter is a positive supermartingale. In this paper, we study the filtering of diffusion signals. Using the Kushner-Stratonovich equations, we decompose the relative entropy supermartingale into decreasing and local martingale terms, and we derive an entropy bound on information and error measures of the difference between conditional distribution and incorrectly initialized filter.

KEYWORDS: Nonlinear filtering; Relative entropy

1 Introduction

Consider the problem of estimating a Markov signal observed in additive white noise. The theory of nonlinear filtering provides equations and formulas for constructing the conditional distribution of the signal given the noise, the most familiar being the Kallianpur-Striebel formula, the Zakai equation, and the Kushner-Stratonovich equation. It is interesting to study the sensitivity of filtering equations to modelling and perturbation errors, and the simplest issue to pose is that of initialization error. Given that the dynamics of the signal-observation pair are fixed and assumed known, the two inputs to any filtering formula are (i) the observation path and (ii) the initial law of the signal. A solution to the filtering equations initialized at a probability law that does not match the initial law of the signal will be called an *incorrectly initialized filter*. The actual conditional distribution of the signal given the observations will be called the *exact filter*, and we are interested in comparing incorrectly initialized and exact filters as time increases.

This paper explores the behavior of relative entropy as a measure of the difference between the exact filter and an incorrectly initialized filter. J.M.C.

Clark observed that the relative entropy between filters is a positive supermartingale so that, in effect, relative entropy is a kind of stochastic Lyapunov function for comparing an incorrectly initialized filter to the exact filter. Clark's observation is developed in Clark, Ocone, and Coumarbatch (1997), where it is shown that the supermartingale property gives rise to an entropy bound on the error between the optimal and suboptimal estimates of the observation function; the precise result is recalled in Proposition 2.1 below. In this paper, we study the case in which the signal is a diffusion process. The aim is to derive the decompositon of the relative entropy into its decreasing and local martingale parts by using the Kushner-Stratonovich equations. This calculation leads to an improved entropy inequality that also bounds the cumulative, average difference between the logarithmic gradients of the exact and incorrectly initialized filters. Our work is related to the paper of Mayer-Wolf and Zakai (1984), who derive a stochastic equation for the relative entropy between the exact filter and the unconditioned signal distribution in filtering problems.

The results here are related to the problem of asymptotic stability of filters. We say that an incorrectly initialized filter is asymptotically stable if the difference between the exact conditional distribution and an incorrectly initialized filter tends to zero in the long run. The Kalman filter provides the classical example; under assumptions of observability and detectability of the linear filtering model, the Kalman filter equations produce asymptotically optimal filters regardless of the initial conditions. It is not yet known in what generality asymptotic stability holds, but there is recent progress on a variety of filtering models. See Delyon and Zeitouni (1991), Atar and Zeitouni (1994), (1997), Atar (1997), Budhiraja and Kushner (1997), Budhiraja and Ocone (1997), Ocone and Pardoux (1996), Cerou (1994) for a sample (by no means complete) of recent work. In Ocone (1996) it is shown that the relative entropy difference measure tends always to 0 in the scalar Beneš filtering model. For other applications of information theory to filtering, see Bobrovsky and Zakai (1976), Bucy (1979), and Duncan (1970).

The paper is organized as follows. In section 2, we develop the background and notation of the paper and state the main result, which is presented in Theorem 2.1. To clarify the presentation, we choose to state the result assuming a priori regularity on the solutions of the Kushner-Stratonovich equation. Later, in section 4, we state conditions on the coefficients of the filtering model under which the assumptions of Theorem 2.1 can be verified. The proof of Theorem 2.1 is presented in section 3. We believe that some of the technical smoothness assumptions can be relaxed.

2 Statement of Main Theorem

We shall study relative entropy for the filtering problem defined by a signal-observation pair (X, Y) satisfying:

$$dX(t) = f(X(t))\,dt + \sigma(X(t))\,dB(t) \tag{1}$$

$$dY(t) = h(X(t))\,dt + dW(t) \tag{2}$$

Here $X(t) \in R^d$, $Y(t) \in R^p$, $B(t) \in R^m$, $W(t) \in R^p$, and B and W are independent standard Brownian motions. Our computations will require some regularity of solutions to the Kushner-Stratonovich equations. Rather than enforce that regularity here through the imposition of conditions on f, σ, and h, we prefer simply to assume it; see hypothesis (H3) below. Then in section 4, we state theorems giving detailed conditions on f, σ, and h under which the regularity required in (H3) does hold. In this section we assume only the following basic restrictions consistent with (H3) and sufficient for existence and uniqueness of solutions to (1).

- $\|\sigma(x)\| \leq K(1 + |x|)$ and $f^{\dagger}(x) \cdot x \leq K(1 + |x|^2)$ for some constant K;
- $f \in C^1$ and $\sigma \in C^2$; (H1)
- $h \in C^0$.

Rather than work on a general probability space, we shall find it convenient to let the signal-observation pair (X, Y) be the canonical process on the path space $\Omega = C([0,\infty) : R^d) \times C_0([0,\infty); R^p)$; thus, if $(\omega, \eta) \in \Omega$, then $(X, Y)(\omega, \eta)(t) = (\omega(t), \eta(t))$. ($C_0([0,\infty); R^p)$ is the space of continuous paths that start at 0.) Let $\mathcal{F}$ denote the σ-algebra generated by $(X, Y)(t)$, $t \geq 0$. Given a probability measure ν on R^d, we let P^ν denote the unique measure on $(\Omega, \mathcal{F})$ that solves the martingale problems associated to the equations (1)-(2) and that gives $(X(0), Y(0))$ the probability law $\nu \times \delta_0$. Since the filter will be a functional of the observation process alone, we will usually want to work just on observation path space $\Omega_o := C_0([0,\infty); R^p)$. We shall again use Y to denote the canonical process on Ω_o; it should always be clear from context when Y is to be considered on Ω or Ω_o. We let $\overset{o}{\mathcal{Y}}_t := \sigma\{Y(s);\ 0 \leq s \leq t\}$ denote the filtration generated by Y, whether on Ω or Ω_o, and we set $\overset{o}{\mathcal{Y}} = \sigma\{Y(s);\ 0 \leq s < \infty\}$. Finally, we use Q^ν to denote the measure induced on observation path space when ν is the law of $X(0)$; that is,

$$Q^\nu(A) = P^\nu\left(Y(\cdot) \in A\right), \qquad A \in \overset{o}{\mathcal{Y}}.$$

Expectation with respect to P^ν or Q^ν shall be denoted by E^ν.

Fix two, distinct probability laws ν and $\overline{\nu}$ on R^d. We want to construct filters when either ν or $\overline{\nu}$ is the law of $X(0)$. To this end we assume the standard finite energy conditions,

$$\left.\begin{array}{ll} E^{\nu}\left[\int_0^T |h(X(s))|^2\, ds\right] < \infty, & \forall T > 0; \\ E^{\overline{\nu}}\left[\int_0^T |h(X(s))|^2\, ds\right] < \infty, & \forall T > 0. \end{array}\right\} (H2)$$

Let Q_W be standard Wiener measure on $(\Omega_o, \overset{o}{\mathcal{Y}})$. Under the hypotheses (H1), it is well-known that Q^{ν}, $Q^{\overline{\nu}}$, and Q_W are all mutually absolutely continuous on $(\Omega_o, \overset{o}{\mathcal{Y}}_T)$, for any finite $T > 0$. In doing stochastic calculus with the observation process, it is necessary to augment the observation filtration with classes of null sets, for the purpose of preserving adaptiveness when choosing versions. In this paper we employ the following convention. If we fix $T < \infty$ and work on the finite time interval $[0, T]$, we let $\mathcal{N}_T$ be the class of Q_W-null (equivalently Q^{ν} or $Q^{\overline{\nu}}$-null) sets in $\overset{o}{\mathcal{Y}}_T$ and define

$$\mathcal{Y}_t := \overset{o}{\mathcal{Y}}_t \vee \mathcal{N}_T. \tag{3}$$

Then $\{\mathcal{Y}_t\}$ will be a right continuous filtration and will serve under any of the measures Q, Q^{ν}, $Q^{\overline{\nu}}$, because of their mutual equivalence. When stating a result on the time interval $[0, \infty)$, $\mathcal{Y}_t$ will be defined instead as $\mathcal{Y}_t := \overset{o}{\mathcal{Y}}_t \vee \mathcal{N}_n$ if $n - 1 \leq t < n$ for positive integers n. We adopt this definition because in general the classes of null sets in $\overset{o}{\mathcal{Y}}$ are not the same for the different measures. In either case, note that for an integrable random variable Z, any versions of $E^{\nu}[Z/ \overset{o}{\mathcal{Y}}_t]$ and $E^{\nu}[Z/\mathcal{Y}_t]$ are equal up to a Q^{ν}-null set.

In this paper, by the filter $\{\pi_t(\eta);\ t \geq 0, \eta \in \Omega_o\}$ we mean a $\{\mathcal{Y}_t\}$-adapted, probability-measure-valued process such that, for each $t \geq 0$, $\pi_t(Y(\cdot))$ is a version of the conditional distribution of $X(t)$ given $\mathcal{Y}_t$ for the process (X, Y) on (Ω, P^{ν}). Likewise, $\{\overline{\pi}_t\}$ is the measure-valued-process such that $\overline{\pi}_t$ is the conditional distribution of $X(t)$ given $\mathcal{Y}_t$ for the process (X, Y) on $(\Omega, P^{\overline{\nu}})$. Because of the equivalence of the measures Q^{ν} and $Q^{\overline{\nu}}$ on $\mathcal{Y}_t$ for any finite t, there is no ambiguity in considering $\overline{\pi}_t$ as a process on (Ω_o, Q^{ν}). Throughout the paper we will use the notation $\pi_t(\phi) = \int \phi(x)\, d\pi_t(x)$, and similarly for $\overline{\pi}_t$. The notation $\langle \psi, \phi \rangle$ will be used for L^2 inner products.

From now on, we shall work on the probability space (Ω_o, Q^{ν}). On this space, π_t is equal in law to the conditional distribution of $X(t)$ given $\mathcal{Y}_t$ when ν is the initial distribution of X; it is thus the "exact" filter. The process $\overline{\pi}_t$, on the other hand, is an incorrectly initialized filter when run on (Ω_o, Q^{ν}), because $\overline{\pi}_0$ equals $\overline{\nu}$ rather than ν. Our goal is to compare π_t

and $\bar{\pi}_t$ *on* (Ω_o, Q^ν) by studying the relative entropy

$$r(t) := \int \ln\left[\frac{d\pi_t}{d\bar{\pi}_t}(x)\right] d\pi_t(x).$$

(One defines the relative entropy to be infinite if π_t is not absolutely continuous with respect to $\bar{\pi}_t$.)
The following statement is a direct consequence of the Theorems 2.3 and 3.4 in Clark, Ocone, and Coumarbatch (1997).

Proposition 2.1 *If* $r(0) < \infty$, *then* $\{r(t),\ t \geq 0\}$ *is a positive* $\{\mathcal{Y}_t\}$-*supermartingale. Moreover,*

$$E^\nu\left[\int_0^\infty |\bar{\pi}_t(h) - \pi_t(h)|^2\, dt\right] \leq 2r(0). \tag{4}$$

Strictly speaking, it is shown that $r(t)$ is a $\{\overset{o}{\mathcal{Y}}_t\}$ supermartingale when π_t and $\bar{\pi}_t$ are defined as $\overset{o}{\mathcal{Y}}_t$-measurable random measures, but this distinction is not important because of the equivalence up to null sets of the augmented and non-augmented filtrations.
Proposition 2.1 is proved in greater generality than the model (1)-(2). But in this paper, we exploit the added structure of the diffusion model to explore r in more detail. In particular, we find the decomposition of $r(t)$ into a decreasing process plus a local martingale and derive an improved relative entropy inequality for diffusion signals.
Let $a(x) = \sigma(x)\sigma^\dagger(x)$, where σ is the diffusion coefficient of the signal in equation (1), and $\sigma^\dagger$ denotes the transpose of σ. Let

$$L = \frac{1}{2}\sum_{i,j=1}^{d} a_{i,j}(x)\frac{\partial^2}{\partial x_i \partial x_j} + \sum_{i=1}^{d} f_i(x)\frac{\partial}{\partial x_i}$$

be the generator of the signal X, and let L^* be its formal adjoint. A jointly measurable process

$$u = \left\{u(t,x,\eta);\ t > 0,\ x \in R^d,\ \eta \in \Omega_o\right\}$$

will be called a classical solution of the Kushner-Stratonovich equation if

(i) For each $t > 0$, $u(s,x,\eta)$ for $(s,x,\eta) \in (0,t) \times R^d \times \Omega_o$ is $\mathcal{B}((0,t)) \times \mathcal{B}(R^d) \times \mathcal{Y}_t$–measurable.

(ii) For any $T > 0$, the function $(t,x) \to u(t,x,\eta)$ is in $C^{0,2}\left((0,T) \times R^d\right)$ except for η in a Q^ν–null set of $\mathcal{Y}_T$.

(iii) For any $T > t > s > 0$, $x \in R^2$,

$$\begin{aligned} u(t,x)-u(s,x) &= \textstyle\int_s^t L^* u(\theta,x)\,d\theta \\ &+ \textstyle\int_s^t u(\theta,x)\left(h(x)-\langle h,u(\theta,\cdot)\rangle\right)\cdot\left(d\eta(\theta)-\langle h^T(\cdot),u(\theta,\cdot)\rangle\,d\theta\right), \end{aligned} \tag{5}$$

except on a Q_W-null set of $\mathcal{Y}_t$.

Implicit in this definition is the assumption that the process $\langle h^T, u(\theta)\rangle$ is well-defined and that the stochastic integrals in which it appears make sense. We agree to work with a version of u that is continuous in (t,x) for all η. Then, because of assumptions (i) and (ii), the map $(x,\theta,\eta)\to u(\theta,x,\eta)$ is indistinguishable from a $\mathcal{B}(R^d)\times\mathcal{P}$ measurable process, where $\mathcal{P}$ is the predictable σ-algebra. Also notice that the the terms in the Kushner-Stratonovich equation other than the stochastic integral are automatically a.s. continuous in (t,x) by assumption (ii). Hence we can define a version (in (t,x)) of the stochastic integral by insisting that the Kushner-Stratonovich equation be valid for all (t,x), almost-surely. We agree always to use a version of the stochastic integral continuous in (x,t). Then we can refine the statement of (5) to say that for every $T>0$, (5) holds for all $(t,x)\in(s,T]\times R^d$ except on a $\mathcal{Y}_T$-null set. Finally, in this definition of classical solution, continuity and differentiability down to the boundary $t=0$ is not required.

The notion of classical solution is used to formulate the final hypothesis.

(H3) There are classical solutions p and $\overline{p}$ of the Kushner-Stratonovich equation such that for every $T>s>0$, both $\inf_{s\le t\le T} p(t,x)>0$ and $\inf_{s\le t\le T}\overline{p}(t,x)>0$ for all x, Q^ν almost surely. In addition, the measure valued processes

$$\begin{aligned} \pi_t(A) &:= \int_A p(t,x)\,dx, \quad t>0, \quad \pi_0=\nu \quad \text{and} \\ \overline{\pi}_t(A) &:= \int_A \overline{p}(t,x)\,dx, \quad t>0, \quad \overline{\pi}_0=\overline{\nu}. \end{aligned}$$

define filters for (X,Y) on the respective spaces (Ω,P^ν) and $(\Omega,P^{\overline{\nu}})$.

In general, a filter process is defined only up to modification on null sets, but henceforth, by π and $\overline{\pi}$ we mean the specific versions defined in (H3). Note that hypothesis (H2) and the simple assumption that π_t admits a jointly measurable density $p(t,x)$ already imply that

$$\begin{aligned} E^\nu\left[\int_0^T |\langle h,p(\theta,\cdot)\rangle|^2\,d\theta\right] &\le E^\nu\left[\int_0^T \pi_\theta(|h|^2)\,d\theta\right] \\ &= E^\nu\left[\int_0^T |h(X(\theta))|^2\,d\theta\right]<\infty. \end{aligned} \tag{6}$$

(The inequality in (6) is a consequence of Jensen's inequality.) Of course, (6) remains true when p is replaced by $\overline{p}$ and ν by $\overline{\nu}$.

We defer until section 4 the question of what conditions on the system coefficients and ν and $\overline{\nu}$ imply (H3).
The statement and proof of the main theorem use the innovations process

$$\hat{Y}(t) := Y(t) - \int_0^t \pi_\theta(h)\,d\theta.$$

As is well-known, the innovations process is a Brownian motion on (Ω_o, Q^ν). The term $\overline{\pi}_t(h) - \pi_t(h)$ occurs frequently in the calcululations. For notational simplicity we shall denote it by $e(t)$. Finally, when it is not necessary to display the x-dependence of $p(t,x)$ or $\overline{p}(t,x)$ explicitly, we write $p(t)$ and $\overline{p}(t)$.
Here is the main result of this section. Recall that $r(t)$ is the entropy of π_t relative to $\overline{\pi}_t$.

Theorem 2.1 *Assume (H1)-(H3). Then*

$$E^\nu\left[\int_0^\infty |e(t)|^2\,dt\right] + E^\nu\left[\int_0^\infty \pi_t\left(\left|\sigma^\dagger\nabla\ln\left[\frac{p(t)}{\overline{p}(t)}\right]\right|^2\right)dt\right] \le 2r(0) \quad (7)$$

If, in addition, $r(0) < \infty$, *and*

$$Q^\nu\left(\int_s^T \pi_t^2\left(|h|\left|\ln\left[\frac{p(t)}{\overline{p}(t)}\right]\right|\right)dt < \infty\right) = 1, \quad (8)$$

then $r(t)$ *admits the following representation for* $0 \le s \le t \le T$:

$$r(t) - r(s) = -\frac{1}{2}\int_s^t |e(\theta)|^2\,d\theta - \frac{1}{2}\int_s^t \pi_\theta\left(\left|\sigma^\dagger\nabla\ln\left[\frac{p(\theta)}{\overline{p}(\theta)}\right]\right|^2\right)d\theta$$
$$+\int_s^t\left[(e(\theta)) + \pi_\theta\left(h\ln\left[\frac{p(\theta)}{\overline{p}(\theta)}\right]\right) - r(\theta)\pi_\theta(h)\right]\cdot d\hat{Y}(\theta). \quad (9)$$

In particular, (8) is true for $s = 0$ *and any* $T > 0$ *in the case that* h *is bounded and* $r(0) < \infty$.

Remarks. Observe that

$$\pi_t\left(\left|\sigma^\dagger\nabla\ln\left[\frac{p(t)}{\overline{p}(t)}\right]\right|^2\right)$$
$$= \int_{R^d}\nabla^T\ln\left[\frac{p(t,x)}{\overline{p}(t,x)}\right]\sigma(x)\sigma^\dagger(x)\nabla\ln\left[\frac{p(t,x)}{\overline{p}(t,x)}\right]p(t,x)\,dx.$$

Suppose that for some $\epsilon > 0$, $\sigma(x)\sigma^T(x) \ge \epsilon I$, for all x. This will be the case in all the examples we consider because it helps guarantee the regularity of solutions to the Kushner-Stratonovich equation. Then we obtain from (7) that

$$E^{\nu}\left[\int_0^{\infty} \pi_t \left(|\nabla \ln \left[\frac{p(t)}{\overline{p}(t)}\right] |^2\right) dt\right] \leq \frac{2r(0)}{\epsilon}.$$

In the case $d = 1$, the expression

$$\pi_t \left(|\nabla \ln \left[\frac{p(t)}{\overline{p}(t)}\right] |^2\right) = \int_{R^d} \left|\frac{\partial}{\partial x} \ln \left[\frac{p(t,x)}{\overline{p}(t,x)}\right]\right|^2 p(t,x)\, dx$$

may be thought of as a kind of relative Fisher information between $p(t,\cdot)$ and $\overline{p}(t,\cdot)$. This should be compared to the similar expression in the dynamics of the relative entropy between the exact filter and the unconditioned filter density, as computed in Mayer-Wolf and Zakai (1984).

3 Proof of Theorem 2.1

To prove (7) we shall prove the stronger inequality

$$E^{\nu} \int_s^T |e(t)|^2\, dt + E^{\nu} \int_s^T \pi_t \left(\left|\sigma^{\dagger} \nabla \ln \left[\frac{p(t)}{\overline{p}(t)}\right]\right|^2\right) dt \leq 2E^{\nu}\left[r(s) - r(T)\right]. \tag{10}$$

for any $0 < s < T < \infty$. Using the supermartingale property and positivity of r, $E[r(s) - r(T)] \leq r(0)$. Thus, letting $s \downarrow 0$ and $T \uparrow \infty$ will lead to (7). Similarly, it will suffice to establish the stochastic integral representation (9) on any finite time interval. Hence, we fix $0 < s < T < \infty$ for the rest of this section and work on the interval $[s, T]$; then $\{\mathcal{Y}_t\}$ shall be defined as in (3), and "a.s." shall mean "almost-surely up to Q^{ν}-null sets of $\mathcal{Y}_T$."
Before beginning the technical details, let us indicate the formal argument leading to Theorem 2.1. To derive (9), first apply Ito's rule to $p(t,x) \ln \left[\frac{p(t,x)}{\overline{p}(t,x)}\right]$ and then integrate both sides with respect to x. The result is exactly (9) with t replaced by T, except that the term

$$\int_s^T \int_{R^d} L^* \left\{p(t,x) \ln \left[\frac{p(t,x)}{\overline{p}(t,x)}\right]\right\} dx\, dt$$

also appears. However, a formal integration by parts shows that this term is 0. To derive the inequality (7) rearrange terms in (9), assuming all terms are defined and finite. Then

$$r(t) + \frac{1}{2}\int_s^t |e(\theta)|^2\, d\theta + \frac{1}{2}\int_s^t \pi_\theta \left(\left|\sigma^{\dagger} \nabla \ln \left[\frac{p(\theta)}{\overline{p}(\theta)}\right]\right|^2\right) d\theta =$$

$$r(s) + \int_s^t \left[e(\theta) + \pi_\theta \left(h \ln \left[\frac{p(\theta)}{\overline{p}(\theta)}\right]\right) - r(\theta)\pi_\theta(h)\right] \cdot d\hat{Y}(\theta).$$

The left-hand side is thus a positive local martingale, hence a supermartingale. Since expectations of a supermartingale are decreasing, we obtain (10) upon taking expectations.
Our rigorous proof proceeds a bit in the opposite direction, in an effort to gain as much generality as possible. By employing cut-off functions to help with the interchange of orders of integration and integration by parts, and then passing to the limit, we first establish inequality (10). Then, under the additional assumption (8), we refine the limiting arguments to establish (9). The details are developed in the next two subsections.

3.1 Preliminary remarks

This section establishes some useful preliminary facts. Throughout the section, hypotheses (H1)-(H3) are in force and p and $\overline{p}$ are as in (H3). Also, we assume

$$r(0) < \infty \tag{11}$$

We have made no assumptions on the sample path regularity of r. In fact, once we derive (9), it will follow that that r is a.s. continuous. However, it is useful for the limiting arguments to establish as much a priori regularity as possible. First, we note

Lemma 3.1 $\int \overline{p}(t,x)\,dx = \int p(t,x)\,dx = 1$ *for all* $0 < t \le T$, *almost surely.*

From the a.s. continuity of $\overline{p}$ and Fatou's lemma it is not hard to show $\int \overline{p}(t,x)\,dx \le 1$ for *all* $0 < t \le T$, almost surely, and the same for p The stronger statement of the lemma is convenient, but not absolutely necessary to use. Actually, it is generally proved in theorems of existence for filtering densities, but it follows readily from (H3). The point is that p satisfies the Kushner-Stratonovich equation in weak form as well. Let ϕ_n be a sequence of non-negative, C^∞ functions of compact support increasing to the constant function 1. (ϕ_n can be defined as in the proof of Theorem 2.1 below.) Then we have $\langle p,\phi_n\rangle = \langle p_0,\phi_n\rangle + \int_0^t \langle p, L\phi_n\rangle\,d\theta + \int_0^t [\langle p, h\phi_n\rangle - \langle p,h\rangle\langle p,\phi_n\rangle]\cdot \hat{Y}(\theta)$ almost surely for all $t \le T$. Since p is the conditional density, and since the moment condition (H2) holds, we can take limits as $n \to \infty$, and pick a subsequence so that the integrals on the right-hand side converges uniformly to 0, almost surely. As a result $\langle p, 1\rangle = 1$ for all t, almost-surely. The exact same proof applies to $\overline{p}$. ◇

Next, we look at $r(t) = \int p(t,x)\ln\left[\frac{p(t,x)}{\overline{p}(t,x)}\right] dx$. If g is a function, g^+ denotes its positive part, as usual. Since $(\ln x)^+ < x$ for $x > 0$,

$$p(t,x)\left|\ln\left[\frac{p(t,x)}{\overline{p}(t,x)}\right]\right| = p(t,x)\left(\ln\left[\frac{p(t,x)}{\overline{p}(t,x)}\right]\right)^+ + p(t,x)\left(\ln\left[\frac{\overline{p}(t,x)}{p(t,x)}\right]\right)^+$$

$$\leq \; p(t,x)\left(\ln\left[\frac{p(t,x)}{\overline{p}(t,x)}\right]\right)^{+} + \overline{p}(t,x). \tag{12}$$

Hence

$$p(t,x)\left|\ln\left[\frac{p(t,x)}{\overline{p}(t,x)}\right]\right| \leq p(t,x)\ln\left[\frac{p(t,x)}{\overline{p}(t,x)}\right] + 2\overline{p}(t,x), \tag{13}$$

and, for any bounded $\psi(t,x)$,

$$\int p(t,x)\left|\psi(t,x)\ln\left[\frac{p(t,x)}{\overline{p}(t,x)}\right]\right| dx \leq \|\psi\|_{\infty}\,(r(t)+2) \quad \forall\, 0<t\leq T \quad \text{a.s.} \tag{14}$$

Lemma 3.2 *The function $t \to E^{\nu}[r(t)]$ is right-continuous. Hence, r admits a right-continuous modification $\tilde{r}$.*

Lemma 3.3 *For any $T > s > 0$, $\sup_{s\leq t\leq T} r(t) \leq \sup_{s\leq t\leq T} \tilde{r}(t)$ a.s, and*

$$Q^{\nu}\left(\sup_{s\leq t\leq T} \tilde{r}(t) < \infty\right) = Q^{\nu}\left(\sup_{s\leq t\leq T} r(t) < \infty\right) = 1.$$

Proof of Lemma 3.2: The process $r(t)$ is actually lower-semicontinuous a.s. To see this, assume $t_n \to t$ and recall

$$r(t_n) = \int p(t_n,x)\left(\ln\left[\frac{p(t_n,x)}{\overline{p}(t_n,x)}\right]\right)^{+} dx - \int p(t_n,x)\left(\ln\left[\frac{\overline{p}(t_n,x)}{p(t_n,x)}\right]\right)^{+} dx. \tag{15}$$

Fatou's lemma can be applied to the first term as $n \to \infty$. As for the second term, by applying Fatou to $\overline{p}(t_n,x) - p(t_n,x)\left(\ln\left[\frac{\overline{p}(t_n,x)}{p(t_n,x)}\right]\right)^{+}$, which we can do because of (12), and to $p(t_n,x)\left(\ln\left[\frac{\overline{p}(t_n,x)}{p(t_n,x)}\right]\right)^{+}$ itself, and using Lemma 3.1, we find that, almost surely, for all $t > 0$ and sequences $t_n \to t$,

$$\int p(t_n,x)\left(\ln\left[\frac{\overline{p}(t_n,x)}{p(t_n,x)}\right]\right)^{+} dx \to \int p(t,x)\left(\ln\left[\frac{\overline{p}(t,x)}{p(t,x)}\right]\right)^{+} dx.$$

As a result, taking limits in (15), shows that, almost surely, $r(t) \leq \lim r(t_n)$ as $t_n \to t$. Now $r(t)$ is a positive process, and so Fatou's lemma applies and shows that $t \to E^{\nu}[r(t)]$ is lower-semicontinuous. But $E^{\nu}[r(t)]$ is also non-increasing in t, because r is a supermartingale. Thus $E[r(t)]$ is right continuous in t. The existence of a right continuous version of r, given the right-continuity of $E[r(t)]$, is well-known; see, for example Karatzas and Shreve (1988). ◇

Proof of Lemma 3.3 The paths of a right-continuous sub or super martingales are a.s. bounded on compacts. From lower-semicontinuity of r and the fact that r and $\tilde{r}$ are modifications of each other, it follows that $r(t) \leq \tilde{r}(t)$ for all $s \leq t \leq T$, a.s., so the sample paths of r are a.s. bounded as well. ◇

3.2 Proof of Theorem 2.1

Let p and $\overline{p}$ be the solutions of the Kushner-Stratonovich equations provided by (H3). An application of Ito's rule show that

$$\begin{aligned}
p(t,x)\ln\left[\frac{p(t,x)}{\overline{p}(t,x)}\right] - p(s,x)\ln\left[\frac{p(s,x)}{\overline{p}(s,x)}\right] &= \int_s^t L^*\left(p(\theta,x)\ln\left[\frac{p(\theta,x)}{\overline{p}(\theta,x)}\right]\right)d\theta \\
&-\frac{1}{2}\int_s^t |e(\theta)|^2 p(\theta,x)\,d\theta - \frac{1}{2}\int_s^t \left|\sigma^\dagger(x)\nabla\ln\left[\frac{p(\theta,x)}{\overline{p}(\theta,x)}\right]\right|^2 p(\theta,x)\,d\theta \\
&+\int_s^t (e(\theta))\cdot(h(x)-\pi_\theta(h))\,p(\theta,x)\,d\theta \qquad (16)\\
&+\int_s^t p(\theta,x)\left[e(\theta)+\ln\left[\frac{p(\theta,x)}{\overline{p}(\theta,x)}\right](h(x)-\pi_\theta(h))\right]\cdot d\hat{Y}(\theta).
\end{aligned}$$

It is this formula that leads to (9) after a formal integration. The rigorous proof employs a cut-off function. Let ψ be a non-increasing, C^∞ function on R_+ such that $\psi(z) = 1$ if $0 \le 1 \le z$ and $\psi(z) = 0$ if $z \ge 2$. For $x \in R^d$, set $\phi_n(x) := \psi(|x|/n)$. Note that each ϕ_n is infinitely differentiable with compact support and that $\phi_n(x) \uparrow 1$ pointwise. Define

$$r_n(t) = \int \phi_n(x)p(t,x)\ln\left[\frac{p(t,x)}{\overline{p}(t,x)}\right]dx.$$

For future reference, we note several properties of r_n. First, the assumption of a.s. continuity in (H3) and the compact support of ϕ_n imply that $r_n(t)$ is continuous in t for each n almost-surely. Second, $\lim_{n\to\infty} r_n(t) = r(t)$ for all $0 \le t \le T$, a.s. Third, for any random time τ, (14) implies

$$|r_n(\tau)| \le r(\tau) + 2. \qquad (17)$$

Finally, using Lemma 3.1 and some obvious bounds

$$r_n(\tau) \ge -\int p(\tau,x)\left(\ln\left[\frac{\overline{p}(\tau,x)}{p(\tau,x)}\right]\right)^+ dx \ge -\int \overline{p}(\tau,x)\,dx = -1, \qquad (18)$$

for any random time τ, a.s.

We shall obtain a stochastic integral equation for r_n by multiplying (16) on both sides by ϕ_n and integrating. The equation is given below in (22). To obtain it, we interchange the order of integration throughout to bring the integrals in dx inside those in $d\theta$ and $d\hat{Y}(\theta)$. We briefly justify that this is allowed. For the terms involving the positive integrands on the right-hand side of (16), one need only apply Tonelli's theorem. To verify the interchange for the other ordinary integrals requires only checking the integrand is a.s. in $L^1(R_+ \times R^d)$. The term $\phi_n(x)L^*\left(p(\theta,x)\ln\left[\frac{p(\theta,x)}{\overline{p}(\theta,x)}\right]\right)$ is easy to handle because it is a.s. continuous by (H3), has compact support, and hence is a.s. bounded. When the interchange is carried out, one obtains $\int_s^t \langle \phi_n L^* \left(p(\theta) \ln \left[\frac{p(\theta)}{\overline{p}(\theta)} \right] \right) \rangle \, d\theta$, and, after integration by parts, this becomes

$$\int_s^t \langle L\phi_n, p(\theta) \ln \left[\frac{p(\theta)}{\overline{p}(\theta)} \right] \rangle \, d\theta.$$

For the fourth integral in (16), we observe the following, which will be useful later:

$$|[e(\theta)] \cdot [\pi_\theta (h\phi_n) - \pi_\theta (\phi_n) \pi_\theta (h)]| \leq 2\pi_\theta (|h|) \, |e(\theta)| \tag{19}$$

and

$$\begin{aligned} E^\nu \int_s^T 2\pi_\theta (|h|) \, |e(\theta)| \, d\theta \\ \leq \; 2 \left(E^\nu \left[\int_0^T |e(\theta)|^2 \, d\theta \right] E^\nu \left[\int_0^T \pi_\theta (|h|^2) \, d\theta \right] \right)^{1/2} < \infty. \end{aligned} \tag{20}$$

The left hand side of (19) is precisely the integrand that would appear after interchange of dx and $d\theta$ integrals, and (20) shows that it is in L^1.

Finally, we want to interchange dx and $d\hat{Y}(\theta)$ integrals in

$$\langle \phi_n, \int_s^t \rho(\theta, x) \cdot d\hat{Y}(\theta) \rangle,$$

where $\rho(\theta, x)$ is the integrand of the stochastic integral in (16). The stochastic Fubini theorem stated in Protter(1990), page 160, allows this to be done if

$$Q^\nu \left(\int_s^T \langle \phi_n, |\rho|^2(\theta, \cdot) \rangle \, d\theta < \infty \right) = 1. \tag{21}$$

But

$$\begin{aligned} \langle \phi_n, |\rho|^2(\theta, \cdot) \rangle \; \leq \; & 3 \Big[|e(\theta)|^2 \langle \phi_n, p^2(\theta) \rangle + \langle \phi_n, p^2(\theta) |h|^2 | \ln \left[\frac{p(\theta)}{\overline{p}(\theta)} \right] |^2 \rangle \\ & + \pi_\theta (|h|^2) \, \langle \phi_n, p^2(\theta) | \ln \left[\frac{p(\theta)}{\overline{p}(\theta)} \right] |^2 \rangle . \Big] \end{aligned}$$

The inner product terms are all bounded and continuous in t, a.s. because of (H3). This and the inequalities (4) of Proposition 2.1 and (6) imply (21). The end result of these interchanges and the integration by parts is the equality:

$$r_n(t) - r_n(s) = \int_s^t \langle L\phi_n, p(\theta, \cdot) \ln \left[\frac{p(\theta)}{\overline{p}(\theta)} \right] \rangle \, d\theta$$

$$-\frac{1}{2}\int_s^t \pi_\theta(\phi_n)\,|e(\theta)|^2\,d\theta$$
$$-\frac{1}{2}\int_s^t \pi_\theta\left(\phi_n\left|\sigma^\dagger\nabla\ln\left[\frac{p(\theta)}{\overline{p}(\theta)}\right]\right|^2\right)d\theta$$
$$+\int_s^t e(\theta)\cdot\left[\pi_\theta(h\phi_n)-\pi_\theta(\phi_n)\pi_\theta(h)\right]d\theta \tag{22}$$
$$+\int_s^t\left[\pi_\theta(\phi_n)e(\theta)+\pi_\theta\left(\phi_n h\ln\left[\frac{p(\theta)}{\overline{p}(\theta)}\right]\right)-\pi_\theta(h)\,r_n(\theta)\right]\cdot d\hat{Y}(\theta).$$

We shall complete the proof of (7) by taking expectations in (22) and then letting $n\to\infty$. The following observations are helpful. The first is elementary, the second uses (14) and the supermartingale property of r, and the third is a straightforward computation using the growth bounds on $f(x)\cdot x$ and $||\sigma(x)||$ in (H1). We have:

$$\left|\langle L\phi_n, p(\theta)\ln\left[\frac{p(\theta)}{\overline{p}(\theta)}\right]\rangle\right|\le\|L\phi_n\|_\infty\int p(\theta,x)\left|\ln\left[\frac{p(\theta,x)}{\overline{p}(\theta,x)}\right]\right|dx; \tag{23}$$

$$\begin{aligned}E^\nu\left[\int_0^T\int p(\theta,x)\left|\ln\left[\frac{p(\theta,x)}{\overline{p}(\theta,x)}\right]\right|dx\,d\theta\right] &\le E^\nu\left[\int_0^T(r(\theta)+2)\,d\theta\right]\\ &\le T(r(0)+2)\ <\ \infty.\end{aligned} \tag{24}$$

There is a constant M independent of n such that

$$|L\phi_n(x)|\le M\mathbf{1}_{\{n\le|x|\le 2n\}}\qquad\text{for all } x. \tag{25}$$

Now we proceed with the proof. Let $g_n(\theta)$ denote the integrand of the stochastic integral in (22). For each positive integer N and n, define the stopping time

$$\tau_n^N=\inf\left\{t>s;\int_s^t|g_n(\theta)|^2\,d\theta\ge N\right\}\wedge T.$$

The stochastic integral is a martingale up to this stopping time, and so, after some rearrangement of terms, taking expectations in (22) yields,

$$\begin{aligned}&\frac{1}{2}E^\nu\int_s^{\tau_n^N}\left[\pi_\theta(\phi_n)\,|e(\theta)|^2+\pi_\theta\left(\phi_n\left|\sigma^\dagger\nabla\ln\left[\frac{p(\theta)}{\overline{p}(\theta)}\right]\right|^2\right)\right]d\theta+E^\nu[r_n(\tau_n^N)]\\ &\quad=E^\nu\left[r_n(s)\right]+E^\nu\int_s^{\tau_n^N}\langle L\phi_n, p(\theta,\cdot)\ln\left[\frac{p(\theta,\cdot)}{\overline{p}(\theta,\cdot)}\right]\rangle\,d\theta\\ &\quad+E^\nu\int_s^{\tau_n^N}e(\theta)\cdot\left[\pi_\theta(h\phi_n)-\pi_\theta(\phi_n)\pi_\theta(h)\right]d\theta.\end{aligned} \tag{26}$$

Now let $N \to \infty$. For each n, the sequence τ_n^N is increasing in N, and $Q^\nu\left(\cup_{N\geq 1}\{\tau_n^N = T\}\right) = 1$. Thus $r_n(\tau_n^N) \to r_n(T)$ as $N \to \infty$ a.s. Because of (18), Fatou's lemma applies to $E^\nu[r_n(\tau_n^N)]$. Also, monotone convergence applies to the other terms on the left-hand side of (26). At the same time, (23), (20), and (24) show that dominated convergence applies to the right-hand side of (26). Hence, after $N \to \infty$, one obtains

$$\frac{1}{2}E^\nu \int_s^T \left[\pi_\theta(\phi_n)\,|e(\theta)|^2 + \pi_\theta\left(\phi_n\left|\sigma^\dagger\nabla\ln\left[\frac{p(\theta)}{\overline{p}(\theta)}\right]\right|^2\right)\right] d\theta + E^\nu[r_n(T)]$$

$$\leq E^\nu[r_n(s)] + E^\nu \int_s^T \langle L\phi_n, p(\theta,\cdot)\ln\left[\frac{p(\theta)}{\overline{p}(\theta)}\right]\rangle\, d\theta$$

$$+E^\nu \int_s^T e(\theta)\cdot\left[\pi_\theta(h\phi_n) - \pi_\theta(\phi_n)\,\pi_\theta(h)\right] d\theta. \tag{27}$$

Now take limits as $n \to \infty$. The integrands on the right-hand side tend to 0 pointwise and are uniformly dominated or increasing as we have seen in (17), (19) and in (25). Also $r_n(T)$ and $r_n(s)$ tend to $r(T)$ and $r(s)$ a.s. and are uniformly dominated because of (17). Thus, (10) is obtained in the limit, completing the proof of (7).

Assume now that the condition (8) stated in Theorem 2.1 holds. We will show that the stochastic integral equation (9) is true by taking limits in (22) as $n \to \infty$. Note first that (22) is true for all $s \leq t \leq T$, a.s., by viture of the a.s. continuity of r_n, if we choose to work with the a.s. continuous versions of the stochastic integrals.

The bounds previously developed show that the ordinary integral terms of (22) converge uniformly for $t \in [s,T]$ to the ordinary integral terms of (9), almost surely. Also we know that $r_n(t)$ converges to $r(t)$ for all $t \in [s,T]$, almost surely. We thus consider the stochastic integral terms. Let $g_n(\theta)$, as above, denote the integrand of the stochastic integral in (22), and let $g(\theta)$ denote the integrand of the stochastic integral in (9). To finish the proof, it suffices to show that

$$\int_s^T |g_n(\theta) - g(\theta)|^2\, d\theta \to 0 \text{ in } Q^\nu\text{-probability} \tag{28}$$

as $n \to \infty$, because this will imply that $\sup_{s\leq t\leq T}|\int_0^T [g(\theta) - g_n(\theta)]\cdot d\hat{Y}(\theta)|$ also converges to 0 in probability as $n \to \infty$; (see for example Protter(1990)). Choosing an a.s. converging subsequence we therefore recover (9) from (22) as we let $n \to \infty$. (In fact, we find that (9) holds for all $s \leq t \leq T$ a.s., and so r is a.s. continuous.) To prove (28) note that

$$\int_s^T |g_n(\theta) - g(\theta)|^2\, d\theta \leq 3\int_s^T \Big[|e(\theta)|^2\pi_\theta(1-\phi_n) +$$

$$+\,\pi_\theta^2(|(1-\phi_n)h\ln\left[\frac{p(\theta)}{\overline{p}(\theta)}\right]| + \pi_\theta(|h|^2)\,|r_n(\theta) - r(\theta)|^2\Big]\, d\theta.$$

As $n \to \infty$ the first two terms tend to 0 by dominated convergence, the error bound (4), and assumption (8). For the last term, one can apply dominated convergence again because $r_n(\theta) \to r(\theta)$ for all θ a.s.,

$$\sup_{s \le \theta \le T} |r_n(\theta)| \le 2 + \sup_{s \le \theta \le T} r(\theta) < \infty$$

almost-surely, and because $E^{\nu}[\int_0^T \pi_\theta\left(|h|^2\right) d\theta] < \infty$.
Finally, assume h is bounded. Then

$$\pi_\theta\left(|h \ln\left[\frac{p(\theta)}{\overline{p}(\theta)}\right]|\right) \le |h|_\infty[r(\theta) + 2],$$

and so Lemma 3.3 implies condition (8). Hence (9) holds for bounded h.

4 Examples

In this section we state various conditons on the coefficients f, g, and h, and on the initial laws ν and $\overline{\nu}$ for which hypothesis (H1)-(H3) hold and to which Theorem 2.1 therefore applies. These conditions are drawn from the literature. We do not aim to give the best possible results.
Throughout the section, we assume that ν and $\overline{\nu}$ admit densities $p_0(x)$ and $\overline{p}_0(x)$, respectively. The notation C_b^m is used to denote the space of functions on R^d with m continuous and bounded derivatives, while H^m is used for the Sobolev space of functions with m generalized derivatives in L^2. We shall need the non-degeneracy assumption: there exist constants $0 < \lambda < \Lambda$ such that

- σ is $d \times d$, symmetric, invertible and $\lambda I \le \sigma\sigma^\dagger \le \Lambda \mathrm{I}$, $\forall x$. (A)

Theorem 4.1 *Assume that f, σ, and h are in C_b^m for some $m > d/2+2$, that p_0 and $\overline{p}_0$ are in H^m, and that (A) is true. Then (H1)-(H3) hold. If $r(0) < \infty$, both the entropy inequality (7) and the Ito equation (9) are valid in this case.*

Proof: (H1) and (H2) are immediate from the regularity and boundedness of the coefficients. Notice also that (8) follows from the boundedness of h once $r(0) < \infty$. The existence of classical solutions p and $\overline{p}$ to the Kushner-Stratonovich equations is a consequence of Corollary 6.2.4, page 237, in Rozovskii (1990), where it is shown also that these solutions are in $C_b^{0,2}([0,T] \times R^d)$, almost-surely. Thus, to prove (H3), it remains only to show the a.s. strict positivity of solutions. It will be shown that for any $0 < s < T$, $p(t,x) > 0$ for all $s \le t \le T$ and for all x, almost surely. Because of the a.s. continuity of p in (t,x), this is enough to imply the slightly stronger positivity statement in (H3). The proof for $\overline{p}$ is precisely the same.

We use the construction of p in Rozovskii (1990) (see Corollary 6.2.1) as

$$p(t,x) = \frac{u(t,x)}{\int u(t,x)\,dx},$$

where u is a classical solution of the Zakai equation,

$$u(t,x) = p_0(x) + \int_0^t L^* u(\theta,x)\,d\theta + \int_0^t u(\theta,x)h(x)\cdot dY(\theta).$$

Consider $v(t,x) = e^{-h(x)\cdot Y(t)}u(t,x)$. This is the well-known transformation for robustifying the Zakai equation. By applying Ito's rule, one finds that, up to a Q^ν-null set of $\mathcal{Y}_T$,

$$v(t,x) = p_0(x) + \int_0^t A_\theta v(\theta,x)\,dx \quad \text{for all } (t,x) \in [0,T]\times R^d, \tag{29}$$

where

$$\begin{aligned} A_\theta &= \frac{1}{2}\sum a_{i,j}(x)\frac{\partial^2}{\partial x_i \partial x_j} + \sum b_i(\theta,x)\frac{\partial}{\partial x_i} + c(\theta,x) \\ \text{with } b(\theta,x) &= a(x)D^\dagger h(x)Y(\theta) + \nabla\cdot a(x) - f(x) \\ \text{and } c(\theta,x) &= -\frac{1}{2}|h(x)|^2 + \frac{1}{2}\left|Y^\dagger(\theta)Dh(x)\sigma(x)\right|^2 \\ &\quad + \frac{1}{2}\sum Y_k(\theta)\mathrm{tr}[a(x)D^2h_k(x)] + (\nabla\cdot a(x))D^\dagger h(x)Y(\theta) \\ &\quad + \frac{1}{2}\mathrm{tr}\nabla^2 a(x) - \nabla\cdot f(x) - Y^\dagger(\theta)Dh(x)f(x). \end{aligned}$$

Here Dh and D^2h_k are the matrices of first and second partial derivatives of h.
Fix Y such that (29) is valid and choose a time t. The solution v to (29) has a Feynman-Kac representation, which we state now. Let $\tilde{B}$ be a Brownian motion, and let Z solve the stochastic differential equation

$$dZ_x(\theta) = \sigma(Z_x(\theta))\,d\tilde{B}(\theta), \quad Z_x(0) = x. \tag{30}$$

Then

$$v(t,x) = E\left[p_0(Z_x(t))L_t e^{\{\int_0^t c(t-\theta, Z_x(\theta))\,d\theta\}}\right], \tag{31}$$

where L_t is the Girsanov transformation factor

$$\exp\left\{\int_0^t \sigma^{-1}b(t-\theta, Z_x(\theta))\,d\tilde{B}(\theta) - \frac{1}{2}\int_0^t \left|\sigma^{-1}b(t-\theta,Z_x(\theta))\right|^2 d\theta\right\}.$$

(In (31), the expectation is taken on a probability space supporting the Brownian motion $\tilde{B}$ and the associated process Z.) By uniform ellipticity of $a = \sigma\sigma^\dagger$, the support of the density of $Z_x(t)$ is all of R^d and so $v(t,x)$ is clearly positive for all (t,x) when $t > 0$. ◇

The representation (31) of the solution to the robust filtering equation is well-known and used, for example, in Picard (1986a) and (1986b) and in Sheu (1983). In fact it extends to filtering models in which the coefficients are unbounded. The following is a slight extension of the results of Sheu (1983).

Theorem 4.2 *Assume (A). Assume also that $\sigma \in C_b^3$ and that f satisfies the assumptions in (H1). Assume in addition that $f, h \in C^3$, that their derivatives up to order 3 admit at most polynomial growth, and that, in particular, for some $m \geq 1$*

- *there are constants c_1, a_1 and c_2, a_2 such that $c_1|x|^{2m} + c_2 \leq |h(x)|^2 \leq a_1|x|^{2m} + a_2$;*
- *$|f(x)| \leq c(1 + |x|^m)$ for some constant c;*
- *All the functions of x in the expression c defined above, except the term $|h(x)|^2$ grow at a polynomial rate of degree strictly less than $2m$.*

Assume also that p_0 and $\overline{p}_0$ are continuous and that

$$\int |x|^k p_0(x)\,dx < \infty \quad \text{and} \quad \int |x|^k \overline{p}_0(x)\,dx < \infty \quad \forall k \geq 1.$$

Then (H1)-(H3) hold and the entropy inequality (7) is valid.
If, in addition, there are constants C and l such that

$$|\ln p_0(x)| + |\ln \overline{p}_0(x)| \leq C(1 + |x|^l), \tag{32}$$

then $r(t)$ satisfies the stochastic Ito equation (9).

Proof (Sketch): Let $||U||_T$ denote the sup norm, $||U||_T = \sup_{[0,T]} |U(t)|$. Hypothesis (H1) and the assumption that the initial densities have moments of all orders imply that $E^\nu\left[||X||_T^p\right] < \infty$ and $E^{\overline{\nu}}\left[||X||_T^p\right] < \infty$ for all positive powers p. Since h admits at most polynomial growth, (H2) easily follows. In fact

$$E^\nu\left[\int_0^T |X(t)|^p\,dt\right] < \infty \quad \text{for all } p \geq 1, \tag{33}$$

and similarly for $E^{\overline{\nu}}$. Also, $E^\nu\left[||Y||_T^p\right] < \infty$ for any positive p.

The conditions on the coeffficents that are imposed in the statement of the theorem are essentially a special case of those in Sheu (1983). In his filtering result, Sheu assumes that the drift f is bounded, but the arguments extend to the case in which f satisfies the conditions on f assumed here. Sheu's result says that if v is defined by (31), it is a classical solution of 29 and

$$p(t,x) = \frac{e^{h(x)\cdot Y(t)}v(t,x)}{\int e^{h(x)\cdot Y(t)}v(t,x)\,dx}$$

is a version of the conditional density $d\pi_t/dx$. Likewise, if $\bar{v}(t,x)$ is defined by (31), but with p_0 replaced by $\bar{p}_0$, and if $\bar{p}(t,x)$ is defined in a similar manner using $\bar{v}$, $\bar{p}$ gives a version of $d\bar{\pi}_t/dx$. These solutions clearly satisfy the positivity condition in hypothesis (H3). To understand the hypotheses, note that they are constructed to guarantee $c(\theta,x)$ is dominated by a term of the form $-c|x|^{2m}$; the existence and regularity properties of v follow from exploitation of this fact.
We next discuss the validity of (8) when condition (32) holds. First we note that the logarithms of the normalizing factors are

$$\begin{aligned}
\ln \int e^{h(x)\cdot Y(t)}v(t,x)\,dx &= \int_0^t \pi_\theta(h)\cdot dY(\theta) - (1/2)\int_0^t |\pi_\theta(h)|^2\,d\theta \\
\ln \int e^{h(x)\cdot Y(t)}\bar{v}(t,x)\,dx &= \int_0^t \bar{\pi}_\theta(h)\cdot dY(\theta) - (1/2)\int_0^t |\bar{\pi}_\theta(h)|^2\,d\theta
\end{aligned}$$

One therefore obtains

$$\ln\left[\frac{p(t,x)}{\bar{p}(t,x)}\right] = \ln\left[\frac{v(t,x)}{\bar{v}(t,x)}\right] + \int_0^t e(\theta)\cdot d\hat{Y}(\theta) - (1/2)\int_0^t |e(\theta)|^2\,d\theta.$$

We shall show that $|\ln v(t,x)|$ and $|\ln \bar{v}(t,x)|$ admit at most polynomial growth. In light of the assumption that $|h(x)|$ admits at most polynomial growth and of the moment bounds in (33), we shall therefore obtain (8) with $s=0$. First, by employing the technique of Theorem 2 in Sheu (1984), keeping track of the dependence on Y, one can show that there are constants α and K depending only on d, m, $\|p_0\|_\infty$, $\|\bar{p}_0\|_\infty$, and the coeficients f, σ, and h, such that

$$v(t,x)+\bar{v}(t,x) \leq \exp\left\{-t\alpha|x|^{m+1} + K(1+\|Y\|_T^4)\right\}.$$

This bound does not require (32), and it shows that $\ln v(t,x)$ and $\ln \bar{v}(t,x)$ are both bounded above by a term of the form $K(1+\|Y\|_T^4)$. We need also a lower bound. For this, apply Jensen's inequality to obtain

$$-\ln v(t,x) \leq E\left[-\ln p_0(Z_x(t)) - \ln L_t - \int_0^t c(t-\theta, Z_x(\theta))\,d\theta\right].$$

Now standard estimates imply that $E[\|Z_x(t)\|_T^p] \leq K_p(1+|x|^p)$ for all positive p. From the explicit expressions for c, $\ln L_t$ and the polynomial bounds on $|\ln p_0(x)|$ it is straightforward to show that there are constants K and k such that

$$-\ln v(t,x) \leq K(1+\|Y\|_T^2)(1+|x|^k)$$

A similar estimate holds for $-\ln \bar{v}(t,x)$. Putting these estimates together with the lower bounds,gives

$$|\ln v(t,x)| \vee |\ln \bar{v}(t,x)| \leq K(1+\|Y\|_T^4)(1+|x|^k)$$

for some (possibly different) constant K. Hence

$$\begin{aligned}\pi_t(|h|^2 \left|\ln\left[\frac{p(t)}{\overline{p}(t)}\right]\right|) &\leq 2K(1+\|Y\|_T^4)^2\pi_t(1+|x|^{2k+2m}) \\ &+2\pi_t(|h|^2)\left|\int_0^t e(\theta)\cdot d\hat{Y}(\theta) - (1/2)\int_0^t |e(\theta)|^2\, d\theta\right|^2.\end{aligned}$$

Using (33), Hölder's inequality, and the boundedness of all positive moments of of $\|X\|_T$ and $\|Y\|_T$, one obtains

$$E^\nu\left[\int_0^T \pi_t(|h|^2|\ln\left[\frac{p(t,\cdot)}{\overline{p}(t,\cdot)}\right]|^2)\,dt\right] < \infty.$$

Then condition (8) follows immediately.

5 Acknowledgements

The observation that relative entropy of filters is a supermartingale is due to J.M.C. Clark. I thank him and Etienne Pardoux for discussions on the topic of this paper.

References

R. Atar, O. Zeitouni (1994). Lyapunov exponents for finite state nonlinear filtering, Technion Department of Electrical Engineering Publication 930, 1994.

R. Atar, O. Zeitouni (1997). Exponential stability for nonlinear filtering, *Annales de l'Institut H. Poincare, Probabilités et Statistiques* **33**.

R. Atar (1997). Exponential stability for nonlinear filtering of diffusion processes in a non-compact domain, to appear in *Annals of Probability.*

A. Budhiraja, D. Ocone (1997). Exponential stability of discret-time filters for bounded observation noise, *Systems & Control Letters* **30**, 185-193.

A. Budhiraja, H.J. Kushner (1997) Robustness of nonlinear filters over the infinite time interval, *SIAM Journal on Control and Optimization*, to appear.

B.Z. Bobrovsky, M. Zakai (1976). A lower bound on the estimation of certain diffusion processes, *IEEE Trans. Infor. Theory* **IT-22**, 45-52.

R.S. Bucy (1979). Information and filtering, *Information Sciences* **18**, 179-187.

F. Cerou (1994) Long time asymptotics for some dynamical noise free non linear filtering problems, preprint.

J.M.C. Clark, D. Ocone, C. Coumarbatch (1997). Relative entropy and filtering bounds for filtering of Markov processes, preprint.

B. Delyon, O. Zeitouni (1991). Lyapunov exponents for filtering problems, in *Applied Stochastic Analysis*, M.H.A. Davis and R.J. Elliott, eds., Gordon and Breach, New York, 511-535.

T. Duncan (1970). On the calculation of mutual information,*SIAM Journal of Applied Mathematics* **19**, 215-220

I. Karatzas, S. Shreve (1988). Brownian Motion and Stochastic Calculus, Springer-Verlag, New York.

E. Mayer-Wolf, M. Zakai (1984), On a formula relating the Shannon to the Fisher information for the filtering problem, *Lecture Notes in Control and Information Sciences*, Vol. **61**, Springer-Verlag, New York, 164-171.

D. Ocone, E. Pardoux (1996). Asymptotic stability of the optimal filter with respect to its initial condition, *SIAM Journal on Control and Optimization* **34**, 226-243.

D. Ocone (1996) Asymptotic Stability of Beneš filters, preprint.

J. Picard (1986a). Filtrage de diffusions vectorielles faiblement bruitées, in *Analysis and Optimization of Systems*, Lecture Notes in Control and Information Science **83**, Springer-Verlag, New York.

J. Picard (1986b). Nonlinear filtering of one-dimensional diffusions in the case of a high signal-to-noise ratio, *SIAM Journal of Applied Mathematics* **46**, 1098-1125.

P. Protter (1990). Stochastic Integration and Differential Equations, Springer-Verlag, New York.

B.L. Rozovskii (1990). Stochastic Evolution Systems, Kluwer Academic Publishers, Boston.

S-J. Sheu (1983). Solution of certain parabolic equations with unbounded coefficients and its application to nonlinear filtering, *Stochastics* **10**, 31-46.

29

Identification for Linear Stochastic Distributed Parameter Systems with Boundary/Point Control

B. Pasik-Duncan

ABSTRACT: Some conditions are given for consistency of a family of least squares estimates of some unknown parameters for a linear stochastic distributed parameter system. The distributed parameter system is described by an analytic semigroup with cylindrical white noise and control that occurs only on the boundary or at discrete points. The consistency of the family of estimates is obtained by relating it to an infinite time horizon control problem and studying the asymptotics of a stationary Riccati equation for the control problem.

KEYWORDS: identification, least squares estimates, stochastic linear distributed parametric systems, consistency of least squares.

1 Introduction

An importatnt family of models for linear distributed parameter systems is the family that is described by analytic semigroups. For controlled linear distributed parameter systems it is often natural to restrict the control to the boundary or to discrete points in the domain. In modeling physical phenomena by these systems, some parameters of the system are typically unknown so it is important to investigate parameter estimation problems for these systems.

In this paper, the consistency of a family of least squares estimates of some unknown parameters for a linear stochastic distributed system with boundary/point control is verified. The consistency is weak or strong depending on the assumptions. The approach is to use the asymptotics of a stationary Riccati equation for a naturally associated control problem. This approach verifies a persistent excitation condition. The least squares estimation scheme does not require switchings or additional excitation. This approach generalizes the results in Pasik-Duncan (1992) and can be used for adaptive control problems such as in Duncan et al. (1994, 1996).

2 Preliminaries

The unknown linear stochastic distributed parameter system with boundary or point control is formally described by the following stochastic differential equation

$$dX(t;\alpha) = (A_0 + A_1(\alpha))X(t;\alpha)dt + BU(t)dt + Q^{1/2}dW(t)$$
$$X(0;\alpha) = x \tag{1}$$

where $X(t;\alpha) \in H$, H is a real, separable Hilbert space, $(W(t),\, t \geq 0)$ is a cylindrical standard Wiener process on H, $Q^{1/2} \in \mathcal{L}(H)$, $\alpha \in \mathcal{A} \subset \mathbb{R}^q$, $\mathcal{A}$ is bounded, $t \geq 0$, $A_0 = A_0^*$ is a infinitesimal generator of an exponentially stable analytic semigroup and $A_1(\alpha)$ is a linear operator of the form (5). Let the filtered probability space be denoted $(\Omega, \mathcal{F}, (\mathcal{F}_t), P)$. Let D_A^γ for $\gamma \in \mathbb{R}$ be the domain of the fractional power $(-A_0)^\gamma$ with the $(-A_0)^\gamma$ graph norm topology. Let $B \in \mathcal{L}(H_1, D_A^{\varepsilon-1})$ where H_1 is a real, separable Hilbert space for some $\varepsilon \in (0,1)$ and $A_1^*(\alpha) \in \mathcal{L}(D_A^\eta, H)$ for some $\eta \in [0,1)$ and all $\alpha \in \mathcal{A}$. For $p \geq 0$, let $M^p(H_1) = \cap_{T>0} M^p(0,T,H_1)$ where

$$M^p(0,T,H_1) = \left\{ U \,|U : [0,T] \times \Omega \to H_1,\ (U(t),\, t \geq 0) \text{ is } (\mathcal{F}_t)\text{-adapted} \right.$$
$$\left. \text{and } E\int_0^T |U(t)|^p dt < \infty \right\}. \tag{2}$$

The control process $(U(t),\, t \geq 0)$ in (1) is assumed to belong to the space $M^p(H_1)$ for some fixed $p > \max(1/\varepsilon, 1/(1-\eta))$ and $p \geq 2$.
The following assumptions are used subsequently.

(A1) $(-A_0)^{-\delta}Q^{1/2}$ is a Hilbert-Schmidt operator for some $\delta \in [0, 1/2)$.

(A2) There is a finite dimensional projection $\tilde{P} : H \to \mathrm{Dom}(A_0)$ such that $(\tilde{P}A_{1i},\, i = 1, \ldots, q)$ is linearly independent.

If (A1) is satisfied and $U \in M^p(H_1)$ then it is known (Proposition 1.1 in Duncan et al. (1996)) that (1) has one and only one mild solution with H-valued continuous sample paths.
To verify the consistency for a family of least squares estimates, some control theory methods are used. Consider the infinite time horizon deterministic control problem where the system is given by (1) with $Q^{1/2} = 0$ and the quadratic cost functional

$$J(x,u,\mu,c) = \int_0^\infty [\langle \tilde{C}(\mu)x(s), \tilde{C}(\mu)x(s)\rangle + c\langle u(s), u(0)\rangle]ds \tag{3}$$

where $c > 0$, $\mu \in \mathbb{R}^q$,

$$\tilde{C}(\mu) = \sum_{i=1}^q \mu_i \tilde{P} A_{1i}, \tag{4}$$

where $\tilde{P}$ is given in (A2), and

$$A_1(\alpha) = \sum_{i=0}^{q} \alpha_i A_{1i} \tag{5}$$

where $A_{1i}^* \in \mathcal{L}(D_A^\eta, H)$. The family of admissible controls is $L^2(0, \infty, H)$. If $(y_1, \ldots, y_k)$ is an orthonormal basis for the range of $\tilde{P}$ then

$$\langle \tilde{C}(\mu)x, \tilde{C}(\mu)x \rangle = \sum_{i=1}^{k} |\langle \tilde{C}^*(\mu)y_i, x \rangle|^2$$

for each $x \in H$ so $\tilde{C}^*(\mu)\tilde{C}(\mu)$ is a symmetric bounded linear operator on H for each $\mu \in \mathbb{R}^q$. It is known (Flandoli (1987) and Lasiecka and Triggiani (1987)) that the infinite time horizon deterministic control problem given by (1) with $Q^{1/2} = 0$ and (3) is well defined and the optimal cost and an optimal feedback control are determined from a weak or equivalently mild solution of the stationary Riccati equation

$$A^*(\alpha)V + VA(\alpha) - c^{-1}VBB^*V + \tilde{C}^*(\mu)\tilde{C}(\mu) = 0 \tag{6}$$

where

$$A(\alpha) = A_0 + A_1(\alpha).$$

To indicate the dependence of V on some terms it is also expressed as $V(\alpha, \mu, c)$. V is the unique, nonnegative, self-adjoint solution of (6) and $V \in \mathcal{L}(H, D_A^{1-\epsilon})$ (Flandoli (1987) and Lasiecka and Triggiani (1987)) where (6) is understood as the inner product equation

$$\langle A(\alpha), Vy \rangle + \langle Vx, A(\alpha)y \rangle - c^{-1}\langle B^*Vx, B^*Vy \rangle + \langle \tilde{C}(\mu)x, \tilde{C}(\mu)y \rangle = 0 \tag{7}$$

for $x, y \in \mathrm{Dom}(A_0)$.

3 Main Results

The estimate of the unknown parameter vector at time t, $\hat{\alpha}(t)$, is a least squares estimates using $\tilde{P}$, that is, it is the minimizer of the functional, $L(t, \alpha)$, with respect to α where

$$\begin{aligned} L(t, \alpha) = & -\int_0^t \langle \tilde{P}A(\alpha)X(s) + \tilde{P}BU(s), d\tilde{P}Q^{1/2}dW(s) \rangle \\ & + \frac{1}{2}\int_0^t |\tilde{P}A(\alpha)X(s) + \tilde{P}BU(s)|^2 ds \end{aligned} \tag{8}$$

The minimizer $\hat{\alpha}(t)$ of (8) satisfies the family of equations

$$\sum_{j=0}^{q}\int_0^t \langle \tilde{P}A_{1i}X(s), \tilde{P}A_{1i}X(s)\rangle ds(\hat{\alpha}_j(t)-\alpha_j^0) = \int_0^t \langle \tilde{P}A_{1i}X(s), d\tilde{P}Q^{1/2}W(s)\rangle \tag{9}$$

for $i = 1, \ldots, q$ where $\hat{\alpha}(t) = (\alpha_1(t), \ldots, \alpha_q(t))$ and $\alpha^0 = (\alpha_1^0, \ldots, \alpha_q^0)$ is the true parameter vector. Multiply both sides of (9) by $\hat{\alpha}_i(t) - \alpha_i^0$ and sum over i to obtain a quadratic form in $\hat{\alpha}(t) - \alpha^0$ as

$$\sum_{i,j=0}^{q}\int_0^t \langle \tilde{P}A_{1i}X(s), \tilde{P}A_{1j}X(s)\rangle ds(\hat{\alpha}_i(t)-\alpha_i^0)(\hat{\alpha}_j(t)-\alpha_j^0) = \sum_{i=0}^{q}\int_0^t \langle \tilde{P}A_{1i}X(s), d\tilde{P}Q^{1/2}W(s)\rangle(\hat{\alpha}_i(t)-\alpha_i^0). \tag{10}$$

To verify the consistency of the family of least squares estimates, $(\hat{\alpha}(t), t > 0)$, it is important to verify an asymptotic property of the solution $V(\alpha^0, \mu, c)$ as $c \downarrow 0$.

Proposition 1 *If* (A1) *and* (A2) *are satisfied then there are real numbers* $r > 0$ *and* $\delta > 0$ *such that if* $c \in (0, \delta)$ *then*

$$\inf_{|\mu|=1} \operatorname{tr}((-A_0)^{\delta} V(\alpha^0, \mu, c) Q(-A_0)^{-\delta}) \geq r c^{1/2} \tag{11}$$

where V *is the solution of* (6) *and* $\alpha = \alpha^0$ *in* (1).

Proof. Let $\bar{V}(\mu, c)$ be defined as

$$\bar{V}(\mu, c) = c^{-1} V(\alpha^0, \mu, c) \tag{12}$$

where $C > 0$ and α^0 is the true parameter vector. Since V is the solution to (6), it follows that $\bar{V}$ is the solution to

$$A^*(\alpha^0)\bar{V} + \bar{V}A(\alpha^0) - \bar{V}B^*B\bar{V} + c^{-1}\tilde{C}^*(\mu)\tilde{C}(\mu) = 0. \tag{13}$$

For $x \in \mathrm{Dom}(A_0)$ it follows from $|a-b| \leq 2(|a|^2 + |b|^2)$ that

$$\langle x, (B\bar{V} - A(\alpha^0))^*(B\bar{V} - A(\alpha^0))x\rangle \leq \langle x, 2(\bar{V}B^*BV + A^*(\alpha^0)A(\alpha^0))x\rangle$$

and using (13) it follows that

$$\langle 2Q\bar{V}\bar{V}Qx, x\rangle \geq -\langle QA^*(\alpha^0)A(\alpha^0)Qx, x\rangle + c^{-1}\langle Q\tilde{C}^*(\mu)\tilde{C}(\mu)Qx, x\rangle. \tag{14}$$

There is a $\tilde{\delta} > 0$ and a family of orthonormal elements $(e_j, j = 1, \ldots, k)$ in $\mathrm{Dom}(A_0)$ such that

$$\inf_{|\mu|=1} \sum_{j=1}^{k} \langle Q\tilde{C}^*(\mu)\tilde{C}(\mu)Qe_j, e_j\rangle > \tilde{\delta}. \tag{15}$$

Using (14) it follows that

$$\begin{aligned}(\mathrm{tr}((-A_0)^{\delta}\bar{V}Q(-A_0)^{-\delta}))^2 &\geq \sum_{j=1}^{k} \langle \bar{V}Qe_j, \bar{V}Qe_j\rangle \\ &\geq -\sum_{j=1}^{k} \langle A(\alpha^0)Qe_j, A(\alpha^0)Qe_j\rangle \\ &\quad + c^{-1}\sum_{j=1}^{k} \langle \tilde{C}(\mu)Qe_j, \tilde{C}(\mu)Qe_j\rangle. \end{aligned} \tag{16}$$

By the inequality (16) and the definition of $\bar{V}(\mu, c)$ it follows that there are real numbers $r > 0$ and $\delta > 0$ such that if $c \in (0, \delta)$ then

$$\inf_{|\mu|=1} \mathrm{tr}((-A_0)^{\delta}V(\alpha^0, \mu, c)Q(-A_0)^{-\delta}) \geq rc^{1/2} \tag{17}$$

□

The lower bound on the trace of the family of solutions of the Riccati equations in the Proposition is used to verify the consistency of a family of least squares estimates as described in the following theorem.

Theorem 1 *Let* $(X(t;\alpha^0), t \geq 0)$ *be the mild solution* (1) *with a control* $(U(t), t \geq 0)$ *from* $M^p(H)$. *Assume that* (A1) *and* (A2) *are satisfied. If the family of random variables*

$$\frac{1}{T}\int_0^T \langle X(t), X(t)\rangle + \langle U(t), U(t)\rangle dt \tag{18}$$

for $T \geq 1$ *is bounded in probability (resp. almost surely) and*

$$\lim_{T\to\infty} \frac{1}{T}\langle X(T), X(T)\rangle = 0 \tag{19}$$

in probability (resp. almost surely), then

$$\lim_{T\to\infty} \hat{\alpha}(T) = \alpha^0 \tag{20}$$

in probability (resp. almost surely) where $(\hat{\alpha}(T), T > 0)$ *is the family of least squares estimates from* (9) *and* α^0 *is the true parameter vector.*

Proof. Assume that (19) and the boundedness of the family (18) are satisfied "almost surely" because the verification "in probability" proceeds in a similar manner. Let $V(\alpha^0, \mu, c)$ be the nonnegative solution of (6). Since V satisfies the stationary Riccati equation (6), it follows by an Itô formula (Corollary 3.5 in Duncan et al. (1994)) that

$$\begin{aligned}\langle V(\alpha^0,\mu,c)X(t),X(t)\rangle &- \langle V(\alpha^0,\mu,c)X(0),X(0)\rangle \\ &= 2\int_0^t \langle V(\alpha^0,\mu,c)X(s), A(\alpha^0)X(s)+BU(s)\rangle ds \\ &+ 2\int_0^t \langle V(\alpha^0,\mu,c)X(s), Q^{1/2}dW(s)\rangle \\ &+ \int_0^t \operatorname{tr}((-A_0)^\delta V(\alpha^0,\mu,c)Q(-A_0)^{-\delta})ds. \end{aligned} \tag{21}$$

Since V determines the optimal cost and an optimal control for (1, 3) where $Q^{1/2} = 0$, it satisfies the stationary Hamilton-Jacobi-Bellman equation, that is, for $z \in \mathrm{Dom}(A_0)$

$$\inf_{u\in H_1} [2\langle V(\alpha^0,\mu,c)(A(\alpha^0)+Bu)z, z\rangle + \langle \tilde{C}(\mu)z, \tilde{C}(\mu)z\rangle + c\langle u,u\rangle] = 0. \tag{22}$$

Using (21) and (22) it follows that

$$\begin{aligned}\langle V(\alpha^0,\mu,c)X(t),X(t)\rangle &- \langle V(\alpha^0,\mu,c)X(0),X(0)\rangle \\ &\geq -\int_0^t [\langle \tilde{C}(\mu)X(s), \tilde{C}(\mu)X(s)\rangle + c\langle U(s),U(s)\rangle]ds \\ &+ 2\int_0^t \langle V(\alpha^0,\mu,c)X(s), Q^{1/2}dW(s)\rangle \\ &+ t\operatorname{tr}((-A_0)^\delta V(\alpha^0,\mu,c)Q(-A_0)^{-\delta})). \end{aligned}$$

Rewriting this inequality it follows that

$$\begin{aligned} C(T,\mu,c) &- T\operatorname{tr}((-A_0)^\delta V(\alpha^0,\mu,c)Q(-A_0)^{-\delta}) + \langle V(\alpha^0,\mu,c)X(T),X(T)\rangle \\ &\leq \langle V(\alpha^0,\mu,c)X(0),X(0)\rangle + 2\int_0^t \langle V(\alpha^0,\mu,c)X(s), Q^{1/2}dW(s)\rangle \end{aligned} \tag{23}$$

where

$$C(T,\mu,c) = \int_0^T [\langle \tilde{C}(\mu)X(s), \tilde{C}(\mu)X(s)\rangle + c\langle U(s),U(s)\rangle]ds.$$

Since $V(\alpha^0,\mu,c)$ is uniformly bounded for $|\mu| = 1$ and $c \in (0,\delta)$ and (18) is bounded almost surely it follows from the Strong Law of Large Numbers for Brownian motion that

$$\lim_{T\to\infty} \frac{1}{T}\int_0^T \langle V(\alpha^0,\mu,c)X(s), Q^{1/2}dW(s)\rangle = 0 \qquad \text{a.s.} \tag{24}$$

It follows directly from (24) that

$$\lim_{T\to\infty} \frac{1}{T} C(T,\mu,c) \geq \mathrm{tr}((-A_0)^{\delta} V(\alpha^0,\mu,c) Q(-A_0)^{-\delta}) \qquad \text{a.s.} \qquad (25)$$

for $|\mu| = 1$. By the Proposition it follows from (25) that

$$\lim_{T\to\infty} \frac{1}{T} C(T,\mu,c) \geq rc^{1/2} \qquad \text{a.s.} \qquad (26)$$

for $c \in (0,\delta)$ and $|\mu| = 1$. The inequality (26) implies a persistent excitation condition, that is, for $\varepsilon > 0$ there is a $\delta > 0$ and a $T_0 > 0$ such that

$$P\left(\frac{1}{T}\int_0^T \langle \tilde{C}(\mu)X(s), \tilde{C}(\mu)X(s)\rangle ds \geq \delta|\mu|^2 \right.$$
$$\left. \text{for all } \mu \in \mathbb{R}^q \text{ and } T \geq T_0 \right) \geq 1 - \varepsilon. \qquad (27)$$

For $\mu \neq 0$ it follows by the uniform boundedness of (18) that

$$\lim_{T\to\infty} \frac{1}{T} \sum_{i=1}^{q} \int_0^t \langle \tilde{P} A_{1j} X(s), \tilde{P} Q^{1/2} dW(s)\rangle \mu_j = 0 \qquad \text{a.s.} \qquad (28)$$

Using (9, 27, 28) it follows that

$$\lim_{T\to\infty} \hat{\alpha}(T) = \alpha^0 \qquad \text{a.s.}$$

□

Acknowledgments: This research was supported in part by NSF Grant DMS 9623439.

References

Duncan, T. E., Maslowski, B., and Pasik-Duncan, B. (1994). Adaptive boundary and point control of linear stochastic distributed parameter systems. *SIAM J. Control Optim.* **32**, 648–672.

Duncan, T. E., Maslowski, B., and Pasik-Duncan, B. (1996). Adaptive boundary control of stochastic linear distributed parameter described by analytic semigoups. *Appl. Math. Optim.* **33**, 107–138.

Flandoli, F. (1987). Algebraic Riccati equation arising in boundary control problems. *SIAM J. Control Optim.* **25**, 612–636.

Lasiecka, I. and Triggiani, R. (1987). The regulator problem for parabolic equations with Dirichlet boundary control.I. *Appl. Math. Optim.* **16**, 147-168.

Pasik-Duncan, B. (1992). On the consistency of least squares identification procedure in linear evolution systems. *Stochastic and Stochastics Reports* **39**, 83–94.

30

Monte Carlo Estimation of Diffusion Distributions at Inter-sampling Times

C. J. Wypasek, J. V. Butera and B. G. Fitzpatrick

ABSTRACT: In this paper we present a Monte Carlo technique for estimating diffusion distributions at inter-sampling times. The technique performs a pathwise analysis of a given diffusion model and initial distribution. The result is a procedure which matches a desired terminal distribution while providing distribution estimation at intermediate times.

KEYWORDS: Diffusion Distributions, Estimation, Markov Chain Monte Carlo

1 Introduction

In using diffusions to model physical phenomena, one is confronted with the problem of fitting models to observed data. In many applications, such data may be available only at a few times, making model comparisons difficult. In this paper, we discuss a Monte Carlo estimation technique for "interpolating" observed diffusion data in time.

Our studies have been motivated primarily from problems in subsurface contaminant transport. In field scale studies (see, e.g., Tennessee Valley Authority (1993) a description of one such experiment), one often performs a tracer test by first injecting a substance into the groundwater and then sampling its concentration at certain times and spatial locations. Within this context, the problem at hand is to infer the contaminant distribution over the spatial and temporal domain of interest.

The algorithm presented here involves a pathwise analysis for a general diffusion. In the simplest case of a random walk on a grid, one can enumerate paths, then back-propagate path probabilities in a way that is reminiscent of neural networks. Although straightforward and complete, the algorithm's run time and storage requirements can quickly exceed practical limits. For large cases or more general diffusions, the algorithm can be executed via Monte Carlo simulation with minimal storage requirements.

In the next section, we provide some background for the diffusions of interest for contaminant transport applications. We then proceed to the algorithm, provide some example computations which illustrate the versatility of the approach. We conclude with some remarks and suggestions for future work.

2 The Advection-Dispersion Model

The basic physical phenomena of interest in contaminant transport through porous media are advection, dispersion, adsorption, and degradation. There are many possible system behaviors, and these depend heavily on the specific contaminants and soil types. We consider here transport of contaminants which dissolve into the groundwater. For simplicity we consider a non-reactive, non-biodegrading contaminant.

Mass-balance considerations lead to a model for a single solute of the form

$$nc_t(t,x) + \nabla \cdot (v(x)c(t,x)) = \nabla \cdot (D\nabla c(t,x)), \tag{1}$$

for $t > 0$, $x \in \Omega$, with initial condition $c(0,x) = c_0(x)$. In this model $c(t,x)$ is the contaminant concentration at time t, position x, $v = v(x)$ is the groundwater velocity vector, D is the diffusion coefficient, n is the medium's porosity.

The porosity n represents the amount of volume which is not taken up by soil. For this quantity to be well-defined, one must choose a representative volume scale (see Bear, 1972, or Freeze and Cherry, 1979). The soil, then takes up the remaining $1-n$ proportion of the volume. In saturated flow, nc represents the volumetric concentration of the contaminant, while c represents the concentration of contaminant in the water.

The groundwater velocity, v, is a crucial component in our model. Typically, v is modeled via Darcy's Law (again, see Bear, 1972, or Freeze and Cherry, 1979):

$$v = -K\nabla h,$$

where K is the hydraulic conductivity of the medium, and h is the hydraulic head (the sum of the pressure head and elevation head). This equation is a generalization of Darcy's empirical observation that the ratio of volumetric fluid flux through a region to the cross sectional area (perpendicular to the flow) is proportional to the head gradient. The constant of proportionality is the hydraulic conductivity, which depends on the type of fluid (usually water in our applications) and the type of soil. In field scale settings, this constant of proportionality is no longer a constant, but a "randomly fluctuating" spatially dependent quantity.

Mass conservation and incompressibility considerations lead to models of the form

$$\nabla \cdot K\nabla h = 0 \qquad \text{in } \Omega. \tag{2}$$

Another important feature of the model is the contaminant dispersion. The dispersion matrix D is often modeled as

$$D = d_0 I + D_1(v), \tag{3}$$

where $d_0 I$ models molecular diffusion and $D_1(v)$ models the hydrodynamic dispersion, due to variations in the velocity field, which are in turn due to variations in the hydraulic conductivity. The following model is commonly used for D_1:

$$(D_1(v))_{ij} = a_T |v| \delta_{ij} + (a_L - a_T) \frac{v_i v_j}{|v|}. \tag{4}$$

Here a_T is called the "transverse dispersivity," and a_L is called the "longitudinal dispersivity." In this dispersion model, the principal axis of dispersion is along the velocity vector (longitudinal), and there is less dispersion in directions orthogonal to v ($a_T < a_L$). Moreover, the dispersion is modeled as (roughly) proportional to speed.

Justifications of this model (see, e.g., Gelhar, 1993) involve some form of "long time" asymptotics: the basic idea is that the highly variable nature of the velocity field is producing plume dispersion that can be mimicked by increasing the diffusion coefficient.

To be specific, we consider a differential equation of the form

$$dx(t) = v(x(t))dt + \varepsilon dW(t), \tag{5}$$

in which x denotes a contaminant particle position at time t, W is a standard Brownian motion, with $(1/2)\varepsilon^2 = d_0$ in the notation above. We take v to be a stochastic process whose statistics are determined by Darcy's law coupled with the statistical model for K. The diffusion term $\varepsilon dW(t)$ is typically of much smaller magnitude (i.e., variance) than the velocity v.

Most stochastic treatments of dispersion approximate the mean and variance of $x(t)$ to determine a Gaussian estimate of the plume. Since the constant coefficient convection-diffusion equation

$$nc_t + \bar{v} \cdot \nabla c = D \Delta c \tag{6}$$

has through the heat kernel a Gaussian density as its solution (with Dirac delta initial condition), the value of D is determined by the variance structure of the velocity field.

Typical tracer test applications involve measurement of the contaminant concentration at (a few) specific times and (a few) specific spatial locations. One major modeling challenge is to "interpolate" these values to other times of interest. Another is to estimate relevant parameters in the diffusion model from data (including even statistical parameters of the velocity field).

We consider now an algorithm for estimating the concentration distribution based on back-propagating contaminant particle trajectories.

3 The Back-Propagation Technique

A basic principle required in this analysis is the approximation of a continuous diffusion by discrete random walks. Given a particular diffusion model (such as the one given above), we may devise (as described in detail in Kushner, 1977) a random walk whose walk probabilities are determined by the diffusion parameters. The idea here is to begin with such an estimate and adjust the transition probabilities so that the random walk matches observed data.

Before examining the case of interest, we motivate the algorithm by considering first the case of a nearest neighbor random walk on an integer grid in 1 dimension. A few randomly chosen sample paths from the same initial point are depicted in Figure 1. It is possible to enumerate all of the sample paths for m time steps using a simple recursive technique for traversing a tree structure. Let $\{\mathcal{P}_1, \ldots, \mathcal{P}_n\}$ denote the set of paths. Here $n = (\#\text{initialpoints})(\#\text{branches})^{m-1}$, but we shall address the issue of large n shortly. Temporarily, also assume that the initial distribution is concentrated at a single point.

For notational purposes for any diffusion we consider, step or *frame* times will be denoted by $t_0, t_1, \ldots, t_f$. Let x denote a location in the domain and $F(t, x)$ will denote the contaminant's mass density function at time t. This physical quantity also has the interpretation as a probabilistic quantity. It is this interpretation that we exploit in our algorithm. Our goal here is to find a distribution $F(t, y)$ satisfying $F(t_f, y) \approx \widehat{F}(t_f, y)$, where $\widehat{F}$ is a distribution from observed data.

Let $x(\mathcal{P}, t)$ denote the location of the path $\mathcal{P}$ at time t. When F is taken on a grid (i.e., we set a predetermined discretization for the diffusion), $x(\mathcal{P}, t)$ will denote the nearest grid point. In particular, when the paths are enumerable, the *natural terminal distribution* is given by

$$F(t_f, y) = \sum_{x(\mathcal{P}_i, t_f) = y} \mathrm{P}(\mathcal{P}_\mathrm{i}).$$

For Markov random walks, the probability, $p_i = \mathrm{P}(\mathcal{P}_\mathrm{i})$, can be calculated as the paths are enumerated. For our nearest neighbor example, let $n_d(\mathcal{P}_i), n_h(\mathcal{P}_i), n_u(\mathcal{P}_i)$ denote the number of steps down, horizontal and up respectively for the given sample path and let $[q_d, q_h, q_u]$ denote the stationary distribution of steps down, horizontal and up. It follows that the path probability for $\mathcal{P}_i$ is

$$p_i = p_i(t_0) = F(t_0, x(\mathcal{P}_i, t_0)) q_d^{n_d(\mathcal{P}_i)} q_h^{n_h(\mathcal{P}_i)} q_u^{n_u(\mathcal{P}_i)}.$$

Of special importance is the ratio,

$$\frac{p_i}{F(t_f, x(\mathcal{P}_i, t_f))}.$$

The ratio represents the conditional probability that a particle chosen at random from the initial distribution follows path $\mathcal{P}_i$ given its terminal location. As a function of the terminal location y,

$$\frac{p_i 1_{\{x(\mathcal{P}_i,t_f)=y\}}}{F(t_f,y)}$$

becomes a probability measure over paths coterminal with y, a measure that is entirely defined by the underlying diffusion.

Now, recall that $\widehat{F}(y)$ denotes a terminal distribution for the diffusion as estimated by data or some other outside information. A good example would be an estimate of a contaminant plume at time t_f generated from well data. Starting from a given model F, we expect that $\widehat{F}$ is not the same as F, due to modeling, measurement or estimation errors. For example, in the case of a contaminant plume, differences between $\widehat{F}$ and F might be attributed to unobservable aspects in the velocity field (due to, e.g., large variations in the soil composition).

If the diffusion is to model the data in a reasonable way, it is clear that $\widehat{F}$ must be somehow *obtainable* by the diffusion model, i.e., $\widehat{F}$ must at least be absolutely continuous with respect to F. In the case of random walks on grids with a non point mass initial distribution, the desired terminal distribution cannot represent mass movement faster than what is possible in the assumed diffusion model. An example demonstrating the difference between a point mass initial distribution and broader initial distribution is presented at the beginning of the next section.

Taking into account matters of the obtainability of $\widehat{F}$ by the diffusion model, the function,

$$\frac{p_i \widehat{F}(x(\mathcal{P}_i,t_f))}{F(t_f,x(\mathcal{P}_i,t_f))},$$

becomes a probability measure on sample paths, defined by and consistent with the underlying diffusion model, that yields the desired terminal distribution. This new probability measure on sample paths has some interesting properties. There are obvious inadequacies in the diffusion model when $\widehat{F}$ differs from F. If the diffusion model is to be acceptable and $\widehat{F}(y) < F(t_f,y)$, paths coterminal with y need to be reduced in likelihood. Where $\widehat{F}(y) > F(t_f,y)$, paths need to be increased in probability.

The dynamic censoring and promotion of sample paths induces a back-propagation technique. Whether by path enumeration or simulation, one can capture frames of F at times, $t_0, t_1, \ldots, t_f$. The set, $\{(x,t_i) : x$, a grid point, t_i, a frame time, can be thought of as a collection of nodes in the evolution of the distribution. The precise algorithm for enumerated paths proceeds as follows. Sequentially, construct each sample path $\mathcal{P}_i$, and record a history of the nodes visited. Calculate the path probability, p_i. Once all paths are constructed, calculate the natural terminal distribution. The new path probabilities, $p_i(t)\widehat{F}(x(\mathcal{P}_i,t_f))/F(t_f,x(\mathcal{P}_i,t_f))$, computed by adjusting the model to meet the terminal distribution, now gives a distribution of the diffusion at time t in terms of the path $\mathcal{P}_i$, analogous to an Eulerian description. To determine the distribution in terms of the nodes (i.e., the Lagrangian description), one must perform the "book-keeping" operation of counting the paths which visit particular nodes at each time of interest.

The term *back-propagation* is intuitive, for paths are constructed forward, natural outputs are compared to desired outputs, then updated path probabilities are back-propagated.

A difficulty with this approach is the increase in computational burden for two or three dimensional problems. For a nearest neighbor walk on the integers on the line, less than 10 frames is practical. For a nearest neighbor walk in 2 dimensions (9 neighbors), only about 5 frames is practical. However, for larger sizes or for more general diffusions, one can still consider the frames as the estimation of the time dependent distribution on a grid. Paths can be taken as the sequence of nodes visited (or at least a neighborhood visit) at frame times. For a node path, $\mathcal{P}'$, the path probability can be estimated via Monte Carlo simulation,

$$p' = \frac{\sum_{i=1}^{N} 1_{\{\mathcal{P}_i = \mathcal{P}'\}}}{N},$$

where N node paths, $\mathcal{P}_i$, are simulated. The terminal distribution, F, will always have to be estimated or have an assumed form. While our example computations are given in terms of the simplest random walk models for F, it is possible (perhaps even desirable) to choose F based on prior modeling considerations (e.g., from the advection-dispersion equation above for contaminant transport modeling). One advantage is that the normalization constant, N, does not have to be carried in each back-propagation. The Monte Carlo version of the algorithm is as follows. Estimate the terminal distribution, F. Initialize all node estimates to 0. Sequentially simulate sample paths and record a history of the nodes visited. Once the terminal location, $x(\mathcal{P}, t_f)$, is known, back-propagate the ratio $\widehat{F}(x(\mathcal{P}, t_f))/F(t_f, x(\mathcal{P}, t_f))$ (recall that the path probabilities are equal initially, from the Monte Carlo simulation) to the nodes estimates along the history of $\mathcal{P}$. One can continue generating sample paths to improve the quality of the approximation. The intermediate time distributions must again be converted from this path form to node form by counting paths.

4 Examples

We present several examples to demonstrate the usefulness of the back-propagation algorithm. In all examples, the desired terminal distributions are contrived to emphasize noticable changes in the distribution frames.

Applying this algorithm to the MAcroDispersion Experiment (MADE2, Tennessee Valley Authority, 1993) site data will appear in a later paper.
Our first example is a 1 dimensional nearest neighbor walk on a grid using the precise algorithm. The stationary step distribution is $[\frac{1}{4}, \frac{1}{2}, \frac{1}{4}]$ for steps up, horizontal and down, respectively. In Figure 2, the first graph depicts the intermediate steps leading to the natural terminal distribution at time 7. The initial distribution at time 1 is uniformly distributed over integers 0, 1 and 5. An important aspect of the back-propagation technique is whether paths from different initial conditions are allowed to interact. With a terminal distribution uniformly distributed over 1, 2, 3 and 4, the second graph of Figure 2 calculation of the updated path probabilities using paths throughout the initial distribution. In this case, there is no guarantee that back-propped initial distribution will be the same as the actual initial distribution. The reason for this behavior is that the back-propagation is attempting to reconstruct an initial distribution that is consistent with the diffusion and the terminal distribution. An exact agreement can be forced by considering the expected behavior with respect to the initial distribution as depicted in the third graph of Figure 2. In this case, the algorithm is repeated for each node in the initial distribution and the natural distribution for each iteration is calculated over only those paths starting at the given node. The per node estimates are then weighted by the initial distribution.
Figure 3 depicts a 2 dimension nearest neighbor walk on a grid with 4 time steps. Each step is a composite of horizontal and vertical steps taken independently with the same distribution as in the first example. Contour plots of the location-wise square root of the distributions are represented in the graphs since there is relatively quick movement away from the point mass initial condition. With the precise algorithm, there is perfect evolution to the the contrived terminal distribution (uniformly distributed over the enclosed coordinates).
Figures 4 and 5 represent 1 and 2 dimensional examples of Gaussian random walks where at each time step, a normal step is taken. In each figure, the first graph represents a Monte Carlo run to estimate the terminal distribution which is normal. The theoretical terminals were then used in a second Monte Carlo run with back-propagation to obtain an excellent match with the contrived terminals. In Figure 4, the contrived terminal is bimodal, a mixture of two normals. In Figure 5, the contrived terminal is bivariate normal with correlation of .5 and smaller variances. Again, square root distributions are represented. It is interesting to note in the terminal of Figure 5 how the desired terminal is approaching the limits of what is obtainable by the diffusion.
Figures 6 and 7 demonstrate application of the back-propagation technique to match field data in one and two dimensional models of groundwater contaminant transport. As discussed above, traditional models for dispersion in groundwater involve a Laplacian term to simulate spreading of a contaminant plume in addition to molecular diffusion. Field scale experiments

indicate a scale (or time) dependence on the dispersion parameters (see, e.g, Gelhar, 1993) Since the additional spreading of the contaminant is due to variation in the groundwater velocity, we attempt to model dispersion by introducing variations in the velocity field directly, rather than using a Laplacian type operator. A thorough description of this technique is contained in Butera (1997).

In both cases for Figures 6 and 7, the natural terminal distribution is approximately normal, whereas the desired terminal distribution is uniform over two rectangular regions. In Figure 7, it is particularly interesting to note that the mass associated with the farther terminal locations is clearly moving faster. Furthermore, as the majority leading edge of the plume, the behavior of this mass can be observed fairly early in the plume evolution.

5 Conclusions and Future Work

In this paper we have presented a back-propagation technique for estimating diffusion distributions given an initial distribution and a desired terminal distribution. The "forward" algorithm is based on a "particle dynamics" (or approximation of the diffusion by random walks) viewpoint, which is consistent with (but more general than) the advection-dispersion models commonly used in practice. The back propagation provides a simple and straightforward means of estimating intermediate and terminal distributions from discrete data. This approach is easier to implement and computationally less intensive than the deterministic inverse problem approach of Fitzpatrick (1995). The included examples clearly demonstrate its utility.

We must also remark that this problem is in a sense analogous to the backward heat equation problem (which is an illposed problem). It is in fact the specification of the model distribution F which stabilizes the inversion. One can view this specification as being Bayesian in spirit: the model F is our prior, and through the data $\widehat{F}$ we compute a posterior. The specification of the prior plays the role of regularization in the backward heat equation problem (see Fitzpatrick (1991) for more discussion on this point).

In future studies, we plan to consider several generalizations and applications. One avenue of interest is to apply the algorithm to field scale data such as the MADE experimental database, detailed in Tennessee Valley Authority (1993). We also plan to use the estimates as nonparametric estimators for the diffusion density, estimators we can then use to compare deterministic parametric models, such as the advection dispersion model discussed in Fitzpatrick (1995) and the model of Butera (1997). It would be of considerable interest to compare this form of density estimation with more commonly used nonparametric density estimators (discussed and analyzed, e.g., in Ibragimov and Hasminskii, 1981). The intermediate information provided by our technique should improve the behavior of parameter estimation schemes. Finally, we plan to examine direct parameterizations of the transition probabilities (such as those discussed, e.g., in Kushner, 1997), which will allow us to compare the estimated probabilities with parameter dependent probabilities, so that determination of physical parameters may be possible by probabilistic means.

Acknowledgments: This research was funded in part by AFOSR grant F49620-95-1-0214. We owe many thanks to Dr. T. Stauffer of AL/EQ, Tyndall AFB, for supplying some very interesting and challenging data to analyze, and for engaging in some very helpful conversations.

References

Bear, J. (1972) *Dynamics of Fluids in Porous Media,* Dover, New York.

Butera, J. V. (1997) *A Random Process Model for Dispersion in Contaminant Transport Through Porous Media: Analysis and Computation,* Ph. D. Dissertation, Department of Mathematics, N. C. State University, Raleigh.

Fitzpatrick, B. (1991) *Bayesian Analysis in Inverse Problems,* Inverse Problems, **7**, no 5, pp. 675-702.

Fitzpatrick, B. (1995) *Analysis and Approximation for Inverse Problems in Contaminant Transport and Biodegradation Models,*, J. Num. Func. Anal. Opt., **16**, no. 7/8, pp. 847-866.

Freeze, R. A., Cherry, J. A. (1979) *Groundwater,* Prentice-Hall, New Jersey, 1979.

Gelhar, L. (1993) *Stochastic Subsurface Hydrology,* Prentice-Hall, New Jersey.

Ibragimov, I. A., and R. Z. Hasminskii. (1981) *Statistical Estimation, Asymptotic Theory,* Springer-Verlag, New York.

Kushner, H. (1977) *Probability Methods for Approximations in Stochastic Control and for Elliptic Equations,* Academic Press, New York.

Tennessee Valley Authority (1993) *Database for the Second Macrodispersion Experiment (MADE-2),* EPRI TR-102072, Project 2485-05, Interim Report.

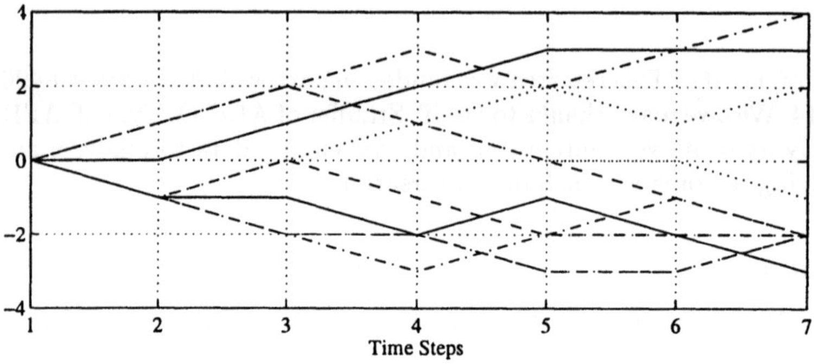

FIGURE 1. Randomly Chosen Paths for a Nearest-Neighbor Walk on the Integers.

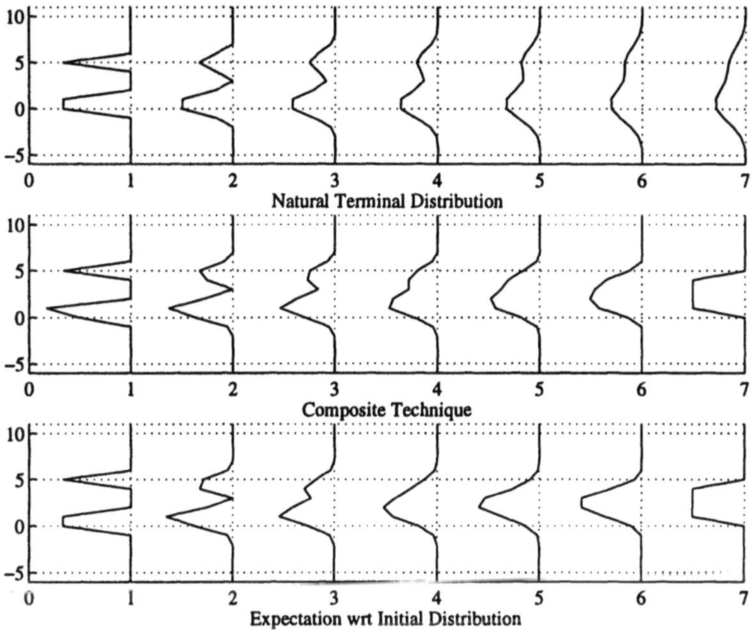

FIGURE 2. Nearest-Neighbor Walk on the Integers in 1 Dimension.

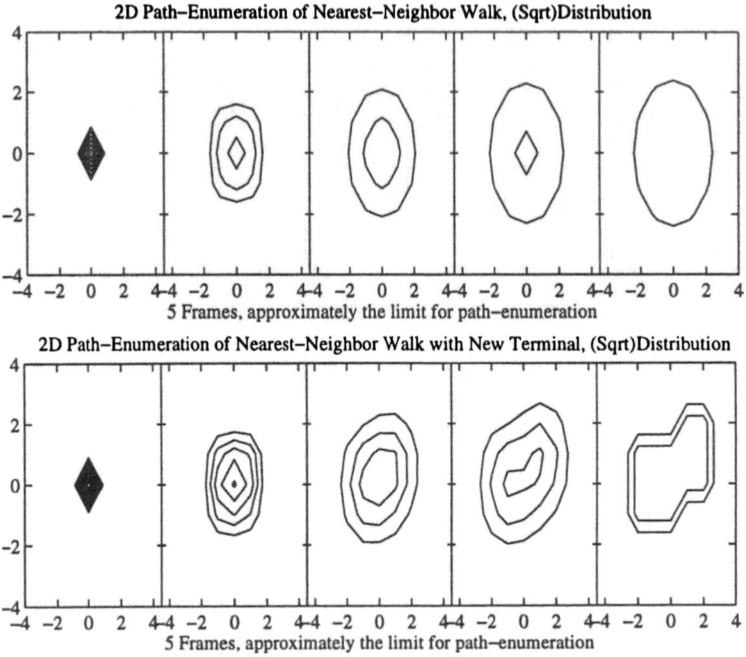

FIGURE 3. Nearest-Neighbor Walk on the Integers in 2 Dimensions.

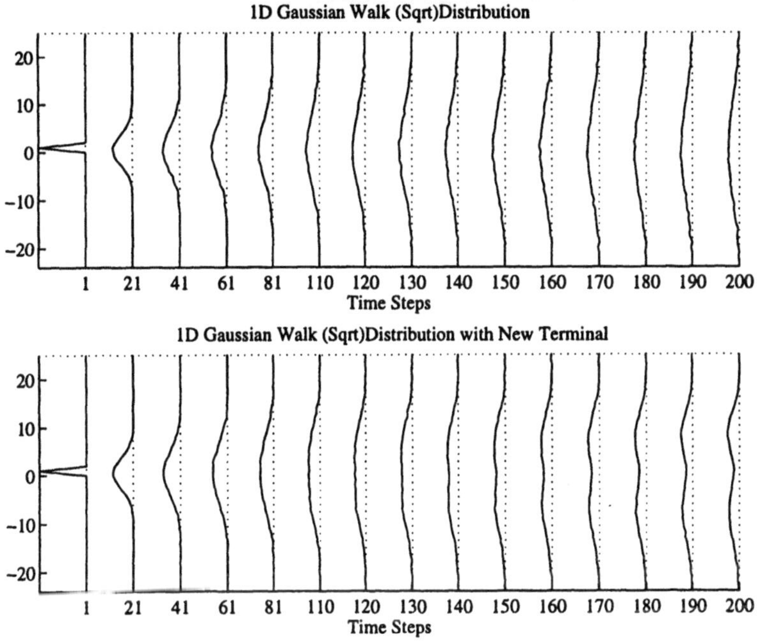

FIGURE 4. Guassian Walk in 1 Dimension.

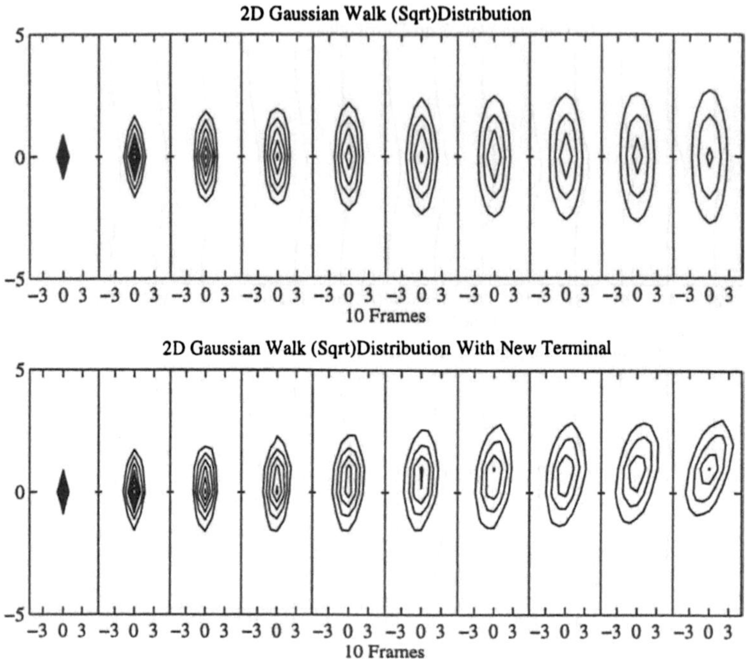

FIGURE 5. Guassian Walk in 2 Dimensions.

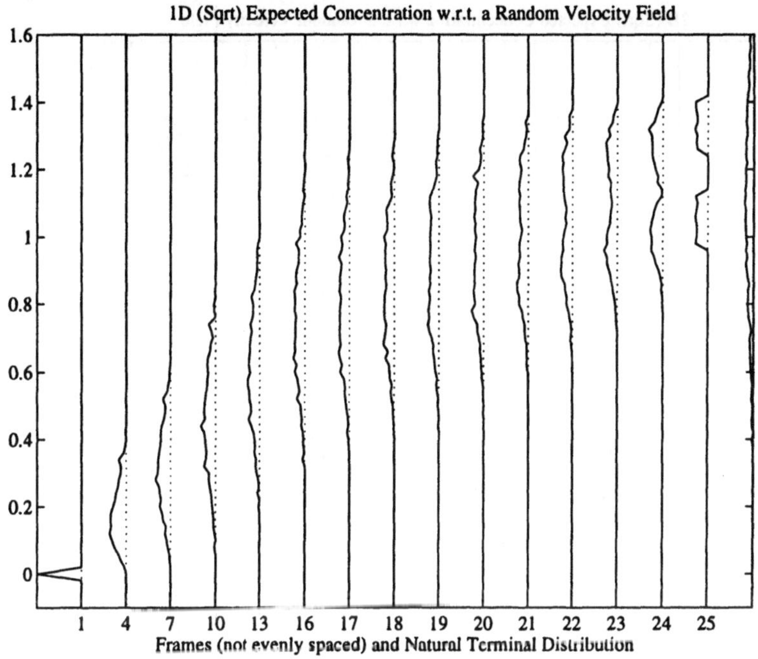

FIGURE 6. Expected Concentration w.r.t. a Random Velocity Field in 1 Dimension.

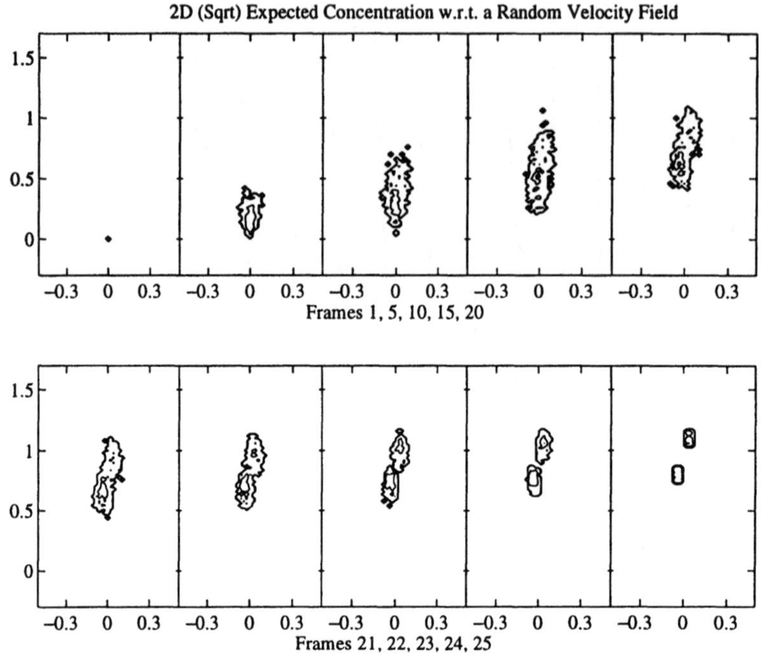

FIGURE 7. Expected Concentration w.r.t. a Random Velocity Field in 2 Dimensions.

Part IV

Mathematical Finance and Other Applications

31

Option Pricing in a Market with Frictions

Alain Bensoussan
Hugues Julien

ABSTRACT: We study in this paper an extension of the Black Scholes approach to take account of frictions in financial markets. The frictions we consider are related to costs linked to holding a portfolio of securities. The friction model is relatively general, the only limitation being that the friction affects the wealth function continuously, and not impulsively, as it is the case when modelling transaction costs. So our model differs from Davis, Panas and Zariphopoulou (1993), but ideas concerning the definition of selling and buying prices of options are taken from their approach. We shall consider European as well as Americans options. Interesting differences will be apparent.

KEYWORDS: frictions, portfolio, Wiener process, wealth function, option pricing

1 The Model

1.1 Notation and Assumptions

The basic securities consist of $n+1$ assets. One of them is a non risky one with price per unit governed by the equation

$$dS^0 = S^0 r(t)dt \tag{1.1}$$

where $r(t)$ is a deterministic bounded function.
In addition there are n risky assets whose price per unit is modelled through linear stochastic differential equations

$$dS^i = S^i(\alpha^i(t)dt + \sum_{j=1}^{n} \sigma^i_j(t)dw^j), i = 1, ..., n \tag{1.2}$$

where $\alpha^i(t), \sigma^i_j(t)$ are deterministic bounded functions. The matrix $\sigma(t) = \sigma^i_j(t)$ satisfies

$$\sigma(t) \text{ is invertible } \sigma^{-1}(t) \text{ is bounded} \tag{1.3}$$

The process $w(t) = w^j(t)$ is a standard Wiener process defined on a probability space $(\Omega, \mathcal{A}, P)$. We denote by F^t the filtration attached to the Wiener process

$$F^t = \sigma(w(s), s \leq t). \tag{1.4}$$

We shall consider portfolios $y(t) = (y^0(t), \ldots, y^n(t)), y^i(t)$ representing the amount of each asset held in the portfolio.
We shall assume that

$$\int_0^T (y^i(t)S^i(t))^2 dt < \infty, \text{ a.s. }, \qquad \forall i = 1, ..., n \tag{1.5}$$

and that

$$y^i(t), \quad i = 1, \ldots n \text{ is } F^t \text{ adapted} \tag{1.6}$$

The wealth is defined by:

$$W(t) = \sum_{i=0}^n y^i(t)S^i(t). \tag{1.7}$$

We consider now a mechanism of evolution of the portfolio which is both self financing and affected by friction. Its "intuitive" and "formal" description is possible, when one uses increments $\Delta y^i(t), i = 0, \ldots n$, writing

$$\sum_{i=0}^n \Delta y^i(t)S^i(t) = -\sum_{i=1}^n S^i(t)y^i(t)f^i(y(t), S(t), t)\Delta t \tag{1.8}$$

where

$$f^i(y, S, t) \text{ is a deterministic bounded function of the arguments } y \epsilon \Re^n, S \epsilon \Re^n, t \tag{1.9}$$

$$\sum_{i=1}^n S^i y^i f^i(y, S, t) \geq 0 \forall y, S, t \tag{1.10}$$

The boundedness assumption is convenient technically, the condition (1.10) expresses the fact that the right hand side of (1.8) is a friction term.
Condition (1.8) is mathematically expressed, as follows

$$\begin{aligned} &W \text{ has a Ito differential} \\ &dW = \sum_{i=0}^n y^i(t)dS^i(t) - \sum_{i=1}^n S^i(t)y^i(t)f^i(y(t), S(t), t)dt \end{aligned} \tag{1.11}$$

or using (1.1), (1.2), (1.7), we have

$$\begin{aligned} dW &= rWdt + \sum_{i=1}^n y^i(t)S^i(t)(\alpha^i(t) - r(t) - f^i(y(t), S(t), t))dt \\ &+ \sum_{i=1}^n \sum_{j=1}^n y^i(t)S^i(t)\sigma_j^i(t)dw^j(t). \end{aligned} \tag{1.12}$$

This expression can also be written as

$$dW = rWdt + \sum_{i=1}^{n}\sum_{j=1}^{n} y^i(t)S^i(t)\sigma_j^i(t)d\hat{w}^j(t) \tag{1.13}$$

where the processes $\hat{w}^j(t)$ are defined by their Ito differentials

$$\sum_{j=1}^{n}\sigma_j^i(t)d\hat{w}^j(t) = (\alpha^i(t)-r(t)-f^i(y(t),S(t),t))dt+\sum_{j=1}^{n}\sigma_j^i(t)dw^j(t) \tag{1.14}$$

which defines uniquely the processes $\hat{w}^j(t)$, since the matrix $\sigma(t)$ is invertible. An important feature is that $\hat{w}(t)$ depends on the portfolio $y(t)$, because of the presence of a friction term. Whenever the friction is absent, the process $\hat{w}(t)$ is independent of the portfolio. From Girsanov transformation, we can define a unique probability Q^y, depending of the portfolio such that $\hat{w}(t)$ becomes a standard Wiener process. For this probability measure, it follows from (1.13) that the discounted wealth is a martingale. Again when the market is frictionless, the probability Q^y does not depend on y, and thus is the unique probability under which the discounted wealth is a martingale.

Moreover using (1.14) in (1.2) yields

$$dS^i = S^i(r(t) + f^i(y,S,t))dt + \sum_{i=1}^{n}\sum_{j=1}^{n} S^i\sigma_j^i(t)d\hat{w}^j(t) \tag{1.15}$$

1.2 Value Functions

We summarize the above. Under the probability system $(\Omega, \mathcal{A}, Q^y, F^t)$, depending on the portfolio $y(t)$ the process $\hat{w}(t)$ is a standard F^t Wiener process, and the wealth and securities prices equations are governed by (1.13) and (1.15). For technical reasons we add the following restriction on the portfolio $y(t)$.

$$E^{Q^y}\int_0^T (y^i(t)S^i(t))^2dt < \infty \tag{1.16}$$

Consider now a utility function U satisfying

$$\begin{aligned} &U(x) \text{ is concave, non decreasing on } \Re \text{ and} \\ &U(0) = 0 \\ &U(x) \le \bar{U} \end{aligned} \tag{1.17}$$

We define the selling and buying value functions:

$$\begin{aligned} V_s(W,S,t) &= \sup_y E^{Q^y}U(W(T) - h(S(T))) \\ V_b(W,S,t) &= \sup_y E^{Q^y}U(W(T) + h(S(T))) \end{aligned} \tag{1.18}$$

where

$$h \text{ deterministic }, 0 \leq h(S) \leq h_0 + h_1 S, S \geq 0 \tag{1.19}$$

Note that from the last assumption (1.17), we allow the possibility

$$E^{Q^y} U(W(T) - h(S(T))) = -\infty$$

for some portfolio y. So the value function is well-defined, provided we accept the value $-\infty$ as possible.
Note also that the definition (1.18) implies also that t represents the initial value of time, in which we have the initial values

$$W(t) = W, \; S(t) = S. \tag{1.20}$$

Following Davis et al.(1993) we define from the value functions V_s, V_b the option prices

$$\begin{aligned} u_s(S,t) &= \inf\{W | V_s(W,S,t) \geq 0\} \\ u_b(S,t) &= \sup\{-W | V_b(W,S,t) \geq 0\} \end{aligned} \tag{1.21}$$

Our objective is to characterize as well as possible, the option prices u_s, u_b. In particular, are they solution of PDE, or can we find easily computable bounds ?

2 Bellman equations and approximation

2.1 Bellman equation for the value functions

We can write for V_s, V_b the corresponding Bellman equation

$$\begin{aligned} &\frac{\partial V_s}{\partial t} + rW\frac{\partial V_s}{\partial W} + \sum_i S^i r \frac{\partial V_s}{\partial S^i} + \frac{1}{2}\sum_{i,j} S^i S^j a_{ij}(t) \frac{\partial^2 V_s}{\partial S^i \partial S^j} \\ &+\frac{1}{2}\max_y \left\{ \sum_{i,j} \left(S^i y^i a_{ij} S^j y^j \frac{\partial^2 V_s}{\partial W^2} + 2 S^j y^j a_{ji} S^i \frac{\partial^2 V_s}{\partial W \partial S^i} \right) \right. \\ &\left. +2\sum_i S^i f^i(y,S,t) \frac{\partial V_s}{\partial S^i} \right\} = 0 \\ &V_s(W,S,T) = U(W - h(S)) \end{aligned} \tag{2.1}$$

$$\begin{aligned} &\frac{\partial V_b}{\partial t} + rW\frac{\partial V_b}{\partial W} + \sum_i S^i r \frac{\partial V_b}{\partial S^i} + \frac{1}{2}\sum_{i,j} S^i S^j a_{ij}(t) \frac{\partial^2 V_b}{\partial S^i \partial S^j} \\ &+\frac{1}{2}\max_y \left\{ \sum_{i,j} \left(S^i y^i a_{ij} S^j y^j \frac{\partial^2 V_b}{\partial W^2} + 2 S^j y^j a_{ji} S^i \frac{\partial^2 V_b}{\partial W \partial S^i} \right) \right. \\ &\left. + 2\sum_i S^i f^i(y,S,t) \frac{\partial V_b}{\partial S^i} \right\} = 0 \\ &V_b(W,S,T) = U(W - h(S)) \end{aligned} \tag{2.2}$$

where we have introduced the matrix

$$a(t) = \sigma(t)\sigma^*(t) \tag{2.3}$$

Although feasible, the global approach through (2.1), (2.2) and (1.21) to characterize the option prices $u_s(S,t), u_b(S,t)$ is quite cumbersome, hence it is important to find alternatives.

2.2 Definition of bounds

We introduce the following PDE

$$\begin{aligned} &-rv_s + \frac{\partial v_s}{\partial t} + r\sum_i S^i \frac{\partial v_s}{\partial S^i} + \frac{1}{2}\sum_{i,j} S^i S^j a_{ij} \frac{\partial^2 v_s}{\partial S^i \partial S^j} \\ &+ \max_y \left\{ \sum_i S^i f^i(y,S,t) \frac{\partial v_s}{\partial S^i} \right\} = 0 \\ &v_s(S,T) = h(S) \end{aligned} \tag{2.4}$$

$$\begin{aligned} &-r\bar{v}_s + \frac{\partial \bar{v}_s}{\partial t} + r\sum_i S^i \frac{\partial \bar{v}_s}{\partial S^i} + \frac{1}{2}\sum_{i,j} S^i S^j a_{ij} \frac{\partial^2 \bar{v}_s}{\partial S^i \partial S^j} \\ &+ \sum_i S^i f^i \left(\frac{\partial \bar{v}_s}{\partial S}, S, t \right) \frac{\partial \bar{v}_s}{\partial S^i} = 0 \\ &\bar{v}_s(S,T) = h(S) \end{aligned} \tag{2.5}$$

and

$$\begin{aligned} &-rv_b + \frac{\partial v_b}{\partial t} + r\sum_i S^i \frac{\partial v_b}{\partial S^i} + \frac{1}{2}\sum_{i,j} S^i S^j a_{ij} \frac{\partial^2 v_b}{\partial S^i \partial S^j} \\ &+ \min_y \left\{ \sum_i S^i f^i(y,S,t) \frac{\partial v_b}{\partial S^i} \right\} = 0 \\ &v_b(S,T) = h(S) \end{aligned} \tag{2.6}$$

$$\begin{aligned} &-r\bar{v}_b + \frac{\partial \bar{v}_b}{\partial t} + r\sum_i S^i \frac{\partial \bar{v}_b}{\partial S^i} + \frac{1}{2}\sum_{i,j} S^i S^j a_{ij} \frac{\partial^2 \bar{v}_b}{\partial S^i \partial S^j} \\ &+ \sum_i S^i f^i \left(-\frac{\partial \bar{v}_b}{\partial S}, S, t \right) \frac{\partial \bar{v}_b}{\partial S^i} = 0 \\ &\bar{v}_b(S,T) = h(S) \end{aligned} \tag{2.7}$$

The interpretation of the above functions is easy. The functions $v_s(S,t)$, $v_b(S,t)$ are expressed as

$$v_s(S,t) = \sup_y E^{Q^y}(h(S(T))) \exp - \int_t^T r(s)ds \tag{2.8}$$

$$v_b(S,t) = \inf_y E^{Q^y}(h(S(T))) \exp - \int_t^T r(s)ds \tag{2.9}$$

Moreover, considering the specific feedback

$$\bar{y}^s(S,t) = \frac{\partial \bar{v}_s}{\partial S}(S,t) \tag{2.10}$$

$$\bar{y}^b(S,t) = \frac{\partial \bar{v}_b}{\partial S}(S,t) \tag{2.11}$$

to which correspond stochastic controls

$$\bar{y}^s(t) = \bar{y}^s(S(t),t) \tag{2.12}$$

$$\bar{y}^b(t) = \bar{y}^b(S(t),t) \tag{2.13}$$

through the SDE

$$dS^i = S^i(r(t) + f^i(\bar{y}^s(S(t),t)S(t),t))dt + \sum_{j=1}^n \sigma_j^i(t)d\hat{w}^j(t), \;\; S^i(t) = S^i \tag{2.14}$$

$$dS^i = S^i(r(t) + f^i(\bar{y}^b(S(t),t), S(t),t))dt + \sum_{j=1}^n \sigma_j^i(t)d\hat{w}^j(t), \;\; S^i(t) = S^i \tag{2.15}$$

then one has the formulas:

$$\bar{v}_s(s,t) = E^{Q^{\bar{y}^s}}(h(S(T)))\exp - \int_t^T r(s)ds \tag{2.16}$$

$$\bar{v}_b(s,t) = E^{Q^{\bar{y}^b}}(h(S(T)))\exp - \int_t^T r(s)ds. \tag{2.17}$$

It is thus clear that one has the following relations

Proposition 2.1 *One has*

$$v_b(S,t) \leq \bar{v}_b(S,t) \leq \bar{v}_s(S,t) \leq v_s(S,t) \tag{2.18}$$

Proof The only thing to be proven is the intermediate relation

$$\bar{v}_b(S,t) \leq \bar{v}_s(S,t) \tag{2.19}$$

Setting $z = \bar{v}_s - \bar{v}_b$. We deduce from (2.5), (2.7)

$$-\frac{\partial z}{\partial t} + rz - r\sum_i S^i \frac{\partial z}{\partial S^i} - \frac{1}{2}\sum_{i,j} a_{ij} S^i S^j \frac{\partial^2 z}{\partial S^i \partial S^j}$$
$$= \sum_i S^i f^i(\frac{\partial \bar{v}_s}{\partial S}, S, t)\frac{\partial \bar{v}_s}{\partial S^i} - \sum_i S^i f^i(\frac{\partial \bar{v}_b}{\partial S}, S, t)\frac{\partial \bar{v}_b}{\partial S^i} \geq 0$$

from the assumption (1.10) on friction. Since $z(s,T) = 0$, the maximum principle implies $z \geq 0$, hence the result. □

2.3 An interval for u_s and u_b

We shall prove in this section the following result

Theorem 2.1 *We assume (1.1), (1.2), (1.3), (1.9), (1.10), (1.11), (1.17), then the option prices $u_s(S,t)$ and $u_b(S,t)$ defined by (1.21) satisfy the relation*

$$v_b(S,t) \le u_s(S,t) \le \bar{v}_s(S,t) \tag{2.20}$$

$$\bar{v}_b(S,t) \le u_b(S,t) \le v_s(S,t) \tag{2.21}$$

Proof We prove (2.20), the proof of (2.21) being similar. We begin with the right inequality. Consider the process (2.14), corresponding to the portfolio $\bar{y}^s(t)$ see (2.12), then from Ito's calculus, one gets easily

$$h(S(T)) = \bar{v}_s(S,t)R_t^T + \int_t^T R_s^T \sum_{i,j} \frac{\partial \bar{v}_s}{\partial S_i} S^i \sigma_j^i d\hat{w}j \tag{2.22}$$

where we have set

$$R_t^T = \exp \int_t^T r(s)ds \tag{2.23}$$

Next, using the same portfolio $\bar{y}^s(t)$ in (1.13) yields

$$W(T) = WR_t^T + \int_t^T R_s^T \sum_{i,j} \frac{\partial \bar{v}_s}{\partial S_i} S^i \sigma_j^i d\hat{w}_j \tag{2.24}$$

hence

$$W(T) - h(S(T)) = (W - \bar{v}(S,t))R_t^T \tag{2.25}$$

therefore

$$E^{Q^{\bar{y}^s}} U(W(T) - h(S(T))) = U((W - \bar{v}(S,t))R_t^T) \tag{2.26}$$

and from the definition of V_s, since $\bar{y}^s$ is a particular control

$$V_s(W,S,t) \ge U((W - \bar{v}(s,t))R_t^T). \tag{2.27}$$

This implies that $V_s(\bar{v}(s,t),S,t) \ge 0$, hence using the definition of $u_s(S,t)$ the right inequality (2.20) holds. To prove the left inequality, we consider an arbitrary portfolio y. Considering the process $S(t)$ defined by (1.15) and computing the Ito differential of $v_b(S(t),t)R_t^T$, we obtain from the PDE (2.6)

$$h(S(T)) \ge v_b(S,t)R_t^T + \int_t^T R_s^T \sum_{i,j} \frac{\partial v_b}{\partial S_i}(S(t),t) S^i \sigma_j^i d\hat{w}_j. \tag{2.28}$$

Moreover from (1.13) we have

$$W(T) = WR_t^T + \int_t^T R_s^T \sum_{i,j} y^i S^i \sigma_j^i d\hat{w}_j. \tag{2.29}$$

hence

$$(W - v_b(S,t))R_t^T + \int_t^T R_s^T \sum_{i,j}(y^i - \frac{\partial v_b}{\partial S_i})S^i\sigma_j^i d\hat{w}_j \geq W(T) - h(S(T))$$

and from the monotonicity of U.

$$E^{Q^y} U\left((W - v_b(S,t))\, R_t^T + \int_t^T R_s^T \sum_{i,j}(y^i - \frac{\partial v_b}{\partial S_i} S^i\sigma_j^i d\hat{w}_j)\right)$$
$$\geq E^{Q^y} U(W(T) - h(S(T))).$$

From the concavity of U, $U((W - v_b(S,t))R_t^T) \geq E^{Q^y}U(W(T) - h(S(T)))$, and since y is arbitrary, we obtain $U((W - v_b(S,t))R_t^T) \geq V_s(W,S,t)$. It follows that if $W \leq v_b(S,t)$, one has $V_s(W,S,t) \leq 0$, which implies $u_s(S,t) \geq v_b(S,t)$. □

Remark 2.1 *Suppose that there is no function $f^i = 0$, then clearly from (2.4), (2.5), (2.6), (2.7), we have:*

$$v_s(S,t) = \bar{v}_s(S,t) = v_b(S,t) = \bar{v}_b(S,t) = u(S,t)$$

where u is the solution of

$$-ru + \frac{\partial u}{\partial t} + r\sum_i S^i \frac{\partial u}{\partial S^i} + \frac{1}{2}\sum_{i,j} S^i S^j a_{ij} \frac{\partial u^2}{\partial S^i \partial S^j} = 0 \qquad (2.30)$$
$$u(S,T) = h(S)$$

and from (2.20), (2.21), it follows also

$$u_s(S,t) = u_b(S,t) = u(S,t) \qquad (2.31)$$

In this case the selling and buying option prices coincide with the usual Black-Scholes price. All probabilities Q^y coincide with a unique probability Q, for which the discounted wealth is a martingale. In this model, the function V_s and V_b are given by

$$V_s(W,S,t) = U((W - u(S,t)R_t^T) \qquad (2.32)$$

$$V_b(W,S,t) = U((W + u(S,t)R_t^T) \qquad (2.33)$$

We are in a situation of complete markets.

3 Option Prices for an agent with a CARA utility

3.1 The model

In our general framework, the function V_s and V_b solutions of (2.1), (2.2) are not given by the formulas

$$V_s(W,S,t) = U((W - u_s(S,t))R_t^T) \qquad (3.1)$$

$$V_b(W,S,t) = U((W - u_b(S,t))R_t^T) \tag{3.2}$$

We have seen that this is the case for complete markets, see Remark 2.1. In the absence of such a property the intervals obtained in Theorem 2.1 are quite useful. We may next wonder whether there are situations going beyond complete markets for which it is possible to obtain the properties (3.1), (3.2). The purpose of this section is to show that this is the case whenever the agent has a CARA (Constant Absolute Risk Aversion) utility, that is U is given by

$$U(x) = K(1 - e^{-\gamma x}) \tag{3.3}$$

which implies the relation

$$U''(x) = -\gamma U'(x) \tag{3.4}$$

and γ is called the risk aversion parameter.

3.2 Option prices in case of CARA utility

Our objective is to obtain the following result

Theorem 3.1 *We make the assumptions of Theorem 2.1, and (3.3), then the functions V_s and V_b solutions of (2.1), (2.2) are given by the formulas (3.1), (3.2) in which $u_s(S,t), u_b(S,t)$ are the solutions of*

$$\begin{aligned} &-\frac{\partial u_s}{\partial t} + ru_s - \sum_i S^i r \frac{\partial u_s}{\partial S^i} - \frac{1}{2}\sum_{i,j} S^i S^j a_{ij} \frac{\partial^2 u_s}{\partial S^i \partial S^j} \\ &+\frac{1}{2}\max_y \Big\{ -\gamma R_t^T \sum_{i,j} S^i S^j (y^i - \frac{\partial u_s}{\partial S^i}) a_{ij} (y^j - \frac{\partial u_s}{\partial S^j}) \\ &- 2\sum_i S^i f^i(y,S,t) \frac{\partial u_s}{\partial S^i} \Big\} = 0 \\ &u_s(S,T) = h(S) \end{aligned} \tag{3.5}$$

$$\begin{aligned} &-\frac{\partial u_b}{\partial t} + ru_b - \sum_i S^i r \frac{\partial u_b}{\partial S^i} - \frac{1}{2}\sum_{i,j} S^i S^j a_{ij} \frac{\partial^2 u_b}{\partial S^i \partial S^j} \\ &+\frac{1}{2}\min_y \Big\{ \gamma R_t^T \sum_{i,j} S^i S^j (y^i + \frac{\partial u_b}{\partial S^i}) a_{ij} (y^j + \frac{\partial u_b}{\partial S^j}) \\ &- 2\sum_i S^i f^i(y,S,t) \frac{\partial u_b}{\partial S^i} \Big\} = 0 \\ &u_b(S,T) = h(S) \end{aligned} \tag{3.6}$$

Proof Note first that the nonlinear terms in the PDE (3.5), (3.6) have linear growth in $\frac{\partial u_b}{\partial S}$ therefore these equations have unique solutions. Simple calculations show that functions defined by (3.1), (3.2) are solutions of Bellman equations (2.1) (2.2). We have not however discussed the uniqueness of the solution of (2.1), (2.2). To avoid this difficulty, we prove directly here that the functions defined by (3.1), (3.2) coincides with those defined in (1.18). We stick to V_s, to fix the ideas. Consider the function V_s defined by (3.1), y, we have by Ito's calculus

$$
\begin{aligned}
&U(W(T) - h(S(T))) - U((W - u_s(s,t))R_t^T) \\
&= \int_t^T U'((W(\tau) - u_s(S(\tau),\tau))R_\tau^T) \sum_{i,j} (y^i(\tau) - \frac{\partial u_s}{\partial S^i}(S(\tau),\tau)) \\
&\times S^i(\tau)\sigma_j^i(\tau)R_\tau^T d\hat{w}_j \\
&-\frac{1}{2}\int_t^T U'(\cdot)R_\tau^T \left\{\gamma R_\tau^T \left((y^i(\tau) - \frac{\partial u_s}{\partial S^i}(S(\tau),\tau)\right)\right) S^i(\tau)S^j(\tau)a_{ij}(\tau)(y^j(\tau) \\
&-\frac{\partial u_s}{\partial S^j}(S(\tau),\tau)) + 2\frac{\partial u_s}{\partial S^i}(S(\tau),\tau)S^i(\tau)f^i(y(\tau),S(\tau),\tau) \\
&-\min_y \Big[\gamma R_\tau^T (y^i(\tau) - \frac{\partial u_s}{\partial S^i}(S(\tau),\tau))S^i(\tau)S^j(\tau)a_{ij}(\tau)(y^j(\tau) - \frac{\partial u_s}{\partial S^j}(S(\tau),\tau)) \\
&+ 2\frac{\partial u_s}{\partial S^i}(S(\tau),\tau)S^i(\tau)f^i(y(\tau),S(\tau),\tau)\Big]\Big\} d\tau
\end{aligned}
\tag{3.7}
$$

Cleary

$$E^{Q^y} U(W(T) - h(S(T))) \leq U((W - u_s(S,t))R_t^T) \tag{3.8}$$

and since y is arbitrary we have

$$V_s(W,S,t) \leq U((W - u_s(S,t))R_t^T). \tag{3.9}$$

On the other hand taking the specific portfolio, which realizes the minimum in the bracket expression of the right hand side of (3.7), calling this portfolio $\hat{y}$ yields $E^{Q^{\hat{y}}} U(W(T) - h(S(T))) = U((W - u_s(S,t))R_t^T)$ and thus $V_s(W,S,t) \geq U((W - u_s(S,t))R_t^T)$ which completes the proof. □

3.3 Behavior with respect to the risk parameter

It is convenient to denote the solution of (3.5) by $u_s^\gamma(S,t)$ and that of (3.6) by $u_b^\gamma(S,t)$. We then have the following property

Theorem 3.2 *We make the assumptions of Theorem 3.1, then $u_s^\gamma(S,t)$ is increasing with respect to γ, whereas $u_b^\gamma(S,t)$ is decreasing with respect to γ.*

Proof We consider only the case of selling prices. Pick $\gamma' > \gamma$ and set $\tilde{u}_s(S,t) = u_s^{\gamma'}(S,t) - u_s^{\gamma}(S,t)$. Then from (3.5) we deduce

$$
\begin{aligned}
&-\frac{\partial \tilde{u}_s}{\partial t} + r\tilde{u}_s - \sum_i rS^i \frac{\partial \tilde{u}_s}{\partial S^i} - \frac{1}{2}\sum_{i,j} S^i S^j a_{ij} \frac{\partial^2 \tilde{u}_s}{\partial S^i \partial S^j} = \\
&-\frac{1}{2}\max_y \Bigg\{ \gamma' R_t^T \sum_{i,j} S^i S^j (y^i - \frac{\partial u_s^{\gamma'}}{\partial S^i}) a_{ij} (y^j + \frac{\partial u_s^{\gamma'}}{\partial S^j}) \\
&- 2\sum_i S^i f^i(y,S,t) \frac{\partial u_s^{\gamma'}}{\partial S^i} \Bigg\} \\
&+\frac{1}{2}\max_y \Bigg\{ -\gamma R_t^T \sum_{i,j} S^i S^j (y^i - \frac{\partial u_s^{\gamma}}{\partial S^i}) a_{ij} (y^j + \frac{\partial u_s^{\gamma}}{\partial S^j}) \\
&- 2\sum_i S^i f^i(y,S,t) \frac{\partial u_s^{\gamma}}{\partial S^i} \Bigg\}
\end{aligned} \tag{3.10}
$$

and since $\gamma' > \gamma$, we have:

$$
\begin{aligned}
&-\frac{\partial \tilde{u}_s}{\partial t} + r\tilde{u}_s - \sum_i rS^i \frac{\partial \tilde{u}_s}{\partial S^i} - \frac{1}{2}\sum_{i,j} S^i S^j a_{ij} \frac{\partial^2 \tilde{u}_s}{\partial S^i \partial S^j} \\
&\geq -\frac{1}{2}\max_y \Bigg\{ -\gamma R_t^T \sum_{i,j} S^i S^j (y^i - \frac{\partial u_s^{\gamma'}}{\partial S^i}) a_{ij} (y^j + \frac{\partial u_s^{\gamma'}}{\partial S^j}) \\
&- 2\sum_i S^i f^i(y,S,t) \frac{\partial u_s^{\gamma'}}{\partial S^i} \Bigg\} \\
&+\frac{1}{2}\max_y \Bigg\{ -\gamma R_t^T \sum_{i,j} S^i S^j (y^i - \frac{\partial u_s^{\gamma}}{\partial S^i}) a_{ij} (y^j + \frac{\partial u_s^{\gamma}}{\partial S^j}) \\
&- 2\sum_i S^i f^i(y,S,t) \frac{\partial u_s^{\gamma}}{\partial S^i} \Bigg\}
\end{aligned} \tag{3.11}
$$

and choosing $\hat{y}$ realizing the first max in the right hand side obtains

$$
\geq \sum_i S^i f^i(\hat{y},S,t) \frac{\partial \tilde{u}_s}{\partial S^i} - \gamma R_t^T \sum_{i,j} (\hat{y}^i - \frac{\partial u_s^{\gamma}}{\partial S^i}) a_{ij} S^i S^j \frac{\partial \tilde{u}_s}{\partial S^j} \tag{3.12}
$$

and the maximum principle implies $\tilde{u}_s \geq 0$. □
We proceed by indicating the following

Proposition 3.1 *One has the properties*

$$
u_s^0(S,t) = v_b(S,t) \tag{3.13}
$$

$$
u_s^{\infty}(S,t) = \bar{v}_s(S,t) \tag{3.14}
$$

$$u_b^0(S,t) = v_s(S,t) \tag{3.15}$$

$$u_b^\infty(S,t) = \bar{v}_b(S,t) \tag{3.16}$$

Proof This is clear from (3.5), (3.6), after taking $\gamma = 0$ or $+\infty$ in these equations. □

Remark 3.1 *Using Proposition 3.1 and Theorem 3.2, we recover the inequalities*

$$v_b(S,t) \leq u_s^\gamma(S,t) \leq \bar{v}_s(S,t), \quad \bar{v}_b(S,t) \leq u_b^\gamma(S,t) \leq v_s(S,t)$$

which was already stated in general in Theorem 2.1. The fact that the bounds of the interval can be attained whenever $\gamma = 0$ or $\gamma = +\infty$ proves that they are sharp.

Remark 3.2 *There is no a priori relation between the buying and the selling prices. We remark from the above that $u_b^0(S,t) \geq u_s^0(S,t)$ and conversely $u_b^\infty(S,t) \leq u_s^\infty(S,t)$. Since $u_s^\gamma(S,t)$ increases with γ, and $u_b^\gamma(s,t)$ decreases with γ there exists a unique $\gamma(S,t)$ such that*

$$\gamma \geq \gamma(S,t) \Rightarrow u_s^\gamma(S,t) \geq u_b^\gamma(S,t)$$
$$\gamma < \gamma(S,t) \Rightarrow u_s^\gamma(S,t) < u_b^\gamma(S,t)$$

Remark 3.3 *An application: We consider the following particular case, $n = 1$ and*

$$f(y,S,t) = -(R_f - r_f)1\!\!I_{y \leq 0}, R_f \geq r_f \tag{3.17}$$

For sake of symmetrization we change r into $r - r_f$, so that

$$r + f(y,S,t) = r - r_f - (R_f - r_f)1\!\!I_{y\leq 0} = \left| \begin{array}{lll} r - r_f & \text{if} & y > 0 \\ r - r_f & \text{if} & y \leq 0 \end{array} \right.$$

Note that

$$\begin{aligned} \max_y f(y,S,t)\frac{\partial v_s}{\partial S} &= \max_y \left[-(R_f - r_f)1\!\!I_{y\leq 0}\tfrac{\partial v_s}{\partial S}\right] \\ &= (R_f - r_f)(\tfrac{\partial v_s}{\partial S})^- \text{ and} \\ f(\frac{\partial v_s}{\partial S}, S, t)\frac{\partial v_s}{\partial S} &= (R_f - r_f)(\tfrac{\partial v_s}{\partial S})^-. \end{aligned}$$

In this example, one has

$$v_s(S,t) = \bar{v}_s(S,t) \tag{3.18}$$

$$v_b(S,t) = \bar{v}_b(S,t) \tag{3.19}$$

and they are solutions of

$$-rv_s + \frac{\partial v_s}{\partial S} + (r - r_f)S\frac{\partial v_s}{\partial S} + \frac{1}{2}S^2\sigma^2\frac{\partial^2 v_s}{\partial S^2} + (R_f - r_f)(\frac{\partial v_s}{\partial S})^- = 0 \tag{3.20}$$

$$-rv_b + \frac{\partial v_b}{\partial t} + (r - r_f)S\frac{\partial v_b}{\partial S} - (R_f - r_f)(\frac{\partial v_b}{\partial S})^+ + \frac{1}{2}S^2\sigma^2\frac{\partial^2 v_b}{\partial S^2} = 0 \tag{3.21}$$

4 American Option Prices

4.1 Definition of value functions

Going back to the basic definitions (1.18) of value functions, we change here (1.18) into:

$$\begin{aligned} V_s(W,S,t) &= \sup_y \inf_\theta E^{Q^y} U((W(\theta) - h(S(\theta)))R_\theta^T) \\ V_b(W,S,t) &= \sup_y \inf_\theta E^{Q^y} U((W(\theta) + h(S(\theta)))R_\theta^T) \end{aligned} \tag{4.1}$$

where θ is an F^t stopping time. We still characterize the option selling and buying prices by (1.21). Using the theory of optimal stopping and variational inequalities, see A. Bensoussan - J.L. Lions (1978), the Bellman inequalities corresponding to the functions (4.1) are written as follows

$$\begin{aligned} V_s(W,S,T) &= U(W - h(S)) \\ V_s(W,S,t) &\le U((W - h(S))R_t^T) \end{aligned} \tag{4.2}$$

$$\begin{aligned} &\frac{\partial V_s}{\partial t} + rW\frac{\partial V_s}{\partial W} + r\sum_i s^i\frac{\partial V_s}{\partial S^i} + \frac{1}{2}\sum_{i,j} S^i S^j a_{ij}\frac{\partial^2 V_s}{\partial S^i \partial S^j} \\ &+\frac{1}{2}\max_y\left\{\sum_{i,j}(S^i S^j y^i y^j a_{ij}\frac{\partial^2 V_s}{\partial W^2} + 2S^i S^j y^j a_{ji}\frac{\partial^2 V_s}{\partial W \partial S^i}\right) \\ &+ 2\sum_i S^i f^i(y,s,t)\frac{\partial V_s}{\partial S^i}\Bigg\} \ge 0 \\ &(V_s(W,S,t) - U((W - h(S))R_t^T))\left[\frac{\partial V_s}{\partial t} + rW\frac{\partial V_s}{\partial W} + r\sum_i S^i\frac{\partial V_s}{\partial S^i}\right. \\ &+\frac{1}{2}\sum_{i,j} S^i S^j a_{ij}\frac{\partial^2 V_s}{\partial S^i \partial S^j} + \frac{1}{2}max_y\left\{\sum_{i,j}(S^i S^j y^i y^j a_{ij}\frac{\partial^2 V_s}{\partial W^2}\right. \\ &\left.\left. + 2S^i S^j y^j a_{ji}\tfrac{\partial^2 V_s}{\partial W \partial S^i}) + 2\sum_i S^i f^i(y,S,t)\frac{\partial V_s}{\partial S^i}\right\}\right] = 0 \end{aligned}$$

and

$$\begin{aligned} V_b(W,S,T) &= U(W + h(S)) \\ V_b(W,S,t) &\le U((W + h(S))R_t^T) \end{aligned} \tag{4.3}$$

$$\frac{\partial V_b}{\partial t} + rW\frac{\partial V_b}{\partial W} + r\sum_i S^i \frac{\partial V_b}{\partial S^i} + \frac{1}{2} S^i S^j a_{ij} \frac{\partial^2 V_b}{\partial S^i \partial S^j}$$
$$+\frac{1}{2}\max_y \left\{ \sum_{i,j} (S^i S^j y^i y^j a_{ij} \frac{\partial^2 V_b}{\partial W^2} + 2S^i S^j y^j a_{ji} \frac{\partial^2 V_b}{\partial W \partial S^i} \right)$$
$$+ 2\sum_i S^i f^i(y,S,t)\frac{\partial V_b}{\partial S^i} \Bigg\} \geq 0$$
$$(V_b(W,S,t) - U((W+h(S))R_t^T)) \left[\frac{\partial V_b}{\partial t} + rW\frac{\partial V_b}{\partial W} + r\sum_i S^i \frac{\partial V_b}{\partial S^i} \right.$$
$$+\frac{1}{2}\sum_{i,j} S^i S^j a_{ij} \frac{\partial^2 V_b}{\partial S^i \partial S^j} + \frac{1}{2} max_y \left\{ \sum_{i,j} (S^i S^j y^i y^j a_{ij} \frac{\partial^2 V_b}{\partial W^2} \right.$$
$$\left. \left. + 2S^i S^j y^j a_{ji} \frac{\partial^2 V_b}{\partial W \partial S^i}) + 2\sum_i S^i f^i(y,S,t)\frac{\partial V_b}{\partial S^i} \right\} \right] = 0$$

4.2 Definition of bounds

We introduce here functions $\hat{v}_s(S,t), \bar{v}_s(S,t), \hat{v}_b(S,t), \bar{v}_b(S,t)$ be bounds for $u_s(S,t), u_b(S,t)$. One has to be careful in treating the stopping time. We define the various functions by the relations

$$-r\hat{v}_s + \frac{\partial \hat{v}_s}{\partial t} + r\sum_i S^i \frac{\partial \hat{v}_s}{\partial S^i} + \frac{1}{2}\sum_{i,j} a_{ij} S^i S^j \frac{\partial^2 \hat{v}_s}{\partial S^i \partial S^j}$$
$$+\min_y \left\{ \sum_i S^i f^i(y,S,t)\frac{\partial \hat{v}_s}{\partial S^i} \right\} \leq 0$$
$$\hat{v}_s(S,t) \geq h(S)$$
$$(\hat{v}_s(S,t) - h(S)) \left[-r\hat{v}_s + \frac{\partial \hat{v}_s}{\partial t} + r\sum_i S^i \frac{\partial \hat{v}_s}{\partial S^i} + \frac{1}{2}\sum_{i,j} a_{ij} S^i S^j \frac{\partial^2 \hat{v}_s}{\partial S^i \partial S^j} \right.$$
$$\left. + min_y \left\{ \sum_i S^i f^i(y,S,t)\frac{\partial \hat{v}_s}{\partial S^i} \right\} \right] = 0$$
$$\hat{v}_s(S,T) = h(S) \tag{4.4}$$

$$
\begin{aligned}
&-r\hat{v}_b + \frac{\partial \hat{v}_b}{\partial t} + r\sum_i S^i \frac{\partial \hat{v}_b}{\partial S^i} + \frac{1}{2}\sum_{i,j} a_{ij} S^i S^j \frac{\partial^2 \hat{v}_b}{\partial S^i \partial S^j} \\
&+ \max_y \left\{ \sum_i S^i f^i(y, S, t) \frac{\partial \hat{v}_b}{\partial S^i} \right\} \geq 0 \\
&\hat{v}_b(S,t) \leq h(S) \\
&(\hat{v}_s(S,t) - h(S)) \left[-r\hat{v}_b + \frac{\partial \hat{v}_b}{\partial t} + r\sum_i S^i \frac{\partial \hat{v}_b}{\partial S^i} + \frac{1}{2}\sum_{i,j} a_{ij} S^i S^j \frac{\partial^2 \hat{v}_b}{\partial S^i \partial S^j} \right. \\
&\left. + \, max_y \left\{ \sum_i S^i f^i(y, S, t) \frac{\partial \hat{v}_b}{\partial S^i} \right\} \right] = 0 \\
&\hat{v}_b(S,T) = h(S)
\end{aligned}
\tag{4.5}
$$

$$
\begin{aligned}
&-r\bar{v}_s + \frac{\partial \bar{v}_s}{\partial t} + r\sum_i S^i \frac{\partial \bar{v}_s}{\partial S^i} + \frac{1}{2}\sum_{i,j} a_{ij} S^i S^j \frac{\partial^2 \bar{v}_s}{\partial S^i \partial S^j} \\
&+ \sum_i S^i f^i(\frac{\partial \bar{v}_s}{\partial S}, S, t) \frac{\partial \bar{v}_s}{\partial S^i} \leq 0 \\
&\bar{v}_s(S,t) \geq h(S) \\
&(\bar{v}_s(S,t) - h(S)) \left[-r\bar{v}_s + \frac{\partial \bar{v}_s}{\partial t} + r\sum_i S^i \frac{\partial \bar{v}_s}{\partial S^i} + \frac{1}{2}\sum_{i,j} a_{ij} S^i S^j \frac{\partial^2 \bar{v}_s}{\partial S^i \partial S^j} \right. \\
&\left. + \sum_i S^i f^i(\frac{\partial \bar{v}_s}{\partial S}, S, t) \frac{\partial \bar{v}_s}{\partial S^i} \right] = 0 \\
&\bar{v}_s(S,T) = h(S)
\end{aligned}
\tag{4.6}
$$

$$
\begin{aligned}
&-r\bar{v}_b + \frac{\partial \bar{v}_b}{\partial t} + r\sum_i S^i \frac{\partial \bar{v}_b}{\partial S^i} + \frac{1}{2}\sum_{i,j} a_{i,j} S^i S^j \frac{\partial^2 \bar{v}_b}{\partial S^i \partial S^j} \\
&+ \sum_i S^i f^i \frac{\partial \bar{v}_b}{\partial S}, S, t) \frac{\partial \bar{v}_b}{\partial S^i} \geq 0 \\
&\bar{v}_b(S,t) \leq h(S) \\
&(\bar{v}_b(S,t) - h(S)) \left[-r\bar{v}_b + \frac{\partial \bar{v}_b}{\partial t} + r\sum_i S^i \frac{\partial \bar{v}_b}{\partial S^i} + \frac{1}{2}\sum_{i,j} a_{ij} S^i S^j \frac{\partial^2 \bar{v}_b}{\partial S^i \partial S^j} \right. \\
&\left. + \sum_i S^i f^i(\frac{\partial \bar{v}_b}{\partial S}, S, t) \frac{\partial \bar{v}_b}{\partial S^i} \right] = 0 \\
&\bar{v}_b(S,T) = h(S)
\end{aligned}
\tag{4.7}
$$

These functions are easily interpretable. We have namely

$$\begin{aligned}
\hat{v}_s(S,t) &= \inf_y \sup_\theta E^{Q^y}(h(S(\theta)) \exp - \int_t^\theta r(s)ds) \\
\hat{v}_b(S,t) &= \sup_y \inf_\theta E^{Q^y}(h(S(\theta)) \exp - \int_t^\theta r(s)ds) \\
\bar{v}_s(S,t) &= \sup_\theta E^{Q^{\bar{y}^s}}(h(S(\theta)) \exp - \int_t^\theta r(s)ds) \\
\bar{v}_b(S,t) &= \inf_\theta E^{Q^{\bar{y}^b}}(h(S(\theta)) \exp - \int_t^\theta r(s)ds)
\end{aligned} \tag{4.8}$$

where the specific control $\bar{y}^s, \bar{y}^b$ are defined as in (2.11), (2.12). One has clearly the relations

Proposition 4.1 *One has*

$$\bar{v}_b(S,t) \leq \hat{v}_b(S,t), \hat{v}_s(S,t) \leq \bar{v}_s(S,t) \tag{4.9}$$

□

Remark 4.1 *In the case $\theta = T$, european options, then $\hat{v}_s(S,t) = v_b(S,t)$, $\hat{v}_b(S,t) = v_s(S,t)$. Note that here the relation $\bar{v}_b(S,t) \leq \bar{v}_s(S,t)$ is trivial.*
□

4.3 Intervals for u_s and u_b

We prove here the analogue of Theorem 2.1, namely

Theorem 4.1 *We make the assumptions of Theorem 2.1, then the option prices $u_s(S,t), u_b(S,t)$ satisfy the relations*

$$\hat{v}_s(S,t) \leq u_s(S,t) \leq \bar{v}_s(S,t) \tag{4.10}$$

$$\hat{v}_b(S,t) \leq u_s(S,t) \leq \hat{v}_b(S,t) \tag{4.11}$$

Proof We prove (4.13). We begin with the right inequality. Consider the process (2.13), corresponding to the portfolio $\bar{y}^s(t)$, see (2.12), where $\bar{v}_s(S,t)$ is of course here the solution of (4.6). Pick θ to be any stopping time. Then by Ito's calculus and (4.6) we have:

$$\bar{v}_s(S(\theta),\theta)R_\theta^T - \bar{v}_s(S,t)R_t^T \leq \int_t^\theta R_s^T \sum_{i,j} \frac{\partial \bar{v}_s}{\partial S^i} S^i \sigma_j^i d\hat{w}_j \tag{4.12}$$

hence also

$$h(S(\theta))R_\theta^T - \bar{v}_s(S,t)R_t^T \leq \int_t^\theta R_s^T \sum_{i,j} \frac{\partial \bar{v}_s}{\partial S^i} S^i \sigma_j^i d\hat{w}_j \tag{4.13}$$

On the other hand, with the same portfolio

$$W(\theta)R_\theta^T - WR_t^T = \int_t^\theta R_s^T \sum_{i,j} \frac{\partial \bar{v}_s}{\partial S^i} S^i \sigma_j^i d\hat{w}_j \tag{4.14}$$

which implies

$$(W(\theta) - h(S(\theta)))R_\theta^T \geq (W - \bar{v}_s(S,t))R_t^T \tag{4.15}$$

and also

$$E^{Q^{y^*}} U((W(\theta) - h(S(\theta)))R_\theta^T \geq U(W - \bar{v}_s(S,t))R_t^T). \tag{4.16}$$

Since θ is arbitrary, we can take the inf in θ, to yield

$$\inf_\theta E^{Q^{y^*}} U((W(\theta) - h(S(\theta)))R_\theta^T \geq U((W - \bar{v}_s(S,t))R_t^T)$$

and thus

$$V_s(W,S,t) \geq U((W - \bar{v}_s(S,t))R_t^T) \tag{4.17}$$

This implies the right inequality (4.13). We prove next the left inequality (4.13). Take any control y, and consider the process $S(t)$, as defined by (1.15). Let $\hat{\theta}$ be the hitting time by $S(t)$ of the boundary of the domain $\{S, t | \hat{v}(S,t) > h(S)\}$. Since in this domain, the differential equality holds, we deduce from (4.4), applying Ito's calculus to $\hat{v}_s(S(t),t)R_t^T$ that

$$\hat{v}_s(S(\hat{\theta}),\hat{\theta})R_{\hat{\theta}}^T - \hat{v}_s(S,t)R_t^T \geq \int_t^{\hat{\theta}} R_s^T \sum_{i,j} \frac{\partial \hat{v}_s}{\partial S^i} S^i \sigma_j^i d\hat{w}_j \tag{4.18}$$

and from the definition of $\hat{\theta}$, also

$$h(S(\hat{\theta}))R_{\hat{\theta}}^T - \hat{v}_s(S,t)R_t^T \geq \int_t^{\hat{\theta}} R_s^T \sum_{i,j} \frac{\partial \hat{v}_s}{\partial S^i} S^i \sigma_j^i d\hat{w}_j. \tag{4.19}$$

On the other hand, we may write also

$$W(\hat{\theta})R_{\hat{\theta}}^T - WR_t^T = \int_t^{\hat{\theta}} R_s^T \sum_{i,j} y^i S^i \sigma_j^i d\hat{w}_j$$

therefore, one has

$$\begin{aligned}(W(\hat{\theta}) - h(S(\hat{\theta})))R_{\hat{\theta}}^T &\leq (W - \hat{v}_s(S,t))R_t^T \\ &+ \int_t^{\hat{\theta}} R_s^T \sum_{i,j} (y^i - \frac{\partial \hat{v}_s}{\partial S^i}) S^i \sigma_j^i d\hat{w}_j.\end{aligned} \tag{4.20}$$

Therefore also

$$E^{Q^y}U((W(\hat\theta)-h(S(\hat\theta)))R^T_{\hat\theta}) \le E^{Q^y}U\Bigg((W-\hat v_s(S,t))R^T_t + \int_t^{\hat\theta} R^T_s \sum_{i,j}(y^i-\frac{\partial \hat v_s}{\partial S^i})S^i\sigma^i_j d\hat w_j\Bigg).$$

Using the concavity of U, it follows

$$E^{Q^y}U((W(\hat\theta)-h(S(\hat\theta)))R^T_{\hat\theta}) \le U((W-\hat v_s(S,t))R^T_t)$$

hence

$$\inf_\theta E^{Q^y}U((W(\theta)-h(S(\theta)))R^T_\theta) \le U((W-\hat v_s(S,t))R^T_t)$$

and since y is arbitrary, we obtain $V_s(W,S,t) \le U((W-\hat v_s(S,t))R^T_t)$ which concludes the result. □

4.4 Equilibrium option prices for an agent with a CARA utility

We consider here, as in section 3.2, the case of an utility of the form (3.3). In this case, we postulate that we have:

$$V_s(W,S,t) = U((W-u_s(S,t))R^T_t) \quad (4.21)$$

$$V_b(W,S,t) = U((W-u_b(S,t))R^T_t) \quad (4.22)$$

with the following variational inequalities for u_s, u_b

$$u_s(S,T)=h(S), \;\; u_s(S,t)\ge h(S)$$

$$-\frac{\partial u_s}{\partial t} + ru_s - \sum_i rS^i\frac{\partial u_s}{\partial S^i} - \frac{1}{2}\sum_{i,j} S^iS^ja_{ij}\frac{\partial^2 u_s}{\partial S^i\partial S^j}$$
$$+\frac{1}{2}\max_y\Bigg\{-\gamma R^T_t\sum_{i,j}S^iS^ja_{ij}(y^i-\frac{\partial u_s}{\partial S^i})(y^j-\frac{\partial u_s}{\partial S^j})$$
$$-2\sum_i S^if^i(y,S,t)\frac{\partial u_s}{\partial S^i}\Bigg\} \ge 0$$

$$(u_s(S,t)-h(S))\Bigg[-\frac{\partial u_s}{\partial t} + ru_s - r\sum_i S^i\frac{\partial u_s}{\partial S^i} - \frac{1}{2}\sum_{i,j} S^iS^ja_{ij}\frac{\partial^2 u_s}{\partial S^i\partial S^j}$$
$$+\frac{1}{2}\max_y\Bigg\{-\gamma R^T_t\sum_{i,j}S^iS^ja_{ij}(y^i-\frac{\partial u_s}{\partial S^i})(y^j-\frac{\partial u_s}{\partial S^j})$$
$$-2\sum_i S^if^i(y,S,t)\frac{\partial u_s}{\partial S^i}\Bigg\}\Bigg] = 0$$

(4.23)

$$
\begin{aligned}
&u_b(S,T) = h(S)\\
&u_b(S,t) \geq h(S)\\
&-\frac{\partial u_b}{\partial t} + ru_b - r\sum_i S^i \frac{\partial u_b}{\partial S^i} - \frac{1}{2}\sum_{i,j} S^i S^j a_{ij} \frac{\partial^2 u_b}{\partial S^i \partial S^j}\\
&+\frac{1}{2}\min_y \left\{ \gamma R_t^T \sum_{i,j} S^i S^j a_{ij}(y^i + \frac{\partial u_b}{\partial S^i})(y^j + \frac{\partial u_b}{\partial S^j}) \right.\\
&\left. -2\sum_i S^i f^i(y,S,t) \frac{\partial u_b}{\partial S^i} \right\} \leq 0\\
&(u_b(S,t) - h(S)) \left[-\frac{\partial u_b}{\partial t} + ru_b - r\sum_i S^i \frac{\partial u_b}{\partial S^i} - \frac{1}{2}\sum_{i,j} S^i S^j a_{ij} \frac{\partial^2 u_b}{\partial S^i \partial S^j} \right.\\
&+\frac{1}{2}\min_y \left\{ \gamma R_t^T \sum_{i,j} S^i S^j a_{ij}(y^i + \frac{\partial u_b}{\partial S^i})(y^j + \frac{\partial u_b}{\partial S^j}) \right.\\
&\left.\left. -2\sum_i S^i f^i(y,S,t) \frac{\partial u_b}{\partial S^i} \right\} \right] = 0
\end{aligned}
\tag{4.24}
$$

Remark 4.2 *Unlike in the case of European options, one has by construction*

$$u_b(S,t) \leq u_s(S,t) \tag{4.25}$$

for any value of γ.

4.5 Behaviour as γ tends to $+\infty$ or 0

Denoting by $u_s^\gamma(S,t), u_b^\gamma(S,t)$ the solutions of (4.26), (4.27), one checks immediately

$$u_s^\infty(S,t) = \bar{v}_s(S,t) \tag{4.26}$$

$$u_s^0(S,t) = \hat{v}_s(S,t) \tag{4.27}$$

$$u_b^\infty(S,t) = \bar{v}_b(S,t) \tag{4.28}$$

$$u_b^0(S,t) = \hat{v}_b(S,t) \tag{4.29}$$

We then proceed with the monotonicity property:

Theorem 4.2 *We make the assumptions of Theorem 3.1, then* $u_s^\gamma(S,t)$ *is increasing with respect to* γ, *whereas* $u_b^\gamma(S,t)$ *is decreasing with respect to* γ.

Proof Pick $\gamma' > \gamma$, and set $\tilde{u}_s(S,t) = u_s^{\gamma'}(S,t) - u_s^\gamma(S,t)$. Consider the domain $\{S,t | u_s^\gamma(S,t) > h(S)\}$, on which the differential inequality in (4.26) is an equality. We can write on this domain

$$-\frac{\partial \tilde{u}_s}{\partial t} + r\tilde{u}_s - \sum_i rS^i \frac{\partial \tilde{u}_s}{\partial S^i} - \frac{1}{2}\sum_{i,j} S^i S^j a_{ij} \frac{\partial^2 \tilde{u}_s}{\partial S^i \partial S^j} \geq$$

$$-\frac{1}{2}\max_y \left\{ -\gamma' R_t^T \sum_{i,j} S^i S^j a_{ij}(y^i - \frac{\partial u_s^{\gamma'}}{\partial S^i})(y^j - \frac{\partial u_s^{\gamma'}}{\partial S^j}) \right.$$

$$\left. -2\sum_i S^i f^i(y,S,t)\frac{\partial u_s^{\gamma'}}{\partial S^i} \right\}$$

$$+\frac{1}{2}\max_y \left\{ -\gamma R_t^T \sum_{i,j} S^i S^j a_{ij}(y^i - \frac{\partial u_s^{\gamma}}{\partial S^i})(y^j - \frac{\partial u_s^{\gamma}}{\partial S^j}) \right.$$

$$\left. -2\sum_i S^i f^i(y,S,t)\frac{\partial u_s^{\gamma}}{\partial S^i} \right\}$$

and since $\gamma' > \gamma$:

$$\geq -\frac{1}{2}\max_y \left\{ -\gamma R_t^T \sum_{i,j} S^i S^j a_{ij}(y^i - \frac{\partial u_s^{\gamma'}}{\partial S^i})(y^j - \frac{\partial u_s^{\gamma'}}{\partial S^j}) \right.$$

$$\left. -2\sum_i S^i f^i(y,S,t)\frac{\partial u_s^{\gamma'}}{\partial S^i} \right\}$$

$$+\frac{1}{2}\max_y \left\{ -\gamma R_t^T \sum_{i,j} S^i S^j a_{ij}(y^i - \frac{\partial u_s^{\gamma}}{\partial S^i})(y^j - \frac{\partial u_s^{\gamma}}{\partial S^j}) - 2\sum_i S^i f^i(y,S,t)\frac{\partial u_s^{\gamma}}{\partial S^i} \right\}.$$

We proceed as in the proof of Theorem 3.2, choosing $\hat{y}$ realizing the first max in the right hand side, to obtain:

$$\geq \sum_i S^i f^i(\hat{y},S,t)\frac{\partial \tilde{u}_s}{\partial S^i} - \gamma R_t^T \sum_{i,j}(\hat{y}^i - \frac{\partial u_s^{\gamma}}{\partial S^i}) a_{ij} S^i S^j \frac{\partial \tilde{u}_s}{\partial S^j}.$$

Outside the set $\{S,t | u_s^{\gamma}(S,t) > h(S)\}$, one has clearly $\tilde{u}_s \geq 0$, and at time T, $\tilde{u}_s(S,T) = 0$. By maximum principle considerations, $\tilde{u}_s(S,t) \geq 0$. □

References

Bensoussan A. and Lions J.L. (1978) *Applications of Variational Inequalities in Stochastic Control*, North Holland.

Davis, M.H.A., Panas, V.G., Zariphopoulou (1993) European option pricing with transaction costs, *SIAM J. Control Optim.*

Fleming W.H, Rishel R.W. (1975) *Deterministic and Stochastic Optimal Control*, Springer-Verlag.

32

Pathwise Comparison of Arithmetic Brownian Motions and Log-normal Processes

G. Ferreyra
P. Sundar

ABSTRACT: Pathwise comparison results for controlled arithmetic Brownian motion processes and for controlled log-normal, also called logarithmic Brownian motion, processes are obtained. Applications of these results to finance are given.

KEYWORDS: Stochastic differential equations, comparison theorem, log-normal, finance, stochastic control.

1 Introduction

Given $x_i \in \mathbb{R}$, $i = 1, 2$, and μ_i, σ_i strictly positive constants, given any reference probability system $\nu = (\Omega, \mathcal{F}, \{\mathcal{F}_t\}, P, B)$ (see Fleming and Soner (1993), p. 160), and any given strictly positive, bounded, progressively measurable control processes u_i with C^1 paths, let $X_i(t)$ be the solutions of the one-dimensional controlled stochastic differential equations with different drift and different dispersion coefficients

$$X_i(t) = x_i + \int_0^t u_i(s)\mu_i f(X_i(s))\,ds + \int_0^t u_i(s)\sigma_i f(X_i(s))\,dB(s), \qquad (1)$$

where f is an appropiate function. The goal of this paper is to obtain pathwise comparison results of the form

$$P\{X_1(t) - x_1 \leq X_2(t) - x_2 : 0 \leq t < \tau\} = 1, \qquad (2)$$

for a certain stopping time τ in each of the following three cases.

1. Proportionally controlled log-normals: $u_1(t) = Ku_2(t)$, $f(x) = x$;
2. Log-normals: $u_1 = u_2 = 1$, $\quad f(x) = x$;
3. Controlled arithmetic Brownian motions: $f(x) = 1$.

To prove the comparison theorems, we consider the transformation

$$F_i(t,x) = \frac{1}{u_i(t)\sigma_i}\int_{x_i}^{x}\frac{d\xi}{f(\xi)}, \tag{3}$$

or a slight variant of it, and apply the Ito rule to the processes

$$Y_i(t) = F_i(t, X_i(t)). \tag{4}$$

The transformation (3) is similar to the one used in Gal'čhuk and Davis (1981/82) and in Ferreyra and Sundar (1997) where pathwise comparison results for other situations are obtained. Pathwise comparison results for solutions of stochastic differential equations with either equal or different dispersion coefficients are numerous. For a list of references, see Ferreyra and Sundar (1997). The novelty here is that (1) depends on control processes and that $F_i(\cdot, x)$ is a random process. By the way, we note that if D_1 denotes the derivative of F_i with respect to the first variable, then

$$D_1 F_i(t, X_i(t)) = -\frac{u_i'(t)}{u_i(t)}Y_i(t). \tag{5}$$

Of the results in the literature, the closest to those contained in this paper appear to be those in O'Brien (1980). But the restriction on the drift coefficients in relation to the dispersion coefficients in O'Brien (1980) makes those results inapplicable to (1).

2 Proportionally controlled log-normals

In the successful Black-Scholes theory, the price of a risky asset is modelled as a log-normal, i.e., an equation of the form of (1) with $u_i = 1$ and $f(x) = x$. In such a model, μ_i is the average return of the asset and σ_i is the volatility of its price X_i. In this and the next section we obtain comparison theorems for log-normal processes.
Assume that

(i) $f(x) = x$,

(ii) The controls u_i are nonanticipating and strictly positive processes with C^1 paths. Moreover, assume that the controls are proportional, i.e., there is a constant $K > 0$ such that $u_2(t) = Ku_1(t)$, for all t,

(iii) $x_1, x_2, \sigma_1, \sigma_2 > 0$,

(iv) $\sigma_1 \leq K\sigma_2$,

(v) $\sigma_1 x_1 = K\sigma_2 x_2$,

(vi) $\frac{\mu_1}{\sigma_1} - \frac{u_1(t)\sigma_1}{2} \leq \frac{\mu_2}{\sigma_2} - \frac{u_2(t)\sigma_2}{2}$ for all t, a.e. $\omega \in \Omega$.

Note that the above assumptions imply

$$X_1, X_2 > 0, \quad \text{and} \quad x_1 \geq x_2. \tag{6}$$

Also note that (v) is a condition on x_1 in terms of x_2 or vice versa.
Let τ be the exit time of the process $X_1 - x_1$ from the interval $(-x_2, \infty)$.

Theorem 1 *Under the assumptions (i)-(vi),*

$$P\{X_1(t) - x_1 \leq X_2(t) - x_2 : 0 \leq t < \tau\} = 1, \tag{7}$$

Proof. Let $0 \leq t < \tau$ and for $i = 1, 2$, let $Z_i = X_i - x_i$. Then, $Z_i > -x_i, Z_1 > -x_2$ and

$$Z_i(t) = \int_0^t u_i(s)(Z_i(s) + x_i)\mu_i \, ds + \int_0^t u_i(s)(Z_i(s) + x_i)\sigma_i \, dB(s). \tag{8}$$

Let

$$F_i(t, x) = \frac{1}{u_i(t)\sigma_i} \int_0^x \frac{d\xi}{\xi + x_i} \qquad x \in (-x_2, \infty). \tag{9}$$

By (v) the lines $\xi \mapsto (\xi + x_1)\sigma_1$ and $\xi \mapsto K(\xi + x_2)\sigma_2$ intersect at $\xi = 0$. This, and the relation for their slopes (iv), implies

$$F_2(t, x) \leq F_1(t, x), \quad t < \tau, x \in (-x_2, \infty). \tag{10}$$

(This inequality is related to the condition on the dispersion coefficients in Gal'čhuk and Davis (1981/82) and Ferreyra and Sundar (1997).)
By the Ito rule (see Protter (1990)), $Y_i(t) = F_i(t, Z_i(t))$ satisfies

$$Y_i(t) = \int_0^t L_i(s, Y_i(s)) \, ds + \int_0^t dB(s), \tag{11}$$

where

$$L_i(t, y) = -\frac{u_i'(t)}{u_i(t)} y + \frac{\mu_i}{\sigma_i} - \frac{u_i(t)\sigma_i}{2}. \tag{12}$$

Since the lines $L_i(t, \cdot), i = 1, 2$, are parallel, (vi) and the standard pathwise comparison theorem for one-dimensional solutions of stochastic differential equations with equal dispersion coefficients (see Ikeda and Watanabe (1977)) imply that

$$P\{Y_1(t) \leq Y_2(t) : 0 \leq t < \tau\} = 1. \tag{13}$$

This, together with (10) implies

$$\begin{aligned}\frac{1}{Ku_1(t)\sigma_2}\int_0^{Z_1(t)}\frac{d\xi}{\xi+x_2} &= F_2(t,Z_1(t)) \le F_1(t,Z_1(t)) = Y_1(t)\\ &\le Y_2(t) = \frac{1}{Ku_1(t)\sigma_2}\int_0^{Z_2(t)}\frac{d\xi}{\xi+x_2}.\end{aligned}$$

Since the integrands are positive, $P\{Z_1(t) \le Z_2(t) : 0 \le t < \tau\} = 1$. This is just (7). ∎

3 Log-normals

Suppose the hypotheses (i) and (iii) of the previous section hold and that $u_1 \equiv u_2 \equiv K = 1$. Then the solution of (1) is a log-normal process and Theorem 1 is a comparison result for two log-normal processes. As an application, we can establish a sufficient condition for increasing gains in the face of increased volatility of asset prices. That is, if higher volatility of asset prices is encountered, what assumption on the mean return does guarantee increased gains in pathwise sense ? Or in other words, if $\sigma_2 > \sigma_1$, what conditions do guarantee that (2) holds ?
Assuming $\sigma_2 > \sigma_1$, $u_1 \equiv u_2 \equiv K = 1$, (i), (iii) and (v) above, we see that to obtain (vi), we need

$$\mu_2 \ge \left(\frac{\mu_1}{\sigma_1} - \frac{\sigma_1}{2} + \frac{\sigma_2}{2}\right)\sigma_2. \tag{14}$$

Since $\sigma_2 > \sigma_1$ implies $\frac{\sigma_2}{2} - \frac{\sigma_1}{2} > 0$, and $\sigma_2/\sigma_1 > 1$, (14) gives a lower bound on the mean return $\mu_2\,(> \mu_1)$ in order to guarantee the pathwise relation (2).

4 Controlled arithmetic Brownian motions

Consider the one-dimensional controlled SDE given by

$$dX(t) = u(t)\mu\, dt + u(t)\sigma\, dB(t). \tag{15}$$

This controlled stochastic differential equation is a model for reinsurance, see Taksar and Zhou (1997) and Hojgaard and Taksar (1998). If $X(t)$ is the reserve at time t, the constant μ is the profit rate, the constant σ is the volatility, and the nonanticipating control $u(t)$ satisfies $0 \le u(t) \le 1$, with $1 - u(t)$ the reinsurance fraction at time t.
The question we explore in this section is the following. Under what conditions does $u_1(t) \le u_2(t)$ for all t, imply that the corresponding reserve processes X_1, X_2 satisfy

$$P\{X_1(t) \le X_2(t) : 0 \le t\} = 1? \tag{16}$$

Consider the stochastic differential equations (1) and assume that

(i) $f(x) \equiv 1$,

(ii) $\mu_1 = \mu_2 > 0, \sigma_1 = \sigma_2 > 0$,

(iii) The controls u_i are nonanticipating and strictly positive processes with C^1 paths,

(iv) $0 < \frac{x_1}{u_1(0)} \le \frac{x_2}{u_2(0)}$, a.e.$\omega \in \Omega$

(v) $\frac{u_1'(t)}{u_1(t)} \ge \frac{u_2'(t)}{u_2(t)}$ for all t, a.e. $\omega \in \Omega$.

For $i = 1, 2$, consider the functions

$$F_i(t, x) = \frac{x}{u_i(t)\sigma}, \tag{17}$$

and define

$$Y_i(t) = F_i(t, X_i(t)). \tag{18}$$

Let τ be the minimum of the exit times of the processes Y_i from the interval $(0, \infty)$.

Theorem 2 *Assume (i)-(v) above. Then*

$$P\Big\{0 \le \frac{X_1(t)}{u_1(t)} \le \frac{X_2(t)}{u_2(t)} : 0 \le t < \tau\Big\} = 1. \tag{19}$$

Proof. The Ito rule implies

$$dY_i(t) = (-\frac{u_i'(t)}{u_i(t)} Y_i(t) + \frac{\mu}{\sigma})\, dt + dB(t), \tag{20}$$

with initial conditions satisfying $Y_1(0) \le Y_2(0)$ by (iv). Then, by (v) and the usual comparison theorem for solutions of one-dimensional stochastic differential equations with equal dispersion coefficients (see Ikeda and Watanabe (1977))

$$P\{Y_1(t) \le Y_2(t) : 0 \le t < \tau\} = 1. \tag{21}$$

Then (19) follows. ∎

Corollary 1 *In addition to (i)-(v) above assume*

(vi) $0 < u_1(t) \leq u_2(t)$ *for all* t.

Then

$$P\{0 < X_1(t) \leq X_2(t) : 0 \leq t < \tau\} = 1. \tag{22}$$

Acknowledgments: The second author is supported by U.S. Army Research Office grant DAAH04-94-G-0249.

References

Ikeda, N. and Watanabe, S. (1977), A comparison theorem for solutions of stochastic differential equations and its applications, *Osaka J. Math.* **14**, 619-633.

Ferreyra, G. and Sundar, P. (1997), Comparison of solutions of stochastic equations and applications, submitted. Available at http://math.lsu.edu/TigerNotes/1998/VOL98.html

Fleming, W. H. and Soner, H. M. (1993), *Controlled Markov Processes and Viscosity Solutions*, Appl. Math., vol 25, Springer-Verlag, New York.

Gal'čhuk, L. I. and Davis, M. H. A. (1981/82), A note on a comparison theorem for equations with different diffusions, *Stochastics* **6**, 147-149.

Hojgaard, B. and Taksar, M. (1998), Controlling risk exposure and dividends pay-out schemes: Insurance company example, preprint.

O'Brien, G. L. (1980), A new comparison theorem for solutions of stochastic differential equations, *Stochastics* **3**, no. 4, pp. 245-249.

Protter, P. (1990), *Stochastic Integration and Differential Equations*, Appl. Math., vol 21, Springer-Verlag, New York.

Taksar, M. and Zhou, Xun Yu, (1997), In *Proceedings of the 36th Conference on Decision and Control*, San Diego, CA, Dec 10-12, 1997, pp. 2791-2794.

33

Critical Power for Asymptotic Connectivity in Wireless Networks

Piyush Gupta
P. R. Kumar

KEYWORDS: Wireless networks, connectivity, critical power, critical range, continuum percolation.

ABSTRACT: In wireless data networks each transmitter's power needs to be high enough to reach the intended receivers, while generating minimum interference on other receivers sharing the same channel. In particular, if the nodes in the network are assumed to cooperate in routing each others' packets, as is the case in *ad hoc* wireless networks, each node should transmit with just enough power to guarantee connectivity in the network. Towards this end, we derive the *critical* power a node in the network needs to transmit in order to ensure that the network is connected with probability one as the number of nodes in the network goes to infinity. It is shown that if n nodes are placed in a disc of unit area in $\Re^2$ and each node transmits at a power level so as to cover an area of $\pi r^2 = (\log n + c(n))/n$, then the resulting network is asymptotically connected with probability one if and only if $c(n) \to +\infty$.

1 Introduction

Wireless communication systems consist of nodes which share a common communication medium: namely, radio. Signals intended for a receiver cause interference at other receiver nodes. This results in reduced signal to noise ratio at the latter receivers, and thus, in the lowering of their information-processing capacity. Hence, it becomes essential to control the transmitter power such that the information signals reach their intended receivers, while causing minimal interference for other receivers sharing the same channel. To achieve this objective, many iterative power control algorithms have been developed (Bambos, Chen and Pottie (1995), Ulukus and Yates (1996) and the references therein).

In this paper we look at the problem from a different perspective. We assume that nodes in the network cooperate in routing each others' data packets. Examples of such networks are mobile *ad hoc* networks (Gupta and

Kumar (1996) and Johnson and Maltz (1996)). They are networks formed by a group of mobile nodes which communicate with each other over a wireless channel and without any centralized control. In such networks, a critical requirement is that each node in the network has a *path* to every other node in the network, i.e., the network is connected. With this in mind, we consider the problem of determining the *critical* power at which each node needs to transmit so as to guarantee asymptotic connectivity of the network.

More precisely, we consider the following problem: Let $\mathcal{D}$ be a disc in $\Re^2$ having unit area. Let $\mathcal{G}(n, r(n))$ be the network (graph) formed when n nodes are placed uniformly and independently in $\mathcal{D}$, and two nodes i and j can communicate with each other if the distance between them is less than $r(n)$. That is, if x_k is the location of node k, nodes i and j can communicate if $\|x_i - x_j\| \leq r(n)$, where the norm used is the Euclidean norm (i.e., L^2-norm). The radius $r(n)$ is usually referred to as the *range* of a node in $\mathcal{G}(n, r(n))$. Then the problem is to determine $r(n)$ which guarantees that $\mathcal{G}(n, r(n))$ is asymptotically connected with probability one, i.e., the probability that $\mathcal{G}(n, r(n))$ is connected, denoted by $P_C(n, r(n))$, goes to one as $n \to \infty$. For this problem, we show that if $\pi r^2(n) = \frac{\log n + c(n)}{n}$, then $P_C(n, r(n)) \to 1$ if and only if $c(n) \to +\infty$.

A related problem that has been considered in the literature is connectivity in *Bernoulli* graphs: Let $\mathcal{B}(n, p(n))$ be a graph consisting of n nodes, in which edges are chosen independently and with probability $p(n)$. Then, it has been shown that if $p(n) = \frac{\log n + c(n)}{n}$, the probability that $\mathcal{B}(n, p(n))$ is connected goes to one if and only if $c(n) \to +\infty$ (Theorem VII.3 in Bollobás (1985)). Even though the asymptotic expression is the same, connectivity in $\mathcal{G}(n, r(n))$ is quite different from connectivity in $\mathcal{B}(n, p(n))$. The event that there are links between i and j, and between j and k, is not independent of the event that there is a link between i and k (as, fixing x_i, the former is true given the latter only if j lies in the intersection of two discs of radius $r(n)$ and centered at i and k, with $\|x_i - x_k\| \leq r(n)$. This has lower probability than the probability $(\pi r^2(n))^2$ of the event that there are links between i and j, and j and k). As it turns out, an entirely different proof technique was needed to prove asymptotic connectivity in $\mathcal{G}(n, r(n))$.

Another closely related problem considered in the literature is the *coverage* problem: Disks of radius a are placed in a unit-area disc $\mathcal{D} \in \Re^2$ at a Poisson intensity of λ, i.e., number of discs having their centers in a set $\mathcal{A} \subset \mathcal{D}$ of area $|\mathcal{A}|$ is Poisson distributed with mean $\lambda|\mathcal{A}|$. Let $\mathcal{V}(\lambda, a)$ denote the vacancy within $\mathcal{D}$, i.e., $\mathcal{V}(\lambda, a)$ is the region of $\mathcal{D}$ not covered by the disks. Then it has been shown in Hall (1988) (Theorem 3.11) that

$$\begin{aligned}\frac{1}{20}\min\left\{1, (1+\pi a^2\lambda^2)e^{-\pi a^2\lambda}\right\} &< P(|\mathcal{V}(\lambda, a)| > 0) \\ &< \min\left\{1, 3(1+\pi a^2\lambda^2)e^{-\pi a^2\lambda}\right\}. \qquad (1)\end{aligned}$$

Note that (1) has more stringent asymptotics on $a(n)$ than our result. If $\lambda = n$ and $\pi a^2(n) = \frac{\log n + \log\log n + c(n)}{n}$, then $\lim_{n\to\infty} P(|\mathcal{V}(n, a(n))| > 0) = 0$ for $c(n) \to +\infty$, and $\lim_{n\to\infty} P(|\mathcal{V}(n, a(n))| > 0) \geq 1/20$ for $c(n) \to -\infty$. Also, note that coverage of $\mathcal{D}$ by discs of radius $a(n) = r(n)$ does not guarantee connectivity in $\mathcal{G}(n, r(n))$ (recall $r(n)$ is the range of nodes in $\mathcal{G}(n, r(n))$)). However, $a(n) = r(n)/2$ does; the corresponding lower bound on $r(n)$ is $\pi r^2(n) = 4\frac{\log n + \log\log n + c(n)}{n}$ for $c(n) \to +\infty$, which is much weaker than the one we obtain. Moreover, since $\mathcal{G}(n, r(n))$ can be connected without $\mathcal{D}$ being entirely covered by n discs of radius $r(n)$, this approach does not lead to any necessary conditions on $r(n)$ for asymptotic connectivity in $\mathcal{G}(n, r(n))$.

Yet another related problem considered is in *continuum percolation* theory (Kesten (1982), Mesteer and Roy (1996)): Nodes are assumed to be distributed with Poisson intensity λ in $\Re^2$, and two nodes are connected to each other if the distance between them is less than r. Then the problem considered is to find a critical value of r such that the origin is connected to an infinite-order component. Of course for this to make sense, the node distribution process is conditioned on the origin having a node. We will, in fact, make use of some results from percolation theory while deriving the sufficient condition on $r(n)$ for asymptotic connectivity in $\mathcal{G}(n, r(n))$ (cf. Section 3).

The rest of the paper is organized as follows. In Section 2 we derive the necessary condition on $r(n)$ for asymptotic connectivity of $\mathcal{G}(n, r(n))$. The sufficiency of this condition is proved in Section 3. We conclude in Section 4 with some comments on extensions of the problem considered.

2 Necessary Condition on $r(n)$ for Connectivity

In this section we derive necessary conditions on the radio range of a node in the network for asymptotic connectivity. In the following, to avoid technicalities which obscure the main ideas, we will neglect edge effects resulting due to a node being close to the boundary of $\mathcal{D}$. The complete proofs which take the edge effects into account are given in the Appendix.

We will frequently use the following bounds.

Lemma 2.1 *(i) For any* $p \in [0, 1]$

$$(1 - p) \leq e^{-p}. \tag{2}$$

(ii) For any given $\theta \geq 1$, *there exists* $p_0 \in [0, 1]$, *such that*

$$e^{-\theta p} \leq (1 - p), \quad \textit{for all } 0 \leq p \leq p_0. \tag{3}$$

If $\theta > 1$, *then* $p_0 > 0$.

Lemma 2.2 *If* $\pi r^2(n) = \frac{\log n + c}{n}$, *then, for any fixed* $\theta < 1$ *and for all sufficiently large* n

$$n(1 - \pi r^2(n))^{n-1} \geq \theta e^{-c}. \tag{4}$$

Proof: Taking the logarithm of the left hand side of (4), we get

$$\log(\text{L.H.S. of (4)}) = \log n + (n-1)\log\left(1 - \pi r^2(n)\right).$$

Using the power series expansion for $\log(1-x)$,

$$\begin{aligned}\log(\text{L.H.S. of (4)}) &= \log n - (n-1)\sum_{i=1}^{\infty}\frac{\left(\pi r^2(n)\right)^i}{i} \\ &= \log n - (n-1)\left(\sum_{i=1}^{2}\frac{(\log n + c)^i}{in^i} + \mathcal{E}(n)\right), \end{aligned}\tag{5}$$

where

$$\begin{aligned}\mathcal{E}(n) &= \sum_{i=3}^{\infty}\frac{(\log n + c)^i}{in^i} \\ &\leq \frac{1}{3}\int_2^{\infty}\left(\frac{\log n + c}{n}\right)^x dx \\ &= \frac{1}{3\log\left(\frac{\log n + c}{n}\right)}\left(\frac{\log n + c}{n}\right)^x\Bigg|_2^{\infty} \\ &\leq \frac{1}{3}\left(\frac{\log n + c}{n}\right)^2, \end{aligned}\tag{6}$$

for all large n. Substituting (6) in (5), we get

$$\begin{aligned}\log(\text{L.H.S. of (4)}) &\geq \log n - (n-1)\left(\frac{\log n + c}{n} + \frac{5(\log n + c)^2}{6n^2}\right) \\ &\geq -c - \frac{(\log n + c)^2 - (\log n + c)}{n} \\ &\geq -c - \epsilon,\end{aligned}$$

for all sufficiently large n. The result follows by taking the exponent of both sides and using $\theta = e^{-\epsilon}$. □

Now, let $P^{(k)}(n, r(n)), k = 1, 2, \ldots$ denote the probability that a graph $\mathcal{G}(n, r(n))$ has at least one order-k component. By an order-k component we mean a set of k nodes which form a connected set, but which are not connected with any other node. Also, let $P_d(n, r(n))$ denote the probability that $\mathcal{G}(n, r(n))$ is disconnected.

Theorem 2.1 *If* $\pi r^2(n) = \frac{\log n + c(n)}{n}$, *then*

$$\liminf_{n\to\infty} P_d(n, r(n)) \geq e^{-c}\left(1 - e^{-c}\right), \tag{7}$$

where $c = \lim_{n\to\infty} c(n)$.

Proof: We first study the case where $\pi r^2(n) = \frac{\log n + c}{n}$ for a fixed c. Consider $P^{(1)}(n, r(n))$, the probability that $\mathcal{G}(n, r(n))$ has *at least* one order-1 component. That is, $P^{(1)}(n, r(n))$ is the probability that $\mathcal{G}(n, r(n))$ has at least one node which does not include any other node in its range. Then

$$\begin{aligned} P^{(1)}(n, r(n)) &\geq \sum_{i=1}^{n} P(\{i \text{ is the only isolated node in } \mathcal{G}(n, r(n))\}) \\ &\geq \sum_{i=1}^{n} \Big(P(\{i \text{ is an isolated node in } \mathcal{G}(n, r(n))\}) \\ &\quad - \sum_{j\neq i} P(\{i \text{ and } j \text{ are isolated nodes in } \mathcal{G}(n, r(n))\}) \Big) \\ &\geq \sum_{i=1}^{n} P(\{i \text{ is isolated in } \mathcal{G}(n, r(n))\}) \\ &\quad - \sum_{i=1}^{n} \sum_{j\neq i} P(\{i \text{ and } j \text{ are isolated in } \mathcal{G}(n, r(n))\}). \end{aligned} \tag{8}$$

Neglecting edge effects, we get

$$\begin{aligned} P(\{i \text{ is isolated in } \mathcal{G}(n, r(n))\}) &\sim (1 - \pi r^2(n))^{n-1}, \\ P(\{i \text{ and } j \text{ isolated in } \mathcal{G}(n, r(n))\}) &\sim (4\pi r^2(n) - \pi r^2(n))(1 - \tfrac{5}{4}\pi r^2(n))^{n-2} \\ &\quad + (1 - 4\pi r^2(n))(1 - 2\pi r^2(n))^{n-2}. \end{aligned} \tag{9}$$

The first term on the RHS above takes into account the case where j is at a distance between $r(n)$ and $2r(n)$ from i. Substituting (9) in (8), we get

$$\begin{aligned} P^{(1)}(n, r(n)) &\geq n(1 - \pi r^2(n))^{n-1} - n(n-1)\Big(3\pi r^2(n)(1 - \tfrac{5}{4}\pi r^2(n))^{n-2} \\ &\quad + (1 - 2\pi r^2(n))^{n-2}\Big). \end{aligned}$$

Using Lemmas 2.1 and 2.2, we get that for $\pi r^2(n) = \frac{\log n + c}{n}$, and for any fixed $\theta < 1$ and $\epsilon > 0$,

$$\begin{aligned} P^{(1)}(n, r(n)) &\geq \theta e^{-c} - n(n-1)\Big(3\pi r^2(n) e^{-\frac{5}{4}(n-2)\pi r^2(n)} + e^{-2(n-2)\pi r^2(n)}\Big) \\ &\geq \theta e^{-c} - (1+\epsilon) e^{-2c}, \end{aligned}$$

for all $n > N(\epsilon, \theta, c)$. Since $P^{(1)}(n, r(n)) \leq P_d(n, r(n))$, we have

$$P_d(n, r(n)) \geq \theta e^{-c} - (1+\epsilon)e^{-2c}, \tag{10}$$

for all $n > N(\epsilon, \theta, c)$. Now, consider the case where c is a function $c(n)$ with $\lim_{n\to\infty} c(n) = \bar{c}$. Then, for any $\epsilon > 0$, $c(n) \leq \bar{c} + \epsilon$ for all $n \geq N'(\epsilon)$. Also, the probability of disconnectedness is monotone decreasing in c. Hence

$$P_d(n, r(n)) \geq \theta e^{-(\bar{c}+\epsilon)} - (1+\epsilon)e^{-2(\bar{c}+\epsilon)}.$$

for $n \geq \max\{N(\epsilon, \theta, \bar{c}+\epsilon), N'(\epsilon)\}$. Taking limits

$$\liminf_{n\to\infty} P_d(n, r(n)) \geq \theta e^{-(\bar{c}+\epsilon)} - (1+\epsilon)e^{-2(\bar{c}+\epsilon)}.$$

Since this holds for all $\epsilon > 0$ and $\theta < 1$, the result follows. □

Corollary 2.1 *Graph $\mathcal{G}(n, r(n))$ is asymptotically disconnected with positive probability if $\pi r^2(n) = \frac{\log n + c(n)}{n}$ and $\limsup_n c(n) < +\infty$.*

3 Sufficient Condition on $r(n)$ for Connectivity

In order to derive a lower bound on $r(n)$ so as to ensure asymptotic connectivity in $\mathcal{G}(n, r(n))$, we make use of some results from *continuum percolation* (Meester and Roy (1996)). In percolation theory, nodes are assumed to be distributed with Poisson intensity λ in $\Re^2$ (results are in fact available for more general cases, see Meester and Roy (1996)). As in $\mathcal{G}(n, r(n))$, two nodes are connected to each other if the distance between them is less than $r(\lambda)$. Let $\mathcal{G}^{\text{Poisson}}(\lambda, r(\lambda))$ denote the resultant (infinite) graph. Also, let $q_k(\lambda, r(\lambda))$ be the probability that the node at the origin is a part of an order-k component. Of course for this to make sense, the node distribution process is conditioned on the origin having a node. Then, $(1 - \sum_{k=1}^{\infty} q_k(\lambda, r(\lambda))) =: q_\infty(\lambda, r(\lambda))$ gives the probability that the origin is connected to an infinite-order component. It can be shown that almost surely $\mathcal{G}^{\text{Poisson}}(\lambda, r(\lambda))$ has at most one infinite-order component for each $\lambda \geq 0$ (Theorem 6.3 of Meester and Roy (1996)). Furthermore, the following is true (Propositions 6.4-6.6 of Meester and Roy (1996))

$$\lim_{\lambda\to\infty} \frac{1}{q_1(\lambda, r(\lambda))} \sum_{k=1}^{\infty} q_k(\lambda, r(\lambda)) = 1. \tag{11}$$

Hence, as $\lambda \to \infty$, almost surely the origin in $\mathcal{G}^{\text{Poisson}}(\lambda, r(\lambda))$ lies in either an infinite-order component or an order-1 component (i.e., it is isolated). Now, our original problem concerning a fixed number of nodes n in the unit-area disc $\mathcal{D}$ can be approximated by regarding that process as the restriction to $\mathcal{D}$ of the Poisson process on $\Re^2$ with $\lambda = n$. Let the graph obtained by restricting $\mathcal{G}^{\text{Poisson}}(n, r(n))$ to $\mathcal{D}$ be denoted by $\mathcal{G}_{\mathcal{D}}^{\text{Poisson}}(n, r(n))$. Then, by the above observation, the probability that $\mathcal{G}_{\mathcal{D}}^{\text{Poisson}}(n, r(n))$ is disconnected, denoted by $P_d^{\text{Poisson}}(n, r(n))$, is asymptotically the same as the probability that it has at least one isolated node, denoted by $P^{\text{Poisson};(1)}$ $(n, r(n))$. Although $\mathcal{G}_{\mathcal{D}}^{\text{Poisson}}(n, r(n))$ has a Poisson(n) number of nodes in $\mathcal{D}$, the difference between $\mathcal{G}_{\mathcal{D}}^{\text{Poisson}}(n, r(n))$ and $\mathcal{G}(n, r(n))$ is negligible for large n. This is made precise below.

Lemma 3.1 *If* $\pi r^2(n) = \frac{\log n + c(n)}{n}$, *then*

$$\limsup_{n\to\infty} P^{\text{Poisson};(1)}(n, r(n)) \leq e^{-c}, \tag{12}$$

where $c = \lim_{n\to\infty} c(n)$.

Proof: Note that since $e^{-n}\frac{n^j}{j!}$ is the probability that $\mathcal{G}_{\mathcal{D}}^{\text{Poisson}}(n, r(n))$ has j nodes, and defining a graph with 0 nodes to be connected, we have

$$P^{\text{Poisson};(1)}(n, r(n)) = \sum_{j=1}^{\infty} P^{(1)}(j, r(n)) e^{-n}\frac{n^j}{j!}. \tag{13}$$

Let $E_1(j, r(n))$ denote the expected number of order-1 components in $\mathcal{G}(j, r(n))$. Then

$$\begin{aligned} P^{(1)}(j, r(n)) &\leq E_1(j, r(n)) \\ &= E[\sum_{i=1}^{j} I(i \text{ is isolated in } \mathcal{G}(j, r(n)))] \\ &= jP(\{j \text{ is isolated in } \mathcal{G}(j, r(n))\}) \\ &\sim j(1-\pi r^2(n))^{j-1}. \end{aligned} \tag{14}$$

Substituting (14) in (13), we get

$$\begin{aligned} P^{\text{Poisson};(1)}(n, r(n)) &\leq \sum_{j=1}^{\infty} j(1-\pi r^2(n))^{j-1} e^{-n}\frac{n^j}{j!} \\ &= n\sum_{j=0}^{\infty}(1-\pi r^2(n))^j e^{-n}\frac{n^j}{j!} \\ &= ne^{-n\pi r^2(n)}, \end{aligned} \tag{15}$$

from which the result follows. □

The following must be a known fact though we are not aware of any reference for it.

Lemma 3.2 *For all $\epsilon > 0$ and sufficiently large n*

$$\sum_{j=1}^{n} e^{-n} \frac{n^j}{j!} \geq (\frac{1}{2} - \epsilon). \tag{16}$$

We are now ready to give a sufficient condition on $r(n)$ for asymptotic connectivity in $\mathcal{G}(n, r(n))$.

Theorem 3.1 *If $\pi r^2(n) = \frac{\log n + c(n)}{n}$ and $\lim_{n\to\infty} c(n) = c$, then*

$$\limsup_{n\to\infty} P_d(n, r(n)) \leq 4e^{-c}. \tag{17}$$

Proof: By (11) and the observation made thereafter, we get that, for any $\epsilon > 0$ and for all sufficiently large n,

$$P_d^{\text{Poisson}}(n, r(n)) \leq (1+\epsilon) P^{\text{Poisson};(1)}(n, r(n)). \tag{18}$$

Note that

$$P_d^{\text{Poisson}}(n, r(n)) = \sum_{j=1}^{\infty} P_d(j, r(n)) e^{-n} \frac{n^j}{j!}. \tag{19}$$

For a fixed range $r = r(n)$, we have

$$P_d(k, r) \leq P(\{\text{node } k \text{ is isolated in } \mathcal{G}(k, r)\}) + P_d(k-1, r).$$

which after recursion gives, that for $0 < j < n$,

$$\begin{aligned} P_d(n, r(n)) &\leq \sum_{k=j+1}^{n} P(\{\text{node } k \text{ is isolated in } \mathcal{G}(k, r(n))\}) + P_d(j, r(n)) \\ &\leq \sum_{k=j+1}^{n} (1 - \pi r^2(n))^{k-1} + P_d(j, r(n)) \\ &\leq \frac{(1-\pi r^2(n))^j}{\pi r^2(n)} + P_d(j, r(n)). \end{aligned} \tag{20}$$

Substituting (20) in (19), we get

$$\begin{aligned} P_d^{\text{Poisson}}(n, r(n)) &\geq P_d(n, r(n)) \sum_{j=1}^{n} e^{-n} \frac{n^j}{j!} - \sum_{j=1}^{n-1} \frac{(1-\pi r^2(n))^{j-1}}{\pi r^2(n)} e^{-n} \frac{n^j}{j!} \\ &\geq P_d(n, r(n)) (\frac{1}{2} - \epsilon) - \frac{e^{-n\pi r^2(n)}}{\pi r^2(n)}, \end{aligned} \tag{21}$$

where we have used Lemma 3.2. Using (18), we get

$$P_d(n, r(n)) \leq 2(1+4\epsilon) \left[P^{\text{Poisson};(1)}(n, r(n)) + \frac{e^{-n\pi r^2(n)}}{\pi r^2(n)} \right]. \tag{22}$$

For the given $\pi r^2(n) = \frac{\log n + c(n)}{n}$, from Lemma 3.1, we get that, for any fixed $\epsilon > 0$, the following holds for all sufficiently large n

$$P_d(n, r(n)) \leq 2(1+4\epsilon)\left[e^{-c(n)} + \frac{e^{-c(n)}}{\log n + c(n)}\right].$$

Thus, since $\epsilon > 0$ is arbitrary,

$$\limsup_{n\to\infty} P_d(n, r(n)) \leq 2e^{-c}. \tag{23}$$

□

The following is an obvious consequence of Theorem 3.1.

Corollary 3.1 *Graph $\mathcal{G}(n, r(n))$ is asymptotically connected with probability one for $\pi r^2(n) = \frac{\log n + c(n)}{n}$ if $c(n) \to +\infty$.*

Combining Corollaries 2.1 and 3.1, we get the main result of the paper.

Theorem 3.2 *Graph $\mathcal{G}(n, r(n))$, with $\pi r^2(n) = \frac{\log n + c(n)}{n}$ is connected with probability one as $n \to \infty$ if and only if $c(n) \to +\infty$.*

4 Concluding Remarks

We have derived the *critical* range of nodes placed randomly in a disc of unit area, for the resulting network to be connected with probability one as the number of nodes tends to infinity (cf. Theorem 3.2). One can consider the following extensions of the problem discussed in this paper:

- Our lower and upper bounds on $P_d(n, r(n))$ are not tight. A more refined argument may lead to bounds which hold for all n. In particular, we believe that for $\pi r^2(n) = \frac{\log n + c(n)}{n}$, $P_d(n, r(n)) \to 1$ if $c(n) \to -\infty$.

- Consider the following generalization of the problem: Even if a node has another node in its range, it can communicate with that node with probability $p(n), 0 \leq p(n) \leq 1$. The quantity $p(n)$ can be regarded as the reliability of a link, and is tantamount to Bernoulli deletion of edges in $\mathcal{G}(n, r(n))$. Our conjecture is that Theorem 3.2 is true with $\pi r^2(n)$ replaced by $\pi r^2(n)p(n)$. This conjecture holds for at least two special cases: $\pi r^2(n) \equiv 4$ (i.e., range of each node includes $\mathcal{D}$) and $p(n)$ arbitrary in $[0, 1]$ (Theorem VII.3 in Bollobás (1985)), and $r(n)$ arbitrary and $p(n) \equiv 1$ (cf. Theorem 3.2). As in the proof of Theorem 3.1, continuum percolation theory results can be used to obtain sufficient conditions on $\pi r^2(n)p(n)$. Clearly, Theorem 2.1 still holds. However, stronger necessary conditions need to be worked out.

- A much harder problem to analyze is when nodes are not placed independently in the disc $\mathcal{D}$. For example, nodes may be placed in clusters, with a specified probability distribution on the size of a cluster.

Acknowledgments: We would like to thank Prof. T. Seidman for some helpful discussions, and Prof. V. Ananthram for pointing out the reference Meester and Roy (1996) to us. The material presented here is based upon work supported in part by the Air Force of Scientific Research under Contract No. AF-DC-5-36128, and in part by the U.S. Army Research Office under Contract No. DAAH-04-95-1-0090. Any opinions, findings, and conclusions or recommendations expressed in this publication are those of the authors and do not necessarily reflect the views of AFOSR or USARO.

References

Bambos, N., Chen, S. and Pottie, G. (1995) " Radio link admission algorithms for wireless networks with power control and active link quality protection," in *Proc. IEEE INFOCOM*, Boston, MA.

Bollobás, B. (1985). *Random Graphs*, Academic Press, Orlando, FL.

Hall, P. (1988). *Introduction to the Theory of Coverage Processes*, John Wiley & Sons, New York.

Kesten, H. (1982). *Percolation Theory for Mathematicians*, Birkhäuser, Boston, MA.

Gupta, P. and Kumar, P. R. (1997). "A system and traffic dependent adaptive routing algorithm for ad hoc networks," in *Proc. IEEE 36th Conf. on Decision and Control*, San Diego, pp2375-2380.

Johnson, D. and Maltz, D. (1996). "Dynamic source routing in ad hoc wireless networks," in *Mobile Computing* (ed. T. Imielinski and H. Korth), Kluwer Academic Publishers, Dordrecht, The Netherlands.

Meester, R. and Roy, R. (1996). *Continuum Percolation*, Cambridge University Press, Cambridge, UK.

Ulukus, S. and Yates, R. (1996). "Stochastic power control for cellular radio systems," Preprint.

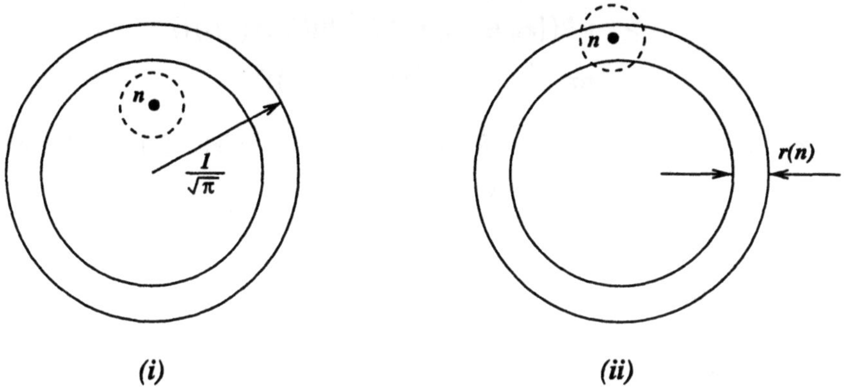

FIGURE 1. Two cases for evaluating $P(\{n \text{ is isolated in } \mathcal{G}(n, r(n)\})$.

1 Appendix

Here we give the complete proofs of the theorems given in the main body of the paper, taking the edge effects into account.

Proof of Theorem 2.1:
As before, we first study the case where $\pi r^2(n) = \frac{\log n + c}{n}$ for a fixed c. Consider $P^{(1)}(n, r(n))$, the probability that $\mathcal{G}(n, r(n))$ has *at least* one order-1 component. Then, as argued in (8), we have

$$\begin{aligned} P^{(1)}(n, r(n)) \geq & \sum_{i=1}^{n} P(\{i \text{ is isolated in } \mathcal{G}(n, r(n))\}) \\ & - \sum_{i=1}^{n} \sum_{j \neq i} P(\{i \text{ and } j \text{ are isolated in } \mathcal{G}(n, r(n))\}). \quad (24) \end{aligned}$$

Now, let us consider each sum in (24) separately. For this purpose, define the notation

$$\begin{aligned} N^{(1)}(\mathcal{G}) &:= \{i \in \mathcal{G} : i \text{ is an isolated node in } \mathcal{G}\}, \\ \mathcal{D}^o &:= \{x \in \mathcal{D} : \|x\| \leq \frac{1}{\sqrt{\pi}} - r(n)\}, \\ \partial\mathcal{D} &:= \mathcal{D} - \mathcal{D}^o. \quad (25) \end{aligned}$$

Then, as illustrated in Figure 1, we need to consider two cases to evaluate the probability that node n is isolated, namely: When $x_n \in \mathcal{D}^o$ (recall that x_n is the position of node n in $\mathcal{D}$), and when $x_n \in \partial\mathcal{D}$. To obtain a lower bound, we consider only the first case, i.e.,

$$\sum_{i=1}^{n} P(\{i \text{ is isolated in } \mathcal{G}(n, r(n))\}) = nP(\{n \text{ is isolated in } \mathcal{G}(n, r(n)\})$$

$$\begin{aligned}
&\geq nP(\{x_n \in \mathcal{D}_o,\ n \in N^{(1)}(\mathcal{G}(n, r(n)))\}) \\
&= n\pi(\frac{1}{\sqrt{\pi}} - r(n))^2(1 - \pi r^2(n))^{n-1}.
\end{aligned}$$

Using Lemma 2.2, we see that for any $\theta < 1$, when n is sufficiently large,

$$\sum_{i=1}^{n} P(\{i \text{ is isolated in } \mathcal{G}(n, r(n)\}) \geq \theta e^{-c}. \tag{26}$$

Next, consider the second sum in (24), which in the notation of (25) can be written as

$$\begin{aligned}
&\sum_{i=1}^{n}\sum_{j\neq i} P(\{i, j \in N^{(1)}(\mathcal{G}(n, r(n)))\}) \\
&= \sum_{i=1}^{n}\sum_{j\neq i} \Big(P(\{i, j \in N^{(1)}(\mathcal{G}(n, r(n))), \text{ and } x_i \text{ or } x_j \in \partial\mathcal{D}\}) \\
&\quad + P(\{i, j \in N^{(1)}(\mathcal{G}(n, r(n))), \text{ and } x_i, x_j \in \mathcal{D}^o\})\Big) \\
&\leq n(n-1)\Big(2P(\{n, n-1 \in N^{(1)}(\mathcal{G}(n, r(n))), \text{ and } x_n \in \partial\mathcal{D}\}) \\
&\quad + P(\{n, n-1 \in N^{(1)}(\mathcal{G}(n, r(n))), \text{ and } x_n, x_{n-1} \in \mathcal{D}^o\})\Big).
\end{aligned} \tag{27}$$

The first term can be written as

$$\begin{aligned}
&2n(n-1)P(\{n, n-1 \in N^{(1)}(\mathcal{G}(n, r(n))), \text{ and } x_n \in \partial\mathcal{D}\}) \\
&\quad = 2n(n-1) \cdot P(\{n \in N^{(1)}(\mathcal{G}(n, r(n))), \text{ and } x_n \in \partial\mathcal{D}\}) \cdot \\
&\quad P(\{n-1 \in N^{(1)}(\mathcal{G}(n, r(n)))\Big| n \in N^{(1)}(\mathcal{G}(n, r(n))), x_n \in \partial\mathcal{D}\}).
\end{aligned} \tag{28}$$

Now $nP(\{n \in N^{(1)}(\mathcal{G}(n, r(n))), \text{ and } x_n \in \partial\mathcal{D}\})$ can be evaluated using Figure 2, to give

$$\begin{aligned}
nP(\{n \in N^{(1)}(\mathcal{G}(n, r(n))), \text{ and } x_n \in \partial\mathcal{D}\}) &\leq n \int_0^{r(n)} \Big(1 - \\
\left(\pi - \cos^{-1}\frac{y}{r(n)}\right) r^2(n) + \mathcal{E}(y)\Big)^{n-1} &2\pi(\frac{1}{\sqrt{\pi}} - y)dy,
\end{aligned} \tag{29}$$

where

$$\begin{aligned}
\mathcal{E}(y) &\leq 2r(n) \cdot \left(\frac{1}{\sqrt{\pi}} - \sqrt{\frac{1}{\pi} - r^2(n)}\right) \\
&\leq 2r(n) \cdot \left(\frac{1}{\sqrt{\pi}} - \frac{1}{\sqrt{\pi}}\left(1 - \pi r^2(n)\right)\right) \\
&= 2\sqrt{\pi} r^3(n).
\end{aligned} \tag{30}$$

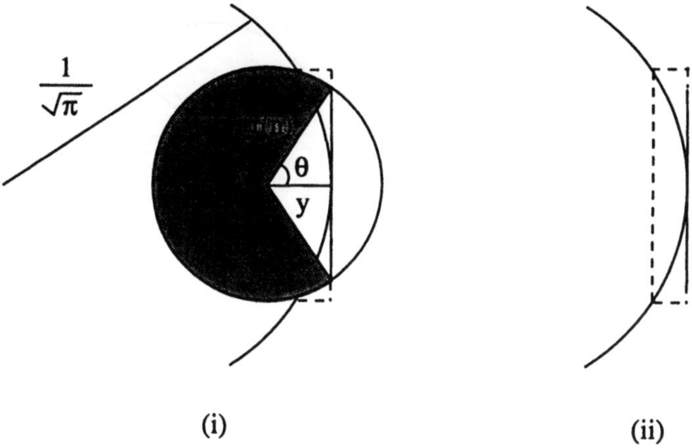

FIGURE 2. Evaluating $nP(\{n \in N^{(1)}(\mathcal{G}(n,r(n))), \text{ and } x_n \in \partial\mathcal{D}\})$: (i) Shaded area gives a lower bound on the area which nodes $1, 2, \ldots, n-1$ should not lie in for node n to be isolated, when n is at a distance of y from the boundary of $\mathcal{D}$, and (ii) Area within the rectangle is an upper bound on the error $\mathcal{E}(y)$ due to approximation of the portion of the disc $\mathcal{D}$ within the range of node n by its tangent.

Substituting (30) in (29), and changing the variable to $\theta = \cos^{-1}\frac{y}{r(n)}$, we get

$$\begin{aligned} & nP(\{n \in N^{(1)}(\mathcal{G}(n,r(n))), \text{ and } x_n \in \partial\mathcal{D}\}) \\ \leq\ & n\int_0^{\frac{\pi}{2}} \left(1-(\pi-\theta)r^2(n)+2\sqrt{\pi}r^3(n)\right)^{n-1} 2\sqrt{\pi}r(n)\sin\theta d\theta \\ \leq\ & n\cdot 2\sqrt{\pi}r(n)\int_0^{\frac{\pi}{2}} e^{-(n-1)((\pi-\theta)r^2(n)-2\sqrt{\pi}r^3(n))}\frac{e^{j\theta}-e^{-j\theta}}{2j}d\theta \\ =\ & 2\sqrt{\pi}nr(n)e^{-(n-1)(\pi r^2(n)-2\sqrt{\pi}r^3(n))}\cdot \\ & \frac{e^{(n-1)\frac{\pi}{2}r^2(n)}(n-1)r^2(n)+1}{((n-1)r^2(n))^2+1}. \end{aligned} \tag{31}$$

For the given $\pi r^2(n) = \frac{\log n + c}{n}$, we thus have

$$\begin{aligned} & nP(\{n \in N^{(1)}(\mathcal{G}(n,r(n))), \text{ and } x_n \in \partial\mathcal{D}\}) \\ \leq\ & \frac{4\pi n e^{-(n-1)(\frac{\pi}{2}r^2(n)-2\sqrt{\pi}r^3(n))}}{(n-1)\sqrt{\pi}r(n)} \\ \leq\ & \frac{4(1+\epsilon)\pi e^{-\frac{c}{2}}}{\sqrt{\log n}}, \end{aligned} \tag{32}$$

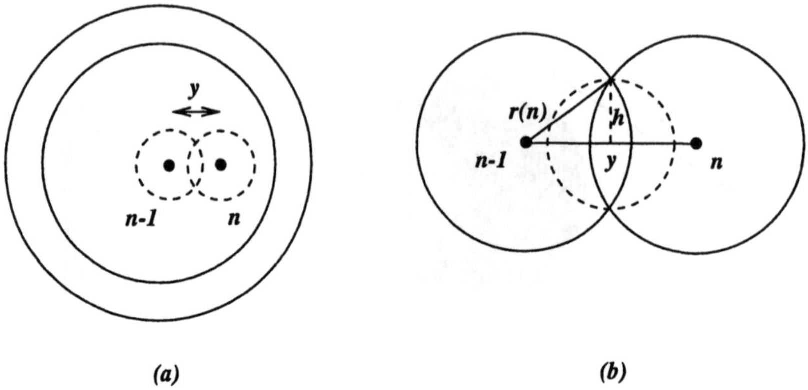

FIGURE 3. Evaluating $P(\{n,(n-1) \in N^{(1)}(\mathcal{G}(n,r(n))), \text{ and } x_n, x_{n-1} \in \mathcal{D}^o\})$. Note that $h^2 = r^2(n) - \left(\frac{y}{2}\right)^2$.

for any $\epsilon > 0$ and sufficiently large n. The remaining factor in (28) can be evaluated as

$$\begin{aligned}
&2(n-1)P(\{n-1 \in N^{(1)}(\mathcal{G}(n,r(n)))\Big| n \in N^{(1)}(\mathcal{G}(n,r(n))), x_n \in \partial\mathcal{D}\}) \\
&= 2(n-1)\Big(P(\{n-1 \in N^{(1)}(\mathcal{G}(n,r(n))), x_{n-1} \in \partial\mathcal{D}\Big| \\
&\quad n \in N^{(1)}(\mathcal{G}(n,r(n))), x_n \in \partial\mathcal{D}\}) + P(\{n-1 \in N^{(1)}(\mathcal{G}(n,r(n))), \\
&\quad x_{n-1} \in \mathcal{D}^o \Big| n \in N^{(1)}(\mathcal{G}(n,r(n))), x_n \in \partial\mathcal{D}\})\Big) \\
&\leq 2(n-1)\Big(2\sqrt{\pi}r(n)(1-\pi r^2(n)+2\mathcal{E}(0))^{n-2} \\
&\quad + (1-\frac{3}{2}\pi r^2(n)+\mathcal{E}(0))^{n-2}\Big), \qquad (33)
\end{aligned}$$

where $\mathcal{E}(\cdot)$ is defined in (30). For the given $\pi r^2(n) = \frac{\log n + c}{n}$, we thus have

$$2(n-1)P(\{n-1 \in N^{(1)}(\mathcal{G}(n,r(n)))\Big| n \in N^{(1)}(\mathcal{G}(n,r(n))), x_n \in \partial\mathcal{D}\})$$

$$\leq 4(1+\epsilon)\sqrt{\frac{\log n}{n}}, \qquad (34)$$

for any $\epsilon > 0$ and all sufficiently large n. Substituting (32) and (34) in (28), we get

$$\begin{aligned}
&2n(n-1)P(\{n, n-1 \in N^{(1)}(\mathcal{G}(n,r(n))), \text{ and } x_n \in \partial\mathcal{D}\}) \\
&\leq \frac{4(1+\epsilon)\pi e^{-\frac{c}{2}}}{\sqrt{\log n}} \cdot 4(1+\epsilon)\sqrt{\frac{\log n}{n}}
\end{aligned}$$

$$\leq \frac{16(1+\epsilon')\pi e^{-\frac{c}{2}}}{\sqrt{n}}, \tag{35}$$

for any $\epsilon' > 0$ and all sufficiently large n. The second term in (27) is (as illustrated in Figure 3),

$$\begin{aligned}
& n(n-1)P(\{n,(n-1) \in N^{(1)}(\mathcal{G}(n,r(n))), \text{ and } x_n, x_{n-1} \in \mathcal{D}^o\}) \\
\leq \;& n(n-1)P(\{x_n \in \mathcal{D}^o\}) \cdot \Big[P(\{r(n) < |x_n - x_{n-1}| \leq 2r(n), |x_i - x_j| \\
& > r(n), 1 \leq i \leq n-2; j = n, n-1\} \Big| x_n \in \mathcal{D}^o) + P(\{2r(n) < |x_n \\
& - x_{n-1}|, |x_i - x_j| > r(n), 1 \leq i \leq n-2; j = n, n-1\} \Big| x_n \in \mathcal{D}^o)\Big] \\
\leq \;& n(n-1)\pi(\frac{1}{\sqrt{\pi}} - r(n))^2 \cdot \\
& \left[\int_{r(n)}^{2r(n)} \left(1 - 2\pi r^2(n) + \pi\left(r^2(n) - \frac{y^2}{4}\right)\right)^{n-2} 2\pi y dy + \right. \\
& \left. + \left(1 - \pi r^2(n)\right)\left(1 - 2\pi r^2(n)\right)^{n-2}\right] \\
\leq \;& n(n-1)\left[\int_{r(n)}^{2r(n)} e^{-(n-2)\pi(r^2(n)+\frac{y^2}{4})} 2\pi y dy + (1 - 2\pi r^2(n))^{n-2}\right] \\
\leq \;& n(n-1)\left[e^{-(n-2)\pi r^2(n)} \frac{4}{n-2} e^{-\frac{(n-2)\pi y^2}{4}}\Big|_{2r(n)}^{r(n)} + e^{-(n-2)2\pi r^2(n)}\right] \\
\leq \;& n(n-1)\left[\frac{4}{n-2} e^{-(n-2)\frac{5}{4}\pi r^2(n)} + e^{-(n-2)2\pi r^2(n)}\right] \\
\leq \;& n(n-1)(1+\epsilon'')e^{-(n-2)2\pi r^2(n)} \\
\leq \;& (1+\epsilon')e^{-2c},
\end{aligned} \tag{36}$$

for any $\epsilon' > 0$, the given $\pi r^2(n) = \frac{\log n + c}{n}$ and all sufficiently large n. Substituting (35) and (36) in (27), we get

$$\begin{aligned}
& \sum_{i=1}^{n} \sum_{j \neq i}^{n} P(\{i \text{ and } j \text{ are isolated in } \mathcal{G}(n,r(n))\}) \\
& \qquad \leq \frac{16\pi(1+\epsilon'')}{\sqrt{n}} e^{-\frac{c}{2}} + (1+\epsilon')e^{-2c} \\
& \qquad \leq (1+\epsilon)e^{-2c}.
\end{aligned} \tag{37}$$

for any $\epsilon > 0$ and all sufficiently large n. Substituting (26) and (37) in (24), we get

$$P^{(1)}(n,r(n)) \geq \theta e^{-c} - (1+\epsilon)e^{-2c},$$

for all $n > N(\epsilon, \theta, c)$. Since $P^{(1)}(n, r(n)) \leq P_d(n, r(n))$, we have

$$P_d(n, r(n)) \geq \theta e^{-c} - (1+\epsilon)e^{-2c}, \tag{38}$$

for all $n > N(\epsilon, \theta, c)$. Now, consider the case where c is a function $c(n)$ with $\lim_{n\to\infty} c(n) = \bar{c}$. Then, for any $\epsilon > 0$, $c(n) \leq \bar{c} + \epsilon$ for all $n \geq N'(\epsilon)$. Also, the probability of disconnectedness is monotone decreasing in c. Hence

$$P_d(n, r(n)) \geq \theta e^{-(\bar{c}+\epsilon)} - (1+\epsilon)e^{-2(\bar{c}+\epsilon)}.$$

for $n \geq \max\{N(\epsilon, \theta, \bar{c}+\epsilon), N'(\epsilon)\}$. Taking limits

$$\liminf_{n\to\infty} P_d(n, r(n)) \geq \theta e^{-(\bar{c}+\epsilon)} - (1+\epsilon)e^{-2(\bar{c}+\epsilon)}.$$

Since this holds for all $\epsilon > 0$ and $\theta < 1$, the result follows. □

Proof of Lemma 3.1:
As before,

$$P^{\text{Poisson};(1)}(n, r(n)) = \sum_{j=1}^{\infty} P^{(1)}(j, r(n))e^{-n}\frac{n^j}{j!}. \tag{39}$$

Let $E_1(j, r(n))$ denote the expected number of order-1 components in $\mathcal{G}(j, r(n))$. Then

$$\begin{aligned} P^{(1)}(j, r(n)) &\leq E_1(j, r(n)) \\ &= E[\sum_{i=1}^{j} I(i \text{ is isolated in } \mathcal{G}(j, r(n)))] \\ &= jP(\{j \text{ is isolated in } \mathcal{G}(j, r(n))\}). \end{aligned} \tag{40}$$

Using the definitions of $N^{(1)}(\mathcal{G}), \mathcal{D}^o$ and $\partial\mathcal{D}$ given in (25), we can write

$$\begin{aligned} &P(\{j \text{ is isolated in } \mathcal{G}(j, r(n))\}) \\ &\quad = P(\{j \in N^{(1)}(\mathcal{G}(j, r(n))) \text{ and } x_j \in \mathcal{D}^o\}) \\ &\quad + P(\{j \in N^{(1)}(\mathcal{G}(j, r(n))) \text{ and } x_j \in \partial\mathcal{D}\}). \end{aligned} \tag{41}$$

From (31) and (41) , we get

$$\begin{aligned} P^{(1)}(j, r(n)) &\leq jP(\{j \text{ is isolated in } \mathcal{G}(j, r(n))\}) \\ &\leq j\pi\left(\frac{1}{\sqrt{\pi}} - r(n)\right)^2 (1 - \pi r^2(n))^{j-1} + 2\sqrt{\pi} j r(n) \cdot \\ &\quad e^{-(j-1)f_1(r(n))}\frac{e^{(j-1)f_2(r(n))}(j-1)r^2(n) + 1}{((j-1)r^2(n))^2 + 1}, \end{aligned} \tag{42}$$

where $f_1(r) = \pi r^2 - 2\sqrt{\pi}r^3$, and $f_2(r) = \pi r^2/2$. From (39) and (42), we get

$$p^{\text{Poisson};(1)}(n, r(n)) \le \sum_{j=1}^{\infty} jP(\{j \text{ is isolated in } \mathcal{G}(j, r(n))\})e^{-n}\frac{n^j}{j!} \tag{43}$$

$$\begin{aligned}
\le\ & \sum_{j=1}^{\infty} j(1 - \pi r^2(n))^{j-1} e^{-n}\frac{n^j}{j!} + 2\sqrt{\pi}r(n)e^{-n}\frac{n}{1!} \\
& + 2\sqrt{\pi}r(n)\sum_{j=2}^{\infty} j\left[\frac{e^{-(j-1)(f_1(r(n))-f_2(r(n)))}}{(j-1)r^2(n)} + \frac{e^{-(j-1)f_1(r(n))}}{((j-1)r^2(n))^2}\right] e^{-n}\frac{n^j}{j!} \\
\le\ & n\sum_{j=0}^{\infty}(1 - \pi r^2(n))^j e^{-n}\frac{n^j}{j!} + 2\sqrt{\pi}r(n)ne^{-n} \\
& + \frac{2\sqrt{\pi}}{r(n)}\sum_{j=2}^{\infty}\frac{j}{j-1}\frac{e^{-(j-1)(f_1(r(n))-f_2(r(n)))}n^j}{j!}e^{-n} \\
& + \frac{2\sqrt{\pi}}{r^3(n)}\sum_{j=2}^{\infty}\frac{j}{j-1}\frac{j+1}{j-1}\frac{e^{-(j-1)f_1(r(n))}n^j}{(j+1)!}e^{-n} \\
\le\ & ne^{-n\pi r^2(n)} + 2\sqrt{\pi}r(n)ne^{-n} + \frac{2\sqrt{\pi}}{r(n)}2e^{f_1(r(n))-f_2(r(n))}\,. \\
& \cdot e^{ne^{-(f_1(r(n))-f_2(r(n)))}}e^{-n} + \frac{2\sqrt{\pi}}{r^3(n)}\frac{2\cdot 3e^{2f_1(r(n))}}{n}c^{ne^{-f_1(r(n))}}e^{-n} \\
\le\ & ne^{-n\pi r^2(n)} + 2\sqrt{\pi}r(n)ne^{-n} + \frac{4\sqrt{\pi}}{r(n)}e^{-(n-1)(f_1(r(n))-f_2(r(n)))}\,. \\
& \cdot e^{\frac{n(f_1(r(n))-f_2(r(n)))^2}{2}} + \frac{12\sqrt{\pi}}{nr^3(n)}e^{-(n-2)f_1(r(n))}e^{\frac{nf_1^2(r(n))}{2}},
\end{aligned} \tag{44}$$

where we have used $e^{-x} \le 1 - x + \frac{x^2}{2}$. For the given $\pi r^2(n) = \frac{\log n + c(n)}{n}$, we thus have

$$\begin{aligned}
p^{\text{Poisson};(1)}(n, r(n)) \ \le\ & e^{-c(n)} + 2\sqrt{n(\log n + c(n))}e^{-n} + \frac{4\pi(1+\epsilon)e^{-c(n)}}{\sqrt{\log n + c(n)}} \\
& + \frac{12\pi^2(1+\epsilon)}{\sqrt{n(\log n + c(n))^3}}e^{-c(n)},
\end{aligned} \tag{45}$$

for any $\epsilon > 0$ and all sufficiently large n. The result follows. □

Proof of Lemma 3.2:
By Chebyshev's inequality, we have that for any α

$$\sum_{j=n+n^{\alpha}+1}^{\infty} e^{-n}\frac{n^j}{j!} \le \frac{n}{n^{2\alpha}}.$$

Let $\alpha = \frac{1}{2} + \epsilon$ for some $\epsilon > 0$, then

$$\sum_{j=n+n^\alpha+1}^{\infty} e^{-n}\frac{n^j}{j!} \leq \frac{1}{n^{2\epsilon}}. \tag{46}$$

Also,

$$\begin{aligned}
\sum_{j=n+1}^{n+n^\alpha} e^{-n}\frac{n^j}{j!} &= \sum_{j=1}^{n^\alpha} e^{-n}\frac{n^{n+j}}{(n+j)!} \\
&= \sum_{j=1}^{n^\alpha} e^{-n}\frac{n^{n-j-1}}{(n-j-1)!}\cdot\frac{1}{\prod_{i=0}^{j}\left(1-\left(\frac{i}{n}\right)^2\right)} \\
&\leq \left(\sum_{j=1}^{n^\alpha} e^{-n}\frac{n^{n-j-1}}{(n-j-1)!}\right)\cdot \\
&\quad \max_{1\leq j\leq n^\alpha}\frac{1}{\prod_{i=0}^{j}\left(1-\left(\frac{i}{n}\right)^2\right)}.
\end{aligned} \tag{47}$$

Now,

$$\begin{aligned}
\max_{1\leq j\leq n^\alpha}\frac{1}{\prod_{i=0}^{j}\left(1-\left(\frac{i}{n}\right)^2\right)} &\leq \max_{1\leq j\leq n^\alpha}\frac{1}{1-\sum_{i=1}^{j}\left(\frac{i}{n}\right)^2} \\
&= \frac{1}{1-\sum_{i=1}^{n^\alpha}\left(\frac{i}{n}\right)^2} \\
&= \frac{1}{1-\frac{n^\alpha(n^\alpha+1)(2n^\alpha+1)}{6n^2}} \\
&\leq \frac{1}{1-\frac{(1+\epsilon')n^{3\epsilon-\frac{1}{2}}}{3}},
\end{aligned} \tag{48}$$

for the chosen $\alpha = \frac{1}{2}+\epsilon$, any $\epsilon' > 0$ and all sufficiently large n. Substituting (48) in (47), we get

$$\begin{aligned}
\sum_{j=n+1}^{n+n^\alpha} e^{-n}\frac{n^j}{j!} &\leq \left(\sum_{j=1}^{n^\alpha} e^{-n}\frac{n^{n-j-1}}{(n-j-1)!}\right)\cdot\left(1+\frac{2(1+\epsilon')n^{3\epsilon-\frac{1}{2}}}{3}\right) \\
&\leq \left(\sum_{j=1}^{n} e^{-n}\frac{n^j}{j!}\right)\cdot\left(1+\frac{2(1+\epsilon')n^{3\epsilon-\frac{1}{2}}}{3}\right).
\end{aligned} \tag{49}$$

From (46), (49) and the fact that $\sum_{j=0}^{\infty} e^{-n} n^j/j! = 1$, we get that

$$\begin{aligned}\sum_{j=1}^{n} e^{-n}\frac{n^j}{j!} &\geq \frac{1 - e^{-n} - \frac{1}{n^{2\epsilon}}}{1 + 1 + \frac{2(1+\epsilon')n^{3\epsilon - \frac{1}{2}}}{3}} \\ &\geq \frac{1}{2} - \epsilon'',\end{aligned}$$

for any $\epsilon < \frac{1}{6}$, $\epsilon'' > 0$ and all sufficiently large n. □

Proof of Theorem 3.1:
By (11) and the observation made thereafter, we get that, for any $\epsilon > 0$ and for all sufficiently large n,

$$P_d^{\text{Poisson}}(n, r(n)) \leq (1 + \epsilon) P^{\text{Poisson};(1)}(n, r(n)). \tag{50}$$

Note that

$$P_d^{\text{Poisson}}(n, r(n)) = \sum_{j=1}^{\infty} P_d(j, r(n)) e^{-n}\frac{n^j}{j!}. \tag{51}$$

For a fixed range $r = r(n)$, we have

$$P_d(k, r) \leq P(\{\text{node } k \text{ is isolated in } \mathcal{G}(k, r)\}) + P_d(k-1, r).$$

which after recursion gives, that for $0 \leq j < n$

$$\begin{aligned}P_d(n, r(n)) \leq& \sum_{k=j+1}^{n} P(\{\text{node } k \text{ is isolated in } \mathcal{G}(k, r(n))\}) \\ &+ P_d(j, r(n)).\end{aligned} \tag{52}$$

Substituting (52) in (51), we get

$$\begin{aligned}&P_d^{\text{Poisson}}(n, r(n)) \\ &\geq P_d(n, r(n))\sum_{j=1}^{n} e^{-n}\frac{n^j}{j!} \\ &\quad - \sum_{j=1}^{n-1}\sum_{k=j+1}^{n} P(\{k \text{ is isolated in } \mathcal{G}(k, r(n))\}) e^{-n}\frac{n^j}{j!} \\ &\geq P_d(n, r(n))(\frac{1}{2} - \epsilon) - \sum_{k=2}^{n} P(\{k \text{ is isolated in } \mathcal{G}(k, r(n))\}) \sum_{j=1}^{k-1} e^{-n}\frac{n^j}{j!} \\ &\geq P_d(n, r(n))(\frac{1}{2} - \epsilon) - \sum_{k=2}^{n} P(\{k \text{ is isolated in } \mathcal{G}(k, r(n))\}) k e^{-n}\frac{n^k}{k!},\end{aligned}$$

where we have used Lemma 3.2 and the fact that $e^{-n}\frac{n^k}{k!}$ increases with k, for $1 \le k \le n$. Using (50), we get

$$\begin{aligned} P_d(n, r(n)) &\le 2(1+6\epsilon)\Big[P^{\text{Poisson};(1)}(n, r(n)) \\ &\quad + \sum_{k=1}^{\infty} kP(\{k \text{ is isolated in } \mathcal{G}(k, r(n))\})e^{-n}\frac{n^k}{k!}\Big]. \end{aligned}$$

For the given $\pi r^2(n) = \frac{\log n + c(n)}{n}$, and from Lemma 3.1, and (43) we get that for any $\epsilon > 0$,

$$P_d(n, r(n)) \le 2(1+6\epsilon)2 \cdot (1+\epsilon')e^{-c(n)}.$$

holds for all sufficiently large n. Thus,

$$\limsup_{n\to\infty} P_d(n, r(n)) \le 4(1+\epsilon'')e^{-c}.$$

Since ϵ'' can be made arbitrarily small, the result follows. □

34
Pricing Models with Transaction Fees

James E. Hodder
Thaleia Zariphopoulou

ABSTRACT: We discuss existing and new results for mathematical finance models with transaction costs, namely models of portfolio management and international asset pricing in the presence of political risk. The main analysis comes from the theories of singular stochastic control and viscosity solutions for first and second order nonlinear partial differential equations.

KEYWORDS: Portfolio management, Asset Pricing, Political Risk, Transaction and shipping fees, Singular control, Viscosity Solutions.

1 Introduction

Transaction costs have always played an important role in modern financial economic theories not only in the derivation of fundamental pricing principles but also for their impact in real-world applications. The purpose of this paper is to discuss existing and new results for financial models which incorporate transaction fees, namely intertemporal models of optimal investment and consumption and models of international asset pricing. These models can be formulated as singular stochastic control problems due to the possibility of singular (lump-sum) transactions. The main tools for studying these problems come from the theories of singular stochastic control and viscosity solutions of nonlinear first and second order partial differential equations.

The investment/consumption models study the optimal trading actions of a utility maximizing consumer who is offered the opportunity to invest in two financial securities, a bond with a deterministic rate of return and a stock whose price is a stochastic process. Two types of models are studied herein: models with the underlying stock price being either a diffusion process or a pseudo-bond price with the deterministic rate of return being replaced by a continuous time Markov-chain process. The essential feature in both types of problems is the penalty that the investor is charged with any time she decides to make a transaction. This penalty is paid in the form of transaction fees which are proportional to the amount transacted. These models are of fundamental importance in optimal portfolio management and also in the area of derivative pricing. In the latter case, it is well known that the classical Black and Scholes theory to price a contingent claim written on the underlying stock, by creating a perfectly replicating portfolio, breaks down due to the presence of transaction costs. One approach to face this difficulty is to employ a utility-maximization criterion for the price claim. This way, the risk which is associated with the claim position, albeit impossible to be hedged, can be measured in terms of the individual risk aversion index of the writer or the buyer of the derivative security.[1]

The goal in the study of these models is to determine the optimal consumption process and trading rules as well as to compute the value function. In subsequent sections we will see that the possibility of lump transactions results in the existence of a "buy" and "sell" region, where action is taken instantaneously and a "consume only" region which separates the above singular transaction sets.

The international asset pricing models study the dynamic equilibrium prices and the real exchange rate across countries. They differ from the aforementioned investment and consumption models in that the two securities, the bond and the stock accounts are replaced by two stochastic processes which model the capital stock of the underlying goods of each country. The role of the consumption process of the single investor is played by the so-called production technology processes which are associated with the individual countries. These production technologies can be consumed locally, reinvested in the corresponding country's production component, or shipped to the other country for either consumption or investment in that location. The shipments of country goods correspond to the financial transactions of the single-agent model and they are charged with proportional shipping costs. In recent years, a lot of attention has been drawn to the factors affecting the returns of the capital stocks across countries. It is widely believed that one of the main factors is the so called "political risk" associated with each country.

At a basic level, political risk can be defined as uncertainty about future government actions which may impact the value of firms and/or the welfare of individuals. Government actions which alter a firm's profitability will invariably impact that firm's owners and will also typically affect other firms and individuals in the economy via changes in equilibrium pricing and consumption.

There are many forms of political risk. In an international context, people sometimes think of expropriation or nationalization with inadequate compensation. However, there are more common and less dramatic risks, such as changes in a country's tax code or in the enforcement of that code. Other

[1]We refer the reader interested in this approach to Hodges and Neuberger (1989), Davis, Panas and Zariphopoulou (1993), Davis and Zariphopoulou (1994), Constantinides and Zariphopoulou (1997,1998), and Barles and Soner (1995).

risks could include uncertainty about pollution restrictions, price controls, tariffs, quotas, safety regulations, labor laws, etc. Governments may target particular firms or groups for favorable or unfavorable treatment. This clearly includes the possibility of using policies which differentiate between domestic and foreign firms and/or individuals in their roles as investors or as consumers.

In what follows, we propose a new asset pricing model in order to explore some of the implications of political risk in an international setting. Capital markets are fully integrated in the sense that individuals from each country can own claims to assets located in either country. These assets serve as both production inputs and consumption goods. To make the model tractable we concentrate on a model with two countries; the production technology is random and differs across the two countries; each country is endowed with its own technology but only one of the countries exhibits political risk. We model political risk via a Markov-chain z_t which alters the drift of the stochastic production process in the politically risky country. We allow z_t to take on two values. The lower state can be interpreted as representing the local government's ability to impose a tax on firms producing in that country. Symmetrically, the high state can be interpreted as either a lower tax rate or even a subsidy for local production.

Goods can be shipped between countries for either investment or consumption purposes. We structure the model with shipping costs in a manner similar to Dumas (1992). This results in a partial market segmentation which influences asset values as well as consumption patterns. If the ratio of the output prices in the two economies is interpreted as a real exchange rate, than that exchange rate will exhibit sustained deviations from its Purchasing Power Parity (PPP) value. In a similar vein, the real interest rate can differ across the two countries due to this partial segmentation.

The paper is structured as follows: in section 2 we discuss existing results on two fundamental singular stochastic control problems associated with transaction costs. The purpose of this section is to offer an exposition of these models to the readers who are not familiar with this class of applications of stochastic control. Section 3 is dedicated to the new model we propose for the study of the political risk factor in asset prices.

2 Consumption and Portfolio Management Models with Transaction Costs: Review on Existing Results

In this section we present the basic optimal investment and consumption models and we discuss some fundamental results for their solutions and the associated optimal policies. These models examine the actions of a single investor who invests between two assets with known returns and he desires to maximize his expected utility payoff that comes from the consumption stream. The first asset has a deterministic rate of return, so one can think of it as a bond with a constant rate of return; the other asset is a stock whose rate of return is governed either by a diffusion process or by a continuous-time Markov chain. The underlying economic feature is that transaction fees are paid any time a financial transaction takes place, in other words when the investor decides to exercise an investment control policy. It turns out that the appropriate way to study these models is by formulating them as singular control problems and analyzing the associated Hamilton-Jacobi-Bellman (HJB) equation. In the diffusion case, the HJB equation turns out to be a second order Variational Inequality as opposed to the Markov-chain case where the HJB equation is actually a system of first order Variational Inequalities coupled through the zero-th order term. In both cases, the Variational Inequalities have gradient constraints related ultimately to the optimal direction of the singular transactions. Moreover, we observe a depletion of the state space to three regions, namely the non-transaction (NT), buy stock (B) and sell stock (S) regions. If the state of the investor belongs to the NT region, the optimal policy is to consume only without trading and if his state belongs to the B (respectively S) region, the optimal policy amounts to buy stock (respectively sell stock) instantaneously, until the state trajectory hits the NT region of trading idleness.

Below, we discuss the two models separately and we provide existing results for the maximized utility (value function). The main mathematical tools come from the theories of singular stochastic control and (constrained) viscosity solutions of nonlinear equations. A selected list of references is presented at the end of each subsection.

2.a *The case of diffusion stock prices.*

We consider an economy with two securities, a bond with price $B(t)$, and a stock with price $S(t)$, at date $t \geq 0$. Prices are denominated in units of a consumption good, say dollars.

The bond pays no coupons, is default free and has price dynamics

$$B_t = e^{rt} B_0, \qquad t \geq 0 \tag{2.1}$$

where r is the *constant rate of interest.*

We denote by $W(t)$ a one-dimensional standard Brownian motion which generates the filtration $\mathcal{F}_t$ on a fixed, complete probability space $(\Omega, \mathcal{F}, P)$. The stock price is the diffusion process

$$\begin{aligned} S_t &= S_0 + \int_0^t \mu S_\tau d\tau + \int_0^t \sigma S_\tau dW_\tau \\ &= S_0 \exp\left\{\left(\mu - \frac{\sigma^2}{2}\right) t + \sigma W_t\right\} \end{aligned} \tag{2.2}$$

where μ is the *mean rate of return* and σ is the *volatility*; μ and σ are constants such that $\mu > r$ and $\sigma \neq 0$.

The investor holds x_t dollars of the bond and y_t dollars of the stock at date t. We consider a pair of right-continuous with left limits (CADLAG), non-decreasing processes (L_t, M_t) such that L_t represents the cumulative dollar amount transferred into the stock account and M_t the cumulative dollar amount transferred out of the stock account. By convention, $L_0 = M_0 = 0$. The stock account process is

$$y_t = y + \int_0^t \mu y_\tau d\tau + \int_0^t \sigma y_\tau dW_\tau + L_t - M_t. \tag{2.3}$$

with $y_0 = y$.

Transfers between the stock and the bond accounts incur *proportional transaction costs*. In particular, the cumulative transfer L_t into the stock account reduces the bond account by βL_t and the cumulative transfer M_t out of the stock account increases the bond account by αM_t, where $0 < \alpha < 1 < \beta$.

The investor consumes at the rate c_t dollars out of the bond account. There are no transaction costs on transfers from the bond account into the consumption good. The bond account process is

$$x_t = x + \int_0^t \{r x_\tau - c_\tau\} d\tau - \beta L_t + \alpha M_t. \tag{2.4}$$

with $x_0 = x$. The integral represents the accumulation of interest and the drain due to consumption. The last two terms represent the cumulative transfers between the stock and bond accounts, net of transaction costs.

A policy is a $\mathcal{F}_t$-progressively measurable triple (c_t, L_t, M_t). We restrict our attention to the set of admissible policies $\mathcal{A}$ such that

$$\begin{cases} c_t \geq 0 \quad \text{and} \quad E\int_0^t c_\tau d\tau < \infty \quad \text{a.s. for} \quad t \geq 0 & (2.5) \\ \text{and} & \\ w_t = x_t + \binom{\alpha}{\beta} y_t \geq 0 \quad \text{a.s. for} \quad t \geq 0 & (2.5)' \end{cases}$$

where we adopt the notation

$$\binom{\alpha}{\beta} z = \begin{cases} \alpha z & \text{if } z \geq 0 \\ \beta z & \text{if } z < 0. \end{cases} \tag{2.6}$$

We refer to w_t as the *net worth*. It represents the investor's bond holdings, if the investor were to transfer the holdings from the stock account into the bond account, incurring in the process the transaction costs.

The investor has von Neumann-Morgenstern preferences

$$E\left[\int_0^{+\infty} e^{-\rho t} U(c_t) dt\right]$$

over the consumption stream $\{c_t, t \geq 0\}$, where ρ is the *subjective discount rate* and the *felicity function* $U : R_0^+ \to R_0^+$ is assumed to be increasing and concave with $U(0) = 0$ and to satisfy: $U \in C([0,+\infty)) \cap C^1((0,+\infty))$ as well as $U(C) \leq M(1+C)^\gamma$ for some constant $M > 0$ and $0 < \gamma < 1$.

Given the initial endowment (x, y) in $D = \left\{(x,y) \in R^2 : x + \binom{\alpha}{\beta} y \geq 0\right\}$, we define the *value function* V as

$$V(x,y) = \sup_{(c,L,M)\in\mathcal{A}} E\left[\int_0^{+\infty} e^{-\rho t} U(c_t) dt \mid x_0 = x, y_0 = y\right]. \tag{2.7}$$

To guarantee that the value function is well defined, we either assume, as in Davis and Norman (1990), that

$$\rho > r\gamma + \frac{\gamma(\mu - r)^2}{2\sigma^2(1-\gamma)} \tag{2.8}$$

or, assume, as in Shreve and Soner (1994), that

$$\rho > r\gamma + \gamma^2(\mu - r)^2 / 2\sigma^2(1-\gamma)^2, \tag{2.9}$$

without an associated upper bound on $\mu - r$. Either the set of conditions (2.8) and (2.9), yield that the value function which corresponds to $\alpha = \beta = 1$ and $U(c) = \frac{1}{\gamma}c^\gamma$ is finite and, therefore, all functions with $0 \leq \alpha < 1, \beta \geq 1$ are finite. We also assume that $\rho \geq \mu$.

A straightforward argument along the lines of Constantinides (1979) shows that the value function is increasing and jointly concave in (x, y); it can also be shown that it is uniformly continuous on D (see Tourin and Zariphopoulou (1994)). Furthermore, the value function is expected to solve the Hamilton-Jacobi-Bellman equation (HJB) associated with the stochastic control problem (2.7). The HJB equation turns out to be a Variational Inequality with gradient constraints given by

$$\min\left[\mathcal{L}V, \beta\frac{\partial V}{\partial x} - \frac{\partial V}{\partial y}, -\alpha\frac{\partial V}{\partial x} + \frac{\partial V}{\partial y}\right] = 0 \tag{2.10}$$

where the differential operator $\mathcal{L}$ is

$$\mathcal{L}V = \rho V - \tfrac{1}{2}\sigma^2 y^2 \frac{\partial^2 V}{\partial y^2} - \mu y \frac{\partial V}{\partial y} - r x \frac{\partial V}{\partial x} - \max_{c \geq 0}\{-c\frac{\partial V}{\partial x} + U(c)\}. \tag{2.11}$$

In the special case of power utility functions, Davis and Norman (1990) obtained a closed form expression for the value function employing the special homogeneity of the problem. They also showed that the optimal policy confines the investor's portfolio to a certain wedge-shaped region in the wealth plane and they provided an algorithm and numerical computations for the optimal investment rules. The same class of utility functions was later further explored by Shreve and Soner (1994) who relaxed some of the technical assumptions on the market parameters of Davis and Norman (1990) related to the finiteness of the value function and the nature of the optimal policies. Shreve and Soner (1994) also provided results related to the regularity of the value function and the location of the exercise boundaries.

In the case of general utility functions the value function is not necessarily smooth and, therefore, it might not satisfy the HJB equation in the classical (strong) sense. It turns out that the appropriate class of weak solutions are the so-called (constrained) *viscosity solutions* and this is the class of solutions we will be using throughout the paper. In models with transaction costs, this class of solutions was first employed by Zariphopoulou (1992) and, subsequently, among others by Davis, Panas and Zariphopoulou (1993), Davis and Zariphopoulou (1994), Tourin and Zariphopoulou (1994), Shreve and Soner (1994), Barles and Soner (1995), and Pichler (1996).

The characterization of V as a constrained solution is natural because of the presence of state constraints given by (2.5)$'$. The following result was proved in Tourin and Zariphopoulou (1994) and Shreve and Soner (1994).

Theorem 2.1: *The value function V is a constrained viscosity solution of (2.10) on $\bar{D}$, in the class of uniformly continuous, concave and increasing functions.*

The notion of *viscosity solutions* was introduced by Crandall and Lions (1983) for first-order equations, and by Lions (1983) for second-order equations. *Constrained viscosity solutions* were introduced by Soner (1986) and Capuzzo-Dolcetta and Lions (1987) for first-order equations (see also Ishii and Lions (1990)). For a general overview of the theory we refer to the *User's Guide* by Crandall, Ishii and Lions (1992) and the book by Fleming and Soner (1993).

We continue by proving that the value function V is the unique constrained viscosity solutions of (2.10). Actually, we will state the uniqueness results in the form of comparison results for constrained solutions.

Theorem 2.2: *Let u be an upper semi-continuous viscosity subsolution of (2.10) on $\bar{D}$ with sublinear growth and v be a bounded from below uniformly continuous viscosity supersolution of (2.10) in D. Then, $u \leq v$ on $\bar{D}$.*

Sketch of the proof: We first construct a positive strict supersolution of (2.10) in D. To this end, let $w(x, y)$ be the value function defined as in (2.7) with U replaced by some U_1 such that $U_1(c) > U(c)$ for $c > 0$, $U_1(0) = U(0)$ and $\alpha = \beta = 1$. This value function is the solution to the classical Merton consumption-portfolio problem in the absence of transaction costs and satisfies $w(x,y) = v(z)$, where $z = x + y$ and v solves

$$\begin{cases} \rho v = -\dfrac{(\mu - r)^2}{2\sigma^2}\dfrac{v'^2}{v''} + rzv' + \max\limits_{c\geq 0}\{-cv' + U_1(c)\} \quad (z > 0) \\ v > 0, v' > 0 \text{ and } v'' < 0, \quad (z > 0). \end{cases} \tag{2.12}$$

We now define, for $(x, y) \in D$

$$W(x,y) = v(x + ky) + x + ky + K \text{ with } \frac{\rho - r}{\rho - \mu}\alpha < k < \beta \text{ and } K > 0$$

and claim that W is a positive strict supersolution of (2.10).
The choice of k, together with the fact that $\frac{\rho - r}{\rho - \mu} > 1$, implies

$$x + ky > 0 \text{ whenever } (x,y) \in D$$

which combined with (2.12) yields $v > 0$ and $v' > 0$. It then follows that

$$\begin{cases} \beta W_x(x,y) - W_y(x,y) = (\beta - k)v'(x + ky) + \beta - k \\ \qquad = f_1(x,y) > \beta - k \\ -\alpha W_x(x,y) + W_y(x,y) = (k - \alpha)v'(x + ky) + k - \alpha \\ \qquad = f_2(x,y,t) + k - \alpha = f_2(x,y,t) > k - \alpha. \end{cases} \tag{2.13}$$

Moreover, using (2.12), we obtain

$$\begin{aligned} &\rho W - \frac{1}{2}\sigma^2 y^2 W_{yy} - \mu y W_y - rxW_x - \max_{c\geq 0}\{-cW_x + U(c)\} = \\ &= \rho[v(x + ky) + (x + ky) + K] - \frac{1}{2}\sigma^2(ky)^2 v'' - \mu y(kv' + k) \\ &\quad -rx(v' + 1) - \max_{c\geq 0}\{-c(v' + 1) + U(c)\} \\ &\geq [\rho v - \frac{1}{2}\sigma^2(ky)^2 v'' - (\mu - r)(ky)v' \\ &\quad - \max_{c\geq 0}\{-cv' + U(c)\} - rzv'] \\ &\quad -k\mu y - rx + \rho(x + ky) + \rho K \end{aligned} \tag{2.14}$$

where we used that $z = x + ky$ and that the function $H(p) = \max\limits_{c\geq 0}\{-cp + U(c)\}$ is decreasing in p. Moreover

$$\Delta = \frac{1}{2}\sigma^2(ky)^2 v'' + (\mu - r)(ky)v' \leq -\frac{(\mu - r)^2}{2\sigma^2}\frac{(v')^2}{v''} \tag{2.15}$$

since the right-hand side of the above inequality is the maximum value of the quadratic expression Δ.
Finally, it follows from the fact that $x + ky > 0$ and the size of ρ that

$$\rho(x+ky) - k\mu y - rx > 0. \tag{2.16}$$

Combining (2.14), (2.15) and (2.16) we get

$$\begin{aligned}
\rho W - &\frac{1}{2}\sigma^2 y^2 W_{yy} - \mu y W_y - rxW_x - \max_{c\geq 0}\{-cW_x + U(c)\} \\
&\geq \rho K + \Big[\rho v + \frac{(\mu - r)^2}{2\sigma^2}\frac{(v')^2}{v''} - rzv' \\
&\quad - \max_{c\geq 0}\{-cu' + U_1(c)\}\Big] + \Big[\max_{c\geq 0}\{-cu' + U_1(c)\} \\
&\quad - \max_{c\geq 0}\{-cu' + U(c)\}\Big] \geq \rho K
\end{aligned} \tag{2.17}$$

where we used that v solves (2.12) and that $U_1 \geq U$.
Now, denote the Hamiltonian

$$\begin{aligned}
H(X, W, DW, D^2W) = \min\Big\{&\rho W - \frac{1}{2}\sigma^2 y^2 W_{yy} - \mu y W_y - rxW_x \\
&- \max_{c\geq 0}\{-cW_x + U(c)\}, \beta W_x - W_y, -\alpha W_x + W_y\Big\}.
\end{aligned}$$

Combining (2.13) and (2.17) yields

$$H(X, W, DW, D^2W) \geq \min\{\beta - k, k - \alpha, \rho K\} = M > 0.$$

To conclude the proof of the theorem we will need the following lemma. Its proof follows along the lines of Theorem VI.5 in Ishii and Lions (1990).

Lemma 2.2: *Let u be upper semi-continuous with sublinear growth viscosity subsolution of (2.10) on D and v be bounded from below uniformly continuous viscosity supersolution of $H(X, u, Du, D^2u) - M$, in D with M being a positive constant. Then $u \leq v$ on D.*

We now conclude the proof of the theorem. We define the function $w^\theta = \theta v + (1-\theta)W$ where $0 < \theta < 1$ and we observe that w^θ is a viscosity supersolution of $H(x, w^\theta, Dw^\theta, D^2w^\theta) - M = 0$. Applying the above lemma to u and w^θ we get $u \leq w^\theta$ on D; sending θ to 1 concludes the proof.

2.b *The case of stock price with Markov-chain parameters.*

We consider a market with the same type of assets as before, i.e. a bond and a stock with prices B_t, given by (2.1) and P_t with satisfies

$$\begin{cases} dP_t = z(t)P_t dt \\ P_0 = P \end{cases}$$

The rate of return z is a finite state continuous time Markov chain, defined on some underlying probability space (Ω, F, Q) with jumping rate $q_{zz'}$, from state z to state z'. The state space is denoted by Z. The associated generator $\mathcal{L}$ of the Markov chain has the form

$$\mathcal{L}v(z) = \sum_{z' \neq z} q_{zz'}[v(z') - v(z)]. \tag{2.18}$$

Let $K = \max_{z \in Z} Z$. A natural assumption is $K \geq r$.
The bond account process x_t satisfies an equation similar to (2.4) and the stock account process y_t solves

$$y_t = y + \int_0^t z(s) y_s ds + L_t - M_t \tag{2.19}$$

A policy is an $\mathcal{F}_t$-progressively measurable triple (c_t, L_t, M_t) where $\mathcal{F}_t$ is the complete σ-algebra generated by the process z_t. The policy (c_t, L_t, M_t) is admissible if it is CADLAG and the state and control constraints (2.5) and (2.5) are satisfied with E being the expectation under the measure Q. The goal of the investor is the same as in the diffusion case, i.e. to maximize the expected discounted utility that comes from the consumption stream. The value function u is given by

$$u(x, y, z) = \sup_{\mathcal{A}} E \int_0^{+\infty} e^{-\rho t} U(c_t) dt \tag{2.20}$$

where $\mathcal{A}$ is the set of admissible policies defined in a similar manner as in the diffusion case.
In the context of transaction costs, this model was first proposed and extensively analyzed and studied in Zariphopoulou (1992).
As it was demonstrated in the previous case, the presence of singular policies results in the fact that the HJB equation is a Variational Inequality with gradient constraints. In fact, when the dynamics of the state process evolve as in (2.19), the Hamilton-Jacobi (HJ) equation turns out to be a system of such Variational Inequalities coupled through the zeroth order term, namely

$$\begin{aligned} &\min\{\rho u - r x u_x - z y u_y - \max_{c \geq 0}[-c u_x + U(c)] - \mathcal{L}u(z), \\ &\beta u_x - u_y, -\alpha u_x + u_y\} = 0 \end{aligned} \tag{2.21}$$

where $\mathcal{L}u(z)$ is given by (2.18).
As in the case of diffusion prices, studied earlier, it turns out that the natural class of solutions is the class of constrained viscosity solutions. The next theorem was presented in Zariphopoulou (1992).

Theorem 2.3. *The value function u is a constrained viscosity solution of the system of Variational Inequalities given in (2.21)* $\forall (x, y) \in D$ and $z \in Z$ with $D = \left\{(x, y) : x + \binom{\alpha}{\beta} y \geq 0\right\}$.

We conclude by stating a comparison result for viscosity solutions of (2.21).

Theorem 2.4. *Let u be an upper semi-continuous viscosity subsolution of (2.21) on D with sublinear growth and v be a bounded from below uniformly continuous viscosity supersolution of (2.21) in D. Then, $u \leq v$ on D.*

Proof. The proof follows along the same lines as the proof of Theorem 2.2. In fact, we only need to show the existence of a positive strict supersolution of (2.21) since the rest of the arguments can be worked in a similar manner. To this end, we define $\phi : D \to R^+$ by $\phi(x,y) = C_1 x + C_2 y + k$, where C_1, C_2, k are positive constants satisfying

$$\alpha C_1 < C_2 < \beta C_1$$

and

$$\rho k > rC_1 + KC_2 + \max_{c \geq 0}[-cC_1 + U(c)].$$

Then, an easy calculation shows that there ϕ is indeed a positive strict supersolution of (2.21).

3 An International Asset Pricing Model with Political Risk

In this section we introduce an asset pricing model to study the effects of political risk on international asset prices, consumption and investment behavior across countries.

We concentrate on a simplified two-country model where capital markets are fully integrated in the sense that individuals from each country can own claims to assets located in either country. These assets serve as production inputs and consumption goods. The production technology is stochastic and differs across the two countries; however there is a single technology in each country. One of the two countries is considered "politically unstable" and exhibits political risk. We model the political risk via a continuous time Markov chain which affects the rate of return of the stochastic production process in the politically risky country. We assume, for simplicity, that there are only two states from the Markov chain, a low and a high state. As it was pointed out in the introduction, the different states can be interpreted, among other things, as representing the local government's ability to alter the tax rate on firms producing in that country or to alter its subsidizing policy on local production. In this context, the low state corresponds to a high tax rate, with the high state representing either a low tax or perhaps a subsidy.

We denote the goods of the two countries by **X** and **Y** and the production technology process by X_t and Y_t in countries $\mathcal{X}$ and $\mathcal{Y}$, respectively.

Country $\mathcal{Y}$ is considered to be politically stable and its production process Y_t is modelled as a diffusion process with drift $b(Y_t)$ and volatility term $\sigma_2(Y_t)$. Country $\mathcal{X}$ has a production technology process with similar diffusion structure — with volatility term $\sigma_1(X_t)$ — but its drift is affected by the two-state Markov chain, say z_t, which represents the effects of the political instability.

Consumption on country $\mathcal{X}$ is denoted by C^x, which includes both consumption of local output and of imports from country $\mathcal{Y}$. Consumption in country $\mathcal{Y}$ is defined in an analogous manner and is denoted by C^y. Cumulative shipments, as of time t, from country $\mathcal{X}$ to country $\mathcal{Y}$ are denoted by L_t; such shipments (exports from country $\mathcal{X}$) incur proportional transaction costs at a rate λ. In a similar manner, cumulative shipments from country $\mathcal{Y}$ (imports by country $\mathcal{X}$) are denoted by M_t with a proportional shipping cost of μ. Without loss of generality, we assume that country $\mathcal{X}$ is charged with the shipping costs.

Using the above definitions, we can write the state processes for the capital stocks in the two countries as

$$dX_t = z_t X_t dt - C_t^x dt + \sigma_1(X_t)dW_t^1 - (1+\lambda)dL_t + (1-\mu)dM_t \quad (3.1)$$

$$dY_t = b(Y_t)dt - C_t^y dt + \sigma_2(Y_t)dW_t^2 + dL_t - dM_t \quad (3.2)$$

with W_t^1 and W_t^2 being Brownian Motions on a probability space $(\Omega, \mathcal{F}, P)$ with correlation $\delta \in (-1,1)$; for this we can take $W_t^2 = \delta W_t^1 + \sqrt{1-\delta^2}\, B_t$ with B_t being a Brownian motion independent of W_t^1.

The volatility coefficients $\sigma_1(x)$ and $\sigma_2(y)$ as well as the drift term $b(y)$ are assumed to be concave functions and locally Lipschitz; we denote by K the generic Lipschitz constant. Moreover it is assumed that $h(0) = 0$ where $h(\cdot)$ is $\sigma_1(\cdot)$, $\sigma_2(\cdot)$ or $b(\cdot)$.

The process z_t is a continuous-time Markov chain with two states z_1 and z_2 with

$$0 < z_1 < z_2 \quad (3.3)$$

As it was discussed above, the low state z_1 is associated with an unfavorable political state (from the perspective of the production process owners) as opposed to the high state z_2 which represents the favorable political state in country $\mathcal{X}$. We denote by p_{ij}, $i,j = 1,2$ the transition probabilities of z_t for the above states.

The *collective* (or integrated) *utility payoff* for consumers of both countries over their consumption rates is

$$E\int_0^{+\infty} e^{-\rho t} U(C_t^x, C_t^y)dt.$$

A policy (C_t^x, C_t^y, L_t, M_t) is admissible if it is $\mathcal{F}_t$-progressively measurable — where $\mathcal{F}_t = \sigma((W_s^1, B_s, Z_s) : 0 \le s \le t)$ — with L_t and M_t being nondecreasing CADLAG processes such that $C_t^x \geq 0$, $C_t^y \geq 0$ a.s. and

$$E\int_0^t e^{-\rho s}(C_s^x + C_s^y)ds < +\infty \qquad t \geq 0 \tag{3.4}$$

and the following state constraints are satisfied

$$x_t \geq 0 \text{ and } y_t \geq 0 \text{ a.e. } t \geq 0. \tag{3.5}$$

The collective consumer function $U : [0,+\infty) \times [0,+\infty) \to [0,+\infty)$ is assumed to be increasing and concave in both arguments with

$$U(0,0) = 0 \quad \text{and} \quad U(C^x, C^y) \leq M(1 + C^x + C^y)^p$$

for some $M > 0$ and $0 < p < 1$.
We define the collective across-countries value function $V(x,y;z)$ as

$$V(x,y;z) = \sup_{\mathcal{A}_z} E\int_0^{+\infty} e^{-\rho t}U(C_t^x, C_t^y)dt \tag{3.6}$$

The set $\mathcal{A}_z$ of admissible policies is defined along the same lines as in the previous section.
From the technical point of view, the stochastic control problem (3.6) is a hybrid extension of the diffusion and the Markov-chain models presented in the previous section. As a matter of fact we can obtain — using a combination of arguments used in Theorems 2.1 to 2.4 — the following result.

Theorem 3.1. *The value function is the unique constrained viscosity solution on* $\{(x,y) = x \geq 0, y \geq 0\}$, *of the system of the variational inequalities*

$$\begin{aligned}\min\{&\rho V(x,y;z_1) - \mathcal{L}V(x,y;z_1) - H(V_x(x,y;z_1),\ V_y(x,y;z_1))\\ &-p_{12}(V(x,y;z_2) - V(x,y;z_1)) - z_1 x V_x(x,y;z_1)\\ &-b(y)V_y(x,y;z_1),\ (1+\lambda)V_x(x,y;z_1) - V_y(x,y;z_1),\\ &-(1-\mu)V_x(x,y;z_1) + V_y(x,y;z_1)\} = 0\end{aligned} \tag{3.7}$$

and

$$\begin{aligned}\min\{&\rho V(x,y;z_2) - \mathcal{L}V(x,y;z_2) - H(V_x(x,y;z_2),\ V_y(x,y;z_2))\\ &-p_{21}(V(x,y;z_1) - V(x,y;z_2)) - z_2 x V_x(x,y;z_2)\\ &-b(y)V_y(x,y;z_2),\ (1+\lambda)V_x(x,y;z_2) - V_y(x,y;z_2),\\ &-(1-\mu)V_x(x,y;z_2) + V_y(x,y;z_2)\} = 0\end{aligned} \tag{3.8}$$

where $\mathcal{L}$ *is the differential operator*

$$\mathcal{L}V = \frac{1}{2}\sigma_1^2(x)V_{xx} + \delta\sigma_1(x)\sigma_2(y)V_{xy} + \frac{1}{2}\sigma_2^2(y)V_{yy} \tag{3.9}$$

and

$$H(q_1,q_2) = \max_{C^x,C^y}\{-q_1C^x - q_2C^y - U(C^x,C^y)\}. \tag{3.10}$$

The proof follows from a combination of standard-albeit tedious arguments – used in the proofs of the previous section. Although these arguments are not presented here, the technically oriented reader can find them in Hodder, Tourin and Zariphopoulou (1998).
As it was explained in previous sections, the presence of singular policies leads to a depletion of the state space into regions of three types, namely the Buy ($\mathcal{B}$), Sell ($\mathcal{S}$) and No Trade ($\mathcal{NT}$) regions. No-closed form solutions exist up to date, for the free boundaries associated with the optimal singular policies. On the other hand, these boundaries can be analyzed and studied numerically. In Hodder, Tourin and Zariphopoulou (1998), we provide an extensive numerical study for the value function and the free boundaries. The main ingredients of the numerical work is the use of fractional-time filtered second-order schemes and the stability properties of viscosity solutions.

Remark 3.1: In the special case of linear diffusion coefficients in the state dynamics (3.1) and (3.2) and a combined utility function of the HARA type, $U(C^x, C^y) = \frac{1}{\gamma}[(C^x)^\gamma + (C^y)^\gamma]$ for $0 < \gamma < 1$, one can show that the value function is homogeneous of degree γ. This fact provides valuable information about the free boundaries which turn out to be straight lines passing through the origin.

We continue this section by presenting some results related to analytic bounds of the value function as well as alternative characterizations of it in terms of a class of "pseudo-collective" value functions. The latter results are expected to enhance our intuition for the economic significance of the proposed pricing model. We only present the main steps of the proofs of these results; the underlying idea is to use the HJB equations (3.7) and (3.8) and interpret them as HJB equations of new pseudo-utility problems. The comparison between the new "pseudo-value functions" and the original value function stems from the uniqueness result in Theorem 3.1 as well as the fact that the pseudo-value functions are viscosity solutions of the associated HJB equations.
To this end, consider the following pairs $(\underline{x}_t, \underline{y}_t)$ and $(\bar{x}_t, \bar{y}_t)$ of state dynamics where $\underline{x}_t$, $\underline{y}_t$, $\bar{x}_t$ and $\bar{y}_t$ solve respectively

$$d\underline{x}_t = z_1 \underline{x}_t dt - \underline{C}_t^x dt + \sigma_1(\underline{x}_t) dW_t^1 - (1+\lambda) d\underline{L}_t + (1-\mu) d\underline{M}_t \quad (3.11)$$

$$d\underline{y}_t = b(\underline{y}_t) dt - \underline{C}_t^y dt + \sigma_2(\underline{y}_t) dW_t^2 + d\underline{L}_t - d\underline{M}_t \quad (3.12)$$

and

$$d\bar{x}_t = z_2 \bar{x}_t dt - \bar{C}_t^{\bar{x}} dt + \sigma_1(\bar{x}_t) dW_t^1 - (1+\lambda) d\bar{L}_t + (1-\mu) d\bar{M}_t \quad (3.13)$$

$$d\bar{y}_t = b(\bar{y}_t) dt - \bar{C}_t^{\bar{y}} dt + \sigma_2(\bar{y}_t) dW_t^2 + d\bar{L}_t - d\bar{M}_t \quad (3.14)$$

where z_1, z_2 are the two states of the process z_t.

We define for (3.11), (3.12) and (3.13), (3.14) the sets of admissible policies $\mathcal{A}_{z_1}$ and $\mathcal{A}_{z_2}$ in a similar manner as $\mathcal{A}_z$.
The following result shows that the original value function V is bounded between $\underline{v}$ and $\bar{v}$ with $\underline{v}$ and $\bar{v}$ being respectively the value functions of two international asset pricing models with the original collective utility but with no political risk. More precisely, $\underline{v}$ (respectively $\bar{v}$) is the collective value function for countries $\underline{\mathcal{X}}$ and $\underline{\mathcal{Y}}$ (respectively $\bar{\mathcal{X}}$ and $\bar{\mathcal{Y}}$) with $\underline{\mathcal{X}}$ (respectively $\bar{\mathcal{X}}$) not exhibiting political instability but with a modified mean rate of return in its capital stock. Models of this type were studied by Dumas (1992) in the case of HARA utilities.

Theorem 3.2. *Consider the value functions $\underline{v}$, $\bar{v} : [0,+\infty) \times [0,+\infty) \to [0,+\infty)$,*

$$\underline{v}(x,y) = \sup_{\mathcal{A}_{z_1}} E \int_0^{+\infty} e^{-\rho t} U(\underline{C}_t^x, \underline{C}_t^y) dt \tag{3.15}$$

and

$$\bar{v}(x,y) = \sup_{\mathcal{A}_{z_2}} E \int_0^{+\infty} e^{-\rho t} U(\bar{C}_t^{\bar{x}}, \bar{C}_t^{\bar{y}}) dt. \tag{3.16}$$

Then

$$\underline{v}(x,y) \le V(x,y;z) \le \bar{v}(x,y)$$

for $(x,y) \in [0,+\infty] \times [0,+\infty)$ and $z = z_1, z_2$.

We finish this section by discussing two collective-utility asset pricing problems without political risk but with different discount factors and "enhanced" collective-bequest functions. It turns out, as it is stated in Theorem 3.3 that their value functions u^1 and u^2 — given below by (3.19) and (3.20) — coincide with the original value function for states z_1 and z_2.
To this end, define two collective-bequest function Φ_1, $\Phi_2 : [0,+\infty) \times [0,+\infty) \to [0,+\infty)$ as

$$\Phi_1(x,y) = p_{12} V(x,y;z_2), \quad \Phi_2(x,y) = p_{21} V(x,y;z_1) \tag{3.17}$$

and the discount factors

$$\rho_1 = \rho + p_{12}, \quad \rho_2 = \rho + p_{21}. \tag{3.18}$$

Also consider the controlled state processes $(\underline{x}_t, \underline{y}_t)$ and $(\bar{x}_t, \bar{y}_t)$ solving (3.11), (3.12) and (3.13), (3.14) and the associated sets of admissible policies $\mathcal{A}_{z_1}$ and $\mathcal{A}_2$.

Theorem 3.3. *Define the value functions u_1, $u_2 : [0,+\infty) \times [0,+\infty) \to [0,+\infty)$ by*

$$u^1(x,y) = \sup_{\mathcal{A}_{z_1}} E \int_0^{+\infty} e^{-\rho_1 t} [U(\underline{C}_t^x, \underline{C}_t^y) + \Phi_1(\underline{x}_t, \underline{y}_t)] dt \tag{3.19}$$

with $\underline{x}_0 = x$, $\underline{y}_0 = y$, *and*

$$u^2(x,y) = \sup_{\mathcal{A}_{z_1}} E \int_0^{+\infty} e^{-\rho_2 t}[U(\overline{C_t^{\overline{x}}}, \overline{C_t^{\overline{y}}}) + \Phi_2(\bar{x}_t, \overline{y}_t)]dt \tag{3.20}$$

with $\bar{x}_0 = x$, $\overline{y}_0 = y$ *and* $\rho_1, \rho_2, \Phi_1, \Phi_2$, *as in (3.18) and (3.17). Then*

$$V(x, y; z_1) = u^1(x, y) \quad \textit{and} \quad V(x, y; z_2) = u^2(x, y) \tag{3.21}$$

Idea of the proof:. Observe that $V(x, y; z_1)$ is the unique solution of the HJB equation (3.7), which can be rewritten as

$$\begin{aligned} \min\{&\rho_1 V(x,y;z_1) - \mathcal{L}V(x,y;z_1) - H(V_x(x,y;z_1),\ V_y(x,y;z_1)) \\ &-z_1 x V_x(x,y;z_1) - b(y)V_y(x,y;z_1) - \Phi_1(x,y), \\ &(1+\lambda)V_x(x,y;z_1) - V_y(x,y;z_1), \\ &-(1-\mu)V_x(x,y;z_1) + V_y(x,y;z_1)\} = 0. \end{aligned} \tag{3.22}$$

We easily get that the above equation can be interpreted as the HJB equation of a new stochastic control problem with value function u^1. The result then follows from the uniqueness of (constrained) viscosity solutions of (3.7) and the fact that u^1 is a (constrained) viscosity solution of (3.22).

We can interpret the ratio of the partial derivatives of $V(x, y; z)$ as the relative price of good **X** in terms of good **Y**. Then, the equilibrium prices for the states z_1 and z_2 will be respectively

$$P_1(x,y;z_1) = \frac{u_x^1(x,y;z_1)}{u_y^1(x,y;z_1)} \quad \text{and} \quad P_2(x,y;z_2) = \frac{u_x^2(x,y;z_2)}{u_y^2(x,y;z_2)}$$

If we further assume that $U(C^x, C^y) = \frac{(C^x)^\gamma}{\gamma} + \frac{(C^y)^\gamma}{\gamma}$, then each of the value functions of the two pseudo-problems (3.19) and (3.20) is homogeneous of degree γ. For the state z_1, there will be a cone with linear boundaries in the x, y plane within which no shipping occurs. Similarly, there will be a "no-shipping cone" for the state z_2; however, these cones will generally be different. This situation is analyzed numerically in Hodder, Tourin and Zariphopoulou (1998). One can observe that for the "politically favorable" state z_2 the expected returns for production in country $\mathcal{X}$ is relatively high as compared with the situation for the politically "unfavorable" state z_1. Consequently, a shift to z_1 makes production in country $\mathcal{X}$ less attractive and may cause a costly shipment to take place between countries. In other words, transitioning from z_2 to z_1 (or vice versa) can cause the no-shipping cone to shift. Hence, jumps in the coefficients of the asset prices can occur when z_t switches values; this contrasts with the smoothing changing prices obtained in Dumas (1992). On the other hand, as in Dumas (1992), the prices are expected to deviate from a parity value of one for potentially substantial periods of time. Both prices are bounded between the $1/1+\lambda$ and $1/1-\mu$ which represent the prices at the cone boundary where shipments take place. However, shifts in the cone as z_t transitions between z_1 and z_2 can result in cone rotation. If the shift is from z_2 to z_1 and this results in cone rotation such that the current x, y pair is no longer an interior (no-shipping) point of the cone, the price of x will drop from a level larger than $\frac{1}{1+\lambda}$ to the minimum level $\frac{1}{1+\lambda}$. In a similar manner, if the x, y pair had been at the upper cone boundary with state z_2, that will be an interior point of the cone, when moving to the state z_1. Hence, the price of **X** will drop from $\frac{1}{1-\mu}$ to a lower value. Thus, the price of **X** jumps downward if a transition takes place from z_2 to z_1 and resulting cone rotation causes a discrete move from an interior point to the boundary or from the boundary to an interior point. We conjecture that a downward jump also occurs if the current (x, y) pair is an interior point of both cones.

References

Capuzzo-Dolcetta, I., P.-L. Lions, P.-L. (1990). *Viscosity solutions of Hamilton-Jacobi equations.* Trans. Amer. Math. Soc., **318**, pp 643–683.

Constantinides, G. M., Zariphopoulou, T. (1998). *Bounds on prices of derivative securities in an intertemporal setting with proportional transaction costs and multiple securities*, working paper.

Constantinides, G. M., Zariphopoulou, T. (1997). *Bounds on prices of contingent claims in an intertemporal economy with proportional transaction costs and general preferences*, submitted for publication.

Crandall, M., Ishii, H., Lions, P.-L. (1992). *User's guide to viscosity solutions of second order partial differential equations.* Bulletin of the American Mathematical Society, **27**, pp 1–67.

Crandall, M., Lions, P.-L. (1983). *Viscosity solutions of Hamilton-Jacobi equations.* Transactions of the American Mathematical Society, **277**, pp 1–42.

Davis, M. H. A., Norman, A. R. (1990). *Portfolio selection with transaction costs.* Mathematics of Operations Research, **15**, pp 676–713.

Davis, M. H. A., Panas, V. G., Zariphopoulou, T. (1993). *European option pricing with transaction costs.* SIAM J. of Control and Optimization, **31**, pp 470–493.

Davis, M. H. A., Zariphopoulou, T. (1995). *American Options and Transaction Fees.* Mathematical Finance, Springer-Verlag.

Dumas, B. (1992). *Dynamic Equilibrium and the Real Exchange Rate in a Spatially Separated World*, Review of Financial Studies, **5**, pp 153–180.

Fleming, W., Soner, H. M. (1993). *Controlled Markov Processes and Viscosity Solutions.* Springer-Verlag, New York.

Hodder, J. E., Tourin, A., Zariphopoulou, T. (1998). *Numerical schemes for variational inequalities arising in international price models,* in preparation.

Hodges, S. D., Neuberger, A. (1989). *Optimal replication of contingent claims under transaction costs.* Review of Futures Markets, **8**, pp 222–239.

Ishii, H., Lions, P.-L. (1990). *Viscosity solutions of fully nonlinear second-order elliptic partial differential equations.* Journal of Differential Equations, **83**, pp 26–78.

Lions, P.-L. (1983). *Optimal control of diffusion processes and Hamilton-Jacobi-Bellman equations.* Part 1: *The dynamic programming principle and applications.* Part 2: *Viscosity solutions and uniqueness.* Communications in PDE, **8**, pp 1101–1174 and pp 1229–1276.

Pichler, A. (1996). *On transaction costs and HJB equations.* Preprint.

Shreve, S. E., Soner, H. M. (1994). *Optimal investment and consumption with transaction costs.* The Annals of Applied Probability, **4**, pp 609–692.

Soner, H. M. (1986). *Optimal control with state space constraints.* SIAM Journal of Control and Optimization, **24**, pp 552–562, pp 1110–1122.

Tourin, A., Zariphopoulou, T. (1994). *Numerical schemes for investment models with singular transactions.* Computational Economics, **7**, pp 287-307.

Tourin, A., Zariphopoulou, T. (1995). *Portfolio Selection with Transaction Costs,* Progress in Probability, **4**.

Zariphopoulou, T. (1992). *Investment-consumption models with transaction fees and Markov-chain parameters.* SIAM Journal of Control and Optimization, **30**, pp 613–636.

Zariphopoulou, T. (1994). *Investment consumption models with constraints,* SIAM Journal of Control and Optimization, **32**, pp 59–84.

35

A Verification Theorem in General Equilibrium Model of Asset Prices

Chi-Fu Huang
Michael Taksar
Steven H. Zhu

ABSTRACT: We consider an investment/consumption model for economic agents each of whom has a HARA utility function and whose objective is maximization of the total expected utility of consumption. There is a diffusion process Y_t which describes the random technology changes and whose state effects the fluctuations of the wealth of each of the agents. This model was considered by Cox, Ingersoll and Ross in the context of the problem of equilibrium asset prices. In this paper we establish the conditions of the existence of the classical solution to the corresponding Bellman equation and consider its dependence on the type of the utility functions the individual agent possesses.

KEYWORDS: Asset pricing, dynamic programming, general equilibrium and verification theorem.

1 Introduction

In the general equilibrium model by Cox, Ingersoll and Ross [3] and [4] the random changes of the state of technology are characterized by a k-dimensional Markov diffusion process

$$dY_s = \mu(Y_s, s)ds + S(Y_s, s)dB_s. \tag{1}$$

where B_s is a standard k-dimensional Brownian motion defined on a probability space $(\Omega, \mathcal{F}, \mathcal{P})$, representing the sources of uncertainties in the economy.

Following the notions of the Arrow-Debreu model, the economy consists of many identical agents, each of whom has a HARA utility function, that is a function of the form

$$U(C,t) = e^{-\lambda t}\frac{C^\gamma - 1}{\gamma}. \tag{2}$$

Each of the agents faces a consumption/investment problem of maximizing the expected cumulative utility of consumption subject to the budgetary constraints. Mathematically the problem of an individual agent consists of finding

$$\text{Max} \qquad E\int_0^T U(C_s, s)ds \tag{3}$$

subject to $W_s \geq 0$ for all $s \geq 0$, where $W.$ is the process whose value at time s is interpreted as the amount of wealth the individual agent possesses at this time. This process is governed by the following stochastic differential equation

$$dW_s = W_s[r + a'(\alpha - r\mathbf{1}) + b'(\beta - r\mathbf{1})]ds - C_s ds + W_s(a'G + b'H)dB_s. \tag{4}$$

In the above equation, α and β are interpreted as the stochastic rates of return to scale and $G(Y,t)$ as the random matrix of volatilities in the linear system of production activities described in above mentioned papers by Cox, Ingersoll and Ross; while a and b are the control "parameters" representing the proportions of wealth invested in the risky assets and the contingent claims, and $C.$ is the rate of current consumption, respectively. Thus the consumption/investment problem the individual agent faces is a dynamic optimization problem, which can be identified as a stochastic optimal control problem (i.e., Fleming/Rishel [5]). Standard dynamic programming principle implies that the value function J for this optimal control problem must satisfy in an appropriate sense the following Bellman equation:

$$\partial_t J(W,Y,t) + \max_{a,b,C}\{\mathcal{L}^{a,b,C}J(W,Y,t) + U(C,t)\} = 0, \tag{5}$$

on $Q_T = R^+ \times R^k \times [0,T)$ with the terminal condition $J(W,Y,T) = 0$. Here, the differential operator $\mathcal{L}^{a,b,C}$ is given by

$$\begin{aligned}
\mathcal{L}^{a,b,C} &= \frac{1}{2}W^2(a'G+b'H)(a'G+b'H)'\frac{\partial^2}{\partial W^2} \\
&+ W(a'G+b'H)S(Y,t)\frac{\partial^2}{\partial W\partial Y} + \frac{1}{2}S(Y,t)S'(Y,t)\frac{\partial^2}{\partial Y^2} \\
&+ W[r + a'(\alpha - r\mathbf{1}) + b'(\beta - r\mathbf{1})]\frac{\partial}{\partial W} + \mu(Y,t)\frac{\partial}{\partial Y} - C\frac{\partial}{\partial W}.
\end{aligned}$$

Furthermore, the boundary condition at $W = 0$ is given by

$$J(0,Y,t) = 0, \qquad \text{if } \gamma > 0, \tag{6}$$

$$J(0,Y,t) = -\infty, \qquad \text{if } \gamma < 0. \tag{7}$$

The solution of the Bellman equation will then give rise to different patterns of the investor's optimal policy. Some related work on this subject can been found in [1], [2], [10] and [16].

The Bellman equation (5) is a 2nd-order nonlinear partial differential equation of a parabolic type. In general, such an equation does not admit a "smooth" classical solution due to singularity at $W = 0$ and a possible degeneracy of the technology process Y_s. The goal of this paper is to obtain conditions under which the value function $J(W, Y, t)$ satisfies the Bellman equation (5) with boundary conditions in a well-defined sense, and to provide a mathematically rigorous proof of the verification theorem in the absence of the usually required regularity properties and polynomial growth conditions on the value function.

2 Classical Solutions: A Verification Theorem

In this section, we present a sufficient condition under which the value function is a classical solution of the Bellman equation (5).
We say that the system of state dynamics (1) is 'nondegenerate' if there exists a constant $q > 0$ such that

$$\xi^T S(Y,t) S^T(Y,t)\xi \geq q|\xi|^2 \tag{8}$$

for any $(Y, t) \in R^k \times [0, T]$ and $\xi \in R^N$. We assume that the drift and diffusion of the state dynamics satisfy the following conditions:

- $\mu(Y, t)$ and $S(Y, t)$ are at most of linear growth, i.e., there exists $M > 0$ such that

$$|\mu(Y,t)| \leq M(1 + |Y|) \tag{9}$$
$$|S(Y,t)| \leq M(1 + |Y|) \tag{10}$$

- $\mu(Y, t)$ and $S(Y, t)$ are continuous in t and locally Lipschitz in Y, i.e., for each $K > 0$ there exists R_K, such that

$$|\mu(Y,t) - \mu(Y',t)| \leq R_K|Y - Y'| \tag{11}$$
$$|S(Y,t) - S(Y',t)| \leq R_K|Y - Y'| \tag{12}$$

  whenever $|Y| \leq R$ and $|Y'| \leq R$.

The control processes of investment/consumption (a, b, C) are chosen from a class of admissible policies defined below.
Definition: A triple (a, b, C) of $\mathcal{F}_t^s$-adapted processes is called an admissible policy if
1). a_s, b_s are square-integrable, that is, $\int_0^t a_s^2 + b_s^2 ds < \infty$, $\mathcal{P}$-a.s. $\forall t \geq 0$.
2).The process C_s is nonnegative with the rate process $C_s/W_s \equiv c_s$ having continuous paths. In addition, $E\int_t^T |U(C_s, s)|ds < \infty$ for all $t \leq T$.
3) The solution W_t to the stochastic differential equation (4) is a.s. nonnegative for all $t \leq T$.

Note that in the above definition, we do not put constraints on a_s, b_s. This implies that borrowing and shortselling are permitted in our model.
Denote by $\mathcal{AC}$ the class of all admissible policies. Below we state without proof the general optimality principle.
Proposition: The optimal value function $J(W,Y,t)$ satisfies the "weak" dynamic programming principle: For any stopping time $\tau \in [t,T]$ with respect to $\mathcal{F}_t^s$,

$$J(W,Y,t) = \max_{a,b,C\in\mathcal{AC}} E_{Y,t}\{\int_t^T U(C_s,Y_s,s)ds + J(W_\tau,Y_\tau,\tau)\}, \qquad (13)$$

For readers interested in the subject, we refer to Chapter V in the recent book by Fleming and Soner [6] for the proofs and various formulations and ramifications of the above principle. In the context of portfolio models, we remark that the dynamic programming principle implies that optimal strategy in the consumption/ investment model of the equilibrium framework must be Markovian. This means that at each moment of time the optimal policy depends only on the current state of the economy, and therefore it cannot be replicated in all states of nature by the policies in the past as well as the future.
Define by $C_p^{2,1}(Q_T)$ the space of functions $\Phi(W,Y,t)$, $W \geq 0, t \leq T, Y \in Q_T$, which are continuous in all arguments, continuously differentiable in t twice continuously differentiable in Y and such that Φ, Φ_t, Φ_Y and Φ_{YY} are subject to a polynomial growth condition.
Theorem: (*Verification*) Suppose there exists a function $V(W,Y,t) \in C_p^{2,1}(Q_T)$ concave in W, such that for all (W,Y,t) with a possible exception for $W=0$ it satisfies the Bellman equation (5) in the classical sense. Then, for each $(W,Y,t) \in Q_T$,

$$V(W,Y,t) \geq E\int_t^T U(C_s,s)ds, \text{for any} \quad (a,b,C) \in \mathcal{AC}.$$

Further suppose there exists an admissible policy (a^*,b^*,C^*) with the corresponding wealth process W_s^* such that

$$\partial_t V(W_s^*,Y_s,s)+\mathcal{L}^{a^*,b^*}V(W_s^*,Y_s,s)+U(C_s^*,s) = 0, \ \forall a.e.s \in [t,T), \ P-a.s. \qquad (14)$$

Then, (a^*,b^*,C^*) is the optimal consumption/investment policy in the interval $[t,T]$ and for this policy $V(W,Y,t) = J(W,Y,t)$.

In the following sections, we will show that $V(W,Y,t) = f(Y,t)U(W,t) + g(Y,t)$ with the functions $f(Y,t)$ and $g(Y,t)$ satisfying a pair of parabolic partial differential equations. Under the assumptions (9)–(12), we provide a rigorous proof of the above theorem for all cases of HARA utility functions $U(C,t)$. Furthermore, in the last section we will demonstrate that the 'uniform ellipticity' condition (8) can be relaxed so that the relaxed condition allows a "generalized" solution to the Bellman equation.

3 Power Utility vs. Logarithmic Utility

It is known that the indirect utility (value) function admits a separation between wealth and state variables

$$J(W,Y,t) = f(Y,t)U(W,t) + g(Y,t) \tag{15}$$

for some functions $f(Y,t)$ and $g(Y,t)$. This observation, along with the fact that $J(W,Y,t)$ is concave in wealth W, was pointed out in CIR's original paper and had been verified in other context by Hakansson [8], Merton [14] and Samuelson [15].

The first result of this section will be the demonstration that the functions $f(Y,t)$ and $g(Y,t)$ in (5) can be determined by solving two separate differential equations of lower dimension. In the following, it will become clear that these functions are of a completely nature in the cases of power and logarithmic utility.

3.1 Case I: Power Utility Function

In this subsection we will consider the case of $U(C,t) = e^{-\lambda t}C^{\gamma}/\gamma$, for $\gamma \neq 0$ with $\gamma \neq 0$, $\gamma < 1$. (Two different subcases of $0 < \gamma < 1$ and $\gamma < 0$ correspond to the power utility with positive and negative exponent respectively). In this case we will seek a solution to the Bellman equation (5) in the form

$$J(W,Y,t) = f(Y,t)e^{-\lambda t}\frac{W^{\gamma}}{\gamma} + g(Y,t). \tag{16}$$

Trivial calculations show that for the function J given by (16)

$$\partial_t J = (\partial_t f - \lambda f)U(W,t) + \partial_t g,$$

and

$$\begin{aligned}
\mathcal{L}^{a,b,C}J &= \frac{1}{2}W^2(a'G+b'H)(a'G+b'H)'fU_{WW} \\
&+ W(a'G+b'H)S'(Y,t)f_Y U_W + \frac{1}{2}S(Y,t)S'(Y,t)[f_{YY}U + g_{YY}] \\
&+ W[r + a'(\alpha - r\mathbf{1}) + b'(\beta - r\mathbf{1})]fU_W \\
&+ \mu(Y,t)(f_Y U + g_Y) - CfU_W.
\end{aligned}$$

By the property of the power function,

$$WU_W(W,t) = \gamma U(W,t), \qquad W^2U_{WW}(W,t) = \gamma(\gamma-1)U(W,t)$$

Substituting U given by (16) into (5) and using the above relations, we see that in order for U to satisfy the Bellman equation (5), the function $g(Y,t)$ must satisfy the following differential equation

$$\partial_t g + \frac{1}{2}S(Y,t)S'(Y,t)g_{YY} + \mu(Y,t)g_Y = 0, \tag{17}$$

with terminal condition $g(Y,T) = 0$, while $f(Y,t)$ must satisfy a quasi-linear differential equation:

$$\partial_t f + \frac{1}{2}S(Y,t)S'(Y,t)f_{YY} + \mu(Y,t)f_Y - \lambda f + \gamma R_\gamma(f, f_Y) = 0, \qquad (18)$$

with terminal condition $f(Y,T) = 0$, where $R_\gamma(f, f_Y)$ is given by

$$\begin{aligned} R_\gamma(f, f_Y) &= \max_{a,b,c}\{(a'G + b'H)S'(Y,t)f_Y + [r + a'(\alpha - r\mathbf{1}) + b'(\beta - r\mathbf{1}) \\ &+ \frac{(\gamma-1)}{2}(a'G + b'H)(a'G + b'H)' - c]f + U(c)\}. \end{aligned}$$

Note that in (18) the maximization is done not with respect to the absolute rate C rather with respect to its ratio to the wealth, namely with respect to the consumption rate $c = C/W$.
Clearly, $g(Y,t) \equiv 0$ is the only solution of (17) and therefore in this case, $J(W,Y,t) = f(Y,t)U(W,t)$ is separable in the multiplicative form, provided the solution to (18) exists.
From (8) it follows that

$$S(Y,t)S'(Y,t) \geq \theta|Y|^2, \qquad \forall Y \in R^k,$$

Combining this with the growth conditions (9)-(12), we can apply the regularity result from the theory of (parabolic) PDEs (e.g., see, Friedman [7]) to obtain existence of a unique $C^{2,1}$-solution $f(Y,t)$ for the 2nd-order quasi-linear equation (18) such that there exist an integer $m > 0$ and a constant $M > 0$,

$$0 < f(Y,t) \leq M(1 + |Y|^m)$$

The proof that $J(W,Y,t) = f(Y,t)U(W,t)$ is a well-defined solution to the Bellman equation (5) will be detailed in the next section.

3.2 Case II: Logarithmic Utility Function

In this case, the utility function is equal to $U(C,t) = e^{-\lambda t}\log C$. We will seek solution to (5) in the form

$$J(W,Y,t) = f(Y,t)e^{-\lambda t}\log W + g(Y,t). \qquad (19)$$

Note that $W^2U_{WW} = -e^{-\lambda t}$ and $WU_W = e^{-\lambda t}$, and for the function $J(W,Y,t)$ given by (19)

$$\partial_t J = (\partial_t f - \lambda f)U(W,t) + \partial_t g.$$

and

$$\mathcal{L}^{a,b,C}J = -\frac{1}{2}(a'G + b'H)(a'G + b'H)'fe^{-\lambda t}$$

$$
\begin{aligned}
&+ \quad (a'G+b'H)S'(Y,t)f_Y e^{-\lambda t}\\
&+ \quad \frac{1}{2}S(Y,t)S'(Y,t)[f_{YY}U(W,t)+g_{YY}]\\
&+ \quad [r+a'(\alpha-r\mathbf{1})+b'(\beta-r\mathbf{1})]fe^{-\lambda t}\\
&+ \quad \mu(Y,t)(f_Y U(W,t)+g_Y)-cfe^{-\lambda t}.
\end{aligned}
$$

Substituting $J(W,Y,t)$ given by (19) into (5) and rearranging terms, we see that $f(Y,t)$ must satisfy the following linear differential equation

$$\partial_t f+\frac{1}{2}S(Y,t)S'(Y,t)f_{YY}+\mu(Y,t)f_Y-\lambda f+1=0 \tag{20}$$

with terminal condition $f(Y,T)=0$.
Similarly, the function $g(Y,t)$ must satisfy a quasi-linear differential equation (cf., **Case I**)

$$\partial_t g+\frac{1}{2}S(Y,t)S'(Y,t)g_{YY}+\mu(Y,t)g_Y+R_0(Y,t)=0 \tag{21}$$

with terminal condition $g(Y,T)=0$, and

$$R_0(Y,t) = \frac{e^{-\lambda t}-e^{-\lambda T}}{\lambda}\max_{a,b,c}\{r+a'(\alpha-r\mathbf{1})+b'(\beta-r\mathbf{1}) \tag{22}$$

$$-\frac{1}{2}(a'G+b'H)(a'G+b'H)'-c\}+U(c)\} \tag{23}$$

where $c=C/W$ as before.
In contrast to **Case I**, the solution to (20) can be found directly

$$f(Y,t)=\frac{1}{\lambda}(1-e^{-\lambda(T-t)}). \tag{24}$$

As a consequence, we can see that this solution is only a function of time t. Further note that $R_0(Y,t)$ is equal to $R_\gamma(e^{-\lambda}f,f_Y)$ with $\gamma=0$ where R_γ is the same as in the **Case I** and $f(Y,t)$ is replaced by the solution (24). Repeating the arguments of CIR's paper one can show that under the the market clearing condition[1], the optimal investment is determined by the formula

$$a^*(Y,t)=(GG')^{-1}\alpha+\frac{1-\mathbf{1}'(GG')^{-1}\alpha}{\mathbf{1}'(GG')^{-1}\mathbf{1}}(GG')^{-1}\mathbf{1}$$

That is, the optimal portfolio decision is a function of the state variables. Moreover, the optimal consumption rate c^* can also be determined explicitly from the expression $R_0(Y,t)$ as

$$c^*=(1-e^{-\lambda(T-t)})/\lambda$$

[1]The market is said to be cleared, if and only if every buy order can be matched with a sell order with no transaction cost incurred.

which is only a function of t, and not of the state Y. Finally, the equilibrium interest rate is given by

$$\begin{aligned} r^*(Y,t) &= a^{*\prime}\alpha - a^{*\prime}GG'a^* \\ &= \frac{\mathbf{1}'(GG')^{-1}\alpha - 1}{\mathbf{1}'(GG')^{-1}\mathbf{1}} \end{aligned}$$

where $\alpha(Y,t)$ is the stochastic rate of return to scale and $G(Y,t)$ is the random matrix of volatilities in the linear system of production activities. The examples below illustrates that specification of the production processes, results in the equilibrium rate r^* being determined very explicitly alongside with the specific nature of optimal investment decision $a^*(Y,t)$.

Example I: A Linear Production with Constant Scale

We consider a special case, discussed in the CIR model, in which the means and variances of the rates of return on the production processes are proportional to the state variable Y, that is,

$$\alpha = \hat{\alpha}Y \qquad \text{and} \qquad GG' = \Omega Y$$

where $\hat{\alpha}$ and Ω are constant, independent of the state Y.

Then, it is easy to verify that the vector of the optimal investments in the portfolio satisfies

$$\begin{aligned} a^*(Y,t) &= (GG')^{-1}\alpha + \frac{1 - \mathbf{1}'(GG')^{-1}\alpha}{\mathbf{1}'(GG')^{-1}\mathbf{1}}(GG')^{-1}\mathbf{1} \\ &= \Omega^{-1}\hat{\alpha} + \frac{1 - \mathbf{1}'\Omega^{-1}\hat{\alpha}}{\mathbf{1}'\Omega^{-1}\mathbf{1}}\Omega^{-1}\mathbf{1} \equiv \text{Constant} \end{aligned}$$

and the equilibrium interest rate can be written as

$$r^*(Y) = \frac{\mathbf{1}'\Omega^{-1}\hat{\alpha} - 1}{\mathbf{1}'\Omega^{-1}\mathbf{1}}Y \tag{25}$$

which is a linear function of Y. Consequently, the rate r^* will be governed by the same stochastic differential equation as that of the process Y.

In the framework of CIR model, the equilibrium interest rate could be either a log-normal or a mean-reverting process as determined by the underlying state process. However, in general the optimal investment rate $a^*(Y,t)$ need not be a constant and the equilibrium rate $r^*(Y,t)$ need not be a linear function of the state Y and the general relation of dependence in Equation(25) can give rise to a rather different pattern of behaviors from that of to the underlying process Y. In particular, the sign of $\mathbf{1}'(GG')^{-1}\mathbf{1}-1$ will determine the fluctuation of the interest rates between the positive and negative levels.

Example I.1: *Log-Normal State Process*

In this example, we consider the case when the process of technological changes follows a geometric Brownian motion

$$dY_s = \mu Y_s dt + \sigma Y_s dB$$

where μ and σ represent the instantaneous rate of return and standard deviation on those changes.
Obviously, the function $g(Y,t)$ must satisfy

$$\partial_t g + \frac{1}{2}\sigma^2 Y^2 g_{YY} + \mu Y g_Y + R_0(Y,t) = 0.$$

Making change of variable $Y = e^Z$, we see that $G(Z,t) = g(e^Z,t)$ satisfies the following linear equation with constant coefficients

$$\partial_t G + \frac{1}{2}\sigma^2 G_{ZZ} + (\mu - \frac{1}{2}\sigma^2)G_Z + \hat{R}_0(Z,t) = 0 \tag{26}$$

which can be solved in a closed form. To this end, we introduce a process Z defined as

$$dZ_s = (\mu - \frac{1}{2}\sigma^2)ds + \sigma dB_t.$$

The solution to(26) can be obtained via the Feynman-Kac formula as

$$\begin{aligned} g(Y,t) &= G(Z,t) \\ &= E\int_t^T \hat{R}_0(Z_s,s)ds \\ &= \int_t^T \int_{-\infty}^{\infty} K(\xi,s;Z,t)\hat{R}_0(\xi,s)d\xi ds \end{aligned}$$

with $Z = \log Y$ being the initial state, $\hat{R}_0(Z,t) = R_0(e^Z,t)$ and the kernel

$$K(\xi,s,Z,t) = \frac{1}{\sqrt{2\pi\sigma^2(s-t)}}\exp\{-\frac{[\xi - Z - (\mu - \frac{1}{2}\sigma^2)]^2}{\sigma^2(s-t)}\}$$

is the transition density of the process Y.. Thus, the equilibrium interest rate r^* in Equation (25) is a logarithmic Brownian motion and the value function is given in the explicit form $J(W,Y,t) = (1 - e^{-\lambda(T-t)})/\lambda +$ the 'Feynman-Kac' integral.
Example I.2: *Mean-Reversion State Process*
In this example, we consider the CIR model with the technology state following a mean-reversion process

$$dY = \mu(\bar{Y} - Y)dt + \sigma\sqrt{Y}dB$$

where $\bar{Y}$ denotes the long-run average of the process Y, μ represents the intensity of mean-reversion and σ is the intensity of volatility of the process Y.
Under the condition of a linear production with constant rates to scale, the optimal allocation of investment $a^*(Y,t)$ is constant and $r^*(Y,t)$ is a linear function of the state Y. In particular, this implies that the equilibrium rate r^* would also follow a mean-reverting process proportional to the process

Y of random technology change. Thus the operator R_0 given by (22) takes on the form

$$\begin{aligned} R_0(Y,t) &= \frac{e^{-\lambda t}-e^{-\lambda T}}{\lambda}[-a^{*\prime}GG'a^* + r^*(Y,t) + a^{*\prime}(\alpha - r\mathbf{1}) - c^*] + U(c^*) \\ &= r_1(t)Y + r_2(t) \end{aligned}$$

and the Equation (21) becomes

$$\partial_t g + \frac{\sigma^2}{2} Y g_{YY} + \mu(\bar{Y} - Y)g_Y + r_1(t)Y + r_2(t) = 0.$$

Straightforward verification shows that the solution to this equation can be expressed as

$$g(Y,t) = m_1(t)Y + m_2(t)$$

with $m_1(t)$ and $m_2(t)$ uniquely determined by solving a pair of ordinary differential equations

$$\begin{aligned} m_1'(t) - \mu m_1(t) + r_1(t) = 0, \\ m_2'(t) + \mu\bar{Y} m_1(t) + r_2(t) = 0, \end{aligned}$$

with the terminal conditions $m_1(T) = 0 = m_2(T)$. Therefore, the value function can be expressed in a closed form

$$J(W,Y,t) = \frac{e^{-\lambda t}-e^{-\lambda T}}{\lambda} \log W + m_1(t)Y + m_2(t).$$

This results in a well-defined problem having an optimal solution with many useful properties.

4 General Verification Procedure

In this section, we provide a rigorous proof for the indirect utility function to satisfy the Bellman equation in the absence of regularity properties and polynomial growth conditions.
We start with formulation of the lemma which estimates the moments of the wealth process

$$dW_s = W_s[r + a'(\alpha - r\mathbf{1}) + b'(\beta - \mathbf{1})]ds - C_s ds + W_s(a'G + b'H)dB_s.$$

Lemma: For any admissible policy $(a, b, C) \in \mathcal{AC}$, we have

$$0 \leq E(W_s^{m\gamma}) \leq W^{m\gamma} \exp\{K(T-t)\}, \qquad \text{for } t \leq s \leq \theta,$$

where $\theta = \inf\{s \geq t : W_s = 0\}$ is the bankruptcy time, $0 \neq \gamma < 1$ is the risk-aversion constant (either positive or negative), $m > 0$ is an integer and

$K > 0$ is a constant depending only on the known parameters. The proof of this lemma will be presented in Appendix.

Theorem: Suppose that the state process (1) satisfies the ellipticity condition (8), and the utility function is either a power function or a logarithmic function in C. Then the value function must take on a separable form

$$J(W,Y,t) = f(Y,t)U(W,t) + g(Y,t)$$

with $f(Y,t)$ and $g(Y,t)$ satisfying a pair of (quasi) linear 2nd-order parabolic differential equations.

Below we outline the proof of this theorem in three different cases.

4.1 Case I: $U(C,t) = e^{-\lambda t}C^{\gamma}/\gamma$, for $0 < \gamma < 1$

In this case,

$$g(Y,t) \equiv 0$$

and $f(Y,t) \in C^{2,1}$ satisfies the differential equation

$$\partial_t f + \frac{1}{2}S(Y,t)S'(Y,t)f_{YY} + \mu(Y,t)f_Y - \lambda f + \gamma R_\gamma(f, f_Y) = 0, \tag{27}$$

with terminal condition $f(Y,T) = 0$. Therefore, we can define the verification function

$$V(W,Y,t) = f(Y,t)U(W,t)$$

and it is easy to show that this function satisfy the Bellman inequality

$$\partial_t V(W,Y,t) + \mathcal{L}^{a,b,C}V(W,Y,t) + U(C,t)\} \le 0, \tag{28}$$

for any (a,b,C).

Next, we recall that $C = cW$ and from the admissibility condition and the Lemma on the moment estimates

$$E_t \int_t^T |U(C_s,s)|ds < \infty \tag{29}$$

for any admissible consumption process C_s.

For a given $(a,b,C) \in \mathcal{AC}$ and a fixed $\tau \in (t,T)$, we introduce

$$M_{\theta\wedge\tau} = \int_t^{\theta\wedge\tau} U(C_s,s)ds + V(W_{\theta\wedge\tau}, Y_{\theta\wedge\tau}, \theta\wedge\tau).$$

where $W_{\theta\wedge\tau}$ and $Y_{\theta\wedge\tau}$ are stopped processes of W_s and Y_s, respectively. From Ito's lemma and inequality (28) follows

$$\begin{aligned} & M_{\theta\wedge\tau} - M_t \\ = & \int_t^{\theta\wedge\tau} U(C_s,s)ds + \int_t^{\theta\wedge\tau} [\partial_t V + \mathcal{L}^{a,b,c}V](W_s,Y_s,s)ds \\ + & \int_t^{\theta\wedge\tau} [W_s V_W(a'G + b'H) + V_Y^T S]dB_s \end{aligned}$$

$$M_{\theta\wedge\tau} - M_t \leq \int_t^{\theta\wedge\tau} [W_s V_W(a'G + b'H) + V_Y^T S] dB_s \tag{30}$$

where $M_t = V(W, Y, t)$.
From the definition

$$\begin{aligned} V_W(W,Y,t) &= f(Y,t)U_W(W,t), \\ V_Y(W,Y,t) &= f_Y(Y,t)U(W,t). \end{aligned}$$

In order to apply the martingale argument, we define a sequence of stopping times

$$\theta_n = \inf\{s \leq \theta\wedge\tau : \int_t^s |W_s V_W(a'G + b'H) + V_Y^T S|^2 ds = n\}$$

for $n = 1, 2, \ldots$. From the growth condition on $f(Y, t)$ and the moment estimates on W_s, we can verify that

$$\begin{aligned} & E\int_t^{\theta\wedge\tau} |W_s V_W(a'G + b'H)|^2 ds \\ = \;& E\int_t^{\theta\wedge\tau} e^{-2\lambda s} W_s^{2\gamma} f^2(Y_s, s)|a'G + b'H|^2 ds < \infty \end{aligned}$$

and

$$\begin{aligned} & E\int_t^{\theta\wedge\tau} |V_Y^T S|^2 ds \\ = \;& E\int_t^{\theta\wedge\tau} e^{-2\lambda s} W_s^{2\gamma} |f_Y^T(Y_s, s)\cdot S|^2 ds < \infty \end{aligned}$$

for any admissible policy (a, b). Hence $\theta_n \nearrow \theta\wedge\tau$, $\mathcal{P}$-a.s. (for instance, see Chapter 4 in Liptser&Shiryayev [12]). Moreover, by virtue of the generalized Ito's formula, the process

$$\int_t^s [W_s V_W(a'G + b'H) + V_Y S] dB_s$$

is a martingale on $[t, \theta_n)$.
Replacing $\theta\wedge\tau$ by θ_n in (30) and taking expectation of both sides, yields

$$E[\int_t^{\theta_n} U(C_s, s)ds + V(W_{\theta_n}, Y_{\theta_n}, \theta_n)] \leq M_t = V(W, Y, t). \tag{31}$$

By condition (29) and due to $\theta_n \nearrow \theta\wedge\tau$, we can apply the Lebesgue convergence theorem and pass to a limit as $n \to \infty$ in the integral below,

$$E_t[\int_t^{\theta_n} U(C_s, s)ds] \to E_t[\int_t^{\theta\wedge\tau} U(C_s, s)ds]. \tag{32}$$

Also, it follows that

$$E_t[V(W_{\theta_n}, Y_{\theta_n}, \theta_n)] \to E_t[V(W_{\theta\wedge\tau}, Y_{\theta\wedge\tau}, \theta\wedge\tau)] \tag{33}$$

thanks to the growth condition on $f(Y,t)$ and the moment estimates for W_s.
From (31), (32) and (33) follows

$$E[\int_t^{\theta\wedge\tau} U(C_s, s)ds + V(W_{\theta\wedge\tau}, Y_{\theta\wedge\tau}, \theta\wedge\tau)] \le V(W,Y,t).$$

In view of the terminal condition on $V(W,Y,t)$ at T and the boundary condition at $W=0$, we can let $\tau \nearrow T$ and take maximum over $(a,b,C) \in \mathcal{AC}$, to obtain

$$E\int_t^T U(C_s, s)ds \le V(W,Y,t).$$

4.2 Case II: $U(C,t) = -e^{-\lambda t}C^{\gamma}/\gamma$, for $\gamma > 0$

In this case, the power utility has a negative exponent and the value function takes on the following form

$$J(W,Y,t) = -f(Y,t)\frac{e^{-\lambda t}}{\gamma W^{\gamma}}. \tag{34}$$

The function $f(Y,t)$ satisfies the same equation (27) as the function f of the previous case (i.e., **Case I**) but with γ replaced by $-\gamma$.
The difficulty here arises from the singularity at $W=0$ where in this case the value function $J(W,Y,t)$ would no longer have the property of polynomial growth in W as required for the application of the martingale method in the standard proof of the verification theorem.
In order to get around this difficulty, we introduce a new variable of reciprocal wealth

$$Z = \frac{1}{W}.$$

Using Ito's formula for the process Z_s, we can see that it satisfies the following stochastic differential equation

$$\begin{aligned} dZ_s &= -Z_s[r + a'(\alpha - r\mathbf{1}) + b'(\beta - r\mathbf{1}) - c - \frac{1}{2}(a'G + b'H)(a'G + b'H)']ds \\ &\quad - Z_s(a'G + b'H)dB_s \end{aligned}$$

Note the similarity of the above equation and that of (4). We rewrite the value function as

$$\ddot{J}(Z,Y,t) = -f(Y,t)e^{-\lambda t}\frac{Z^{\gamma}}{\gamma}$$

and the corresponding Bellman equation becomes

$$\partial_t \hat{J}(Z,Y,t) + \max_{a,b,c}\{\mathcal{D}^{a,b,c}\hat{J}(Z,Y,t) + U(C,t)\} = 0,$$

on $Q_T = R^+ \times R^k \times [0,T)$ with terminal/boundary condition

$$\hat{J}(W,Y,T) = 0 \qquad \text{and} \qquad \hat{J}(0,Y,t) = 0.$$

The operator $\mathcal{D}^{a,b,c}$ is given by

$$\begin{aligned}
\mathcal{D}^{a,b,c} &= \frac{1}{2}Z^2(a'G+b'H)(a'G+b'H)'\frac{\partial^2}{\partial Z^2} \\
&- Z(a'G+b'H)S(Y,t)\frac{\partial^2}{\partial Z\partial Y} + \frac{1}{2}S(Y,t)S'(Y,t)\frac{\partial^2}{\partial Y^2} \\
&- Z[r + a'(\alpha - r\mathbf{1}) + b'(\beta - r\mathbf{1}) \\
&- c - \frac{1}{2}(a'G+b'H)(a'G+b'H)']\frac{\partial}{\partial Z} + \mu(Y,t)\frac{\partial}{\partial Y}.
\end{aligned}$$

Therefore, the problem at hand is reduced to the Case I with the same zero terminal/boundary conditions as before.

4.3 Case III: $U(C,t) = e^{-\lambda t}\log C$

From the results of Section 2, we know that

$$f(Y,t) = \frac{1}{\lambda}(1 - e^{-\lambda(T-t)})$$

and the function $g(Y,t) \in C^{2,1}$ satisfies the following differential equation

$$\partial_t g + \frac{1}{2}S(Y,t)S'(Y,t)g_{YY} + \mu(Y,t)g_Y + R(Y,t) = 0$$

with terminal condition $g(Y,T) = 0$.
The difficulty here is again the singularity of $\log W$ at $W = 0$. To get around this difficulty, we again introduce a new variable $X = \log W$. From Ito's formula,

$$\begin{aligned}
dX_s &= [r + a'(\alpha - r\mathbf{1}) + b'(\beta - r\mathbf{1}) - c - \frac{1}{2}(a'G+b'H)(a'G+b'H)']ds \\
&\quad +(a'G+b'H)dB_s.
\end{aligned}$$

Thus, the value function can be rewritten as

$$V(X,Y,t) = \frac{e^{-\lambda t} - e^{-\lambda T}}{\lambda}X + g(Y,t)$$

and the corresponding Bellman equation becomes

$$\partial_t \hat{J}(X,Y,t) + \max_{a,b,c}\{\mathcal{D}^{a,b,c}\hat{J}(W,Y,t) + U(C,t)\} = 0,$$

on $Q_T = R \times R^k \times [0,T)$ with the only terminal condition $\hat{J}(X,Y,T) = 0$. The operator $\mathcal{D}^{a,b,c}$ is now generated by the logarithmic process X and the state process Y

$$\begin{aligned}
\mathcal{D}^{a,b,c} &= \frac{1}{2}(a'G+b'H)(a'G+b'H)'\frac{\partial^2}{\partial X^2} \\
&+ (a'G+b'H)S(Y,t)\frac{\partial^2}{\partial X \partial Y} + \frac{1}{2}S(Y,t)S'(Y,t)\frac{\partial^2}{\partial Y^2} \\
&+ [r + a'(\alpha - r\mathbf{1}) + b'(\beta - r\mathbf{1}) \\
&- c - \frac{1}{2}(a'G+b'H)(a'G+b'H)']\frac{\partial}{\partial X} + \mu(Y,t)\frac{\partial}{\partial Y}.
\end{aligned}$$

Therefore, in this case the verification theorem is also being reduced to the standard argument which employs the same polynomial growth condition as that of in the CIR's original paper [3] and [4].

5 Viscosity Solution

In the previous sections, we assumed that the state of technology is a non-degenerate stochastic process, i.e. (8) holds. In this section, we relax this restriction using a perturbation analysis. More importantly, we show that the value maximization mechanism is preserved under a random perturbation of the economy with uncertainty specified by an exogenous state of the technology process.

Suppose that (8) does not hold. We introduce a small perturbation into the technology processes

$$dY_s^\epsilon = \mu(Y_s^\epsilon, s)ds + S_\epsilon(Y_s^\epsilon, s)dB_s \tag{35}$$

where $S_\epsilon S_\epsilon^T = SS' + \epsilon^2 I_k$ and I_k is an $k * k$-identity matrix. The conditions (9–12) guarantee existence of a solution Y_s^ϵ to this system. Intuitively, such a perturbation uncovers 'hidden' sources of uncertainty which are not captured by original system due to random degeneracy.

The following standard estimate of perturbation can be found in Krylov [13].

Lemma: For any $T > 0$, there exists a constant $K_T > 0$ such that

$$E|Y_s^\epsilon - Y_s|^2 \leq K_T\sqrt{\epsilon}, \qquad \forall t \leq s \leq T \tag{36}$$

That is, the solution Y^ϵ converges to the solution Y uniformly on any finite time interval.

We can define the maximization problem for the perturbed system

$$J^\epsilon(W,Y,t) = \max_{(a,b,c)\in\mathcal{AC}} E\int_t^T U(C_s,s)ds,$$

with initial state $W_t = W$ and $Y_t^\epsilon = Y$. The Bellman equation for this perturbed problem is

$$\partial_t J^\epsilon(W,Y,t) + \max_{a,b,C}\{\mathcal{L}_\epsilon^{a,b,C} J^\epsilon(W,Y,t) + U(C,t)\} = 0, \qquad (37)$$

on $Q_T = R^+ \times R^k \times [0,T)$ with the same boundary condition at $W = 0$ as for the original problem. However, the operator $\mathcal{L}_\epsilon^{a,b,C} = \epsilon^2/2 + \mathcal{L}^{a,b,C}$ is uniformly elliptic as a result of the perturbation. From the results of the previous section follows

$$J^\epsilon(W,Y,t) = f^\epsilon(Y,t)U(W,t) + g^\epsilon(Y,t)$$

with the functions $f^\epsilon(Y,t)$ and $g^\epsilon(Y,t)$ satisfying a pair of parabolic differential equations (17)-(18) and (20)-(21), in the Case I of the power utility and in the Case II of the log utility respectively.
In order to analyze the Bellman equation for the perturbed system we use the theory of viscosity (weak) solutions to 2nd-order nonlinear differential equations. The notion of a weak solution provides a convenient framework within which one can characterize solutions to the Bellman equations, and can prove continuous dependence of these solutions on the problem data. Following Ishii-Lions [9], we define $x = (W,Y)$ and consider the equation

$$\partial_t u + H(u_{xx}, u_x, t) = 0, \qquad \text{on } Q_T \qquad (38)$$

where $H(A,p,x,t) = \max\{\mathcal{L}^{a,b,C}u + U(C)\}$ in the case of our Belllman equation (5).
Definition: Let $u(Y,t) : \bar{R}_T \to R$ be a locally bounded, continuous function.

- The function u is a (viscosity) subsolution of (38) if for any $\phi \in C^{2,1}(Q_T)$ and any local maximum $(x_0,t_0) \in Q_T$ of $u - \phi$,

$$\partial_t \phi(x_0,t_0) + H(u_{xx}(x_0,t_0), u_x(x_0,t_0), x_0, t_0) \le 0, \qquad (39)$$

- The function u is a (viscosity) supersolution of (38) if for any $\phi \in C^{2,1}(Q_T)$ and any local minimum $(x_0,t_0) \in Q_T$ of $u - \phi$,

$$\partial_t \phi(Y_0,t_0) + H(u_{xx}(x_0,t_0), u_x(x_0,t_0), x_0, t_0) \ge 0. \qquad (40)$$

The function u is a viscosity (or 'weak') solution of (38) if it is both a subsolution and a supersolution of (38).
Finally, we present the verification theorem in the general case and outline briefly the proof for the value function (i.e., the solution to the Bellman equation (5)) to be a'weak' limit of the classical solutions to the perturbed Bellman equations (37).
Theorem: The indirect utility function $J(W,Y,t)$ in the CIR model is a viscosity solution to the Bellman equation (5).

The proof of this theorem is based on the following proposition, often referred as the maximum principle (see Fleming/Soner [6]) and the 'weak limit' argument in the proof of convergence of the viscosity solutions. (see P.L. Lions [11]).

Proposition: Let J_1 and J_2 be respectively the subsolution and supersolution of

$$\partial_t J^\epsilon(W,Y,t) + \max_{a,b,C}\{\mathcal{L}_\epsilon^{a,b,C} J^\epsilon(W,Y,t) + U(C,t)\} = 0,$$

on $Q_T = R^+ \times R^k \times [0,T)$. If $J_1(W,Y,T) \le J_2(W,Y,T)$, then $J_1 \le J_2$ on Q_T.

For a proof, see P.L. Lions [11].

Using the 'weak limit' argument, we define the functions

$$\bar{J}(W,Y,t) = \overline{\lim}_{\epsilon\to 0} J^\epsilon(W',Y',t') \tag{41}$$

$$\hat{J}(W,Y,t) = \underline{\lim}_{\epsilon\to 0} J^\epsilon(W',Y',t) \tag{42}$$

and one can show that $\bar{J}$ is a subsolution while $\hat{J}$ is a supersolution. Therefore, $\bar{J} = \hat{J}$.

6 Appendix: Proof of Lemma

Let $\theta_s = a_s' G + b_s' H$. The admissibility of (a,b) implies that the process θ_s is $\mathcal{L}^2$-integrable. Therefore we can define an equivalent martingale measure

$$\frac{dQ}{dP} = \exp\{m\gamma \int_t^s \theta_r dB_r - \frac{m\gamma}{2}\int_t^s |\theta_r|^2 dr\}. \tag{43}$$

where $m > 0$ is an integer, and $\gamma < 1$ is the risk-aversion parameter. Applying Ito's formula to $\log(W_s^{m\gamma})$, we obtain

$$\begin{aligned} &\log(W_s^{m\gamma}) - \log(W_s^{m\gamma}) \\ = \; & m\gamma \int_t^s [r + a_v'(\alpha - r\mathbf{1}) + b_v'(\beta - r\mathbf{1})] - c_v dv \\ &+ m\gamma \int_t^s \theta_v dB_v - \frac{m\gamma}{2}\int_t^s |\theta_v|^2 dv. \end{aligned} \tag{44}$$

Taking expectation of both sides of (44), we get

$$EW_s^{m\gamma} = W^{m\gamma} E_Q \exp\{m\gamma \int_t^s [r + a_v'(\alpha - r\mathbf{1}) + b_v'(\beta - r\mathbf{1})] - c_v dv\} \tag{45}$$

where $E = E_P$ and E_Q are the expectation under the original measure P and the equivalent martingale measure Q respectively.

Let

$$Z_s = \int_t^s [r + a_v'(\alpha - r\mathbf{1}) + b_v'(\beta - r\mathbf{1})]dv - c_v dv.$$

The integral rule of calculus implies

$$\begin{aligned} & \exp(\frac{1}{2}m\gamma Z_s) \\ = \; & 1 + \frac{1}{2}m\gamma \int_t^s [r + a_v'(\alpha - r\mathbf{1}) + b_v'(\beta - r\mathbf{1})] \exp(\frac{1}{2}m\gamma Z_v)dv \\ \leq \; & 2 + \frac{K}{2}m^2\gamma^2 \int_t^s \exp(m\gamma Z_v)dv \end{aligned}$$

for some constant $K > 0$. The last inequality in the above expression follows from Holder's inequality, nonegativity of c_t and the fact that (a, b) is square-integrable. Finally, we take expectations of both sides and apply Gronwall's inequality to obtain

$$E_Q \exp(m\gamma Z_s) \leq 2\exp\{\frac{K}{2}m^2\gamma^2(T - t)\}. \tag{46}$$

Substitution of (46) into (45) yields (36).

Acknowledgments: This research is supported by the National Science Foundation Grant DMS-9705017.

Bibliography

[1] J. Cox and C.F. Huang (1989). Optimal consumption and portfolio policies when asset prices follow a diffusion process. J. Econ. Theory 49, pp33-83.

[2] J. Cox and C.F. Huang (1991). A variational problem arising from a financial economics model, Mathematical Finance, pp152-203.

[3] J. Cox, J.E. Ingersoll and S.A. Ross (1985). An intertemporal general equilibrium model of asset prices, Econometrica, 53, pp363-384.

[4] J. Cox, J.E. Ingersoll and S.A. Ross (1985). A theory of the term structure of interest rates, Econometrica, 53, pp385-407.

[5] W.H. Fleming and R.W. Rishel (1975). *Deterministic and Stochastic Optimal Control*, Springer-Verlag, New York.

[6] W.H. Fleming and H.M.Soner (1992). *Controlled Markov Processes and Viscosity Solutions*, Springer-Verlag, New York.

[7] A. Friedman (1975). *Stochastic Differential Equations and Applications* Vol. I & II, Academic Press, Cambridge.

[8] N.H. Hakansson (1970). Optimal investment and consumption strategies under risk for a class of utility functions, Econometrica 38, pp587-607.

[9] H.Ishii and P.L. Lions (1992). User's Guide on viscosity solutions of fully nonlinear 2nd-order partial differential equations, Comm. on PDE's, 27, pp. 1-67.

[10] I. Karatzas, J. Lehoczky and S. Shreve (1987). Optimal portfolio and consumption decisions for a 'small investor' on a finite horizon, SIAM J. on Control 25, pp1557-1586.

[11] P.L. Lions (1983). Optimal control of diffusion processes and Hamilton-Jacobi-Bellman equations, Part 1 and Part 2, Comm. on PDE's 8, pp1101-1174 and pp1229-1276.

[12] D. Liptser and N. Shiryayev (1977). *Statistics of Random Processes*, Springer-Verlag, New York.

[13] N. Krylov (1980). *Controlled Diffusion Processes*, Springer-Verlag, New York.

[14] R. Merton (1971). Optimal consumption and portfolio rules in a continuous time model, J. Econ. Theory 3, pp373-413.

[15] P.A. Samuelson (1969). Lifetime portfolio selection by dynamic stochastic programming, Rev. of Economics and Statistics 51, pp239-246.

[16] D. Duffie, W. Fleming, H. M. Soner and T. Zariphopoulou (1996) Hedging in an incomplete market with HARA utility, J. on Economic Dynamics and Control, 21, pp. 753-782.

36

Optimal Portfolio Management with Partial Observations and Power Utility Function

Raymond Rishel

1 Introduction

A common model for a stock price is that it is a "geometric Brownian motion". That is the price $s(t)$ of a stock satisfies a stochastic differential equation of the type

$$ds(t) = s(t)(\bar{r}\, dt + \sigma d\beta(t)) \tag{1}$$

in which $\beta(t)$ is a Brownian motion process and $\bar{r}$ and σ are constants. For this model the growth rate of the stock has stochastic differential

$$\bar{r}\, dt + \sigma d\, \beta(t) \tag{2}$$

with $\bar{r}\, dt$ being the infinitesmal mean growth rate and $\sigma d\, \beta(t)$ the infinitesmal random fluctuations about this mean growth rate. Since the process

$$\bar{r}t + \sigma\beta(t) \tag{3}$$

has stationary increments, the growth rate of the stock over any two periods of equal length has the same statistical properties. This appears to rule out bull and bear markets.

A model for a stock price which can have features of bull and bear markets is a doubly stochastic model of the form

$$d\, s(t) = s(t)(r(t)\, dt + \sigma d\, \beta(t)) \tag{4}$$

in which $\beta(t)$ is a Brownian motion and $r(t)$ is also a random process. For this model the infinitesmal mean growth rate $r(t)dt$ is random as well as the random fluctuations $\sigma d\, \beta(t)$ about this mean. If $\bar{r}$ is a target mean growth rate for the stock and there is a period for which $r(t)$ is substantially larger than $\bar{r}$ this could be considered as a bull market. Conversely when $r(t)$ is substantially less than $\bar{r}$ this could be considered as a bear market.

Optimal portfolio management problems with this type of doubly stochastic stock price model have been considered by a number of authors [2] [4] [5] [6]. Explicit optimal policies for these models have been obtained only for the case of logarithmic utility functions. For logarithmic utility functions a separation principle holds. That is the optimal policy for $r(t)$ random is obtained from the optimal policy when $r(t)$ is a constant $\bar{r}$ by substituting the best estimate of $r(t)$, that is the conditional mean of $r(t)$ given past observations of the prices, for the constant $\bar{r}$ in the optimal policy for constant mean growth rate.
This paper investigates the problem of investment for long term growth with this type of model with utility of wealth given by

$$U(w) = w^{\gamma} \quad \text{where } 0 < \gamma < 1, \tag{5}$$

and the random mean growth rate given by

$$r(t) = \bar{r} + m(t) \tag{6}$$

where $\bar{r}$ is a known constant and $m(t)$ is the solution of a linear Gaussian system.
This problem has many interesting features. Surprisingly for some situations it has no finite solution. When it does have a solution it does not satisfy the separation principle. It's solution is intimately related to an auxiliary linear quadratic control problem.
To give a simple exposition of the results only a market of a bond and a single stock and a scalar linear system for the mean growth rate is considered. Most of the results obtained can be extended to markets with finitely many stocks with mean growth rates given by vector linear systems.

2 The Portfolio Problem

Consider a market formed from a bond whose price evolves as

$$d\,b(t) = r\,b(t)\,dt; \quad b(0) = b_0 \tag{7}$$

and a single stock whose price evolves as

$$d\,s(t) = s(t)(r(t)\,dt + \sigma\,d\,\beta(t)); \quad s(0) = s_0 \tag{8}$$

where

$$r(t) = \bar{r} + m(t) \tag{9}$$

and $m(t)$ is a solution of

$$d\,m(t) = a\,m(t)\,dt + b\,d\alpha(t); \quad m(0) = m_0 \tag{10}$$

In the above $\alpha(t)$ and $\beta(t)$ are independent Brownian motions, m_0 is a Gaussian random variable independent of $\alpha(t)$ and $\beta(t)$ with mean 0 and

variance Γ. The quantities $r, b_0, \sigma, s_0, \bar{r}, a, b$ are known constants, all positive except perhaps a and b.
An investor invests a "fraction" $u(t)$ of his wealth in the stock and $1-u(t)$ in the bond. His wealth evolves as

$$d\,w(t) = (1-u(t))w(t)r\,dt + u(t)w(t)((\bar{r}+m(t))\,dt + \sigma d\,\beta(t));\;\; w(0) = w_0 \tag{11}$$

where w_0 is a positive constant. He observes the bond price $b(t)$ and stock price $s(t)$ but cannot observe $m(t)$ or the Brownian motions $\alpha(t)$ and $\beta(t)$. He invests for the period $0 \le t \le T$. He chooses the portfolio policy $u(t)$ based on his observation of prices up to the current time. His objective is to choose the policy to maximize his utility of terminal wealth

$$E[U(w(T))] = E[w(T)^\gamma] \tag{12}$$

Short selling and borrowing are allowed so the "fraction" $u(t)$ is allowed to take on arbitrarily real values. Let

$$\mathcal{S}_t = \sigma\{b(s), s(s);\;\; 0 \le s \le t\} \tag{13}$$

denote the sigmafield of stock and bond prices up to time t. By choosing $u(t)$ based on prices up to time t we shall mean that each policy $u(t)$ is an $\mathcal{S}_t$ adapted stochastic process. That is $u(t)$ is a measurable stochastic process such that for each t, $u(t)$ is $\mathcal{S}_t$ measurable. We shall impose the further technical condition

$$E\left[\int_0^T [u(t)w(t)]^2\,dt\right] < \infty \tag{14}$$

on the policies.

3 Reduction to a Completely Observed Problem

It may be verified using Ito's stochastic differentiation rule that solutions $b(t)$ and $s(t)$ of equations (7) and (8) together with (9) are given by

$$b(t) = b_0\,e^{rt} \tag{15}$$

and

$$\begin{aligned} s(t) &= s_0\,e^{-\frac{\sigma^2 t}{2}}\,e^{\int_0^t r(s)\,ds + \int_0^t \sigma\,d\,\beta(s)} \\ &= s_0\,e^{-(\frac{\sigma^2}{2}-\bar{r})t}\,e^{\int_0^t m(s)\,ds + \int_0^t d\,\beta(s)} \end{aligned} \tag{16}$$

Since $b(t)$ and $s(t)$ have this form and b_0, s_0, r, and $\bar{r}$ are known constants observing

$$(b(s), s(s);\;\; 0 \le s \le t) \tag{17}$$

is equivalent to observing

$$(y(s);\quad 0 \le s \le t) \tag{18}$$

where

$$y(t) = \int_0^t m(s)\, ds + \int_0^t \sigma\, d\beta(s). \tag{19}$$

In fact it follows from (15) (16) and (19) that if we define the sigmafields $\mathcal{Y}_t$ by

$$\mathcal{Y}_t = \sigma\{y(s);\quad 0 \le s \le t\} \tag{20}$$

that the sigmafields $\mathcal{S}_t$ and $\mathcal{Y}_t$ are the same. Thus requiring $u(t)$ to be $\mathcal{S}_t$ adapted is equivalent to requiring $u(t)$ to be $\mathcal{Y}_t$ adapted.
Since the sigmafields $\mathcal{S}_t$ and $\mathcal{Y}_t$ are the same, when taking conditional expectations with respect to the past measurements, we may take conditional expectations with respect to the sigmafields $\mathcal{Y}_t$. The theory of Kalman filtering is concerned with taking conditional expectations of a process of the form (10) with respect to measurements of the type (19). If we define $\hat{m}(t)$ by

$$\hat{m}(t) = E[m(t)|\mathcal{Y}_t] \tag{21}$$

the theory of Kalman filtering, [3] p. 135, implies that $\hat{m}(t)$ is a solution of the equations

$$d\,\hat{m}(t) = a\hat{m}(t)\, dt + \frac{R(t)}{\sigma}\, d\nu(t);\quad \hat{m}(0) = 0 \tag{22}$$

where $R(t)$ is the solution of the Riccati equation

$$\frac{dR}{dt} = 2aR - \frac{R^2}{\sigma^2} + b^2;\quad R(0) = \Gamma \tag{23}$$

and $\nu(t)$ satisfies

$$d\nu(t) = \sigma^{-1}[dy_t - \hat{m}(t)dt] = \sigma^{-1}[(m(t) - \hat{m}(t))dt + \sigma d\beta(t)];\quad \nu(0) = 0 \tag{24}$$

The process $\nu(t)$ is called the innovations process and is a $\mathcal{Y}_t$ adapted Brownian motion process. Since from (24)

$$m(t)\, dt + \sigma\, d\,\beta(t) = \hat{m}(t)\, dt + \sigma\, d\,\nu(t) \tag{25}$$

the wealth equation (11) can be rewritten in terms of the innovations Brownian motion as

$$d\,w(t) = (1 - u(t))w(t)r\, dt + u(t)w(t)((\bar{r} + \hat{m}(t))\, dt + \sigma\, d\,\nu(t));\quad w(0) = w_0 \tag{26}$$

Thus our portfolio management problem can be expressed in terms of $\mathcal{Y}_t$ adapted quantities as:

Maximize

$$E[w(T)^\gamma] \tag{27}$$

subject to

$$d\,w(t) = (1-u(t))w(t)r\,dt + u(t)w(t)((\bar{r}+\hat{m}(t))\,dt + \sigma\,d\,\nu(t)); \quad w(0) = w_0 \tag{28}$$

$$d\,\hat{m}(t) = a\,\hat{m}(t)\,dt + R(t)\sigma^{-1}\,d\,\nu(t); \quad \hat{m}(0) = 0 \tag{29}$$

over all $\mathcal{Y}_t$ adapted policies $u(t)$ for which

$$E\int_0^T [u(t)w(t)]^2\,dt < \infty. \tag{30}$$

In the terminology of, [3] p. 162, this is a nonanticipative control problem. Theorem 4.1 and Corollary 4.2 of [3] imply that the following dynamic programming theorem gives a sufficient condition for a policy to be optimal.[1]

Theorem 1: *A sufficient condition that there be an optimal policy of the form $u(t) = u^*(t, w(t), \hat{m}(t))$ is that there is a twice continuously differentiable solution of the dynamic programming partial differential equation*

$$\operatorname*{Max}_u \; [V_t(t,w,\hat{m}) + V_w(t,w,\hat{m})w[r + u(\bar{r}+\hat{m}-r)] + V_{\hat{m}}(t,w,\hat{m})a\hat{m}$$
$$+\tfrac{1}{2}u^2w^2\sigma^2\,V_{ww}(t,w,\hat{m}) + u\,wR(t)V_{\hat{m}w} + \tfrac{1}{2}\frac{R(t)^2}{\sigma^2}\,V_{\hat{m}\hat{m}}] = 0 \tag{31}$$

with boundary condition

$$V(T,w,\hat{m}) = w^\gamma \tag{32}$$

and that the function $u^*(t,w,\hat{m})$ attains the maximum over u in (31).

4 Reduction of the Solution of the Dynamic Programming PDE to the Solution of a Related PDE

Let's look for a solution of (31) (32) of the form

$$V(t,w,\hat{m}) = e^{g(t,\hat{m})}w^\gamma \tag{33}$$

[1] Condition (2) p. 162 of [3] for an admissible nonanticipative control is not satisfied, but Corollary 4.2 follows in the current case using the condition (30) instead of condition (2) of [3].

substituting (33) into the dynamic programming PDE (31) gives (using an abbreviated notation)

$$\begin{aligned}\operatorname*{Max}_{u}\ & [e^g g_t w^\gamma + \gamma e^g w^\gamma (r + u(\bar{r} + \hat{m} - r)) + e^g g_{\hat{m}} w^\gamma a\hat{m} \\ & + \tfrac{1}{2} u^2 \sigma^2 \gamma(\gamma - 1) e^g w^\gamma + uR(t) e^g g_{\hat{m}} \gamma w^\gamma \\ & + \tfrac{1}{2} \frac{R(t)^2}{\sigma^2} e^g g_{\hat{m}}^2 w^\gamma + \tfrac{1}{2} \frac{R(t)^2}{\sigma^2} e^g g_{\hat{m}\hat{m}} w^\gamma \Big] = 0.\end{aligned} \tag{34}$$

Dividing out $e^g w^\gamma$ in (34) we see that (33) will be a solution of (31) if $g(t, \hat{m})$ is a solution of

$$\begin{aligned}\operatorname*{Max}_{u}\ & [g_t + \gamma r + \gamma u(\bar{r} + \hat{m} - r) + g_{\hat{m}} a\hat{m} \\ & + \tfrac{1}{2} u^2 \sigma^2 \gamma(\gamma - 1) + uR(t)\gamma g_{\hat{m}} \\ & + \tfrac{1}{2} \frac{R(t)^2}{\sigma^2} g_{\hat{m}}^2 + \tfrac{1}{2} \frac{R(t)^2}{\sigma^2} g_{\hat{m}\hat{m}} \Big] = 0.\end{aligned} \tag{35}$$

The terms involving u in (35) are

$$\gamma u(\bar{r} + \hat{m} - r) + \tfrac{1}{2} u^2 \sigma^2 \gamma(\gamma - 1) + uR(t)\gamma g_{\hat{m}} \tag{36}$$

The maximum of (36) over u is attained at

$$u = \frac{1}{(1-\gamma)\sigma^2} [\bar{r} + \hat{m} - r + R(t) g_{\hat{m}}] \tag{37}$$

the value of (36) at this maximum is

$$\tfrac{1}{2} \frac{\gamma}{(1-\gamma)\sigma^2} [\bar{r} + \hat{m} - r + R(t) g_{\hat{m}}]^2 \tag{38}$$

substituting (38) for the terms (36) in (35) we see, after a little simplification that (33) will be a solution of (31) if $g(t, \hat{m})$ is a solution of

$$\begin{aligned}& g_t + \gamma r + g_{\hat{m}} a\hat{m} + \tfrac{1}{2} \frac{\gamma}{(1-\gamma)\sigma^2} (\bar{r} + \hat{m} - r)^2 \\ & + \frac{\gamma R(t)}{(1-\gamma)\sigma^2} (\bar{r} + \hat{m} - r) g_{\hat{m}} + \tfrac{1}{2} \frac{R(t)^2}{(1-\gamma)\sigma^2} g_{\hat{m}}^2 + \tfrac{1}{2} \frac{R(t)^2}{\sigma^2} g_{\hat{m}\hat{m}} = 0\end{aligned} \tag{39}$$

The boundary condition (32) will be satisfied by (33) if $g(t, \hat{m})$ satisfies the boundary condition

$$g(T, \hat{m}) = 0. \tag{40}$$

5 The Auxiliary Control Problem

Notice that

$$\operatorname*{Max}_{v} \left[\frac{R(t)}{(1-\gamma)^{1/2}} g_{\hat{m}} v - \frac{\sigma^2}{2} v^2 \right] = \tfrac{1}{2} \frac{R(t)^2}{(1-\gamma)\sigma^2} g_{\hat{m}}^2. \tag{41}$$

Now (41) implies that (39) is equivalent to the PDE

$$\begin{aligned}\operatorname*{Max}_{v} \Big[& g_t + \gamma r + \tfrac{1}{2}\frac{\gamma}{(1-\gamma)\sigma^2}(\bar{r}+\hat{m}-r)^2 - \frac{\sigma^2}{2}v^2 \\ & + g_{\hat{m}}\left(a\hat{m} + \frac{\gamma R(t)}{(1-\gamma)\sigma^2}(\bar{r}+\hat{m}-r) + \frac{R(t)}{(1-\gamma)^{1/2}}v\right) \\ & + \tfrac{1}{2}\frac{R(t)^2}{\sigma^2}g_{\hat{m}\hat{m}}\Big] = 0\end{aligned} \tag{42}$$

Notice that (42) with boundary condition (40) are the dynamic programming PDE and boundary condition for the stochastic control problem with state $\hat{m}(t)$, control $v(t)$, and Brownian motion $\xi(t)$ given by
Maximize

$$E\int_0^T \left[\gamma r + \tfrac{1}{2}\frac{\gamma}{2(1-\gamma)\sigma^2}(\bar{r}+\hat{m}(t)-r)^2 - \frac{\sigma^2}{2}v(t)^2\right] dt \tag{43}$$

subject to:

$$d\hat{m}(t) = \left[a\hat{m}(t) + \frac{\gamma R(t)}{(1-\gamma)\sigma^2}(\bar{r}+\hat{m}(t)-r) + \frac{R(t)}{(1-\gamma)^{1/2}}v(t)\right]dt + \frac{R(t)}{\sigma}d\xi(t) \tag{44}$$

over admissible controls $v(t)$.
This auxiliary control problem is a linear quadratic control problem whose structure is well known. However since the quadratic forms in the state and the control in the criteria to be optimized have opposite signs it is also well known that this control problem may not have a solution for arbitrarily large time intervals.
The following theorem governs the solution of the Dynamic Programming PDE for this control problem.

Theorem 2: *The PDE (39) and terminal condition (40) will have a solution of the form*

$$g(t,\hat{m}) = K(t)\hat{m}^2 + J(t)\hat{m} + L(t) \tag{45}$$

if there is a solution of

$$\dot{K}(t) + \frac{2R(t)^2}{(1-\gamma)\sigma^2}K(t)^2 + 2\left(a + \frac{\gamma R(t)}{(1-\gamma)\sigma^2}\right)K(t) + \tfrac{1}{2}\frac{\gamma}{(1-\gamma)\sigma^2} = 0 \tag{46}$$

with terminal condition

$$K(T) = 0 \tag{47}$$

defined on all of $[0,T]$, *where* $J(t)$ *and* $L(t)$ *are solutions of*

$$\begin{aligned}&\dot{J}(t) + J(t)\left(a + \frac{1}{(1-\gamma)\sigma^2}\left(\gamma R(t) + 2R(t)^2K(t)\right)\right) \\ &\quad + \frac{\gamma}{(1-\gamma)\sigma^2}(2R(t)K(t)+1)(\bar{r}-r) = 0\end{aligned} \tag{48}$$

with terminal condition

$$J(T) = 0 \tag{49}$$

and

$$\begin{aligned}\dot{L}(t) + \gamma r + \frac{\gamma}{(1-\gamma)\sigma^2} J(t)R(t)(\bar{r} - r) + \tfrac{1}{2}\frac{R(t)^2 J(t)^2}{(1-\gamma)\sigma^2} \\ + \tfrac{1}{2}\frac{\gamma}{(1-\gamma)\sigma^2}(\bar{r} - r)^2 + \frac{R(t)^2}{\sigma^2} K(t) = 0\end{aligned} \tag{50}$$

with terminal condition

$$L(T) = 0. \tag{51}$$

The proof of this theorem is obtained by the standard method of substituting (45) in the left-hand side of (39) and gathering together the coefficients of $\hat{m}^2, \hat{m}$ and the constant terms. These are given respectively by the left hand sides of (46), (48) and (50). When these are zero, (45) will be a solution of (39). The boundary condition (40) and (45) imply (47), (49) and (51).

However this method will give a solution of the PDE (39) with boundary condition (40), only if the ordinary differential equations (46), (48), (50) with respective boundary conditions (47), (49), (51) have solutions on $[0, T]$. Notice that if (46), (47) have a solution for $K(t)$ on $[0, T]$, then (48) (49) for $J(t)$ and (50) (51) for $L(t)$ will have solutions on $[0, T]$, since (48) (49) is a linear ordinary differential equation and (50) (51) is just an integration. Since the explicit form of equations (46) (48) (50) will be important for later work, this standard derivation of Theorem 2 is carried out in the appendix.

6 Solution to the Portfolio Problem

Theorem 2 and (34)-(39) imply the following results for the portfolio problem.

Theorem 3: *If (46) and (47) have a solution $K(t)$ defined on all of $[0, T]$, then the dynamic programming PDE (31) and boundary condition (32) have the solution*

$$V(t, w, \hat{m}) = e^{[K(t)\hat{m}^2 + J(t)\hat{m} + L(t)]} w^{\gamma} \tag{52}$$

where $J(t)$ is the solution of (48) (49) and $L(t)$ the solution of (50) (51). The function

$$u^*(t, w, \hat{m}) = \frac{1}{(1-\gamma)\sigma^2}\left(\bar{r} - r + \hat{m} + R(t)[2K(t)\hat{m} + J(t)]\right) \tag{53}$$

attains the maximum over u in the dynamic programming PDE (31).

Thus the optimal policy for the portfolio management problem is given by

$$u^*(t) = \frac{1}{(1-\gamma)\sigma^2}\left(\bar{r} - r + \hat{m}(t) + R(t)[2K(t)\hat{m}(t) + J(t)]\right) \tag{54}$$

where $\hat{m}(t)$ is the solution of (22) in which $\nu(t)$ is given by (24).

7 Non Separation of Control and Estimation

When (46) and (47) have a solution $K(t)$ defined on all of $[0, T]$, the solution of (48) (49) for $J(t)$ can be expressed by

$$J(t) = I(t)(\bar{r} - r) \tag{55}$$

where

$$I(t) = \int_t^T \frac{\gamma}{(1-\gamma)\sigma^2}(2R(s)K(s) + 1)e^{\int_t^s (a + \frac{1}{(1-\gamma)\sigma^2}(\gamma R(v) + 2R(v)^2 K(v))\, dv}\, ds \tag{56}$$

Thus the optimal policy (53) can be expressed as

$$u^*(t) = \frac{(1 + R(t)I(t))(\bar{r} - r) + (1 + 2R(t)K(t))\hat{m}(t)}{(1-\gamma)\sigma^2} \tag{57}$$

For the portfolio problem in which the infinitesmal mean growth rate $r(t)$ is just a constant $\bar{r}$, it is well known, [1], p. 151, that the optimal policy is given by

$$u(t) = \frac{\bar{r} - r}{(1-\gamma)\sigma^2} \tag{58}$$

The separated policy obtained by substituting the conditional expectation of $r(t)$ given the past measurements for the constant $\bar{r}$ in (58) is

$$u(t) = \frac{\bar{r} + \hat{m}(t) - r}{(1-\gamma)\sigma^2} \tag{59}$$

This differs from the optimal policy (57) which has the additional non-zero terms

$$R(t)I(t)(\bar{r} - r) + 2R(t)K(t)\hat{m}(t) \tag{60}$$

in the numerator. Thus the separation principle does not hold for this problem.

8 Non Existance

What can be said about the portfolio management problem when the solution $K(t)$ of (46) (47) becomes infinite at some $\tau > 0$. Let us show if $\bar{r} > r$

that there are policies which make

$$E[w(T)^\gamma] \tag{61}$$

arbitrarily large in this case. Thus a solution of the problem does not exist in this case.

Consider the policy $u^\epsilon(t)$ defined by

$$u^\epsilon(t) = \begin{cases} 0 \quad \text{if } 0 \le t < \tau + \epsilon \\ \dfrac{1}{(1-\gamma)\sigma^2}\left(\bar{r} + \hat{m}(t) - r + R(t)[2K(t)\hat{m}(t) + J(t)]\right) \\ \quad \text{if } \tau + \epsilon \le t \le T \end{cases} \tag{62}$$

and the wealth process $w^\epsilon(t)$ which is a solution of the wealth equation (11) with policy $u^\epsilon(t)$.

Since $K(t), J(t), L(t)$ satisfying (56)-(51) will be defined on $\tau + \epsilon \le t \le T$ our previous work implies that for $\tau + \epsilon \le t \le T$

$$V(t, w, \hat{m}) = e^{[K(t)\hat{m}^2 + J(t)\hat{m} + L(t)]} w^\gamma \tag{63}$$

will be a solution of the PDE (31) with boundary condition (32) and $u^*(t, w, \hat{m})$ given by (53) will attain the maximum over u in this PDE. Since this is the case differentiating

$$V(t, w^\epsilon(t), \hat{m}(t)) = e^{[K(t)\hat{m}(t)^2 + J(t)\hat{m}(t) + L(t)]}(w^\epsilon(t))^\gamma \tag{64}$$

using Ito's differential rule, it will follow from the PDE (31) that the dt terms in the stochastic differential are zero, and hence on $\tau + \epsilon \le t \le T$ that

$$\begin{aligned} d\,V(t, w^\epsilon(t), \hat{m}(t)) &= V_w(t, w^\epsilon(t), \hat{m}(t)) u^\epsilon(t) w^\epsilon(t) \sigma \, d\nu(t) \\ &\quad + V_{\hat{m}}(t, w^\epsilon(t), \hat{m}(t)) R(t) \sigma^{-1} \, d\nu(t). \end{aligned} \tag{65}$$

This implies for $\tau + \epsilon \le t \le T$ that

$$\begin{aligned} w^\epsilon(T)^\gamma &= V(t, w^\epsilon(t), \hat{m}(s)) \\ &+ \int_t^T [V_w(s, w^\epsilon(s), \hat{m}(s)) u^\epsilon(s) w^\epsilon(s) \sigma + V_{\hat{m}}(s, w^\epsilon(s), \hat{m}(s)) R(s) \sigma^{-1}] \, d\nu(s). \end{aligned} \tag{66}$$

Now since the expected value of the stochastic integral will be zero, we have

$$\begin{aligned} E[w^\epsilon(T)^\gamma] &= E[V(\tau + \epsilon, w^\epsilon(\tau + \epsilon), m(\tau + \epsilon))] \\ &= E[e^{K(\tau+\epsilon)\hat{m}(\tau+\epsilon)^2 + J(\tau+\epsilon)\hat{m}(\tau+\epsilon) + L(\tau+\epsilon)}(w^\epsilon(\tau + \epsilon))^\gamma] \end{aligned} \tag{67}$$

Now the solution $K(t)$ of (46) (47) will be nonnegative and for $\bar{r} \geq r$, the solutions $J(t)$ and $L(t)$ of (48) (49) and (50) (51) will be nonnegative. From the definition (62) of $u^\epsilon(t)$ and (11) it follows that

$$w^\epsilon(\tau+\epsilon) = w_0\, e^{r(\tau+\epsilon)}. \tag{68}$$

Hence from (67) (68) and the nonnegativity of $J(t)$ and $L(t)$ we obtain the estimate

$$E[w^\epsilon(T)^\gamma] \geq e^{K(\tau+\epsilon)}\, pr[\hat{m}(\tau+\epsilon) > 1]\ \ (w_0\, e^{r(\tau+\epsilon)})^\gamma. \tag{69}$$

Since

$$\lim_{\epsilon \downarrow 0}\ K(\tau+\epsilon) = +\infty \tag{70}$$

it follows that

$$\lim_{\epsilon \downarrow 0}\ E[(w^\epsilon(T)^\gamma] = +\infty. \tag{71}$$

Thus the policies $u^\epsilon(t)$ are a sequence of policies for which the expected terminal utility of wealth becomes infinite.

9 Solutions of the Riccati equation for $K(t)$

When will solutions of

$$\dot{K}(t) + \frac{2R(t)^2}{(1-\gamma)\sigma^2}\, K(t)^2 + 2\left(a + \frac{\gamma R(t)}{(1-\gamma)\sigma^2}\right) K(t) + \tfrac{1}{2}\frac{\gamma}{(1-\gamma)\sigma^2} = 0 \tag{72}$$

with boundary condition

$$K(T) = 0 \tag{73}$$

satisfy

$$\lim_{t \downarrow \tau}\ K(t) = +\infty \tag{74}$$

for some τ for which $0 < \tau < T$?

Elementary computations show that solutions $R(t)$, of (23), traverse in a monotone fashion between Γ and the positive root of the right hand side of (23) set equal to zero. Thus $R(t)$ is positive, bounded, and bounded away from zero. That is there are positive numbers $\underline{R}$ and $\bar{R}$ so that

$$\underline{R} \leq R(t) \leq \bar{R} \tag{75}$$

Consider the function $F(t,K)$ given by

$$F(t,K) = \frac{2R(t)^2}{(1-\gamma)\sigma^2}\, K^2 + 2\left(a + \frac{R(t)}{(1-\gamma)\sigma^2}\right) K + \tfrac{1}{2}\frac{\gamma}{(1-\gamma)\sigma^2} \tag{76}$$

and define $\underline{F}(K)$ and $\bar{F}(K)$ respectively by replacing $R(t)$ in (76) by $\underline{R}$ or $\bar{R}$. For positive values of K, (75) implies

$$\underline{F}(K) \leq F(t, K) \leq \bar{F}(K) \tag{77}$$

Thus as long as they all exist, the solution of (46) (47) on $[0, T]$ is bounded above by the solution of

$$\dot{K}(t) + \bar{F}(K(t)) = 0; \quad K(T) = 0 \tag{78}$$

and below by the solution of

$$\dot{K}(t) + \underline{F}(K(t)) = 0; \quad K(T) = 0. \tag{79}$$

If $B^2 - 4AC > 0$, the solution of the differential equation and boundary condition

$$\dot{K}(t) + A\,K(t)^2 + B\,K(t) + C = 0; \quad K(T) = 0 \tag{80}$$

is given by

$$\begin{aligned} K(t) = \frac{\sqrt{B^2 - 4AC}}{2A} \tanh & \left[\tfrac{1}{2}\sqrt{B^2 - 4AC}(t - T) \right. \\ & \left. + \operatorname{arctanh} \left(\frac{B}{\sqrt{B^2 - 4AC}} \right) \right] - \frac{B}{2A} \end{aligned} \tag{81}$$

If $4AC - B^2 > 0$, the solution of (80) is given by

$$\begin{aligned} K(t) = \frac{\sqrt{4AC - B^2}}{2A} \tan & \left[\tfrac{1}{2}\sqrt{4AC - B^2}(T - t) \right. \\ & \left. + \arctan \left(\frac{B}{\sqrt{4AC - B^2}} \right) \right] - \frac{B}{2A}. \end{aligned} \tag{82}$$

Of course (81) is bounded and (82) is not. The solution (82) becomes infinite at a time τ somewhere in the interval

$$\left[T - \frac{2\pi}{\sqrt{4AC - B^2}}, T \right].$$

For the equation (78)

$$\begin{aligned} B^2 - 4AC &= 4\left(a + \frac{\gamma \bar{R}}{(1-\gamma)\sigma^2} \right)^2 - 4\frac{\bar{R}^2 \gamma}{(1-\gamma)^2 \sigma^4} \\ &= 4\left[a^2 + \frac{2a\gamma \bar{R}}{(1-\gamma)\sigma^2} - \frac{\gamma \bar{R}^2}{(1-\gamma)\sigma^4} \right] \end{aligned} \tag{83}$$

Similarly for the equation (79)

$$4AC - B^2 = 4\left[\frac{\gamma \underline{R}^2}{(1-\gamma)\sigma^4} - a^2 - \frac{2a\gamma \underline{R}}{(1-\gamma)\sigma^2} \right] \tag{84}$$

Since the solution of (78) bounds above the solution of (46) (47), the solution of (46) (47) will exist on all of $[0, T]$ if

$$(1-\gamma)\sigma^4 a^2 + 2\sigma^2 a\gamma\bar{R} - \gamma\bar{R}^2 > 0 \tag{85}$$

Since the solution of (79) bounds below the solution of (46) (47), if

$$\gamma\underline{R}^2 - (1-\gamma)\sigma^4 a^2 - 2\sigma^2 a\underline{R} > 0 \tag{86}$$

and

$$T > \frac{2\pi}{\sqrt{4AC - B^2}} \tag{87}$$

where $4AC - B^2$ is given by (84), the solution of (79) satisfies

$$\lim_{t \downarrow \tau} K(t) = +\infty \tag{88}$$

for some τ satisfying

$$\tau \geq T - \frac{2\pi}{\sqrt{4AC - B^2}} > 0. \tag{89}$$

10. Conclusions

A doubly stochastic model for a stock price in which both the infinitesmal mean growth rate and the fluctuations about this mean growth rate are random was considered. When the mean growth rate is given by a linear system, a now standard argument reduces the portfolio management problem with this type of stock price model to a completely observed stochastic control problem.

Computing optimal policies for this problem involves solution of the dynamic programming partial differential equation for the value function. By making an appropriate guess for the functional form of the value function, solution of the dynamic programming PDE equation for the portfolio management problem was reduced to the solution of a partial differential equation for a related linear quadratic control problem. This linear quadratic control problem does not have a positive definite criterion, so there is the possibility on non existence of a solution. However when it does have a solution, solution of the original portfolio management problem can be expressed in terms of it. In cases in which it did not have a solution it was shown that there could be nonexistence of the solution of the original portfolio management problem.

10 Appendix: Reduction of the solution of the PDE (39) to the solution of ordinary differential equations

The partial differential equation (39) is

(39) $$g_t + \gamma r + \tfrac{1}{2}\frac{\gamma}{(1-\gamma)\sigma^2}(\bar{r} + \hat{m} - r)^2$$
$$+ \left(a\hat{m} + \frac{\gamma R}{(1-\gamma)\sigma^2}(\bar{r} + \hat{m} - r)\right) g_{\hat{m}} +$$
$$\tfrac{1}{2}\frac{R^2}{(1-\gamma)\sigma^2} g_{\hat{m}}^2 + \tfrac{1}{2}\frac{R^2}{\sigma^2} g_{\hat{m}\hat{m}} = 0.$$

If it has a solution of the form

(45) $$g(t,\hat{m}) = K(t)\hat{m}^2 + J(t)\hat{m} + L(t)$$

substituting this in the PDE requires that

(90) $$\dot{K}(t)\hat{m}^2 + \dot{J}(t)\hat{m} + \dot{L}(t) + \gamma r + \tfrac{1}{2}\frac{\gamma}{(1-\gamma)\sigma^2}(\bar{r} + \hat{m} - r)^2$$
$$+ \left(a\hat{m} + \frac{\gamma R(t)}{(1-\gamma)\sigma^2}(\bar{r} + \hat{m} - r)\right)(2K(t)\hat{m} + J(t))$$
$$+ \tfrac{1}{2}\frac{R(t)^2}{(1-\gamma)\sigma^2}(2K(t)\hat{m} + J(t))^2 + \tfrac{1}{2}\frac{R(t)^2}{\sigma^2}2K(t) = 0.$$

Multiplying out terms in (90) gives

(91) $$\dot{K}(t)\hat{m}^2 + \dot{J}(t)\hat{m} + \dot{L}(t) + \gamma r$$
$$+ \tfrac{1}{2}\frac{\gamma}{(1-\gamma)\sigma^2}(\bar{r} - r)^2 + \frac{\gamma}{(1-\gamma)\sigma^2}(\bar{r} - r)\hat{m} + \tfrac{1}{2}\frac{\gamma}{(1-\gamma)\sigma^2}\hat{m}^2$$
$$+ 2\left(a + \frac{\gamma R(t)}{(1-\gamma)\sigma^2}\right) K(t)\hat{m}^2$$
$$+ \frac{2\gamma R(t)}{(1-\gamma)\sigma^2}(\bar{r} - r)K(t)\hat{m} + \left(a + \frac{\gamma R(t)}{(1-\gamma)\sigma^2}\right) J(t)\hat{m}$$
$$+ \frac{\gamma R(t)}{(1-\gamma)\sigma^2}(\bar{r} - r)J(t)$$
$$+ \frac{2R(t)^2}{(1-\gamma)\sigma^2}K(t)^2\hat{m}^2 + \frac{2R(t)^2}{(1-\gamma)\sigma^2}K(t)J(t)\hat{m} + \tfrac{1}{2}\frac{R(t)^2 J(t)^2}{(1-\gamma)\sigma^2}$$
$$+ \frac{R(t)^2 K(t)}{\sigma^2} = 0.$$

Gathering together coefficients of powers of $\hat{m}$

$$(92)\quad \left[\dot{K}(t)+\tfrac{1}{2}\frac{\gamma}{(1-\gamma)\sigma^2}+2\left(a+\frac{\gamma R(t)}{(1-\gamma)\sigma^2}\right)K(t)+\frac{2R(t)^2}{(1-\gamma)\sigma^2}K(t)^2\right]\hat{m}^2$$

$$+\left[\dot{J}(t)+\frac{\gamma}{(1-\gamma)\sigma^2}(\bar{r}-r)+\frac{2\gamma R(t)}{(1-\gamma)\sigma^2}(\bar{r}-r)K(t)\right.$$

$$\left.+\left(a+\frac{\gamma R(t)}{(1-\gamma)\sigma^2}\right)J(t)+\frac{2R(t)^2}{(1-\gamma)\sigma^2}K(t)J(t)\right]\hat{m}$$

$$\dot{L}(t)+\gamma r+\tfrac{1}{2}\frac{\gamma}{(1-\gamma)\sigma^2}(\bar{r}-r)^2+\frac{\gamma R(t)}{(1-\gamma)\sigma^2}(\bar{r}-r)J(t)$$

$$+\tfrac{1}{2}\frac{R(t)^2J(t)^2}{(1-\gamma)\sigma^2}+\frac{R(t)^2K(t)}{\sigma^2}=0.$$

Thus (45) will be a solution of the PDE (39) if $K(t), J(t), L(t)$ are solutions of the differential equations (46), (48), (50).

References

[1] D. Duffie, *"Dynamic Asset Pricing Theory"*, Princeton University Press (1992).

[2] M.V. Dothan, D. Feldman, *Equilibrium Interest Rates and Multiperiod Bonds in a Partially Observed Economy"*, J. Finance, **41** (1986) pp. 369-382.

[3] W.H. Fleming, R.W. Rishel, *"Deterministic and Stochastic Optimal Control"*, Springer-Verlag (1975).

[4] G. Gennotte, *"Optimal Portfolio Choice Under Incomplete Information"*, J. Finance, **41** (1986), pp. 733-746.

[5] I. Karatzas, X. Xue, *"A Note on Utility Maximization Under Partial Observations"*, Math. Finance, **1** (1991) pp. 57-70.

[6] Y. Kuwana, *"Certainty Equivalence and Logarithmic Utilities in Comsumption/Investment Problems"*, Math. Finance, **5** (1995) pp. 297-309.

37

Hierarchical Production Controls for a Stochastic Manufacturing System with Long-Run Average Cost: Asymptotic Optimality

Suresh P. Sethi
Hanqin Zhang

ABSTRACT: This paper presents an extension of earlier research on hierarchical control of stochastic manufacturing systems with long-run average cost in which a positive inventory deterioration/cancellation rate for each product is assumed. Here we drop the assumption of the positive inventory deterioration/cancellation rate for each product, and give an asymptotic analysis of the manufacturing systems as the rates of change of the machine states approach infinity. We obtain a limiting problem in which the stochastic machine availability is replaced by its equilibrium mean availability. We use a near optimal control of the limiting problem to construct nearly asymptotically optimal open-loop piecewise deterministic controls for the original problem.

KEYWORDS: Stochastic manufacturing system; hierarchical control; dynamic programming equation; viscosity solution.

1 Introduction

We consider with the problem of controlling the production rate of a multi-product stochastic manufacturing system in order to meet a constant demand for each product type at a minimum long-run average expected cost of inventory/backlog. The stochastic nature of the system is due to the machines that are subject to random breakdown and repair. Bielecki and Kumar (1988) obtained an explicit optimal solution for a single machine (with two states, up and down) single product problem with linear holding and backlog costs. They showed that the optimal policy is a hedging point policy. Presman *et al.* (1998) extended their result to the case of two products, when the average machine capacity is sufficiently large.
A number of other attempts (see Presman *et al.* (1998) and Sethi *et al.*

(1997b) for references) have been made to extend the Bielecki-Kumar model in different directions. But these attempts are heuristic in nature, since they do not rigorously prove the optimality of the policies for their extensions of the Bielecki-Kumar Model.

The difficulty in proving the optimality lies in the fact when the problem is generalized to include convex costs, multiple products or multiple machine capacity levels, it becomes too complex for an explicit solution to be obtained. As a result, the approaches of Bielecki and Kumar and Presman *et al.* of proving optimality based on the explicitness of optimal solutions are no longer applicable.

Ghosh *et al.* (1997) considered the convex cost problem for a single product with a nondegenerate diffusion term in the dynamics of the manufacturing system. The convex cost problem for single and multiple products without diffusion was studied by Sethi *et al.* (1997a, 1998) with the use of the vanishing discount method. These papers develop appropriate dynamic programming(DP) equations, establish the existence of optimal solutions, and obtain verification theorems for optimality.

Since optimal solutions of the stochastic manufacturing problems are difficult to obtain in all but the simplest of the cases, researchers have attempted to obtain suboptimal or near-optimal solutions of the problems. Of particular importance to us is the so-called hierarchical control approach, which is based on the reduction of a given complex problem into simpler approximate problems or subproblems, and subsequent construction of a satisfactory solution for the given problem from the solutions of the simpler problems. Moreover, in the case of stochastic systems where fluctuation rates of some processes are much faster than the fluctuation rates of other processes, the hierarchical control approach can provide solutions that are asymptotically optimal as the fluctuation rates of the faster processes become large. This is a powerful idea and there is now a considerable literature devoted to asymptotic analyses of hierarchical manufacturing systems consisting of failure-prone machines with discounted cost criteria; see Sethi and Zhang (1994) and Yin and Zhang (1997) for references and details.

For an asymptotic analysis of hierarchical manufacturing systems with the long-run average cost criterion, Sethi *et al.* (1997b) treated a multiproduct manufacturing system consisting of a single or a number of parallel machines subject to breakdown and repair and facing constant demand for its products. Their treatment required the assumption of a positive inventory deterioration/cancellation rate for each product. The assumption corresponds to a stability condition typically imposed on infinite horizon problems (cf. Fleming and McEneaney (1995)). Under this assumption, they obtained a limiting problem as the rates of machine breakdown and repair go to infinity, and used the optimal control of the limiting problem to construct asymptotically optimal controls for the original problem.

It is the purpose of this paper to perform an asymptotic analysis of the multiproduct system of Sethi *et al.* (1997b) *without a positive inventory deterioration/cancellation rate for each product.* We obtain a limiting problem as the rates of machine breakdown and repair go to infinity, and use the optimal control of the limiting problem to construct open-loop piecewise deterministic controls for the original problem. Furthermore, It is shown that these controls are nearly asymptotically optimal under certain assumptions on the cost functions involved.

In Section 2 we introduce the problem and specify the required assumptions. Section 3 is devoted to the study of the value function associated with the corresponding control problem with a discounted cost. It establishes the results needed for the so-called *vanishing discount approach* often used for analysis of average cost minimization problems. In Section 4, we establish the convergence of the minimum long-run average expected cost for the original problem to the minimum long-run average cost for the limiting problem. Finally, nearly asymptotically optimal open-loop controls are constructed from the solution of the limiting problem in Section 5, and the convergence behavior of the associated average cost is examined.

2 Problem Formulation

Consider a manufacturing system that produces n distinct part types using m identical machines in parallel. With the production rate $u(t) := (u_1(t), ..., u_n(t)) \in R^n$, $u(t) \geq 0$, the total inventory/backlog or surplus $x(t) := (x_1(t), ..., x_n(t)) \in R^n$, and the demand rate $d := (d_1, ..., d_n) \in R^n$, the system dynamics is

$$\dot{x}(t) = u(t) - d, \ x(0) = x. \tag{2.1}$$

Assume that the machines are failure-prone and repairable and that each machine has a unit production capacity when not broken down. We can therefore let $\mathcal{M} = \{0, 1, ..., m\}$ denote the set of machine capacity states and let a random processes $\alpha(\epsilon, t) \in \mathcal{M}$, defined on the standard probability space $(\Omega, \mathcal{F}, P)$, denote the total capacity process for the manufacturing system. We assume that $\alpha(\epsilon, t)$ is a finite-state Markov process governed by a generator

$$Q^\epsilon = \epsilon^{-1} Q, \tag{2.2}$$

where Q is an $(m+1) \times (m+1)$ matrix such that $Q = (q_{ij}, q_{ij} \geq 0, \ i \neq j$ and $q_{ii} = -\sum_{j \neq i} q_{ij}$. Now we define the set of admissible control $u(\cdot) = (u_1(\cdot), ..., u_n(\cdot))$ as follows:

Definition 2.1 We say that a control $\mathbf{u}(\cdot) = \{u(t) : \ t \geq 0\}$ is admissible if (i) $u(\cdot)$ is an $\mathcal{F}_t^\epsilon = \sigma\{\alpha(\epsilon, s), \ s \leq t\}$ adapted measurable process and (ii) $u(t) \in U_{\alpha(\epsilon,t)}$ for all $t \geq 0$, where $U_i = \{p = (p_1, ..., p_n) \geq 0 : \sum_{j=1}^n p_j \leq i\}$. We use $\mathcal{A}^\epsilon(i)$ to denote the set of all admissible controls with the initial condition $\alpha(\epsilon, 0) = i$.

Definition 2.2 A function $w(\boldsymbol{x}, j)$ defined on $R^n \times \mathcal{M}$ is called an admissible feedback control, or simply a feedback control, if (i) for any given initial surplus $\boldsymbol{x}$ and production capacity i, the equation

$$\dot{\boldsymbol{x}}(t) = w(\boldsymbol{x}(t), \alpha(t)) - \boldsymbol{d}$$

has a unique solution and (ii) the control defined by $\boldsymbol{u}(\cdot) = \{\boldsymbol{u}(t) = w(\boldsymbol{x}(t), \alpha(t)),\ t \geq 0\} \in \mathcal{A}^\epsilon(i)$. With a slight abuse of notation, we simply call $w(\boldsymbol{x}, i)$ a feedback control when no ambiguity arises.
Let $h(\cdot)\ :\ R^n \to R_+$ and $c(\cdot)\ :\ R^n_+ \to R_+$ denote the surplus cost and the production cost, respectively. For any $\boldsymbol{u}(\cdot) \in \mathcal{A}^\epsilon(i)$, define the cost objective

$$J^\epsilon(\boldsymbol{x}, i, \boldsymbol{u}(\cdot)) = \limsup_{T\to\infty} \frac{1}{T} E \int_0^T (h(\boldsymbol{x}(t)) + c(\boldsymbol{u}(t)))\, dt, \tag{2.3}$$

where $\boldsymbol{x}(\cdot)$ is the surplus process corresponding to the production rate $\boldsymbol{u}(\cdot)$. Our goal is to choose $\boldsymbol{u}(\cdot) \in \mathcal{A}^\epsilon(i)$ so as to minimize $J(\boldsymbol{x}, i, \boldsymbol{u}(\cdot))$. We formally summarize our control problem as follows:

$$\mathcal{P}^\epsilon : \begin{cases} \min J^\epsilon(\boldsymbol{x}, i, \boldsymbol{u}(\cdot)) = \limsup_{T\to\infty} \frac{1}{T} E \int_0^T (h(\boldsymbol{x}(t)) + c(\boldsymbol{u}(t)))\, dt, \\ \text{s.t. } \dot{\boldsymbol{x}}(t) = \boldsymbol{u}(t) - \boldsymbol{d},\ \boldsymbol{x}(0) = \boldsymbol{x},\ \boldsymbol{u}(\cdot) \in \mathcal{A}^\epsilon(i), \\ \text{minimum average cost } \ \lambda^\epsilon(\boldsymbol{x}, i) = \inf_{\boldsymbol{u}(\cdot)\in\mathcal{A}^\epsilon(i)} J^\epsilon(\boldsymbol{x}, i, \boldsymbol{u}(\cdot)). \end{cases} \tag{2.4}$$

We assume that $h(\cdot)$, $c(\cdot)$, and $\alpha(\epsilon, \cdot)$ satisfy the following assumptions:
(A.1) $h(\cdot)$ and $c(\cdot)$ are nonnegative convex functions with $h(0) = 0$ and $c(0) = 0$.
(A.2) For all $\boldsymbol{x}$, $\boldsymbol{x}'$, $\boldsymbol{y}$ and $\boldsymbol{y}'$, there exist positive constants C_{21} and p such that

$$0 \leq h(\boldsymbol{x}) \leq C_{21}(1 + |\boldsymbol{x}|^p),$$
$$|h(\boldsymbol{x}) - h(\boldsymbol{x}')| \leq C_{21}(1 + |\boldsymbol{x}|^p + |\boldsymbol{x}'|^p)|\boldsymbol{x} - \boldsymbol{x}'|,$$
$$|c(\boldsymbol{y}) - c(\boldsymbol{y}')| \leq C_{21}(1 + |\boldsymbol{y}|^p + |\boldsymbol{y}'|^p)|\boldsymbol{y} - \boldsymbol{y}'|.$$

(A.3) Q is weakly irreducible in the following sense: the equations

$$\boldsymbol{q}Q = 0 \text{ and } \sum_{j=0}^m q_j = 1$$

have a unique solution $\boldsymbol{q} = (q_0, q_1, ..., q_m)$ with $q_j \geq 0\ j = 0, 1, ..., m$. The vector $\boldsymbol{q}$ is called the equilibrium distribution vector of the Markov process $\alpha(\epsilon, t)$.
(A.4) The average capacity $\bar{\alpha} \equiv \sum_{j=0}^m jq_j > \sum_{j=1}^n d_j$.
Remark 2.1 Convexity assumptions in (A.1) are standard in the convex production planning literature. The restrictions on the growth rate of the

surplus cost function $h(\cdot)$ in (A.2) are usual assumptions to ensure the existence of a solution to the dynamic programming equation to be considered later.

Remark 2.2 Assumption (A.4), i.e., the average capacity of the machine is greater than demand, is necessary in the sense that if it were not true, then even if the system could always be producing at its maximum capacity, it still would not meet the demand, and the backlog could build up without bound over time with the consequence that no optimal solution with a finite cost would exist.

Remark 2.3 In Sethi *et al.* (1997b), they consider the same manufacturing system but with a positive inventory deterioration/cancellation rate for each product type. That is, instead of (2.1), their system dynamics is

$$\dot{\boldsymbol{x}}(t) = -\boldsymbol{a}\boldsymbol{x}(t) + \boldsymbol{u}(t) - \boldsymbol{d}, \ \boldsymbol{x}(0) = \boldsymbol{x}, \tag{2.5}$$

with $\boldsymbol{a} = (a_1, ..., a_n)$, a positive constant vector. In view of the fact that control $\boldsymbol{u}(\cdot)$ is bounded between 0 and m, (2.5) implies that the system state $\boldsymbol{x}(\cdot)$ remains bounded. Consequently, the presence of a positive inventory deterioration/cancellation rate gives some smooth properties of $J^\epsilon(\boldsymbol{x}, i, \boldsymbol{u}(\cdot))$; see Sethi *et al.* (1997b). In the present paper, with (2.1), instead of (2.5), the system state $\boldsymbol{x}(\cdot)$ may no longer remains bounded.

Remark 2.4 By Assumption (A.4) we know that $\lambda^\epsilon(\boldsymbol{x}, i)$ is independent of initial condition $\boldsymbol{x}$ and i. So we will simply write $\lambda^\epsilon(\boldsymbol{x}, i)$ as λ^ϵ in the remainder of this paper.

Since we will use the vanishing discount approach to study our problem, we provide next the required analysis of the discounted problem.

3 Analysis of the Discounted Cost Problem

In order to derive the DP equation for $\mathcal{P}^\epsilon$ and carry out an asymptotic analysis of the minimum long-run average expected cost λ^ϵ, we introduce a corresponding control problem $\mathcal{P}^{\epsilon,\rho}$ with the cost discounted at a rate $\rho > 0$:

$$\mathcal{P}^{\epsilon,\rho} : \begin{cases} \min \ J^{\epsilon,\rho}(\boldsymbol{x}, i, \boldsymbol{u}(\cdot)) = \int_0^\infty e^{-\rho t} \left(h(\boldsymbol{x}(t)) + c(\boldsymbol{u}(t))\right) dt \\ \text{s.t. } \dot{\boldsymbol{x}}(t) = \boldsymbol{u}(t) - \boldsymbol{d}, \ \boldsymbol{x}(0) = \boldsymbol{x}, \ \boldsymbol{u}(\cdot) \in \mathcal{A}^\epsilon(i) \\ V^{\epsilon,\rho}(\boldsymbol{x}, i) = \inf_{\boldsymbol{u}(\cdot)\in\mathcal{A}^\epsilon(i)} J^{\epsilon,\rho}(\boldsymbol{x}, i, \boldsymbol{u}(\cdot)) \end{cases} \tag{3.1}$$

From Sethi and Zhang (1994), the value function $V^{\epsilon,\rho}(\boldsymbol{x}, i)$ is the only viscosity solution to the DP equation

$$\begin{aligned} \rho V^{\epsilon,\rho}(\boldsymbol{x}, i) = \quad & \min_{\boldsymbol{u}\in U_i} \left[(\boldsymbol{u} - \boldsymbol{d}) \cdot \frac{\partial V^{\epsilon,\rho}(\boldsymbol{x}, i)}{\partial \boldsymbol{x}} + c(\boldsymbol{u}) \right] \\ & + h(\boldsymbol{x}) + \sum_{j=0}^{m} q_{ij}^\epsilon V^{\epsilon,\rho}(\boldsymbol{x}, j) \end{aligned} \tag{3.2}$$

for any $i \in \mathcal{M}$, where q^{ϵ}_{ij} is the (i,j)-element of the generator Q^{ϵ}, $V^{\epsilon,\rho}(\cdot,i)$ is convex, and it satisfies the growth condition

$$|V^{\epsilon,\rho}(\boldsymbol{x},i) - V^{\epsilon,\rho}(\boldsymbol{x}',i)| \leq \quad C_{31}(\rho)(1+|\boldsymbol{x}|^{p}+|\boldsymbol{x}'|^{p})|\boldsymbol{x}-\boldsymbol{x}'|, \quad \boldsymbol{x},\boldsymbol{x}' \in R^{n} \text{ and } i \in \mathcal{M} \tag{3.3}$$

for some positive constant $C_{31}(\rho)$ dependent on ρ, where p is specified in Assumption (A.2).

In order to study the long-run average cost control problem using the vanishing discount approach, we must first obtain some estimates for the value function $V^{\epsilon,\rho}(\boldsymbol{x},i)$.

Lemma 3.1 *For any $v > 0$, let*

$$\tau^{\epsilon}(v) = \inf\left\{t \ : \ \int_0^t \left(\alpha(\epsilon,s) - \sum_{j=1}^{n} d_j\right) ds = v\right\}.$$

If Assumptions (A.3)-(A.4) *hold, then for any $r \geq 1$, there are constants C_{32}, independent of v and ϵ, and ϵ_0 such that for all $\epsilon \in (0,\epsilon_0]$,*

$$E\left(\tau^{\epsilon}(v)\right)^{r} \leq C_{32}(1+v^{r}).$$

Proof. Note that from Corollary C.2 of Sethi and Zhang (1994), there exist $C^{(1)}_{32}$ and ϵ_0 such that for $\epsilon \in (0,\epsilon_0]$ and any $t > 0$,

$$E\exp\left\{\frac{1}{(t+1)^{1/2}}\left|\int_0^t (\alpha(\epsilon,s)-\bar{\alpha})ds\right|\right\} \leq C^{(1)}_{32}, \tag{3.4}$$

where $\bar{\alpha}$ denotes the average capacity as defined in Assumption (A.4). Then, using (3.4), along the lines of the proof of Lemma 7.1 in Sethi *et al.* (1998), we can complete the proof. □

Lemma 3.2 *Let Assumptions* (A.3)-(A.4) *hold. For any $(\boldsymbol{x},i),(\boldsymbol{x}',i') \in R^{n} \times \mathcal{M}$, there exist constants ϵ_0, C_{33}, independent of ϵ, $\boldsymbol{x}$ and $\boldsymbol{x}'$, and a control policy $u^{\epsilon}(\cdot) \in \mathcal{A}^{\epsilon}(i)$ such that for any $r \geq 1$ and $\epsilon \in (0,\epsilon_0]$*

$$E\left(\tau^{\epsilon}(i,i',\boldsymbol{x},\boldsymbol{x}')\right)^{r} \leq C_{33}\left(1+\sum_{j=1}^{n}|x_j - x'_j|^{r}\right),$$

where

$$\tau^{\epsilon}(i,i',\boldsymbol{x},\boldsymbol{x}') = \inf\{t \geq 0 \ : \ (\boldsymbol{x}(t),\alpha(\epsilon,t)) = (\boldsymbol{x}',i')\}$$

and $\boldsymbol{x}(\cdot)$ is the surplus process corresponding to the control policy $\boldsymbol{u}^{\epsilon}(\cdot)$ and the initial condition $(\boldsymbol{x}(0),\alpha(\epsilon,0)) = (\boldsymbol{x},i)$.

Proof. From Lemma C.3 of Sethi and Zhang (1994) and Assumption (A.3), there exist constants T and ϵ_0 such that for $t \geq T$ and $\epsilon \in (0,\epsilon_0]$,

$$P(\alpha(\epsilon,t) = j|\alpha(\epsilon,0) = i) \geq q_j/2, \tag{3.5}$$

for any $i, j \in \mathcal{M}$. The remainder of the proof is similar to the proof of Theorem 3.2 of Sethi *et al.* (1998), and it is therefore omitted. □

Remark 3.1 From Meyn and Tweedie (1994), we can see that the Markov process $x(t)$ is positive Harris recurrent under Assumptions (A.3)-(A.4). Therefore, we have $E\tau^\epsilon(i, i', x, x') < \infty$, which is not as strong as Lemma 3.2 required for this paper.

With Lemma 3.2 in hand, we can follow the corresponding proof of Theorem 3.3 in Sethi *et al.* (1998) to prove the following theorem.

Theorem 3.3 *Under Assumptions* (A.1)-(A.4), *there exist constants* ρ_0 *and* ϵ_0 *such that*

(i) $\{\rho V^{\epsilon,\rho}(0,0): \ 0 < \rho \leq \rho_0, \ 0 < \epsilon \leq \epsilon_0\}$ *is bounded;*

(ii) *For* $\epsilon \in (0, \epsilon_0]$ *and* $\rho \in (0, \rho_0]$, *the function*

$$W^{\epsilon,\rho}(x, j) \equiv V^{\epsilon,\rho}(x, j) - V^{\epsilon,\rho}(0, 0)$$

is convex in x;

(iii) *For* $\epsilon \in (0, \epsilon_0]$ *and* $\rho \in (0, \rho_0]$, $W^{\epsilon,\rho}(x, j)$ *is locally Lipschitz continuous in* x, *i.e., there exists a constant* C_{34}, *independent of* ρ *and* ϵ, *such that*

$$|W^{\epsilon,\rho}(x, j) - W^{\epsilon,\rho}(x', j)| \leq C_{34}(1 + |x|^p + |x'|^p)|x - x'|$$

for $j \in \mathcal{M}$ *and all* $x, x' \in R^n$, *where* p *is given in* Assumption (A.2).

With the help of Theorem 3.3, we are ready to deal with the convergence of λ^ϵ the minimum average cost of the problem $\mathcal{P}^\epsilon$, with respect to ϵ.

4 Convergence of the Minimum Average Cost

We obtain the limiting control problem as $\epsilon \to 0$ and show that λ^ϵ, defined in (2.4) converges to the minimum average cost associated with the limiting control problem. Corresponding to the discounted cost DP equation (3.2), we can formally write the DP equation associated with the long-run average cost case as

$$\hat{\lambda}^\epsilon = \inf_{u \in U_i} \left\{ (u - d) \cdot \frac{\partial \hat{W}^\epsilon(x, i)}{\partial x} + c(u) \right\} + h(x) + \sum_{j=0}^{m} q_{ij}^\epsilon \hat{W}^\epsilon(x, j), \quad (4.1)$$

for any $i \in \mathcal{M}$ and $x \in R^n$, where $\hat{\lambda}^\epsilon$ is a constant and $\hat{W}^\epsilon(x, i)$ is a real-valued function on $R^n \times \mathcal{M}$. Since the partial derivative $\frac{\partial \hat{W}^\epsilon(x,i)}{\partial x}$ may not exist in general, we give a precise interpretation of (4.1) (see Fleming and Soner (1992));

Definition 4.1 We say that $(\hat{\lambda}^\epsilon, \hat{W}^\epsilon(x, i))$ is a viscosity solution to (4.1), if

(i) $\hat{W}^\epsilon(x, i)$ is continuous in x and $|\hat{W}^\epsilon(x, i)| \leq C_{41}(1 + |x|^{p_1})$, where C_{41} and p_1 are positive constants;

(ii) For all $r \in D^+\hat{W}^\epsilon(x,i) = \{r \in R^n : \limsup_{|h|\to 0}[\hat{W}^\epsilon(x+h,i) - \hat{W}^\epsilon(x,i) - r\cdot h]|h|^{-1} \le 0\}$,

$$\lambda^\epsilon \le \inf_{u\in U_i}\{(u-d)\cdot r + c(u)\} + h(x) + \sum_{j=0}^m q_{ij}^\epsilon \hat{W}^\epsilon(x,j);$$

(iii) For all $r \in D^-\hat{W}^\epsilon(x,i) = \{r \in R^n : \liminf_{|h|\to 0}[\hat{W}^\epsilon(x+h,i) - \hat{W}^\epsilon(x,i) - r\cdot h]|h|^{-1} \ge 0\}$,

$$\hat{\lambda}^\epsilon \ge \inf_{u\in U_i}\{(u-d)\cdot r + c(u)\} + h(x) + \sum_{j=0}^m q_{ij}^\epsilon \hat{W}^\epsilon(x,j).$$

Lemma 4.1 *If Assumptions* (A.1)-(A.4) *hold, then the DP equation* (4.1) *has a unique solution in the following sense: If* $(\hat{\lambda}_1^\epsilon, \hat{W}_1^\epsilon(x,i))$ *and* $(\hat{\lambda}_2^\epsilon, \hat{W}_2^\epsilon(x,i))$ *are viscosity solutions to* (4.1), *then* $\hat{\lambda}_1^\epsilon = \hat{\lambda}_2^\epsilon$.

Proof. See Theorem G.1 in Sethi and Zhang (1994). □

Remark 4.1 While the proof of Theorem G.1 in Sethi and Zhang (1994) works for the uniqueness of $\hat{\lambda}^\epsilon$, it cannot be adapted to prove whether or not $\hat{W}_1^\epsilon(x,i)$ and $\hat{W}_2^\epsilon(x,i)$ are equal. Clearly, if $(\hat{\lambda}^\epsilon, \hat{W}^\epsilon(x,i))$ is a viscosity solution to (4.1), then for any constant C_{42}, $(\hat{\lambda}^\epsilon, C_{42} + \hat{W}^\epsilon(x,i))$ is also a viscosity solution to (4.1).

According to (i) and (iii) of Theorem 3.3, for a fixed $\epsilon \in (0,\epsilon_0]$ and any subsequence of $\{\rho\}$, denoted by $\{\rho'\}$, there is a further subsequence of $\{\rho'\}$, still denoted by $\{\rho'\}$, such that

$$\lambda^{\epsilon,*} = \lim_{\rho'\to 0} \rho' V^{\epsilon,\rho'}(0,0) \tag{4.2}$$

and

$$W^\epsilon(x,j) = \lim_{\rho'\to 0} W^{\epsilon,\rho'}(x,j) \tag{4.3}$$

for all $(x,j) \in R^n \times \mathcal{M}$. Furthermore, we have the following lemma.

Lemma 4.2 *There exists a constant* ϵ_0 *such that for* $\epsilon \in (0,\epsilon_0]$, *we have the following.*

(i) *The function* $W^\epsilon(x,j)$ *given in* (4.3) *is convex and is locally Lipschitz continuous in* x, *i.e., there is a constant* C_{43} *such that*

$$|W^\epsilon(x,j) - W^\epsilon(x',j)| \le C_{43}(1 + |x|^p + |x'|^p)|x - x'|$$

for all $x, x' \in R^n$ *and* $j \in \mathcal{M}$;

(ii) $(\lambda^{\epsilon,*}, W^\epsilon(x,j))$ *is a viscosity solution to the DP equation* (4.1);

(iii) *There exists a constant* C_{44} *such that for all* $x \in R^n$ *and* $i,j \in \mathcal{M}$

$$\frac{1}{\epsilon}|W^\epsilon(x,i) - W^\epsilon(x,j)| \le C_{44}(1 + |x|^{p+1}).$$

Proof. (i) From the convexity of $V^{\epsilon,\rho}(\boldsymbol{x},j)$, we get the convexity of $W^{\epsilon}(\boldsymbol{x},j)$. According to (4.3), it follows from (iii) of Theorem 3.3 that the function $W^{\epsilon}(\boldsymbol{x},j)$ is locally Lipschitz continuous in $\boldsymbol{x}$.
(ii) To justify $(\lambda^{\epsilon,*}, W^{\epsilon}(\boldsymbol{x},j))$ to be a viscosity solution to (4.1), let $\boldsymbol{x}_0 \in R^n$ and $\boldsymbol{r} \in D^{-}W^{\epsilon}(\boldsymbol{x},j)$ for any fixed $j \in \mathcal{M}$. Then there exists a continuously differential function ϕ and a neighborhood $S(\boldsymbol{x}_0)$ of $\boldsymbol{x}_0$ (cf. Fleming and Soner (1993)) such that

$$\left.\frac{\partial\phi(\boldsymbol{x})}{\partial\boldsymbol{x}}\right|_{\boldsymbol{x}=\boldsymbol{x}_0} = \boldsymbol{r},$$
$$0 = W^{\epsilon}(\boldsymbol{x}_0,j) - \phi(\boldsymbol{x}_0) = \inf_{\boldsymbol{x}\in S(\boldsymbol{x}_0)} \{W^{\epsilon}(\boldsymbol{x},j) - \phi(\boldsymbol{x})\},$$

and $\boldsymbol{x}_0$ is the only minimum point of $W^{\epsilon}(\boldsymbol{x},j)-\phi(\boldsymbol{x})$ on $S(\boldsymbol{x}_0)$. Now choose $\boldsymbol{x}_0^{\rho'} \in S(\boldsymbol{x}_0)$ such that

$$V^{\epsilon,\rho'}(\boldsymbol{x}_0^{\rho'},j) - \phi(\boldsymbol{x}_0^{\rho'}) = \inf_{\boldsymbol{x}\in S(\boldsymbol{x}_0)} \left\{V^{\epsilon,\rho'}(\boldsymbol{x},j) - \phi(\boldsymbol{x})\right\}.$$

Then, as $\rho' \to 0$,

$$\boldsymbol{x}_0^{\rho'} \to \boldsymbol{x}_0. \tag{4.4}$$

Moreover,

$$\left.\frac{\partial\phi(\boldsymbol{x})}{\partial\boldsymbol{x}}\right|_{\boldsymbol{x}=\boldsymbol{x}_0^{\rho'}} \in D^{-}V^{\epsilon,\rho'}(\boldsymbol{x}_0^{\rho'},j). \tag{4.5}$$

Note that $V^{\epsilon,\rho'}(\boldsymbol{x},j)$ is a unique viscosity solution of the DP equation (3.2). Then, (4.5) implies

$$\begin{aligned} \rho' V^{\epsilon,\rho'}(\boldsymbol{x},j) \geq \quad & \inf_{\boldsymbol{u}\in U_j} \left\{ (\boldsymbol{u}-\boldsymbol{d}) \cdot \left.\frac{\partial\phi(\boldsymbol{x})}{\partial\boldsymbol{x}}\right|_{\boldsymbol{x}=\boldsymbol{x}_0^{\rho'}} + c(\boldsymbol{u}) \right\} \\ & + h(\boldsymbol{x}_0^{\rho'}) + \sum_{i=0}^{m} q_{ji}^{\epsilon} V^{\epsilon,\rho'}(\boldsymbol{x}_0^{\rho'},i). \end{aligned}$$

Hence,

$$\begin{aligned} & \rho' V^{\epsilon,\rho'}(0,0) + \rho' \left[V^{\epsilon,\rho'}(\boldsymbol{x},j) - V^{\epsilon,\rho'}(0,0) \right] \\ & \geq \inf_{\boldsymbol{u}\in U_j} \left\{ (\boldsymbol{u}-\boldsymbol{d}) \cdot \left.\frac{\partial\phi(\boldsymbol{x})}{\partial\boldsymbol{x}}\right|_{\boldsymbol{x}=\boldsymbol{x}_0^{\rho'}} + c(\boldsymbol{u}) \right\} \\ & \quad + h(\boldsymbol{x}_0^{\rho'}) + \sum_{i=0}^{m} q_{ji}^{\epsilon} \left[V^{\epsilon,\rho'}(\boldsymbol{x},i) - V^{\epsilon,\rho'}(0,0) \right], \end{aligned}$$

where $\sum_{i=0}^m q_{ji}^\epsilon V^{\epsilon,\rho'}(0,0) = 0$ is applied. Letting $\rho' \to 0$ and using (4.4), we obtain

$$\lambda^{\epsilon,*} \geq \inf_{\boldsymbol{u}\in U_j} \{(\boldsymbol{u}-\boldsymbol{d})\cdot \boldsymbol{r} + c(\boldsymbol{u})\} + h(\boldsymbol{x}_0) + \sum_{i=0}^m q_{ji}^\epsilon W^\epsilon(\boldsymbol{x},i),$$

for all $\boldsymbol{r} \in D^- W^\epsilon(\boldsymbol{x},j)$. Similarly, we can show that the opposite inequality for $\boldsymbol{r} \in D^+ W^\epsilon(\boldsymbol{x},j)$ holds, and thus prove that (ii) holds.
(iii) First we consider the point of differentiability of $W^\epsilon(\boldsymbol{x},i)$ for each i. Let $\boldsymbol{x}_0$ be the differentiable point of $W^\epsilon(\boldsymbol{x},i)$ for each i. Then by (ii), equation (4.1) holds, and we have $\sum_{j=0}^m q_{ij} W^\epsilon(\boldsymbol{x}_0,j) = \epsilon l(\boldsymbol{x}_0,i)$, where

$$l(\boldsymbol{x}_0,i) = \lambda^{\epsilon,*} - \inf_{\boldsymbol{u}\in U_i} \left\{ (\boldsymbol{u}-\boldsymbol{d}) \cdot \left.\frac{\partial W^\epsilon(\boldsymbol{x},i)}{\partial \boldsymbol{x}}\right|_{\boldsymbol{x}=\boldsymbol{x}_0} + c(\boldsymbol{u}) \right\} - h(\boldsymbol{x}_0) - \sum_{j=0}^m q_{ij}^{(1)} W^\epsilon(\boldsymbol{x}_0,j).$$

Observe that due to (i) and Assumptions (A.1)-(A.2),

$$|l(\boldsymbol{x}_0,i)| \leq C_{44}^{(1)}(1+|\boldsymbol{x}_0|^{p+1}), \text{ for each } i, \ \epsilon \text{ and a suitable } C_{44}^{(1)}. \qquad (4.6)$$

Also, the irreducibility of Q implies that the kernel of Q is the one-dimensional subspace spanned by the vector $\mathbf{1} = (1,...,1) \in R^{m+1}$. Hence, for any $i,j \in \mathcal{M}$ we have

$$\frac{1}{\epsilon}|W^\epsilon(\boldsymbol{x},i) - W^\epsilon(\boldsymbol{x},j)| \leq \sup_{j'\in\mathcal{M}} |l(\boldsymbol{x}_0,j')| \leq C_{44}^{(1)}(1+|\boldsymbol{x}_0|^{p+1}). \qquad (4.7)$$

Recall that $\boldsymbol{x}_0$ is a point of differentiability of $W^\epsilon(\boldsymbol{x},i)$ for each i. But such points are dense because of the local Lipschitz continuity of $W^\epsilon(\boldsymbol{x},i)$. Hence, (4.7) holds everywhere. Thus we have (iii). □
From (ii) of Lemma 4.2, for each subsequence of $\{\rho\}$, denoted by $\{\rho'\}$, the limit of $\{\rho' V^{\epsilon,\rho'}(0,0)\}$ together with some function $W^\epsilon(\boldsymbol{x},i)$ is a viscosity solution to (4.1). With the help of Lemma 4.1, we get that $\lambda^{\epsilon,*}$, the limit of $\{\rho' V^{\epsilon,\rho'}(0,0)\}$, does not depend on the choice of the subsequence $\{\rho'\}$. Consequently, we have the following result.
Lemma 4.3 *Under Assumptions* (A.1)-(A.4), *we have*

$$\lim_{\rho\to 0} \rho V^{\epsilon,\rho}(0,0) = \lambda^\epsilon.$$

□

We now study the limiting behavior as $\epsilon \to 0$. By multiplying both sides of (4.1) at i by q_i (see Assumption (A.3)) and summing over i, we have

$$\lambda^\epsilon = \sum_{i=0}^m q_i \inf_{\boldsymbol{u}\in U_i} \left\{ (\boldsymbol{u}-\boldsymbol{d}) \cdot \frac{\partial W^\epsilon(\boldsymbol{x},i)}{\partial \boldsymbol{x}} + c(\boldsymbol{u}) \right\} + h(\boldsymbol{x}),$$

where $qQ = 0$ is applied. Then we let $\epsilon \to 0$ to obtain

$$\hat{\lambda} = \sum_{i=0}^{m} q_i \inf_{\boldsymbol{u} \in U_i} \left\{ (\boldsymbol{u} - \boldsymbol{d}) \cdot \frac{\partial \hat{W}(\boldsymbol{x})}{\partial \boldsymbol{x}} + c(\boldsymbol{u}) \right\} + h(\boldsymbol{x}). \tag{4.8}$$

Such a procedure will be justified in the next theorem. Note that

$$\sum_{i=0}^{m} q_i \inf_{\boldsymbol{u} \in U_i} \left\{ (\boldsymbol{u} - \boldsymbol{d}) \cdot \frac{\partial \hat{W}(\boldsymbol{x})}{\partial \boldsymbol{x}} + c(\boldsymbol{u}) \right\} = \inf_{\boldsymbol{u} \in U_{\bar{\alpha}}} \left\{ (\boldsymbol{u} - \boldsymbol{d}) \cdot \frac{\partial \hat{W}(\boldsymbol{x})}{\partial \boldsymbol{x}} + \bar{c}(\boldsymbol{u}) \right\},$$

where $\bar{\alpha}$ is given in Assumption (A.4) and

$$\bar{c}(\boldsymbol{u}) = \inf \left\{ \sum_{i=0}^{m} q_i c(\boldsymbol{u}^{(i)}) \ : \ \boldsymbol{u}^{(i)} \in U_i \text{ and } \sum_{i=0}^{m} q_i \boldsymbol{u}^{(i)} = \boldsymbol{u} \right\}.$$

Thus, (4.8) is equivalent to

$$\hat{\lambda} = \inf_{\boldsymbol{u} \in U_{\bar{\alpha}}} \left\{ (\mathbf{u} - \boldsymbol{d}) \cdot \frac{\partial \hat{W}(\boldsymbol{x})}{\partial \boldsymbol{x}} + \bar{c}(\boldsymbol{u}) \right\} + h(\boldsymbol{x}), \tag{4.9}$$

which is the DP equation of the following deterministic problem:

$$\mathcal{P} : \begin{cases} \min J(\boldsymbol{x}, \hat{\boldsymbol{u}}(\cdot)) = \limsup_{T \to \infty} \dfrac{1}{T} \displaystyle\int_0^T (h(\hat{\boldsymbol{x}}(t)) + \bar{c}(\hat{\boldsymbol{u}}(t)))dt, \\ \text{s.t. } \dot{\hat{\boldsymbol{x}}}(t) = \hat{\boldsymbol{u}}(t) - \boldsymbol{d}, \ \hat{\boldsymbol{x}}(0) = 0, \ \hat{\boldsymbol{u}}(\cdot) \in \hat{\mathcal{A}}, \\ \text{minimum average cost } \lambda = \inf_{\hat{\mathbf{u}}(\cdot) \in \hat{\mathcal{A}}} J(\boldsymbol{x}, \hat{\boldsymbol{u}}(\cdot)), \end{cases} \tag{4.10}$$

where $\hat{\mathcal{A}} = \{\hat{\boldsymbol{u}}(\cdot) : \hat{\boldsymbol{u}}(t) \in U_{\bar{\alpha}}$ and $\hat{\boldsymbol{u}}(\cdot)$ is a deterministic measurable process$\}$.

Theorem 4.4 *Under Assumptions* (A.1)-(A.4), $\lambda^\epsilon \to \lambda$ *as* $\epsilon \to 0$.

Proof. Similar to Lemma 4.1, (4.9) has a unique viscosity solution in the same sense as defined in Lemma 4.1. Now using (i) of Theorem 3.3 and (i) of Lemma 4.2 and Lemma 4.3, we have for any subsequence of $\{\epsilon\}$, denoted by $\{\epsilon'\}$, a further subsequence of $\{\epsilon'\}$, still denoted by $\{\epsilon'\}$, such that

$$\lambda^* = \lim_{\epsilon' \to 0} \lambda^{\epsilon'} \tag{4.11}$$

and

$$W(\boldsymbol{x}, j) = \lim_{\epsilon' \to 0} W^{\epsilon'}(\boldsymbol{x}, j) \tag{4.12}$$

for all $(\boldsymbol{x}, j) \in R^n \times \mathcal{M}$, where λ^ϵ and $W^\epsilon(\boldsymbol{x}, j)$ are given in Lemma 4.3 and (4.3), respectively. From (iii) of Lemma 4.2, we know that $W(\boldsymbol{x}, j)$ is independent of j. We therefore write $W(\boldsymbol{x}, j)$ as $W(\boldsymbol{x})$. The proof that $(\lambda^*, W(\boldsymbol{x}))$ is a viscosity solution of (4.9) is similar to the proof of (ii) of Lemma 4.2. □

In the rest of this section, we present another deterministic control problem $\bar{\mathcal{P}}$, which is equivalent to $\mathcal{P}$. It is $\bar{\mathcal{P}}$ that is used in the next section. Observing the equality

$$\sum_{j=0}^{m} q_j \inf_{\boldsymbol{u}\in U_j}\left\{(\boldsymbol{u}-\boldsymbol{d})\cdot\frac{\partial \hat{W}(\boldsymbol{x})}{\partial \boldsymbol{x}}+c(\boldsymbol{u})\right\}$$
$$=\inf_{(\boldsymbol{u}^{(0)},\boldsymbol{u}^{(1)},\ldots,\boldsymbol{u}^{(m)})\in U_0\times U_1\times\cdots\times U_m}\left\{\left(\sum_{j=0}^{m} q_j\boldsymbol{u}^{(j)}-\boldsymbol{d}\right)\cdot\frac{\partial \hat{W}(\boldsymbol{x})}{\partial \boldsymbol{x}}\right.$$
$$\left.+\sum_{j=0}^{m} q_j c(\boldsymbol{u}^{(j)})\right\},$$

we get that (4.8) is also equivalent to

$$\hat{\lambda}=\inf_{(\boldsymbol{u}^{(0)},\boldsymbol{u}^{(1)},\ldots,\boldsymbol{u}^{(m)})\in U_0\times U_1\times\cdots\times U_m}\left\{\left(\sum_{j=0}^{m} q_j\boldsymbol{u}^{(j)}-\boldsymbol{d}\right)\cdot\frac{\partial \hat{W}(\boldsymbol{x})}{\partial \boldsymbol{x}}\right.$$
$$\left.+\sum_{j=0}^{m} q_j c(\boldsymbol{u}^{(j)})\right\}+h(\boldsymbol{x}). \quad (4.13)$$

This motivates us to define another limiting control problem whose DP equation is precisely (4.13). Consider the enlarged control space

$$\bar{\mathcal{A}}=\{\quad \bar{\boldsymbol{u}}=(\bar{\boldsymbol{u}}^{(0)}(\cdot),\bar{\boldsymbol{u}}^{(1)}(\cdot),\cdots,\bar{\boldsymbol{u}}^{(m)}(\cdot)):\ \boldsymbol{u}^{(i)}(t)\in U_i$$
$$\text{and } \boldsymbol{u}^{(i)}(\cdot) \text{ is a deterministic measurable process}\}.$$

Then we define the control problem:

$$\bar{\mathcal{P}}:\begin{cases}\min \bar{J}(\boldsymbol{x},\bar{\boldsymbol{u}}(\cdot))=\limsup_{T\to\infty}\frac{1}{T}\int_0^T\left(h(\bar{\boldsymbol{x}}(t))+\sum_{j=0}^{m} q_j c(\boldsymbol{u}^{(j)}(t))\right)dt\\ \text{s.t. } \dot{\bar{\boldsymbol{x}}}(t)=\sum_{j=0}^{m} q_j\boldsymbol{u}^{(j)}(t)-\boldsymbol{d},\ \bar{\boldsymbol{x}}(0)=\boldsymbol{x},\ \bar{\boldsymbol{u}}(\cdot)\in\bar{\mathcal{A}}\\ \lambda=\inf_{\bar{\boldsymbol{u}}(\cdot)\in\bar{\mathcal{A}}}\bar{J}(\boldsymbol{x},\bar{\boldsymbol{u}}(\cdot)).\end{cases} \quad (4.14)$$

It should be noted that the problems $\mathcal{P}$ and $\bar{\mathcal{P}}$ are equivalent in view of the facts that they have the same DP equation (4.8) and that the equation has a unique viscosity solution; see also Chapter 3 in Sethi and Zhang (1994).

5 Nearly Asymptotically Optimal Open Loop Controls

We use a near optimal open-loop control policy associated with the limiting problem $\bar{\mathcal{P}}$ given in (4.14) to derive an open-loop control policy for the original problem $\mathcal{P}^\epsilon$ given in (2.4). We study the asymptotics of the derived control policy. To this end, we give several technical lemmas concerning the asymptotic properties of $\alpha(\epsilon, t)$.

Lemma 5.1 *Under Assumption* (A.3) , *there exists constants* $C_{51} > 0$ *and* $k_0 > 0$ *such that for all* $t \geq 0$

$$|P(\alpha(\epsilon,t) = j) - q_j| \leq C_{51} e^{-k_0 t/\epsilon}.$$

Proof. We refer the reader to Lemma C.1 in Sethi and Zhang (1994).

Lemma 5.2 *Under Assumption* (A.3), *for any bounded deterministic process* $\{\beta(t) : 0 \leq t < \infty\}$, *there exists a constant* C_{52} *such that for each* $j \in \mathcal{M}$ *and* $T_1, T_2 \in [0, \infty)$ *with* $T_1 < T_2$,

$$E\left|\int_{T_1}^{T_2} \left(I_{\{\alpha(\epsilon,t)=j\}} - q_j\right) \beta(t)dt\right|^2 \leq C_{52}\epsilon(T_2 - T_1).$$

Proof. We refer the reader to Lemma C.4 in Sethi and Zhang (1994). □

Corollary 5.3 *Let Assumption* (A.3) *hold. For any bounded deterministic process* $\{\beta(t) : 0 \leq t < \infty\}$, $j \in \mathcal{M}$, *and* $T > 0$,

$$E\left|\int_0^T \left(I_{\{\alpha(\epsilon,t)=j\}} - q_j\right) \beta(t)dt\right| \leq (C_{52}\epsilon T)^{1/2},$$

where C_{52} *is given in Lemma* 5.2.

Proof. Note that from Jensen's inequality,

$$\begin{aligned} &\left(E\left|\int_0^T \left(I_{\{\alpha(\epsilon,t)=j\}} - q_j\right) \beta(t)dt\right|^2\right)^{\frac{1}{2}} \\ &\geq E\left|\int_0^T \left(I_{\{\alpha(\epsilon,t)=j\}} - q_j\right) \beta(t)dt\right|. \end{aligned} \tag{5.1}$$

Thus, the corollary follows from (5.4) and Lemma 5.1. □

Now, let $\bar{\boldsymbol{u}}^\delta(\cdot) = (\bar{\boldsymbol{u}}^{(0),\delta}(\cdot), \bar{\boldsymbol{u}}^{(1),\delta}(\cdot), ..., \bar{\boldsymbol{u}}^{(m),\delta}(\cdot)) \in \bar{\mathcal{A}}$, for $\delta > 0$, be an δ-optimal control for the limiting control problem $\bar{\mathcal{P}}$. We construct a control

$$\boldsymbol{u}^{\epsilon,\delta}(t) := \sum_{i=0}^{m} I_{\{\alpha(\epsilon,t)=i\}} \bar{\boldsymbol{u}}^{(i),\delta}(t). \tag{5.2}$$

Then, we have $\boldsymbol{u}^{\epsilon,\delta}(t) \in \mathcal{A}^\epsilon$. Moreover, let $\boldsymbol{x}^{\epsilon,\delta}(t)$ and $\bar{\boldsymbol{x}}^\delta(t)$ denote the corresponding states of the systems $\mathcal{P}^\epsilon$ and $\bar{\mathcal{P}}$, respectively. Then

$$\dot{\boldsymbol{x}}^{\epsilon,\delta}(t) = \boldsymbol{u}^{\epsilon,\delta}(t) - \boldsymbol{d}, \quad \boldsymbol{x}^{\epsilon,\delta}(0) = \boldsymbol{x}, \tag{5.3}$$

$$\dot{\bar{\boldsymbol{x}}}^\delta(t) = \sum_{i=0}^{m} q_i \bar{\boldsymbol{u}}^{(i),\delta}(t) - \boldsymbol{d}, \quad \bar{\boldsymbol{x}}^\delta(0) = \boldsymbol{x}. \tag{5.4}$$

We can now derive the following asymptotic relationship between $\bar{\boldsymbol{x}}^{\delta}(t)$ and $\boldsymbol{x}^{\epsilon,\delta}(t)$.

Lemma 5.4. *Let Assumption* (A.3) *hold and let* $\liminf_{t\to\infty} \bar{x}_i^{\delta}(t) > 0$ $(i = 1, ..., n)$, *then*

$$\liminf_{t\to\infty} x_i^{\epsilon,\delta}(t) > 0 \text{ with probability one.}$$

Proof. Let $\tau_1^{\epsilon} = \inf\{t : \alpha(\epsilon, t) \neq \alpha(\epsilon, 0)\}$ and $\tau_i^{\epsilon} = \inf\{t \geq \tau_{i-1}^{\epsilon} : \alpha(\epsilon, t) \neq \alpha(\epsilon, \tau_{i-1}^{\epsilon})\}$ $(i = 2, 3....)$. Define $N^{\epsilon}(t) = \max\{i : \tau_i^{\epsilon} \leq t\}$. Then

$$\begin{aligned}
\boldsymbol{x}^{\epsilon,\delta}(t) &= \sum_{k=1}^{N^{\epsilon}(t)} \int_{\tau_{k-1}^{\epsilon}}^{\tau_k^{\epsilon}} \left(\sum_{i=0}^{m} I_{\{\alpha(\epsilon,s)=i\}} \bar{\boldsymbol{u}}^{(i),\delta}(s) - \boldsymbol{d} \right) ds \\
&+ \int_{\tau_{N^{\epsilon}(t)}^{\epsilon}}^{t} \left(\sum_{i=0}^{m} I_{\{\alpha(\epsilon,s)=i\}} \bar{\boldsymbol{u}}^{(i),\delta}(s) - \boldsymbol{d} \right) ds.
\end{aligned}$$

Note that $\{\int_{\tau_{k-1}^{\epsilon}}^{\tau_k^{\epsilon}} (\sum_{i=0}^{m} I_{\{\alpha(\epsilon,s)=i\}} \bar{\boldsymbol{u}}^{(i),\delta}(s) - \boldsymbol{d})\, ds, k \geq 1\}$ is a sequence of independent random variables. Therefore, by the Strong Law of Large Numbers, we have the lemma. □

In order to prove that $\boldsymbol{u}^{\epsilon,\delta}(t)$ given in (5.2) is a δ-asymptotically optimal open-loop control, we assume the following.

(A.5) $h(\boldsymbol{x}) = \sum_{i=1}^{n} c_i^{+} x_i^{+} + c_i^{-} x_i^{-}$, where $x^{+} = \max\{0, x\}$ and $x^{-} = \{0, -x\}$.

We can now prove the following theorem.

Theorem 5.5. *Let Assumptions* (A.1)-(A.5) *hold and let* $\liminf_{t\to\infty} x_i^{\delta}(t) > 0$ $(i = 1, ..., n)$. *Then*

$$\lim_{\epsilon\to 0} \left| J^{\epsilon}(\boldsymbol{x}, i, \boldsymbol{u}^{\epsilon,\delta}(\cdot)) - \lambda \right| \leq \delta. \tag{5.5}$$

Moreover, $u^{\epsilon,\delta}(\cdot)$ *is δ-asymptotically optimal, i.e.,*

$$\lim_{\epsilon\to 0} \left| J^{\epsilon}(\boldsymbol{x}, i, \boldsymbol{u}^{\epsilon,\delta}(\cdot)) - \lambda^{\epsilon} \right| \leq \delta. \tag{5.6}$$

Proof. From Theorem 4.4 and δ-optimality of $\bar{\boldsymbol{u}}^{\delta}(\cdot)$, it suffices to show that for a large enough $T > 0$,

$$\begin{aligned}
\frac{1}{T} \Bigg| & \int_0^T E\left(h(\boldsymbol{x}^{\epsilon,\delta}(t)) + c(\boldsymbol{u}^{\epsilon,\delta}(t)) \right) dt \\
& - \int_0^T \left(h(\bar{\boldsymbol{x}}^{\delta}(t)) + \sum_{i=0}^{m} q_i c(\bar{\boldsymbol{u}}^{(i),\delta}(t)) \right) dt \Bigg| \leq C_{53} \epsilon^{1/2},
\end{aligned} \tag{5.7}$$

for some C_{53}. Note that from Assumption (A.2), $c(\bar{\boldsymbol{u}}^{(i),\delta}(t))$ $(i \in \mathcal{M})$ are bounded on $[0, \infty)$. Thus, by Corollary 5.3,

$$\frac{1}{T} \left| \int_0^T Ec(\boldsymbol{u}^{\epsilon,\delta}(t))dt - \int_0^T \sum_{i=0}^{m} q_i c(\bar{\boldsymbol{u}}^{(i),\delta}(t))dt \right|$$

$$= \frac{1}{T}\left| \int_0^T Ec\left(\sum_{i=0}^m I_{\{\alpha(\epsilon,t)=i\}} \bar{u}^{(i),\delta}(t) \right) dt - \int_0^t \sum_{i=0}^m q_i c(\bar{u}^{(i),\delta}(t))dt \right|$$
$$\le \frac{1}{T}\sum_{i=0}^m E \left| \int_0^T \left(I_{\{\alpha(\epsilon,t)=i\}} - q_i \right) c(\bar{u}^{(i),\delta}(t))dt \right|$$
$$\le C_{53}^{(1)} \epsilon^{1/2} \tag{5.8}$$

for some $C_{53}^{(1)}$. If we can show

$$\frac{1}{T}\left| \int_0^T Eh(x^{\epsilon,\delta}(t))dt - \int_0^T h(\bar{x}^{\delta}(t))dt \right| \le C_{53}^{(2)}\epsilon \tag{5.9}$$

for some $C_{53}^{(2)}$, then from (5.8) we know that (5.7) holds. By Lemma 5.4, there exist a T_1 and $B \in \mathcal{F}$ with $P(B) > 1-\epsilon$ such that for $t \ge T_1$

$$E|x^{\epsilon,\delta}(t) I_{\{B^c\}}| < \epsilon, \text{ and } x^{\epsilon,\delta}(t,\omega) > 0, \ \omega \in B.$$

Hence,

$$\frac{1}{T}\left| \int_0^T Eh(x^{\epsilon,\delta}(t))dt - \int_0^T h(\bar{x}^{\delta}(t))dt \right|$$
$$\le \frac{1}{T}\int_0^T \left| Eh(x^{\epsilon,\delta}(t)) - h(\bar{x}^{\delta}(t)) \right| dt \tag{5.10}$$
$$\le \frac{1}{T}\left(\int_0^{T_1} E|x^{\epsilon,\delta}(t) - \bar{x}^{\delta}(t)|dt + \int_{T_1}^T |Ex^{\epsilon,\delta}(t) I_{\{B\}} - \bar{x}^{\delta}(t)|dt \right.$$
$$\left. + \int_{T_1}^T E|x^{\epsilon,\delta}(t)| I_{\{B^c\}} dt \right)$$
$$\le \frac{C_{53}^{(3)}}{T}\left(\int_0^{T_1} E|x^{\epsilon,\delta}(t) - \bar{x}^{\delta}(t)|dt \right.$$
$$\left. + \int_{T_1}^T |Ex^{\epsilon,\delta}(t) - \bar{x}^{\delta}(t)|dt + \epsilon(T-T_1) \right),$$

for some $C_{53}^{(3)}$. By Lemma 5.1 and the boundedness of $|\bar{u}^{(i),\delta}(s)|$,

$$\int_{T_1}^T |Ex^{\epsilon,\delta}(t) - \bar{x}^{\delta}(t)|dt$$
$$= \int_{T_1}^T \left| E\left(\int_0^t \sum_{i=0}^m I_{\{\alpha(\epsilon,s)=i\}} \bar{u}^{(i),\delta}(s) - d \right) ds \right.$$
$$\left. - \int_0^t \left(\sum_{i=0}^m q_i \bar{u}^{(i),\delta}(s) - d \right) ds \right| dt$$

$$
\begin{aligned}
&= \int_{T_1}^{T} \left| \sum_{i=0}^{m} \int_0^t (P(\alpha(\epsilon,s)=i)-q_i)\,\bar{u}^{(i),\delta}(s)ds \right| dt \\
&\le C_{53}^{(4)} \int_{T_1}^{T} \left(\sum_{i=0}^{m} \int_0^t |P(\alpha(\epsilon,s)=i)-q_i|\,ds \right) dt \qquad (5.11)\\
&\le (m+1)C_{53}^{(4)} \int_{T_1}^{T} \left(\int_0^t e^{-k_0 s/\epsilon} ds \right) dt \\
&= (m+1)C_{53}^{(4)} \int_{T_1}^{T} \frac{\epsilon}{k_0} \left(1 - e^{-k_0 t/\epsilon} \right) dt \\
&\le C_{53}^{(5)} (T-T_1)\epsilon.
\end{aligned}
$$

for some $C_{53}^{(5)}$. This and (5.10) imply (5.9). □

Acknowledgments: This work was done while the second author was visiting School of Management, University of Texas at Dallas. This research was supported in part by NSERC Grant A4619, a grant from Chinese Natural Science Foundation, and the Faculty Research Grants from University of Texas at Dallas. The authors are grateful to Qing Zhang for his valuable suggestions and comments.

References

T. Bielecki and P. R. Kumar. (1988). Optimality of zero-inventory policies for unreliable manufacturing systems, *Operations Research*, Vol. 36, 532-546.

W. H. Fleming and W. M. McEneaney. (1995). Risk sensitive control on an infinite horizon, *SIAM J. on Control and Optimization*, Vol. 33, 1881-1921.

W. H. Fleming and M. H. Soner. (1992). *Controlled Markov Processes and Viscosity Solutions*, Springer-Verlag, New York.

M. K. Ghosh, A. Arapostathis and S. Marcus. (1997). Ergodic control of switching diffusion, *SIAM J. on Control and Optimization*, Vol. 35, 1952-1988.

S. P. Meyn and R. L. Tweedie. (1994) *Markov Chains and Stochastic Stability*, Springer-Verlag, London.

E. Presman, S. P. Sethi, H. Zhang and Q. Zhang. (1998). Optimal control policies in unreliable two-part-type manufacturing systems, *Dynamics of Continuous, Discrete and Impulsive Systems*, to appear.

S. P. Sethi, W. Suo, M. I. Taksar and H. Yan. (1998). Optimal production planning in a multi-product stochastic manufacturing system with long-run average cost, *Discrete Event Dynamic Systems: Theory and Applications,* Vol. 8.

S. P. Sethi, W. Suo, M. I. Taksar and Q. Zhang. (1997a). Optimal production planning in a stochastic manufacturing system with long-run average cost, *Journal of Optimization Theory and Applications*, Vol. 92, 161-188.

S. P. Sethi, H. Zhang and Q. Zhang. (1997b). Hierarchical production control in a stochastic manufacturing system with long-run average cost, *Journal of Mathematical Analysis and Applications,* Vol. 214, 151-172.

S. P. Sethi and Q. Zhang. (1994). *Hierarchical Decision Making in Stochastic Manufacturing Systems*, Birkhäuser Boston, Cambridge, MA.

G. Yin and Q. Zhang. (1998). *Continuous-Time Markov Chains and Applications: A Singular Perturbation Approach*, Springer-Verlag, New York.

MIX
Papier aus verantwortungsvollen Quellen
Paper from responsible sources
FSC® C105338

If you have any concerns about our products,
you can contact us on
ProductSafety@springernature.com

In case Publisher is established outside the EU,
the EU authorized representative is:
Springer Nature Customer Service Center GmbH
Europaplatz 3, 69115 Heidelberg, Germany

Printed by Libri Plureos GmbH
in Hamburg, Germany